S0-CAH-347

ALLE ZEIT WACH
1842

Wendy A. Warr (Ed.)

Chemical Structures

The International Language of Chemistry

With 213 Figures and 18 Tables

Springer-Verlag
Berlin Heidelberg New York London Paris Tokyo

Dr. Wendy A. Warr
Information Services Section
ICI Pharmaceuticals Division
Mereside, Alderley Park
Macclesfield, Cheshire SK 10 4TG, UK

ISBN 3-540-50143-6 Springer-Verlag Berlin Heidelberg New York
ISBN 0-387-50143-6 Springer-Verlag New York Berlin Heidelberg

Typesetting: Hope Services (Abingdon) Ltd. Printing: Weihert-Druck GmbH, Darmstadt.
Binding: J. Schäffer GmbH & Co. KG., Grünstadt
2151/3140 - 543210 - Printed on acid-free paper

PREFACE

This book constitutes the Proceedings of the conference 'Chemical Structures: The International Language of Chemistry' which was held at Leeuwenhorst Congress Centre, Noordwijkerhout in the Netherlands, between May 31 and June 4, 1987. The conference was jointly sponsored by the Chemical Structure Association, the American Chemical Society Division of Chemical Information, and the Chemical Information Groups of the Royal Society of Chemistry and the German Chemical Society.

The purpose of the conference was to bring together experts and an international professional audience to discuss and to further basic and applied research and development in the processing, storage, retrieval and use of chemical structures, to focus international attention on the importance of chemical information and the vital research being carried out in chemical information science and to foster co-operation among major chemical information organisations in North America and Europe. Subjects covered included integrated in-house databases, substructure searching methodology, spectral databanks, new technologies (microcomputers, CD-ROM, parallel processing and expert systems) and chemical reactions.

The keynote address was given by Mike Lynch of the University of Sheffield. In this, the opening chapter of the book, Mike discusses progress made in chemical information science in the last fifteen years and describes his own approach to research.

In a plenary session, Myra Williams of Merck, Sharp and Dohme considered future trends from the point of view of the information manager and strategic planner in industry. She emphasises the need for integration, open architecture and a uniform user interface.

The next group of chapters is concerned with in-house chemical structure databases and related property data. Tom Hagadone has carried out a survey of current practice and emerging trends in the integration of in-house databases and he presents his results. Mike Allen describes Glaxo's Chemical and Biological Information System, CBIS, with particular emphasis on registration and stereochemistry. Deibel and de Jong explain Akzo's integration of DARC structure handling software with the database management system ORACLE. Fisons, however, have integrated OSAC with their System 1032-based data handling system ABACUS. Dave Magrill gives details. Pfizer have developed SOCRATES to store and retrieve chemical structures, reactions and related data. SOCRATES, which exploits System 1032 for database management, but as a whole is proprietary to Pfizer, is described in Trevor Devon and David Bawden's paper.

The next four chapters in this group all involve MACCS to a greater or lesser extent. Arnold Lurie describes both Kodak's private registry system with Chemical Abstracts Service and the use of MACCS for a separate chemical structure handling system. The considerable advantages of integrating these two systems are not addressed and remain the subject of interesting speculation. Harold Schlevin and his co-workers at Ciba-Geigy integrate MACCS with a variety of commercially available software tools for handling and analysing data. They have written an in-house system, ChemSketch, to integrate text and chemical structures. Bill Henckler and his co-workers at Merck have converted their out-dated Chemical Structure Information System, CSIS, with its intricate database updating, to a new MACCS-based system. Finally, Erick Ahrens describes how customisable chemical database systems, and in particular MACCS-II, can serve the diverse information needs of the chemical and pharmaceutical industries.

Substructure searching methodology is covered in the next set of chapters. John Barnard's chapter gives an historic overview and outlines areas of current research interest. Peter Bruck and his co-workers demonstrate how fast very large files can be searched when the database is tree-structured. Peter Willett's team at Sheffield University have developed techniques for substructure searching in files of three-dimensional chemical structures. Their paper also describes hardware and software techniques for identifying the maximal common substructure in a set of three-dimensional chemical structures. David Bawden of Pfizer (in conjunction with Sheffield University) has developed techniques for measuring similarity among structures. He describes associated ranking and clustering procedures and the concept of browsing rather than exact searching.

The next three chapters concern generic structure search. Mike Lynch reviews progress in the Sheffield University Generic Chemical Structures project and the problems that his team are at present tackling. Kathy Shenton and co-authors describe the construction of a database of generic structures and the 'Markush DARC' search software developed collaboratively by Derwent, Tlsystémes and the French Patent Office, INPI. Yoshihiro Kudo's poster paper illustrated a system for searching using both specific and generic structural formulae.

Three chapters about online databases follow. Clemens Jochum's concerns the construction of the Beilstein Online database. Of particular interest are the complete representation of stereochemistry and the replacement of bond orders by electronic information, allowing a global, and unique, representation of tautomers. Ole Norager describes the ECDIN project, in which data has been collected on environmental chemicals and made accessible online. Gerry vander Stouw shows how the Chemical Abstracts Registry System Enhancements Feasibility Study project identified aspects of the CAS structuring conventions which create problems for users and he considers the improvements which may be recommended.

Two separate teams, those of Wolfgang Bremser and Henk van't Klooster, were chosen to cover the subject of spectral databases, computer-aided library search systems and expert systems for structure analysis.

The chapter by Ernst Meyer and his colleague describes techniques for detecting classes of substances which might prove to be biologically active. Bench chemists at BASF can apply the system to a large collection of biological test data in their quest for 'lead' compounds.

The next group of chapters arises from a session on new technologies. Bill Town's chapter covers hardware and software developments in the microcomputer area and, in particular, the use of microcomputers for chemical structure input and display, for managing personal databases, for offline structural query negotiation and for downloading of chemical structures. Dan Meyer defines five major categories of microcomputer software for handling chemical structures and then compares a number of packages for managing databases of chemical structures. Patrick Gibbins outlines the benefits and limitations of publishing on CD-ROM. Todd Wipke and Mathew Hahn describe a symbolic, non-numerical approach to molecular model building and conformational analysis. The last two chapters in the new technology group both cover parallel processing techniques. Peter Jochum and his colleague describe an architecture consisting of a search machine with many independent parallel processing units administered by a master processor. Christian Zeidner and co-workers report on the completely new hardware and software for Chemical Abstracts' second generation, parallel structure searching system, which will involve inverted file searching at the screening stage (instead of the present serial search method).

Eleven chapters on the subject of chemical reactions follow. Peter Johnson gives an overview of current approaches to reaction indexing with emphasis on the Leeds University system ORAC. A chapter by Tom Moock *et al.* addresses newer features of the REACCS system, especially atom-atom mapping in reaction substructure search. A poster by Guenther Grethe and colleagues was also centred on REACCS. David Elrod's poster gave an example of the integrated use of CHIRON and REACCS. CHIRON (a program developed by Steve Hanessian of Montreal University) was used to predict chiral precursors for an anthracycline and then REACCS was used to work out a synthetic pathway starting from those chiral precursors.

Johnny Gasteiger and his co-workers seek to predict the course of complex organic reactions using empirical methods to quantify the electronic and thermochemical factors which influence chemical reactivity.

Willi Sieber describes a method of integrating the complementary approaches of reaction retrieval and synthesis planning.

The University of Nijmegen houses a Dutch national facility for both computer-assisted organic synthesis and computer-assisted molecular modelling. Jan Noordik describes the tools it makes available to the academic bench chemist.

George Vladutz's contribution, co-authored by Scott Gould, examines whether a file organisation consisting of superimposed structures ('hypergraphs') and superimposed reaction graphs has potential for improved retrieval capabilities. A second hyperstructure of superimposed reaction skeleton graphs, could be cross- referenced to the first hyperstructure.

Rainer Herges has developed algorithms to search systematically for new reactions. With the computer system IGOR, he and his co-workers have predicted hitherto unknown reactions and realised them in the laboratory.

Bob Dana and colleagues at Chemical Abstracts Service presented a poster on the machine generation of multi-step reactions for the CASREACT system on STN International. The online version of FIZ's ChemInform database will be available on STN International also. Alex Parlow gave a poster presentation on the creation of ChemInform both as an electronic database (for online and in-house use) and as a printed product.

The book concludes with two 'miscellaneous' chapters, Kurt Loening's on chemical nomenclature and Robert Fugmann's on grammar in chemical indexing languages.

In assembling the Conference programme, the organising committee hoped to cover most areas of chemical information science. Nearly all the 44 contributions were specifically invited by the committee. In the few cases where a speaker did not supply a written version of his paper, I, the editor, have written a summary. We hope, therefore, that this book truly reflects the state-of-the-art in chemical information science.

I am deeply indebted to the organising committee, without whose enthusiasm and hard work this book would not have been possible. The committee members were Charles Citroen (CID-TNO, The Netherlands), David Johnson (Exxon Research and Engineering, USA), Reiner Luckenbach (Beilstein Institute, Federal Republic of Germany), Peter Nichols (ISI, UK) and Bill Town (Hampden Data Services, UK). I myself had the pleasure of chairing the committee.

The interest and support of the four sponsoring organisations is gratefully acknowledged.

I am also extremely grateful to my secretary, Mary Burgess, who has patiently and accurately word-processed every word of all forty-four chapters.

The book was typeset from the floppy disks by Hope Services Ltd., of Clifton Hampden, Oxfordshire. Their helpfulness and efficiency has been much appreciated.

Two members of ICI Pharmaceuticals' Information Services Section, Madeline Gray and Pat Holohan, have helped with online searching for literature references, standard journal abbreviations and key subject terms, and Frank Loftus, of the same Section, redrafted a number of diagrams.

I would also like to make special mention of David Johnson of Exxon, USA, who spent much time and care preparing the index.

Finally I particularly want to thank Bill Town of Hampden Data Services for managing the finances, for suggesting and implementing the electronic typesetting exercise and for supplying constant advice and support when needed.

May 1988

Wendy A Warr
ICI Pharmaceuticals
Macclesfield
Cheshire UK

CONTENTS

KEYNOTE ADDRESS

R & D IN CHEMICAL INFORMATION SCIENCE: RETROSPECT AND PROSPECT

Michael F Lynch
Department of Information Studies, University of Sheffield, Western Bank, Sheffield, S10 2TN, U.K.

It is a great honour for me to be invited to give the keynote address at this conference, not least since it gives every appearance of being a worthy successor to the earlier conference held in 1973 here at Noordwijkerhout, which was quite seminal in its effects on R&D in chemical information sciences. It brought together, for the first time, a number of chemists and chemical information scientists who were working in different cultures and with different orientations to discover that they shared substantial interests, not merely in the objects of their research, but also in common methods and models.

I congratulate the International Organising Committee and the Netherlands organiser, Charles Citroen, on their success; they can feel very gratified that their efforts have been rewarded so fully by the very large attendance here this week, as well as by the calibre of this most promising programme of speakers and papers.

We had to thank Todd Wipke, Steve Heller, Richard Feldmann and Ernie Hyde for their wisdom and foresight in organising the 1973 conference at such a timely juncture, and with such valuable results – not least through publication of the papers given at it in the form of the proceedings which have become a working source for everyone in the field since then. That conference and its record provide me with a baseline in time, to review what has developed in the intervening time, so that we can also look forward to see what is in prospect.

It is a particular pleasure for me to see so many here who have an association with the Department of Information Studies in Sheffield. There are current and former students or researchers, collaborators, and others who are simply good friends. Janet Ash, my first research student in 1965, is here, still involved in the field. Bill Town, George Adamson, David Bawden, Peter Willett, George Vladutz, and John Barnard are here, and so too are all the current members of our chemical information research group. Peter Willett and I are grateful to the conference organisers and to the committee of the Chemical Structure Association, in particular, for the award of studentships which enable all of these to participate this week.

The main themes in my discussion of R & D in chemical information science will be, first, to look backward to identify the seminal contributions which formed the foundation on which current systems are based, and identify common factors in the developments. Next, I wish to assess the health of current developments, and then to try to identify factors of which we need to be aware in order to ensure healthy and productive continuing progress. This is necessary in order to harness a technology, the capabilities of which continue to maintain rapid performance growth rates such as are attained by no other technology past or current.

W. A. Warr (Ed.)
Chemical Structures

Doubts have been voiced over whether this continued growth in performance can be maintained, as fundamental physical limitations are reached. We can take hope, it appears, from a discovery which was made close to here at Leiden University in 1911. I refer, of course, to the discovery of superconductivity by Kamerlingh-Onnes; the prospect of the room-temperature superconducting switch now seems to assure continuing advances in processing power for our needs.

In my review, I shall be looking particularly at questions of the models and formalisms which are available to us, and asking what other formalisms we may need to understand for research purposes, the better to meet the demands posed by the needs of the disciplines. Equally, I want to look at the new challenges which now face us or loom ahead.

Many workers have taken similar paths. Many common topics were evident early. In part, this was due to the need to replace functions served by existing, conventional information and documentation systems, and their extensions as made possible by the vastly more flexible technology. Chemical information science has also seen the highly imaginative opening up of novel possibilities, not previously conceived of within classical information systems.

A welcome feature of research in chemical information science is the probability that it will attract early implementation if successful and useful. This contrasts strongly with the position in the information retrieval field at large, where the rate of innovation, certainly by the large host organisations, is slow, where useful research results seem slow to be acknowledged, and where the user market appears very much less well-informed than is the case in our community. Systems operators there are more inclined to dabble in improvements based on cosmetic factors, rather than on the evaluation and possible application of potentially useful research results. It is thus rewarding for many researchers in this field, with shorter and more effective lines of communication between researchers and possible implementers, to see research results reaching meaningful applications in the short term.

The technology provides its own impetus too. In the case of some of the earliest contributions, some ideas were not fully implementable, because the facilities were not commonly available at reasonable costs. Thus, we sometimes fail to appreciate that the concern for graphic representations dates from the earliest days, e.g., the attention given to graphics displays of structural information by Ascher Opler in the late 1950s, also the work of Ernst Meyer at BASF and IDC on the design and use of the Formellesegeraet – the optical formula reading machine for graphic input at attainable cost.

Comparing the 1973 and the 1987 conference topics, we find substantial degrees of commonality between them. While methods and techniques have moved ahead strongly, the preoccupations are not overly dissimilar. The chemical structure diagram and its associated data are the paramount focus of interest, then as now. There is now, however, a much greater emphasis on accomplishment in installed and operating systems, particularly those operated as in-house systems, and taking full account of matters of systems integration, both within user organisations and between them and public database suppliers. Tomorrow morning's sessions will provide a full overview of the health of these systems – though I note with regret the relative absence of similar contributions from university departments in regard to the integration of such facilities into teaching programmes. World-wide, these are not good years for academics, a song which I could sing at some length.

The record of the 1973 conference was incomplete in one important respect, in that it did little to reflect the fact that at the user level in industry the emphasis was still principally on low technology rather than on high technology. The ICI paper on CROSSBOW described work based on the WLN notation, although in other respects it was advanced for its time. Most people using computers for chemical structure applications at that time were still using paper-based displays of permuted notations.

Graphics display methods were available only to the most richly funded researchers in 1973, but are now commonplace, thanks to the extraordinary rate of developments in the professional microcomputer workstation. It is hard to think of another factor which has so strongly influenced what chemists do in relation to storage, display and manipulation of chemical structures than this development, as Bill Town will be telling us. This appears still to continue at undiminished rates, soon to be further enhanced by the advent of yet more powerful workstations, and communications networks to support them.

I can recollect very little sense, in 1973, that this revolution was imminent. Georges Anderla, well-known to us Europeans as one of the early visionaries about the impact of information technology on scientific communication, prepared a report for Organisation for Economic Cooperation and Development in 1972, in which he forecast the wholesale mushrooming of computer-based information systems for the sciences. He recently revisited his report, and noted that the future impact of very large scale integration (VLSI) was evident only to a very few at that time, so that he quite properly based his projections on the characteristics of the mainframe and minicomputer types then extant. As a result, he noted, he quite failed to see, first, that the PC would result in expertise becoming vastly more widely disseminated, with power passing out of the hands of the small priesthood of computer experts, thus tapping a huge reservoir of innovative thinking, and, second, that the workstation, rather than the dumb terminal, would become standard. On another front, however, the bandwidth of the co-axial cable between processor and screen will not readily be attained for general networking purposes.

Substructure searching methods, then as now, retain pride of place in interest. Here, the advances are especially strongly evident; many of the papers this week report on the operations of in-house and public structure and data retrieval systems, now very widespread in industry. Fourteen years ago, the ICI and BASF systems were two of a few operational search systems in industry, with Richard Feldmann's forerunner of the Chemical Information System(CIS) at the National Institutes of Health. The ICI CROSSBOW system, though, depended for some of its functions on the use of notations at the user interface, which restricted its use somewhat, as chemists wereunwilling to learn to encode structures for input. The performance of the BASF system was remarkable for the time, yet had already been in operation for over five years, and even included some generic structure storage and search capabilities.

Today, the focus of interest has become much wider. The large public substructure search services will soon have been in operation for a decade. The concern is now with second generation facilities, as also with the improved organisation of output, or with improved organisation of the output of a search. The reflection of stereochemistry was present only in Todd Wipke's experimental synthesis planning routines and in Jacques-Emile Dubois' DARC system in 1973. Today, in addition to packages which represent stereochemistry, we move on to focus on true 3-D searching, scarcely even thought of at that earlier time.

The horizon is widening in other senses too, to the challenging problems of searching Markush formulae, with work currently in progress variously at Derwent and

Telesystemes, IDC, Chemical Abstracts, and Sheffield University, which are partly reflected in contributions here during the coming week.

The widening of the focus of interest from the individual structures themselves to their interrelations in chemical reactions was much in evidence 14 years ago. It was the subject of Jacques Valls' paper, which reviewed the existing range of methods, and the application of simple computer methods to them; reaction indexing, using more innovative methods, was also discussed by Ernie Hyde and Diane Eakin, and by myself. In the meantime, we have novel systems such as ORAC and REACCS, operating on highly selective reaction databases, but still no chemistry-wide database, so that progress in that quarter has been slower than one might have imagined.

Perhaps the most significant innovative thinking evident at the 1973 meeting was the strong presence of contributors working within chemistry departments on synthesis planning systems – particularly the laboratories of Todd Wipke and of Ivar Ugi. These and related approaches have seen strong growth in the meantime.

The interrelation of reaction databases and synthesis design systems is brought into close focus this week by a variety of presentations, not least from the originator of many of the ideas on automatic reaction analysis and synthesis design, George Vladutz.

The link with structure-associated data is as strong today as a decade and a half ago. First, there is the welcome start to building the Beilstein database, with a wealth of organised data, second, the increasing provision of analytical and interpretive methods as they relate to spectral, property and other data. Then as now, this is one of the major bridgeheads between our field and the artificial intelligence area.

One concern which was marked in 1973 but is less evident here this week, though burgeoning in other respects, is molecular modelling and associated developments in molecular biology, and I shall be referring to these later.

Some observations about the intervening period are in order here. First, I note the continuity both of many themes and of the researchers involved in them. Most of the researchers reporting work in 1973 are still active, and some are here today to report on their current work.

Secondly, there is the substantial scale of research effort, in our experience – although the resources required at the research stage, when the basic principles are being elaborated, are nothing like those required to implement them for operational purposes in secure and robust systems with good interfaces. Our work on reaction indexing, in total, ran to something like 15 man-years in all, while our generic structures studies have involved almost 20 man-years on a time-scale of 7 years.

What can be said to have improved as regards research methods over the intervening period? First, I think, there is much wider understanding of the nature of algorithms, their formal description, and their properties. Many of the algorithms we need deal with problems which are NP-complete in nature. Many of the most interesting are combinatorially explosive, rapidly generating very large tree-structured spaces, hence a canny look at algorithms is appropriate. Second, we have improved working environments, from operating systems to languages, and from fast processors to larger backing stores. For most workers in the early 1970s, only batch processing with serial storage media was available. Networking, at various bandwidths, is vastly more convenient today too. In 1973, the transatlantic telephone rate was $240/hour.

So much for the retrospective view. What now by way of prospect? I have tended not to spend much time in front of crystal balls, rather, I have been a jobbing researcher, often seeing a function which was clearly in need of upgrading, and setting about the means of attaining it for others to use. This is at a certain cost, too, in that staying close to the front requires one to dig deep narrow holes, as it were, and the consequent degree of specialisation can and does lead to ignorance of the work of others. Not that ignorance is all bad – a little of it at the right time can be beneficial. Where would the inventor of the Mettler balance have got to, if he had not been ignorant of all of the problems of conventional balance design, as is reputed to be the case?

For the prospect, I also want to look a little at the methods of research, at the well-springs of ideas and their interrelations, as well as at the tools we need to apply increasingly in the future.

In common with many other areas of science, the goals of research may be either phenomenon-driven or applications-driven, i.e., they may derive either from observations of an interesting and potentially valuable phenomenon which is capable of yielding new knowledge, and, in turn, new opportunities for practical applications, or they may be based on the identification of roles or functions which may be served if algorithms and data structures which support the function can be developed – the technology replacement approach which we have often seen applied. In my experience, tough problems help to evoke strong and innovative solutions.

As elsewhere, it is desirable that research should be timely and have goals which are attainable within reasonable time – certainly if it is to enjoy the support of research sponsors.

The human dimension is mostly lacking from published accounts of research, both as regards the well-springs of ideas and the pain and pleasure. The pleasure comes in the sighting of an opportunity for a breakthrough, in its elaboration through discussions with colleagues, and in the achievement of objectives. The pain, much more frequent, it seems, comes in the long and often tedious working up to proving the point, or in recognising that what seemed to be a good idea will not fly.

Where, then, does a jobbing researcher like myself get his ideas from, what are the influences which affect the course of research once embarked upon, and, equally important, what are the tools which can be applied to solutions of the problems, or elucidation and understanding of the phenomena as they crop up?

The truth to tell, many of the influences, apart from those of organisations and mentors, are often seemingly trivial observations, noted and tucked away in long-term memory against a rainy day when there is time to develop them further. Organisations and mentors are immensely formative factors, we would all recognise. I had the good fortune to work closely with Malcolm Dyson during my apprenticeship years in this area, also with Fred Tate – characters as different as you could wish, each contributing in major ways to information science, Dyson through his vision of what the emergent technology could mean for a chemical information database spanning all of the years, Tate's preoccupation being with using the technology to get the show on the road, to continue to provide services which had outgrown printer's ink, leading, in time to a cumulative database. Each was right in his own vision of the future, differing mainly about the means by which they were to be achieved.

Dyson and Tate stood above the technology of their time – the fact that they started with an 8K IBM 1401 did not limit their thinking – rather, they were confident that when the need arose, the technology would rise to the occasion. There is a close parallel in this with an anecdote from the history of the old Austrian Empire in the middle of the 19th century. Austria needed to build a rail link to join Vienna to Trieste, its port on the Mediterranean, but a minor range of the Alps stood in the way. The builders who began construction of the Semmering mountain railway in 1848 – the world's first – were faced with gradients which no locomotive could then attempt. Their confidence in the future of the technology was exactly similar to that which Dyson and Tate showed, so that when they completed the track six years later in 1854, lo, there were the locomotives which could take the grades. Some of our research themes have had a similar underlying assumption.

One example of the kind of observation I mentioned, which was to evince itself in a number of subsequent lines of research was the following: examine the middle pages of a molecular formula index, and you will note that the listings of substance names under particular values for molecular formulae differ greatly in number. Some patterns are evident. These are, most notably, that common combinations of carbon and hydrogen atoms, together with one to a couple of nitrogen or oxygen atoms, on their own or together, show large numbers of entries – these are the domains of organic chemistry familiar to us all, which continue to enthral us by their incredible variety and structural diversity. Thus $C_{15}H_{24}O$, in a CA volume index, accounts for around 100 compound entries. Add to one of these popular formulae a less common atom, say, a chlorine or a phosphorus atom, or a couple of them, and the numbers of structures listed there will plummet. $C_{15}H_{54}OS$ accounts for only 5 entries. There is manifestly good reason for this – the probability that a chemist, somewhere, has looked at a substance with an unusual combination of atoms is low. Seen through the mirror of a representation, however, this offered the first glimpse of a strategy on which chemical screen search methods might be based, and on which a more general principle might be elucidated. Mooers, indeed, had had related ideas.

This notion re-emerged as important in a very different context. Look at a well-known printed chemical subject index, and you will note that the number of entries under different headings is very disparate there too. This effect is due to the structure of the subject in question, and to the vocabulary used. It turns out that indexers' description of the content of a particular document differs in ways influenced directly by their estimation of the number of postings expected at particular subject headings – and that one might therefore be able to capture decisions made by indexers which would reveal the microstructure of conceptual fields – what today would be called the knowledge structure of the domain. In order to capture these relations it was necessary to compare two entries, and identify variant subject descriptions by first isolating the common parts of the two entries.

This thought itself has not led further. It did, however, give rise to the notion of applying an algorithm to pairs of organic structures, so as to identify maximal common substructures (MCS's), so as to classify or to cluster these in a meaningful way. This was totally unrealistic for the day, and for the tools available to us – a Mercury computer with no internal memory, but in its place, as Janet Ash will remember, one and a half magnetic drums – and a very simple list processing language with which molecules containing just a few atoms could be compared. We depended on the remote ATLAS computer at Harwell for back-up – and when the compiler was lost, our first attempts foundered.

The ground had been prepared, however; George Vladutz, a few years earlier, had made the suggestion that automatic identification of the MCS would give a means of identifying skeletal reaction schemes from records of substances involved in reactions. Here was a means of doing just that, by focussing on the unchanged parts of molecules. As early as 1972, we had, for instance, the first machine-readable chemical reaction database, compiled from the pages of Current Abstracts of Chemistry and Index Chemicus, in the form of WLN records of reactions – and with it the possibility of identifying the individual stages of multi-stage reactions, a topic still of interest and to be the subject of a paper later this week.

Our first attempts were over-ambitious for the day, and our early work only modestly successful. Real success came only 10 or so years later, and that because George Vladutz held a Fellowship in our Department in the year that Peter Willett came to us as a student. George encouraged Peter to look afresh at the problem, with results which are known to all of you. Peter, more recently, has successfully extended the MCS work to 3-D MCS identification for pharmacophore identification.

Since the earliest MCS work was impracticable, we began instead to look at a more readily attained and more practically useful objective – indexing of chemical structures in order to provide screens for substructure search. Note, however, that the stimulus came from an analogous linguistic problem context – a topic to which I will return again.

This proved to be a fertile vehicle for both applicable research results, and for theory development – indeed, it was on this theme that I talked here 14 years ago.

The question was, given the very great variety of structural features characteristic of databases of specific chemical structures, how might one detect and characterise these in order to achieve efficient and effective screenout of the great majority of non-candidate structures. The methodology which we devised, and which was later to be elaborated further elsewhere, derived from the observation I mentioned earlier relating to the relative abundances of molecules under molecular formula entries with combinations of less common atoms. We counted the frequencies of a variety of hierarchically related species of fragments from samples of chemical structure databases, and studied the regular rank-frequency distributions which they showed. We coined such names as *augmented atoms* and *augmented pairs* for these species, and developed methods for selecting some fixed number of these so that, for that number of fragments, they showed up more or less equally frequently both in database structures and in queries, and thus tended towards optimal screening power when applied in serial searches of large databases. While our work examined certain larger fragments of up to eight atoms, it was left to others to use atom, bond and connectivity sequences. Our early work thus laid the basis of a methodology developed and extended by others, and in wide use today. A recent instance of its use in another context is in Peter Willett's approach to 3-D search.

This work in fact extended the classical approach of Calvin Mooers in applying information theory to questions of searching databases – although his were recorded on edge-notched cards. He had argued for codings which utilised the fixed capacity of the edge-notched card by superimposed coding reflecting term posting frequencies, an idea successfully applied by Ernst Meyer in his superimposed screens for fast substructure search. We were able to extend this in another sense, by considering the relations between the variety and types of screens, and their selection in order to attain optimal distributions. We showed that the entropy of these distributions was an appropriate measure for assessing the value of a screen set of a particular size, and indicated methods which were applicable to achieving relatively high values for the entropy of screen sets.

The idea of using symbol frequencies to guide the selection of an indexing vocabulary for search in the two-dimensional domain of chemical structure representations then prompted us to look at the applicability of these ideas in the text processing context – an example of phenomenon – driven research. This led to a series of studies of the microstructure of text, which were a delight to carry out, though they have seen little in the way of application. They provided insights into the microstructure of text databases, and led to a reinterpretation of Shannon's communication theory. This preoccupation with language and language structure was invaluable to us later in tackling the problems presented by generic chemical structures, in that it laid the groundwork for an approach with firm foundations in modern theories of language to this very refractory problem.

I will not say much about this, since I will be talking about our work on it later this week. An anecdote will perhaps illustrate why we had not attempted anything on these lines earlier. About 1964 or so, I attended a meeting in Washington called by the US Patent Office and the National Bureau of Standards to announce details of their research on the chemical patents question. The meeting was attended by the distinguished Princeton mathematician, Julian Bigelow. His reaction, when the project staff had described their work, was to observe that there was a certain fundamental and insoluble difficulty in what they were attempting. They were dealing with infinite sets of molecules and proposing to search them with a finite search algorithm. He was told sharply that he didn't understand what he was talking about, and his advice disregarded, but, *verb. sap.*, it was enough to deter me from making any attempt on the problem until I was fairly sure that there was a way forward. By the late 1970s, some decades of work on formal language theory was available. Professor E. V. Krishnamurthy of Bangalore in South India pointed out to me that this theory, which, for example, underpins the design of modern programming languages, could be applied in several ways to assist in the solution of this refractory problem. Even with the aid of such strong formalisms, finding general solutions has not been easy, and there are aspects still in need of resolution.

More generally, however, it is often the case that tough problems engender strong solutions, and I hope, interesting work. Thus, the design of GENSAL has provided the basis for several very similar approaches to the problem of generic structure representation – and could provide a rational basis for the design of a more conventional chemical notation system, if anyone felt that that was still a sufficiently important goal. Equally, TOPOGRAM and its possible extensions have provided a very powerful tool with which to overcome some of the consequences of unbounded generic terms. The notion of reduced graphs is central to our work on structure and substructure searching of generic structures. This too is based on the notion of a higher-level linguistic description of the gross characteristics of chemical structures. Even though it has long been implicit in existing methods of structure description, as in nomenclature and notations, it may still prove to be important as a method for substructure screening and in other respects too.

The insights I offer here are based on extremely simple notions. They were based initially on simply observation and intuition, if rationalised and given firmer foundations later. They are often sufficient to deal with problems at the levels to which we have taken them thus far. I suggest, however, that in many respects we are still climbing in the foothills of larger problem ranges. In many aspects of our current work we are constantly aware that the formalisms we are using are insufficiently strong, certainly inadequately open to us, if they exist, even for our present purposes. By way of an example, let me mention the matter of ring system description, which is crucial for a fully adequate treatment of generically described ring systems in patents. What areas of graph theory deal adequately with the variety of ring systems known to chemists, or indeed, capable of stable existence? While the DENDRAL project does include ring system generation – just the aspect which Dennis Smith spoke about here on the occasion of the last conference – it is not clear to me that the tools of graph theory are sufficiently well developed to enable us to talk in a sufficiently abstract way about characteristics of ring systems.

What I am saying is that the formalisms which we incorporate in the methods we apply today in chemical information systems are at an early stage of development, that we need to ensure that the tools of both classical and modern mathematics are more readily available to us, and that our younger researchers need to be better versed in these matters. This, I am sure, is an issue on which Ivar Ugi would support me, in that, in applying the earlier studies of Dugundji, he and his group have sought to investigate reaction prediction at a difference level from those of the library-based approaches.

We, for instance, have drawn on aspects of graph, grammar, set, and automata theory. Others, in the area of protein chemistry, are applying the predicate calculus, as exemplified by PROLOG, in unravelling the relations between secondary and tertiary protein structures, and, of course, the predicate calculus is also implicit in much work on expert systems. I am absolutely sure that we need to be more alert to the possibilities of applying the abstract tools available to us through symbolic representations in mathematics, to open up new opportunities, and develop new tools which can heighten the capabilities of the systems we devise. Many of the simpler problems have now been examined and successfully dealt with. We move on to more difficult problem areas, where we need input from other disciplines, for instance, from clever graph theoreticians whose work we may only half understand.

I am indebted to Jacques-Emile Dubois for a quotation from Lavoisier which in certain respects sums up what I am trying to say in relation to applying linguistic models to some aspects of handling representations of chemical structures. Lavoisier observed once that to change a science, one needs to change the language of the science. This is as true today as it was exactly two centuries ago.

I do wish to draw your attention to the very major challenges and opportunities which derive from the requirements for advanced computational support for molecular biology and its supporting disciplines, and for its applications in protein engineering. Common ground between interests in low and high molecular weight chemical species has long been provided by X-ray crystallography. Structure elucidation and modelling in the nucleic acid and protein engineering areas have grown apace in the demands they make on computational methods. Progress in those and related areas is unthinkable today without such support through the maintenance of databases of great complexity, through quite different, often ill–defined and vastly more demanding search requirements, and through the need to reason about sequence and structure and activities, and to deduce optimal experiment designs. There are many supporting activities, including the design of intelligent laboratory instruments, superfast detectors for faster, more adequate data collection and reduction, and telecommunication networks to enable investigators to collaborate more effectively, which also demand attention. Sir Hans Kornberg has rightly drawn attention recently to the need for all concerned with low molecular weight chemistry to recognise the challenges which exist in these emergent areas with which we share much common ground.

We have gained quite major insights into fundamentals of structure representation and manipulation, and good experience in building systems to meet the needs of industry and academia in relation to the needs of low molecular weight species. There are many ways in which they can be extended to meet these new challenges too. My remarks about the adequacy of the formalisms, models and tools which we use or need relate, *a fortiori*, to this burgeoning field, and I commend it to you for your attention.

Many of you are practical people, getting on with important work with the best tools available. I may appear to you to have placed too much emphasis on seemingly arcane and obscure theory. Let me end by quoting Lord Rayleigh to you: 'Nothing is more practical than good theory'.

FUTURE DIRECTIONS IN INTEGRATED INFORMATION SYSTEMS: IS THERE A STRATEGIC ADVANTAGE?

Myra Williams and Gary Franklin
Merck, Sharp and Dohme Research Laboratories, Building 80K, P O Box 2000, Rahway, NJ 07065, USA

ABSTRACT

Major advances in hardware, software, and connectivity of devices during the past decade are making new approaches to information management feasible. Although current software and hardware can meet specific information needs, generalised integrated solutions are not yet available. A number of companies have had the capability of merging chemical and biological data for years. In addition, software is available that permits text and graphics to be merged in a single document. In fact, most specific information needs can be met today by building specialised bridges between existing databases. However, problems of redundancy, incompatible systems, paucity of standards, numerous user interfaces and closed architectures argue that better solutions must be found. An immediate option is to use a windowing approach that facilitates the movement of information from one system to another. While this provides excellent functionality to the computer expert, it does not meet the broad based needs of most scientific and management staff. The increasingly competitive environment of the pharmaceutical industry mandates that information be readily available for effective decision making at all levels within the corporation. In response to this need, future information systems will utilise a common interface and hence one set of commands for accessing numerous databases. Software will adopt an open architecture so that only one graphics editor, one text editor, etc., will be required. The system will be capable of handling numerous different data types including text, chemical structures, spectra, graphs, restriction maps, among others. Moreover, even when integrated in a report, the data types will maintain their integrity; hence, when the report is distributed electronically, the underlying data can stay associated with it. And, finally, future information systems will exist in a multivendor, distributed environment. The driving force for these changes is the need to improve the productivity of research. Moreover, the emphasis reflects the fact that information is now a crucial component of strategic management.

INTRODUCTION

The world pharmaceutical market is undergoing a transition that will make it increasingly difficult for many companies to compete effectively. For example, the issue of cost containment is assuming greater urgency as governments throughout the world strive to slow the growth of health care costs. In response to this concern, some governments have encouraged the use of generics, others have limited price increases or even mandated price decreases for drugs, while still others have restricted

W. A. Warr (Ed.)
Chemical Structures

reimbursements for drug purchases to a limited list of agents. At the same time, regulatory requirements for drug approval are continuing to intensify, increasing both the time and cost of drug development. The relative profitability of the pharmaceutical industry is also attracting new entrants such as DuPont, Procter and Gamble, and Kodak, resulting in greater competition in an already highly fragmented market. When these events are combined with growing safety and environmental concerns, decreased exclusivity, a trend toward self-medication, and a shift from primary care physicians to health maintenance organisations, a turbulent future for the industry can be foreseen.

These conditions are of concern; however, they are offset at least in part by new opportunities in therapy and an expanding market-place due to the aging population. The most important factor is that existing drug therapy is not optimal. With the possible exception of vaccines and antibiotics, major improvements are needed in almost every area of therapy. Moreover, the expansion of basic knowledge in fields such as immunology, molecular biology, receptor technology, and disease mechanisms is opening new and more rational approaches to treatment. These same advances lead the way to earlier introduction of second generation compounds since the basis of the first discovery is now frequently understood – an advantage to the company that is poised to move swiftly, but a disadvantage to the original innovator. Further changes can also be anticipated from significant improvements in drug delivery systems and diagnostics that will gradually alter present concepts of medical practice.

It is clear that the most successful companies will be those that can rapidly translate research advances into marketed products with significant advantages over competitive agents, and can communicate these advantages persuasively. A key ingredient for success in this objective is improved use of information and computer resources for enhancing the overall productivity and effectiveness of R&D.

INFORMATION REQUIREMENTS

The expressions 'strategic information systems' and the 'use of computer resources to obtain a strategic advantage' have been among the most frequently discussed topics in the computer and information literature during the past year. While the expressions have perhaps been over-utilised, it is important to understand how information is used to develop research strategy, and how information can contribute to improving the effectiveness of R&D. Such knowledge provides the foundation for future information systems; without it, systems will continue to be designed to meet isolated needs and major opportunities will be missed.

While numerous approaches can be taken to describe categories of information used in pharmaceutical R&D, the one we have chosen here reflects the types of information required for the development of research strategy. These are displayed in Figure 1 and are described below.

Project Information

As used here, project information spans all aspects of data collection, analysis, and storage throughout R&D. The information ranges from screening data at the basic research level to adverse experience reports in Regulatory Affairs. The requirements at this level include the integration of data of various types (i.e., chemical, biological, text,

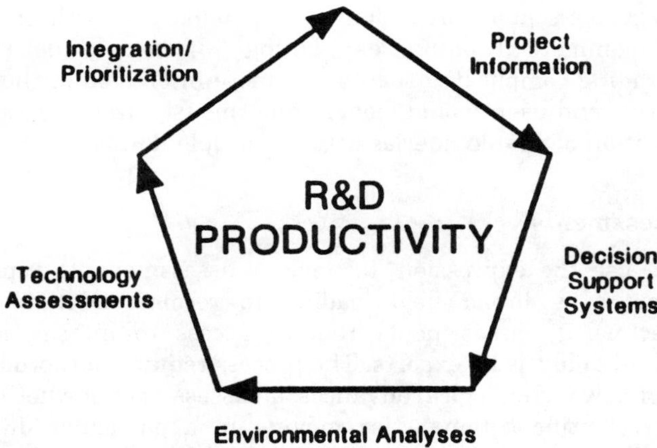

Figure 1. Information requirements.

graphics, etc.), the sharing of information between groups, appropriate security control, documentation to prove invention, and support for regulations such as Good Laboratory Practices or Good Clinical Practices. Project data consist of individual files that might include raw data as well as analysed data, shared files that combine input from several sources, or central files spanning projects in R&D. Project information incorporates the most extensive use of computerisation.

Decision Support Systems

Decision support systems depend upon chemical and biological information, critical path dates, human and financial resource allocation, historical data, probabilities of success, and sales data. These systems tend to use refined data, but it is still important for these systems to be able to integrate information across various data types and databases. Since decision support systems become most valuable for managers and executives, individuals who may not be very experienced with computers, such systems must have a uniform user interface. Key to the success of such systems is the ability to experiment with different scenarios to model likely outcomes both in timing and financial impact. Project management and scheduling tools, and financial modelling to determine the intrinsic value of projects, are frequently used applications in this area. The addition of knowledge-based software would be a real advantage in improving the utility of decision support systems.

Environmental Analyses

It is impossible to plan for the future without understanding the external trends that can have an impact on the success of the program. Information on legislative activities, regulatory actions, industry trends, consumer trends, demographic shifts, and global economics is vital to the development of meaningful strategy. Most of the sources for such data are outside the corporation, with published literature, activities of liaison groups, and reports from experts being among the most critical. The ability to search the scientific and business literature online is very important with end-user searching of in-house and commercial databases gaining support. While the information professional performing complicated searches is more interested in thoroughness than in ease of use, the end-user would benefit enormously from systems that would facilitate the execution of simple queries across multiple databases.

Technology Assessments

In this paper, we use the expression 'technology assessment' to represent a multi-disciplinary review of a broad field leading to recommendations on significant opportunities. Technology assessments require access to information from many different sources and analysis by experts. The process requires a thorough understanding of the impact of new technological advances, an assessment of whether the advances provide a strong scientific rationale for moving in a particular direction, and an evaluation of medical and marketing needs. The final recommendations have to be made within the framework of the competitive environment. Scientific literature, searching of online databases, consultants' reports, and trip reports are useful in broadening the perspective of the participants. The objective of the review is to examine a field critically while minimising restraints due to internal biases. Proprietary as well as published information can be used in a very provocative way in this process.

Integration/Prioritisation

At each level of information (from project information to technology assessments), computers are utilised in the acquisition, processing, and storage of data but the ultimate relevance of information systems depends upon the integration of information from numerous different sources and a high degree of judgment in the interpretation of data and prioritising of opportunities.

CURRENT INFORMATION SYSTEMS

While not all information systems make extensive use of computers, computerised information systems are no longer adjuncts to the way we conduct science and business; they are integral components of our entire organisation. All scientists and administrators as well as clerical employees should be utilising computers in their routine operations. This fact places considerable pressure on the way computer systems are designed today.

The rapid proliferation of computer-based tools, while providing numerous benefits in many areas, and while promising exciting developments for the future, presents unique new challenges and problems to the organisation. The exponential increase of hardware, software and networking alternatives has resulted in a plethora of heterogeneous computer-based solutions to users' needs. In most large organisations, many incompatible, poorly co-ordinated, and redundant systems exist. This state of affairs is the natural outgrowth of a technology still in its infancy with attendant rapid growth, change, and uncertainty. Often, inconsistent solutions were and are still necessary in order to fulfil a diverse set of requirements within reasonable time and cost constraints. However, continued proliferation of disparate systems will have long-lasting financial implications, will threaten the effectiveness of operations, and will diminish the capacity to accommodate new developments.

A number of systems are currently in use to meet the information needs of scientists. While bridges are sometimes built between systems, the requirements are generally met by specially designed, closed systems or by simple file systems. For example, several programs are used throughout the industry to store and manipulate chemical structures and related information. Separate software packages are used for managing biological data with some bridges being implemented to associated chemical structures. Additional programs are used for proteins and nucleic acid sequence data and other supportive information for molecular biology research. Various word processors are employed for document handling and still additional software for chemical reaction storage and retrieval. Although these systems provide valuable functionality, they present major obstacles to maximising the use of information in improving the productivity of the laboratory.

Vendors and software specialists strive to overcome the limitations of their products by integrating additional functionality in their closed systems. These systems tend to force a separation between the primary data they were designed to handle and other related databases. Such separations are often cumbersome and artificial. Realising that users require combinations of functionality, some vendors have provided external hooks which are often insufficient. Such closed systems have significant limitations with some of the following far-reaching consequences:

1. *They make information sharing difficult.* As a result, islands of uncontrolled information threaten one of our greatest assets – corporate databases. Moreover, it is difficult and costly to extract and merge data from multiple closed systems, a frequently requested need.
2. *They promote inconsistent user interfaces.* Users are forced to learn numerous systems in order to access different applications. Such a requirement is intimidating to the novice user, inhibiting to the typical user, and a waste of time for the expert. Multiple systems dramatically increase the effort required to realise the benefits of computer-based tools. Hence, they distract from scientific work and adversely affect productivity.
3. *They produce redundancy and incompatibility.* In addition to redundant and incompatible user interfaces, closed systems foster data redundancy, hardware and software incompatibility, duplication of programs and duplication of security schemes. In some cases, the user or programmer finds it easier to re-enter data that already reside in an existing database than to build a bridge to extract the information. Clearly, such duplication makes maintenance of consistency between data in various databases a major challenge. Each level of redundancy compounds the problem. Not only are direct costs increased unnecessarily, but the integrity, accuracy, and security of information are threatened.
4. *They stagnate development.* Integrating new applications can be time-consuming and costly. Closed commercial packages and vendor dependence hinder the production of new functionality, handicap innovation, and hence, imperil long-term competitive ability.
5. *They complicate training.* Even if the users were willing to invest a substantial percentage of their time mastering new computer application programs, training alone becomes prohibitive. In a small organisation, there is not a sufficient critical mass to justify a training and support group that can cover all of the various applications programs being used. And, in a large organisation, one is faced with teaching the same course hundreds of times as well as providing daily response to

questions that arise because of differences between systems. Even if such support were feasible, normal turnover of employees combined with staff growth can result in almost 25% of the staff being new after only three years. Such numbers would result if the normal turnover rate from departures and retirements is 5% and staff growth also 5%. Thus, for a large, growing organisation, the requirement of training people to use numerous applications packages becomes almost an impossible challenge.

Although today's systems meet essential application requirements, serious deficiencies exist. Integration of systems is required in order to facilitate evolution and to realise fully the benefits of investments. However, starting again by designing new integrated systems would necessitate reproducing all of the functionality that now exists and would be extremely expensive and unproductive. Fortunately, industry standards are emerging in many areas including communications, graphics, database management and operating systems. In addition, new products are surfacing which specifically address areas of systems integration. Hence, changing technology now enables us to eliminate some of the deficiencies of current systems.

FUTURE DIRECTIONS

Information itself does not have an intrinsic value; it is only the use of information that contributes value to the corporation. As described in the earlier sections of this paper, computer-based information systems must change substantially if they are to be used effectively by clerks, scientists, and administrators at all levels. In response to this need, future systems will be far more friendly, augmenting the human-computer interface in capability, response time, context-sensitive help, consistency and other aids to improve use. Information storage and retrieval will evolve from today's database to tomorrow's knowledge-based technology as front-ends, and progressively the systems themselves, incorporate greater intelligence. Systems will adopt an open architecture to facilitate extension, customisation and integration. The use of the object-oriented paradigm will support the numerous datatypes necessary for handling scientific information. In short, modular components and reusable tools will bring optimal flexibility and function to both the user and the applications programmer while simultaneously accommodating the evolution of systems. The specific properties of the systems are discussed below from the users' perspective, the systems design considerations, and the impact on integration.

The user will have the enormous power of information immediately available with the following functional attributes.

Minimal Training

Time invested in learning new software packages or understanding major upgrades to old systems is not necessarily well spent. Systems in the future will be so easy to use that most of the functionality can be obtained with minimal assistance. The achievement of this objective requires that systems be designed with a uniform user interface. This will be accomplished with flexible tools for employing menus, command dialogues, windowing, graphics and online help across applications.

Laboratory Automation

While automated data aquisition from laboratory instruments, control of experiments through robotics, automated data reduction and summary analysis have increased productivity in many laboratories, much work remains to be done. Automation from data acquisition through data presentation will be expanded. The manual keying of data will be minimised regardless of whether the source of the data is a laboratory instrument or an online database. Further, experimental design tools will be developed.

Analytical Tools

Tools for the manipulation and analysis of results will be readily available. Some of the standard tools will be a database management system, software for statistical analysis, and graphics capabilities.

Office Management

Future information systems will provide easy access to electronic mail capabilities, calendaring/scheduling, electronic forms, budget and capital information, and human resources data. Software for salary and expense planning will also be available. These facilities will cover inter- as well as intra-departmental needs.

Complex Documents

Current systems require many steps to include graphical objects and other non-textual entities into documents. The new systems will support easy incorporation of multiple data types (e.g., chemical structures, reaction schemes, graphs, charts, data tables, etc.) while retaining the characteristics of the objects. One will be able to manipulate the object in its native mode and append ancillary data to the document if desired.

Information Sharing

Information sharing will be facilitated with appropriate security to cover a wide range of functions. These include shared database access, multiple database integration, electronic filing systems, electronic mail, and fully editable document transfer between different word processing packages.

The systems design will provide optimal flexibility to the applications programmer, place significant computer power under the control of the end-user, and improve communications as follows.

Software. One of the dominant requirements in the future generations of software is that the software adopt an open architecture. Traditionally, the word 'architecture' has been associated with hardware, but its use in connection with software reflects the fact that a single vendor can neither anticipate nor supply all the needs of users. The central concept of open architecture entails well-documented, modular interfaces to system components. If vendors would adopt an open architecture, the benefits of purchasing commercial products could be realised and the user would retain the advantage of being able to customise and extend the systems according to individual needs.

A second requirement is that the software be able to accommodate multiple data types such as chemical structures, DNA sequences, restriction maps, NMR spectra, etc. Object-oriented programming combined with a database management system (DBMS) would provide an elegant solution to this objective since object-oriented languages permit complex objects to be handled in a uniform manner. Other properties of object-oriented languages would also facilitate program development and maintenance. The evolution of current database management facilities from hierarchical and network models to relational models, with the SQL query language being adopted as a standard, is another major asset in future systems designs. When combined with the power of knowledge-based systems, these advances in software will enable the production of far more meaningful information systems.

A final point on software is the distribution of information. Most current DBMS's are centralised with both the data and related programs residing on one machine at a single geographical location. The trend towards functional distribution has created a demand for distributed database management support in which data are managed in a networked environment. With a distributed DBMS, users and application programmers can access remote databases just as if they were centralised. In a sense, the entire network becomes a potential resource. With transparent operations, the user only needs to know the name of a database in order to access it. When the software support for this technology is mature, distribution of databases will increase rapidly.

Hardware. As in other areas of computer technology, there is going to be a re-structuring of the concepts and distinctions between system architectural components. As the overlap of communication, I/O, software, and hardware elements increases, the traditional distinction between centralised computer complexes and distributed elements will fade. Functional distribution is going to be enabled by changing technology on all fronts. Although there is great progress in new technology such as parallel architectures and massive memory machines, these will not become mainstream systems in the near future; existing hardware components and vendor architectures will remain important. The most important developments will occur in the distribution of processing and in the interconnection of components.

Technology will place substantial power on users' desktops. Personal computers and workstations will replace today's terminals. The substantial processing power and high bandwidth displays will enable user interface improvements and new functions such as text and graphics integration. Optical disk technology will offer substantial storage capabilities at all levels of computing. Further, there will be dramatic improvements in the local interconnection of machines.

Telecommunications. As the networking of distributed, frequently heterogeneous elements becomes increasingly important, the establishment of and adherence to industry standards becomes the key factor. Organisations such as ISO (International Standards Organisation), IEEE (Institute for Electrical and Electronics Engineers) and CCITT (International Telegraph and Telephone Consultative Committee) will be major players in establishing networking models such as the OSI (Open Systems Interconnect) seven layer model and the associated protocols for the various layers.

Most of the protocols are expected to be completed by the early 1990s. Vendor compliance with these protocols will be critical in this networking strategy. However, this compliance will be dictated by financial and political factors and not simply by the availability of these protocols. Local area networking standards such as Ethernet and

Token Ring will be important. During this period of 'non-compliance', bridges and gateways will provide the necessary functionality between heterogeneous systems.

As workstation technology and functional distribution take hold, communication demands will increase. The bandwidth of today's asynchronous communications will be insufficient for meeting future application requirements. Providing the medium for higher bandwidth communications will be necessary. Typically, network support is standard with today's workstations. Many organisations will need to upgrade or replace existing equipment in order to network. Networking heightens the importance of planning and standardisation. Indiscriminate heterogeneity will prove costly. Moreover, software support for network management, security and control must be considered.

While the above paragraphs describe many of the properties of future information systems, there is one underlying theme that deserves specific discussion – that of integration.

Integration

Integration is the most frequently utilised term in the description of future information systems reflecting the fact that knowledge results from the assimilation of numerous different inputs. Integration pertains to all of the above-described functional areas as well as to general system features that are common to all applications. However, the term 'integration' has diverse meanings with varying degrees being possible. Levels of integration can be considered conceptually in three dimensions (Figure 2).

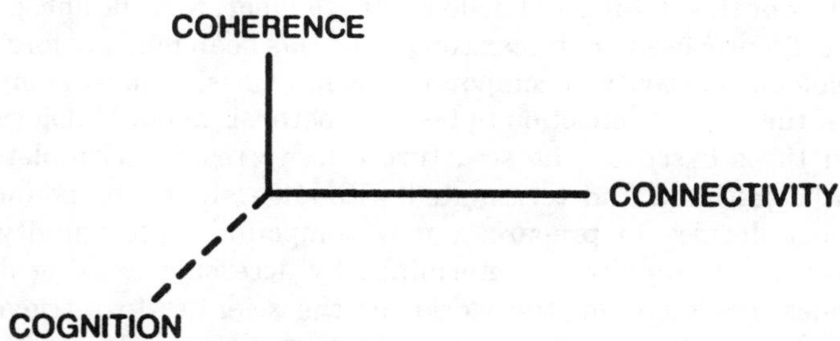

Figure 2. Integration.

Connectivity represents the lowest level of integration with connectivity implying access. Software that supports multiple functions or that permits access to data residing in different systems is sometimes referred to as an 'integrated software package'. However, in many cases, only the lowest level of connectivity is supported.

When the users talk about integration, what they are frequently requesting is not just connectivity, but coherence. Coherence provides uniformity across services in a connected environment. Coherence requires consistent user interfaces, standard networking protocols, and agreed upon data representations – hence, greater standardisation. While coherence is a worthwhile objective in future systems development, there is a large installed base of effective software that needs to be better utilised.

The users wish to gain access to the full capabilities of current applications software without having to learn multiple command languages. They would benefit from an expert system overlay that guides them through the appropriate use of the tools. These goals lead us to the third dimension of integration, the cognitive dimension. The tools of artificial intelligence will enable us to move from the current focus on connectivity and coherence to systems that will provide more value to individuals in all levels of the organisation.

BENEFITS/VALUE

Future information systems will require a major investment in software, hardware, telecommunications, and training. A large laboratory will be facing an investment of millions of dollars, not just thousands. For those of us who spend weeks getting approval to purchase a terminal, months to acquire a PC, and a year or more to obtain a new minicomputer, the challenge we face in persuasion is apparent. The justification has to be built on the importance of the new systems to improving the overall effectiveness of research – not a trivial goal. It is easy to demonstrate improvements in productivity such as the increase in assay output of a laboratory due to use of automation, or the filing of an NDA in a shorter period of time. However, the only parameter that really counts is the flow of significant new products from the laboratory to the market place, and years are required before an impact can be observed in this area.

There are numerous ways that future information systems will increase the productivity of the laboratory. The first is by simply reducing unnecessary duplication of work. Some companies have claimed substantial expense savings simply from setting up an online inventory of chemicals available in-house and providing substructure search capabilities of this database. Facile access to laboratory databases would permit a scientist to see if a proposed or similar compound has been made before, check what is known about biological activity of compounds of that class, request samples of existing compounds from the sample collection to test a hypothesis about biological activity, and model new structures based on the structure-activity results. Complete programmes can sometimes be initiated and terminated without a single compound being made. Once the chemist decides to prepare a new compound, patentability and possible synthetic pathways can rapidly be determined by accessing existing databases. The computer provides a window on the world for the scientist to retrieve and analyse information rapidly, conduct simulations, and report results. Moreover, computerised information systems provide a framework for rapid sharing of project results regardless of the physical location of the scientists.

On the development side, major advances could be made through the remote entry of clinical data with validation criteria applied when the data are initially keyed, the automated encoding of these data, and expanded use of graphics and relational technology for analysis. The complex document capabilities described earlier in this paper could also revolutionise the current approach to the generation of NDA's. For example, an electronic version of the NDA could be submitted along with the hard copy. While the hard copy would make heavy use of graphics and summary data, the underlying data could be retrieved through the electronic version. This approach differs significantly from the current electronic NDA pilot projects in which access to the raw data is provided through a menu-driven interface. The two approaches would be complementary.

Direct savings would also result from the establishment of an electronic filing system that would eliminate the countless paper copies stored today. Moreover, the associated automated search and retrieval with remote access capabilities would be useful for applications ranging from project reports and NDA's to the storage of NMR data. While such systems exist today, optical storage technology will greatly expand the utility of such systems once standards are established.

The applications described above are not novel. They are essentially extensions of systems that have already been implemented. Future information systems will simply facilitate the full utilisation of such systems by incorporating some of the attributes described earlier in this paper.

While productivity improvements are important (i.e., generating higher quality information and throughput in reduced time), the most important goal for information systems is qualitative. The real value of information is in improving decision making and hence enhancing the overall effectiveness of R&D. The success of the laboratory depends first of all on having the best people in place. The second ingredient for success is the selection of the right projects for support. While both recruitment and project selection involve considerable judgment, information is inherent throughout the process.

The long-term success of any plan for computer and information resources will be based on its impact in stimulating innovation while improving the quality of decision making. The combination of these ingredients, people, projects, and decision support, will designate those companies that will be most successful long-term in this rapidly changing environment.

CURRENT APPROACHES AND NEW DIRECTIONS IN THE MANAGEMENT OF IN-HOUSE CHEMICAL STRUCTURE DATABASES

Tom R Hagadone
The Upjohn Company, Kalamazoo, Michigan 49001, U.S.A.

ABSTRACT

The period of the 1970s saw a great deal of emphasis placed on the design and implementation of database systems for the management of collections of chemical structures. Techniques were developed for creating canonical connection table representations, efficient substructure search systems were devised, and effective computer-graphics-based structure drawing software was written. Since that time, most in-house developed systems have been replaced with commercially available standardised software for managing chemical structure databases. The emphasis today is on improving integration of existing in-house databases and the addition of new databases to provide better access to a wide variety of compound-related information. This article will provide a survey of current practice and emerging trends in the management of in-house collections of structures, reactions, compound property data, and documents. Topics to be covered include the types of databases currently being maintained and the level of integration achieved, software and hardware environments, user training and support issues, and new approaches to improving the level of integration and accessibility of in-house databases.

INTRODUCTION

Information systems for managing proprietary research information are an important component of all research programmes. Pharmaceutical companies devote significant resources to the development and maintenance of internal information systems for use in all phases of the research process. Information management in the lead finding and lead optimisation phases of research is a topic of particular importance that has received considerable attention during the last decade.[1–3].

Recent interest has centred on the concept of integrated systems that combine information from a wide variety of disciplines into a single system where it can be accessed in a uniform manner. Figure 1 is a diagram of such an ideally integrated system in which each discipline contributes its own data to the central pool where it can be shared with people in other disciplines. Although this simple diagram may be an attractive way of looking at the information management problem, it is not a very good description of current reality. A more accurate description of existing information systems is shown in Figure 2. Instead of all information being collected in a monolithic central storage area, information is distributed among a number of varied types of databases and accessed through a number of different software systems.

W. A. Warr (Ed.)
Chemical Structures

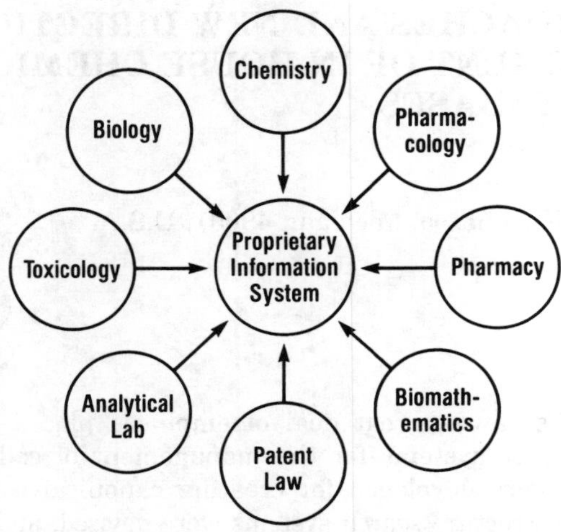

Figure 1. An idealised research information system.

It is the goal of the information system designer to integrate this fragmented scene to provide the greatest possible utility for information retrieval and correlation for the end-user scientist. This challenge has been met in a variety of ways within the pharmaceutical industry. Traditionally, chemistry-related research information has been organised in the form of a chemical information system as shown in Figure 3. A central store of company registry numbers, structures, and nomenclature is

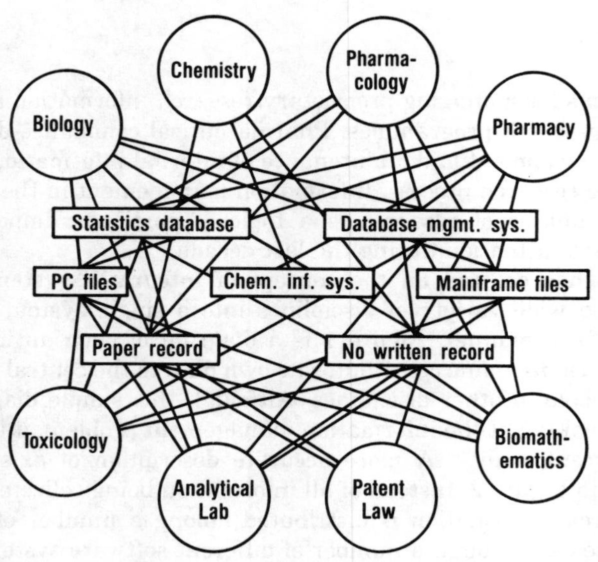

Figure 2. A more realistic view of research information.

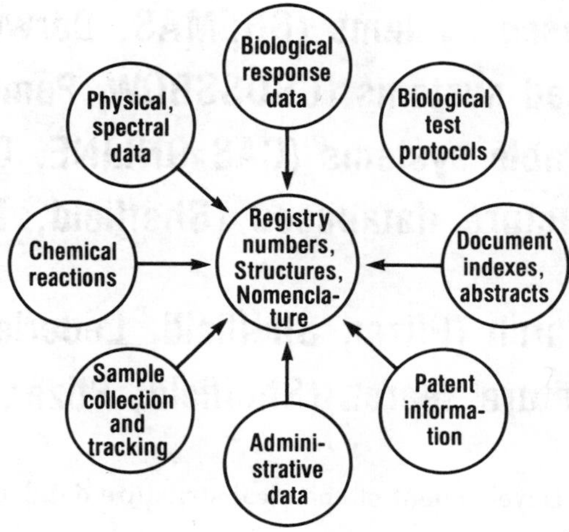

Figure 3. Components of a chemical information system.

surrounded by collections of information that can be linked by registry number. In the 1970s chemical information systems were based largely on software written in-house and responsibility for improvements lay directly with the in-house software development group. The approach has changed in the 1980s however, and internal databases are now managed by a variety of commercial software packages.[4] Although each package performs well in its area of speciality, integration of the individual components into a homogenous system with a single user interface is a critical concern. The important question for chemical information management has become: how can the databases and functionality of the various commercial packages in use be combined to create better integrated information systems?

This article will explore the answers to this question that have been arrived at to date and will discuss trends that are emerging in the direction of greater integration and functionality. We will begin with a brief discussion of the types of information in each component of a chemical information system and the associated methods used for data collection, analysis, storage and retrieval. Hardware and software environments will be examined along with the increasingly important role of the microcomputer. Then, a number of current approaches to providing better levels of integration will be described and compared. Usage levels and training and security issues will be considered, and finally a brief look will be taken at emerging trends.

STRUCTURES

The development of chemical information systems, as shown in Figure 4, has centred around the development of chemical structure database management systems (DBMS). This in turn has closely followed advances in computer hardware technology. Thus in the 1950s and 1960s retrieval systems were built based on the chemical fragments present in a molecule.[5] Although mostly obsolete now, fragment-based systems are still in use for databases of generic patent structures. In the 1960s and 1970s fragment-

1. Fragment based systems (GREMAS, Derwent)
2. Notation based systems (CROSSBOW, Pomona MedChem)
3. Connection table systems (CAS ONLINE, DARC, MACCS)
4. Generic structure databases (Sheffield, Derwent-DARC, CAS)
5. Similarity search (Pfizer, Sheffield, Lederle)
6. 3-D substructure search (Sheffield, Pfizer, MDL)

Figure 4. Development of chemical structure database software.

based systems were replaced with notation-based systems, most notably those employing Wiswesser Line Notation (WLN).[6] Notation is still used as the storage form in a few systems such as the Pomona College MedChem software which is based on the SMILES notation.

The availability of relatively inexpensive graphics terminals paved the way for the connection table-based systems of the 1970s and 1980s that provide the current standard.[7–9] Work continues to be done to enhance structure searching software with improvements in the handling of tautomerism, stereochemistry, and graphical query languages. Beyond this, the University of Sheffield,[10] Chemical Abstracts Service (CAS), and Télésystèmes and Derwent[11] are actively developing techniques and software for searching databases of generic chemical structures associated with patent claims. CAS has recently received a patent for a method of generic structure searching based on a hierarchical group classification scheme.[12] The University of Sheffield and Pfizer[13] and Lederle[14] have pioneered the concepts of structural similarity searching as a complement to substructure searching. Sheffield, Pfizer, and others have explored pharmacophoric pattern matching techniques in files of three-dimensional structures.[15]

In the 1970s most companies developed their own structure management software or adapted outside software for in-house use. With the commercial availability of effective graphics and connection table-based software in the early 1980s, most companies replaced their in-house systems with commercial systems. A necessary step in this process was the conversion of in-house structure databases, which range in size from a few thousand to over 500,000 structures, to the required standard connection table formats. Conversion was easiest for those companies already using connection table based software. Software conversion of WLN files was more difficult, with successful automatic conversion occurring for about 80–90 per cent of the collection depending on the quality of WLN. Companies with fragment-based or paper files had to go through the painful process of manually entering each structure. These conversions are now largely complete except for a few companies that have chosen to continue to maintain their in-house developed systems because of advantages they perceive in functionality and/or adaptability.[16]

Although there has been much development and consolidation of chemical structure DBMS's, there are still a number of systems with which a chemist may need to deal, some of which are shown in Figure 5. Interestingly, basic substructure search software is still being designed and implemented. The Cambridge Crystallographic Data Centre is working on improved graphics-based substructure search software for the Cambridge Structural Database,[17] the Pomona College Medicinal Chemistry Project group is working on substructure search software for their medicinal chemistry structure databases,[18] the Beilstein Institute is designing substructure search software and hardware for online access to their database,[19] and a number of chemical information systems with substructure search capabilities have been introduced for use on microcomputers.[20] The continuing emergence of new substructure search systems results from the need of database providers to control distribution of their databases as well as to provide specialised retrieval and analysis software specific to their respective databases. Database specific software may be necessary; however, the proliferation of substructure search systems creates additional complexity for users since each system presents a different interface to the user which must be learned. As a partial solution to this problem a number of companies have converted structure databases to the standard form used in-house. For example, the Cambridge Structural Database has been converted by ICI, Lederle, and Upjohn, and the Pomona College databases have been converted by Smith Kline & French Laboratories and others. Of great service to chemists would be the ability to search both in-house and external databases through the use of a single graphical query language. Unfortunately, there has been very little movement in this direction. Real progress will require an organised effort by the pharmaceutical and chemical industries to set appropriate standards and to lobby the key chemical software vendors for the necessary co-operation.

REACTION DATABASES

An area closely related to structure database software is that of chemical reaction indexing. Although reaction indexing systems have a less full history than structure database software, they are now being very actively developed and used.[21] Nearly all pharmaceutical companies have now acquired commercial in-house software for reaction searching. The emphasis in reaction searching is on databases of published reactions in contrast to the proprietary focus of structure management software. Exceptions to this are Kodak, McNeil Pharmaceuticals, and a small group of other

In-house proprietary databases (MACCS, DARC)
Online databases (CAS ONLINE, DARC, Derwent)
Special purpose databases (Pomona MedChem, Cambridge Structural Database)
Microcomputer systems (ChemBase, ChemSmart, PSIDOM)

Figure 5. Substructure search systems currently in use.

companies that have developed databases of proprietary process-related reactions. The concept of proprietary databases of research reactions has not gained acceptance at this time. The large effort necessary to build and maintain research reaction databases is difficult to justify and chemists have been reluctant to participate in the database building and maintenance processes.

One criticism of reaction indexes has been the lack of completeness in the databases of reactions. This criticism is being answered by the introduction of databases such as the ISI Current Chemical Reactions database and the CAS reaction database now under development. As was previously mentioned for structure databases, a single graphical query interface for searching both in-house and external reaction databases would be of great value to the end-user chemist.

BIOLOGICAL RESPONSE DATA

The notion of a chemical structure database is a well-defined concept and a single database can be used to store a company's entire collection of structures representing a wide range of therapeutic areas. Second in importance to a chemical structure database is a collection of biological response data. Biological response data however, comes in many different forms by nature and must be dealt with on a case by case basis. This diversity has resulted in a variety of approaches currently being used to collect, store, and retrieve biological data. Data may be collected directly off an instrument, entered at a terminal by the biologist in the laboratory, or written on paper forms for keyboarding by a central data entry group. The trend is clearly toward online data capture and away from the paper forms approach. Of the companies interviewed, approximately 20 per cent of the biology laboratories are currently capturing data directly from instruments. Most other data are entered directly in the laboratory and a small and diminishing amount of biological data are stillentered on paper forms for later keyboarding.

After being collected, multiple dose and multiple observation data are typically subjected to statistical analysis before being stored. The final resting place of the evaluated data can be a statistics database, an operating system file, a chemical structure database management system, or a general database management system (DBMS). For large collections of data, storage in a general DBMS has many advantages to offer including effective data entry procedures, efficient searching and retrieval, and sophisticated report generation. A disadvantage of general DBMS storage is that users may need to learn two systems, the chemical structure DBMS and the general DBMS. For production of routine reports containing structures, biology, or both this disadvantage is largely circumvented by providing a set of pre-programmed procedures that can be invoked by simply selecting the desired report from a menu of available reports. For creation of non-routine reports however, the user is generally forced to deal with the command languages of the underlying structural and general DBMS's.

Biological data is stored in two basic ways. A single file may be used to contain the results of all tests or a separate file may be used for each test. The single file approach has the advantage of collecting all results for a compound in a single place, but the disadvantage of difficult searching and reporting of test specific data such as ED_{50} and IC_{50}. The separate file per test approach has the advantage of providing for relatively easy searching and reporting of detailed results but makes it difficult to find all the data for a given compound. This problem is often solved by providing a summary file containing only test names, dates, and a qualitative activity indication for all test results in addition to the individual table per test for detailed data.

A useful complement to the test results themselves is a file of documents describing test protocols and interpretation guidelines. For most effective use, these documents should be available online from within the chemical information system.

OTHER DATA

Beyond structures, reactions, and biological data there is great variety in the level of integration achieved for the other types of chemistry-related information shown in Figure 6. To date, success in integrating spectral databases has been limited. A few companies include references and/or peak values of spectra in their chemical information system. Integration of full spectra, however, must await better solutions to the problems of providing integrated spectra searching capabilities and efficient management of large files of full spectral images. Many companies maintain sample inventory and tracking systems. Online access to sample inventories is particularly useful since scientists often need to know if a sample is available for testing for those compounds found in a substructure or other type of search.

Most companies maintain the Fine Chemicals Directory (FCD) for in-house searching. In a few cases the FCD has been linked by internal registry number or catalogue number to an in-house inventory of fine chemicals available at a central location or in individual laboratories. Such inventory systems have been responsible for more than doubling access to the FCD in some cases.

Other compound-related information such as internal document indexes and patent status information is normally kept in separate information systems and is not usually linked to the in-house chemical information system. For the few companies that have integrated these additional types of information, the convenience of having it readily available with other chemical information has been of benefit to users.

HARDWARE AND SOFTWARE ENVIRONMENTS

The hardware and software environments used to support chemical information systems have undergone a great deal of change in recent years. Two important changes

1. **Physical, spectral data**
2. **Sample inventory and tracking**
3. **Commercial compounds available in-house**
4. **Internal document indexes, abstracts**
5. **Patent information**
6. **Miscellaneous administrative data**

Figure 6. Other types of proprietary research data.

are the emergence of standard commercial software packages specialised for managing particular types of research information and the large scale introduction of microcomputers into the laboratory as chemists' workstations. Figure 7 shows the most popular hardware and software combinations in use today.

The DEC VAX series of computers has become dominant for use with chemical information and molecular modelling systems. IBM mainframes are used for these applications in a number of companies however, due mainly to historical corporate influences.

Each of the mainframe software systems shown in Figure 7 is a powerful tool for managing its specialised type of information. However, in general, each system is supplied by a different vendor and comes with its own user interface. This independence of vendors and interfaces makes it difficult to achieve a well-integrated chemical information system; however, the benefits of active competition between vendors are of positive value to the industry.

The survey of pharmaceutical research organisations found that the Apple Macintosh and IBM PC were being acquired in approximately equal numbers, although each company generally preferred one over the other. The Macintosh is selected for its structure drawing and graphics capabilities while the PC is chosen for laboratory automation applications and the wealth of available software. However, changes in the hardware and software options for both machines are bringing them closer together in functionality. The ChemDraw chemical structure drawing program has been largely responsible for the introduction of the Macintosh into the chemists' laboratory;

Mainframe hardware and software
— Mainframes and superminis (DEC, IBM)
— Structure database management (MACCS, DARC)
— Reaction databases (REACCS, ORAC, SYNLIB)
— General database management (ORACLE, INGRES, System 1032, INQUIRE)
— Statistical software (RS/1, SAS)
— Text management (BASIS, STAIRS, INQUIRE)

Microcomputer hardware and software
— Microcomputers (IBM PC, Apple Macintosh)
— Structure drawing (ChemDraw)
— Chemical wordprocessing (ChemText)
— Chemical databases (ChemBase, ChemSmart, PSIDOM)
— Terminal emulation (TGRAF, EMU-TEK, Versaterm Pro)

Figure 7. Hardware and software environments.

however, most of the recently developed chemical microcomputer software has been designed for the IBM PC.[22]

Chemical structure drawing and graphics terminal emulation for access to mainframe chemical software account for a large portion of microcomputer use. In addition, the traditional microcomputer applications of wordprocessing, spreadsheet, and database management receive frequent use. Often however, the microcomputer is viewed as a tool for integrating text, data, structures, and reactions from a variety of internal and external information sources. Microcomputers can be used to manage collections of personal chemical information; however, most large organisations support a policy of uploading information of common interest to the research mainframe to facilitate information sharing and maintenance of large databases.

INTEGRATION ISSUES

An ideal chemical information system would provide a single easy-to-use interface for accessing all chemistry-related research information. Current systems however, can only approximate this ideal with varying degrees of success. Lack of integration results primarily from the fact that current systems employ a number of major software packages to manage the different pieces of the chemical information puzzle. Each of packages shown in Figure 7 is usually supplied by a separate vendor. Each vendor has a different view of the software market and provides his own unique user interface. Given this diversity of software, how can integration be achieved? Three general approaches to the solution of this problem can be identified; the integrated data approach, the integrated user interface approach, and the user-customised software approach.

The philosophy of the integrated data approach is to store as much information as possible in the database format of a single software package such as a structure DBMS or a general DBMS. This approach has the advantage of providing a single interface for all information in the chosen system but the disadvantage of inappropriate retrieval and analysis facilities for data types for which the package has not been designed.

The integrated interface approach attempts to provide a single interface for accessing the data and retrieval capabilities of a number of software packages. This approach minimises the need for training and allows information to be combinedand compared in useful ways but has the disadvantage of limiting access to a subset of the capabilities of the underlying systems. The success of this technique is also very dependent on the quality of the 'hooks' provided by the underlying packages.

The user-customised software approach places the responsibility on the customer for the creation of an integrated system through the availability of software tools such as subroutine libraries, built-in procedural programming languages, and standard formats for file export and inport. The success of this approach is dependent on the software development resources available locally and the quality of the customisation tools provided with each package to be integrated.

Within these general approaches, five current strategies for chemical information system integration can be identified and are shown in Figures 8 to 12.

No Integration (Figure 8).

Ignoring the integration issue is the easiest way outfor a research information systems group but provides the least service to the user. Unfortunately, a number of companies

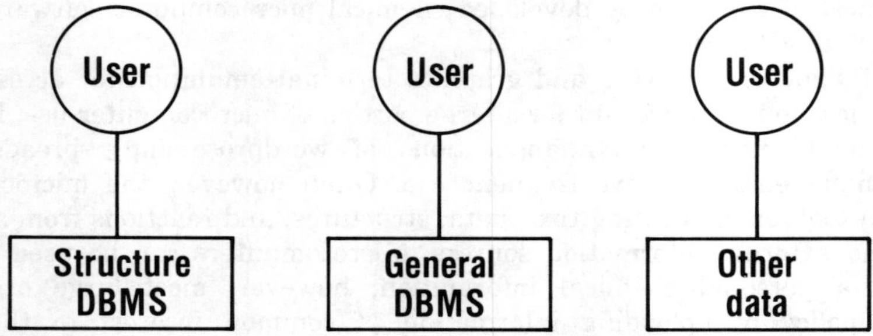

Figure 8. No integration.

are still following this approach in which several unconnected software packages are used to manage different subsets of the total collection of research information. This approach creates obvious training problems and difficulties in combining and correlating information contained in different components of the system.

All Data in the Structure DBMS (Figure 9).

The next step up in integration places as much information as possible in the chemical structure or reaction DBMS. For example, Molecular Design Limited's MACCS system has been used to store structures, reactions, biology, spectra, inventory, and document index information.[23] This approach provides a good degree of data integration but suffers from poor performance for large databases. Additionally, the retrieval capabilities of the structure DBMS are inadequate for certain types of information.

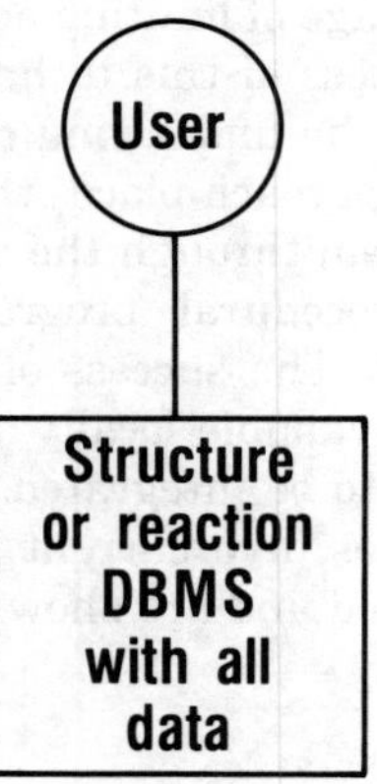

Figure 9. Integration by storage of all data in a structure DBMS.

Integration Through a Custom Supervisor Program (Figure 10).

For larger collections of biological data a popular approach is the use of a structure DBMS for storing structures and a general DBMS for storing biology and other information. An upper level supervisor program is created to provide the primary user interface that controls access to the underlying systems. This approach provides a good method for combining structures with biology and related information and partially insulates the user from the underlying systems for routine queries. For non-routine queries however, the user is forced to deal with the details of the individual DBMS languages or to seek out expert help.

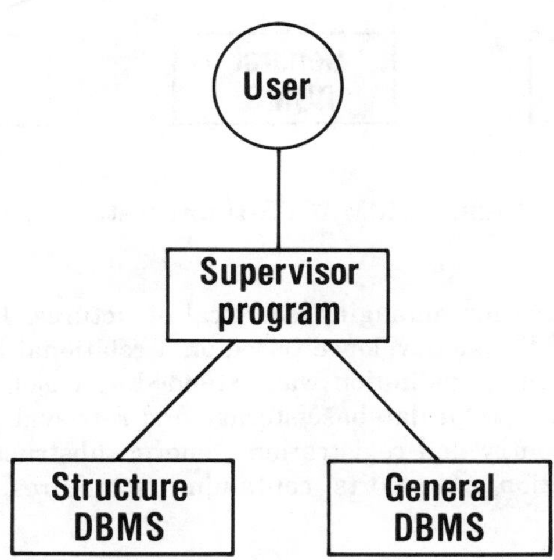

Figure 10. Integration through a custom supervisor program.

MACCS-II Integration (Figure 11).

A new approach to the integration problem has been introduced by Molecular Design Limited in their MACCS-II software.[24] MACCS-II provides a common interface for accessing structures, general DBMS data, graphics images, and other research data. The system, which may be customised by users through a built-in procedural language, is being actively explored by a number of companies.[25] Some limitations of this approach are: graphics terminals are required for all interaction even if only textual information is being retrieved; the general form of each report must be defined in advance by a system specialist; the general DBMS interface does not provide full access to the power and flexibility of the underlying DBMS.

Structurally Extended Relational DBMS Integration (Figure 12).

Relational database management systems[26,27] are becoming increasingly popular for managing research information as well as business data. However, existing relational

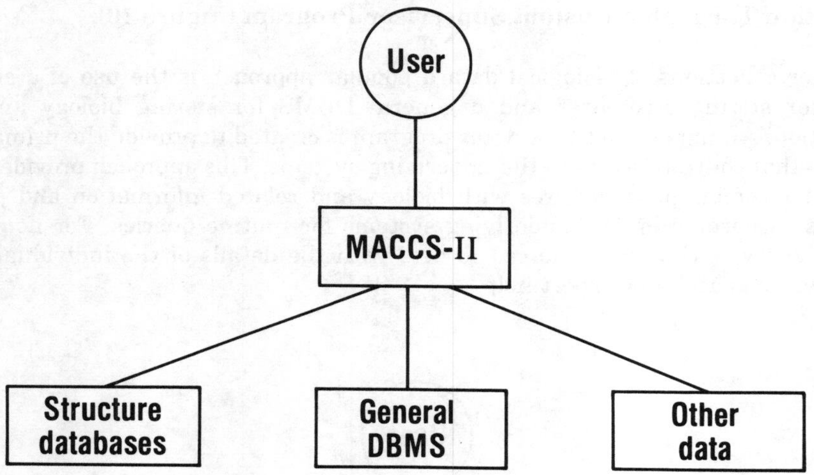

Figure 11. MACCS-II integration.

systems have no facilities for managing chemical structures. In the late 1970s the Upjohn Cousin system[28,29] was developed based on a relational DBMS using the SQL language. The SQL language definition was extended at Upjohn to include chemical structure as a new data type for database storage and retrieval. Facilities were added for graphical structure entry and registration, generic substructure query entry and searching, and generation of reports containing structures, biology, and other

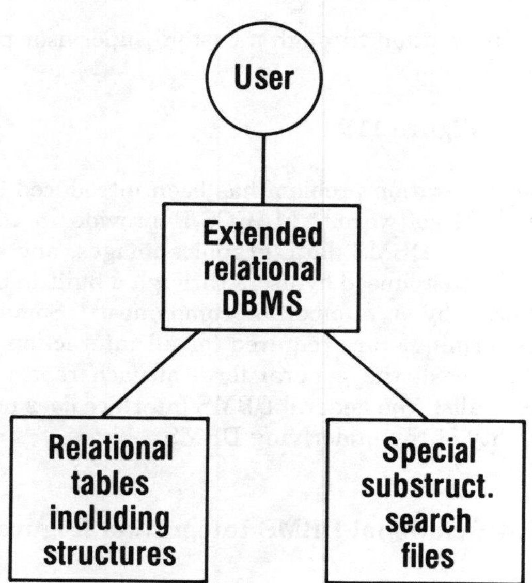

Figure 12. Structurally extended relational DBMS integration.

information from the relational database. Cousin was introduced to the Upjohn research community in 1981 and has seen extensive use since that time. The main advantage of the extended relational DBMS approach is its clean integration of structures and associated information. A homogenous view of all chemical information is provided through the relational DBMS model and the extended SQL language provides a single uniform method of information retrieval. Limitations of the Cousin system are its lack of access to information outside the relational DBMS environment and its dependence on in-house support for software maintenance and enhancement.

USAGE PATTERNS AND LEVELS

One measure of the success of a chemical information system is the level of end-user access enjoyed by the system. Unfortunately, most companies do not keep detailed records of chemical information system usage levels and patterns and it is therefore difficult to directly compare the successes of the various approaches to integration. Usage data from the Upjohn Cousin system however, provide one example of the usage levels that can be expected from a well integrated end-user oriented chemical information system. At Upjohn a continuing effort has been made to integrate as much research information as possible into the system and to provide easy end-user access to this information. The system includes three large structure databases, the collection of Upjohn research compounds, the Fine Chemicals Directory (FCD), and the Cambridge Structural Database as well as a number of smaller project-related databases. Data for new biological screens and assays is actively sought out and added to the database whenever feasible. Sample inventory and tracking information and a technical report index are maintained as well. Figure 13 shows a plot of monthly usage levels of Cousin

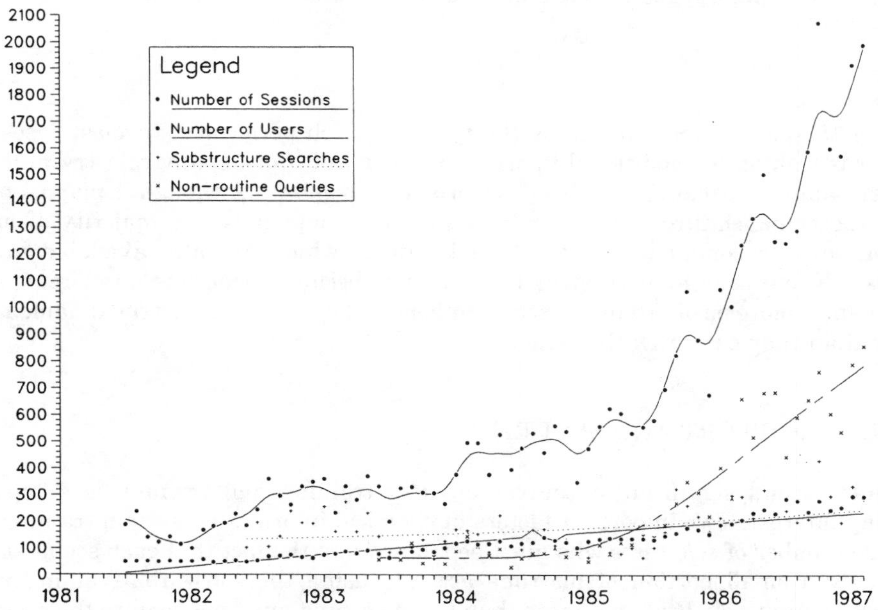

Figure 13. Upjohn Cousin system monthly usage levels, support staff excluded.

for the past six years excluding support staff activities such as database updating, development work, and consulting.

While the number of users increased by a factor of six over the time period, the number of sessions increased by a factor of ten. The average session length has remained at a fairly constant average of fifteen minutes per session. The number of sessions has increased faster than the user population because of the increase in the amount and type of information available in the database as well as the acquisition of microcomputer-based graphics terminals by many individual chemistry laboratories. Although substructure searching is considered an important function of the system, the number of substructure searches has only doubled since the introduction of the system.

Figure 14 shows usage patterns for various types of data in the Cousin database for all users. Although chemical structures are clearly the most desired piece of information, other components of the database are frequently retrieved as well. An inventory of fine chemicals ordered by each laboratory was created and linked to the FCD by supplier catalogue number to provide a method of locating chemicals available in-house. This capability was recently introduced but has already more than doubled use of the FCD.

Registry number..........................2356 (est)
Name search 772
Substructure search.......................375
Molecular formula search..................175
Other types of searches...................???

Figure 14. Usage patterns for Upjohn Cousin data. Average retrievals per month for 10/86 – 3/87.

Figure 15 shows Cousin usage by the type of search performed. In most cases the searcher is looking for additional information for a known company registry number; however, name searching has been surprisingly popular. Although Upjohn enters systematic nomenclature into the database for all compounds, the majority of name searches are for common or other trivial names, which are also available in the database. Name searching receives frequent use because scientists often remember trivial names more easily than registry numbers and can type in the trivial name more quickly than they can draw the structure.

TRAINING, SUPPORT AND SECURITY

Companies spend significant resources on user training and training is a topic of recurring interest. Two important factors in chemical information system training are the large number of scientists who must be trained and the fact that each scientist will spend only a small portion of his time actually using the software. Training must therefore be relatively brief, not more than two days, and must not require the scientist to memorise lists of commands. Voluminous user guides and reference manuals will not

Structures, names, inventory	**3671**
Detailed biological data	**725**
Biological summary data	**309**
Sample tracking data	**210**
Technical report index	**75**
Fine Chemicals Directory	**251**
Lab. fine chemicals inventory	**324**

Figure 15. Usage patterns for Upjohn Cousin data. Average searches per month for 10/86 – 3/87.

be used by the typical scientist although a small pocket reference will be. A menu-driven interface is therefore a requirement for the occasional user although a way of by-passing the menus will be appreciated by the few expert users. Although many queries can be phrased by the end-users themselves, frequently there will be cases in which expert advice is required. A clearly identified support person needs to be available to users and can help identify weaknesses in the system needing correction as well as provide expert consultation.

Information security is an important issue that is being dealt with in a variety of ways. At one extreme some companies have discouraged integration as a means of maintaining information security. Others allow access only on a demonstrated need-to-know basis and do not allow access from terminals at home. At the other end of the spectrum are companies that allow full database access for all research employees as well as access from home. The balance between the need to share research information and the need to provide security is a difficult one at which to arrive. As chemical information systems become more powerful and companies become more security-conscious this balance will need to be adjusted in the direction of greater security control.

TRENDS

Predicting trends in the chemical software and database arena is a risky business; however, a few currently emerging directions seem reasonably clear. Further integration of functionality from the user interface point of view is a high priority need voiced by many organisations. Software packages for managing structures, reactions, spectroscopic data, general databases, text databases, statistical data, and document creation have evolved independently into powerful tools. Now, there is a strong desire to integrate these components under a single common interface. The difficult question to answer is how can such integration be achieved and who will do it?

A commercial provider of a top level interface is in a very powerful position since the interface controls the level of functionality of the underlying packages that is passed through to the user. A successful interface will require the co-operation of the vendors of the underlying packages to provide the appropriate 'hooks' to allow each package to be effectively integrated. Vendors are unlikely to co-operate unless they can see increased profitability or security for themselves through such co-operation. Better organised and more powerful user groups from the pharmaceutical and chemical industries will be necessary to push vendors towards a more unified approach. Additionally, because of the potential rewards for the provider of the 'universal interface', as well as the many technical and organisational problems that must be overcome, we are likely to see such interfaces offered by a number of vendors.

Although better integration is important, competition is as well. It has been and will continue to be to the advantage of the pharmaceutical and chemical industries to have many companies vying to provide software and database services. Most companies are wary of becoming too dependent on a single vendor for access to chemical information. Competition spurs innovation, makes vendors more responsive to users' needs, and keeps prices down. However, competition in the chemical software arena may lead in the direction of more fragmented rather than better integrated systems.

Another likely trend is the availability of new chemical databases for in-house use. We are already seeing this. Examples are additional reaction libraries, chemical structure, biological response, and bibliographic databases tailored to particular areas of interest, as well as additional collections of physico- chemical property data such as log P,pK_a, and spectra libraries.[30]

A final trend will be stronger links between chemical information systems and drug design tools such as quantitative structure-activity relationships (QSAR) and molecular modelling techniques. Information generated by these techniques such as physico-chemical property estimates, three-dimensional structures from X-ray and NMR studies, as well as conformational and electronic information from modelling studies has the potential for being integrated into chemical information systems. This information in turn will provide new clues concerning the types of structures that should be synthesised as well as which compounds in valuable existing research collections should be tested for possible additional activities.

Whichever trends actually emerge into accepted practice, it is clear that the rapid changes occurring in computer hardware and software, in information services and regulatory requirements, and in the underlying science will continue to provide new challenges to those responsible for managing proprietary research information.

ACKNOWLEDGMENTS

This article is based on published reports and interviews with a representative sample of companies in the pharmaceutical industry. A survey of this type would not have been possible without the co-operation of those involved in proprietary research information management at the companies interviewed. The following people are to be especially thanked for very open and helpful discussions of their current methods and future plans for managing in-house chemical information: Brian Bergner, Squibb; Tom Bromstrup, Eli Lilly; François Choplin, Rhône-Poulenc; John Chu, Wyeth; Walter Gall and Peter Gund, Merck Research Laboratories; Kevin Haraki, Lederle Laboratories; Tricia Johns, Searle; Ted Legatt, Schering; George Rosko, Consultant; Alan Salzman, Schering; Steve Schmidt, Abbott Laboratories; John Vinson and Roger Westland, Warner Lambert; Merrie Wise, Smith-Kline & French Laboratories.

REFERENCES AND NOTES

1. Brown, H.D. 'A Drug is Born: Its Information Facets in Pharmaceutical Research and Development'. *J. Chem. Inf. Comput. Sci.* 1985, *25*, 218–224.
2. Brown, H.D. 'Information Requirements for Chemists in the Pharmaceutical Industry'. *J. Chem. Inf. Comput. Sci.* 1984, *24*, 155–158.
3. *Communication, Storage and Retrieval of Chemical Information*; Ash, J.E.; Chubb, P.; Ward, S.E.; Welford, S.M.; Willett, P., Eds; Ellis Horwood: Chichester, 1985.
4. Bowman, C.M.;Moses, P.B. 'Evolution of Industrial Chemical Information Systems'. *J. Chem. Inf. Comput. Sci.* 1985, *25*, 197–202.
5. Fugmann, R. 'The IDC System'. In *Chemical Information Systems*; Ash, J.E.; Hyde, E., Eds.; Ellis Horwood: Chichester, 1975.
6. Eakin, D.R. 'The ICI CROSSBOW System'. In *Chemical Information Systems*; Ash, J.E.; Hyde, E., Eds; Ellis Horwood: Chichester, 1975.
7. Dittmar, P.G. et al. 'The CAS ONLINE Search System. I. General System Design and Selection, Generation, and Use of Search Screens'. *J. Chem. Inf. Comput. Sci.* 1983, *23*, 93–102.
8. Dubois, J.E.; Sobel, Y. 'DARC System for Documentation and Artificial Intelligence in Chemistry'. *J. Chem. Inf. Comput. Sci.* 1985, *25*, 326–333.
9. Anderson, S. 'Graphical Representation of Molecules and Substructure Search Queries in MACCS'. *J. Mol. Graphics* 1984, *2*, 83–90.
10. Lynch, M.F. et al. 'Generic Structure Storage and Retrieval'. *J. Chem. Inf. Comput. Sci.* 1985, *25*, 264–270.
11. Shenton, K.E.; Norton, P.; Ferns, E.A. 'Generic Searching of Patent Information'. In these Proceedings.
12. Fisanick, W. 'Storage and Retrieval of Generic Chemical Structure Representations by Computer'. U.S. patent US 4,642,762A, 10 Feb 1987.
13. Bawden, D. 'Browsing and Clustering of Chemical Structures'. In these Proceedings.
14. Carhart, R.E. et al. 'Atom Pairs as Molecular Features in Structure-Activity Studies: Definition and Applications'. *J. Chem. Inf. Comput. Sci.* 1985, *25*, 64–73.
15. Brint, A.T.; Mitchell, E.; Willett, P. 'Substructure Searching in Files of Three-Dimensional Chemical Structures'. In these Proceedings.
16. Upjohn, Pfizer, and Monsanto are examples of companies that have decided to continue to support their in-house developed chemical structure database management systems.
17. Kennard, O. et al. 'The Cambridge Structural Database in Molecular Modelling and Drug Design', paper presented at the Cambridge Structural Database Workshop, Hilton Head Island, South Carolina, 1987.
18. Plans for and progress on the Merlin substructure search system were presented at the Pomona College Medicinal Chemistry Project User's Group meeting in Claremont, California, 1987, and described in the Spring 1987 users' group newsletter (MUGWUMP).
19. Jochum, C.J. 'Building a Structure-oriented, Numerical, Factual Database'. In these Proceedings.
20. Town, W.G. 'The Impact of Microcomputers on Chemical Information Systems'. In these Proceedings.
21. Johnson, A.P. 'Reaction Indexing: An Overview of Current Approaches'. In these Proceedings.
22. Meyer, D.E. 'Use of Microcomputer Software to Access and Handle Chemical Data'. In these Proceedings.
23. Barcza, S. 'Integrated Chemical-Biological-Spectroscopy-Inventory-Reactions Preclinical Database'. *J. Chem. Inf. Comput. Sci.* 1986, *26*, 198–204.
24. Ahrens, E.K.F. 'Customisation for Chemical Database Applications'. In these Proceedings.
25. Abbott Laboratories and Dow Chemical have been leaders in exploring MACCS-II. Many other pharmaceutical and chemical companies have recently started investigating MACCS-II's capabilities and limitations.
26. Date, C.J. 'An Introduction to Database Systems'. Addison-Wesley: Reading, Massachusetts, 1982.
27. ORACLE from Oracle Corporation and INGRES from Relational Technology Inc. are two popular relational systems for the DEC VAX series of computers. SQL/DS and DB2 from IBM are popular for IBM mainframes. A large number of relational systems are available for the

IBM PC and a SQL-based relational DBMS will be an integral part of the announced OS/2 PC operating system.

28. Howe, W.J.; Hagadone, T.R. 'Molecular Substructure Searching: Computer Graphics and Query Entry Methodology'. *J. Chem. Inf. Comput. Sci.* 1982, *22*, 8–15.
29. Hagadone, T.R.; Howe, W.J. 'Molecular Substructure Searching: Minicomputer-Based Query Execution'. *J. Chem. Inf. Comput. Sci.* 1982, *22*, 182–186.
30. Examples are ISI's Current Chemical Reactions database and Index Chemicus subset databases as well as the Rhône-Poulenc pK database.

SOFTWARE AND DATABASES

BASIS – Text management system, hardware supported: IBM mainframe, DEC VAX, etc., available from: Information Dimensions, Inc., 655 Metro Place, Dublin, OH43017, U.S.A.

CAS ONLINE – Online chemical information system, service available from: Chemical Abstracts Service, P.O. Box 3012, Columbus, Ohio 43210, U.S.A.

Cambridge Structural Database System – Crystallographic database with associated retrieval, analysis and display software, hardware supported: DEC VAX and limited generic computer support, available from: Cambridge Crystallographic Data Centre, University Chemical Laboratory, Lensfield Road, Cambridge CB2 1EW, U.K.

ChemText – Microcomputer chemical information system and chemical word processor, hardware supported: IBM PC, available from: Molecular Design Limited, 2132 Farallon Drive, San Leandro, CA 94577, U.S.A.

ChemDraw – Microcomputer chemical structure drawing program, hardware supported: Apple Macintosh, available from: Cambridge Scientific Computing, Inc., P.O. Box 2123, Cambridge, MA 02238, U.S.A.

ChemSmart – Microcomputer chemical information system, hardware supported: IBM PC, available from: ISI Software, 3501 Market Street, Philadelphia, PA 19104, U.S.A.

Current Chemical Reactions In-house Database – Growing database of 30,000 recently published reactions, available from: ISI Software, 3501 Market Street, Philadelphia, PA 19104, U.S.A.

DARC – Chemical information system for online and in-house use, available from: Télésystèmes Questel, 83–85, bd Vincent-Auriol, 75013 Paris, France.

EMU-TEK – Tektronix terminal emulator, hardware supported: IBM PC, available from: FTG Data Systems, 10801 Dale Street, Suite J-2, PO Box 615, Stanton, CA 90680, U.S.A.

Fine Chemicals Directory – Database of commercially available chemicals, available from: Fraser Williams (Scientific Systems) Ltd., London House, London Rd. Sth., Poynton, Cheshire SK12 1YP, U.K. and Molecular Design Limited, 2132 Farallon Drive, San Leandro, CA 94577, U.S.A.

INGRES – Relational database management system, hardware supported: DEC VAX, IBM mainframe, etc., available from: Relational Technology, Inc., 1080 Marina Village Parkway, Alameda, CA 94510, U.S.A.

INQUIRE – General and text database management system, hardware supported: IBM mainframe, available from: Infodata Systems, Inc., 5205 Leesburg Pike, Falls Church, VA 22041, U.S.A.

MACCS – Chemical information system for in-house use, hardware supported: DECVAX, IBM mainframe, and Prime, available from: Molecular Design Limited, 2132 Farallon Drive, San Leandro, CA 94577, U.S.A.

MedChem – Chemical information and modelling system, hardware supported: DEC VAX, available from: Pomona College Medicinal Chemistry Project, Seaver Chemistry Laboratory, Claremont, CA 91711, U.S.A.

ORAC – Reaction indexing system, hardware supported: DEC VAX, available from: Wolfson CADOS Unit, Department of Organic Chemistry, University of Leeds, Leeds LS2 9JT, U.K.

ORACLE – Relational database management system, hardware supported: DEC VAX, IBM, etc., available from: Oracle Corporation, 20 Davis Drive, Belmont, CA 94002, U.S.A.

PSIDOM – Microcomputer chemical information system, hardware supported: IBM PC, available from: Hampden Data Services Ltd, Hampden Cottage, Abingdon Road, Clifton Hampden, Abingdon, Oxfordshire OX14 3EG, U.K.

RS/1 – Data analysis system, hardware supported: DEC VAX, IBM mainframe & PC, etc., available from: BBN Software Products Corp., 10 Fawcett St., Cambridge, MA 02238, U.S.A.

REACCS – Reaction indexing system for in-house use, hardware supported, DEC VAX, IBM mainframe, & Prime, available from: Molecular Design Limited, 2132 Farallon Drive, San Leandro, CA 94577, U.S.A.

SAS – Data analysis system, hardware supported: IBM mainframe and PC, DEC VAX, etc., available from: SAS Institute, Inc. (NC), PO Box 8000, Cary, NC 27511, U.S.A.

STAIRS – Text management system, hardware supported: IBM mainframe, available from IBM, Old Orchard Road, Armonk, NY 10504, U.S.A.

System 1032 – General database management system, hardware supported: DEC VAX, available from: Software House, 1000 Massachusetts Ave., Cambridge MA 02138, U.S.A.

TGRAF – Tektronix terminal emulator, hardware supported: IBM PC, available from: Grafpoint, 4340 Stevens Creek Blvd., San Jose, CA 95129, U.S.A.

VersaTerm-PRO – Tektronix terminal emulator, hardware supported: Apple Macintosh, available from: Peripherals, Computers & Supplies, Inc., 2457 Perkiomen Ave., Mt. Penn, PA 19606, U.S.A.

DEVELOPMENT OF AN INTEGRATED SYSTEM FOR HANDLING CHEMICAL STRUCTURES AND ASSOCIATED DATA

Michael J Allen
Glaxo Group Research Ltd, Greenford, Middlesex UB6 0HE, U.K.

SUMMARY

Glaxo Group Research began developing a system, CBIS (Chemical and Biological Information System), for handling chemical structures and related data, about three years ago.

The system architecture is as follows. Chemical structure information is handled in a graphics based system. A physicochemical data module, not yet fully operational, links analytical data to chemical structures. At a level below the structures are held 'compound data' e.g. sample data such as solubility, and tracking data/test history. Under this are a number of biological project databases. Eventually there will be more than thirty of these, one for each project.

Of the 120,000 compounds in Glaxo's registry, about 38% have stereocentres. It was essential to design in-house conventions for handling stereochemistry if Glaxo were to convert their backlog of chemical structures and move to graphics registration. Conventions were devised to handle E/Z, mixed and unknown configurations about C=C, C=N and N=N; relative, absolute and unknown configurations about C, N, S and P asymmetric centres; partially resolved asymmetric centres; and unknown isomeric composition. Flags and labels are used: flags for atom centres and bonds, labels ('absolute' or 'isomer n') on the whole structure.

Glaxo's backlog of 120,000 compounds was converted to the new graphics system and all the stereochemistry was rechecked. This whole exercise took 100 man-months.

The system places much emphasis on registration tasks; many other companies have ignored graphics chemical registration. Every structure is registered at three levels: parent, version and preparation levels. There are thus three databases. First the structure is drawn and oriented. (Glaxo have fairly strict rules about preferred orientations for compounds). The system then detects stereocentres and the user can enter the desired stereochemistry at each centre. A 2-D chemical structure novelty check takes place before the user decides to register the compound and the system automatically assigns a parent compound number. Salt codes are entered as numbers (which are converted to a chemical structure display) before the version is registered. Finally, the preparation is registered.

ACKNOWLEDGMENTS

This summary of Mike Allen's paper has been prepared by the Editor from tape recordings and notes made during the Conference, with the approval of the author.

W. A. Warr (Ed.)
Chemical Structures

(INTER)FACING DARC – ORACLE

A J C M (Juus) de Jong
Organon International, P O Box 20, 5340 BH Oss, The Netherlands
A M C (Twan) Deibel
Akzo Pharma, P O Box 20, 5340 BH Oss, The Netherlands

ABSTRACT

To improve the interfacing of DARC with external systems a set of programs has been planned for development by a team of Télésystèmes and Akzo staff. A new open-ended communication protocol has been written initialising a subprocess controlled by transferred logical names. *Via* this protocol the following options are achieved:

1. Data selection from external systems (ORACLE) by:

   Host Language Interface controlled dynamical queries. This function is triggered by the 'display select' option. Results can be integrally displayed with the chemical structure.

   Introducing 'fill boxes' in the standard report made through RDS2 (Report Definition System). These fill boxes are 1:1 related with RDBMS (relational database management system) column names and will be filled at display time by interactive commands.

   In both cases pre-selected data from DARC are transferred to the external system.

2. From an external system (RDBMS) a chemical structure can be selected from the DARC database. This structure will be graphically displayed while remaining in the original (RDBMS approach) process.

INTRODUCTION

Computers have been used for the storage and retrieval of scientific datasince the late 1960s. At first only punched cards and printers with the alphanumerical character set were available for a reasonable price. These facilities were insufficient for a good presentation of chemical structures. The Wiswesser line notation (WLN) was a workable substitute for structures, but it has never become a common language for the chemists. CROSSBOW and a few other programs became available but these batch-oriented programs are not user-friendly for bench chemists. They found them barriers instead of aids in communication. Therefore these systems were used mainly by intermediaries or information scientists.

A new generation of software for in-house processing of chemical structures arrived with topological programs like DARC and MACCS in the early 1980s. It became possible to store a chemical structure directly in a database, to search for structures and

W. A. Warr (Ed.)
Chemical Structures

substructures and to present the results on a graphic screen. There was no chemical code, the structure was drawn graphically with a light pen or mouse and the commands were simple. This type of software was easy to use by our bench chemists. After a test period of several months DARC appeared to be suitable for the management of our chemical structures. By the end of last year Organon decided to use this program.

For data related to the structures such as chemical, physical and biological data the relational database management system (RDBMS) ORACLE is generally used. After the successful building of a system for toxicological data with ORACLE it became a standard tool for developing research information systems. Integration of information systems is relatively easy with ORACLE as the central database. But the relational concept is the strongest argument for using a system like ORACLE. The search process for related data in DARC is a time consuming sequential process. On the other hand searches in ORACLE *via* an index are very fast. The decision to use DARC was under the condition that efficient links with ORACLE would soon become available.

FACING DARC

At the moment it can be said that in our company DARC-SMS (Structure Management System) on a local VAX computer cluster fulfils most of our current needs. These are unique identification of compounds and phrasing queries by drawing a structure including stereochemical aspects, simple commands and fast responses. Shortcomings exist for large molecules like peptides and sugars, but the program will certainly be improved by Télésystèmes. We are using only the SMS part of DARC that covers the structure, a structure number or compound number (CN), the molecular weight and the molecular formula. All other data will be stored in ORACLE and not in the DMS (Data Management System) module of DARC.

FACING ORACLE

ORACLE is available in our company also on the VAX computer cluster. Figure 1 presents the major relevant functions. The kernel comprises all software for the management of the database and other parts of the system. Data are stored in a collection of tables described by a dictionary. A table consists of columns and rows. SQL (Structured Query Language), a standard query language, forms a shell round the kernel. Instructions for the kernel are passed down *via* SQL.

In the outside shell ORACLE has many tools and only the relevant ones have been mentioned. For batch loading there is the ODL (Oracle Data Loader) program. For input from keyboard the IAP (Interactive Application Processor) program is available. The lay-out and controls of the screens are being generated by the IAG (Interactive Application Generator) program. IAP gives a way of access to the database for input and output *via* user-defined screens for every application. Another way of interrogating the database is *via* the UFI (User Friendly Interface) program. This gives a flexible entry to the data and possibilities for short *ad hoc* reports. Despite this name we think that UFI is not really user-friendly and it is therefore reserved for experienced users only. The HLI (Host Language Interface) program finally makes it possible to enter ORACLE from a user program written in a language like FORTRAN that has a 'call' mechanism.

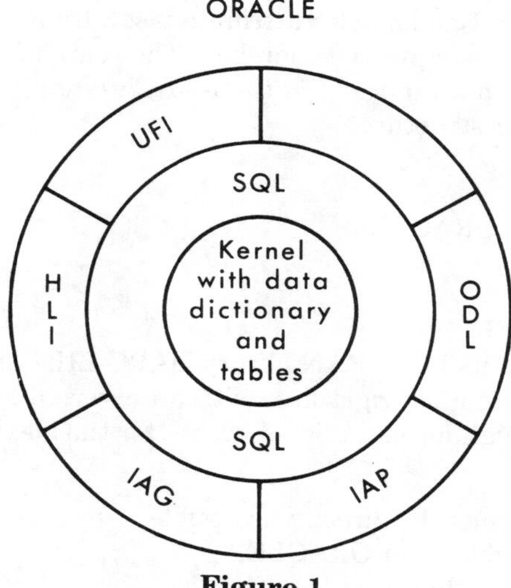

Figure 1

Several collections of research data are planned. At present most of these data are stored in other computer systems and must be converted into ORACLE tables. For this batch loading process the ODL program will be used. Input from old card systems and new data will be processed *via* the IAP program.

The data in ORACLE are kept in tables, where a row in a table consists of a key and a number of attributes. The key identifies this logical amount of information in the row. The relational concept of ORACLE gives the possibility of fast responses for queries where data from one table are related to data from another table. This is possible for all tables as long as the compound number is the key. The data are searched *via* the content of the value and not *via* numerous predefined relations.

A simplified example, Figure 2, can make this more clear.

Chemical/physical data are stored in one table and there is a second table for pharmacological data. In one query you can ask for compounds with a certain value for

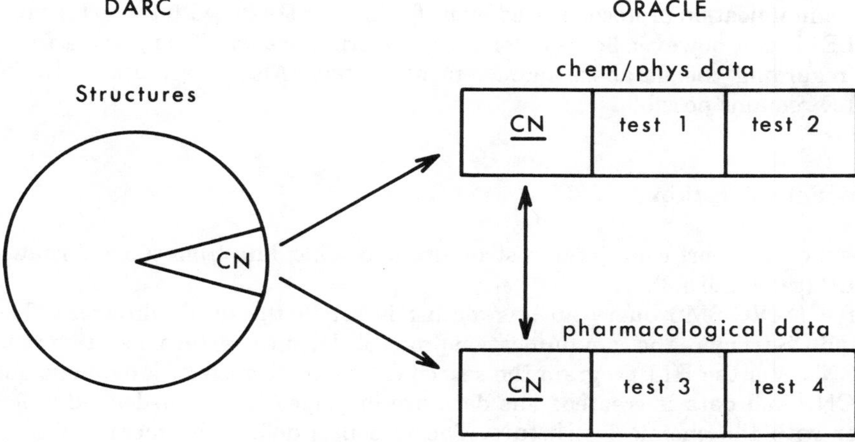

Figure 2

attribute test 1 from table 1 and one for attribute test 3 from table 2. Tables are linked by the compound number as a pre-selection key. The search is performed very quickly because the keys are set in an index. *Via* the DARC system this compound number is 1:1 related to a particular structure.

INTERFACING DARC - ORACLE

Requirements

Standard DARC has with its FORTRAN library DARC-LIB1 and the Report Definition System module, RDS2, a simple rigid link with an external system. This facility was clearly not sufficient enough for us. A list of wishes for the ideal linking program shows the following points:

1. DARC and ORACLE must be directly accessible.
2. DARC must be accessible from ORACLE.
3. ORACLE must be accessible from DARC.
4. Both systems should be linked only when it is necessary.
5. It should be independent from software vendors as much as possible.

To realise our wishes it was decided to develop a new communication protocol and three new linking facilities: display select, fill and structure display as a joint activity.

Communication Protocol

An open-ended communication protocol has been written as the heart of the system. In orderto standardise communication between DARC and other processes, this communication will be performed with the following characteristics:

1. Actual data will always be transmitted within logical names.
2. All data will be transmitted in a character string data type
3. When a call is made to a subprocess, it transfers control to a command procedure.

The communication protocol is used by us for linking DARC with the external system ORACLE. It can however be used for every external system. This gives a freedom of choice regarding the database management system. Also direct use of DARC and ORACLE remains possible.

Display Select Option

This displays a report containing a structure and some data from the external system ORACLE (see Figure 3).

After a DARC-SMS query an answer list is set in the user's directory. The new command DS triggers the communication protocol. Then the external system reads this file of CN's. Via the HLI program the search can be performed by a dynamic query. A list of CN's and data is created. The data are displayed in a user-defined 'select box' together with the chemical structure. The screen is defined by means of the RDS2 program. Only one box can be defined but the size is only limited by the screen. After

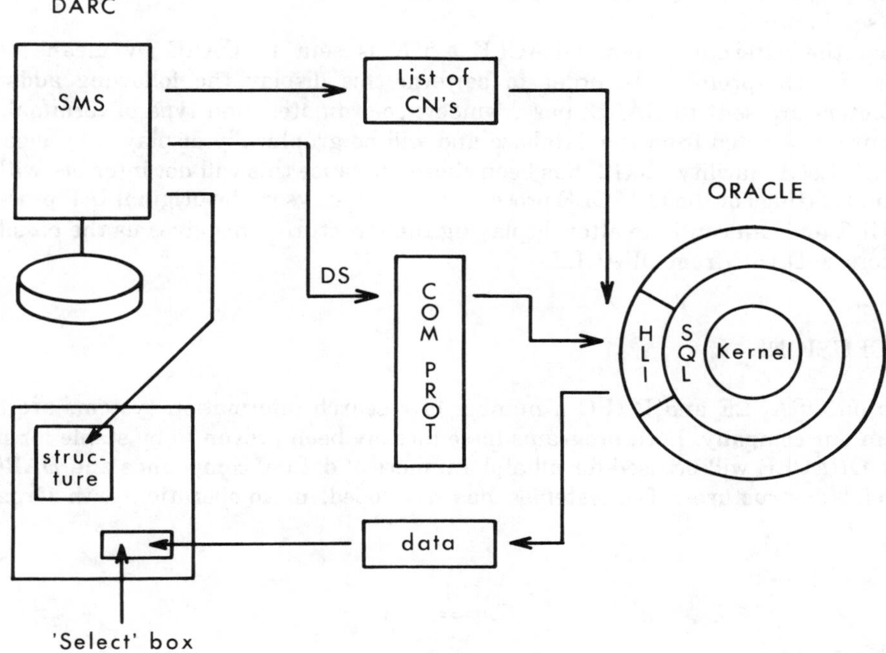

Figure 3

the display the DARC process can be continued. This facility gives the needed flexible link from DARC to ORACLE.

Fill Option

This displays reports containing structure(s) and a standard set of data from the external system ORACLE (see Figure 4).

New types of boxes, 'FILL-boxes', can be defined with RDS2. These 'FILL-boxes' are 1:1 related to RDBMS column names. They are similar to the current field boxes for an external file. The only difference is that they are to be filled with information at display time by an interactive command.

After a DARC query a new display option activated by the command 'OD' triggers the communication protocol and sends the CN and column names to the HLI program. This HLI program collects the data from ORACLE and sends them back to DARC. DARC will write the data on the screen. This option gives only a facility for the execution of a predefined query resulting in a standard report. It will give an easy link from DARC to ORACLE.

Structure Display Option

This displays a single chemical structure on an ORACLE screen with alphanumerical data (see Figure 5).

From the external system ORACLE a CN is sent to DARC by means of the communication protocol. In order to perform this display the following additional parameters are sent to DARC: base, window co-ordinates, and type of terminal. The structure is selected from the database and will be graphically displayed by means of the DEC ReGIS facility. ReGIS has been chosen because this will not interfere with the terminal settings of the ORACLE process. The user stays in the original IAP process of ORACLE and can continue after displaying the structure. This gives us the possibility of access to DARC from ORACLE.

CONCLUSIONS

Based on ORACLE and DARC a number of research information systems are being built in our company. Both programs have already been proven to be stable for many years. ORACLE will be used for all alphanumerical data of compounds and DARC for the related structures. Télésystèmes has developed, in co-operation with Organon,

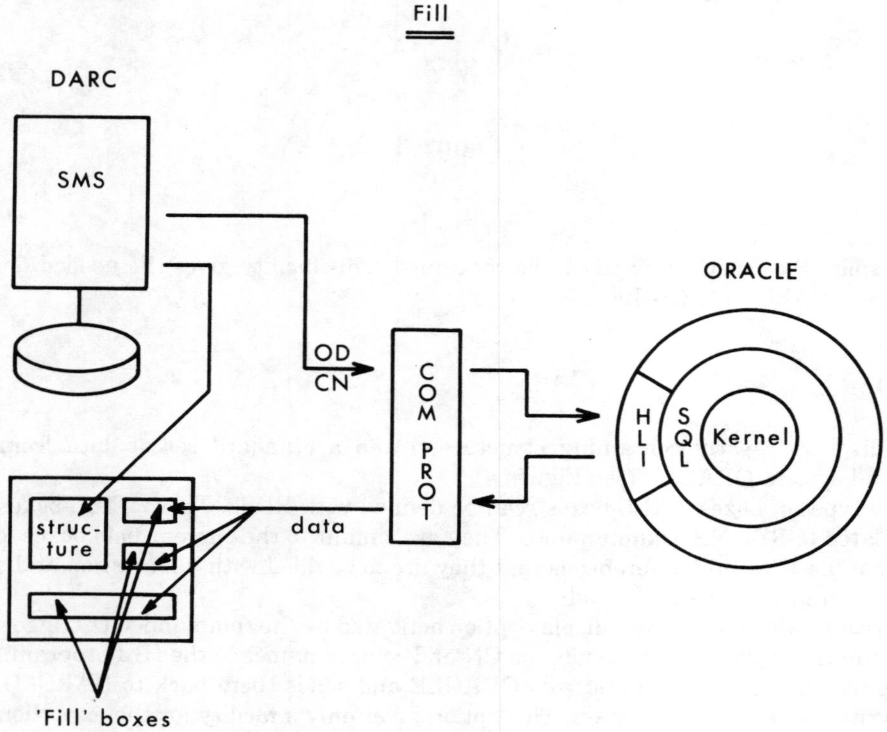

Figure 4

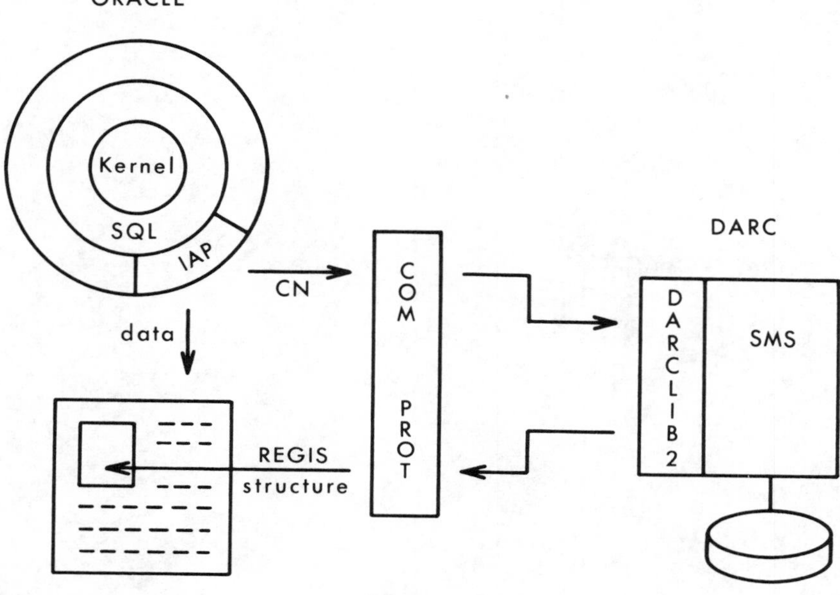

Figure 5

a new communication protocol. *Via* this protocol DARC can be linked directly to an external system. Three different options for the linking have been planned for development. These options will give us great flexibility for a reasonable price. The fact that the communication protocol calls a command procedure makes us independent of a software supplier. We expect the options will soon meet all our defined requirements.

USING OSAC TO KEEP STRUCTURES IN THEIR PLACE

David S Magrill
Fisons PLC, Pharmaceutical Division, Bakewell Road, Loughborough, Leicestershire LE11 0RH, U.K.

INTRODUCTION

Fisons Pharmaceutical Division is a British ethical pharmaceutical company based in the East Midlands. Although modest by the standards of major international drug companies, it has an active Research Group working on the discovery of new chemical entities as potential drugs and like most other drug companies it must manage information on the compounds made and on their performance in biological and other tests.

In 1983 we undertook a major overhaul of our existing system and, for reasons which I shall briefly discuss, we put early emphasis on the module for storing and retrieving biological test data. The new system, which we called ABACUS (Advanced Biological And Chemical Unified System) has been operational for about 18 months although it is still evolving.

Our next major challenge was to integrate a chemical structure handling component into ABACUS. Our progress in using OSAC (Organic Structures Accessed by Computer) for this purpose is the principal subject of this paper.

BACKGROUND

The objectives of ABACUS were firstly, to maximise the exploitation of research results by providing facilities to record, retrieve and correlate information on the compounds tested and on the test results and related data; and secondly, to help streamline and manage the whole drug discovery process.

Fisons' original biological and chemical records system was set up in 1973 on a remote ICL computer and was subject to the hardware and software limitations of the time. Results were recorded by biologists on datasheets and largely in free format. Some codes were added by information scientists before the data was sent for input. Chemical structures were stored as Wiswesser Line Notations (WLN).

By 1983 the biological component was so rarely used that there was no opposition to the suggestion that it be abandoned. Many biologists had ceased to fill in datasheets with the reasonable rationale that they were unable to retrieve useful data from the system. The outdated computing facilities, operated at second hand, made it difficult to obtain print-outs or to check and correct data that was added to the system.

The same problems had affected the WLN file: no checker program was available and even a cursory inspection revealed a host of errors which could be logged but not easily corrected. The ponderous procedures available made it possible, at best, to print the file in reference number or WLN order but no substructure search was possible beyond that provided by initial WLN strings.

W. A. Warr (Ed.)
Chemical Structures

The installation of a VAX 11/780 computer in 1983 provided hardware for a redesigned system. We decided on the following priorities:

1. Transfer the WLN file to the VAX, cleaning it up as far as possible.
2. Design and implement a new system for the storage and retrieval of biological test data. This was regarded as very urgent since the data was not being captured electronically at the time.
3. Add chemical graphics facilities to the biological test system.

The Information Department, which undertook to upgrade the system, had only slender resources to devote to the project. At best only one staff member could be spared to work on the project and for long periods it was done as a 'spare-time' activity. This situation forced us to take pragmatic approaches to problems and to look for solutions capable of easy and rapid implementation.

WLN FILE

We wrote a program to remove in-house idiosyncrasies from the WLN file and to reformat it for CROSSBOW. Programs from this suite were then used to remove outdated conventions, check for what errors could be detected algorithmically and generate molecular formulae. This was achieved and the file, then of some 13,000 'FPL Compounds', was subsequently maintained by CROSSBOW.

A search facility was provided using the CROSSBOW 'Bit and String' search program, which is sufficient for small files but clumsy to use. However our primary aim was to obtain clean and useful WLN files, ready for conversion to a graphics system in due course, so the extra expense of connection table based searching seemed unjustified.

ABACUS VERSION 1

The biological testing at Fisons operates at relatively low volume but with a high level of complexity:

1. Chemists synthesise about 800 new compounds a year for testing.
2. Each compound is tested in several screens (typically 2–6) and usually more than once in each screen. In total about 40 different screens are currently available.
3. The screens are all different in nature and generate different types of data. The set of screens in active use is in constant flux as tests are developed, refined or dropped.
4. Most screens are project-specific but some, such as simple measures of metabolic clearance rate, are available to all projects.
5. Some cross-screening of compounds takes place – although very little of this is random in nature.

The new system, initially for handling only biological data, was called ABACUS. We used the System 1032 database management system to build it, calling modules written in 1032's command language from a BASIC program. The implementation was very rapid. The principal features of this first version of ABACUS were:

1. Fully menu-driven.
2. Bespoke forms for data input on each screen.

3. Updated directly by users.
4. Simple search system with test-specific display and/or print facilities in a variety of formats.
5. Easy to maintain and to adapt to new requirements.
6. Commonality of tests exploited without affecting their individuality.

ABACUS VERSION 2

For Version 2 of ABACUS we needed a chemical structure handling system to add the following facilities to ABACUS.

1. Graphics-based chemical substructure search system.
2. Integrated chemical and biological search capability.
3. Display of structures with biological results.
4. Printing of structures with biological data.

By 1986, when we were ready to buy chemical structure software, there was a number of systems on offer. A year earlier the position might have been very different. The main contenders were: MACCS, DARC and OSAC, a program derived from the reaction storage and searching program, ORAC. In addition, HTSS and SABRE were of interest although we had doubts about their state of development at the time.

By this time the basic standards for such a program, in terms of structure input, query input, substructure search and structure display capabilities were quite well established and, by and large, we were satisfied that all programs could meet these requirements. We therefore focused on the aspect that would have the heaviest bearing on our small development effort: the ease with which a program might be integrated into ABACUS.

ABACUS is written almost entirely in 1032 code so we sought a system which, as well as providing the normal range of integrated capabilities, would provide all the necessary functions as separate routines callable from 1032. The functions in question included finding structures; displaying them to order with suitable scaling and positioning; printing them as required; supplying query sketchpads; and carrying out searches. Without such facilities it was difficult to see how the graphics system could be integrated smoothly with ABACUS.

It was, in fact, this requirement which suggested the title of this paper, 'Using OSAC to Keep Structures in their Place.' In-house structure handling systems, notably MACCS, provided such a technological leap forward that for their first few years it was understandable if the chemical system was the dominant software. However, now that we are used to the idea, we ought to be seeing the chemical structures for what they are, more pieces of data (admittedly with some distinctive characteristics), and to be losing our awe of the software that handles them. With this philosophy, what we wanted was software that, when necessary, would handle the structures as we dictated, under the direction of 1032, BASIC, FORTRAN or any other language.

We selected OSAC primarily because it came closest to meeting this need. A valued bonus was that it was fully compatible with ORAC (Organic Reactions Accessed by Computer) and that upgrade plans of its originator, Dr A P Johnson of the University of Leeds Industrial Services (ULIS) called for the full integration of the two programs to handle both structure and reaction data (ODAC). We were therefore able to provide ORAC for our chemists and to use this as a training vehicle for later OSAC use.

EXPERIENCE WITH OSAC

We have now had OSAC in-house for about five months. Our first task was to load the FPL Compound file, now containing over 14,000 compounds, by converting the records from WLN's.

In our WLN file salts were recorded as parent compounds with a WLN suffix to specify the salt modification. As we wanted our OSAC file to show salts in their ionised form, and as we proposed to generate the OSAC structures directly from WLN's, we first wrote a program to convert our records to true salt WLN's where appropriate. We then used DARING to convert the WLN file to connection tables. A batch of ferrocenes which are of no current interest was first set aside for later consideration.

The DARING conversion proceeded with only a handful of errors. The resulting file was then added to OSAC which registered the compounds and supplied co-ordinates for display purposes. The program failed to add only 11 compounds and the co-ordinates generated were, for the most part, very acceptable. It made no attempt to deal with bridged structures, of which we had 57 on file, and it had problems with compounds like *ortho*, *ortho'*-disubstituted biphenyls where superposition of substitutents cannot be avoided without using unorthodox bond angles or awkward bond lengths.

We plan to organise a systematic review of all our structures, to check them for correctness (WLN errors cannot be excluded) and clarity and to add any available stereochemical information. In the meantime, however, we have a useful and generally accurate file. Once we had the WLN file in the appropriate form it was passed through DARING, added to OSAC and screened inside a week. We added OSAC as an option on the ABACUS menu but at this stage there was, of course, no real integration.

We had previously organised a number of ORAC training courses, conducted by the experts at Leeds, and these served as OSAC courses for many of our chemists. However, we backed them up with a series of short OSAC courses aimed at explaining the conventions used in our file and one or two traps for the unwary. We have no systematic data but anecdotal evidence suggests that the system has been very well accepted by Fisons chemists.

STRUCTURE DISPLAY IN ABACUS

Our next target was to display structures alongside biological or other data within ABACUS. Since ABACUS was almost totally 1032 driven and our philosophy was to minimise changes our plan was to use terminals which could overlay a graphics plane, containing the structure, on to a VT100 text plane controlled by 1032.

This system offered many advantages. It was easy to develop from the ABACUS standpoint since, in principle, all we needed to do was add calls to OSAC to draw structures on the graphics plane and make sure we left a space for them on the text plane – a simple job for 1032 and one already anticipated in many of our procedures.

There were one or two drawbacks. Obviously, the terminals supported are limited but as Pericom terminals had been widely adopted within Fisons, with our encouragement, this was not a real problem. For users with PC's we have undertaken to support EmuTek also although alternative emulators are now being considered.

One problem we failed to anticipate was that text and graphics screens do not exactly coincide and the correlation varies from model to model. Thus the Pericom MG200, MG400 and MG600 all require different settings to overlay a structure on the same text screen gap. Fortunately the MG600 settings seem all right for the MG700 and MG800 but we had to calibrate each of the first three models and make corrections accordingly.

ULIS were very helpful and responsive in providing the functions we needed and in helping us to use them. Figure 1 shows schematically the link between ABACUS and ODAC (the generalised form of OSAC/ORAC) and VDU-PAC (a set of device independent driver calls provided with ODAC).

The ODAC and VDU-PAC procedures are represented by the bottom boxes but they cannot be called directly by ABACUS 1032 procedures because 1032 can only call integer functions and doesnot support FORTRAN's return codes. It was therefore necessary to write for each procedure a 1032 external procedure and a FORTRAN jacket procedure to ensure correct communications. The ABACUS-ODAC Macro Library is not strictly necessary but it saves much duplication at the 1032 level by, for example, combining into one procedure all the calls needed to find an OSAC record, to manipulate the two screens and to display the structure as required.

Using this method we can now readily display structures on screen forms (used both for input and confirmation of the biological records and as the most comprehensive form of results display) and on other standard displays of ABACUS data. The method can be very readily applied to any display required.

INTEGRATED SEARCH SYSTEM

As part of the design of our integrated search system we compared our needs with

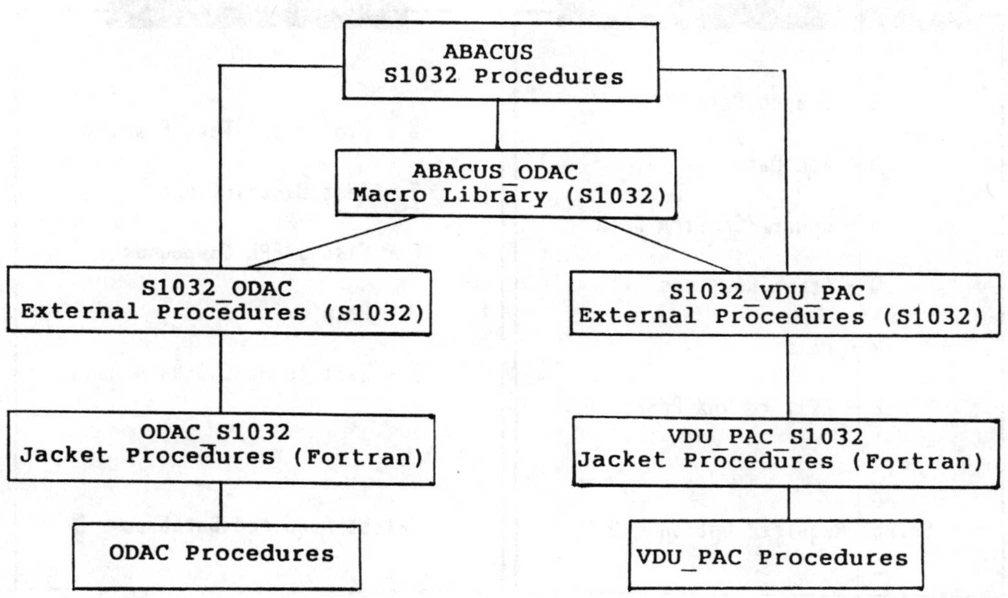

Figure 1

models formalised by Dr A P Cook of ULIS. In this analysis it is recognised that, for example, to the searcher of biological records the chemical structure is just another property of the test and it is merely a matter of processing convenience that causes the chemical record to be kept in a distinct file. In a Type 1 system the user is aware that there are two different files and can transfer hits between them at will. In a Type 2 system the file structure is transparent to the user who searches on chemical and/or non-chemical criteria at will.

The Type 2 system is elegant in its approach and provides an attractive facility but in some respects it does not meet our needs. Since searches are conducted in terms of biological test records any search actually carried out on the compound file has to be mapped back to biological test records with corresponding compound identifiers. In this process the hits for any particular compound will vary from 0 to possibly 60 or 70, depending upon how much it was tested. This is fine if the test records are what matter. If you want to know about the compounds themselves it will not do.

We therefore opted for a combination of the best features of both types. This option will give a Type 2 interface to the searcher who wishes to search biological records and a Type 1 interface to the user who wants to treat the two files as separate and to search only one type of record at a time. Using this system, it will be possible to browse the chemical compound file in terms of chemical records and, if desired, to transfer hits (as compounds identifiers) to corresponding hits (as biological records with the same compound identifiers) in the non-chemical search system.

We have now built a prototype of the Type 2 component of this system, taking the opportunity to upgrade our ABACUS search system at the same time. As before, the work was done with the close co-operation of our ULIS colleagues.

When 'S' is selected from the Main Menu (Figure 2) the database menu is offered on

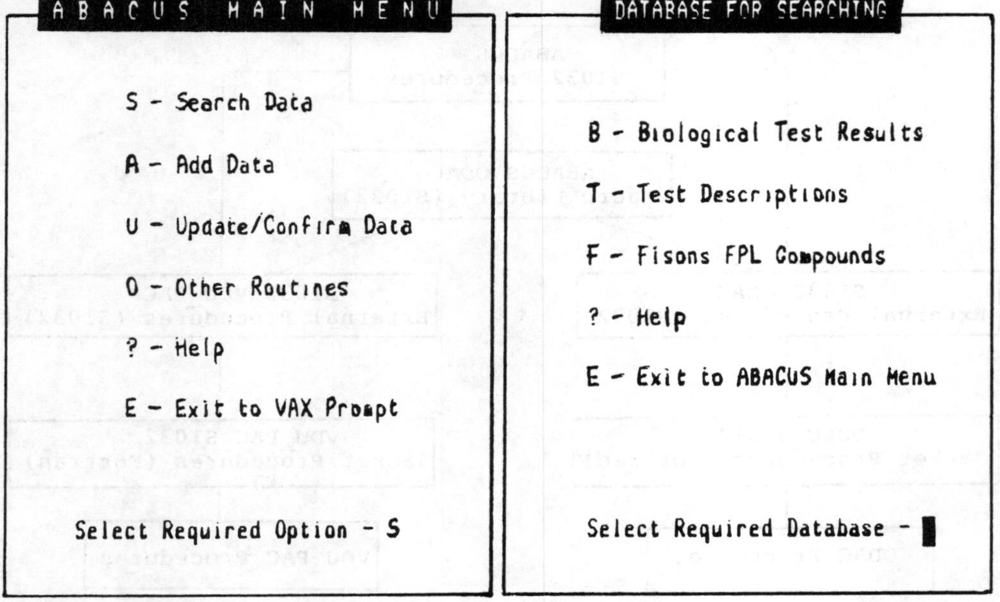

Figure 2

the right of the screen. Selection of 'B' leads to the Biological Test Results - Main Search Menu (Figure 3). The record of hits on the right of the screen remains in place throughout the subsequent operations.

Selection of 'D' enables the user to define a search (Figure 4) in terms of all attributes of the wider biological record, now perceived as including the structure (Option C) and values which are specific to particular tests (Option T). The search may include multiple terms, earlier sets and Boolean logic.

The user who selects 'C' is offered a sketchpad (Figure 5) to input the query structure which is stored as part of the overall question. On exiting from the Search Definition Menu the user is returned to the Main Search Menu. Selection of 'S' then effects the search. The results can be viewed or printed using options in 'V'.

The search is effected in the order specified. If desired, a chemical search may follow a non-chemical one and thus be restricted to a relevant subset of compounds. By this means a query like 'What compounds active in Test 1234 contain ester groups?' becomes simple although in isolation the substructure aspect of such a search would rarely be feasible.

Although this work is not yet complete the prototype system works well and we expect no difficulty in copying it for the chemical arm of the system and for searching other databases.

INTEGRATION OF STRUCTURES AND DATA IN HARDCOPY

In accordance with our plans to exploit the versatility of 1032 we have adopted a 'mark-up language' approach to printing structures and text together. We simply use the

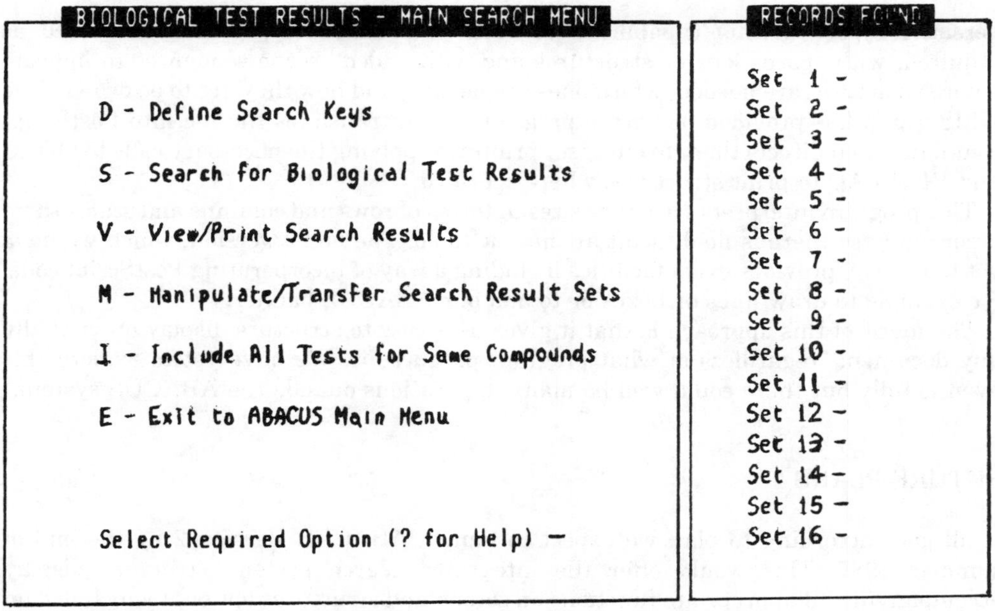

BIOLOGICAL TEST RESULTS - MAIN SEARCH MENU

D - Define Search Keys

S - Search for Biological Test Results

V - View/Print Search Results

M - Manipulate/Transfer Search Result Sets

I - Include All Tests for Same Compounds

E - Exit to ABACUS Main Menu

Select Required Option (? for Help) -

RECORDS FOUND

Set 1 -
Set 2 -
Set 3 -
Set 4 -
Set 5 -
Set 6 -
Set 7 -
Set 8 -
Set 9 -
Set 10 -
Set 11 -
Set 12 -
Set 13 -
Set 14 -
Set 15 -
Set 16 -

Figure 3

```
BIOLOGICAL TEST RESULTS - SEARCH DEFINITION                RECORDS FOUND

        F - FPL Number & Batch                              ) Set  1 -  158
      ) C - Chemical Structure                                Set  2 -
      ) B - Biological Test Number                            Set  3 -
      ) A - Activity Code (Biological)                        Set  4 -
        T - Test-Specific Attribute                           Set  5 -
        P - Project Number                                    Set  6 -
        N - Notebook Reference                                Set  7 -
        I - Initials of Tester                                Set  8 -
        D - Date of Test - Range                              Set  9 -
        R - Record Status                                     Set 10 -
        S - Search Set Records                                Set 11 -
                                                              Set 12 -
   AND) L - Logical Connection of the Search Keys             Set 13 -
                                                              Set 14 -
        E - End Search Key Definition                         Set 15 -
                                                              Set 16 -
        Select Required Option (? for Help) -

Enter Set Number : 1
```

Figure 4

versatile report writing capabilities of 1032 to produce output files formatted as required, with spaces left for structures and with added escape sequences to indicate where structures are needed, which ones are needed, and how they are to be drawn. The ULIS group has provided us with a program which translates the file into PostScript commands and directs them to our laser printer, supplying the necessary calls to ODAC and VDU-PAC to print structures where specified.

This program interprets structure sizes in terms of rows and columns and scales them accordingly so there is no difficult arithmetic to do. The latest version, which we have yet to receive, provides extra facilities including a way of incorporating PostScript code, for example to draw lines or boxes or to add our ABACUS banner.

The merit of this approach is that it gives us access to structure display on virtually any document, regardless of what program produces it. We have yet to explore this avenue fully but there could well be many applications outside the ABACUS system.

FUTURE PLANS

It all goes according to plan we expect to launch ABACUS Version 2 by the end of summer 1987. This would offer the integrated search system, structure display accompanying all appropriate functions on screen and structures optionally included on a range of standard hard copy outputs. Next we shall concentrate on developing Version 3, which will provide direct structure registration by the chemists, stock control

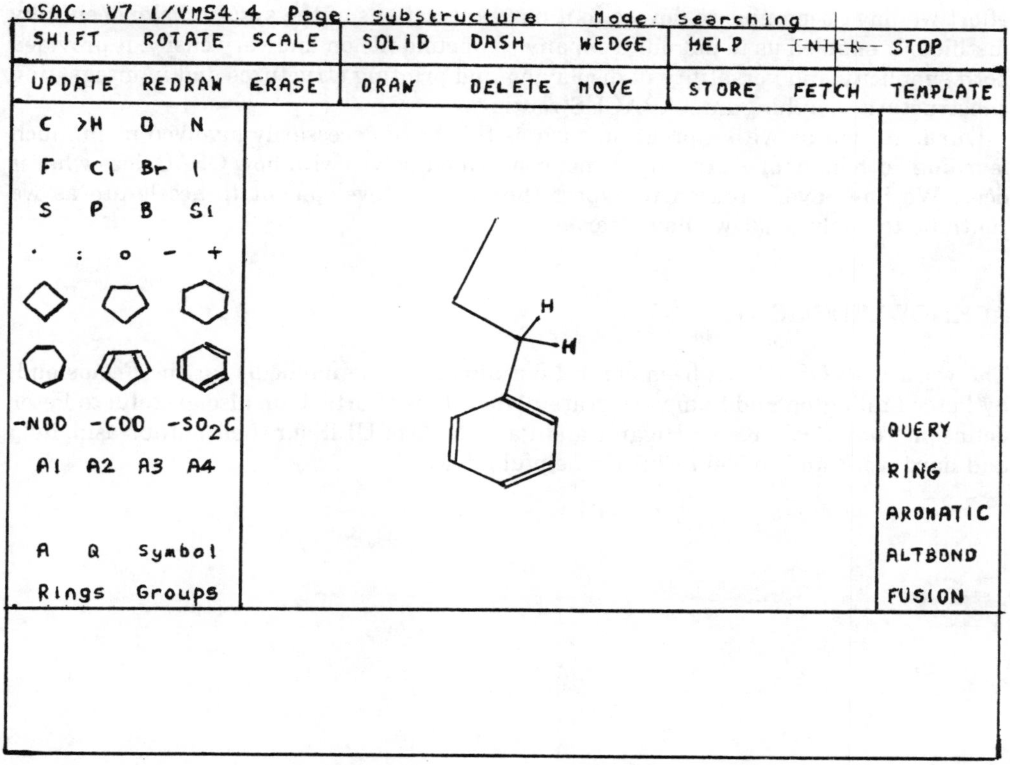

Figure 5

facilities and procedures to control the requesting of tests, the monitoring of progress and the notification of the results.

At this point, which we hope to reach by the end of 1987 or soon after, our major objectives will be met. However, we still have manyplans for ABACUS if the effort can be spared. We are making the Fine Chemicals Directory available on OSAC and expect to be adding several other substantial databases over the next year. We hope to integrate our analytical sample request system with ABACUS to provide substructure access to analytical and physico-chemical data. We are adding procedures to summarise data and, if needed, to interface with RS/1. And we plan to extend the scope of ABACUS to include data other than routine screening data on compounds.

CONCLUSION

We chose OSAC as the chemical component of ABACUS because we judged it the system best capable of meeting our prime need for an efficient chemical database system which could provide a high level of integration flexibility.

Despite the relatively short time we have been working with OSAC and the sparse effort we have been able to devote to it our expectations of the system have been fully justified. It enabled us to put our company compound file on line very easily; it provided both simplicity and versatility in displaying and printing structures and in integrating substructure searching into ABACUS routines.

Our first months with a program such as this have necessarily involved us in much learning, although, of course, we do not concern ourselves with how OSAC does what it does. We have every reason to expect the pace of development to accelerate as we continue to apply what we have learned.

ACKNOWLEDGMENTS

The work with OSAC has been carried out almost single-handedly, at the Fisons end, by Peter Millington and I am very grateful for all his efforts. I am also grateful to Peter Johnson, Tony Cook, Kevin Higgins and Paul Hoyle of ULIS for their enthusiasm, help and hard work and to Neill Clift for helpful advice.

DEVELOPMENT OF THE PFIZER INTEGRATED RESEARCH DATA SYSTEM SOCRATES

David Bawden, Trevor K Devon*, D Tony Faulkner, Jeremy D Fisher, John M Leach, Robert J Reeves and Frank E Woodward
Pfizer Central Research, Sandwich, Kent, U.K.

ABSTRACT

This paper describes the development of SOCRATES, a state-of-the-art user-oriented interactive research data management system in regular use in Pfizer Central Research, Sandwich. The chief attribute of SOCRATES is a high degree of integration of both data and functionality, achieved primarily through the exploitation of a database management system, System 1032, and the effective use of bit screen arrays. These technical features are described and illustrated with examples of SOCRATES in use.

INTRODUCTION

SOCRATES is a full data management system for chemical and biological data generated in-house, comprising subsystems for online compound registration; compound submission for biological testing; direct biological data entry from the laboratories; display, searching and report generation. In addition to in-house data, other computer-readable chemical compound-based files (e.g., Fine Chemicals Directory, Cambridge Crystal Structure Databank) have been integrated into the search subsystem. Other special features of SOCRATES include the incorporation of novel 'browsing' software for chemical structure analogue searching, and the extension of SOCRATES to cater for chemical reaction searching (the CONTRAST subsystem) and 3-D structure searching.

SOCRATES is used widely by Pfizer scientists in a VAX cluster computer environment at Sandwich and is well regarded both for its ease of use and its range of integrated search capabilities, allowing combinations of a variety of chemical and biological search parameters.

THE DEVELOPMENT OF SOCRATES

The SOCRATES system was developed in-house through collaboration of the Pfizer Research Computational Sciences and Research Information Services departments, in consultation with our scientists. Work formally began on the system in early 1983. It was agreed with our users to adopt a 'system prototyping' development approach, this having the benefit of involving our users in providing continuous feedback as system features become available. Thus the first display and search capabilities in SOCRATES became available to our scientists within one year of starting the project, culminating in a fully functional system in mid-1986.

W. A. Warr (Ed.)
Chemical Structures

SOCRATES

- accumulated database of Research compound synthesis and biological testing knowledge
- integrated system for collection, storage, retrieval and distribution of these data
- administration, monitoring and control of the biological testing of Research compounds
- used directly by scientists for compound selection, idea generation, hypothesis testing and synthesis planning
- provides direct access to bought-in databases - eg. crystal structures; commercially available chemicals; drug compendia; chemical reactions

Figure 1.

At Sandwich in 1983 there already existed an internal databanks system for chemical and biological data (INDABS) which possessed a limited degree of integration through the use of chained files. The chemical structure files were coded using the Wiswesser Line Notation (WLN) and INDABS included our own fairly powerful chemical retrieval system, using both WLN string searching and atom-by-atom searching of connection tables generated from the WLN. A fast novelty search capability was also provided using a topological code.[1] Our chemists at Sandwich were trained in coding and understanding WLN, and frequently carried out manual searches of permuted WLN indexes themselves.

Although our scientists had some online interactive access to a variety of in-house information, on most occasions they would seek the assistance of the information staff for their searches. Our target was thus a completely user-oriented interface to the substantial data potentially available. In attempting to achieve a more user-oriented system, a number of basic requirements were established:

1. graphical structure input and output of chemical structures
2. menu-driven access to user options

SOCRATES FEATURES

- menu driven
- graphical structure i/o
- use of System 1032 DBMS
- integrated database
- integrated programs

Figure 2.

3. rapid retrieval of data and structures
4. integration with other research systems and databanks.

When these were examined against our existing system (INDABS), it became clear that a radical re-design of our database structure was going to be necessary and that a graphical structure handling module had to be acquired or written in-house. Our substructure search system also had to be rewritten because structures and chemical searches entered graphically would be stored as connection tables instead of WLN.

THE CHEMICAL STRUCTURE SUBSYSTEM

The components of the chemical structure subsystem of SOCRATES are:

1. graphical structure input for file compounds and enquiries
2. connection table files
3. screen fragments files

4. screen fragments search
5. atom-by-atom connectivity search
6. graphical structure output display/print.

There are a number of related files (e.g., some WLN files have been retained) and functions (e.g., compound registration) that will not be discussed here.

Our first efforts were directed at providing a graphical input for enquiry definition for structure or substructure searches. In this we were greatly aided by the acquisition from John Figueras of his SENTRY program.[2] The FORTRAN code and graphics calls were adapted for the VAX and VT100 Retrographics terminal with light-pen respectively, and some modifications to extend the enquiry input for substructure search were introduced.

It was also fortuitous for us that one of John Figueras' colleagues, Craig Shelley, had written a graphical structure output program that he was prepared to let Pfizer have for use in SOCRATES. This program,[3] also written in FORTRAN, was integrated into our system with only minor modification.

With the graphical output program in place, it was possible to provide our scientists with the first fruits of the early SOCRATES system: graphical structure display at a terminal of compounds from the company files (generated from the stored WLN's via connection tables). The next step was to provide a structure match capability. As mentioned earlier, within INDABS there already existed a fast chemical novelty search which employed an index of topological codes.[1] This was easily integrated into SOCRATES to accept graphically drawn structures from SENTRY instead of WLN's.

The substructure search module required the design and implementation of a fragment screening subsystem. The algorithmically generated screens include atom-, bond- and ring-centred fragments in addition to a number of atom sequence fragments; a set of 1315 screens was chosen on a frequency basis. The screen generation procedure, based on the work of Adamson and Lynch's group at Sheffield,[4] allows appropriate screen sets to be adopted as the structural composition of a file changes with time. Of importance was the decision not to use serial fragment screens, but 'invert' the fragment-compound pointer matrix instead (see Figure 3). This has the particular advantage, at search time, of requiring look-up of only those column vectors that correspond to fragments in the enquiry, instead of serially searching all the compound row vectors. This approach was also to prove subsequently of great use in providing a route for convenient integration of chemical searches with other data in SOCRATES.

The final component of the chemical substructure search system was the atom-by-atom connectivity search which operated on the potential hits from the fragment screening step. An atom-by-atom topological search algorithm was developed in-house along standard lines (our earlier connection table search program proved too specific to WLN-based features to be usable in SOCRATES). One design feature of the search system that has proved popular with our users is the use of 'real time' searching, whereby search hits are displayed as they are found, rather than upon the completion of the search. This also offers the user the opportunity of abandoning a search if it becomes apparent that it is not satisfactory (e.g., to refine a broad search that is retrieving too many hits).

A feature of the SOCRATES connection table design that merits special mention is our inclusion of aromatic bonds: thus an algorithm has been developed for detecting all aromatic bonds in a ring or ring system which more closely accords with the chemists' concept of aromaticity than that found in other systems (e.g., MACCS does not consider

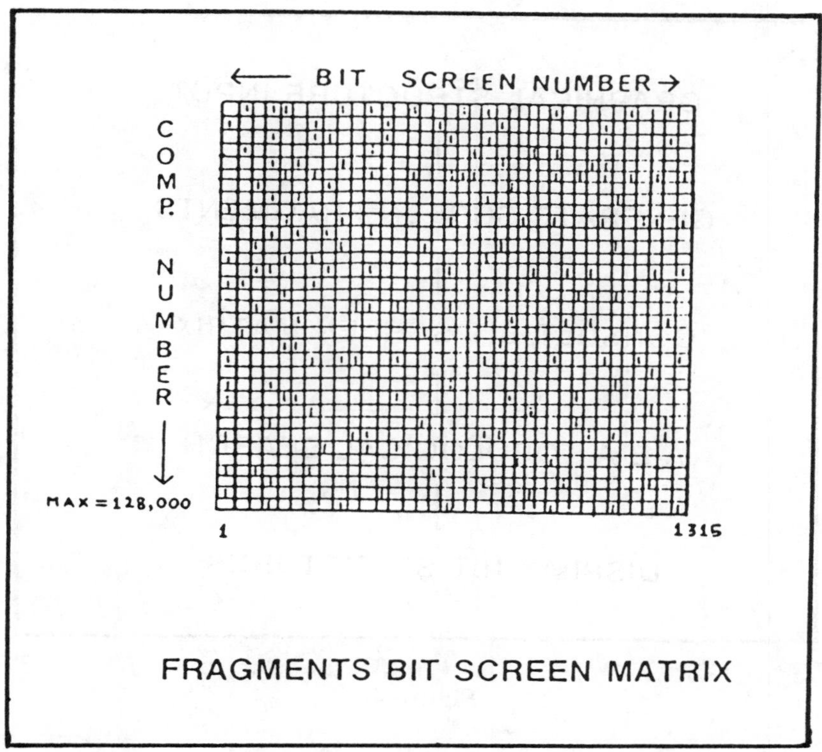

Figure 3.

odd-sized rings suchas thiophene or imidazole as containing aromatic bonds). This requirement arose out of an analysis of a large number of earlier searches carried out in INDABS where the 'aryl' concept featured prominently (see, for example, the anthranilic acid analogues search illustrated in Figure 6 below).

Within the space of one year of starting the project, the prototype chemical structure subsystem of SOCRATES was ready for use by our scientists. Subsequent modifications and extensions to the chemical module were largely a matter of detail (e.g., multifragment search capability; templates) and implementation of different chemical files (in-house company collections; Fine Chemicals Directory; Cambridge Crystal Structure Databank). The substructure search system is shown schematically in Figure 4 below along with examples of the sort of substructural enquiries that can be entered in SOCRATES (Figures 5 and 6). Of particular note is the degree of generic search capability that can be expressed in our system. A major issue in the next phase of development was the nature of the file structures and the database management system (DBMS).

THE DATABASE MANAGEMENT SYSTEM

At an early stage it became evident that the file structures employed in the INDABS integrated system would limit the ultimate flexibility of SOCRATES, and therefore an

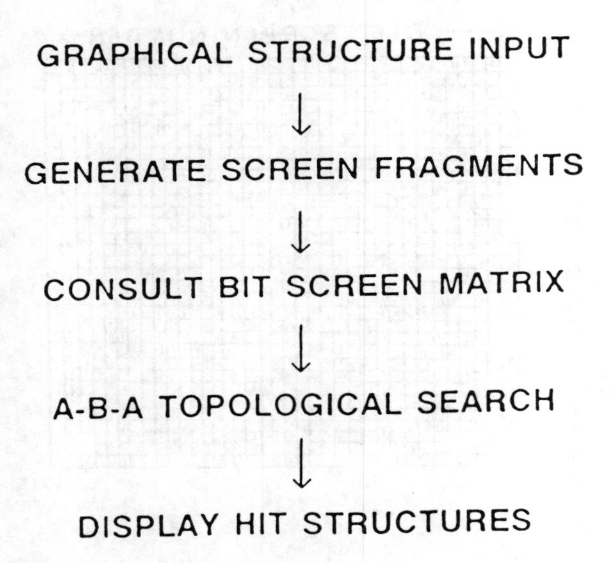

Figure 4.

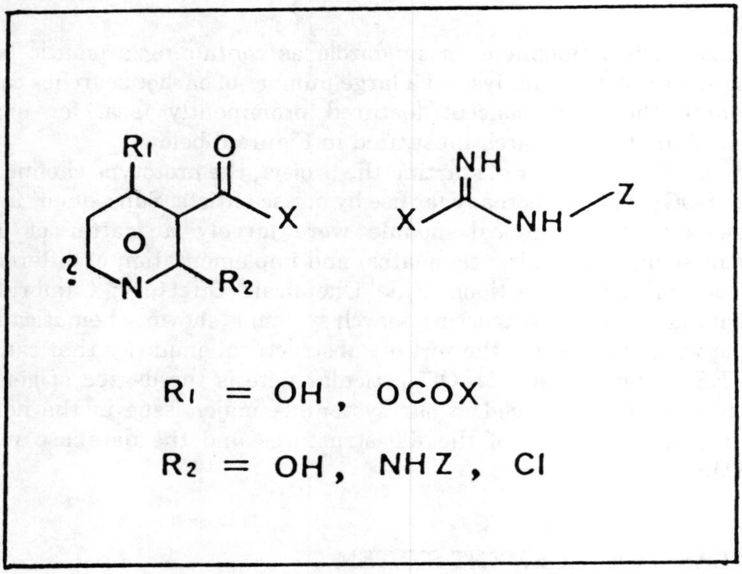

R_1 = OH, OCOX

R_2 = OH, NHZ, Cl

Figure 5.

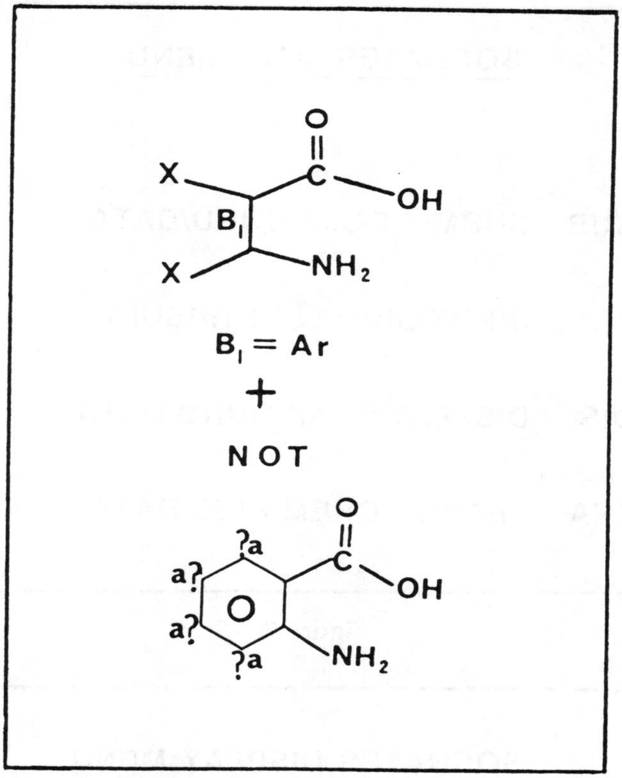

Figure 6.

approach which allowed us to take advantages of new advances in the VAX computing environment was encouraged. At about this time (1983), Software House Inc. had announced System 1032, the VAX version of their DBMS. Our Pfizer colleagues in the USA had started to use S1032 for their biological data, and so an evaluation was undertaken at Sandwich in which S1032 was compared with ORACLE, promoted at the time as the state-of-the-art relational DBMS. S1032 proved most suited to our needs, because of its functionality, operating speed and ease of implementation (especially its host language interface and having its own procedural language).

S1032 was thus selected to handle all the SOCRATES files, with the sole exception of the chemical screening fragments matrix which is stored as a VAX ISAM file. The relative ease of implementation of S1032 in building data-sets from existing computer-readable files enabled us to convert a number of our INDABS files directly into SOCRATES, for example, our screening manager's database, which details all the biological testing that has been carried out for individual compounds. The SOCRATES menu-driven interface was implemented using Forms Management System (FMS) software which had already been successfully employed in an in-house clinical data handling system. Examples of some of the SOCRATES menus are illustrated in Figures 7–9 below. SOCRATES had thus become a FORTRAN shell with calls to FMS and S1032, the latter *via* the powerful host language interface.

SOCRATES MAIN MENU

SUB SUBMIT COMPOUND/DATA

INP INPUT BIOL TEST RESULTS

DIS DISPLAY COMPOUND DATA

SEA SEARCH CHEM/BIOL DATA

Figure 7.

SOCRATES DISPLAY MENU

STR CHEMICAL STRUCTURE
HIS SCREENING HISTORY
BIO BIOLOGICAL TESTING DATA
REG REGISTRATION DETAILS
SYN ROUTE OF SYNTHESIS

Figure 8.

It is not proposed here to describe in detail the implementation of S1032 within SOCRATES, especially since the use of S1022 (earlier, analogous DEC-10 DBMS) for this type of database has already been more than adequately described by Carhart[5] for the Cyanamid MRIDIR system. Suffice it to say that each datafile (e.g., connection tables, screening history, compound registration file) is stored as an individual data-set in S1032. The biological testing data are also stored as individual data-sets for each test. S1032 provides facilities (MAP and JOIN) for linking data from one data-set to

SOCRATES SEARCH MENU

CHE	CHEMICAL STRUCTURE FILES
NOV	NOVELTY CHECK
BRO	BROWSE: SIMILARITY SEARCH
QSM	SCREENING HISTORIES
DAT	BIOLOGICAL TESTING DATA
QFC	FCD TEXT SEARCH
QCD	CSDB TEXT SEARCH

Figure 9.

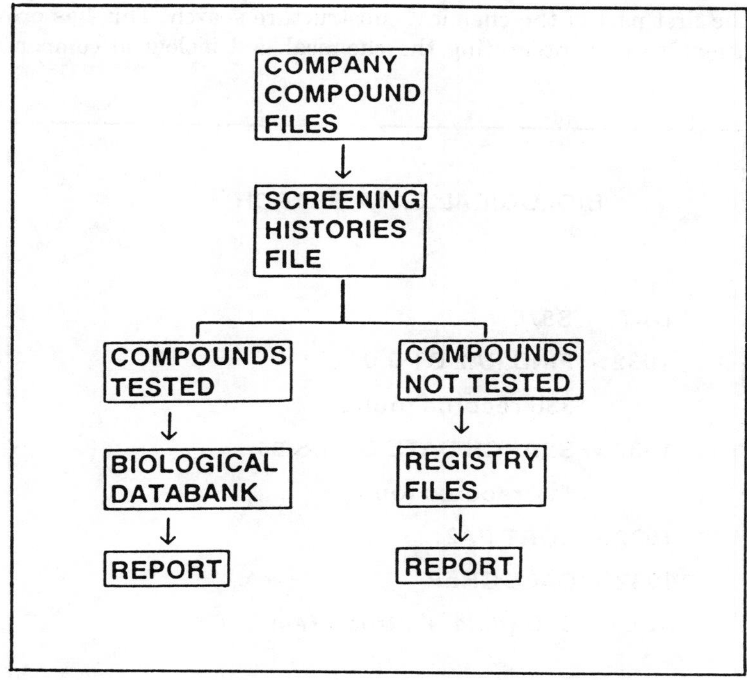

Figure 10.

another, both at the search level and report generation stage. This affords a number of useful integration features.

These sorts of features (e.g., 'MAP' as shown in Figure 10) are presented to the users as options on the various menus provided in SOCRATES, and require no knowledge of S1032 itself. The exception to this is with the raw biological data, where upon selection of a particular test, the scientist uses S1032 commands to interrogate the data-set. This rarely requires the knowledge of more than a few commands (e.g., Find; Search; Print; Sort) and a menu of stored routines (e.g., for printing records out in a standard format) is provided in addition (an example of a biological databank search using S1032 is provided in Figure 11).

INTEGRATED SEARCH FEATURES

Although S1032 affords considerable powers of integration of data-sets, the use of the bit-screen approach has been developed to combine substructure searching with other search parameters, such as biological data, compound availability, dates, and testing histories. Two types of bit-screen are used for this purpose, 'stored' and 'dynamic': the stored screens are for those attributes associated 'permanently' with a given compound (e.g., compound availability) and therefore analogous to the chemical fragments bit-screens; the dynamic bit-screens are vectors generated at search time from a S1032 data-set created in a search, e.g., from a search of biological data. The bit-screen vectors, either stored or dynamic, are simply 'ANDed' to the fragment screen vectors selected as the first part of the chemical substructure search. This has proved a very fast and efficient way of integrating the chemical and biological components of the

```
BIOLOGICAL DATA SEARCH

DAT:   S876
1032>  FIND PA2 GT 9.0
       336 records found
1032>  SEARCH DATE GE 1/6/84
       117 records found
1032>  SORT PA2
1032>  CALL QREP
SOC:   STD (data + structures)
```

Figure 11.

SOCRATES search system (as shown schematically in Figure 12). It is also used within the chemical search subsystem itself to provide 'offspring' searches and 'NOT' logic capabilities (see searches illustrated in Figures 6 & 13).

The chemical fragment bit-screens have also been employed to good effect to offer some novel facilities using similarity or dissimilarity ranking of chemical structures. Thus a calculation of similarity coefficient is applied to the individual compound bit-screen vector, e.g., compared to some target compound, and the coefficients can be used to rank or cluster the compound collection. This has been successfully implemented in SOCRATES as both an alternative to conventional substructure searching (called BROWSE) and for producing sets of chemical compounds for biological testing that are dissimilar amongst themselves and/or some target structure(s). This methodology was developed out of a collaboration with Peter Willett's group at Sheffield and has already been the subject of two publications.[6,7]

An extension of SOCRATES to handle chemical reactions, called CONTRAST, has also been developed at Sandwich, again with some collaboration from Peter Willett at Sheffield.[8,9] In SOCRATES the precursors of our company compounds are stored in the compound registry file, and so application of a reaction site detection algorithm to the starting material and product generates a set of modified connection tables and fragment screens, and a set ofreaction bit-screens. The substructure search graphical structure input menu was modified to enable the chemist to identify the atoms involved in the reaction, but otherwise is essentially the same interface as in SOCRATES.

Other software that has been recently integrated into SOCRATES includes FLATLAND, a package designed to incorporate chemical structures and schemes into word processing systems,[10] and 3D-SEARCH, which extends our searching capability into databases with 3-D structure coordinates (e.g., the Cambridge Crystal Structure

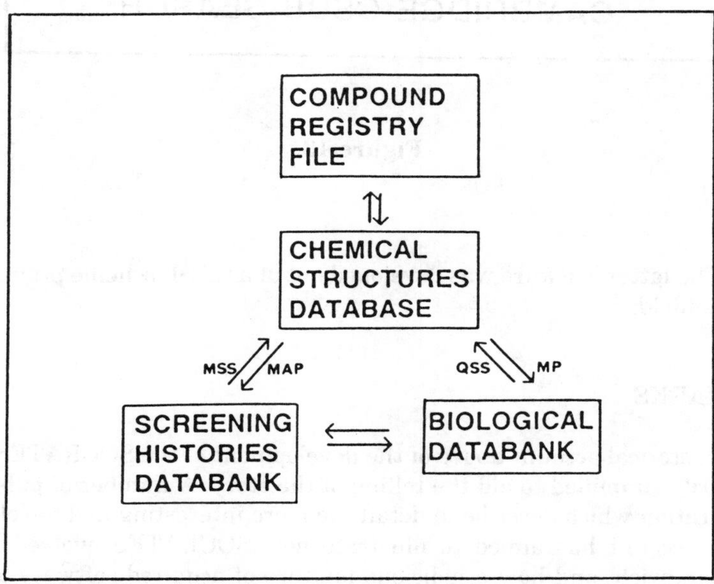

Figure 12.

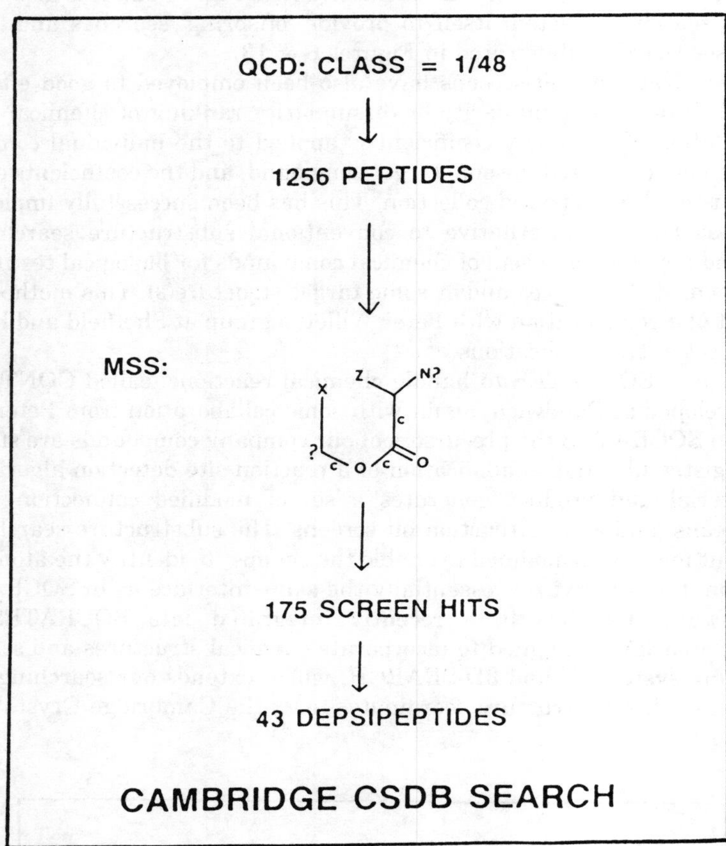

Figure 13.

Databank). The latter software was developed out of a CASE scheme project with Peter Willett at Sheffield.[11]

FINAL REMARKS

The broadly historical account above of the development of the SOCRATES system has been necessarily simplified to aid the telling of the story. A number of publications are now in preparation which describe in detail the more interesting and novel parts of the system. This account has aimed to illustrate how SOCRATES evolved from earlier systems at Sandwich, and how a judicious mixture of acquired software and in-house development has led to a particularly flexible, highly integrated research data management system.

REFERENCES

1. Bawden, D.; Catlow, J.T.; Devon, T.K.; Dalton, J.M.; Lynch, M.F; Willett, P. 'Evaluation and Implementation of Topological Codes for Online Compound Search and Registration'. *J. Chem. Inf. Comput. Sci.* 1981, *21*, 83–86.
2. Figueras, J. 'Chemical Symbol String Parser'. *J. Chem. Inf. Comput. Sci.* 1983, *23*, 48–52.
3. Shelley, C.A. 'Heuristic Approach for Displaying Chemical Structures'. *J. Chem. Inf. Comput. Sci.* 1983, *23*, 61–65.
4. Adamson, G.W.; Cowell, J.; Lynch, M.F.; McLure, A.H.W.; Town, W.G.; Yapp, A.M. 'Strategic Considerations in the Design of a Screening System for Substructure Searches of Chemical Structure Files'. *J. Chem. Doc.* 1973, *13*, 153–157.
5. Carhart, R.E. 'MRIDIR – A System for Integrating Research Databases'. *Drug Inf. J.* 1984, *18*, 153–166.
6. Willett, P.; Winterman, V.; Bawden, D. 'Implementation of Nearest-Neighbour Searching in an Online Chemical Structure Search System'.*J. Chem. Inf. Comput. Sci.* 1986, *26*, 36–41.
7. Willett, P.; Winterman, V.; Bawden, D. 'Implementation of Non-hierarchic Cluster Analysis Methods in Chemical Information Systems: Selection of Compounds for Biological Testing and Clustering of Substructure Search Output'.*J. Chem. Inf. Comput. Sci.* 1986, *26*, 109–118.
8. Bawden, D.; Wood, S.I. 'Design, Implementation and Evaluation of the CONTRAST Reaction Retrieval System'. Chapter 7 in *Modern Approaches to Chemical Reaction Searching*; Willett, P.Ed.; Gower: Aldershot, 1986.
9. Lynch, M.F.; Willett, P. 'The Automatic Detection of Chemical Reaction Sites'. *J. Chem. Inf. Comput. Sci.* 1978, *18*, 154–159.
10. Hanson, R.M., 'The Threefold Challenge of Integrating Text with Chemical Graphics'. Chapter 11 in *Graphics for Chemical Structures*; Warr, W. Ed.; ACS Symposium Series 341; American Chemical Society: Washington, 1987.
11. Jakes, S.E.; Watts, N.; Willett, P.; Bawden, D.; Fisher, J.D. 'Pharmacophoric Pattern Matching in Files of 3-D Chemical Structures: Evaluation of Search Performance'. *J. Mol. Graphics* 1987, *5*, 41–48.

CAS AND MDL REGISTRY SYSTEMS AT EASTMAN KODAK COMPANY

Arnold P Lurie
Department of Information Services, Eastman Kodak Company, Rochester, NY 14650, U.S.A.

SUMMARY

Eastman Kodak has many laboratories worldwide, all with their own information departments, but the Information and Computer Technology Division in Rochester, and in particular the Application and Data Resources Unit, is responsible for the 555,000 compounds in the Chemical Registry System (under both MACCS and CAS Registry System software). There are also on file 17000 reactions under REACCS, 190,000 reports and 65,000 patents in a photographic patents index. The Chemical Information Centre holds 1,100,000 index cards, half of them in accession number order and half in molecular formula sequence.

Kodak installed a CAS Registry II system in 1970 and devised their own system for handling polymers.

The first attempt to integrate the chemical system with the external literature took the form of an Information Imaging Technology Patent Information System, ITPAIS. This is a collaborative venture of Kodak with Agfa, Fuji and the Rochester Institute of Technology. ITPAIS uses a photoscience thesaurus and contains 95,000 chemical accession numbers. Microfilm indexes to text and chemical structures are disseminated to the collaborating companies and DIALOG provides text access to the participants.

In 1975, Kodak asked Chemical Abstracts Service to do a design study on maintaining the Kodak Registry at Columbus, providing the following features. Kodak needed daily registration from data sent by facsimile to CAS. The information needed to be available online to Kodak within 24 hours. Support for Kodak's polymer system was necessary. Access to CAS chemists' expertise was needed, for help with structure conventions, and so on. The Auxiliary Data Management File (ADMF) had to be handled and substructure search was a necessity. Kodak wanted CAS Registry Number overlap detection and the production of a weekly bulletin of the most recent compound updates. CAS had to supply magnetic tapes from which Kodak could generate microfilm indexes, and magnetic tapes of Kodak connection tables. ADMF data had to be returned to Kodak in machine readable form. In addition CAS had to produce guarantees that they would do any vital enhancements needed by Kodak.

A contract was signed in 1978 and the new system was implemented during 1980.

Now that 170 people can logon to the CAS system there is no more need for microfilm indexes. None has been produced since 1985. Chemists can limit their substructure searches to any one of 3 proprietary data bases or to ITPAIS or they can search all four data collections. Eventually Kodak would like to add flags and pointers for reports, for spectral data, for safety data and for the Stores File.

The CAS system has given Kodak the following benefits. There is consistency with CAS conventions, CAS Registry Numbers are available and there is a searching system

W. A. Warr (Ed.)
Chemical Structures

in common with CAS. Kodak have reduced staffing and costs. They are able to support their structure-spectra correlation system using the connection table tapes from CAS. Construction of a REACCS database was facilitated by use of CAS connection tables. The construction of Kodak's MACCS database was done by converting CAS connection tables into MACCS using a program supplied by Molecular Design Limited.

The Kodak chemist can now find out whether a compound is known (using either CAS or MACCS systems); what data is held on a known compound; whether a sample is available; whether a compound is commercially available; and how to synthesise a compound (since multiple REACCS databases are searchable).

The Kodak Registry at Columbus will continue despite Kodak's purchase of MACCS. Certain features of the CAS system are essential. At the moment the MACCS system is updated once a month with CAS input. Kodak is negotiating with CAS for electronic daily transfer of connection tables and their system from Columbus to Rochester. Kodak propose to add more subset indicators to their Registry ADMS and link in more data files once MACCS-II and a relational database management system have been implemented. They will continue training users. Various Kodak/CAS/MDL enhancements and the implementation of new technologies are planned.

ACKNOWLEDGMENTS

This summary of Arnold Lurie's paper has been prepared by the Editor from tape recordings and notes made during the Conference, with the approval of the author.

INTEGRATION OF CHEMICAL STRUCTURES WITH INFORMATION IN SUPPORT OF BUSINESS NEEDS

Harold H Shlevin, Marc M Graham, David F Pennington and Werner von Wartburg*
Research Department, Pharmaceuticals Division, Ciba-Geigy Corporation, Summit, New Jersey 07901, U.S.A.and *Ciba-Geigy Ltd., Basel, Switzerland

ABSTRACT

Co-ordinated, integrated management of critical data in multinational corporations with dispersed research centres can afford significant competitive advantage. We describe the systems we have used to achieve online integration of chemical, biological, archival, textual and other forms of information. A basic strategy has been to use the best commercially available software tool(s) suited to the needs of a particular data domain: Molecular Design Limited's MACCS for structures, REACCS for reactions, ORACLE for efficacy data, BASIS for textual data in document and bibliographic information management systems, RS/1 and SAS for data analysis, Tell-A-Graf for presentation in quality graphics, and All-in-1 for full function scientific word processing. A critical success factor has been the ability to integrate data and graphics from multiple domains in queries, reports, etc. Integration of structural and efficacy data, consistent with strategy, has used MDL's MACCS-ORACLE link, while cross integration of other information has relied on commercial products such as DECpage and ORACLE-SAS, RS/1-Tell-A-Graf and ORACLE-RS/1 links. A weakness has been the unavailability of a commercial system for integration of text and structures which allowed complete data sharing, did not duplicate already available information, embodied full function word processing and maximised user-friendliness. ChemSketch (Copyright of Ciba-Geigy Corporation) meets these needs and provides a powerful unified system that scientists can use to produce shared, high quality documents and reports containing merged structures, text, data and other graphics.

INTRODUCTION

The Drug Discovery Process

Large multinational pharmaceutical and chemical companies synthesise many compounds and intermediates annually. These are typically subjected to biological testing using *in vitro* and *in vivo* models. Of those found active in biological models, the most attractive candidates are selected for follow-up toxicological testing.

The traditional Pharmaceutical Research and Development process is much like an inverted funnel (Figure 1). The historical trends show that approximately 10,000 compounds are synthesised in the basic research stage for every compound that is realised as an approved drug. Of every 10,000 initial chemical entities, approximately 10 (0.01per cent) clear preclinical testing hurdles to enter toxicological testing. Only

W. A. Warr (Ed.)
Chemical Structures

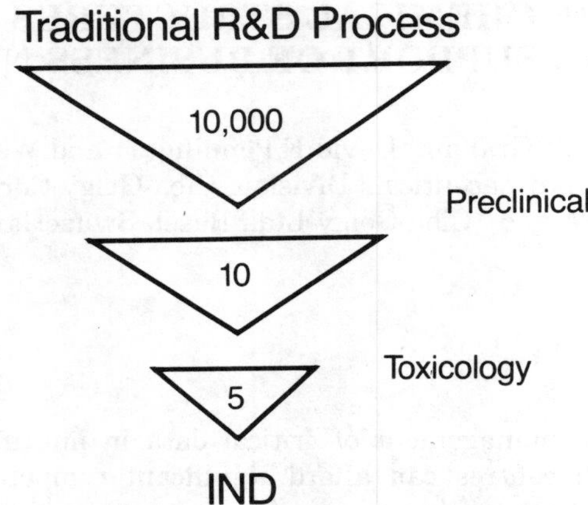

Figure 1. Traditional Research and Development Process: The traditional pharmaceutical R&D process is like a funnel. Of every 10,000 initial new chemical entities, only 5 compounds clear the necessary hurdles to enter human clinical trials. A single weak link at a critical stage in this process will stall all progress.

about 5 (0.05 percent) of every 10,000 tested compounds clear toxicology and enter initial human clinical trials. Considering these odds of success, a close multidisciplinary, interdepartmental co-ordination is required in the drug discovery process. A single weak link at a critical stage will stall progress.

A common denominator or end product of this process is *paper* which is generated to support patents and regulatory approvals. In the preclinical stages this paper takes the form of laboratory notebooks containing chemical, biological and toxicological data records, as well as formal scientific progress and final reports. When a compound clears these major hurdles, the preclinical results are reported in the form of an IND (Investigational New Drug application).

The pharmaceutical industry is confronted by low odds of success, intense competitive pressures, and increasing cost pressures. Yet, the pharmaceutical industry is the world's most research and development intensive major business in terms of investment as a percent of cash flow. In 1985, the health care industry spent approximately $11 billion on research and development. To place this in perspective, the pharmaceutical industry spending on research and development is about 3 times as high (on a percent of sales basis) as manufacturing industries and 45 times higher than spending by the food and beverage industry. Despite this high spending level, several external trends are converging to jeopardise the pay-back on most health care research and development investments.

This has resulted in increased attention to strategies and tactics which can help maximise the productivity of research and development. We describe here, in part, how the Pharmaceuticals Division of Ciba-Geigy has applied strategic management of basic information technology to support better and increase the productivity of its world-wide scientific research efforts. We believe that effective application of such technologies can be a critical success factor in realising longer term business success.

AN INFORMATION MANAGEMENT PERSPECTIVE

In order to manage effectively this massive amount of documentation or paper, the industry initially established paper archives. In the most progressive centres, this paper was often microfilmed for long term storage and retrieval. The ability to identify, locate and retrieve stored printed information was always a difficult challenge. After all, the real objective of an archive is not to input information but rather to retrieve it in the proper form in an effective and efficient manner in relationship to other selected information. Considerable effort in the past was wasted in manually attempting to retrieve such information. Chemical information, because of its graphical nature, posed a unique problem for archival storage and retrieval. Duplicate compounds were sometimes synthesised because the chemists did not have effective means to identify their prior existence.

An increased attention to the overall productivity of research and development, in part, provided the justification to access more effectively the scientific information present within the company. Although this was not a new request and although it was quite simple in principle, realisation had to wait until chemical information, expressed as graphical structural information, was accessible by computer. Similar justification can be used to provide support for the retrieval of biological data and other forms of basic scientific information in a knowledge-based industry.

Challenges of Chemical Information Management

We focus here primarily on chemical information and the key challenges posed in integrating it with other forms of information. These challenges must be met to support the principal business objectives of our organisation.

A key problem in handling chemical structures by the computer is the perception of the unique identity of a chemical structure. It is obvious to the chemist that the real chemistry is a phenomenon which is more than the graphical appearance of the formula. The principal break-through allowing machine processing of structural formulae was the Morgan algorithm which was later extended to include stereochemistry. The stereochemically extended Morgan algorithm, sometimes called the SEMA name, generates a connection table which is unique to a chemical compound. This ability to provide a unique identification of structure, independent of graphical presentation and the reliability of the algorithms, provided the basis to position the computer to understand and perceive the language of the chemist.

General Strategies for Management of Chemical Information

To optimise the use of chemical structures stored in computer databases, Ciba-Geigy's Pharmaceuticals Division follows the following general strategies: it maintains a corporate file of all compounds synthesised by all divisions and it maintains division-specific databases containing all compounds ever synthesised and tested. These are accessible by all research centres world-wide. Associated with the chemical structures which are stored in MACCS, is a minimal set of non-relational information. Additional batch-specific data such as physico-chemical properties are stored in a linked ORACLE database. A principal objective has been to have the structural data available at each research centre and to link it with local biological efficacy databases. In essence, this provides each individual scientist with access to the basic structure-activity knowledge of the company in support of a basic functional operation in the drug discovery process.

The corporate file is currently run on an IBM mainframe and data are stored in Chemical Abstracts Service Registry-II format. An in-house developed graphical interface program allows searching in a very friendly manner. The technology of this software package will be replaced in the near future. This file currently contains approximately 500,000 compounds.

The Pharmaceuticals Division database is implemented in MACCS and ORACLE on DEC VAX computers. It contains approximately 175,000 compounds. Work is still in progress to enter a backlog of about 25,000 very old compounds. Identical databases are maintained at both Summit, New Jersey and Basel, Switzerland. Each of these two major centres enters compounds of its own origin. Updates to the databases routinely occur by automated procedures using DECnet transfer protocols over Ciba-Geigy's private international data communication network. Thus, each centre serves as a 'disaster site back-up' for the other. In Switzerland, individual bench chemists are registering compounds into a temporary MACCS database which after appropriate automated and manual integrity checks are directly registered to the permanent database.

INTEGRATION OF CHEMICAL DATA

The merging of MACCS chemical structure information databases with other information databases has been a strategic goal at Ciba-Geigy. This objective recognised that the majority of data developed in the drug discovery process are functionally linked to a chemical entity. We describe here two specific examples of integration of chemical information with other types of information in support of business needs. These are integration of structural data with relational data contained in ORACLE and integration of structural data with text. The first example utilises commercially available products while the second example utilises proprietary software developed by Ciba-Geigy.

The 'Tool Kit'

MACCS-II is the tool now being used to link data contained in ORACLE databases with the structural information contained in MACCS databases. This 'link' recently created by Molecular Design Limited (MDL), provides a bidirectional data path between MACCS databases and our relational databases (e.g., biological efficacy data, physico-chemical data, chemical stockroom data, etc). It allows controlled retrieval of ORACLE data into a MACCS graphical display and printed graphics.

Therefore, we can obtain data from ORACLE based upon structural retrievals. It also allows us to write information to ORACLE databases from a MACCS application and temporarily switch to an ORACLE 'relational' application if we desire. This latter capability provides a link to ORACLE tools from MACCS. An additional aspect of this link is to use the MACCS interactive display to do *ad hoc* queries of the ORACLE databases. This allows one to generate lists of chemical entities based upon the properties of data outside MACCS. These lists, in turn, can be used to retrieve any related MACCS and ORACLE information into a MACCS display.

In thinking about this 'link', and its application, it is useful to remember the individual strengths of each piece of application software. The strength of MACCS lies with the fact that it is optimised as a chemical structural database. With MACCS, one can draw, display, print, save and retrieve chemical structures based upon the graphical representation of the structure. MACCS in essence 'thinks' of chemical entities in graphical form just as the chemist does. The strength of ORACLE lies in the fact that it is a relational database. This allows the database to maintain the inherent relational characteristics of the experiment itself. Using ORACLE, one can construct multiple tables with multiple records of related data that can be merged upon retrieval based upon various data interrelationships. These relationships can be varied amongst many different applications without changing the physical structure of the database.

The MACCS-II link has its advantages and disadvantages at the moment. To use it, you must always be in MACCS. This problem occurs because you cannot get information from a MACCS database from outside MACCS by any direct method. ORACLE does not have this problem. For example, it is possible to retrieve ORACLE database information into a FORTRAN procedural language program for further manipulation and display. What MACCS-II does provide is a strong link to the outside. This overcomes most of the problems associated with this 'missing' link to MACCS.

We currently use the link in two different ways. In one case the link takes place totally using the MACCS programming interface. The other uses the ability to 'suspend' and exit MACCS, enter an outside application and then return to MACCS. Both of these methods have their advantages and disadvantages.

Our 'Chemical Datasheet', which contains the set of chemical data for a particular synthesis of a compound, is a pure 'MACCS interface' application. The data sheet is a report and because of this, it lends itself well to a pure MACCS application. It retrieves MACCS structural and ORACLE-related data and places it on a sheet of paper or a terminal display. It does not require manipulation of relational data by the person using it. This is not to say that the MACCS programming interface does not allow one to do this. It does, but with a great deal of program development.

The MACCS 'programming tool' consists of building an application out of 'graphical display boxes'. One electronically 'paints' boxes on a graphic computer terminal and each box can then be programmed to read and write an individual piece of information (i.e., a field). This tool easily places one 'record' of information (i.e., a set of records) in one or more tables. It also easily retrieves them. What is difficult to create is 'multiple' record reads and with even more difficulty, multiple record writes. The interface is still basically a MACCS interface: 'One structure, One record'. With the advent of our relational design it is actually: 'One structure, many records'.

Displaying (reading) multiple records of information is basically a problem of making a 'pretty' display or report. Storing (writing) such information requires the creation of a carefully controlled environment for the person who is going to use it. In creation of information, the problems of data integrity and providing guidance to the person who is creating it, become the basic concerns of the application developer. This is why we tend to use ORACLE relational database application development tools for 'relational data entry'. MACCS-II does not let us access its data from these applications, but MACCS-II does let us access these applications from MACCS-II. The key is the ability of MACCS-II to do two things. The first is the MACCS-II ability to 'suspend' MACCS literally and begin running an outside application. The second is the passing of the 'linking' information (i.e., the internal identifier for the chemical entity or formulation) to an application *via* the MACCS-ORACLE link.

Application of MACCS-II

There is a variety of information that needs to be stored and retrieved during the drug discovery and development process. Most of this information is linked to chemical entities recorded in MACCS. There are two major applications of the MACCS-ORACLE link at Ciba-Geigy. The first is the MACCS-ORACLE Chemical Information Database (MOCID), the other is the linking of biological test data with chemical structural data.

Chemical Information Database

The Chemical Information Database contains general information about a chemical entity or drug formulation. Originally, all this information was contained in just a MACCS database. The difficulty with keeping all the data only in MACCS is that there is 'multiple' information about a chemical entity. One of the most obvious examples of this is the data associated with each synthesis (batch) of the compound. There is also multiple information such as synonyms for the name of the compound that also must be stored. Since MACCS is not relational, the best one can do with such data in MACCS is to 'string' data into fields (i.e., MACCS datatypes) and then do a string search of this field when trying to retrieve the information.

The primary purpose of the Chemical Information Database is to take advantage of both the relational power of ORACLE and the chemical structural retrieval power of MACCS. If one needs to retrieve information based upon structure, it is best done in MACCS. The link to ORACLE allows one to control the retrieval of batch information so that each batch can be considered a separate display or page of information. This is possible because each batch in ORACLE is considered a separate set of information. The biological efficacy information is stored in an ORACLE relational database. This information existed before the MACCS-ORACLE link. Because of the link, chemical information in MACCS can be directly linked relationally to biological information in ORACLE. This ability to link biological and chemical information is an important and fundamental tool for drug discovery scientists. This application supports a primary business objective by linking structure-activity data.

The user interface is designed using a menu technique; the implementation today is in DEC's Forms Management System (FMS). It consists of a sequence of menus which allows the user to select different input and query functions. A function is selected by hitting only one key. Pre-selections on each level are available as well as a header line, providing information on the position of the current screen in the menu hierarchy. Another feature to facilitate usage of the system is the help facility: for menu screens, where some fields have to be entered to specify query parameters, help is available for each field.

Those parts of the pharmaceutical database which handle input and output of structural formulae, i.e., the registration of compounds, the data sheets, and part of the result output are implemented using MACCS-II. A typical data sheet output is shown in Figure 2. One of the features of MACCS-II which we heavily use is the interface to the ORACLE database management system. The example data sheet shown in Figure 2

CIBA-GEIGY
Pharmaceuticals Division Basle

CGP002175

Producer	Supplier	Date	Batch
PRICKETT	PRICKETT	01-JAN-78	
		C-Project	Case
		Notebook	MP
		OR	BP
		Mol.Wt. 267.4	Batch Wt.
		Mol Formula C15 H25 N O3	
		Remarks	
		Solubility	
		Location of Batch	
		Gal. Formul.	
Comment	CAS-No.	N or K	Origin PRICKETT
Research Proj.	Compare with	Ref.	

HO
NH
O
O

Figure 2. Typical Data Sheet: Structural formula and the molecular formula are retrieved from MACCS. All the rest of the information would be retrieved from different ORACLE tables. Proprietary information is represented as blank fields in this example.

is produced by retrieving the structural formula and the molecular formula from the MACCS database. All the rest of the information is retrieved from different ORACLE tables.

The MACCS-II sequence language is also used to support the registration process in doing comprehensive, automated checks on parents and salts of the molecule. During registration of a new chemical entity even the division's internal code number is assigned by MACCS. Based on the experience with the paper documentation used over decades we place high emphasis on the consistency of the database. During registration of the 175,000 compounds into the pharmaceutical database we spent several man-months 'cleaning' the information. This involved discussions with the research centres from which the compound originated, and linking to laboratory notebook entries and to biological results. We have still not finished but we have operational tools now as well as tools to meet the future challenges and needs as they arise.

The registration of chemical information is an application where we suspend MACCS. The MACCS-ORACLE application allows the person using it to add and correct the structure information for a given chemical entity or formulation in MACCS-II. When chemical and batch information needs to be manipulated, the MACCS display is switched to an ORACLE SQL*FORMS display. This display carefully controls the addition and updating of compound and batch-related information. It allows them to scan this information quickly using multiple page, multiple line (i.e., record) scrolling displays. Extensive edit checking, database duplication checks, database dictionary checks, field level messages and help are built into this display. Such features are a part of ORACLE'S SQL*FORMS and do not have to be developed. In MACCS, such displays would require considerable development time and soon become 'maintenance nightmares'.

The MACCS-ORACLE Link: The Future

As the biology database becomes more extensive and more historical, a greater need for more complex 'structure-activity' searches will arise. These would allow a scientist to enquire for compounds that have specific structure and biological activity limits. This draws on both MACCS and ORACLE simultaneously, and involves correlating the data retrieval from two different databases.Basically questions such as 'What compounds with this substructure have activity greater than X in result Y of test Z?' would be asked. These questions will require 'intersects' and 'joins' of lists of 'hits' from ORACLE and MACCS. This has now been implemented with 'the link'. Future work involves optimisation and refinement of the report formats to handle radically different biological experimental designs better.

INTEGRATION OF STRUCTURES AND TEXT (CHEMSKETCH)

Purpose

Any chemist who has ever synthesised a compound has wanted to be able to share his findings easily with others. Usually this sharing could only be accomplished by drawing the structure on a piece of paper and distributing copies to all parties of interest. Although this proved to be quite adequate for internal distribution, there was always the problem of how to improve the quality of these structures so that they could be shared externally. Yet, many times the structure did not tell the whole story. The scientist also needed to be able to integrate these structures with text from modern word processing systems. Unfortunately there was no off-the-shelf system available for merging structures and text. Therefore we developed a system that allows a scientist to generate high quality documents containing both chemical structures and text.

This system currently consists of: ChemSketch (written in-house), REACCS and MACCS (from MDL)and DECpage and All-in-1 (from Digital Equipment Corporation, DEC).

Possibilities

In the past, users had to 'cut and paste' documents. This usually meant getting hard copy of a structure from MACCS, cutting it to size, and then taping it in the space left for it in the document. This was certainly a tedious endeavour which consumed many resources. Therefore other possibilities were investigated.

As computers became more widely accepted, computer solutions were evaluated for the production of composite documents (i.e., documents containing both text and graphics). Several commercial software packages were evaluated. Most were limited by the duplicate effort they required in redrawing structures which already existed in MACCS, requiring the user to learn two graphical interfaces (*viz.*, the MACCS DRAW mode plus the drawing peculiarities of the other commercial program). In addition, we were further limited in not being able to share these composite documents with others in the corporation located at geographically dispersed sites. It is often most efficient and expeditious to send documents using VMS or All-in-1 mail over our network. If the document were created using one of the incompatible commercial packages you clearly would not have this option even with file transfer software because of differences between VMS files and those of the other device. Finally, most lacked a full function word processor. Incompatibilities also limited the longer term potential for machine storage of documents containing integrated structures, text and data.

Another solution we considered was using DECslide, a general freehand drawing system developed by DEC. Because DECslide was not created specifically for drawing chemical structures, many of the users' needs were cumbersome to implement. Although this route did solve our problem of distributing documents throughout the corporation it did not address the problem of having to draw the structure twice, once in MACCS and once in DECslide. There was also the problem of learning the commands and switches of DECslide and those of MACCS DRAW mode.

Another problem which was specific to our environment was our lack of ReGIS terminals. DECslide would operate only on ReGIS terminals (VT-240, VT-241, and compatibles). Unfortunately we already had a significant investment in Tektronix 401x and VT-640 emulation terminals since MACCS would not run on ReGIS terminals at the time we purchased our graphics devices. Therefore further evaluation of DECslide was suspended.

In retrospect, we had the majority of the pieces we needed to fulfil our requirements. Our users were already using the WPS+ word processor provided with All-in-1 to create simple documents so it became the natural choice for the word processor we would use to create composite documents. In addition, All-in-1 allows users to send these word processing documents anywhere on the computer network. DECpage as of Version 2.1 of All-in-1 came completely integrated within the word processing (WP) menus of All-in-1. DECpage is a software package developed by DEC that combines graphics files containing ReGIS or Sixel graphics protocol commands with WPS+ documents. Therefore we had everything we needed to produce composite documents. All we needed was a way to get chemical structures into a file containing ReGIS or Sixel protocol commands. To complete this last step we developed ChemSketch.

ChemSketch is a software program developed in-house to draw chemical structures and to retrieve structures from MACCS. It is written in VAX C and will run on any of the VAX CPUs as long as they are running the VMS operating system. ChemSketch currently supports Tektronix 401x, VT-640, and VT-240 terminal emulation.

Description

ChemSketch is an interactive graphics program for drawing chemical structures and reactions. Its user interface is icon-based so that the user can point to the object to be drawn with the mouse (called a button) and then position the cross-hairs where the object is to be drawn. There are also keyboard switches that can be used when drawing chemical structures or reactions to modify the object being drawn. These keyboard switches are implemented to mimic as closely as possible the DRAW mode of MACCS. The buttons and switches allow the user to:

1. Draw chemical structures freehand or with templates.
2. Bring in MACCSstructures *via* a MDL metafile conversion.
3. Save output for further revision at a later date.
4. Convert all output to ReGIS graphics format for inclusion into an All-in-1 document via DECpage.

ChemSketch also has a 'user-friendly' interface (Figure 3) which not only displays errors but also suggests what should be done to remedy the error. The ChemSketch system is integrated into a menu created using a DCL (Digital CommandLanguage) command procedure so that it may be incorporated into our already existing Research Menu System. This menu system allows the scientist to run all of the software on our VAX's by making a choice from the menu. All of these menus mimic as much as possible the menus of All-in-1.

One choice on this menu allows the chemist to pull a structure or reaction from MACCS. It creates a file containing the MDL metafile commands which can later be

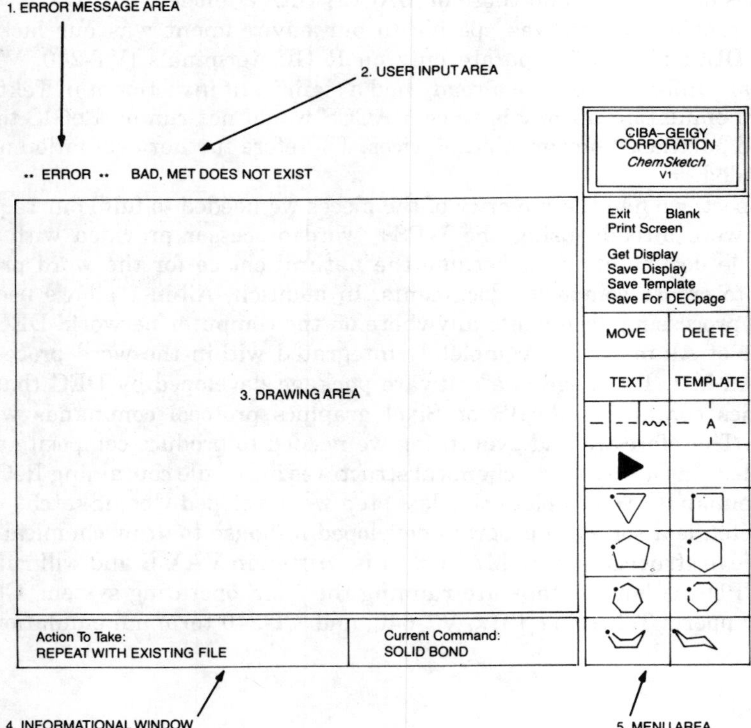

Figure 3. Example of ChemSketch screen: The ChemSketch system provides a variety of user-friendly interfaces. The example shown here was selected to indicate the various fields present on the typical form. Additionally, this is supported by an easily accessed help system.

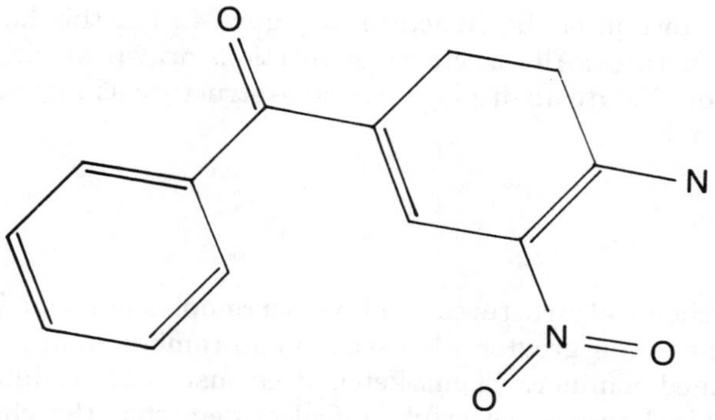

Figure 4a. An example structure extracted from the Fine Chemicals Database of MACCS and placed into a document. Note that there is some distortion to this structure.

converted into a ChemSketch display file using the 'M' switch. The other menu choice is for running the ChemSketch program. Once a command is issued the program prompts the user for the type of graphics terminal currently being utilised. Then the ChemSketch screen is displayed. It is as simple as pressing the button on the mouse once the cross-hairs are positioned over the icon and then pressing the button on the mouse again once the cross-hairs are positioned within the drawing area. All the keyboard switches are the same as the MACCS DRAW mode therefore reducing the amount of training required to bring the user up to speed.

Two key features incorporated into ChemSketch allow the creation of composite documents. The first is the MACCS interface. The user, after extracting an MDL metafile can press the 'M' keyboard switch and quickly review or modify the structure or reaction on the screen. The second key feature is the 'Save for DECpage' command. This command converts the currently displayed structure into a file which contains the ReGIS commands required by DECpage to produce the composite document. Figure 4a is an example of a structure that was extracted from the Fine Chemicals Database of MACCS and placed into this document.

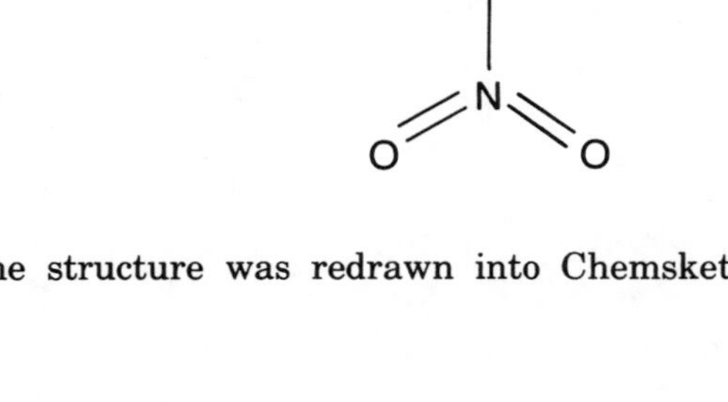

Figure 4b. The same structure was redrawn into Chemsketch and is free of all distortion.

There is some distortion of the structure in Figure 4a but this has to do with how MACCS 'cleans' structures. All structures or reactions drawn within ChemSketch are free of this distortion. Figure 4b displays the same structure as Figure 4a but this time drawn in ChemSketch.

CONCLUSION

The integration of chemical structures and text is a complex problem. The complexity of the problem becomes even greater when you try to combine many different systems together in a unified manner. ChemSketch does just that. It integrates MACCS, DECpage, and All-in-1 into a powerful unified system that the chemist can use to produce high quality documents and reports. We are utilising ChemSketch to extend our abilities to integrate chemical structural data with other primary systems (e.g., BASIS for textual data, SAS and RS/1 for various formsof analysed data, etc.). The combination of tools (*viz.* DECpage, MACCS–II, and ChemSketch) today provide a powerful armamentarium allowing integration of virtually all scientific information.

POSTER SESSION: MOVING TO AN ONLINE ENVIRONMENT FOR CHEMICAL INFORMATION AT MERCK AND COMPANY.

William L Henckler, Debra L Allison, Patricia L Combs,
Gary S Franklin, Walter B Gall and Susan J Sallamack
Merck Sharp and Dohme Research Laboratories, Rahway, NJ 07065,
U.S.A.

ABSTRACT

In 1972 the Merck Sharp and Dohme Research Laboratories (MSDRL) implemented the in-house developed Chemical Structure Information System (CSIS). The system is batch-oriented and the expertise of an information specialist is required to search the database. The software has an intricate database updating process. After more than ten years of use, an extensive study was begun to evaluate alternatives for either replacing or upgrading CSIS. The end result was to have an online interactive system. At the conclusion of the study and beta-testing, the MACCS software from Molecular Design Limited was selected to replace CSIS. A technical staff was organised and trained. The CSIS database conversion was begun. End-user training programs were developed and classes made available to the MSDRL staff. Work continues with the resolution of online updating deficiencies in MACCS versus CSIS and interfacing chemical structures with chemical and biological data.

MOVING TO AN ONLINE ENVIRONMENT FOR CHEMICAL INFORMATION

Electronic data processing of chemical structures at Merck and Co., Inc. began in 1972. Until that time, a manual card system, based on a hierarchical ordering of elements and functional groups developed by Louis R Wiselogle[1] was used to organise chemical structure records. The file was used to check for uniqueness prior to registering new compounds and for very limited substructure searching. After nearly three years of development, beginning in 1969, the Chemical Structure Information System[2] (CSIS) became operational on the IBM mainframe.

The system, still operating to this date, is based on the approach used at the Walter Reed Army Institute of Research in Washington, D.C. The system was extensively enhanced to meet the specific needs of the Merck Sharp and Dohme Research Laboratories (MSDRL) and the Chemical Data Department (CDD) which is responsible for processing chemical structure information at Merck and Co., Inc.

A wide variety of substructure search parameters are available. Among the more noted parameters is the ability to allow for variable substituents. This function is similar to the 'R' group used to represent radical groups in Markush-type structures. Taking the function a step further, CSIS allows for variable positions to be designated on the main structure of the query. Although more flexible, this function is similar to the use of the slashed ring. Both of these functions can be combined, thus substantially

W. A. Warr (Ed.)
Chemical Structures

increasing the power of CSIS to execute generic substructure searches. Other structure-related parameters include indefinite bonds, bond declarations (acyclic or ring) and nondescript atoms. Boolean logic (AND, OR, and NOT) can also be incorporated in a search strategy.

Non-structural parameters such as molecular formula, accession numbers (L-number), registrant's name, and date of registration can all be used to refine the search query. These data are included in the CSIS record. Ancillary data contained in separate databases can also be used to restrict or supplement CSIS output. Examples of such data are sample availability, references to compound sources, and audit trails of sample distribution for biological evaluation.

CSIS has a very sophisticated batch-oriented updating process. Input structures are validated by edit programs that check for syntax rule violations, validate structure against molecular formula, and verify L-number sequence by using a check-letter algorithm. The update programs isolate the base component, if the structure is multi-component (salt, solvate, or formulation); determine if the base is unique to the database; and then determine the appropriate L-number to assign. All records having a common base component have the same root L-number. A hardcopy update report is issued showing what action was taken on the input records.

Although the system was originally designed so questions could be phrased by the users, accessing it does require the interface of an information specialist. Familiarity with the system's capabilities is essential to ensure that no valid answers are missed. It is also necessary to avoid large retrievals that are cluttered with false drops. Even the most frequent users are not aware of the wide range of search strategies that are available. Thus the information specialist is an integral part of CSIS.

Keyboarding chemical structures using the IBM Selectric typewriter initially, and then the IBM 3279 Colour Graphics terminal beginning in 1981, is difficult. It requires the expertise of data entry clerks who perform this task daily. The system operated in batch mode with hardcopy output received within a few hours or the next day. Both of these factors were not in step with the growing interest on the part of the MSDRL staff for online interactive information systems. A move toward such a system came in 1982 when the CSIS edit programs were moved to run under IBM/CICS. Search output was also redirected to a VSAM file that allowed for online viewing using CICS.

The implementation of RS/1 and ORACLE had introduced the MSDRL scientists to online interactive systems for storing and analysing their text and numeric data. The logical next step was the incorporation of chemical structures. Recognising this need, a committee was formed which included the Chemical Data Department, the MSDRL and Corporate computer support groups, and a select group of MSDRL scientists. The object of the committee was to investigate alternatives for either replacing or upgrading CSIS. The result of their work was to recommend an online interactive system for handling chemical structures that would interface with ancillary text and numeric data.

The first task was to define the desired online environment. The scientists wanted the capability to execute online substructure searches, review the results and, if necessary, refine and re-submit the query. All acknowledged the tremendous capabilities of CSIS but all agreed that they were best left for use by the information specialists. These advanced capabilities were classified as 'wants' rather than 'needs' on the part of the scientists and their absence from a system would not completely rule it out from consideration. Other 'needs' were defined as 'user-friendly' graphics input, quick response time (5–7 seconds), the ability to build private project files, and the ability to interface to ancillary data stored in RS/1 or ORACLE.

The Chemical Data Department's criteria were that the system provide the same updating process as CSIS now uses; the same range of generic searching capabilities; the ability to interface with the department's ancillary data files; and large volume production plotting. Another criterion was that the CSIS database be at least 80% machine convertible. This was a shared requirement with the Corporate computer support group who would write the conversion programs. The final requirement was that implementation and support for the system not adversely impact on the overall operations of the department.

The MSDRL and Corporate computer support groups were concerned with the ease of software installation, its operational efficiency on the mainframe, data security provisions, access to an established vendor technical support group, and the availability of vendor supported terminals and plotters. Since a major commitment had been made toward an MSDRL VAX environment, a vendor supported VAX version was a significant requirement.

A number of commercial systems that were available at the time were evaluated. Several failed to meet all criteria, due to a lack of U.S. installations and/or U.S.-based system support groups. Those that failed were DARC/Questel, Télésystèmes, Paris, France; the NIH-EPA Chemical Information System; CHEMPIX, developed by Moreaux of Roussel-Uclaf and marketed by Chemical Information Management Inc.; and SYNLIB, marketed by Smith Kline Beckman. The Upjohn COUSIN system, developed by Dr J Howe, was not commercially available. Chemical Abstracts Service (CAS) private file support was considered too expensive. Security with the CAS file was also a concern of MSDRL since it could not be brought in-house.

At this point only two alternatives remained: MACCS from Molecular Design Limited (MDL) and enhancing CSIS. A comprehensive study by the Corporate computer support group concluded that 36 man-months were needed to complete the work begun in 1982, to enhance CSIS to an online interactive system. MSDRL management felt that a commercial system would be able to meet the scientists' needs in much less time. Also there were only a few IBM compatible terminals in MSDRL, in contrast to the many VAX compatible terminals. Acquiring more IBM terminals presented an additional cost factor to enhancing CSIS. The decision was made to drop CSIS from further consideration.

Much was known about MACCS from demonstrations and extensive discussions with the MDL technical staff. However, many questions remained. An agreement was negotiated with MDL to provide Merck and Co., Inc. with a VAX version of MACCS for a ninety-day evaluation period. Concurrent with these negotiations were discussions on the design of MACCS data format comparable to CSIS format and a methodology for converting the CSIS database. The CDD and MSDRL staffs also received training in theuse of MACCS.

The Fine Chemicals Directory and a subset of CSIS, generated while testing the conversion programs, were the databases used during the evaluation. The CSIS subset contained structures of interest to MSDRL and CDD. This MACCS database provided opportunities for verifying output from a MACCS query with the more familiar output from a CSIS query. In this way much was learned on how MACCS search strategies should be designed to ensure complete results. The CDD staff also used the database to evaluate updating procedures. Knowing the exact content of the database enabled the staff to predict what structures would be found or not.

At the conclusion of the ninety days each group member of the committee reported his results. MACCS was generally acceptable to most scientists. The software was found to be 'user-friendly', offering an easy means of entering chemical structures and associated data into the scientists' private project databases. They found substructure searching to be accurate and fast, although test searching of fixed and flexible datatypes was slow. They recognised the most significant benefit of MACCS was the potential it offered for accessing the CSIS database.

The Corporate computer support group reported that their conversion programs would generate connection tables from the CSIS picture file for an estimated 90% of the database. Writing these conversion programs took significant effort because the specific bonding information, not saved by CSIS, had to be regenerated. With additional time, refinements could be made to improve the conversion rate. However, MDL's LAYOUT programme did cause serious distortion to a few structural types. These would require manual 'cosmetic' editing.

The MSDRL computer support group reported that the software performed well on the VAX, although it was not optimised for that hardware. They recommended that MDL supply a plotting 'metafile' which would address the need for local and large scale production plotting.

The CDD staff concluded that MACCS could not replicate the registration process of CSIS. It also did not provide the sophisticated generic substructure searching capabilities which are used in a least 45% of the searches executed by the CDD staff. The absence of an interface with the department's ancillary databases was another serious deficiency of MACCS.

Although there were reservations, the general consensus was to recommend that MACCS be leased to support the scientists' private project files and that a MACCS version of the CSIS database be built. CSIS would remain operational to support new compound registration, generic substructure searching, and integration with departmental ancillary data. Further, it was recommended that Merck and Co., Inc. begin negotiations with MDL to provide the necessary enhancements to MACCS that would enable CSIS to be retired.

Lease negotiations were concluded and MACCS was installed on the MSDRL VAX in November, 1984. A few project files were generated from CSIS. These provided opportunities to refine the CSIS conversion programmes and develop quality control procedures. Four chemists were hired and trained for the quality control work, and the database conversion began in the second quarter of 1985. CSIS records were converted in blocks of 6,000 records each and in L-number order. The L-numbers from each batch that failed to generate a connection table were saved in a separate file. A CSIS structure print and the corresponding MACCS plot were manually compared by the quality control staff for each record that was converted. Approximately 30% of these required some correction to the structure diagram and/or the text data.

The first pass through the CSIS database of more than 100,000 records required six weeks, and significant IBM and VAX batch time, to complete. The daily updates to CSIS during this time were batched and processed at the end, after the first 100,000+ records had been passed through the conversion programs. Then all structures that failed to generate a connection table the first time were reviewed and appropriate refinements made to the conversion programs. After processing these records the overall rate was calculated to be 98%. Each month the updates to CSIS were converted. This was lowered to every two weeks and is now being done weekly. The quality control phase of the file conversion project was concluded near the end of 1985.

While the conversion was being done, a training program for the MSDRL community was developed. Course material was outlined, 35 mm slides prepared and a reference manual which condensed the 'MDL MACCS Reference' manual was written. Training classes began in June, 1985. The FCD and the CSIS subset were the databases used during the training program.

Security issues were also addressed at this time. A program was written and layered over the MACCS database. A user's logon identification is granted a 'key' which allows access to the database. Only one individual has authorisation to grant this 'key'. Before the 'key' is granted, a security application must be completed and signed by the appropriate managerial levels and department security administrator. With security in place, the Merck Chemical Structure Data Base (MCSDB) was opened to the authorised MSDRL scientists in November, 1985, approximately one year from MACCS installation.

MACCS has been a success at Merck and Co., Inc. Presently more than 300 people have been trained to use it, with additional classes scheduled through 1987. An average of 80 different users per month access MACCS through a variety of terminals: Tektronix, Envision, VT640, IBM and MacIntosh PC's with T-Graf, EMU-TEK, and VersaTerm-PRO emulators, respectively. Over 120 scientists have access to MCSDB. The number of substructure search requests sent to the Chemical Data Department has been reduced by two-thirds. Over 30 project files have been generated by the CDD staff to be used by individual groups of scientists in specific areas of research. The metafile, coupled with an in-house developed program, has enabled the IBM 3800 laser printer to be used for large scale production plotting. However, progress in the implementation of requested enhancements to MACCS required for the discontinuation of CSIS has been very disappointing.

The recent installation of MACCS-II with the Customisation and DBMS Interface modules has provided the capability to interface chemical structures with ancillary data. After receiving training from Molecular Design Limited, the CDD staff and Corporate computer support staff developed screens and sequences to customise MACCS-II for the MSDRL environment, and to replace the current version of MACCS. An orientation program was prepared to acquaint the user community with the new facilities and commands. Technical assistance was given to convert, where necessary, the scientists' project files.

MACCS-II screens and sequences have been generated for one large project biological data ORACLE table; response to this prototype application has been positive. Requests for modifications have been made that reflect the different needs of those who use the database. Other biological data ORACLE tables are being built and will also be interfaced with MCSDB using MACCS-II. The migration of CDD ancillary data from IBM to VAX/ORACLE has begun. A menu sequence will be provided so the user will be presented with options for selecting CDD ancillary data at the conclusion of the substructure search.

Future plans call for investigating new technologies that will further integrate not only diverse databases, but hardware and software as well. Ultimately, terminals and PC's will be replaced by multiapplication workstations that will then provide the MSDRL scientists with greater potential for retrieving and analysing their data.

REFERENCES

1. Buhle, E.L.; Hartnell, E.D.; Moore, A.M.; Wiselogle, L.R.; Wiselogle, F.Y. 'A New System for the Classification of Compounds'. *J. Chem. Ed.* 1946, August, *23*, 375–91.
2. Brown, H.D.; Costlow, M.; Cutler, F.A. Jr.; DeMott, A.N.; Gall, W.B.; Jacobus, D.P.; Miller, C.J. 'The Computer-Based Chemical Structure Information System of Merck Sharp and Dohme Research Laboratories'. *J. Chem. Inf. Comput. Sci.* 1976, *16*, 5–10.

CUSTOMISATION FOR CHEMICAL DATABASE APPLICATIONS

Erick K F Ahrens
Molecular Design Limited, 2132 Farallon Drive, San Leandro, California 94577, U.S.A.

ABSTRACT

An important need in the chemical/pharmaceutical industry is for a chemical database system that will serve diverse information needs. From the chemist's viewpoint, structures are central to these needs, but others involved with biological screening or inventory control have a different, though related, viewpoint. The applications of these diverse users demand flexibility in a chemical database system. Further, new technology in networks, distributed databases, and workstations will lead to additional differences in chemical database systems among companies and their departments. To meet these diverse needs, a chemical database system must be customisable. This paper reviews a system that solves many of these chemical database applications today, Molecular Design's MACCS-II with its Customisation and DBMS Interface Modules, and discusses how the chemical/pharmaceutical industry can take advantage of new technologies through customisable chemical database systems in the future.

INTRODUCTION

On reading the title of this chapter many readers will have at least three questions: 'What is meant by customisation?', 'What is a chemical database?', and 'How does this chapter apply to my company?'.

Customisation is a term applied to the alterability of commercial computer software to the specifications of individual companies and of individual users.

Chemical database refers to a database management system (DBMS) with special features for treating molecules.

This chapter is useful to chemical/pharmaceutical companies that utilise computers for chemical and scientific information management.

Today, most chemical/pharmaceutical companies take advantage of computers to increase the productivity of their research. An important use of research computers is for chemical database applications that permit scientists to access, manipulate, and create reports containing chemical structures. As solutions to these applications have emerged, researchers have turned their attention to broader applications. Scientists now envision chemical-scientific database applications that include structures and related scientific information such as biological data, inventory data, and instrument information, depending on their research interests. An application goal can be imagined to be computerised scientific information management for multidepartment, multisite, multinational companies. Solutions for these applications are immediately complicated by the fact that the various data and hardware of the information system

W. A. Warr (Ed.)
Chemical Structures

differ among companies and their departments. This complication can be addressed through customisation and is the subject of this chapter.

TECHNOLOGY AND NEEDS 1984

Chemical Database Technology

By 1984, the use of chemical database systems was well understood and wide-spread in the chemical/pharmaceutical industry.[1,2] Their use provided researchers with rapid access to company-wide, and in some cases industry-wide, chemical information. A few companies had developed proprietary systems for this purpose such as Upjohn's COUSIN[3] and ICI's CROSSBOW.[4] However, most companies were using commercial chemical database systems such as Molecular Design's (MDL's) MACCS.[5,6]

MACCS. The features of MDL's Molecular Access System, MACCS, began with the ability to draw molecules or fragments of molecules with a computer graphics terminal. Once drawn, the molecule could be registered into a chemical database, or used as a query to search a database. A search could look for an identical molecule, or it could treat the query as a substructure and look for all molecules that contained it. Further, the search algorithms would treat stereochemistry, parent status, tautomerism, isomerism, formula, etc. Along with these drawing, registration, and searching features were plotting and data manipulation features that rounded out the package. MACCS operations were menu-oriented, making the system 'chemist-friendly'.[7]

Most chemical research groups using MACCS created one or more databases of proprietary chemical structures and associated property data.[8] They also utilised the FCD database that was available from MDL in MACCS database format. That database was derived from the Fine Chemicals Directory published by Fraser Williams (Scientific Systems) Limited.

DATACCS. In 1984, MDL augmented the MACCS system with the introduction of the Data Access System, DATACCS. This program simplified data transfer and report generation between various files and MACCS databases. Since the files could be from other computer programs or DBMS's on the same or other computers, DATACCS was an important step in data integration. It also had a simple programming language to allow repetitive tasks to be automated and to provide for data transfer and report generation on a large scale. DATACCS operations were command-oriented, for use by scientific programmers.

General DBMS's. Many chemical/pharmaceutical companies utilised general DBMS's for information related to chemical structures where chemical structure access was not needed directly. Applications included database storage for reports on the biological activity of compounds and for toxicology information. The data integration provided by DATACCS was useful for these applications, though disk file oriented and indirect, since it could merge data from these DBMS's with chemical graphics and property data. Among DBMS's used by chemical/pharmaceutical companies for research, Oracle Corporation's ORACLE was most popular followed by CompuServ/Software House's System 1032 and Infodata's INQUIRE.

Many DBMS vendors claimed that their product was '4th generation'. This label followed the reasoning that computer languages such as FORTRAN and COBOL were 3rd generation, assembly languages were 2nd generation, and machine languages were 1st generation. DBMS's purporting to be 4th generation typically had a high-level language for programming database applications whose use did not require knowledge of low-level data structures and related algorithms. Also, many of these 4th generation languages could drive menus for friendlier applications as compared to command-driven languages. The QUEL language developed at the University of California at Berkeley[9] and the SQL language developed at International Business Machines (IBM)[10] were examples of 4th generation languages designed for DBMS's.

Some DBMS vendors also claimed that their product was 'relational'. This label was properly applied to DBMS's that viewed data storage as tables that could be manipulated by relational operators.[11] These operators included 'select' to search for rows in a table, 'project' to combine rows from tables, and 'join' to combine columns from tables. The QUEL and SQL 4th generation languages were developed for use with relational DBMS's. The ORACLE DBMS used SQL and was claimed to be both 4th generation and relational.

Note that many features of a chemical database system could not be reasonably duplicated with a commercial DBMS. One reason was that general DBMS's were designed to treat numbers, text, dates and money data types, but not abstract data types such as chemical structures. This allowed DBMS vendors to tune the performance of their product to the needs of the general business community, a vast market place, and to avoid specialisation in narrower markets such as the chemical/pharmaceutical industry.

Chemical Database Needs

The chemical database technology described above is fragmented. Scientists had to know the details of several systems, some of which utilised dissimilar hardware, in order to utilise them for their research. The chemical/pharmaceutical industry needed and was asking for an integrated chemical-scientific database system that treated chemical and biological database needs and that adapted easily to their applications.[12]

Chemical Database Considerations. Most researchers found the menu-driven chemical database operations of MACCS friendly and with a good selection of features for their work. They also appreciated the data integration that DATACCS provided between MACCS databases and data from other programs, but the separation of the programs made some applications cumbersome. What was needed was a system that combined the capabilities of MACCS and DATACCS and that interfaced to other programs more directly.

Biological Database Considerations. Since MACCS was oriented to the needs of the research chemist, biological test data and inventory information were easier to treat externally. Many chemical/pharmaceutical companies utilised a general DBMS for data related to chemical structures. These systems served their purpose well, except when chemical structures needed to be inserted into reports or otherwise keyed directly. What was needed was a system that could key both structures and general DBMS data. This required a system that interfaced MACCS with general DBMS's directly.

Applications Adaptability. In 1984, many chemists appreciated 4th generation capabilities and wanted a system that allowed scientific programmers to create menu-driven interfaces for their applications. In addition, they wanted to include the menu-driven chemical database operations of MACCS as part of their applications packages. This would minimise the need to learn multiple systems such as MACCS, DATACCS and DBMS's, and improve their productivity. In summary, scientists needed a system with the adaptability to meet the research application needs of their multidepartment chemical/pharmaceutical companies.

STATE-OF-THE-ART 1987

MACCS-II is MDL's solution to the needs expressed by the chemical/pharmaceutical industry in 1984. It is a 4th generation system that represents the state-of-the-art for chemical-scientific database systems in 1987. The standard version of MACCS-II combines and enhances the features of the MACCS and DATACCS programs. The optional Customisation Module addition to MACCS-II provides users with the ability to design customised menus; accommodate foreign language conversions and the terminology of different scientific disciplines; automate the performance of routine procedures by using an enhanced command sequence language and invoke external procedures.

The optional DBMS Interface Module addition to the Customisation Module expands data access capabilities beyond chemical data to include related scientific information in general DBMS's.

The capabilities of MACCS-II with its optional Customisation and DBMS Interface Modules are described below in terms of its facilities, MACCS menus, and application possibilities.

MACCS-II Facilities

MACCS-II commands are grouped into facilities, each of which represents a chemical-scientific database function. These functions interact to perform the data retrieval, transfer, and reporting activities that are central to the system. Scientific programmers utilise the facilities to create command- and menu-driven applications for use by end-user scientists. The facilities are summarised as follows:

Program Operations. This facility performs operations necessary to start-up, command processing, and exiting MACCS-II, as well as allowing adjustment of some of the terminal characteristics. Additionally, a command is provided to invoke external processes such as MDL's reaction indexing system, REACCS or BBN Software Products Corporation's statistics program, RS/1.

Help. This facility is used to access information on program commands and topics. In addition to the online help provided with MACCS-II, user-defined help pages may be added for use with custom menus and command procedures particular to the research site.

Editors. There are three types of editing common to chemical-scientific database operations: data form and menu, text, and molecule. A separate facility is provided for each in MACCS-II.

Form Edit creates the forms that route data for display and database operations. A form is built of boxes, each of which has properties controlling data format and flow. The data format is illustrated in Figure 1. In that figure, the input device property, IDEV, is shown and indicates the many devices that may simultaneously supply information to the form. Also in that figure, the commands are shown that route information from the form to the various devices. In addition to form editing, this editor may be used to create menus for use in custom applications.

General Edit allows text editing of box labels and data, command sequences, and disk files in an interactive edit mode or from within an automated sequence.

Chemical Edit allows drawing and editing of molecular structures for input to a form or database. Drawing may be free-hand or by use of templates provided by the program. Operation of this mode may be controlled through an interactive menu, through a command-driven molecule editor that may be used as an interactive mode, or from within an automated sequence.

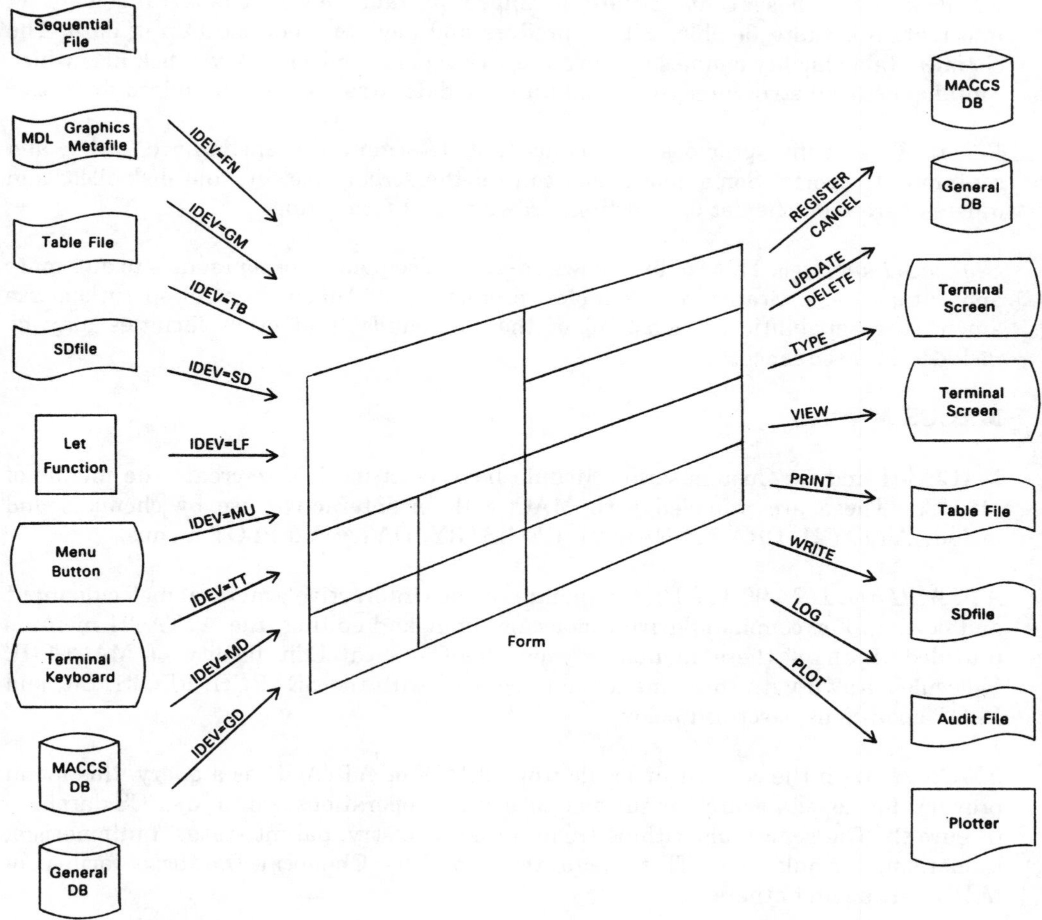

Figure 1. MACCS-II form data flow

Database Operations. There are two classes of database operations common to chemical-scientific database applications: general and chemical. A separate facility is provided for each in MACCS-II.

The General Database operation controls storage, retrieval, searching, and deletion of data in general databases that interface to MACCS-II. ORACLE or System 1032 DBMS interfaces are available from MDL. Other DBMS's that have interface architecture similar to ORACLE's may be user-programmed for use with this facility.

The Chemical Database operation controls storage, retrieval, searching, and deletion of data in MACCS databases. The chemistry treated by MACCS database operations, such as substructure searching, is included in this facility.

File Handlers. There are two classes of table files handled for chemical-scientific database operations: general and chemical. A separate facility is provided for each in MACCS-II.

The general table file facility enables reading and writing of text and numeric data via disk files whose contents are arranged in table form. The arrangement of information is quite flexible, with delimiters and keys to guide location of data. The chemical table facility enables reading and writing of chemical data via disk files which contain chemical structures, text, and numeric data arranged in a standard format.

Report. This facility provides commands that determine the appearance, order, and contents of reports. Some commands control the screen, plotter, and disk files, and others operate on the data by sorting, drawing, and formatting.

Sequence Language. This facility allows creation of sequences of commands to automate application tasks. Branching, variables, prompting, and menu control operations are among the capabilities. Almost all of the commands from other facilities may be included in a sequence.

MACCS Menus

MACCS-II and its Customisation Module have been used to re-create the menus of MACCS. These are provided with MACCS-II for interactive use by chemists and include ATTACH, DRAW, SEARCH, REGISTRY, DATA and PLOT menus.

ATTACH and DRAW. The DRAW menu provides interactive graphical molecule input and editing. For command-driven molecule input and editing, the ATTACH menu is provided. Both of these menus rely on the Chemical Edit facility of MACCS-II. Molecules drawn with these menus may be used with the SEARCH, REGISTER, and PLOT operations described below.

SEARCH. With the current molecule from DRAW or ATTACH as a query, this menu provides for novelty search or substructure search operations over a MACCS database (Figure 2). The search algorithms treat stereochemistry, parent status, tautomerism, isomerism, formula, etc. This menu relies on the Chemical Database facility of MACCS-II among others.

REGISTER. This menu provides for storage of chemical structures in a MACCS database. Itrelies on the Chemical Database facility of MACCS-II among others.

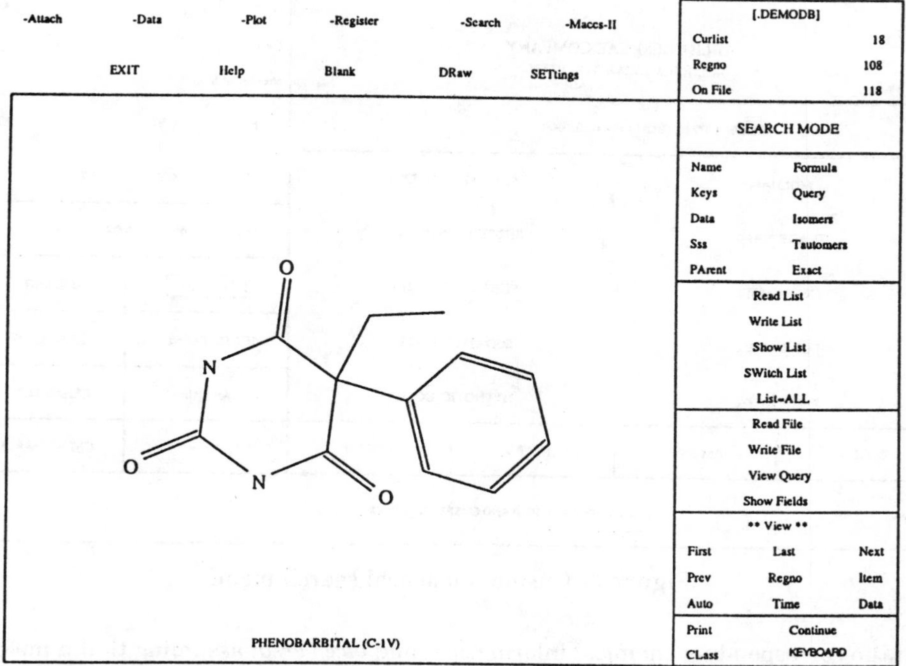

Figure 2. MACCS-II chemical search menu

DATA. This menu provides for bulk storage of text and numeric data in a MACCS database. It relies on the Chemical Database facility of MACCS-II among others.

PLOT. This menu provides for report needs that include chemical graphics. It relies on the Report facility of MACCS-II among others.

Application Possibilities

MACCS-II with its Customisation and DBMS Interface modules can be tailored to the chemical-scientific database needs of entire organisations or individual research groups. Customisation features allow MACCS-II to treat applications expressed in 1984 and many applications that go beyond those. These applications include interfaces with database systems, scientific programs, and scientific instruments that are used in regulatory reporting, laboratory information, and inventory systems. Examples of some applications are as follows.

Database Interfaces. A company that uses a general DBMS for its biological information and a MACCS database for its chemical information might create a biological search menu and a biological-chemical viewing menu for use in conjunction with the chemical search menu. Then the biological database could be searched and the results viewed

FABLE CHEMICAL COMPANY
BIOLOGICAL SEARCH SCREEN

GDB: BARBS
On File: 21
Number of hits: 18

FIELDS 1 TO 10, FROM 13 IN BARBINFO

EQ GE LE WL
NE GT LT
AND OR NOT ( )

MOLNAME SUBSTITUENT_R5
TRADE_NAME SUBSTITUENT2_R5
CRCNUMBER SUBSTITUENT_R1
UP_DATE SEDATIVE_DOSE
BARB_DATA HYPNOTIC_DOSE

SEARCH | USE LIST
SET DATASET | SAVE LIST
CLEAR QUERY | CLEAR LIST

PAGE + | PAGE - | HELP | MACCS-II | VIEW | CHEM SEARCH

Query = HYPNOTIC DOSE LE 300.0 AND ONSET_OF_ACTION GT 25.0

Figure 3. Custom biological search menu

with the corresponding chemical information, and *vice versa*; assuming that a molecule identifier is available in both databases. The sample biological search menu shown in Figure 3 allows the user to construct a relational query, and the sample biological-chemical viewing menu shown in Figure 4 will view information based on a biological search or on a chemical search via the menu shown in Figure 2. A paging button on the viewing menu allows the user to cycle through lengthy biological text information. Both of these biological menus rely on the General Database facility of MACCS-II among others.

Scientific Programs. In some cases, a MACCS-II application will need to access an outside scientific program for data. Examples are MDL's REACCS and BBN's RS/1. One approach to this need is to have MACCS-II generate or prompt the user for input to that program, then it can be run as a background process. Another approach is to execute the other program as a separate interactive process. In either case, when it finishes, MACCS-II can display the results to the user as shown in Figure 5. In that figure, the user provided input to the REACCS program via menu buttons, and MACCS-II displayed the resulting reaction in its form. This menu relies on the Program Operations facility of MACCS-II among others.

Scientific Instruments. In other cases, a MACCS-II application will need to access an outside scientific instrument for data. An example is a spectrograph workstation. An approach to this need is to have MACCS-II generate or prompt the user for input to that instrument, then it can be activated as a background process. When it finishes, MACCS-II can display the results to the user. This capability is shown in Figure 6. For that figure a Hewlett-Packard workstation returned a spectrograph that MACCS-II then displayed in its form. This application relies on the Program Operations facility of MACCS-II among others.

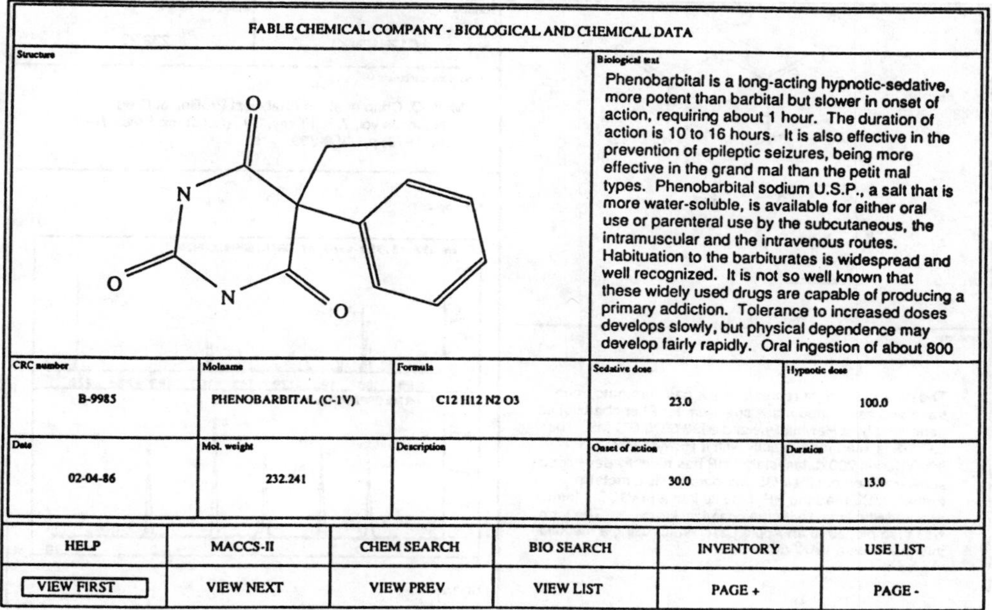

FABLE CHEMICAL COMPANY - BIOLOGICAL AND CHEMICAL DATA
Structure
O
N
O
N
O
Biological text
Phenobarbital is a long-acting hypnotic-sedative, more potent than barbital but slower in onset of action, requiring about 1 hour. The duration of action is 10 to 16 hours. It is also effective in the prevention of epileptic seizures, being more effective in the grand mal than the petit mal types. Phenobarbital sodium U.S.P., a salt that is more water-soluble, is available for either oral use or parenteral use by the subcutaneous, the intramuscular and the intravenous routes. Habituation to the barbiturates is widespread and well recognized. It is not so well known that these widely used drugs are capable of producing a primary addiction. Tolerance to increased doses develops slowly, but physical dependence may develop fairly rapidly. Oral ingestion of about 800
CRC number
B-9985
Molname
PHENOBARBITAL (C-1V)
Formula
C12 H12 N2 O3
Sedative dose
23.0
Hypnotic dose
100.0
Date
02-04-86
Mol. weight
232.241
Description
Onset of action
30.0
Duration
13.0
HELP
MACCS-II
CHEM SEARCH
BIO SEARCH
INVENTORY
USE LIST
VIEW FIRST
VIEW NEXT
VIEW PREV
VIEW LIST
PAGE +
PAGE -

Figure 4. Custom biological-chemical viewing menu

REACTION SEARCH MENU

+

→

extreg	
39985	SRCH RXN
catalyst1	
NaOC2H5	SRCH DAT
catalyst2	
	FIND RXN
solvent1	
EtOH	SAVE QRY
solvent2	
MeOH	DRAW
yield(%)	
0075.0	HELP
time(hrs)	
001.500	MAIN

litext	comments
C.O. Wilson, Ph.D., et. al. Textbook of Organic Med. and Pharm. Chemistry	Mg(OCH3)2 can be used as a catalyst instead of NaOC2H5

Figure 5. Custom menu to invoke REACCS

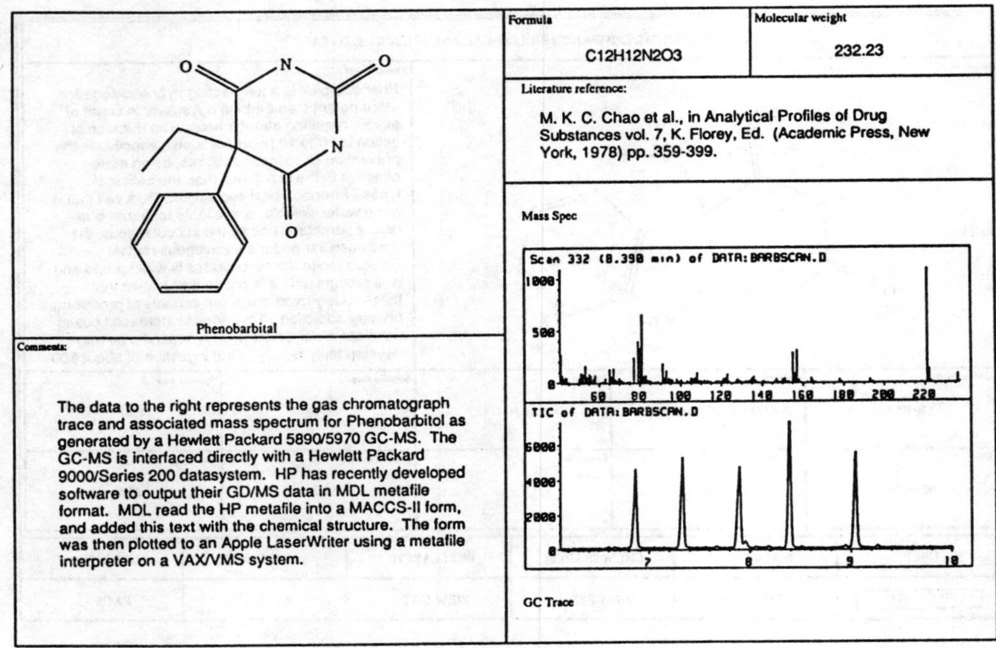

Formula
C12H12N2O3
Molecular weight
232.23
Literature reference:
M. K. C. Chao et al., in Analytical Profiles of Drug Substances vol. 7, K. Florey, Ed. (Academic Press, New York, 1978) pp. 359-399.
Phenobarbital
Comments:
The data to the right represents the gas chromatograph trace and associated mass spectrum for Phenobarbitol as generated by a Hewlett Packard 5890/5970 GC-MS. The GC-MS is interfaced directly with a Hewlett Packard 9000/Series 200 datasystem. HP has recently developed software to output their GD/MS data in MDL metafile format. MDL read the HP metafile into a MACCS-II form, and added this text with the chemical structure. The form was then plotted to an Apple LaserWriter using a metafile interpreter on a VAX/VMS system.
Mass Spec
Scan 332 (8.398 min) of DATA:BARBSCAN.D
TIC of DATA:BARBSCAN.D
GC Trace

Figure 6. Custom form with spectrograph display

NEW TECHNOLOGY 1987

By the time of writing, trends within the chemical/pharmaceutical industry and the computer industry have led to standards and *de facto* standards for software and hardware. These are influencing the new technology appropriate for the chemical/pharmaceutical industry. The primary areas of this new technology are in computer hardware and operating systems, computer software environments, relational database management systems (RDBMS), computer networks and distributed systems, and chemistry.

Hardware and Operating Systems

In the area of computer hardware and operating systems the chemical/pharmaceutical industry increased its use of Digital Equipment Corporation (DEC) VAX computers with the VMS operating system, increased its use of IBM PC computers, and began to talk about more exotic workstations and parallel processors.

VAX VMS. DEC VAX computers with the VMS operating system have become the *de facto* standard hardware/operating system mini/mainframe computing environment for the chemical/pharmaceutical industry. In July of 1985, over 60% of those chemical/pharmaceutical companies using MACCS for research chose VAX with VMS. Other computers, such as Prime and IBM, were each less than 20%. By the time of writing, the percentage for VAX with VMS has risen to over 80% and no other system has made even modest gains in this area. The competitive combination of VAX features and the wide availability of scientific software designed for VMS explains this trend.

Personal Computers. IBM PC's (personal computers) and clones with the MS-DOS operating system have become a popular microcomputing environment for the chemical/pharmaceutical industry. Many software products are available for this system including MDL's ChemText chemical word processor, ChemBase chemical database, and ChemTalk terminal program, ORACLE's RDBMS, and BBN's RS/1 statistics program.

Apple Computer Corporation's Macintosh personal computer is also gaining popularity in the chemical/pharmaceutical industry. However, the availability of software is not as great as for the IBM PC.

Workstations. Workstations, such as the DEC and Sun Microsystem products, are available at ever lower prices. They connect to mini/mainframe computers via networks. These systems typically have more throughput and graphics capabilities than a personal computer, but less processing power than a mini/mainframe computer. They have features that suit them for use in interactive and routine tasks such as editing and graphics processing. Many workstations use the UNIX operating system, so programs common to the chemical/pharmaceutical industry must be converted for use on them.

Back-end Processors. An increasing number of hardware vendors supply back-end and parallel processors. These processors usually connect to mini/mainframe computers and increase the throughput of the system for critical operations. Examples are back-end processors that serve as database machines and parallel processors for use with certain numeric-computation algorithms. Many of these processors use the UNIX operating system, so programs common to the chemical/pharmaceutical industry must be converted for use on them.

Software Environments. It is generally agreed that 4th generation environments, such as ORACLE and MACCS-II, have proven themselves more productive than older systems for most application development needs of the chemical/pharmaceutical industry. Many lower level environments have a computer execution speed advantage, but applications can be created in higher level environments much more quickly. Further, computer scientists specialising in software engineering have embraced 4th generation environments and are further improving the benefits of them.[13] Coupled with the fact that 4th generation systems are being tuned for the needs of their markets (for example, MACCS-II is being tuned for the needs of the chemical/pharmaceutical industry) the trend toward 4th generation software environments is expected to continue.

Relational Database Management Systems

The SQL RDBMS language has become the American National Standards Institute (ANSI) standard database language.[14] Applications written in ANSI SQL should now be transportable to other RDBMS's that support that language. This is both because of and to the benefit of the ORACLE RDBMS and IBM's DB2 RDBMS which both use SQL. Companies whose systems use the QUEL RDBMS language, such as Relational Technology Incorporated's (RTI) INGRES are now creating versions of their systems that use SQL.

ORACLE could become the *de facto* standard RDBMS for the chemical/pharmaceutical industry. In February of 1985, nearly 20% of those chemical/pharmaceutical companies using MACCS and a DBMS for research preferred ORACLE, while INQUIRE and System 1032 tied for second place with slightly more than 10%. By the time of writing, the figure for ORACLE has risen to nearly 30% and the other systems are each less than 10%. The only system besides ORACLE that made even modest gains in this area is RTI's INGRES.

Networks and Distributed Systems

Networks are used to connect computer hardware, often dissimilar hardware, so that data can be transferred and programs can communicate. The confusing array of available networks makes network communication for the diverse chemical/pharmaceutical industry difficult. Recently, standard network protocols have emerged and the difficulties are diminishing. These networks are being used for local area networks (LAN's) and for wide area networks (WAN's). LAN's are useful in multidepartment applications and WAN's are useful in multisite applications.

With standard networks and moderately priced workstations now available, distributed systems could become popular research environments in the chemical/pharmaceutical industry. Distributed systems typically allow the user to do interactive and routine tasks such as editing and report preparation on a workstation with its quick response time. Further, they allow the user to send computational or database operations across a network to a mini/mainframe that acts as a server. This architecture offers the end-user scientist the best of all worlds, immediate response for common operations and quick access to centralised chemical-scientific database operations.

Versions of the ORACLE and INGRES RDBMSs have been announced for distributed use. These will utilise PCs, workstations, or mini/mainframes as nodes. Work done at the University of California, Berkeley on the academic version of INGRES is indicative of the features of these RDBMS's.[9] A user may submit a query at his node that requests information that is distributed over several other nodes. The retrieval algorithm polls the nodes for availability, optimises the query for retrieval speed, and retrieves the requested data across the network. When mature, this technology should be useful in chemical/pharmaceutical companies that are multidepartment, multisite, and multinational organisations.

Chemistry

In addition to the computer software and hardware trends that will impact chemical-scientific database systems in the future, trends in chemistry will also impact them. The disciplines of genetic engineering and polymer chemistry have special needs for chemical database technology, as do computational chemistry and patent applications. The chemical/pharmaceutical industry will enjoy important gains in productivity when these needs are better provided for.

FUTURE-OF-THE-ART – CUSTOMISATION

This section discusses how a customisable chemical-scientific database system can provide for competitive chemical/pharmaceutical research as technology advances. The areas explored include the differing information needs of various research departments, the differing computer configurations among departments and sites, and the possible architecture of the database system.

Data Viewpoints

An important need in the chemical/pharmaceutical industry is for a chemical-scientific database system that will serve diverse information needs. From the chemist's viewpoint, structures are central to these needs. From the biologist's viewpoint, biological screening reports are central, and chemical structures are peripheral. Other departments, such as toxicology and spectroscopy have additional points of view. These viewpoints can be represented with a Venn diagram as in Figure 7.

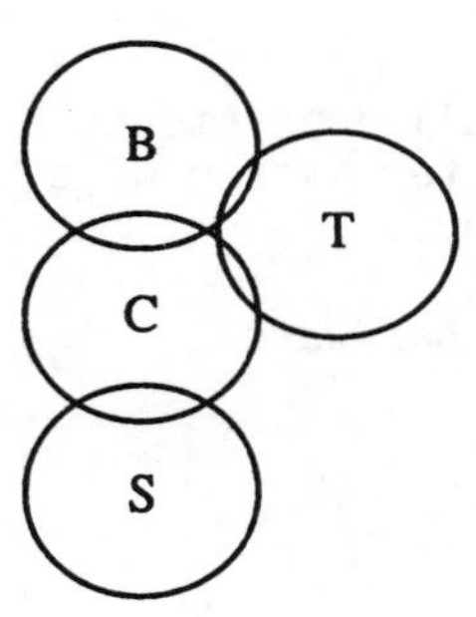

Each circle represents the research database of a department in a chemical/pharmaceutical company:

B - Biology
C - Chemistry
S - Spectroscopy
T - Toxicology

The union of all circles represents the chemical-scientific database of the company.

Intersecting circles indicate data sharing and communication between departments.

Figure 7. Database viewpoints

In Figure 7, each circle depicts the research database of a department in a chemical/pharmaceutical company. The union of all circles represents the chemical-scientific database of the company. Intersecting circles indicate data sharing and communication between departments. Since each department views its data in special ways and may select particular software, it is difficult to design a chemical-scientific database system that treats all unions and intersections. A chemical-scientific database system that can be customised for the various data views is a flexible solution to this difficulty.

Computer Configurations

Another important need in the chemical/pharmaceutical industry is for a chemical-scientific database system that will operate in diverse computer configurations. The computers, workstations, PC's, terminals, and networks of a company may differ by department and site. The hardware configuration for a department can be represented with a Venn diagram as in Figure 8.

In Figure 8, each circle depicts a hardware component of a departmental or site-wide computer system. The union of all circles for all departments and sites represents the computer system of a chemical/pharmaceutical company. Intersecting circles indicate hardware component interactions. Since each department and site may have a unique hardware configuration, it is difficult to design a chemical-scientific database system that treats all unions and intersections. A chemical-scientific database system that is adaptable a wide range of computers, workstations, PC's, terminals, and networks is a good solution to this difficulty.

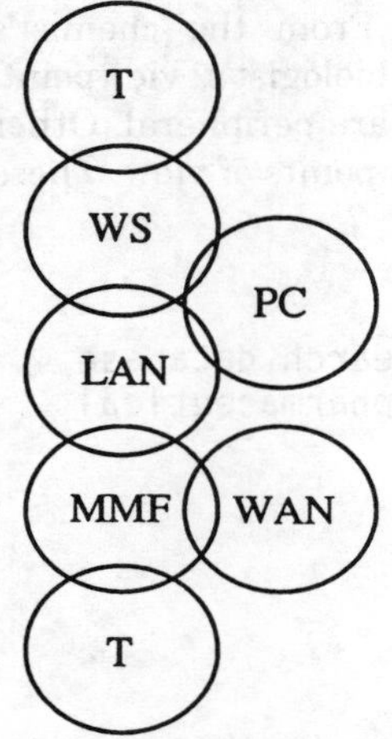

Each circle represents a hardware component of a computer system:

MMF - Mini/mainframe Computer
WS - Workstation
PC - Personal Computer
T - Computer Terminal
LAN - Local Area Network
WAN - Wide Area Network

The union of all circles for all departments and sites represents the computer system of a chemical/pharmaceutical company.

Intersecting circles indicate hardware component interactions.

Figure 8. Hardware configuration

Customisation

A chemical-scientific database system will serve both the database viewpoints among research departments, and the computer hardware configurations among research departments and sites, of a chemical/pharmaceutical company if it is customisable and provides flexible support for standard and emerging technology. The standard, customisation, and interface features of MACCS-II serve as a model. Flexible chemical-scientific database support for standard and emerging technology should include operating systems (VAX VMS in particular), personal computers (IBM PC's in particular), workstations (possibly with UNIX), two-way interaction with 4th generation environments (SQL in particular), communication over networks, and distributed system interaction. Provision for trends in chemistry will complete the future-of-the-art chemical-scientific database system. Chemical/pharmaceutical companies using a customisable chemical-scientific database system for their research will enjoy productivity gains afforded by the system and by emerging technology.

ACKNOWLEDGMENTS

Much of this report discusses the MACCS-II program. That project was managed by the author who gratefully acknowledges the contributions of the development team. Team members included R. Briggs Jr., R. Greenberg, G. Hazen, J. Mikesell, D. Pirkle, D. Raich, B. Reid, S. Shaw, R. Shier, M. Van Duyne and B. Van Vliet. The team in turn thanks the many people in the chemical/pharmaceutical industry and at MDL who contributed to the development of this chemical-scientific database system.

REFERENCES

1. *Communication, Storage and Retrieval of Chemical Information*; Ash, J.E.; Chubb, P.; Ward, S.E.; Welford, S.M.; Willett, P., Eds.; Ellis Horwood: Chichester, 1985.
2. *Aster Guide to Computer Applications in the Pharmaceutical Industry: Molecular Design's Integrated System for Drug Design*; Wipke, W.T. Ed.; Aster Publishing: Springfield, Oregon, 1984.
3. Howe, W.J. 'Molecular Substructure Searching: Computer Graphics and Query Entry Methodology'. *J. Chem. Inf. Comput. Sci.* 1982, *22* (1), 8–15.
4. *Chemical Information Systems*; Ash, J.E.; Hyde, E., Eds.; Ellis Horwood: Chichester: 1975.
5. Barcza, S.; Kelly, L.A.; Wahrman, S.S.; Kirschenbaum, R.E. 'Structured Biological Data in the Molecular Access System'. *J. Chem. Inf. Comput. Sci.* 1985, *25*, 55–59.
6. Adamson, G.W.; Bird, J.M.; Palmer, G.; Warr, W.A. 'Use of MACCS within ICI'. *J. Chem. Inf. Comput. Sci.* 1985, *25*, 90–92.
7. Dill, J.D.; Hounshell, W.D.; Marson, S.; Peacock, S.; Wipke, W.T. 'Search and Retrieval Using an Automated Molecular Access System', paper presented at 182nd National Meeting of the American Chemical Society, New York, August, 1981.
8. Adamson, G.W.; Bird, J.M.; Palmer, G.; Warr, W.A. 'In-house Chemical Databases at Imperial Chemical Industries'. *J. Mol. Graphics* 1986, *4* (3), 165–169.
9. *The INGRES Papers*; Stonebraker, M. Ed.; Addison-Wesley: Menlo Park, California, 1986.
10. *An Introduction to Database Systems*; Date, C.J. Ed.; Addison-Wesley: Menlo Park, California, 1982.
11. Codd, E. 'A Relational Model of Data for Large Shared Databanks'. *Commun. ACM* 1970, *13* (6), 377–387.
12. Kos, A. 'Future Trends in Software', paper presented at the Second European Seminar on Computer-Aided Molecular Design, Basel, Switzerland, October, 1985.
13. Ahrens, E.K.F. 'Applications Prototyping with the MACCS-II Customisation Module', paper presented at the Molecular Design Software Users Group Meeting, San Francisco, California, March, 1987.
14. *Database Language – SQL*; Technical Committee on Database, X3H2; American National Standards Institute, New York, 1986.

PROBLEMS OF SUBSTRUCTURE SEARCH AND THEIR SOLUTION

John M Barnard
Barnard Chemical Information Ltd, 43 Nethergreen Road, Sheffield S11 7EH, U.K.

ABSTRACT

This introductory paper describes some of the problems (both theoretical and practical) of substructure searching. An historical overview is given of the various solutions which have been devised, and some areas of current research interest are outlined.

INTRODUCTION

Substructure searching is the process of searching a file of chemical structure representations, not to find a particular compound, but to find any and all compounds which contain the same specified query substructure.

In the past, systems were developed which allowed such searching by using special fragment codes, or linear notations and ciphers. These had the disadvantage that the query had to be framed in terms of the fragments or line notation symbols that were available in the particular system being used. These systems are not discussed here, though many are still in active use; this paper concentrates on connection-table based substructure search systems, where the query can be expressed as a partial structure diagram.

SUBSTRUCTURE SEARCHING AND SUBGRAPH ISOMORPHISM

Chemical structure diagrams can be regarded as *topological graphs*, in which the atoms are *nodes* or *vertices* and the bonds are *edges* or *branches*. Extending the jargon, the nodes and edges may be *coloured* to distinguish different elements and bond orders, resulting in a *chromatic graph*.[1–2].

The problem of substructure searching may then be considered as the problem of comparing two chromatic graphs – one for the query structure and one for the file structure. The aim is to establish a correspondence or *mapping* between the nodes and edges of the query and a subset of the nodes and edges of the file structure, which preserves the *adjacency* of the nodes (i.e., if two nodes are adjacent in one structure, the nodes of the other structure onto which they map must also be adjacent). Such a mapping is called a *subgraph isomorphism*, and an attempt to find one must be made for all the structures in the file.

Figure 1 illustrates the mapping between the nodes of a query substructure and a file structure; in this case there are two possible correspondents for the nitrogen atom in the query – this is an example of a *multiple isomorphism*.

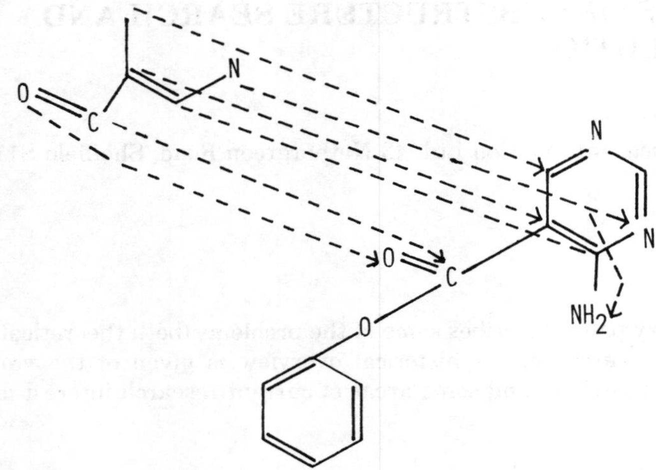

Figure 1. Mapping between the nodes of a query substructure and those of a file structure.

A 'brute force' approach to finding an isomorphism is to try all possible permutations of mapping the query graph onto the file structure graph until either a successful one is found, or all of them have been tried. Obviously there are rather a lot to try, and whilst a successful one might be found straight away, the 'worst case' situation is that the time taken to test a structure is exponential in the number of nodes.

Ideally, what is required is an algorithm which tests for the subgraph isomorphism in *polynomial* time (i.e., the time taken is a direct function of the number of nodes, or its square or cube etc.). Unfortunately, no such algorithm is known, and it is strongly suspected by mathematicians that no such algorithm exists, though this has not been proven.

The subgraph isomorphism problem is one of a class of mathematical problems known as *NP-complete* (another is the well-known 'travelling salesman' problem) and it has been proved that if a polynomial-time algorithm exists for one of these, such algorithms must exist for all of them. But none of them yet has a polynomial time algorithm.

Designers of substructure search systems therefore seem to be stuck with exponential time algorithms, and the normal approach has been to devise heuristics which allow the average time to compare structures to remain quite reasonable in practice, even if the 'worst-case' situation is still exponential in the number of nodes.

SCREENING SYSTEMS

Most of the heuristics used follow the principle of not bothering to make an exhaustive test of a possible isomorphism when it can be determined at an early stage that it will not be successful. The most obvious example of this is screening systems.

At the simplest level, this means that if the query has a tellurium atom in it, there is not much point in looking at any file structures that do not have tellurium atoms in them.

Whilst a screening system based on molecular formula works very well if the query contains a lot of unusual elements, the unfortunate fact remains that most organic compounds found in databases, and hence most queries submitted to them, contain carbon, probably nitrogen or oxygen, and perhaps a halogen or two.

Limited Environment Fragments

As a result, screening systems based on automatically-generated structural fragments have been developed. These fragments consist of small groups of connected atoms and bonds, and if the file structure does not contain all the fragments found in the query, there is no point in looking at it further.

Lynch and his colleagues at Sheffield University in the early 1970s looked at a number of different types of fragment,[3–4] including atom-centred, bond-centred and certain ring descriptor fragments; Figure 2 shows examples of some of those used. Two aspects studied were the frequency distributions of the fragments in a large file of structures, and the degree of independence of the different fragments. It was found that the frequency distributions were very skew (a few fragments occurred in almost all structures, and a lot occurred in only one or two structures each), and there was significant co-occurrence of certain fragments.

It is a general principle of indexing systems that middle-frequency keys are the most useful: the very common ones are no use because they do not allow any structures to be eliminated, the very uncommon ones are no use because it is unlikely that they will ever occur in queries (though they are very useful if they ever do). The indexing keys should also as far as possible be independent (if one fragment only and always occurs in the presence of another, searching for it contributes nothing to the retrieval performance).[5] Using these principles, the BASIC group of companies in Basel compiled a dictionary of fragments for use in a screening system.[6] This dictionary was taken over by Chemical Abstracts Service and now forms the basis of the structure search system on STN International.[7]

The main publicly available structure search system competing with STN International is the DARC system, operated by Télésystèmes.[8–9] This system also uses 'limited-environment' fragments as keys for a screening search, though in this case the set of available fragments (the FREL's) is 'open-ended', rather than controlled by a dictionary. The FREL's (Fragment Reduced to an Environment which is Limited) are generated automatically, and each describes two concentric layers of atoms around a 'focus', which is an atom with a connectivity of at least 3; Figure 3 shows an example. In addition to the FREL's, the DARC system also uses a special bit-screen search, which is mainly concerned with the characteristics of ring systems.

Other Fragments

Other types of fragment can also be employed in screening searches. In many cases fragment codes originally developed for manual encoding of structures are used, and computer programs have been written which enable them to be generated automatically. A good example is the GREMAS code, developed by IDC, which is generated automatically for structures in the Chemical Abstracts Registry File.[10] Though the GREMAS screen is normally the only search step, it can also be used as a preliminary screen before an atom-by-atom search. Work is currently in progress at IDC on the

ATOM-CENTERED	SIMPLE ATOM	C
	CO-ORDINATED ATOM	C4
	BONDED ATOM	-C- (with vertical bonds)
	AUGMENTED ATOM	C - C - CL (with C above and below the central C)
BOND-CENTERED	SIMPLE BOND	-
	SIMPLE PAIR	C - C
	AUGMENTED PAIR	2C - C1
	BONDED PAIR	C - C = (with a single and a double bond on the first C)
	OCTUPLET	C - C = C (with C and =O on the first C)

Figure 2. Some of the fragment types studied by Lynch *et al.*

automatic generation of GREMAS fragments from generic structures in patents (at present encoded manually), which will allow the current GREMAS searching of such structures to continue (with cheaper input), and may also form a screening stage in a more sophisticated system.[11]

Screen Searching

There are various ways of actually doing the screening search. Bit-strings can be set up in which each bit represents one (or perhaps more than one) fragment from the dictionary. The bit-string for the query can then be matched against the bit-string for each file structure in turn. Alternatively, the file structure bit-strings can be 'inverted' with the result that only the bits for the fragments actually present in the query need to be examined. Other more sophisticated tree-structured search methods are also used, especially for open-ended fragment sets such as the DARC FREL's; this type of search is further discussed below.

Meyer[12] has suggested another way of speeding up the screen search – the so-called *superimposition principle*, in which a long bit-string is effectively folded onto a shorter one. Each bit in the long bit-string gives rise to two or more bits in the shorter one, which being shorter is much quicker to do comparisons on. On the other hand, unless

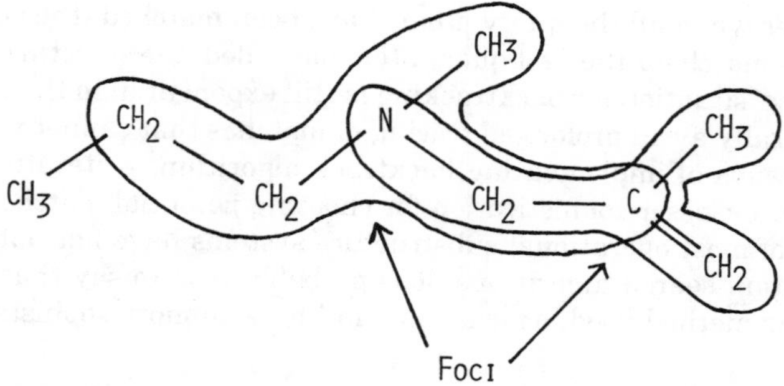

Figure 3. The DARC FREL's generated for a simple acyclic structure.

the original bit-string is fairly sparsely filled, there can be a loss of screen-out performance.

Reduced Graphs

A new type of screening search which has been suggested recently uses the concept of so-called *Reduced Graphs*. In a reduced graph, some of the nodes are collapsed together, thus forming a simpler graph, which may be searched using any of the normal methods available; because the reduced graph is smaller, it is quicker to search. Though this work has so far been aimed primarily at generic structure searching,[13–15] it is equally applicable to searching files of specific structures. A number of different bases may be used to collapse the nodes: in the Sheffield work these include distinction between cyclic and acyclic parts of the structure, or between carbon and heteroatoms; in the Chemical Abstracts work a hierarchy of mutually-exclusive generic group nodes is used. Figure 4 shows how a structure may be collapsed to a reduced graph by using the distinction between cyclic and acyclic nodes.

BACKTRACK SEARCH

Once the screening stage has been completed, a lot of structures have been eliminated from consideration because they cannot contain the specified substructure. On a small file, if there is reasonably good screening performance, this may leave sufficiently few structures for it not to be worth proceeding further. In fact, when the CAS ONLINE system first became available, only the fragment screening stage was implemented.

However, in most cases it will be necessary to test more rigorously for subgraph isomorphism at the atom-by-atom level; many ways have been suggested for doing this as rapidly as possible.

The 'brute force' approach of trying every possible correspondence between the query atoms and the file structure atoms is clearly a non-starter, at least on conventional computer architecture. Beyond this, the simplest approach is *backtracking*, which was first used by Ray and Kirsch in 1957.[16]

This basically involves matching one pair of atoms, then trying to establish matches for their neighbours, and so on. If at any stage, it is not possible to establish a match, the algorithm 'undoes' the last successful match, and tries another one. The algorithm terminates either when all the query atoms have been matched (the structures match) or every possible match for the first query atom has failed (the structures do not match).

The 'worst case' situation for backtracking is still exponential in the number of nodes, but it does generally avoid prolonged tracing of matches that cannot work. There are a lot of different ways of implementing backtrack algorithms – iteratively, recursively, and with greater or lesser sophistication for choosing potential matches to try. Though the originators of most operational substructure systems have not published details of their atom-by-atom search algorithms, it is probably true to say that backtracking is the most popular method used, at least as a fall-back to more sophisticated methods.

RELAXATION METHODS

Though the term has only recently become current amongst chemical information scientists, the technique of *relaxation* has been used by them for several decades.

Relaxation is a numerical technique originally developed in the 1930s by mathematicians, for solving simultaneous equations by iteratively refining initial 'guesses' for the variables; a textbook on the subject was published in 1954.[17]

In the context of chemical structure handling, relaxation basically involves assigning a value to each atom, and then iteratively refining that value by examining the values of the neighbouring nodes. Undoubtedly the best-known example of relaxation is the Morgan algorithm[18] for canonical numbering of the atoms in a structure. Other examples are provided by the Augmented Connectivity Molecular Formula[19] used in the Chemical Abstracts Registry System III, and Lynch and Willett's procedure for detecting chemical reaction sites.[20]

Though a number of earlier algorithms are in fact based on the same principle, the use of relaxation as a substructure search method was first suggested explicitly in the

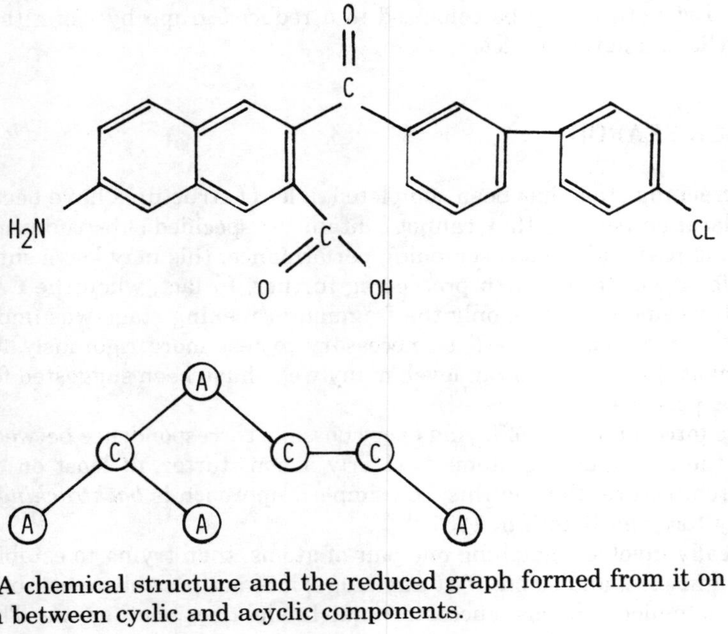

Figure 4. A chemical structure and the reduced graph formed from it on the basis of the division between cyclic and acyclic components.

early 1980s, by Kitchen and Krishnamurthy,[21] and was implemented in a form also capable of handling generic structures (both in the query and the database) by von Scholley.[22]

Von Scholley's algorithm involves assigning to each file structure node the labels of query nodes to which it could correspond (initially on the basis of element type). Each iteration of the algorithm then examines the neighbours of each node and eliminates possible correspondences if there is an inconsistency between the query and file structure node neighbours. The algorithm is iterated either until a particular query node is eliminated from all the possible correspondence lists of all the file structure nodes (in which case, there is no match) or until no further eliminations can be made (in which case there is a match).

Because all the nodes are refined at each iteration, each iteration for any one node takes into account successively more distant nodes, and thus more and more of the node's environment. This is one of the things that distinguishes von Scholley's algorithm from Figueras' set reduction approach,[23] published in 1972. Although Figueras' method also involves assigning label sets, and iteratively refining them, he explicitly uses 2nd and higher order connection tables (i.e., tables which show for each atom, not the immediately adjacent atoms, but the atoms 2, 3 etc. bonds distant).

It is interesting to note that neither von Scholley's algorithm nor Figueras' is guaranteed to give 100% precision. In both cases certain structures are indicated as hits when they do not in fact contain the desired substructure. An example is shown in Figure 5, where in both the query and file structure the 'local environment' of each atom is the same, however many neighbours distant are considered. However, such query types are rare, and Figueras reported that in practice his algorithm gave no false drops.

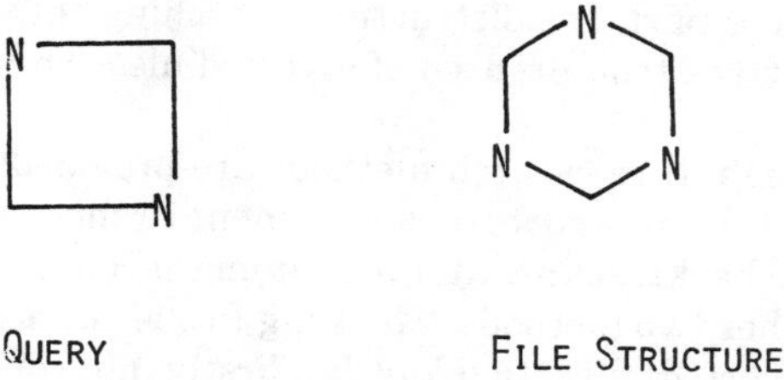

Figure 5. A query/file structure pair which would be incorrectly shown to match by von Scholley's and Figueras' algorithms.

In contrast, Sussenguth's earlier graph-theoretical algorithm[24] does give 100% precision, because, unlike those of Figueras and von Scholley, it uses backtracking as a fall-back technique. Before doing any backtracking, however, Sussenguth's algorithm eliminates as many potential mappings as possible by a logical partitioning procedure, and by application of a 'connectivity property'. This latter operation, which involves examination of the neighbours of the query and file structure nodes, is an example of relaxation.

A subgraph isomorphism algorithm published by Ullmann[25] in 1976 also contains a refinement step, which is essentially an application of relaxation.

PARALLEL PROCESSING

The development of new computer architectures, and in particular of parallel processors, is leading to some new ideas for substructure searching. Because each node in von Scholley's algorithm can be treated independently during each iteration of the algorithm, there is obvious scope for parallelism. This is under active investigation at Sheffield: one paper has already been published,[26] and further results are presented elsewhere in these proceedings.[27]

Wipke and Rogers[28] have also simulated the operation of a parallel substructure search algorithm on a conventional computer. Their approach uses parallelism to get round the problems of the 'brute force' approach, but it has not so far been tried on a 'real' parallel machine.

The use of pairs of minicomputers in parallel for the substructure search system of STN International is well-known, and is discussed elsewhere in these proceedings,[29] as is another proposal for a substructure search system based on a multiprocessor architecture.[30]

TREE-STRUCTURED INVERTED FILES

A rather different approach to substructure searching is used by at least two systems. One of these is the Chemical Information System (CIS) developed by Richard Feldmann and his colleagues at the National Cancer Institute,[31] and its various off-shoots which include the Chemical Databank System (CDS) at the SERC Laboratory in Daresbury. The other is the Hierarchic Tree Substructure Search (HTSS) developed in Hungary.[32]

What these systems have in common is that they attempt to put a large part of the computationally-intensive part of substructure searching into a pre-processing of the database, generating a tree-structured set of inverted files. The search itself can then be relatively rapid.

In Feldmann's approach, three search methods are provided: a Ring Probe, based on the ring systems present in the structure, a Fragment Probe, based on augmented-atom fragments, and a simple backtracking atom-by-atom search, which can be used to refine the hits found by the other two methods. The Ring Probe operates by searching through the tree-structured inverted files to identify, firstly all compounds containing the specified ring pattern (ignoring atom types), then within those, the atom types present in the ring system, the exact positions of those atom types, and finally the positions of attachment of any ring substituents. At each level the search can be 'Exact' or 'Imbed' (e.g., exactly the substituent positions specified, or at least those specified).

The Fragment Probe operates in a similar way, with a search first for the central atom type, then for the number of neighbours, and then in turn for the atom type and bond order for each neighbour. Under the control of the user, the sets of structures retrieved by applying the Fragment Probe to each atom can be ANDed together, along with those retrieved by the Ring Probe, and the atom-by-atom search applied as a final filter. The tree-structured search files used for the FREL search in the DARC system[8–9] are conceptually similar to Feldmann's Fragment Probe.

HTSS is also rather similar, except that in building the hierarchic tree structure, once all the immediate neighbours for the central atom have been processed, what amounts to a relaxation procedure is applied to extend the tree search file, taking account of successively more distant atoms. This effectively obviates the need for an atom-by-atom search stage.

OTHER DEVELOPMENTS

Though substructure search techniques have been under development for a long period, the field is certainly far from played out, and there is a lot of important and interesting work being done, and new systems are being developed.

Generic Structures

Many systems, including the two publicly-available ones, now allow generic query structures, at least as permutations of specific alternatives for R-groups. In addition, the development of systems for searching files of generic structures from patents is a very exciting area at present:[11] both Chemical Abstracts Service[14] and Tèlèsystémes (working in conjunction with Derwent Publications Ltd and the French Patent Office)[33] are developing new systems, and research on the topic is continuing at Sheffield University.[27,34]

3-D Substructure Searching

There is a growing interest in 3-dimensional substructure searching, which considers the topography of the molecule (the spatial arrangements of its atoms and bonds), instead of merely its 2-dimensional topology (the way the atoms and bonds are connected together). Searches for pharmacophoric patterns (the 3-D structural features of molecules which interact with drug receptor sites) mayhave important applications in drug design.[35–36]

Approximate Structure Searching

Another topic of interest is approximate structure matching,[37–38] which can have the advantage of allowing the user to 'browse' in a database, as well as avoiding the computational intensiveness of exact substructure searching. Such work may also allow the ranking of the output from a substructure search.

Stereospecific Searching

Searching for particular stereochemical configurations is in principle just an extension of the methods for searching without stereochemistry: additional conditions of bond orientation etc. need to be satisfied before a match is identified. A growing number of substructure search systems are now offering stereospecific searching.

A number of practical problems are involved, however, most particularly in the interconversion of absolute and relative stereochemical descriptors; because a substructure query does not show all the atoms of a molecule, it is not normally possible to assign an absolute configuration to a chiral centre on the basis of the Cahn-Ingold-Prelog sequence rule, whereas this is likely to be the preferred method for giving stereochemical information in the database structures.

Macromolecular Searching

Biological macromolecules present special problems for substructure searching. In many cases the feature of interest is the one-dimensional sequence of amino-acid or nucleotide residues, and a number of databases of such sequences are now available for searching.[39] Searching of the three-dimensional structure of such macromolecules is also of interest.[40]

QUERY FORMULATION

Whatever algorithms are used, the usefulness of any substructure search system is limited by the types of query which it is possible to put to it, and the interpretation placed on them.

Free Sites

Whilst a substructure search system generally operates by finding all structures in the database which contain the specified query structure embedded within them, a number of different ways are used to specify the exact manner in which the query may be embedded. In some cases (e.g., DARC) this involves specifying as 'free sites' those atoms which can have connections other than those shown in the query; in other cases (e.g., STN International) it involves specifying the permissible number of connections for certain atoms. It may also be possible to specify whether or not a ring system can be embedded in a larger ring system, or have only acyclic substituents.

One possibly controversial question is whether the envelope ring of an ortho-fused ring system should properly be regarded as a substructure of that ring system. In the example shown in Figure 6, two atoms in the query are shown as free sites, and in the case of the first file structure, the only feature which is not in the query is a bond between the two free site atoms of the query. It is arguable whether this accords with the definition of subgraph isomorphism as being a mapping which preserves the adjacency of the nodes in the two structures; however, a rigorous definition of subgraph isomorphism may not be relevant to the practical problems of chemical substructure search. Nevertheless, the fact that different systems do give different answers to this query suggests that the question does need further consideration.

Tautomerism and Alternating Bonds

Two of the great practical problems of substructure search are tautomerism and alternating bonds in rings. It is obviously essential to ensure that structures are not missed simply because the query has been drawn in a different tautomeric form to the file structure. The simplest approach to the problem is to establish conventions for

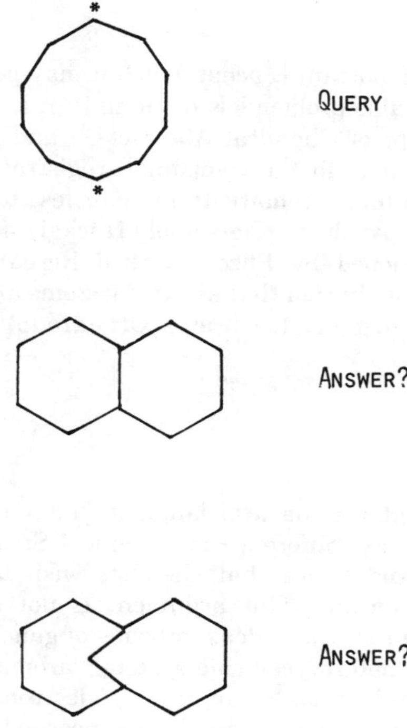

Figure 6. A query substructure with free sites marked, and two possible answers (discussion in text).

which tautomeric form should be used, and to apply these unerringly; this is often easier said than done.

In the Chemical Abstracts Registry System, tautomers and alternating bond systems in file structures are identified automatically,[41] and the bonds in these systems are replaced by 'normalised' bonds; a 'denormalisation' algorithm is also available to return them to conventional single and double bonds. When searching, the user is able to specify in the query that a bond is 'single exact' or 'single or normalised' etc. The rules used by Chemical Abstracts for the definition of tautomerism exclude keto-enol tautomers, and these are therefore registered as separate compounds, and must be searched for separately. It is, however, easily possible to extend the rules to allow the identification of keto-enol tautomerism.

A different method of handling the problem of alternating bonds has been suggested by Gasteiger.[42] In this system, explicit bond orders are not used, but a 'pi electron count' is associated with the atoms at each end of multiple bonds. Thus all the atoms in a phenyl ring have a pi electron count of 1. Gasteiger's approach has recently been extended to tautomeric systems (though the identification of tautomeric groups is essentially the same as Chemical Abstracts'), and will be used in the forthcoming Beilstein Online system.[43]

Aromaticity

Aromaticity is also a very important concept which it may perhaps be desirable to use in queries. One of its particular problems is defining it: it is certainly not the same as the alternating bond concept of Chemical Abstracts,[41] and so their approach to it is basically to ignore it (which, for a compound registration system, is probably appropriate). Some systems limit aromaticity more or less to 6-membered alternating bond rings like benzene and pyridine; others apply Hückel's 4n+2 pi electrons rule. The SOCRATES system[44] developed by Pfizer Central Research in the UK uses the interesting, if rather broad, definition that all ring systems are assumed to be aromatic unless they contain a ring member (such as a CH_2 group) which would destroy the aromaticity.

Truly Generic Queries

It is frequently claimed that the 'natural language' of the chemist is the structure diagram (the very title of this conference is 'Chemical Structures: the International Language of Chemistry') and hence that chemists wish to formulate their search queries using structure diagrams. This argument is not entirely convincing since chemists also frequently express their ideas in terms of generic concepts which cannot actually be drawn, such as 'heterocyclic ring system', 'aromatic group', and with even less reference to a structural basis, 'polar group', 'electron-withdrawing group', etc. These concepts have had to be addressed by those concerned with Markush structures from patents, because they occur in database structures, but they may also be of use in generic query systems.

Ideally, a substructure query language should allow the user to specify the structural features which are desired in the answers to the query, but without having to give them explicitly in terms of structure diagrams. Examples might include specifying the number and size of rings, and the heteroatoms present, or a range of lengths and allowable substituents for carbon chains. Several approaches to this sort of *intensional* description of chemical entities are under development, and may find their way into substructure search systems over the next few years: the *superatom* concept of Markush DARC,[33] the generic group nodes of Chemical Abstracts,[14] the *parameter lists* and chemical grammar (TOPOGRAM) of the Sheffield group,[45] and some recent work in Japan.[46]

REFERENCES

1. Tarjan, R.E. 'Graph Algorithms in Chemical Computation'. In *Algorithms for Chemical Computations*; Christoffersen, R.E. Ed; ACS Symposium Series 46. American Chemical Society: Washington, 1977; pp. 1–19.
2. *Chemical Graph Theory*; Trinajstic, N. Ed; CRC Press, 1983, 2 vols.
3. Lynch, M.F. 'The Microstructure of Chemical Databases and the Choice of Representation for Retrieval'. In *Computer Representation and Manipulation of Chemical Information*; Wipke, W.T.; Heller, S.R.; Feldmann, R.J.; Hyde, E., Eds;Wiley: New York, 1974.
4. Lynch, M.F. 'Screening Large Chemical Files'. In *Chemical Information Systems*; Ash, J.E.; Hyde, E., Eds; Ellis Horwood: Chichester, 1974, 177–94.
5. Mooers, C.N. 'Zatocoding Applied to Mechanical Organisation of Knowledge'. *Am. Doc.* 1951, *2*, 20–32.

6. Graf, W.; Kaindl, H.K.; Kniess, H.; Warszawski, R. 'The Third BASIC Fragment Search Dictionary'. *J. Chem. Inf. Comput. Sci.* 1982, *22*, 177–81.
7. Dittmar, P.G.; Farmer, N.A.; Fisanick, W.; Haines, R.C.; Mockus, J. 'The CAS Online Search System. 1. General System Design and Selection, Generation and Use of Search Screens'. *J. Chem. Inf. Comput. Sci.* 1983, *23*, 93–102.
8. Attias, R. 'DARC Substructure Search System: A New Approach to Chemical Information'. *J. Chem. Inf. Comput. Sci.* 1983, *23*, 102–8.
9. Watson, D. 'Some Experiences with the DARC System'. In *Proceedings of the CNA(UK) Seminar on Chemical Structure Searching of the Published Literature; Daresbury, U.K., 17–19 March 1980*. Chemical Structure Association, 1982.
10. Rssler, S.; Kolb, A. 'The GREMAS System, an Integral Part of the IDC System for Chemical Documentation'. *J. Chem. Doc.* 1970, *10*, 128–34.
11. Barnard, J.M. 'Online Graphical Searching of Markush Structures in Patents'. *Database* 1987, *10* (3), 27–34.
12. Meyer, E. 'Superimposed Screens for the GREMAS System'. In *Mechanised Information Storage Retrieval and Dissemination, Proceedings of FID-IFP Conference, Rome, June 14–17 1967*; Samuelson, K. Ed.; North Holland: Amsterdam, 1968.
13. Gillet, V.J.; Downs, G.M.; Ling, A.(B).; Lynch, M.F.; Venkataram, P.; Wood, J.V.; Dethlefsen, W. 'Computer Storage and Retrieval of Generic Chemical Structures in Patents. 8. Reduced Chemical Graphs, and their Application in Generic Chemical Structure Retrieval'. *J. Chem. Inf. Comput. Sci.* 1987, *27*, 126–37.
14. Fisanick, W. 'Storage and Retrieval of Generic Chemical Structure Representations', U.S. Patent 4,642,762 assigned to American Chemical Society (Feb 10 1987).
15. Nakayama, T.; Fujiwara, Y. 'Computer Representation of Generic Chemical Structures by an Extended Block-Cutpoint Tree'. *J. Chem. Inf. Comput. Sci.* 1983, *23*, 80–7.
16. Ray, L.C.; Kirsch, R.A. 'Finding Chemical Records by Digital Computers'. *Science* 1957, *126*, 814–19.
17. *Relaxation Methods*; Allen, D.N. de G. Ed; McGraw Hill, 1954.
18. Morgan, H.L. 'The Generation of a Unique Machine Description for Chemical Structures – A Technique Developed at Chemical Abstracts Service'. *J. Chem. Doc.* 1965, *5*, 107–13.
19. Freeland, R.G.; Funk, S.A.; O'Korn, L.J.; Wilson, G.A. 'The Chemical Abstracts Service Chemical Registry System. 2. Augmented Connectivity Molecular Formula'. *J. Chem. Inf. Comput. Sci.* 1979, *19*, 94–8.
20. Lynch, M.F.; Willett, P. 'The Automatic Detection of Chemical Reaction Sites'. *J. Chem. Inf. Comput. Sci.* 1978, *18*, 154–59.
21. Kitchen, L.; Krishnamurthy, E.V. 'Fast, Parallel Relaxation Screening for Chemical Patent Database Search'. *J. Chem. Inf. Comput. Sci.* 1982, *22*, 44–8.
22. von Scholley, A. 'A Relaxation Algorithm for Generic Chemical Structure Screening'. *J. Chem. Inf. Comput. Sci.* 1984, *24*, 235–41.
23. Figueras, J. 'Substructure Search by Set Reduction'. *J. Chem. Doc.* 1972, *12*, 237–44.
24. Sussenguth, E.H. 'A Graph-Theoretic Algorithm for Matching Chemical Structures'. *J. Chem. Doc.* 1965, *5*, 36–43.
25. Ullmann, J.R. 'An Algorithm for Subgraph Isomorphism'. *J. Assoc. Comput. Mach.* 1976, *23*, 31–42.
26. Gillet, V.J.; Welford, S.M.; Lynch, M.F.; Willett, P.; Barnard, J.M.; Downs, G.M.; Manson, G.; Thompson, J. 'Computer Storage and Retrieval of Generic Chemical Structures in Patents. 7. Parallel Simulation of a Relaxation Algorithm for Chemical Substructure Search'. *J. Chem. Inf. Comput. Sci.* 1986, *26*, 118–26.
27. Downs, G.M.; Gillet, V.J.; Holliday, J.; Lynch, M.F. 'The Sheffield University Generic Chemical Structures Project – a Review of Progress and of Outstanding Problems'. In these Proceedings.
28. Wipke, W.T.; Rogers, D. 'Rapid Subgraph Search Using Parallelism'. *J. Chem. Inf. Comput. Sci.* 1984, *24*, 255–62.
29. Farmer, N.; Amoss, J.; Farel, W.; Fehribach, J.; Zeidner, C. 'The Evolution of the CAS Parallel Structure Searching Architecture'. In these Proceedings.
30. Jochum, P. 'A Multiprocessor Architecture for Substructure Search'. In these Proceedings.
31. Feldmann, R.J.; Milne, G.W.A.; Heller, S.R.; Fein, A.; Miller, J.A.; Koch, B. 'An Interactive Substructure Search System'. *J. Chem. Inf. Comput. Sci.* 1977, *17*, 157–63.
32. Nagy, M.Z.; Kozics, S.; Veszpremi, T.; Bruck, P. 'Substructure Search on Very Large Files Using Tree-Structured Databases'. In these Proceedings.

33. Shenton, K.E.; Norton, P.; Ferns, E.A. 'Generic Searching of Patent Information'. In these Proceedings.
34. Lynch, M.F.; 'Generic Chemical Structures in Patents (Markush Structures): the Research Project at the University of Sheffield'. *World Patent Information* 1986, *8*, 85–91.
35. Gund, P. 'Three-dimensional Pharmacophoric Pattern Searching'. *Prog. Mol. Subcell. Biol.* 1977, *5*, 117–43.
36. Brint, A.T.; Mitchell, E.; Willett, P. 'Substructure Searching in Files of Three-Dimensional Chemical Structures'. In these Proceedings.
37. Willett, P.; Winterman, V.; Bawden, D. 'Implementation of Nearest-neighbour Searching in an Online Chemical Structure Search System'. *J. Chem. Inf. Comput. Sci.*1986, *26*, 36–41.
38. Bawden, D. 'Browsing and Clustering of Chemical Structures'. In these Proceedings.
39. Kneale, G.G.; Bishop, M.J. 'Nucleic Acid and Protein Sequence Databases'. *Comp. Appl. Biosci.* 1985, *1*, 11–17.
40. Lesk, A.M. 'Detection of Three-dimensional Patterns of Atoms in Chemical Structures'. *Commun. ACM* 1979, *22*, 219–24.
41. Mockus, J.; Stobaugh, R.E. 'The Chemical Abstracts Registry System. VII. Tautomerism and Alternating Bonds'. *J. Chem. Inf. Comput. Sci.* 1980, *20*, 18–22.
42. Gasteiger, J. 'A Representation of Pi-systems for Efficient Computer Manipulation'. *J. Chem. Inf. Comput. Sci.* 1979, *19*, 111–15.
43. Jochum, C. 'Building a Structure-oriented, Numerical, Factual Database'. In these Proceedings.
44. Bawden, D.; Devon, T.K.; Faulkner, D.A.; Fisher, J.D.; Reeves, R.J.; Woodward, F.E. 'Development of the Pfizer Integrated Research Data System SOCRATES'. In these Proceedings.
45. Welford, S.M.; Lynch, M.F.; Barnard, J.M. 'Computer Storage and Retrieval of Generic Chemical Structures in Patents. 3. Chemical Grammars and their Role in the Manipulation of Chemical Structures'. *J. Chem. Inf. Comput. Sci.* 1981, *21*, 161–8.
46. Tokizane, S.; Monjoh, J.; Chihara,H. 'Computer Storage and Retrieval of Generic Chemical Structures Using Structure Attributes'. *J. Chem. Inf. Comput. Sci.* 1987, *27*, 177–87.

SUBSTRUCTURE SEARCH ON VERY LARGE FILES USING TREE-STRUCTURED DATABASES

M Z Nagy, S Kozics, T Veszpremi and P Bruck
HTSS, 1136 Budapest, Balzac u 43, Hungary

ABSTRACT

Substructure search on large databases often exceeds the limits of interactive computer systems. To avoid this difficulty the majority of operations should be performed during the generation of the database rather than during the retrieval procedure. The better the nature of the problem can be reflected by the data structure of the database, the faster will be the retrieval.

In the HTSS (heirarchical tree substructure search) system a single rooted decision tree will be generated from the connectivity tables of the molecules by the gradually refining classification of the atoms. The atoms are classified by their type, by number of their neighbours, by the type of their bonds, and by their position in chains and rings. This primary colouring will be modified by taking into account the colour of their neighbours and this procedure is repeated step by step until its effect reaches the furthermost atoms.

Substructure search in this tree-like database is a tree-walk through the relevant branches from the root toward the leaves. If no leaf can be reached, molecules with the most similar graph are retrieved.

A graphic-oriented interpreter has been designed to control the input of the query structures, the retrieval procedure and the presentation of the answers. This software technology enables the user to customise the menu and to introduce new commands. A hierarchic multiwindow technique facilitates the construction of complex Markush queries.

Substructure searches can be limited to a subset of the database which has a common property (to study structure-property relationships).

The tree-based search technology can offer very good performance on large files for a broad range of computers from IBM-PC to IBM mainframes.

INTRODUCTION

Traditional substructure search systems work in two stages: the first stage is a preliminary filter using fragment codes; the second stage applies atom-by-atom search to check the isomorphism of the subgraphs.

The latter is mathematically an NP-complete problem, therefore its size has to be kept minimal.

Even using the most sophisticated atom-by-atom search algorithms care should be taken because the number of structures remaining after the first filtering stage should not exceed twenty thousand.

W. A. Warr (Ed.)
Chemical Structures

Therefore traditional systems cannot handle a given category of problems for very large databases due to time limitations. However, the exact solution of the problem is unavoidable in many important cases, e.g., to calculate statistical data for structure-property relationship studies on large databases.

The HTSS system avoids this problem by making the atom-by-atom search phase unnecessary.

This system deviates from other systems in two main characteristics: in the description method of the atomic environments, and in the data structure used for the storage of structural information.

CONCENTRIC ENVIRONMENTS

The concept of concentric environment has been introduced by Dubois in the DARC system as an alternative to the traditional fragment concept. The radius of perception was two, i.e., the first and the second neighbours of the atom in question were taken into account. Features used for characterisation were the type of the atom, the type of its bonds and the number of neighbours. To limit the number of fragments only the environment of tertiary and quaternary atoms have been used for structure descriptions in the DARC system.

The HOSE system, developed by Bremser and his co-workers at BASF, used slightly smaller fragments: the second neighbours are not always taken into account, but the concentric environment of each atom is used for structure description of the molecule.

In the case of the HTSS system the radius of the environment can be selected at the generation of the database, according to the problems to be answered. The usual radius, which can be used for a very broad range of problems, is three, i.e., the effects of the first, second and third neighbours of each and every atom are taken into account.

Besides the three traditional characteristics (number of neighbours, atom type and bond types) three further characteristics are used: the position of the atomin rings and chains and its distance from specific sites in the molecule are also taken into account.

Since the size of the fragments is larger and the description of the structural features is more detailed, it is obvious that the fragments tend to become unique: forexample, in a database containing ten million atoms, more than one million environments are different and nearly one million environments are contained only in a single molecule.

In traditional search systems no direct information is available on the relative position of the concentric environments within the molecules. In the HTSS system this additional connectivity information is also taken into account due to the new method of information storage.

It is obvious that atom-by-atom search is not necessary if an adequately precise description of the environments is used. Therefore the problems and limitations related to atom-by-atom search can be avoided in the HTSS system.

THE DECISION TREE

To store such a large amount of information a new data structure has been introduced: the decision tree of the database.

First the atoms are classified according to the number of their neighbours, in the next tree level the type of the atom is used for characterisation, and at the following level the bond types of each atom are used.

We can further characterise an atom by the 'colour' of its neighbours (and repeat this classification at the next levels). However the neighbour of an atom at this level already reflects the characteristics of its first neighbours, therefore at the second level of iteration we can perceive the characteristics of the second neighbours of an atom. So the further we go with iterations, the further we can perceive the environment of an atom within a molecule.

The tree data structure can provide the description of the structural features with the required precision, and has an additional benefit: repetitive structural features are stored in this tree only once.

SUBSTRUCTURE SEARCH

Generally speaking, a substructure has been found in our tree-structured database, if a leaf can be reached during the tree-walk for each atom in the query structure. However, if no leaf can be reached for a few atoms, we still obtain important structural information: which molecules in the database have the most similar structural features to the query structure.

Similarity in this case means the agreement of those structural features which have been used during the generation of the decision tree. (Often, but not always, this is in coincidence with the feelings of organic chemists.)

Since in the decision tree related structures are located in non-distant parts of the tree, the system is well suited to the simultaneous retrieval of a set of similar structures, therefore the HTSS system is able to retrieve very complex Markush queries with good response times. A biographical multiwindow technique has been implemented in the user interface of the HTSS system to enable the easy entry of Markush queries. Functional groups can be defined by the user as special atom types.

The multiwindow technique is very practical for displaying the query and the search results side by side.

The user interface of the HTSS system is controlled by a graphic-oriented interpreter language. This enables the user to customise the HTSS menu and to introduce new commands.

SUBTREES

Working on large databases it is often desirable to restrict searches to a subset of the data. The HTSS system can perform operations on such subsets, which are defined by the serial numbers of the compounds to be included in a 'subtree'. If these numbers are the result of a previous substructure search, all molecules in the subtree will contain this structural feature; if the numbers are the results of a mass spectral retrieval system, the compounds in that subtree have a common spectral property. This feature of the HTSS system facilitates correlation studies between chemical structure and other properties.

Subtrees physically do not exist, the irrelevant part of the tree is simply deselected during tree-walk.

PERFORMANCE

The performance of the system has been studied both on microcomputers as well as on mainframes. On IBM-PC/AT using a database of 15,000 compounds the average retrieval time for a substructure was 15 seconds; the system is CPU-bound.

On an IBM mainframe (3090/150) the average retrieval time for a substructure was under 10 seconds on a database of 1,200,000 compounds. The effect of database size on retrieval time has been studied and the results are very encouraging: increasing the size of the database from 150,000 to 1,200,000 resulted in an increase in retrieval time of only 50 per cent.

These data show an agreement with another study: the retrieval time requirement normalised for one hit decreases as a function of database size for various substructures. These results show that the structure of big trees does not change significantly, even when the number of compounds is quadrupled; therefore the number of disk seeks, which is the most time-consuming operation in database management, will increase only slightly.

ACKNOWLEDGMENTS

The authors are grateful for the cooperation of Dr Clemens Jochum and his co-workers at the Beilstein Institut in the testing of HTSS on large databases and gratefully acknowledge the advice of Dr S R Heller for a period of more than a decade.

SUBSTRUCTURE SEARCHING IN FILES OF THREE-DIMENSIONAL CHEMICAL STRUCTURES

Andrew T. Brint, Eleanor Mitchell and Peter Willett
Department of Information Studies, University of Sheffield, Western Bank, Sheffield S10 2TN, U.K.

ABSTRACT

This paper discusses techniques for chemical substructure searching in files of three-dimensional (3-D) chemical structures. A methodology is presented for the selection of search screens which are based upon interatomic distance information; the use of these screens on a file of 3-D chemical structures taken from the Cambridge Crystallographic Data Bank (CCDB) shows that they allow searches for typical pharmacophoric patterns to be carried out with high efficiency. A range of subgraph isomorphism algorithms is described for geometric searching, the 3-D equivalent of atom-by-atom searching in conventional substructure search systems. Current research involves the development of hardware and software techniques for the identification of the maximal substructures common to sets of 3-D structures, and the extension of the work on substructure searching in the CCDB to the macromolecular structures in the Protein Data Bank.

INTRODUCTION

Computer-based chemical information systems are now widely used for the storage and retrieval of chemical structure information. Efficient searching algorithms are available which allow structure-based searches to be carried out on databases containing many thousands, or even millions, of chemical compounds. In addition, the ease with which machine-readable structural data can be manipulated has led to a wide range of related activities, such as computer-aided synthesis design and studies of quantitative structure-activity relationships (QSAR).[1,2]

A limitation of many current systems is that they are designed in large part for the storage and retrieval of two-dimensional (2-D) chemical compounds, i.e., chemical structure diagrams, and take little account of the three-dimensional (3-D) characteristics of molecules. Information about the stereochemistry of a molecule can be stored by means of suitable annotations to its connection table, the internal machine-readable representation of a chemical structure, and it is possible to use this information in *structure searching* (where the system must carry out a graph isomorphism search for an exact match with an input query structure).[3] It is not, however, generally possible to carry out 3-D *substructure searching* (where the system must carry out a subgraph isomorphism search for all of the molecules in the database which contain an input query substructure).

A complete record of the topography of a molecule requires the availability of the appropriate 3-D co-ordinate data, these coming from molecular mechanics or X-ray crystallographic studies. If such data are available, they can be used to display

W. A. Warr (Ed.)
Chemical Structures

structures in 3-D on a graphics terminal and such *molecular graphics* systems are becoming increasingly widely used in QSAR studies.[4,5] However, the co-ordinate data are used only for display purposes: they cannot be used to provide access to structures in response to queries which specify 3-D patterns. There is an increasing need to provide such facilities since one of the major applications of molecular graphics systems is in the analysis of *pharmacophoric patterns*, i.e., the geometrical arrangement of structural features which causes a biological response at a receptor site. Several methods are available for the identification of pharmacophoric patterns.[4–7]

In this paper, we describe some of the work which has been carried out in Sheffield over the last few years to develop searching techniques which will allow a chemist to retrieve all of those molecules from a file of 3-D structures which contain a query pharmacophoric pattern. The next section outlines the pharmacophoric pattern matching problem and discusses an algorithmic technique which we have developed for the selection of interatomic distances; these distance can be used as first-level screens in pharmacophoric pattern searches. This is followed, in the next section, by an evaluation of the retrieval performance of the screens when they are used in an experimental, pharmacophoric pattern matching system. The next section discusses algorithms which can be used for *geometric searching*, the 3-D equivalent of atom-by-atom searching, to determine whether a molecule contains the precise arrangement of atoms specified in the query pattern. During the course of this work, we have become interested in two related problems, these being the identification of the largest substructure common to a set of 3-D molecules and the extension of our work on substructure searching to macromolecular species: these two ongoing research areas are described in the last two sections.

SELECTION OF DISTANCE SCREENS FOR PHARMACOPHORIC PATTERN MATCHING

Gund[6] has noted that pharmacophoric pattern matching is closely related to the substructure search problem which has been mentioned in the Introduction. Substructure search systems generally involve a two-level retrieval mechanism in which an initial *screening* search is used to eliminate from further consideration large numbers of molecules that cannot possibly contain the query as a substructure; only those few molecules that pass this initial fragment match go onto the second-level, *atom-by-atom* search in which the exact topologies of the molecules are considered. The screening search involves a comparison of fragment substructures present in the query with corresponding sets of fragments which are used to represent the compounds in the file that is being searched. The atom-by-atom search is necessary because even if a compound contains all of the fragments associated with the query, they may be interconnected in many different ways, and thus a match in the screening search is a necessary, but not sufficient, condition for a molecule to contain the query as a substructure. Atom-by-atom searching corresponds to the classical graph theoretic problem of subgraph isomorphism, which is known to be NP-complete, and thus extremely demanding of computational resources if large numbers of structures need to be matched against the query. Accordingly, the overall efficiency of a substructure search system will be crucially dependent upon the ability of the fragment screens to discriminate between potential matches for the query, and those molecules that cannot possibly match it.

The establishment of a *screen set*, that is a set of descriptors which may be assigned to a file of structures to increase search efficiency, requires two things: firstly, some explicit definition of the type, or types, of fragment that are to be considered; secondly, a means of identifying some small number of the possible fragments of the selected type(s), the aim being to ensure a high level of screen-out whilst limiting the size of the screen set to manageable proportions.[1,2]

Many types of substructural feature have been used as screens for 2-D screening: as an example, the CAS ONLINE system involves the use of no less than 12 distinct types of feature, these including atom and ring counts, sequences of atoms or bonds, and a range of atom-centred fragments.[8] For our studies of 3-D screening, we have chosen to use screens based on the distances that separate the (non-hydrogen) atoms in molecules. More specifically, the screens consist of a pair of atoms, possibly augmented by connectivity information, together with a *distance range*. Thus, one of the screens identified by our selection procedure consists of a nitrogen atom separated from an oxygen atom by a distance of between 2.32 and 2.87 Angstroms (all subsequent distances are assumed to be in this unit). Thus a molecule that contained these two atoms separated by a distance of 2.50 Angstroms would be assigned this screen amongst many others, whereas a distance, D, such that $D<2.32$ or $D>2.87$, would result in the assignment of one of the other oxygen nitrogen screens that are available for assignment purposes. We have chosen to use screens based upon interatomic distances primarily because published pharmacophores are generally expressed in such terms, thus making it relatively easy to set up a file of potential queries (see, e.g., [9,10]).

The screens are identified by a statistical analysis of the distribution of interatomic distances in the molecules which are to be searched. The distance information is encoded as a 10-character string of the form *aaxbbycccc* where *aa* and *bb* are the atomic numbers of the two atoms under consideration, *x* and *y* are their connectivities, that is the number of non-hydrogen atoms attached to each of them, and *cccc* is the interatomic distance in units of 0.01; a rather similar type of notation has been described by Crandell and Smith.[11] Each of these 10-character descriptors may be canonicalised by ensuring that $aax \geqslant bby$: thus a molecule which contained a divalent nitrogen 7.13 Angstroms away from a trivalent carbon atom would result in the generation of the descriptor 0720630713.

Studies of screening systems for 2-D substructure searching have shown that the substructural features which are chosen for use as screens should be of intermediate frequency of occurrence in the database which is to be searched; moreover, all of these frequencies should be approximately equal if facile query encoding and good levels of screen-out are to be obtained. Accordingly, there has been considerable interest in methodologies for the generation of equifrequent screen sets,[12–14] and this work forms the basis for the selection procedure used here.

The database that we have used is the Cambridge Crystallographic Data Bank (CCDB)[15] which contains connectivity, co-ordinate and bibliographic data relating to all carbon-containing compounds for which an X-ray structure has been reported. A sub-file of 2,337 molecules was extracted from the CCDB after eliminating records that contained more than a single residue, referred to polymeric species, or contained atoms other than B, Br, C, Cl, F, H, I, N, O, P and S. For each of the chosen molecules, the unit cell co-ordinates were transformed into Cartesian co-ordinates. These data were then analysed to provide the descriptor frequency information that was to form the basis for the selection of the screens. A structure containing NS (non-hydrogen) atoms will give rise to $NS(NS-1)/2$ distinct interatomic distances and descriptors were generated for all such distances in each of the 2,337 molecules in the sample file. In all, a total of 743,749 descriptors was generated, this representing an average of 318 descriptors per compound.

In the actual implementation, which is discussed below, two types of screen set were used and it was decided that each set should contain *circa* 750 screens. Given a screen set of size 750 and a total of 743,749 descriptor occurrences, it would seem that each screen should correspond to a total of about 1000 occurrences in the sample file if an equifrequent screen set is to be obtained. This may be achieved by taking the set of descriptors corresponding to some atom-pair, dividing the observed occurrence frequency by 1000, and then rounding to obtain the number of screens appropriate to this pair of atoms. For example, the sorted file of 743,749 descriptors contains 16,159 distances involving the O-O atom pair; thus, 16,159/1000, i.e., about 16, screens need to be created for the characterisation of O-O distances in the database.

The distance ranges corresponding to such a number of screens may then be obtained by taking the descriptors for that atom-pair, sorting them into increasing distance order and then assigning the first screen to the distance range that includes the first 1000 occurrences, assigning the second screen to the corresponding range for the next 1000 descriptors and so on. In essence, this was the procedure that was used, although some modifications are needed to ensure that all occurrences of a given distance are assigned the same screen and that no distance range terminates in the middle of the occurrences of a particular descriptor; the detailed workings of the selection algorithm are described by Jakes and Willett.[16]

EVALUATION OF SCREEN-OUT

A prototype pharmacophoric pattern matching system has been installed in the Research Information Services Department, Pfizer Central Research (U.K.), the implementation being in FORTRAN 77 on a VAX 8600 running under VMS.[17]

Two types of screen can be generated by the selection algorithm, these being the AA screens and the AX screens. An AA screen is one in which the elemental types of both atoms in a pair are specified, while an AX screen is one in which only one of the elemental types is specified: thus typical AA and AX screens are of the form (O C 2.27–2.30) and (O X 2.29–2.32) respectively, where X represents any non-hydrogen atom and where the second half of each bracketed string corresponds to the chosen distance range. Within both screen sets, some of the screens involving highfrequency atoms (CNO) additionally contain connectivity information for added discrimination. The retrieval experiments used a subset of the CCDB containing some 12,728 structures, with a total of 1471 AA and AX screens available for the characterisation of the structures in the database. The same screens are used to characterise both the structures in the file and query pharmacophoric patterns; the assignment procedures are described in detail by Jakes *et al.*[17]

The file of screened structures is organised so as to allow the query to be matched against the structures as efficiently as possible. Since each of the M screens in the screen set denotes the presence or absence of some specific interatomic distance range within a structure, each structure can be described by a bit string of length M bits. A file of N molecules can hence be represented for search as a bit map, B, containing N rows and M columns, in which the assignment of the J'th screen to the I'th compound ($1 \leq I \leq N$, $1 \leq J \leq M$) is denoted by switching on the bit B[I,J]. Thus the I'th row is a bit string representing the assignment of the M screens to the I'th compound, while the J'th column is a bit string representing the assignment of the J'th screen to the N molecules.

The bit map is set up on disk so that direct access is available to each of the M columns. When a query pattern is submitted, screens are assigned to the query; molecules matching the query at the screen level may then be rapidly identified by carrying out intersection (AND) and union (OR) operations on the bit string columns from B corresponding to the query screens. Matching structures are then candidates for the more precise *distance search* and the *geometric search*. These two levels of search are required since a structure that matches the query at the screen level may well not be an exact match for one of two reasons. Firstly, the screens refer to distance ranges, and not to precise distances (so that a structure may well be assigned a query screen even though it does not contain the precise query distance). Secondly, the screen search identifies merely the presence or absence of various distances, without regard to the exact way in which they are interrelated in 3-D space.

Many of the false matches are eliminated by the distance search. The co-ordinate data for each structure which passes the screen search are retrieved from a disk file, the exact interatomic distances are calculated, and then compared with the set of query distances to determine whether all of the latter are present within the structure. The co-ordinate data are very bulky, and require access to backing storage for each of the molecules involved in the distance search. In addition, there is the computation associated with the calculation of the distances and their subsequent comparison with the set of query distances; thus the distance search is much more time-consuming than the screen search, despite the much smaller number of structures that needs to be processed. While the distance search permits the identification of those structures which contain the precise distances in the query specification, it does not involve consideration of the overall topography of the structures, which can only be established by the final geometric search (which is discussed in detail in a later section).

The performance of the search system was tested using the set of ten published pharmacophoric patterns listed in Table 1; further results are presented by Jakes *et al.*[17] Patterns in the literature are usually specified by a series of distance ranges, and in such cases the published distance ranges formed the basis for the search that was carried out. In a few cases, such as the anti-cholinergic pattern, no range was specified and in such cases the pharmacophore that was searched contained the exact distances plus or minus 0.10 Angstrom, so as to ensure the retrieval of at least some structures. Table 1 contains the results of searching the ten query patterns against the file of 12,728 3-D structures; the table contains the numbers of structures retrieved in response to each query in the screen (SS), distance (DS) and geometric (GS) searches. An inspection of the figures in the table shows that the screen-out and precision vary considerably from query to query. This is hardly surprising given that some of the patterns involve several precise distances, while others are much broader in scope, involving only a few distance ranges. Despite this, the screen-out is consistently high; over the set of queries, the screen searches retrieved a total of 5363 structures, this representing a mean screen-out of 95.8%. A total of 1502 molecules passed the distance search, representing a screen-out of 72% of the molecules passing the screen search. Taken together, the screen search and the distance search thus succeeded in eliminating 99.2% of the structures prior to the geometric search. The geometric search identified a total of 609 structures that exactly matched the query specification, representing precision figures of 11.4% and 40.5% for the sets of molecules passing the

Table 1 Numbers of structures matching in the screen search (SS), distance search (DS) and geometric search (GS) for ten published pharmacophoric patterns.

Pattern	SS	DS	GS
Anti-cholinergic	368	80	4
Adrenergic	325	24	0
Ant-leukemic	485	370	171
Anti-malarial	379	78	0
Anti-neoplastic	283	56	0
Hallucinogenic	542	191	69
Serotonic	519	1	0
Prostaglandin-like	690	55	4
Steroid hormonal	1106	83	102
Analgesic	666	414	259

screen and distance searches respectively. We feel that these results fully justify the screening methodology which we have developed, since the screens clearly allow substantial reductions in the computational requirements of pharmacophoric pattern searching.

GEOMETRIC SEARCHING ALGORITHMS

In this section, we discuss algorithms that can be used for geometric searching, where we wish to identify whether the precise pattern of distances characterising a query is present in a structure in the search file. There is an obvious, brute-force algorithm for geometric searching. Given a query pattern and a structure containing NQ and NS atoms respectively, generate all NS!/(NQ!(NS−NQ)!)) combinations of NQ atoms from the structure and determine whether any of these combinations is an exact match for the query pattern, this match involving a check on the geometric arrangement of the NQ(NQ−1)/2 distinct interatomic distances. Such a procedure is clearly impracticable for all but the smallest structures, and there is accordingly a need for sophisticated techniques which can reduce the number of combinations that need to be tested to an acceptable level. We have recently carried out a comparative evaluation of algorithms which can be used for this purpose, and in this section, we give a brief overview of their *modus operandi*: detailed descriptions are presented by Brint and Willett.[18]

Lesk[19] has described an algorithm primarily for pattern searching in proteins but which can also be used with the smaller molecules present in the CCDB. The algorithm assigns a structure atom, S_x, as a candidate match for a query atom, Q_y, if S_x has other structure atoms of the appropriate atomic types at the same distances from it as does Q_y. All of the structure atoms which are not matched to any query atom are removed from consideration and the candidate matches are checked again to ensure that the deletions have not affected any of the current structure atom-to-query atom matches. This process is repeated until no more eliminations can be made; thus, the algorithm reduces the number of possible structure atoms that can match each of the query atoms. Once this has been done, the brute-force algorithm is invoked to generate all of the allowed combinations of NQ structure atoms, and these combinations are then tested to see whether they match the pharmacophore by trying to rotate them onto the query pattern.

Set reduction involves the successive elimination of candidate structure atoms from sets corresponding to each pattern atom on the basis of an analysis of neighbourhood and connectivity information. The technique has been widely used as a component of 2-D substructure searching systems[1,2] and we have developed a modification of the technique which can be used for geometric searching. The first stage of the algorithm involves the creation of a *distance table*. The NQ pattern atoms are labelled from 1 to NQ and for each of the NQ(NQ−1)/2 distinct interatomic distances in the pattern (or less if not all of the query distances are specified for the search), a list of pairs of atoms from the structure is produced. The distance between the atoms in these pairs is equal to that between the pattern atoms (to within any specified tolerances), and the atom type of the first atom corresponds with the type of the first query atom (and similarly for the second atom).

Once the distance table has been calculated, the main stage consists of taking each pattern atom Q_i in turn and finding the smallest list of pairs in the table that is associated with it. For each pair, S_i–S_j, in this list, check that the atom in correspondence with Q_i also corresponds with it in the other NQ−2 lists corresponding to distances which involve Q_i. If it does not, the pair is removed from the list. When the list has been processed, the pairs in the other lists corresponding to Q_i are checked to see whether the atom which corresponds to Q_i does so in the list which was processed first. If it does not, the pair is again removed. The main stage is repeated until no further eliminations can be made, when the remaining pair lists are passed onto a highly efficient backtracking, depth-first search procedure which allows the identification of those which are present.[18]

Any 3-D molecule can be regarded as a graph in which both the nodes and the edges have labels, these corresponding to the atomic types and the interatomic distances respectively. This being so, it is possible to use established graph theoretic procedures for pharmacophoric pattern matching, and we have tested two such algorithms.

The first approach is based on the *maximal common subgraph* (MCS) algorithm of Levi[20] and Barrow *et al.*,[21,22] where an MCS is defined as the largest subgraph that is common to a pair of graphs;[23] such algorithms have been widely used in chemical information systems as the basis for automatic reaction indexing systems.[24] Given a pair of graphs, Q and S, an MCS algorithm can be used to determine whether Q is a subgraph of S simply by determining whether the MCS is equivalent to Q. The algorithm of Levi and Barrow *et al.* involves the generation of a *correspondence graph*, C, which can be formed from the two original graphs Q and S as follows:

1. Create the set of all pairs of nodes from the two graphs such that the nodes of each pair are of the same type.
2. Form a graph whose nodes are the pairs from Step 1. Two nodes (S_x,Q_m), (S_y,Q_n) are connected if the values of the arcs from S_x to S_y and Q_m to Q_n are the same.

Maximal common subgraphs then correspond to the *cliques* of the correspondence graph, where a clique is a subgraph in which every node is connected to every other node and which is not contained in any larger subgraph with this property. Examples of the use of this technique for 3-D geometric searching have been described by Golender and Rozenblit[25] and by Kuhl *et al.*[26] The former authors have used it to find out if a pharmacophore occurs in a molecule by looking for cliques in the correspondence graph whose size is the same as that of the pattern, and it was this method which was coded, using the well-known clique detection algorithm of Bron and Kerbosch.[27]

The final algorithm tested was the subgraph isomorphism algorithm due to Ullman.[28] This begins with an array M_O of size NQ∗NS, the elements of which have the value one if the relevant pattern and structure nodes could match each other, and zero otherwise. The basic algorithm works by forming a series of matrices M_D (D = 1, 2, . . . NQ) each one being created from its predecessor M_{D-1} by systematically changing all but one of the ones in a row to zero. The final matrix is checked to see whether each row contains a single one and each column contains no more than a single one; if it does not, then backtracking occurs. Ullman modifies this naive tree search by adding a refinement procedure which is based on the fact that, for a subgraph isomorphism, if Q_x is a neighbour of Q_w in the pattern and S_z in the structure matches with Q_w, then there must exist a neighbour, S_y, of S_z which matches with Q_x (and the relevant entry for Q_x–S_y in M_{NQ} must be one); in the context of 3-D chemical graphs, this is equivalent to a node having neighbours at the appropriate interatomic distances.

These four algorithms were implemented in FORTRAN 77 on a Prime 9950 minicomputer. One set of experiments[18] used a subfile of 250 molecules selected from the start of the CCDB. Every 25th molecule was used to generate patterns of size 3, 5 and 7 atoms by randomly selecting the appropriate number of carbon atoms from the structure; these patterns were then run against the subfile to see how many times they occurred. The means and standard deviations of the CPU times for these searches are given in Table 2. No figures are listed for Lesk's algorithm owing to the fact that all of the runs had to be aborted after using more than 2400 CPU seconds, this occurrence arising from occasions where the partitioning step was not sufficient to reduce the number of combinations to an acceptable level (one of the size-7 query patterns, indeed, involved a final stage match involving over 200 million possible combinations). A detailed discussion of these, and other, results is given by Brint and Willett; however, it is clear from the figures presented here that the Ullman algorithm is by far the quickest in operation and its relative advantage increases with an increase in the size of the query pattern. Thus, this algorithm would seem to be well suited to the provision of efficient geometric searching facilities.

Table 2 Mean standard deviation of search times (in CPU seconds on a Prime 9950) for searching a file of 250 molecules from the CCDB.

Algorithm	Pattern Size					
	3		5		7	
Lesk	*	*	*	*	*	*
Set reduction	66	18	131	69	146	82
Clique finding	44	1	109	15	235	24
Ullman	31	2	37	10	33	10

IDENTIFICATION OF MAXIMAL COMMON SUBSTRUCTURES

The work described in the previous sections has provided efficient retrieval techniques when there is a need to search a database for a pharmacophoric pattern of interest. An alternative, but related, problem is that of identifying what is the pharmacophoric pattern which is responsible for activity. One possible approach to the automatic identification of such patterns involves the comparison of a set of molecules, all of which exhibit the activity of interest; then, if it is supposed that there is only a single mode of action present in the data-set, the substructural features which are responsible for the observed activity are likely to correspond to those patterns of atoms which are common to all of the structures. Thus, an MCS algorithm provides an obvious approach to the automatic identification of pharmacophoric patterns.

The MCS algorithm of Crandell and Smith[11] involves taking all the common substructures of size N atoms associated with each molecule in the data-set and adding an extra atom to each such substructure. These enlarged substructures are then canonically named, so as to allow them to be compared rapidly with the enlarged substructures associated with all of the other molecules in the data-set. If a substructure is not found in all of the other molecules' lists, it is deleted from consideration. The surviving substructures form the common substructures of size N+1.

The selection of an atom to add to a substructure in the growing step is done by consulting a distance matrix associated with each molecule. This matrix contains the distances between all atoms in the molecule. Negative distances in the matrix indicate those distances which are not included in the current set of common substructures, while atoms associated with positive distances are available for addition to the common substructures in the next growth phase of the algorithm. Thus, the algorithm consists of the following basic steps:

1. Create the initial distance tables; set N := 1.
2. Grow substructures of size N+1 atoms from the common substructures of size N atoms identified in the previous iteration.
3. Name the substructures.
4. Compare the substructures to identify those which are still in common; if there are none then stop.
5. Amend the distance tables and return to Step 2.

Crandell and Smith[11] describe modifications to this basic procedure to allow for a common starting substructure to be specified, which must be contained in any substructures produced by the algorithm, and to cater for stereochemistry; these modifications are not considered here.

An alternative technique for the identification of 3-D MCS's is the clique detection procedure described above, where it was noted that MCS's correspond to the cliques of the correspondence graph linking a pair of 3-D structures.We have recently carried out an extended comparison of these two approaches to the identification of 3-D MCS's using structures from the CCDB.[29] A typical set of results is shown in Table 3, which gives the CPU times (for FORTRAN 77 programs on a Prime 9950 machine) for MCS searching using pairs of structures containing (a) 14 (b) 15 and (c) 20 non-hydrogen atoms. The pairs of structurally similar molecules were obtained by extracting a structure from the CCDB and then randomly distorting some of the co-ordinate data so as to alter some of the interatomic distances; thus, the extent of the overlap between the structures can be controlled by the amount of distortion imposed on the chosen compound. It will be seen from Table 3 that the clique detection algorithm is substantially faster in operation than the Crandell-Smith algorithm, and similar results were obtained in all of the other comparative tests which were carried out in the course of this evaluation.[29]

Table 3 Search time (in CPU seconds on a Prime 9950) for 3-D MCS identification of pairs of structurally related molecules from the CCDB.

(a)

Algorithm	Size of the largest common substructure 6	7	10	12	14
C-S	4.5	5.3	12.0	47.9	266.3
Clique	2.1	2.3	2.0	2.3	2.4

(b)

Algorithm	Size of the largest common substructure 6	7	9	11	13	15
C-S	2.6	3.2	4.7	19.1	70.6	382.0
Clique	1.5	1.5	1.5	2.3	1.4	1.4

(c)

Algorithm	Size of the largest common substructure 6	7	9	11	14
C-S	8.6	14.6	19.0	59.8	*
Clique	4.1	4.0	3.6	3.8	3.8

One characteristic of the Crandell-Smith algorithm is that it embodies a considerable degree of parallelism, thus making it potentially suitable for implementation in a multiprocessor environment. Computers embodying some degree of parallelism have been widely used in chemistry, e.g., for molecular mechanics and protein crystallography studies; however, these applications typically involve extensive numeric calculations which can be implemented efficiently on vector processors, such as those manufactured by Cray and Floating Point Systems. Wipke and Rogers[30] and Gillet *et al.*[31] have shown that 2-D chemical subgraph matching is a non-numeric task which is well suited to implementation in a multiprocessor environment, and we are currently evaluating whether this is also the case with 3-D MCS matching. In the Crandell-Smith algorithm, the process of generating the size (N+1)-atom substructures (from the N-atom substructures common to all of the molecules which are to be compared) for each of the molecules can be carried out independently of the others; this task can hence be executed in parallel (if sufficient processors are available) as can the updating of the distance matrix in Step 5 of the algorithm. However, there are problems of synchronisation since the processors need to communicate with each other to determine which of the new set of size (N+1)-atom substructures are common and can be input to the next iteration of the algorithm.

The processors which we are using in this work are INMOS transputers[32,33] A transputer is a 10 MIP RISC processor (a 10 million instructions per second reduced instruction set computer) with a limited amount of local storage (4 Kbytes in the model T414 transputer), links to large internal and external memory devices and four high-speed communication channels which allow it to be connected to other transputers in whatever configuration seems appropriate to the task in hand. A language, occam, is available for the programming of transputer systems; the language has been designed specifically for the implementation of concurrent systems and can be run on a single transputer, which time-slices between the various processes in a concurrent program, or on a whole network of transputers, where the processing is spread over the network.

Some initial simulation experiments suggested that the Crandell-Smith algorithm can be speeded-up by about five times (when compared to the processing rate of a single processor) if a pool of sixteen transputers is used to compare a pair of 3-D structures from the CCDB; the degree of speed-up increases with an increase in the number of molecules which are being processed. We are now implementing the algorithm on a ternary tree arrangement of transputers to see whether this increase in efficiency can be obtained in practice.

IDENTIFICATION OF SECONDARY STRUCTURES IN PROTEINS

Our other area of current interest is in the development of 3-D substructure searching techniques for protein molecules, a facility which could be of considerable importance for the emerging technology of protein engineering. There are three main levels of structural organisation in proteins, these being the *primary structure* which describes the linear sequence of amino acid residues comprising the protein, the *secondary structure* which describes the way in which the alpha-carbon atoms of the amino acids group themselves together in a small number of commonly occurring structural types, and the *tertiary structure* which describes the complete 3-D arrangement of the atoms in the structure. There are several sequence, i.e., primary structure, databases available, together with a range of sophisticated searching techniques which allow, for example, the identification of all occurrences of a query subsequence or of the sequence least dissimilar from a query sequence.[34,35] The only available tertiary structure database is the Protein Data Bank (PDB)[36] which contains X-ray crystallographic data for *circa* three hundred macromolecular species. The co-ordinate data are used primarily for display purposes, allowing portions of a protein to be viewed on a graphics terminal for studies in protein folding, protein evolution and drug receptor studies *inter alia*.

Secondary structures are important in defining protein conformations and in attempts to relate structure and function, and it is hence of some importance to be able to identify the secondary structures which are presentin a protein molecule. However, the identification of the secondary structure components is a rather subjective task, the residues comprising each such component being identified by visual inspection of a graphics terminal display, and different workers may assign different secondary structures to a given protein. A pioneering attempt to overcome this problem was made by Levitt and Greer[37] who described algorithmic techniques which could be used to identify several types of secondary structure using hydrogen bonding, torsion angle and interatomic distance information. We are collaborating with colleagues in the Department of Biochemistry at the University of Sheffield to extend this work by making an extended study of the use of interatomic distance information for the automatic identification of secondary structures, specifically the alpha-helix, beta-strand, 3–10 stretch and turn structures. The main body of our work has involved the creation of model secondary structures, using published torsion angle and distance data, and then searching for these, or characteristics derived from them, in structures from the PDB. The secondary structures identified by our programs can then be compared with those which have been assigned by the workers who deposited the original co-ordinate data. Thus, the assignments in the databank are accepted as being correct, and we seek to identify techniques which can correctly mirror these expert assignments.

To date, we have used three major approaches, all of them being based upon just the pattern of alpha-carbon atoms in the protein backbone, and without regard to the corresponding side-chains. The first of these is a modified geometric searching routine which takes the set of interatomic distances characterising a model secondary structure and then moves along the backbone to determine whether a match is present; an error range is specified to allow for the variations in interatomic distance which are known to occur due to slight bends in the axis of the structure. The second approach is based on the fact that it is not necessary to compare all of the interatomic distances to identify some of the secondary structures. For example, the 1:3 distance (i.e. the distance between the alpha-carbons of the first and the third residue in the secondary structure) in a beta-strand is much longer than in the other secondary structures which we have considered; accordingly, it is necessary only to identify this characteristic distance to confirm the presence of this structural feature. Similar characteristics can be derived for the other secondary structures, and the resulting set of discriminating features built into a binary decision tree which can be used to classify an unknown secondary structure into one of the known structural classes. The final approach uses a linear discriminant analysis routine to carry out this assignment task.

The work is still in progress. However, it seems clear that the variability in secondary structure assignments in the PDB is so great that it will probably be impossible to reproduce completely the assignments which have been made by previous workers. That said, our results to date show that it is possible to identify all of the alpha-helix and beta-strand structures with a precision of about 75% and 35% respectively; the results for the 3–10 stretch and turn secondary structures are very much worse, with the precision typically being less than 20%. Full details of this work will be reported elsewhere.

ACKNOWLEDGMENTS

We thank the British Library Research and Development Department, Pfizer Central Research and the Science and Engineering Research Council for funding this research, and David Bawden, Jeremy Fisher, Susan Jakes and Nicola Watts for their contributions to the work.

REFERENCES

1. *Communication, Storage and Retrieval of Chemical Information*; Ash, J.E.; Chubb, P.; Ward, S.E.; Welford, S.M.; Willett, P., Eds; Ellis Horwood: Chichester, 1985.
2. Willett, P. 'A Review of Chemical Structure Retrieval Systems'. *J. Chemometrics*. In the press.
3. Wipke, W.T.; Dyott, T.M. 'Stereochemically Unique Naming Algorithm'. *J. Am. Chem. Soc.* 1974, *96*, 4834–4842.

4. Gund, P.; Andose, J.D.; Rhodes, J.B; Smith, G.M. 'Three-dimensional Molecular Modelling and Drug Design'. *Science* 1980, *208*, 1425–1431.
5. Humblet, C.; Marshall, G.R. 'Three-dimensional Computer Modelling as an Aid to Drug Design'. *Drug Dev. Res.* 1981, *1*, 409–434
6. Gund, P. 'Three-dimensional Pharmacophoric Pattern Searching'. *Prog. Mol. Subcell. Biol.* 1977, *5*, 117–143.
7. Gund, P. 'Pharmacophoric Pattern Searching and Receptor Mapping'. *Annu. Rep. Med. Chem.* 1979, *14*, 299–308.
8. Dittmar, P.G.; Farmer, N.A.; Fisanick, W.; Haines, R.C;Mockus, J. 'The CAS ONLINE Search System. I. General System Design and Selection, Generation and Use of Search Screens'. *J. Chem. Inf. Comput. Sci.* 1983, *23*, 93–102.
9. Kier, L.B. 'The Prediction of Molecular Conformation as a Biologically Significant Property'. *Pure Appl. Chem.* 1973, *35*, 509–520.
10. Kelley, J.M.; Adamson, R.H. 'A Comparison of Common Interatomic Distancesin Serotonin and Some Halluginogenic Drugs'. *Pharmacology* 1973, *10*, 28–31.
11. Crandell, C.W.; Smith, D.H. 'Computer-assisted Examination of Compounds for Common Three-dimensional Substructures'. *J. Chem. Inf. Comput. Sci.* 1983, *23*, 186–197.
12. Adamson, G.W.; Cowell, J.; Lynch, M.F.; McLure, A.H.W.; Town, W.G.; Yapp, A.M. 'Strategic Considerations in the Design of a Screening System for Substructure Searches of Chemical Structure Files'. *J. Chem. Doc.* 1973, *13*, 153–157.
13. Hodes, L. 'Selection of Descriptors According to Discrimination and Redundancy. Application to Chemical Structure Searching'. *J. Chem. Inf. Comput. Sci.* 1976, *16*, 88–93.
14. Willett, P. 'A Screen Set Generation Algorithm'. *J. Chem. Inf. Comput. Sci.* 1979, *19*, 159–162.
15. Allen, F.H.; Bellard, S.; Brice, M.D.; Cartwright, B.A.; Doubleday, A.; Higgs, H.; Hummelink, T.; Hummelink-Peters, B.G.; Kennard, O.; Motherwell, W.D.S.; Rogers, J.R.; Watson, D.G. 'The Cambridge Crystallographic Data Centre: Computer-based Search, Retrieval, Analysis and Display of Information'.*Acta Crystallogr. B35*, 2331–2339
16. Jakes, S.E.; Willett, P. 'Pharmacophoric Pattern Matching in Files of Three-dimensional Chemical Structures. Selection of Interatomic Distance Screens. *J. Mol. Graphics* 1986, *4*, 12–20.
17. Jakes, S.E.; Watts, N.; Willett, P.; Bawden D.; Fisher, J.D. 'Pharmacophoric Pattern Matching in Files of 3-D Chemical Structures: Evaluation of Search Performance'. *J. Mol. Graphics*. In the press.
18. Brint, A.T.; Willett, P. 'Pharmacophoric Pattern Matching in Files of 3-D Chemical Structures: Comparison of Geometric Searching Algorithms'. *J. Mol. Graphics*. In the press.
19. Lesk, A.M. 'Detection of 3-D Patterns of Atoms in Chemical Structures'. *Commun. ACM* 1979, *22*, 219–224.
20. Levi, G. 'A Note on the Derivation of Maximal Common Subgraphs of Two Directed or Undirected Graphs'. *Calcolo* 1972, *9*, 341–352.
21. Barrow, H.G.; Burstall, R.M. 'Subgraph Isomorphism, Matching Relational Structures and Maximal Cliques'. *Inf. Process. Letts.* 1976, *4*, 83–84.
22. Barrow, H.G.; Tenebaum, J.M. 'Computational Vision'. *Proc. IEEE* 1981, *69*, 572–595.
23. McGregor, J.J. 'Backtrack Search Algorithms and the Maximal Common Subgraph Problem'. *Software Pract. Exper.*1982, *12*, 23–24.
24. *Modern Approaches to Chemical Reaction Searching*; Willett, P. Ed.; Gower: Aldershot, 1986.
25. *Logical and Combinatorial Algorithms for Drug Design*; Golender, V.; Rosenblit, A., Eds.; Research Studies Press: Letchworth, 1983.
26. Kuhl, F.S.; Crippen, G.M.; Friesen, D.K. 'A Combinatorial Algorithm for Calculating Ligand Binding'. *J. Comput. Chem.* 1984, *5*, 24–34.
27. Bron, C.; Kerbosch, J. 'Algorithm 457. Finding all Cliques of an Undirected Graph'. *Commun. ACM,* 1973, *16*, 575--577.
28. Ullman, J.R. 'An Algorithm for Subgraph Isomorphism'. *J. Assoc. Comput. Mach.* 1976, *23*, 31–42.
29. Brint, A.T.; Willett, P. 'Algorithms for the Identification of Three-dimensional Maximal Common Substructures'. Submitted for publication.
30. Wipke, W.T.; Rogers, D. 'Rapid Subgraph Search Using Parallelism'. *J. Chem. Inf. Comput. Sci.* 1984, *24*, 255–262.
31. Gillet, V.J.; Welford, S.M.; Lynch, M.F.; Willett, P.; Manson, G.; Thompson, J. 'Computer Storage and Retrieval of Generic Chemical Structures in Patents. 7. Parallel Simulation of a

Relaxation Algorithm for Chemical Substructure Search'. *J. Chem. Inf. Comput. Sci.* 1986, *26*, 118–126.

32. Barron, I.M. 'The Transputer and Occam'. In *Information Proceedings 86*; Kugler, H.J. Ed.; Elsevier: Amsterdam, 1986.
33. May, D.; Taylor, R. 'Occam – An Overview'. *Microprocessors and Microsystems* 1984, *8*, 73–79.
34. Kneale, G.G.; Bishop, M.J. 'Nucleic Acid and Protein Sequence Databases'. *Comput. Appl. Biosci.* 1985, *1*, 11–17.
35. *Time Warps, String Edits and Macromolecules: The Theory and Practice of Sequence Comparison*; Sankoff, D.; Krusaki, J.B., Eds.; Addison–Wesley: Reading, Massachusetts, 1983.
36. Abola, E.E.; Bernstein, F.C.; Koetzle, T.F. 'The Protein Data'. In *The Role of Data in Scientific Progress*; Glaeser, P.J. Ed.; Elsevier: New York, 1985.
37. Levitt, M.; Greer, J. 'Automatic Identification of Secondary Structure in Globular Proteins'. *J. Mol. Biol.* 1977, *114*, 181–203.

BROWSING AND CLUSTERING OF CHEMICAL STRUCTURES

David Bawden
Pfizer Central Research, Sandwich, Kent, U.K.

ABSTRACT

Techniques for browsing and clustering of chemical structures, as implemented in the SOCRATES chemical databank system, are described. Both are based on a quantitative calculation of intermolecular similarity. Browsing, i.e., nearest-neighbour searching, offers easy access to chemical structure files, without the necessity of learning any query language. It also provides a means for identifying analogies, based on overall molecular similarity, thus avoiding bias based on the user's preconceptions. Clustering allows an overview of structural variation within a file, and gives a means of selecting representative subsets of compounds. Subsets can also be chosen, e.g., for biological testing, so that the compounds selected are as dissimilar as possible from each other.

INTRODUCTION

This paper describes techniques, based on quantitative measures of intermolecular similarity, for browsing and clustering within computerised chemical information systems. In particular, their implementation within the SOCRATES chemical databank system at Pfizer Central Research, Sandwich, is described; much of this work was carried out by Vivienne Winterman, a CASE student in the Department of Information Studies, Sheffield University. Although these methods have only recently been introduced into operational information systems, the basic ideas are those pioneered by George Adamson's group at Sheffield.[1]

Two main concepts are involved here. *Browsing*, implying nearest neighbour searching or ranking, has two functions. It provides an easy way of identifying relevant structures within large databanks, without any need for the user to learn a query language. It is also a powerful means of identifying analogies, based on general structural similarity and uninfluenced by the preconceptions of the user.

Clustering also has two functions: to give an easily assimilated overview of the structural variation in a large file, and to allow systematic selection of representative subsets from large sets of structures.

Taken together, browsing and clustering techniques can make structure files much more accessible, particularly to infrequent and non-expert users. Further, they can act as a great stimulus for creativity, thereby magnifying the value of the information resource.[2] They also have a considerable part to play in pattern recognition and structure-activity studies,[3] but this is not discussed here.

There are three main aspects to the implementation of these techniques within SOCRATES, which will now be discussed:

1. Similarity searching/ranking.

W. A. Warr (Ed.)
Chemical Structures

2. Clustering of a file.
3. Selection of dissimilar structures.

All of these are based on the concept of structural similarity, which, although it may be expressed quantitatively, remains a highly subjective concept. It is not possible to speak of a 'correct' ranking or clustering; only a 'useful', 'interesting', or 'intuitively sensible' one. Experience shows that chemists of equivalent expertise will frequently disagree on relative similarities.

Similarity between pairs of structures, in the work described here, is calculated as a numerical index, which is a function of structural features in common, and not in common, between the two structures. It is obviously convenient if the structural features used are common to other functions of the chemical information system, e.g., fragment screens.

SIMILARITY SEARCHING/RANKING

This is used within SOCRATES in two ways: either as a browsing search, an alternative to a 'conventional' substructure search, on a full file, or as a means of making a large output from a substructure search more intelligible, by ranking the retrieved structures in order of similarity to an 'ideal' target structure.

The procedure in both cases is the same, and has been described fully elsewhere.[4] The 'target' or 'ideal' structure is input, using the standard SOCRATES graphics front-end, or simply specifying a name, laboratory code number etc. The procedure thereafter is fully automatic, so that this search method is immediately accessible to the least experienced user.

The system identifies the structures in the file which have more than a threshold number of structural fragments (screens) in common with the target, and from these chooses a cut-off number of screens in common so as to give a 'browsing set' of appropriate size, typically about 50 structures. A measure of similarity between each of these structures and the target is calculated, and they are ranked in order of similarity to the target for display to the user.

The first structure seen by the user is the most similar, and will of course be identical to the target structure, if that is present in the file. This browsing search is therefore a useful alternative, or addition, to an exact structure match procedure, since it will show the closest structural analogues, if the required structure is not found.

The user then inspects the list of structures, in decreasing order of similarity. At any time the set can be broadened (to bring in a greater variety of structural types) or narrowed (to restrict the set to those having more in common with the target). The user can also select any structure seen as the target for a new browsing search, thereby 'wandering' through a variety of related structural types. This is therefore a genuinely interactive browsing system, which operates very rapidly by virtue of the SOCRATES inverted bit-screens.[5]

Several parameters of the procedure can be varied, and all have been exhaustively evaluated and compared;[4,6] the methods adopted in the operational system reflect the best solutions to emerge from these evaluations, with some compromises to improve efficiency. The standard set of 1315 fragment screens, chosen on frequency grounds for use in SOCRATES substructure searching, is used as the structural features, and only presence/absence information, rather than occurrence counts, is used. The Tanimoto coefficient is used as the quantitative measure of intermolecular similarity. This has the form :

$$\frac{\Sigma\ N\ [A,J]\ N\ [B,J]}{\Sigma\ N\ [A,J]^2 + \Sigma\ [B,J]^2 - \Sigma\ N\ [A,J]\ N\ [B,J]}$$

for two structures denoted as A and B where N [A,J] and N [B,J] are the number of occurrences of the Jth fragment in structures A and B respectively, the summation being over the entire set of fragments characterising the set of compounds. When only presence/absence information is considered, the expression reduces to:

$$\frac{C}{A + B - C}$$

where C is the number of fragments in common, and A and B the total number of fragments in structures A and B respectively.

The magnitude of the coefficient therefore varies from 1 (structures containing identical sets of fragments) to 0 (no fragments in common). This coefficient was chosen, out of the many possible alternatives, since it was found to give optimal results both in terms of property prediction and intuitive assessment of similarity.[4,6]

Weighting of structural features is not carried out as a general rule, since there are no *a priori* reasons for choosing a weighting function. It would be possible to allow a user of the system to choose to weight, i.e., to emphasise in the determination of similarity, particular parts of the structure, or general features, e.g., ring systems, functionalities, or overall size and shape.

This structure browsing procedure has proved very popular with users at Pfizer, both as a simple alternative to substructure search, and as a means of identifying analogous structures ('lateral thinking'). Examples of the sort of 'similar structures' produced from browse searches are shown in Figures 1 and 2 (not ranked in order); other examples are given in reference 4. Some of these would clearly be retrieved by 'conventional' substructure searches, while others would not.

CLUSTERING

In a similar fashion to structure browsing, this is used in two ways within SOCRATES. A complete file of structures may be divided into clusters, systematically reflecting the structural variation within the file, and one compound chosen from within each cluster, so as to create a subset of 'structural representatives', e.g., for use in large-scale screening programmes. Alternatively, a substantial output from a substructure search may be clustered, and one 'exemplifying' structure from each cluster examined. In this way, the structural variation within the output may be readily appreciated.

The procedures used have been fully described,[7] and are again derived from extensive evaluation of alternative methods. On the basis of a matrix of intermolecular similarity coefficients, the set of nearest neighbours for each structure is generated. Clusters are then formed, by the Jarvis-Patrick method, which assigns compounds to the same cluster if they have several nearest neighbours in common. This method gives clusterings which are more acceptable, intuitively and for property prediction, than other clustering techniques. The pattern of clusters produced is non-overlapping (i.e., each structure occurs in one, and only one cluster) and non-hierarchical (i.e., the

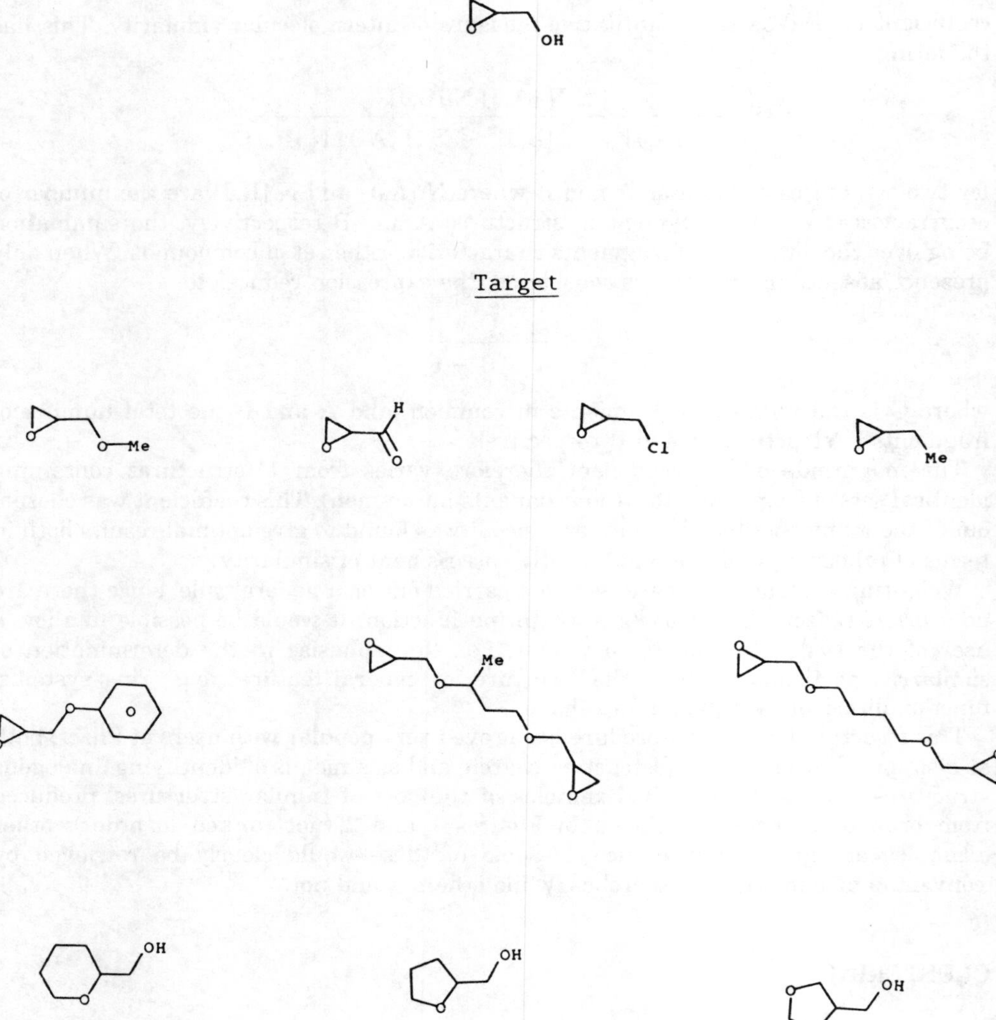

Figure 1. Example of 'similar' structures from a similarity search (unranked).

classification is 'flat', with all clusters equivalent). Although relatively computationally demanding, this procedure can be feasibly carried out on large in-house files in a VAX cluster environment, taking advantage of the SOCRATES inverted bit-screens, and rapid nearest neighbour searching algorithms.

The parameters of the algorithms may be adjusted, so as to vary, for example, the number and size of clusters produced. This means that different clusterings may be produced for a given file, for different purposes, e.g., to produce representative sets of varying size for different biological testing requirements. As with browsing, weighting functions may be used to give clusterings dependent on various structural attributes (e.g., rings or functionality).

SELECTION OF DISSIMILAR STRUCTURES

This procedure is used to generate a subset of a structure file, this subset being of specified size and comprising structures as dissimilar from each other as possible, and also from any other predefined set of structures. For example, it may be desirable to choose, for biological testing, a representative subset of a compound collection, in which the structures are as varied as possible, and also differ from those compounds already tested.

This procedure differs from file clustering, which can also be used to generate a represeniative subset, in allowing the introduction of the predefined set of structures to be avoided. If this is not used, then it would be expected that clustering and

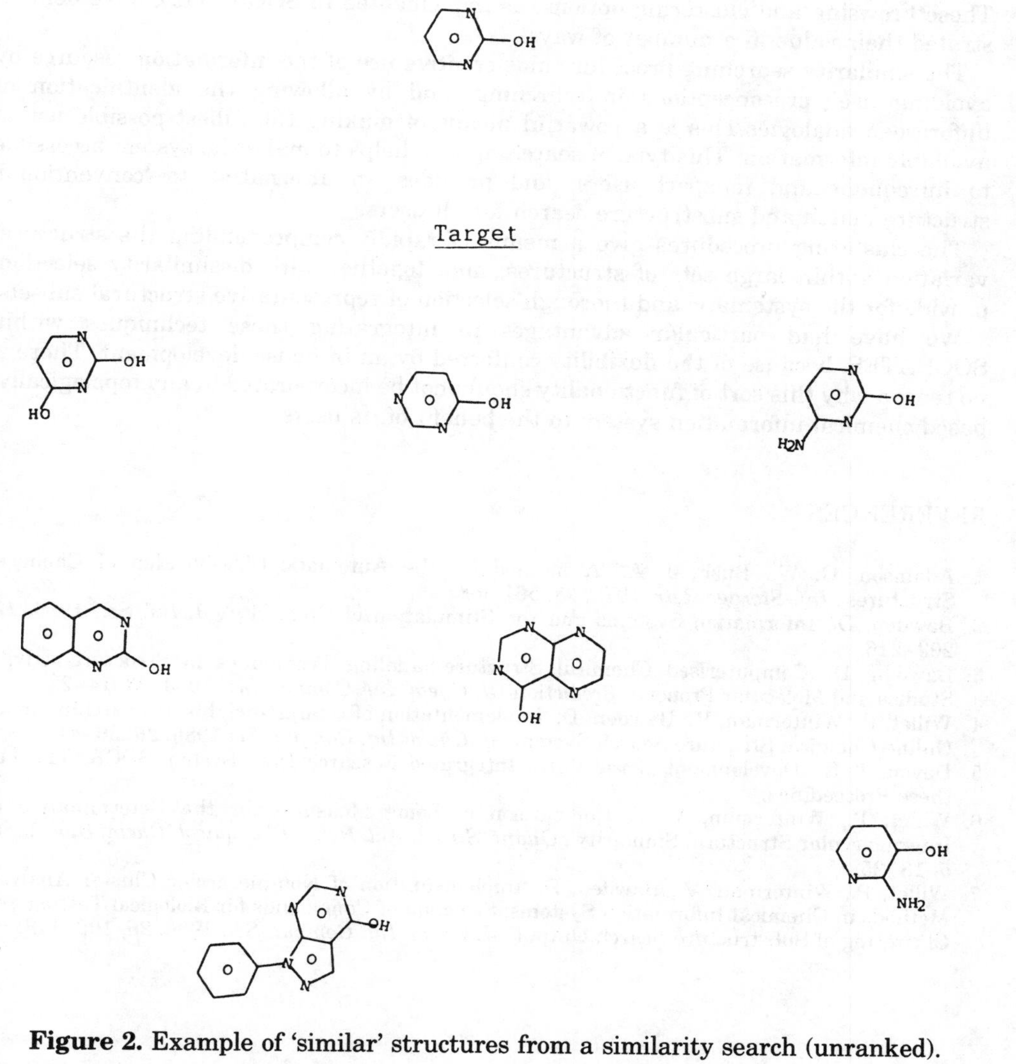

Target

Figure 2. Example of 'similar' structures from a similarity search (unranked).

dissimilarity selection, all other things being equal, would produce the same sort of subsets, though not necessarily containing the same compounds.

Initially three sets of structures are defined: a *candidate* set (from which the subset is to be chosen), a *negation* set (those from which the chosen structures must differ), which may initially be empty, and a *selection* set which will initially be empty. Then, for as many structures as are required to be chosen, the structure in the candidate set with the lowest similarity to any in the negation set is identified, added to the selection and negation sets, and removed from the candidate set. As with clustering, this procedure is relatively computationally demanding, but feasible.

CONCLUSIONS

These browsing and clustering options, as implemented in SOCRATES, have demonstrated their value in a number of ways.

The similarity searching procedure aids creative use of the information resource by avoiding user preconceptions in searching, and by allowing the identification of unforeseen analogies.This is a powerful means of making the fullest possible use of available information. This type of searching also helps to make the system accessible to infrequent and inexpert users, and provides an alternative to 'conventional' structure match and substructure search for all users.

The clustering procedures give a means of rapidly comprehending the structural variation within large sets of structures, and, together with dissimilarity selection, provide for the systematic and thorough selection of representative structural subsets.

We have had particular advantages in integrating these techniques within SOCRATES, because of the flexibility conferred by an in-house development. There is no reason why this sort of functionality should not be incorporated in any topologically-based chemical information system to the benefit of its users.

REFERENCES

1. Adamson, G. W., Bush, J. A. 'A method for the Automatic Classification of Chemical Structures'. *Inf. Storage Retr.* 1973, *98*, 561-568.
2. Bawden, D. 'Information Systems and the Stimulation of Creativity'. *J. Inf. Sci.* 1986, *12*, 203–216.
3. Bawden, D. 'Computerised Chemical Structure-handling Techniques in Structure-activity Studies and Molecular Property Prediction'. *J. Chem. Inf. Comput. Sci.* 1983, *23*, 14–22.
4. Willet, P., Winterman, V., Bawden, D. 'Implementation of Nearest-neighbour Searching in an Online Chemical Structure Search System'. *J. Chem. Inf. Comput. Sci.*1986, *26*, 36–41.
5. Devon, T. K. 'Development of the Pfizer Integrated Research Data System SOCRATES'. In these Proceedings.
6. Willet, P., Winterman, V. 'A Comparison of Some Measures for the Determination of Intermolecular Structural Similarity'. *Quant. Struct.-Act. Relat. Pharmacol. Chem. Biol.* 1986, *5*, 18–25.
7. Willet, P., Winterman, V., Bawden, D. 'Implementation of Non-hierarchic Cluster Analysis Methods in Chemical Information Systems: Selection of Compounds for Biological Testing and Clustering of Substructure Search Output'. *J. Chem. Inf. Comput. Sci.* 1986, *26*, 109–118.

THE SHEFFIELD UNIVERSITY GENERIC CHEMICAL STRUCTURES PROJECT – A REVIEW OF PROGRESS AND OF OUTSTANDING PROBLEMS

Geoffrey M Downs, Valerie J Gillet, John Holliday and Michael F Lynch
Department of Information Studies, University of Sheffield, Sheffield S10 2TN, UK

ABSTRACT

The achievements of the project are reviewed briefly. They include:

a) the design, implementation and testing of a representation which closely mirrors the characteristics of generic structures in patents, and which enables an internal representation to be created,
b) the application of formal grammar theory to some of the problems caused by generic radical terms, and
c) the development of a number of search representations including both fragments and ring descriptors, as well as reduced graph representations, together with appropriate search algorithms, for screening and atom- and bond-level searching.

Work currently in hand or needed for solution of the problem is described. The overall objective is to provide general facilities for full structure and for substructure searching.

The problem is difficult in many respects; it has called for the development of strong solutions, certain of which are also applicable to searches of databases of specific chemical structures.

INTRODUCTION

Many will be familiar with aspects of our work on generic chemical structures in progress at Sheffield for a number of years now; much of it has been published[1–11] and the Chemical Structure Association held an international conference on the topic in Sheffield three years ago, organised by John Barnard, which attracted a substantial international attendance. Even at that stage, a number of other organisations in the UK, France, Germany and the US had also become active in the area, including some which already offered chemical patent structure and substructure retrieval systems based on fragmentation codes. Since then, several new services have been announced, to be offered this year or next.

As a result, this introduction and the review of accomplishments on the Sheffield project to date can be brief, so that current work and the way in which it seeks to solve the outstanding problems are emphasised.

Generic structures pose problems which far outstretch the capabilities of topologically-based systems for substructure search of specific chemical substances. As the example of Figure 1 shows, they are framed in terms of characteristics which include the following:

R1 R2 N (R3)M1 (R5)M2 R6 S R7 R4

```
R1 = H/Cl/PHENYL OSB (Me/MeO/[4] Br);

R2 = H/ALKYL <1-2>;

R1 + R2 = METHYLENEDIOXY/ETHYLENEDIOXY;

R6 = Me/Cl/Br/MeO OSB <1-2>Cl;


R3 = METHYLENE;

M1 = <1-2>;

R4 = H/Et/N-PROPYL/PHENYL SB (Cl/Br/Et)/HETEROCYCLE RC <1>;

R7 = CARBONYL/METHYLENE;

R5 = H/Cl/Br/Me/MeO/EtO;

M2 = <1-3>.
```

Figure 1. A composite generic structure expressed in GENSAL.

1. lists of alternative substituents at fixed points of attachment;
2. variable positions of attachment for substituents, e.g., around the periphery of a ring or ring system;
3. group multiplicity, which may relate to the repetition of singly or doubly connected substituents;
4. substituents which optionally form a new ring or ring-system;
5. groups which are defined intensionally, rather than extensionally, particularly general radical terms;
6. groups defined in terms of properties, e.g., 'pharmaceutically-accepted protecting group';
7. uncertainties about the precise denotation of terms.

Certain of these features, notably the first type, together with 'don't care' bond types, have already appeared in operational substructure search systems, as amplifications to the query language, to simplify the users' task. They represent a small subset of the features which occur in generics in patents.

It is conceivable that extensions to existing substructure search systems, and a good deal more computing power, would be sufficient to deal with searches of generics which include the first four of these features. Extensions not involving new knowledge to deal with the other types can be ruled out.

Our strategy has been one of tackling the problem at a high level of generality from the start. Thus, we designed a general-purpose representation with which the widest range of generic structures typical of patents can be captured, and from this have derived a hierarchy of equivalent searchable representations, in which simplifications and generalisations enable simpler search algorithms to be used, in order to accelerate search speed. This strategy has meant that results have inevitably been a long time coming, but we have not had to scale slopes of increasing gradient one after the other.

The solutions – solutions in principle, still lacking the detailed features of the user interface, etc. – are now much clearer, and it is these that we want to outline in this paper, with particular emphasis on current work.

The design of a fiducial representation was our first concern. The result was GENSAL, an algorithmic language based on similar principles to those which underlie modern block-structured programming languages, that is, on formal grammar theory. As a result, GENSAL is a very powerful and flexible mode of expression, which enables the representation to be as close as possible to the forms in which generic structures are described in patents. It suffers from none of the syntactic ambiguities which bedevilled earlier chemical line notations, in part because of the lack of a firm formal foundation for their definition. It has been thoroughly tested through application to over 2000 patents of recent date.

GENSAL incorporates not only the usual partial structure diagrams, but also descriptions of radicals as names or chemical line formulae. These account for the substituents which are described extensionally, i.e., by recitation of their members. For the groups that are described intensionally, i.e., generic radical terms, we designed parameter-based designations which describe the features of the radicals which are present or absent. These include features such as branching and unsaturation in acyclic radicals, and number, size and types of rings, heteroatom presence or absence, etc., in cyclic radicals. John Barnard and Steve Welford laid these foundations from the outset of the project, and defined the internal representation, the Extended Connection Table Representation (ECTR). John Barnard was responsible for the interpreter which transforms GENSAL expressions into the ECTR.

The problems of search, and hence of searchable representations and of applicable algorithms, can be summarised as follows. Both file structures and queries comprise components which are described in either extensional or intensional terms. Thus either a file or a query structure may include expressions such as 'a branched alkyl radical' or 'a bicyclic nitrogen-containing ring system', and the task of the search is successfully to identify, say, an isobutyl radical with the former and a quinoline ring system with the latter. Further, it is necessary to identify subgraph correspondences too, as with a pyridine or pyrrole ring as a substructure of the class defined by the expression 'bicyclic nitrogen-containing ring system'. There may be more complex requirements, such as, that the intensionally-described radical may or may not match a description in the same terms. Basically, however, what is required is that whatever transformations the file structure or the query is subjected to, it should remain invariant with respect to the other, or should at least be identifiable as corresponding, or potentially corresponding to the other. As shown later, the task is especially difficult when 'don't care' bond situations are specified.

Our efforts have been directed in several distinct directions. From the outset, we were concerned to devise a series of representations, of which the above conditions are true, and which are degenerate forms of the full description, and, as a result, smaller and more rapidly searched, although the full representation, or at least its internal equivalent, the ECTR, remains as the ultimate in exactitude insearch. Additionally, working in parallel with us, Annette von Scholley devised her algorithm for atom-by-atom level searches on a representation derivable from the ECTR – more about this later.

The first objective was to devise a method for fast screen searching based on the hierarchy of fragment types with which we had had much success earlier. For this purpose, non-trivial and very demanding extensions of conventional screen generation routines were devised hand-in-hand with a novel departure, the development and application of TOPOGRAM. Steve Welford carried particular responsibility for these aspects. Like the theoretical basis of GENSAL, TOPOGRAM derives from formal grammar theory. For large areas of structural chemistry, it enables us, from the practical viewpoint, to generate those fragment screens which would characterise the radicals implied by the generic radical term if all of its members were generated exhaustively.

For generic structures, which encompass such large numbers of individual specific structures within each family, a fragment screen is clearly a weak search tool in many circumstances, even when it is two-part, the first part of the screen vector containing those fragments which appear in the invariant part of the structure, and the second part also including those which are variable. Accordingly, our attention has also been directed at other representations which are intermediate in detail between the ECTR and the highly degenerate fragment screens. The chief of these are the reduced graph representation, which describes the gross topological features of the entire generic structure, and the ring descriptions, which, as can readily be imagined, are highly complex for the case of ring systems in which great variation may be present. Val Gillet and Geoff Downs have pushed the work forward in these two difficult areas which are described in greater detail below, while John Holliday is now looking at the generation of ring characterisations for generic radical terms.

GRAPH REDUCTION METHODS FOR SEARCH

The notion of reduced graphs is simple. In one form, it is already widely reflected in nomenclature and in notations, in the sense that in both of these a natural differentiation is usually made between ring systems and acyclic components – consider the case of phenyl naphthyl ether. There are also exceptions, such as benzyl phenyl ether.

In the ring/non-ring reduced graph, therefore, the components of the structure are reduced to nodes of which there are two basic types, the ring node and the non-ring node. This is readily illustrated for a specific structure in Figure 2.

Substantial latitude is available in determining what detail can be given in the nodes in order to extend the descriptive power of the representation for search purposes. Thus, one might further differentiate by giving the count of atoms in the nodes, or by detailing the molecular formula, and so on.

A second and complementary basis for chemical graph reduction is provided by differentiation between contiguous assemblies of carbon atoms and heteroatoms. This is illustrated in Figure 3 for the same structure. Once again, it is possible to ring the changes on the detail given.

These methods result in substantial reductions in the numbers of nodes handled in the reduced representation. In the ring/non-ring reduced graphs, this averages about a

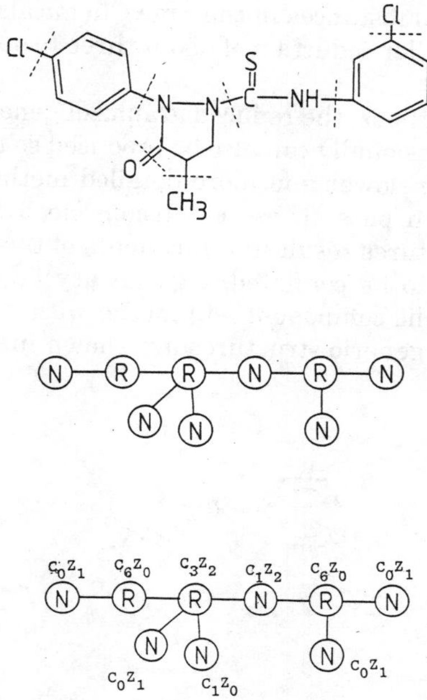

Figure 2. A specific structure and two forms of its reduced chemical graph based on ring/non-ring components.

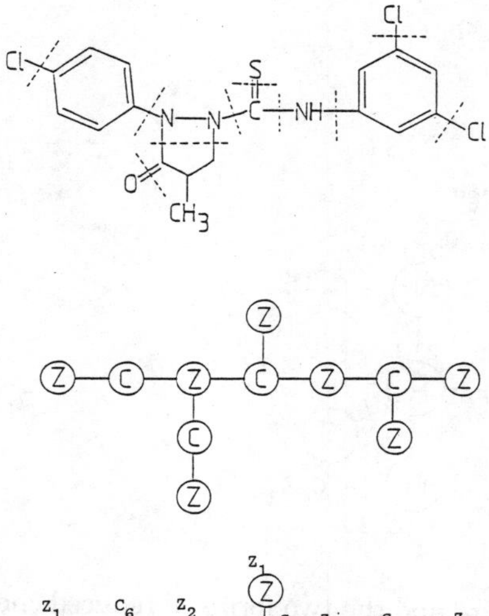

Figure 3. A specific structure and two forms of its reduced chemical graph based on carbon/heteroatom components.

five to one reduction for the substances in the Fine Chemicals Directory. In the carbon/heteroatom reduction, a smaller reduction of about three to one is achieved for the same file.

By extensions to these methods, the reduced graphs of generic structures in which all variables are described extensionally can also be produced so that simpler, faster search methods can be applied, the slower and more detailed methods being invoked for the minority of structures which pass. These extensions need to ensure that, when the descriptions of generic structures result in components of the same type being adjacent to one another, they need to be conflated, e.g., an acyclic component attached as a substituent on another acyclic component will merge with it. Instances of the reduced graphs produced from such generic structures are shown in Figure 4.

R1 = PHENYLENE;
R2 = H / [3] (METHYL / CL / ETHYL);
R3 = H / [4,5] CL / [4] METHYL / [5] ETHYL / [3] BR.

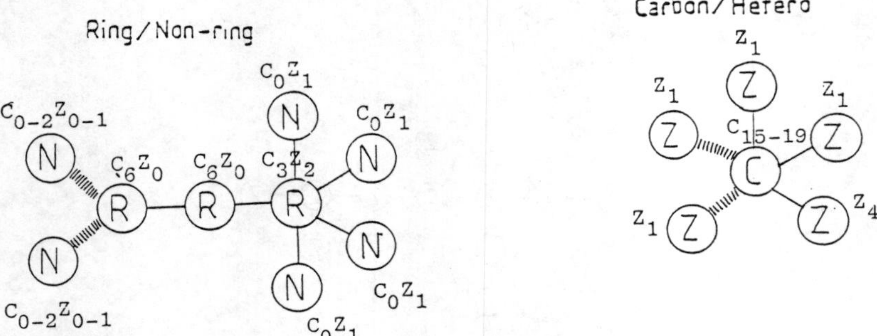

Figure 4. A generic structure and the two forms of reduced chemical graphs describing it.

Some model experiments have recently been carried out by Val Gillet on a small database of generic structures to test the resolving power of these representations.[11] The database comprises generics from 77 patents, in which the variable components are specific partial structures. It was searched using von Scholley's method. The searches involve whole structure matching, and the queries used were of two types: individual specific structures, and generic structures.

The first types of query are specific structures, one each from each of the 77 patents; each should thus retrieve at least the generic from the patent from which it is derived, and possibly others, as a result of degeneracy in the representation. Figure 5 shows the performance obtained for the ring/non-ring representation, where the level of detail in the nodes includes the number or range of numbers of carbon and heteroatoms associated with each node. Apart from three query/structure pairs which are not matched because of the presence of multiplier terms, which are not yet fully handled, 46 of the queries retrieve only their associated generic, and, in the worst cases, two queries each retrieve only 3 additional generics.

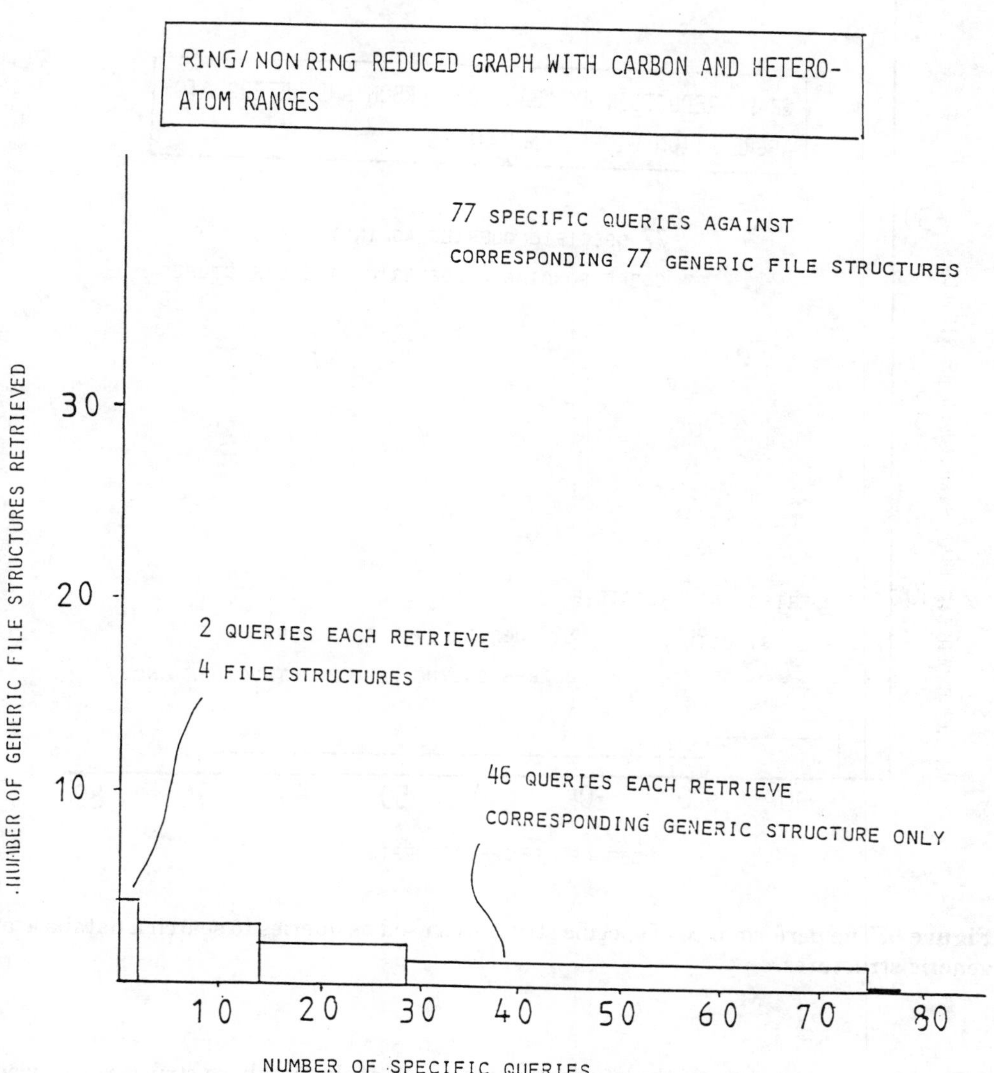

Figure 5. The performance of specific structures used as queries to search a database of generic structures.

The performance of the carbon/heteroatom reduced graph in the same search is slightly better for this database and these queries; 51 queries each retrieve only the associated generic. In the worst cases, five queries retrieve only two additional generics. Figure 6 illustrates these results.

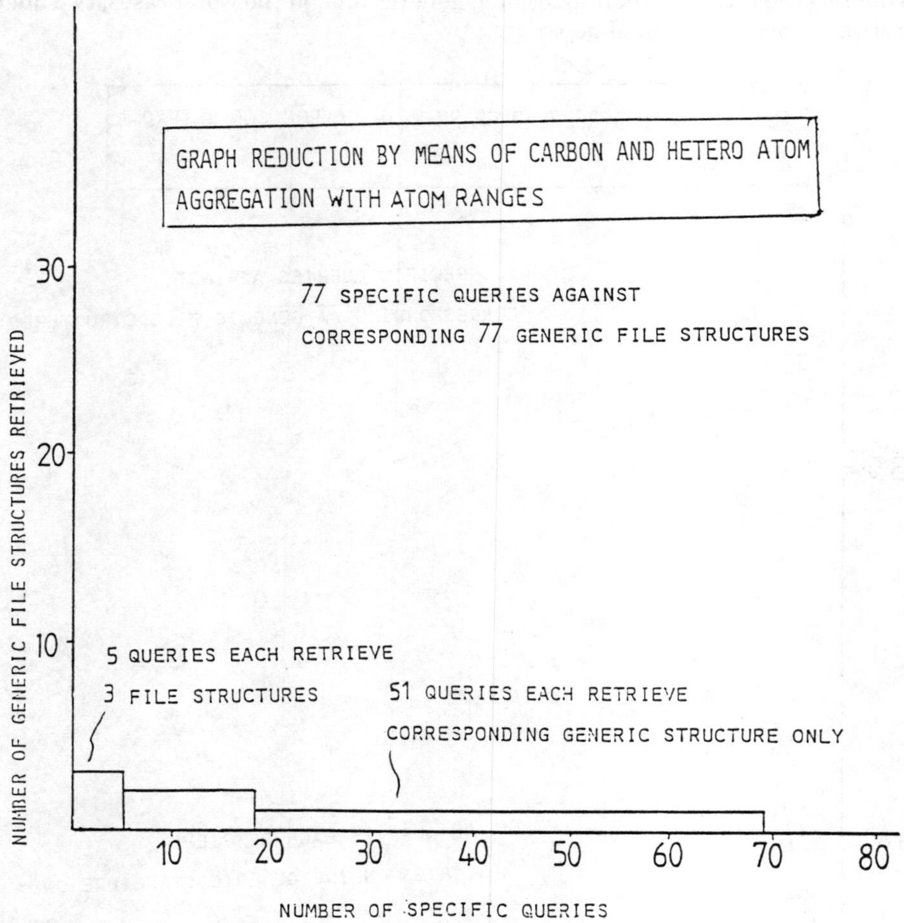

Figure 6. The performance of specific structures used as queries to search a database of generic structures.

The second query type is the generic query – each of the database structures is used in turn as a query to search the database. For this search, Figure 7 shows the result of using the ring/non-ring reduction method, and Figure 8 the carbon/heteroatom reduction.

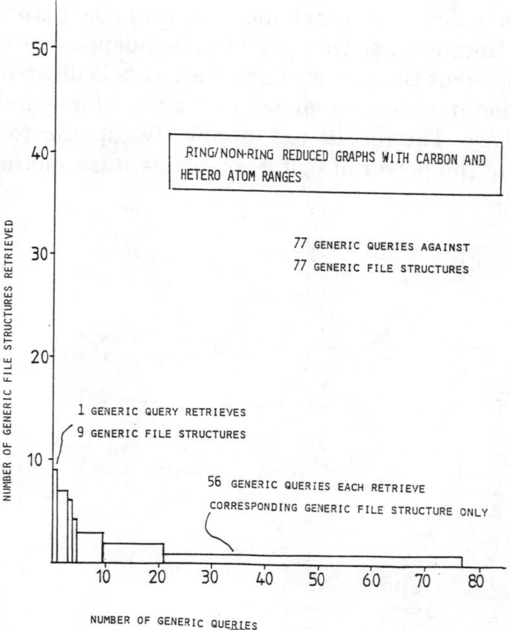

Figure 7. The performance of generic structures used as queries to search a database of generic structures.

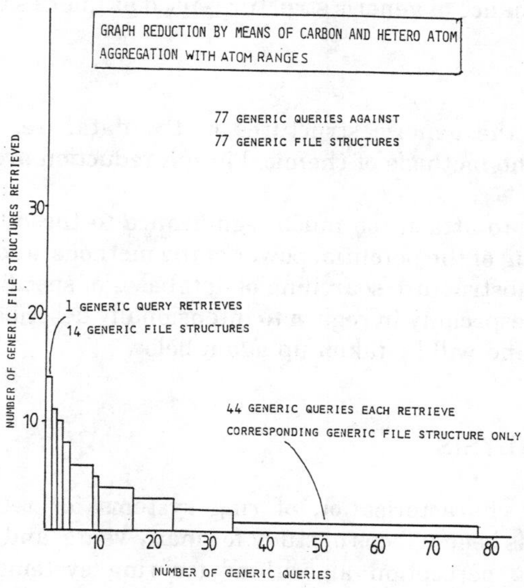

Figure 8. The performance of generic structures used as queries to search a database of generic structures.

One of our purposes in devising search methods is to use features of structures which are orthogonal to one another, i.e., they are largely independent of one another, so that one feature can complement the power of another. This is illustrated when we combine the two graph reduction methods, as shown in Figure 9 for searches of generic queries against the test database. The results are manifestly superior to either of the methods used alone, illustrating the power of using complementary methods.

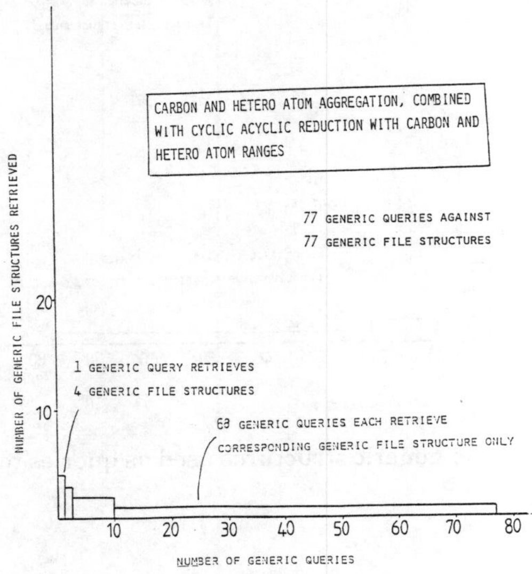

Figure 9. The performance of generic structures used as queries to search a database of generic structures.

The complexity of the generic structures in the database, and the comparative strengths of each of the methods of chemical graph reduction are illustrated in Figure 10.

It would be unwise to attach too much significance to these test results. However, they suggest something of the potential power of the methods, and they may well have a bearing on keys for substructure searching of databases of specific substances. The real problems of generics, especially in regard to intensionally described ring systems, are of much greater order, and will be taken up again below.

RING SYSTEM ANALYSIS

The question of the characterisation of ring systems for retrieval as well as for synthesis planning has been a focus of study for many years, and has evoked more than twenty different ring perception algorithms for ring systems in specific chemical structures. The question of ring characterisation is, if anything, more important for generic structures, and has engaged our efforts, particularly those of Geoff Downs, for over two years.

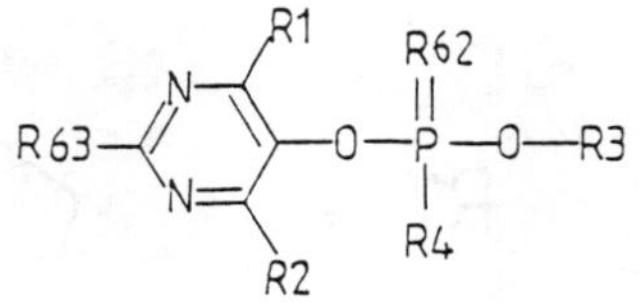

```
R62 = O / S;
R3  = METHYL / ETHYL;
R4  = METHOXY / ETHOXY / S SB N-PROPYL SB [2] METHYL;
R63 = T-BU / I-PR / ETHYL / S SB I-PR / H / METHYL / ETHYL / N-PR SB [2]
      METHYL / N-BUTYL / N-PENTYL / N-OCTYL / PHENYL / METHYL SB ((S /
      SULPHINYL) SB (METHYL / ETHYL) / SULPHONYL SB ETHYL) / S SB (I-PR /
      T-BU) / (S / SULPHINYL / SULPHONYL) SB (METHYL / ETHYL / I-PROPYL) /
      (S / SULPHINYL) SB N-BUTYL / S SB (T-BUTYL / PROPENYL [1,3]) / S SB
      (ETHYL SB METHYLTHIO / PHENYL);
R1  = H / METHYL / METHYLTHIO / S SB ETHYL;
R2  = H / METHYL.
```

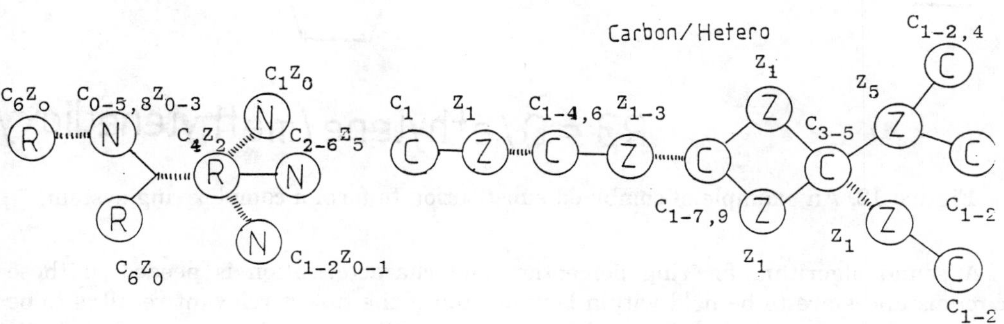

Figure 10. A generic structure and the two forms of reduced graph describing it.

Consider generic ring systems which are extensionally described, i.e., their definitions are solely in terms of structure diagrams rather than of generic radicals. The principal cases to be taken into account are:

1. Doubly connected variable substituent within a ring system, as shown in Figure 11. The substituent may be symmetrical or asymmetrical; it may itself contain a ring system. It may also be defined in terms of repeating groups or of further nested variables.
2. Single or multiple ring annelation, or the combined substituent situation, i.e., R1 and R2 combine to form a ring, as shown in Figure 12. Again, the variable substituent may be symmetrical or otherwise. It may also contain repeating groups or further nested variables.
3. Arbitrary combinations of these.

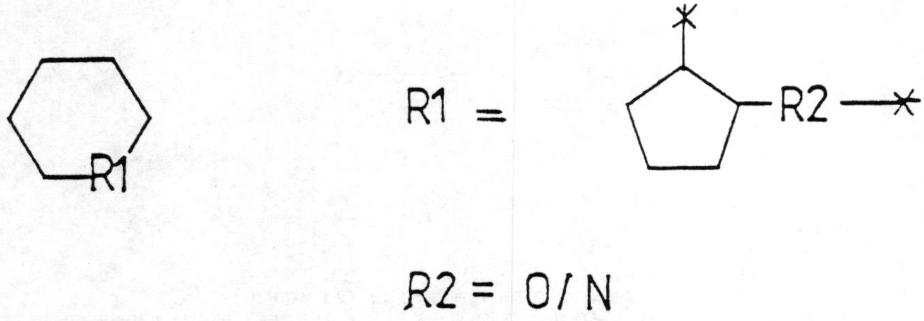

Figure 11. An example of a doubly connected ring variable and its alternative values.

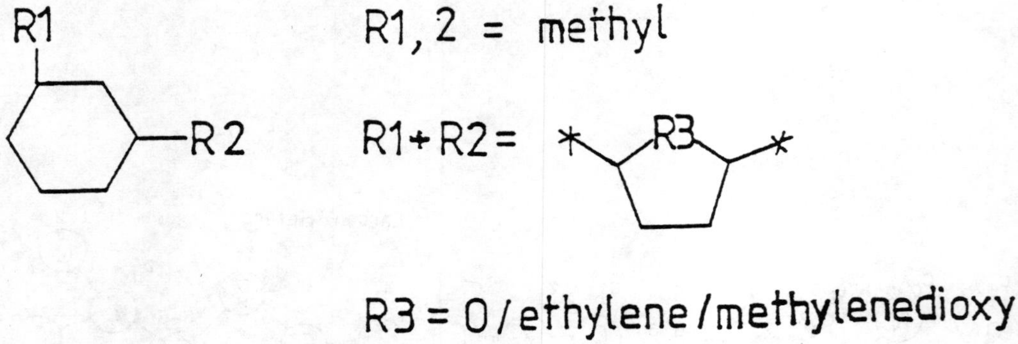

Figure 12. An example of combined substitution to form a complex ring system.

A sound algorithm for ring perception and characterisation is needed, if these circumstances are to be held within bounds, and if the fullest relevant recall is to be attained. Hence, we analysed the existing ring perception algorithms to see what choice was open to us. We were forced to conclude that none was equal to the requirements for optimal retrieval performance, and we needed therefore to develop a sounder theoretical basis for the description of the rings in specific ring systems before applying it to these extensionally described ring systems.

The review of existing ring perception algorithms under the following seven categories showed the following characteristics:

1. All cycles – the set which guarantees total recall, but at the cost of low precision and exceptionally long processing times.
2. All simple cycles – no repeated nodes, and no neighbouring points occurring non-adjacently in the cycle. They give low precision and long processing times.
3. Beta-rings – a subset of all simple cycles in which only those rings which are not combinations of three or more simple cycles are retained. These result in higher precision, but still call for long processing times.
4. The fundamental set of rings – the set of rings grown from the chords of a spanning tree – problematical, since many different spanning trees may exist and the chords depend on them.

5. Smallest set of smallest rings (SSSR) – a special case of category 4, which may result in symmetrically equivalent rings being omitted, and may therefore lead to recall failures.
6. K-Rings – the combination of all possible SSSR's, which can also lead to recall failures.
7. Heuristics – one of the above sets is used and processed to add 'chemically important' rings. At best, this is a compromise solution.

Turning now to the extensionally described generic ring systems, there are additional complexities. First, the ring description should not require the exhaustive generation of all of the specific forms of the class. Secondly, it should not depend on the orientation or form of definition of the structure. The example of Figure 13 underlines these points.

The result of our considerations was the definition of the Extended Set of Smallest Rings (ESSR), which proved adequate when tested against the most difficult specific ring systems used in testing other algorithms, and against a range of extensionally described generic ring systems designed to present worst case difficulties. Its derivation, which will be described in detail elsewhere, depends on the notion of treating a ring system as a three-dimensional solid, when the degree of complexity

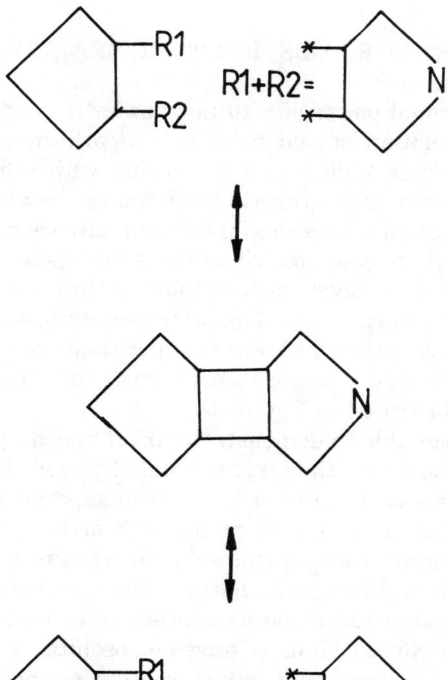

Figure 13. An example of alternative definitions of the same ring system.

necessitates this, and including all simple faces of this solid. A simple face is a simple cycle whose nodes and vertices all lie on the same external plane or surface of the solid three-dimensional structure. It is necessary to consider all possible projections, and there are cases where symmetrical equivalents to those simple faces which lie wholly or partly on an internal plane of the projection must also be included.

We feel confident that the means developed here for extensionally described ring systems can deal entirely adequately with the worst instances that inventors and Patent Offices can produce.

When we consider intensionally described ring systems, we face problems similar to those handled within the DENDRAL work, but greater in degree, in that, within DENDRAL, a fixed or bounded range of atoms is available for construction of the ring systems.[12] Intensionally described ring systems in generics are frequently unbounded with regard to the number of atoms involved, and the worst case circumstance is usual rather than unusual. Nonetheless, for those cases with relatively restricted parameter ranges, there appears to be hope of restricting the outcome of searches. Using generalised descriptions of vertex-reduced graphs, John Holliday is developing relationships which relate both to full structure and to substructure search, on which we will be reporting in due course.

POOLED PROCESSORS FOR SUBSTRUCTURE SEARCH

The need for computational power of a suitable order is much in our minds. Val Gillet has carried out a simulation of von Scholley's algorithm which took account of its internal parallelism.[7] Peter Willett and I, working with colleagues in our Computer Science Department, can now report preliminary results of implementing the algorithm on a pool of actual processors for substructure searches of specific substances – but in a different way, as yet, to that of the simulation, in that the parallelism is expressed at the database level rather than within the algorithm itself.[13] The processors chosen for this purpose are INMOS transputers, which incorporate up to four fast interprocessor links, so that different configurations can readily be constructed.

The configurations studied are the ternary tree, the singly linked chain, and the doubly linked chain, illustrated in Figure 14.

We have recently been able to demonstrate linear speed-up of the relaxation search method on up to 11 processors, the largest number tested thus far, in the case of the doubly linked chain. This configuration has proved superior to either of the others for this purpose, in that speed-up fell off at about 5 or 6 transputers with these configurations, due to communications bottlenecks; the results are illustrated in Figure 15.

These results are still preliminary in nature. The structures in the sample database serve both as database structures and as queries, since we have no readily adaptable screening system for specifics. Again, we have no absolute yardstick for the speed of the relaxation method, as compared with other existing methods. The work is currently being extended in a number of directions to provide a firmer basis for generalisation.

A word of warning. The matter of building and harnessing configurations of transputers for such purposes is no simple matter; without the work of our Computer Science colleagues and their insights into aspects such as communications and programming in occam, we could have made little headway.

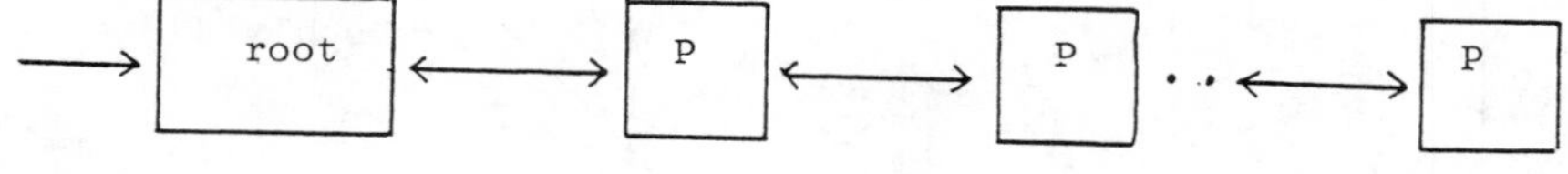

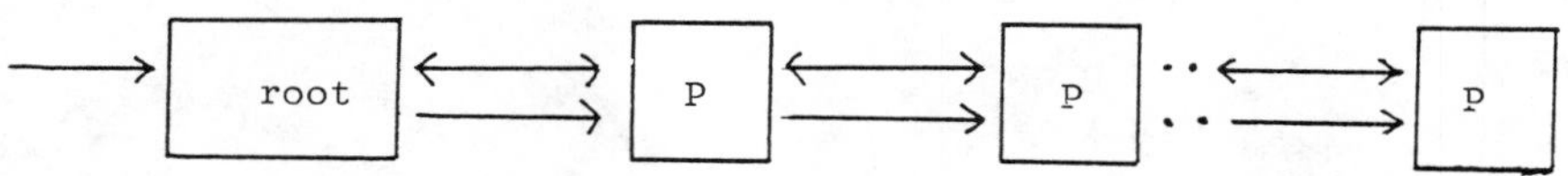

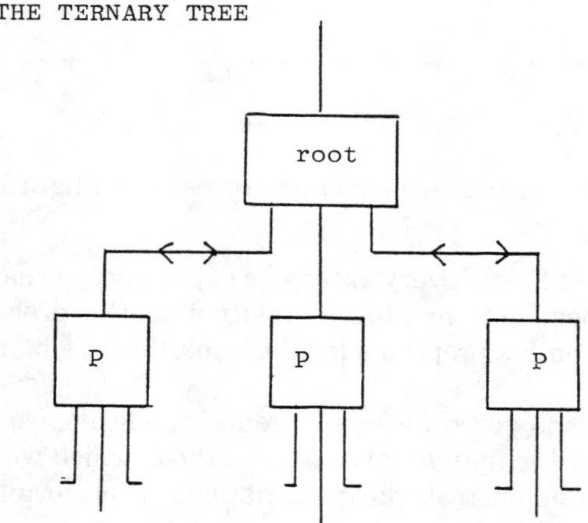

Figure 14. The configurations of pooled processors evaluated for relaxation searching.

OUTSTANDING PROBLEMS

There is a further dimension to the representation and search of generic structures, if the user requirements are to be met fully. Many features of generic structures admit uncertainty, for instance, the term 'aryl', which, without further qualification or exhaustive instantiation, admits interpretation either as a carbocyclic or as a heterocyclic radical. In consequence, in the case of the ring/non-ring graph reduction detailed previously, the term 'radical' is indefinite, while in the carbon/heteroatom reduction, a ring system with, say, two or more heteroatoms, may give rise to different representations depending on whether the heteroatoms neighbour on one another or not.

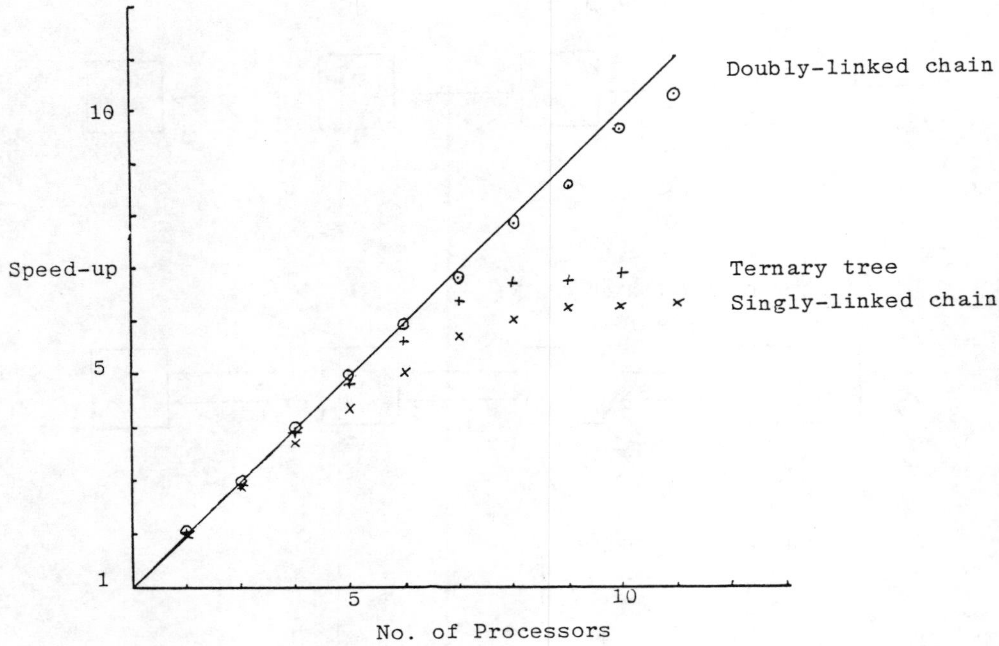

Figure 15. The performance of the three pooled processor configurations in search.

Winfried Dethlefsen of BASF, with very extensive experience of searching on the IDC and BASF systems, has been our mentor in relation to the problems which these uncertainties give rise to, and his proposals for their solution are being reflected in our current work.

Given the extreme problems posed by highly generic radicals, especially as regards ring systems, it is likely that the individual search methods which we have documented thus far will, in isolation, demonstrate large variations in performance; it is only in combination that they are likely to perform reasonably. Thus, fragment screening alone will show least discriminatory power when ring systems are the particular focus of interest in queries, since the definitions of generic ring systems tend to involve the greatest degrees of latitude. We therefore envisage reasonable performance levels being attainable in three stages of increasing degrees of difficulty:

1. Whole structure queries, where the requirement, as in the reduced graph work detailed above, involves the determination of inclusion of the file structure by the query, or the reverse, or of common membership between the two.
2. Substructure queries in which the differentiation between ring and chain bonds is retained, thus enabling the simple application of reduced graph representations at the stage after fragment screening. It seems probable that the power of these representations, as suggested in the results described above, will be adequate to the task.
3. Substructure queries in which the differentiation between ring and chain bonds is wholly or partly neglected; these pose the most difficult problems, and are likely to be the last to be attained at reasonable performance levels.

The definition of the outstanding problems is thus already very much clearer, and work towards their solution is now well in hand, although they must be regarded as posing difficulties which are still substantial in magnitude. Moreover, the integration of the range of facilities researched thus far to produce a unified system, and its implementation in terms of appropriate database access structures on suitable hardware also present tasks which will keep us busy for some time yet.

ACKNOWLEDGMENTS

We gratefully acknowledge financial support of this work by IDC (International Documentation in Chemistry) and by the British Library Research and Development Department.

REFERENCES

1. Lynch, M.F.; Barnard, J.M.; Welford, S.M. 'Computer Storage and Retrieval of Generic Chemical Structures in Patents, Part 1, Introduction and General Strategy'. *J. Chem. Inf. Comput. Sci.* 1981, **21**, 148–150.
2. Barnard, J.M.; Lynch, M.F.; Welford, S.M. 'Computer Storage and Retrieval of Generic Chemical Structures in Patents, Part 2, GENSAL, A Formal Language for the Description of Generic Chemical Structures'. *J. Chem. Inf. Comput. Sci.* 1981, **21**, 151–161.
3. Welford, S.M.; Lynch, M.F.; Barnard, J.M. 'Computer Storage and Retrieval of Generic Chemical Structures in Patents, Part 3, Chemical Grammars and Their Role in the Manipulation of Generic Structures'. *J. Chem. Inf. Comput. Sci.* 1981, **21**, 161–168.
4. Barnard, J.M.; Lynch, M.F.; Welford, S.M. 'Computer Storage and Retrieval of Generic Chemical Structures in Patents, Part 4, An Extended Connection Table Representation for Generic Structures'. *J. Chem. Inf. Comput. Sci.* 1982, **22**, 160–164.
5. Barnard, J.M.; Lynch, M.F.; Welford, S.M. 'Computer Storage and Retrieval of Generic Chemical Structures in Patents, Part 5, Algorithmic Generation of Fragment Descriptors for Generic Structure Screening'. *J. Chem. Inf. Comput. Sci.* 1984, **24**, 57–66.
6. Barnard, J.M.; Lynch, M.F.; Welford, S.M. 'Computer Storage and Retrieval of Generic Chemical Structures in Patents, Part 6, An Interpreter Program for the Generic Structure Language GENSAL'. *J. Chem. Inf. Comput. Sci.* 1984, **24**, 66–70.
7. Gillet, V.J.; Welford, S.M.; Lynch, M.F.; Willett, P.; Barnard, J.M.; Downs, G.M.; Manson, G.A.; Thompson, J. 'Computer Storage and Retrieval of Generic Chemical Structures in Patents, Part 7, Parallel Simulation of a Relaxation Algorithm for Chemical Substructure Search'. *J. Chem. Inf. Comput. Sci.* 1986, **26**, 118–126.
8. Welford, S.M.; Ash, S.; Barnard, J.M.; Carruthers, L.; Lynch, M.F.; Scholley, von A. 'The Sheffield University Generic Chemical Structures Research Project'. In *Computer Handling of Generic Chemical Structures*; Barnard, J.M. Ed.; Gower: Aldershot, 1984.
9. Scholley, von A. 'A Relaxation Algorithm for Generic Chemical Structure Screening'. *J. Chem. Inf. Comput. Sci.* 1984, **24**, 235–241.
10. Lynch, M.F. 'Generic Chemical Structures in Patents – The Research Project at the University of Sheffield'. *World Patent Information* 1986, **8**, 85–91.
11. Gillet, V.J.; Downs, G.M.; Ling, A.; Lynch, M.F.; Venkataram, P.; Wood, J.V. 'Computer Storage and Retrieval of Generic Chemical Structures in Patents, Part 8. Reduced Chemical Graphs and Their Applications in Generic Chemical Structure Retrieval'. *J. Chem. Inf. Comput. Sci.* In the press.
12. *Computer-Assisted Structure Elucidation*; Gray, N.A.B. Ed.; Wiley: New York, 1986, 325–398.
13. *The Application of Reconfigurable Microprocessors to Information Retrieval Problems*; Lynch, M.F.; Manson, G.A.; Willett, P.; Wilson, G.A., Eds; British Library: London, 1987.

GENERIC SEARCHING OF PATENT INFORMATION

Kathleen E Shenton, Peter Norton, E Anthony Ferns
Derwent Publications Limited, Rochdale House, 128 Theobalds Road, London WC1X 8RP, U.K.

ABSTRACT

Over the past twenty-five years, Derwent has been a leader in the indexing of generic structures found in patents. Though traditionally this indexing has been by a fragmentation code, for the past five years Derwent has embarked on a study of the possibilities for graphic indexing of generic (Markush) structures. As a result of this study, Derwent has teamed with Télésystèmes and INPI (the French patent office) in the development of Markush DARC, an extension of the commercially available generic DARC. This paper reviews the progress of this co-operative effort, outlining the indexing techniques used in the creation, for the first time, of a database of Markush structures, and an up-to-date review of the operational Markush DARC search software.

INTRODUCTION

The indexing of structures found in chemically-related patents has always been a challenging problem. Derwent had initially attacked this problem by indexing structures using a fragmentation code. Initially, it was expected that this code would be an interim measure; it was designed to last two to three years. However, the capabilities of computerised systems were not sufficient to provide an alternative. Though the code has been enhanced and improved over the years, Derwent still indexes chemical compounds found in patents using the fragmentation code, twenty-five years after its inception.

Over the past five years, Derwent has been active in the promotion of developments designed for the provision of graphics indexing and retrieval of generic, or Markush, structures. After a detailed survey of organisations active in the area of graphic chemical retrieval, including analyses of programs created by Chemical Abstracts Service (CAS), Molecular Design Limited, the Sheffield University generic chemical structures team, and the Télésystèmes DARC team, Derwent has chosen to collaborate with Télésystèmes in the creation of a product designed for the indexing and the retrieval of generic structures found in patents. Derwent has been fortunate to also collaborate with INPI (the French Patent Office) in this work, in both the development of the software and the creation of generic databases.

Today, in June, 1987, we have finally achieved a working system for the indexing and retrieval of generic structures found in patents. It has not been an easy development, nor has it been without compromises. In this paper, our purpose is to report the progress

W. A. Warr (Ed.)
Chemical Structures

made at this time, to describe the developmental steps and decisions taken over the last few years, and to provide an indication of what tomorrow might bring in this important area.

THE PROBLEM OF INDEXING GENERIC STRUCTURES

The ability to index a specific chemical structure graphically has been available for quite some time. Télésystèmes-DARC launched a commercial system for this purpose in 1979; CAS has been doing this type of indexing in-house since the late 1960s. However, the structures reported in patents present some unique problems: chemicals are described not as individual entities, but as families of compounds represented by a base structure with variable groups; and some possibilities of the compounds protected by the patent are represented as generic terms only.

The problem is simply exemplified by the extension of an imaginary patent covering aspirin and its derivatives. Indexing aspirin is not a difficult job – it can be done by structure, Registry Number, or name. However, if one replaces the alkyl group of this compound with a short list of other possibilities, the number of compounds covered can increase substantially. A further generalisation of the structure to cover *any* alkyl would mean that the patent covers an infinite number of structures.

This is a very simplistic example. In practice, Derwent consistently encounters documents that cover an infinite range of compounds, representing potential variations on multiple sites on a base compound.

WHAT THE USER REQUIRES

The users of patent information have some major problems; their companies expect them to be able to find information that will indicate the validity of a particular invention, any infringements on existing inventions, and potential licensing opportunities. With the breadth of information covered by patent documents today, it is easy to become mired in arguments of what is truly covered by a given disclosure. Derwent realises that if our service is to be valuable, we must provide all the information available in a patent to our users in an unambiguous way – they can then make a determination of the value of a particular document retrieved.

Derwent has also recognised the problem of variability in requirements among users. Though it is not difficult to define patent searching in a broad sense, it is a fact of life that users have varying skills, requirements, and interests. Derwent's task is to provide chemical searching systems that will satisfy *all* of our users, from those requiring a brief glimpse into the wealth of information available within patents, to those who wish everything that they can possibly find for a given structure, no matter how 'derivatised' the information.

Over the years, the availability of graphic search systems has also changed the user community. While previous generations of users have been satisfied in learning complex search systems of codes or nomenclature, the newest users expect systems to provide the capability to define chemicals by structure, 'the language of the chemist'. The next generation of Derwent chemical indexing must, therefore, offer this type of searching, in order to gain universal acceptance and wide usage.

DOWN TO 'BRASS TACKS'

There are, of course, at least two ways to look at patent information. One way is to look at patent information in the same way as information found in literature; in a patent, after all, one is reporting a development of general interest to the chemical community, just as in a journal article. A second way is to look at patents as legal documents, protecting a given area of technology and giving a limited monopoly to the assignee. Derwent feels that it is important to provide information about patent documents in a fashion that satisfies either view of patent information.

When Derwent looked at the elements of development and creation of a graphics chemical database, both of the traditional ways of looking at a patent document were of high consideration. In this regard, the GENSAL team's approach appealed to Derwent – philosophically, here was a way to provide the user with infinite flexibility in retrieval. However, though computer systems have advanced over the past two decades at a rapid rate, Derwent did not feel that the technology was advanced *enough* to provide a working GENSAL-like system in the short term. Our users needed a better way to access chemical information, and they required it *now*. For this reason, the DARC approach was appealing – though the system initially would not offer *everything* that was ultimately required for all of our patent users, it would offer a vast improvement over our current fragmentation code. Markush DARC is a pragmatic approach to solving the problem of searching a generic database.

This approach is illustrated by a simple comparison of Generic DARC and the new Markush DARC. Though Generic DARC offered the ability to form generic queries on only a single compound database, the roots of Markush DARC are in this earlier development. In many ways, Markush DARC's generic database operations mirror the generic query capabilities of Generic DARC. Hence, the DARC team could achieve an operational substructure search against a generic database within a relatively short time-frame.

Whatever the search software offered, however, the database was of utmost importance. The DARC system not only offered a means to search generic structures in the short term, it also offered a way to record *all* of the information about structures found in patents, through a new and flexible input system. This capability was of incredible import to Derwent; we could build a database that would not only serve the users today, but in future could be of more utility as search software was further developed. As opposed to the fragmentation code, there should never be a question of time-ranging. Moreover, though we assume that Markush DARC is a system not only of today but of tomorrow, we are building a database that can be converted and adapted to other search systems that may be offered by other organisations.

THE PROBLEMS OF CREATING A GENERIC DATABASE – AND SOME SOLUTIONS

The problems in the creation of a generic database can be broken up into three main areas, classification of chemicals, expression of generic groups and standardisation of conventional representations.

In addition to work in defining the input and search software required from Télésystèmes, Derwent has invested a great amount of effort in the research of these problems. As a result, Derwent has been able to index chemicals found in New Compound specifications from the beginning of the Derwent Year 1986 (Week 8601). In devising solutions for the indexing of generic structures, Derwent studied all existing systems of chemical indexing, including those of CAS, the Pharma Documentation Group (PDG), the Institute for Scientific Information (ISI), IFI/Plenum Data Corporation, the Sheffield University team, the American Petroleum Institute (API) and others. With 1986 as a test year, we can now look to the indexing of documents issued in 1987, for which we will perform complete (new compound plus known compound) indexing.

Classification of Chemicals

Derwent has classified chemicals by means of two elements of indexing – Role Qualifiers and File Segments. The Role Qualifiers indicate the purpose of the chemical in the given invention. These qualifiers will not be available in the graphics database, but will be available for searching in the bibliographic records of the World Patent Index Latest (WPIL) file. Examples of some Role Qualifiers used by Derwent are as follows.

(N) New Compound
(P) Known Compound Produced
(S) Starting Material
(C) Catalyst
(M) Component of a Mixture

File Segments are a type of screen used to limit the answer set retrieved to a particular class of compounds. Using them, it will be possible to select only records for documents in certain Central Patents Index (CPI) sections, or to limit the search to coordination complexes or other types of structures. These segments will be searchable in the commercial release of the system, and Boolean logic capabilities will be subsequently available.

File Segments are also displayed in the graphics records. An explanation of all of them is given below.

CPI Sections

A Derwent Section A
B Derwent Section B or C
E Derwent Section E

General

Y Mixture
Z Salt
1 Specific Compound List
8 Small Molecules

The use of the CPI section segments is self-evident. Section Y is used when a mixture is being indexed. Components of mixtures are put on separate records – when this is done, each of the records is given the File Segment ‘Y’, plus whatever other file segment(s) is appropriate. Segment 'Z' indicates that the substance on the record is a salt, *whether or not the counter-ion is specifically listed*. This allows us to indicate that a patent covers a compound plus all appropriate salts, for example, in cases where a patent discloses a compound and 'all pharmaceutically acceptable salts'.

Segment 1 denotes that a specific structure has been claimed in a patent in isolation, or as one of a very small group of specific compounds. Derwent is using Segment 1 as an efficiency factor in indexing, and in presentation to the user. For example, if there are 1,000 patents on production of aniline, the user would only wish to see one graphic aniline record, along with any representations of aniline as a possibility of a Markush.

Segment 8 is an important part of the improved treatment of structures containing small molecules, and will be applied to records for small molecule single compounds and to records for Markush formulae which imply or encompass small molecules (i.e., equal to or less than 12 non-hydrogen atoms).

Special Segments

C Co-ordination Compounds/Complexes
L Oligomers
P Polypeptides

C indicates the presence of dative or D- or F-pi bonding, which requires special representation conventions.

Non-Polymer Compounds

M Metals or Alloys
W Extended Structures (Most commonly, inorganic solids)
V Ordinary Chemicals

File Segment V simply denotes that the compound has no special features. These three segments are mutually exclusive, and are also excluded if file segment C has been applied.

Polymers

F Any Polymer (excluding polypeptides and oligomers)

This File Segment allows the searcher to select all polymers, or to exclude all polymers, by the use of a single File Segment. In anticipation of the indexing of polymers in Section A, Plasdoc, Derwent has allowed for the following additional File Segments:

Backbone Type

H Homopolymer or Homocondensate
S Simple Binary Condensate
J Alternating Copolymer or Cocondensate
K Block Copolymer or Cocondensate
R Random Copolymer or Cocondensate
Q Star-block Copolymer
T Tri-block Copolymer
N Natural/other polymer (i.e., backbone not elsewhere classified)

Modification

X Cross-linked Polymer
D Derivatised Polymer
G Grafted Polymer
U Unmodified Polymer

Expression of Generic Groups

Markush DARC includes the capability to express generic groups as 'Superatoms'. In the short term, DARC will allow users to search generic groups as specific Superatom terms; eventually, we hope to allow users to search these terms as specific terms or in translation from specific groups to generic terms or from generic terms to specific groups, as the search requires. At the moment, Superatoms are handled in the same way as atoms of the natural elements in the periodic table. All Superatoms are represented by two or three character codes beginning with a letter.

Derwent expended a great deal of time and effort in the definition of these Superatoms. As a part of this study, Derwent's Product Development team reviewed the work done by the Sheffield University team in the identification of generic terms found in patents, working closely with Derwent's Patents Coding Department. The aim of Derwent in defining Superatoms was the identification of small number of unambiguous terms that could be used to represent the concepts expressed universally in patents. At the conclusion of our study, the following Superatoms were defined:

Acyclic Hydrocarbyl

CHK Alkyl, Alkylene
CHE Alkenyl, Alkenylene
CHY Alkynyl, Alkynylene

Cyclic Hydrocarbyl

CYC Monocyclic or fused carbocycle containing no benzene rings, optionally substituted by acyclic hydrocarbyl, i.e., any alkyl, alkenyl or alkynyl groups that might be present are included within this Superatom.
ARY Monocyclic or fused carbocyclic system containing at least one benzene ring.

Heterocyclic

HEA Single ring heteroaryl, i.e., a ring system having five ring members and either (1) 2 formal double bonds, i.e., the minimum possible degree of hydrogenation, or equivalent structures in which each of the notional double bonds has been replaced by one or two normalised bonds, OR a ring system having six ring members and six normalised bonds.
HET Any single ring heterocyclic, irrespective of ring size, not covered by HEA above, i.e., five or six members partially or fully saturated, or any single ring system of 3, 4 or more than 6 members with any degree of unsaturation.
HEF Fused heterocycle, including all ring sizes and degrees of unsaturation.

Metals

AMX Alkali and alkaline earth metals.
A35 Group IIIa, IVa, Va.
TRM Transition Metals, excluding Lanthanum.
ACT Actinides.
MX Any Metal (not used by Derwent).

Other Concepts

HAL Halogen
ACY Acyl (i.e., residue left after removal of -OH from an acid)
DYE Dye residue, chromophore, fluorescent group, etc.
POL Polymeric or polypeptide residue.
PRT Protecting group (e.g., protected amino = PRT-NH-).
XX Any atom or group excluding hydrogen.
PEG Polymer End Group.
UNK Unknown Group.

Derwent was able to reduce the number of concepts required as Superatoms by the use of further qualifiers, or Attributes, for Superatoms and atoms generally. Markush DARC afforded Derwent the capability not only to index 'normal' types of attributes, such as charge, abnormal mass, abnormal valency, and presence of deuterium or tritium, but also to indicate special attributes on the Superatom terms. These attributes are consistently applied to all occurrences of Ring or Chain Superatoms.

For the chain superatoms, CHK, CHE, and CHY, there are three attributes to define length:

LO 1–6 C
MID 7–10 C
HI >10 C

and two additional attributes to cover branching:

STR Straight
BRA Branched

For the carbocyclic Superatoms CYC and ARY, there are two to indicate whether the system is monocyclic or polycyclic:

MON Monocyclic
FU Fused

CYC and HET have two additional attributes to indicate presence or absence or unsaturation:

SAT Fully saturated
UNS At least one double or triple bond in ring, or in the case of CYC, in a carbon chain attached to the ring.

Derwent also adds textual notes to the input structures to further define Superatom terms and other concepts in a standardised way. This capability completes our requirement for full coverage of all of the information about a Markush structure, and in defining this capability we have drawn on the GENSAL definitions of generic term parameters. Storage of this information will allow the search system to grow in capabilities over time, by providing the information necessary to fully interpret generic terms.

STANDARDISATION OF CONVENTIONAL REPRESENTATIONS

Many compounds can be drawn in more than one way. There are two causes of this. Firstly, chemists have several ways of expressing chemical structures, each tailored to a certain class of compound. For some cases, more than one of these approaches may seem appropriate. Secondly, even using the most appropriate way of depicting a structure, there may be more than one equally valid representation. The most important example of this problem is tautomerism.

A further problem with a graphical indexing system such as DARC is that certain methods of representing structures are not possible. An example is the depiction of pi bonding in metal complexes, where a bond is often shown joining a metal atom to the middle of a double bond in an organic ligand, or to the middle of a ring.

Conventions have therefore been established to ensure uniformity in the representations of chemical structures forming the graphics database, and to define methods for handling structures such as those with pi bonds which cannot be input directly. In arriving at these conventions Derwent's aims have been:

1. To ensure that as far as possible structures input will appear as chemists would expect them to appear.
2. To make the rules as few in number and as simple as possible.
3. To alert users and indexers to the few remaining areas which cannot be resolved by the dogmatic application of conventions.

In devising conventions, we have always had to bear in mind the generic nature of the input, which results in an enormous volume of information. Because many of the input structures are generic, and in complex generics the permutations are so numerous, we cannot rely, as CAS does, on computer perception of structural features such as tautomerism, even if the algorithms for this could work reliably.

The CAS input programs include routines for identifying structures which can be represented in more than one way, and subsequently for standardising these structures. A major difficulty is prototropic tautomerism, in which two or more representations differ only in the position of a hydrogen and the disposition of double or triple and single bonds. An example is the free monothio carboxy group, which could be drawn as –(C=S)OH or as –(C=O)SH, and in fact probably exists as an equilibrium mixture of the two, the balance being determined by factors such as solvent or the nature of the rest of the molecule. The CAS programs deal with this by replacing the double and single bonds in either of these representations by 'tautomeric' bonds, of the order 1.5, so that whichever appears in the source document, only the unambiguous tautomeric form is recorded.

Computer standardisation of structures requires that the structure be completely defined, without variable sites; a change in a variable may completely alter the way that the machine interprets the structure as a whole. Working through all of the specifics implied by a Markush, and subjecting each in turn to structural feature perception routines may seem to be a perfect task for a computer, but is not feasible, as the number of separate specifics generated by some generic formulae is too large. Therefore, these operations must be carried out intellectually by the technical staff responsible for input, and the conventions must be fairly simple for this to be done reliably. A set of rules for preferred representations of structures is now in operation by the Derwent staff. This will also help the user; the understanding of only a small number of rules will allow the user to effectively search the Derwent Markush database.

In drawing up these rules, Derwent has followed the basic principles outlined below.

1. Although tautomeric bonds will be used freely to avoid alternative equivalent representations in cyclic systems, we have tried to avoid using them in structures where it is not necessary to avoid multiple representations. For example, we do not regard a representation of a simple amide group having tautomeric C-O and C-N bonds as helpful to either indexers or searchers, so we have fixed priorities for prototropic structures in most cases.
2. The use of abnormal valencies is avoided if possible. This principle especially must be tempered according to circumstances. For example, it does not seem unreasonable to show a metal atom in an octahedral co-ordination complex with a valency of six, particularly if by doing so, we satisfy the valencies on some or all of the ligand atoms. However, to represent ferrocene with a ten-valent iron atom and ten five-valent carbon atoms (as in CAS) seems confusing and inaccurate. In cases such as this, we will use multifragment representations.
3. The number and size of charges shown in indexed structures should be as small possible. If there are two or more representations which differ in the number or magnitude of charges on atoms, preference is given to the one with the fewest, smallest charges.
4. Charge is localised if possible, on the atom which would 'conventionally' be regarded as carrying it.
5. Bonds which cannot be represented as links between specific atoms, such as pi bonds and electrovalent bonding between ions, will still be handled by representations.
6. Because chemical structure representation in diagrammatic form has inconsistencies and contradictions built into it, it is impossible to anticipate every case. Most of the problem areas can be dealt with by a few general rules, but to cover all eventualities would require a huge system of regulations, with many provisos. Therefore, the aim of our rules is to cover the commonest problems and to ensure that only one representation will need to be assigned in most cases. Even when the rules fail to resolve a choice of structure completely, at least the number of possible structures should be reduced. In cases not covered by our rules, the structures given in the document will be indexed.

THE SYSTEM TODAY

Markush DARC, with a test database of 1321 compounds drawn from INPI and Derwent indexing, went into beta-test on 26 May 1987. We estimate that these 1321 compounds represent, conservatively, four million individual chemical compounds. Considering that the test database represents only five weeks of input data, the possibilities for the future are astounding, if not a bit unnerving.

The future system that will be offered to the chemical searching community represents the high level of investment and effort expended by both Derwent and INPI, in the creation of standards by which chemicals in patents are indexed. However, no database is effective if there is not a system by which it can be searched, and here is where full credit is due to the Télésystèmes DARC team. Markush DARC is available today, a testament to the excellence and ingenuity of Jean-Claude Roussel, Technical Manager of the DARC team, his current staff, and the people that created DARC as a flexible piece of software in the past.

The envisaged commercial release of Markush DARC will offer many of the same capabilities that users expect from Generic DARC; we have intentionally made the search system as similar as possible to the existing Generic DARC in order to minimise problems in training, documentation, and user acceptance. Though there are differences in the capabilities of the two systems, such as the possibility of the use of Superatoms and Attributes, the striking difference is that Markush DARC operates against a *generic* database, not a database of specific compounds.

Markush DARC has a very simple database structure. Variable groups can be nested into other variable groups. In conducting a search, of course, it is not necessary for the software to generate all of the individual compounds represented by the consolidated connection table, as one would expect of any good generic system. Searchers can expect specific retrieval, based on their query structure, wherever this structure appears in the input – in a main group (G0), in a subsidiary group (Gi), or across main and subsidiary groups.

THE SYSTEM – TOMORROW

In 1972, commercial online bibliographic systems were first offered to the information community. Compared to the systems that are available today, such services were primitive. For example, databases were offered for only half the day. Over the past fifteen years, we have seen an explosion in online technology. It is difficult to remember what systems were like in 1972, but it is essential that we not forget that there must be a beginning to everything, from which we can progress.

Markush DARC represents a system from which we can progress. Derwent and INPI have built in essential information into the database that allows for future expansion of capabilities. These capabilities include generic search, translation of generic expressions to specifics and specifics to generic expressions, and sophisticated limitations in search results. Derwent believes that the Markush DARC system should be viewed as not only a product of today, but also as an embryonic representation of what the future holds.

Although it is now a quarter of a century since the start of Farmdoc, Derwent's Pharmaceutical Patents Indexing Service, this is just the beginning.

POSTER SESSION: A SEARCH SYSTEM OF GENERIC NAMES IN THE LIST OF EXISTING CHEMICAL SUBSTANCES OF JAPAN WITH GENERIC STRUCTURES

Yoshihiro Kudo
Yamagata University, Yonezawa, Yamagata 992, Japan

ABSTRACT

A prototype system is designed to treat specific and generic names in the List of Existing Chemical Substances gazetted by the Japanese Government on the basis of a policy for searching such mixed types of representations for a particular substance or a group of substances in terms of structural formulae. Some functions of the system have been developed separately. The system is designed under a policy that a name in the List should be interpreted as widely as possible since it is legal rather than scientific. To cover broad names, two tactics are combined: to describe a name with a set of generic structural formulae described with such generic terms as alkyl, and to draw a structural formula of a broad name by merging more than one specific structural formula on the basis the principle of the coloured complete graph. For easy operation, the present system provides a function of automated derivation of a set of structural formulae as a query from a structural formula drawn by a user. The unique function releases user from reluctant study of a complicated set of *ad hoc* rules and jargons that hold only within a local system. Satisfactory results have been obtained.

INTRODUCTION

Tables 1 to 3 show examples of about 20,000 Existing Chemical Substances (ECS) registered in the List which was gazetted as the attachment of 'The Law of Examination and Regulation of Manufacture, etc. of Chemical Substances' legislated in 1973 by the Japanese Government, and the supplement of the List has been updated several times by addition of New Chemical Substances whose total number was 1742 up to April 14, 1988. The List also has been published in the form of handbooks both in Japanese and in English, in which names are arranged according to structural features for easier searching. None the less, it is still difficult or very tedious even for chemical information experts in Industry to search all names for a particular substance or a group of substances, because the List contains many generic names. Generic names in ECS haveto be interpreted also from a legal viewpoint. For example, octane may be called n-octane, alkane(C8) many other names.

Arbitrariness and ambiguity of generic forms have prevented the application of many techniques established for specific substance systems. It is not practical to try to describe all types of generic names in terms of specific names, because the number of them is frequently unreasonably excessive as seen in Markush formulae, e.g., substituted benzene, used in patent documents.

W. A. Warr (Ed.)
Chemical Structures

Table 1. Examples of ECS

HN	*GN*	*Name*
2-148	2-185	N-Alkylor alkenyl(C16-28)-N,N-dialkyl(C1-5 or H) amine
2-214	9-1971	Aliphatic alkyl (at least one of the alkyl groups is C 8-24, others are C1-5) quaternary ammonium salt
3-4	3-22	Branched alkylbenzene(C3-36)
3-2660	3-2254	Cyclic(7-12 membered ring) brominated or chlorinated hydrocarbons (C1 or Br:4-12)
3-2749	3-2339	Menthyl alkanoate
4-196	4-166	Poly(di-penta)benzylphenylphenol
5-363	9-1986	2-Alkyl (or alkenyl)(C7-23)-4,4-dialkyl(C 1-5 or hydroxymethyl) oxazoline
6-286	6-82	Vinyl chloride Vinyl alkanoate copolymer

GN : The ID number assigned in the gazetted List.
HN : Another number given in the Handbooks.
In both Numbers, the first digits suggest respective structural classes.

The present paper focuses only on searching ECS for a name with structural formulae.

TWO TACTICS FOR REPRESENTING BROAD NAMES AND FRAGMENT CODE

Essentially, in the system, a structural formula is described by a combination of atoms or atomic groups with the ranges of those atoms or groups, for example, alkyl (C3-5) and carbocyclic (3-8 membered).

Aiming at exhaustive retrieval of relevant and/or similar names to be interpreted as widely as possible from a legal viewpoint, the system adopts two tactics: firstly, to describe a specific or generic name with a set of many generic structural formulae which are chosen to be complementary to each other (Figure 1), and secondly, to draw a structural formula by merging more than one usual structural formula (Figure 2).

Figure 1 illustrates the idea of a set of structural formulas with examples of vinyl chloride and vinyl bromide. In this pair, the first (as vinyl halide), third (as alkenyl halide) and fourth (as halogenated carbon group) representations are the same. Figure 2 shows two examples of merged generic formulae: (a) from those of vinyl chloride and vinyl bromide, and (b) from those of 1,1-dibromo-1,2,2,2-tetrachloroethane and 1,2-dibromo-1,1,2,2-tetrachloroethane. Figure 3 shows the difference between 1-bromo-1-chloroethylene and a merged one from vinyl bromide and vinyl chloride. This method can be applied to branched alkane(C5) but not to branched alkane(C30), because of too large a number of member hydrocarbons in the latter case. In such cases, however, the fragment code method is very useful if an index for C-branched is set. The absence is indicated with zero value of the index. Table 4 shows a part of fragment codes.

Table 2. Relationships Among Substances, GN, and HN.

There are various aspects of the relationships.

HN	GN	Name
	(a) One substance : three numbers	
2-70	9-370	1-Chloro-3-bromopropane
2-71	9-1247	1-Bromo-3-chloropropane
2-72	9-2007	1-Bromo-3-chloropropane
	(b) One substance : two groups	
3-9	3-3427	Trialkyl(C1-4)benzene
3-10	3-7	Tri- or tetramethylbenzene
	(c) One H : five GNs	
3-2	3-60	Mono-(or di-)methyl(ethyl,bromoallyl,bromopropyloxycarbonyl or chloropropyloxycarbonyl)benzene
	3-2	Toluene
	3-28	Ethylbenzene
	3-3	Xylene
	3-13	Diethylbenzene
	(d) Nine HNs : One GN	
2-2603	2-2142	Acetylacetone complex salt (Fe,Cr,Al,Co,Mn,Ti,Ce,Zr,Cu)
2-2608		
2-2615		
2-1619		
2-2624		
2-2630		
2-2639		
2-2683		
2-2707		

STRUCTURAL FORMULA IN THE DATABASE AND FOR QUERY

A set of structural formulae is stored in database files as shown in Figures 1 and 3. 0f these files, STF1, STF2, STF3, STF4, and STF5 are for structural formulae and STF9 is for the fragment codes.

Because the system should not impose on users any rules and operations which are arbitrary and hold only within a local system, the method explained above cannot be applied to input a query. The system provides a different way to construct a query because it takes skill to handle queries.

The system derives a query set for a specific or generic name in question from a specific structural formula drawn by a user almost automatically. Each range, defined by figures of minumum and maximum of the number of carbons of alkyl group, the size of ring etc., which is set according to the formula by the user can be easily changed.

n in STFn	Vinyl bromide	Vinyl chloride
1	C=C—Br (1)	C=C—Cl (1)
2	C=C—Hal (1)	C=C—Hal (1)
3	DB—AK—Br (1) (2) (1)	DB—AK—Cl (1) (2) (1)
4	DB—AK—Hal (1) (2) (1)	DB—AK—Hal (1) (2) (1)
5	AK—Hal	AK—Hal

Hal : Halogen

AK (n) : Group of n carbon atoms

DB : Double bond

Figure 1. A set of structural formulae corresponding to a name.

Table 3. Image of fragment code method

HN	*GN*	*Name*	*Na*	*Fe*	*COO*	*Halogen*	*Alkali*
2-22	2-29	1,1-Dialkyl(C4-20)vinylidene	0	0	0	0	0
2-948	2-859	Adipic acid salt(Na,K,Ca)	1	0	2	0	1
2-1337	2-2657	Pentachlorostearic acid salt (Na,K)	1	0	2	2	2
2-702	2-625	Aliphatic monocarboxylic acid (C7-23)salt (Al,Fe)	0	1	2	0	0
3-2660	3-2254	Cyclic(7-12 membered ring) brominated or chlorinated hydrocarbons(Cl or Br:3-12)					

HN	*GN*	*Name*	*C2*	*C4*	*C6*	*C8*	*C10*	*Me*	*C-Branch*
2-8	2-8	Octane	0	0	0	2	0	2	1
2-173	2-155	N,N,N′,N′-Tetramethylalkylene (C2-4)diamine	1	1	0	0	0	2	0
2-238	2-224	t-Alkyl(C4-8) hydroxyperoxide	0	1	1	1	0	2	2
2-883	2-827	Fatty acid(C8-24) mono- or dialkanol(C2-3) amide	1	0	0	1	1	2	1
3-1458	3-73	Telephthalic acid salt(Na,K)	0	0	0	0	0	0	0

2 : present 1 : possible 0 : absent

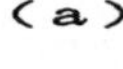

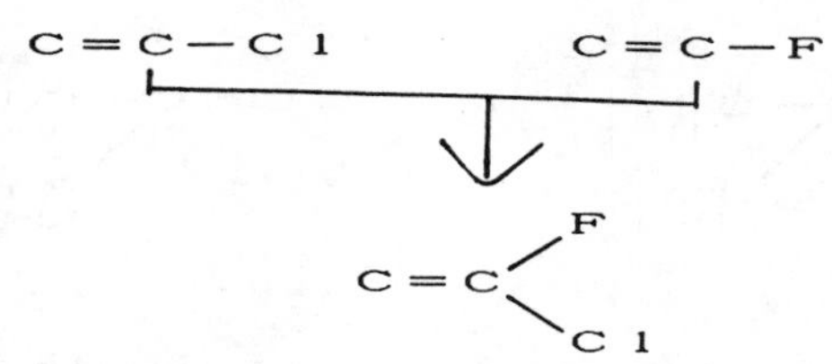

(b)

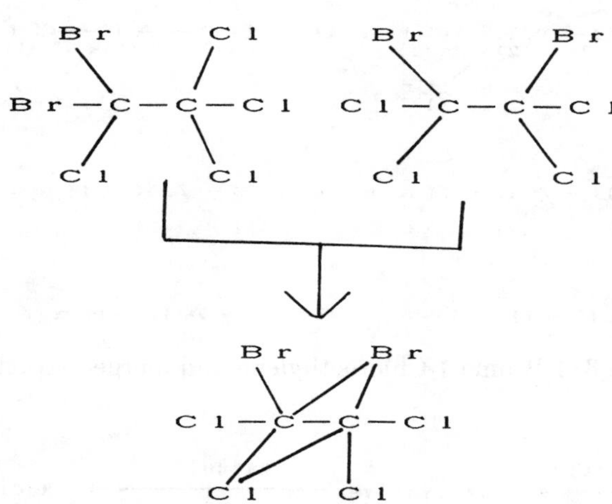

Figure 2. Merging of structural formulae.

Table 4. Examples of fragment codes (See also Table 3)

1	CH3−	11	−CHO	21	O	31	Ca	41	Co
2	−CH2−	12	>C=O	22	S	32	Al	42	Mn
3	CH2=	13	=C=O	23	N	33	Light m.	43	AK3
4	>CH−	14	R CH2	24	O/S/N	34	Heavy m.	44	AK4
5	=CH−	15	R CH	25	the sextet metals	35	Alkali	45	AK5
6		16		28		36		46	

Figure 4 shows that, for example, alkyl(C2-5) can be derived by drawing a structure of any one of alkyl groups and by modification of the range: if the user draws methyl, amyl and decyl, alkyl(C1-1), alkyl(C5-5) and alkyl(10- 10) are derived, respectively, whose ranges can be easily changed. Even if all of the three original alkyl groups are allowed, alkyl(C5-5) would be the best because of maximum (5 in this example) is the same between the original (C5-5) and the target(C2-5).

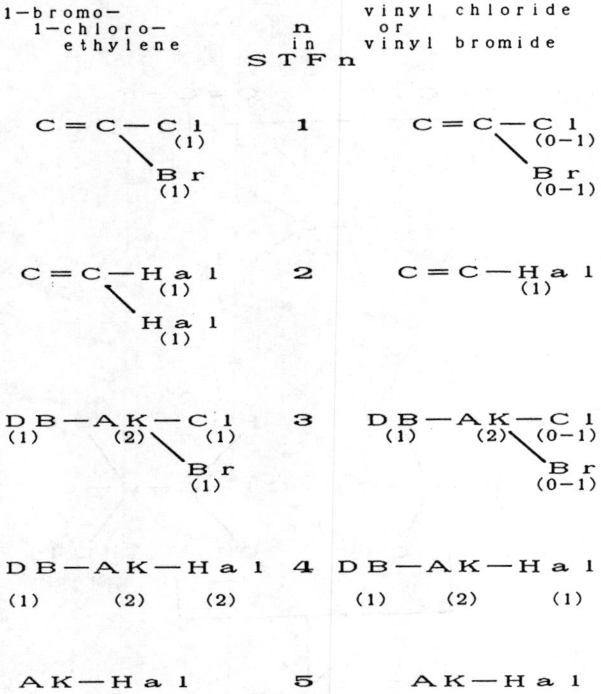

1-bromo-1-chloro-ethylene	n in STFn	vinyl chloride or vinyl bromide
C=C—Cl(1), Br(1)	1	C=C—Cl(0-1), Br(0-1)
C=C—Hal(1), Hal(1)	2	C=C—Hal(1)
DB(1)—AK(2)—Cl(1), Br(1)	3	DB(1)—AK(2)—Cl(0-1), Br(0-1)
DB(1)—AK(2)—Hal(2)	4	DB(1)—AK(2)—Hal(1)
AK—Hal	5	AK—Hal

Figure 3. 1-Bromo-1-Chloroethylene and merged structure.

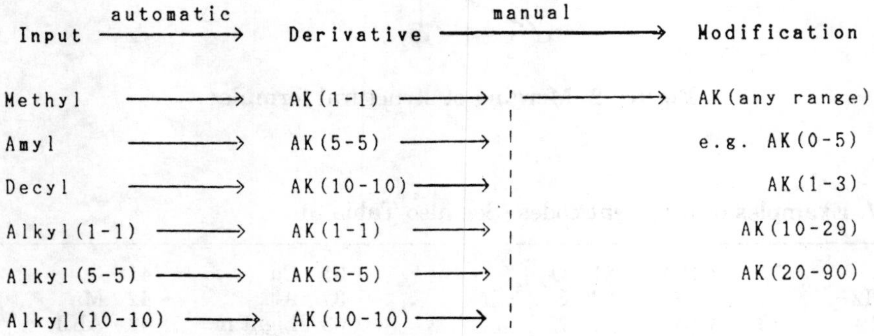

Input	automatic →	Derivative	manual →	Modification
Methyl	→	AK(1-1)	→	AK(any range)
Amyl	→	AK(5-5)	→	e.g. AK(0-5)
Decyl	→	AK(10-10)	→	AK(1-3)
Alkyl(1-1)	→	AK(1-1)	→	AK(10-29)
Alkyl(5-5)	→	AK(5-5)	→	AK(20-90)
Alkyl(10-10)	→	AK(10-10)	→	

Figure 4. Automated derivation and manual modification.

DISCUSSION

The policy of the system would be very reliable for searching of generic representations by means of a set of generic structural formulas which can be built-up without any *ad hoc* technique.

Users treating patent documents would demand a large set consisting of many generic structural formulae to completely cover complicated Markush formulae. The larger the number of generic structural formulae is, the more powerful the system would be and the heavier a load to manage the system would be. So a set would fall into proper size according to a situation.

One of the other features is the ability to accomplish the complete relevancy. So it is in reproducibilities and efficiencies that the difference between experts and beginners would appear.

Although a method in which a generic name is covered with more than one structural formula may be thought to be too large and unwieldy, it would be extremely simple in comparison with an application of artificial intelligence, and, at least at the level of the prototype system, size has not yet proved to be a problem.

Searching of ECS is a kind of legal operation. So the system is almost playing a role as a consultant for users both in the Government and in Industry.

REFERENCES

1. Chemical Products Safety Division, Basic Industries Bureau, Ministry of International Trade and Industry, *Handbook of Existing and New Chemical Substances*, The Chemical Daily Co: Tokyo, Japan, 1986.
2. Kudo, Y., Chihara, H.'Chemical Substance Retrieval System for Searching Generic Representations. I. A Prototype System for the Gazetted List of Existing Chemical Substances of Japan', *ibid*, 1983, *23*, 109–117.
3. Kudo, Y. 'Polymer Retrieval Systems for Generic Polymer Names Consisting of Both Generic and Specific Monomer Names. A Prototype System for the Gazetted List of Existing Chemical Substances of Japan'. *Bull. Yamagata Univ. (Engl.)*, 1987, *19*, 147–156.

BUILDING A STRUCTURE-ORIENTED NUMERICAL FACTUAL DATABASE

Clemens Jochum
Beilstein Institute, Varrentrappstr. 40–42, 6000 Frankfurt 90, West Germany

ABSTRACT

The projected, numerical factual database BEILSTEIN-ONLINE represents a natural extension of the Beilstein Handbook of Organic Chemistry which has been published for more than 100 years in more than 300 volumes. The Handbook together with the structure file cards of the literature period 1960–1980 (E-V Period) contain factual and structural information on more than 4 million compounds. The computer representation of organic chemical structures therefore constitutes the centre part of the database design. The chemical structure is described in a post-topological format. Not only constitutional information is included but also a complete stereochemical representation. Bond orders are replaced by electronic information. This allows a much more global and unique representation of tautomers. This paper gives an overview not only of the data structure of chemical compounds but also of its associated physical numerical data and keywords.

CURRENT PRINTED INFORMATION SYSTEM

The projected, numerical factual database BEILSTEIN-ONLINE represents a natural extension of the Beilstein Handbook of Organic Chemistry which has been published for more than 100 years in more than 340 volumes.

The present Beilstein Information Pool contains Handbook and Registry data of the literature time-frame from 1830 to 1960. Since the handbook has primarily been mostly published using conventional non-computerised typesetting methods, this information pool is for the most part available in only printed form. However, the most recent 25% of the published volumes has been printed using electronic typesetting methods and is therefore available in a computer-readable form.

In addition the primary literature from 1960 to 1980 has been completely abstracted. Over 7 million abstract cards contain the structural formula, numerical physical data, reaction pathways and original literature citations.

This information pool will gradually be transferred into a computer-readable form and extended by additional new sources of information (see below). All future Beilstein information products will be generated from this information pool which will be organised in an internal database.

INFORMATION SOURCES OF THE DATABASE

The Beilstein database will be generated from four sources of information (Figure 1).

W. A. Warr (Ed.)
Chemical Structures

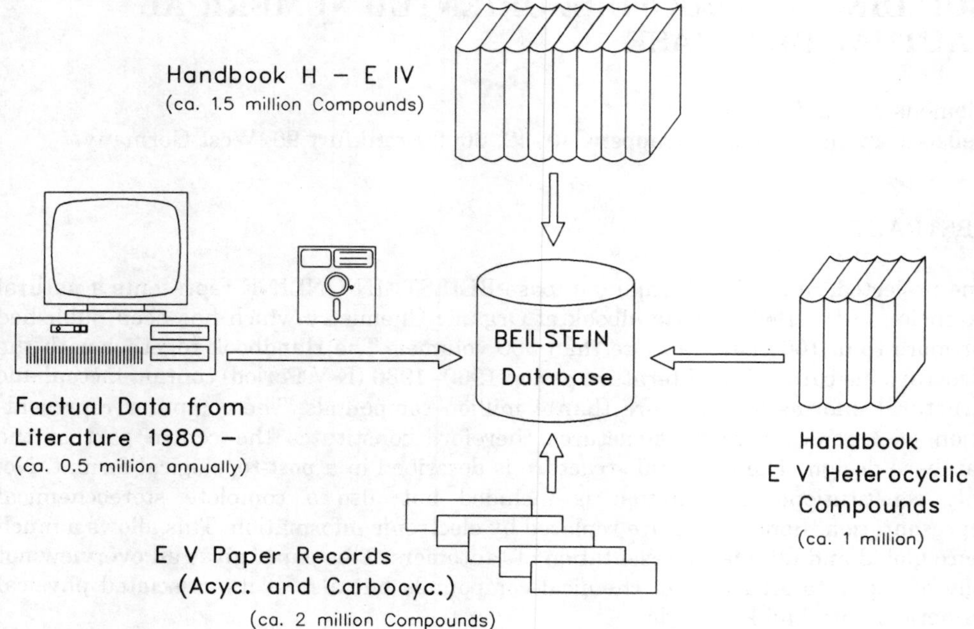

Figure 1

1. The printed Handbook series from the Basic Series up to the fourth Supplementary Series (H to E IV). These series are almost completed and contain the literature time-frame from 1830 to 1960. These volumes contain the factual data of more than 1 million organic compounds, which singles out this database as unique world-wide and thus presents a significant marketing advantage against other numerical and bibliographical databases.
2. The printed Handbook material of the fifth Supplementary Series which contains the literature from 1960 to 1980.As in the previous handbook series, these data have been thoroughly checked for errors and redundancies.
3. The publishing of the Fifth Supplementary Series will probably continue for more than one decade. On the other hand as described above, the primary literature of this time period (1960–1980) has been completely abstracted in more than 7 million excerpts. Some of the excerpts will be added to the database in their original non-checked form in order to give scientists limited access (see below) to these data as soon as possible.
4. Beginning in 1985 the factual data of the primary literature from 1980 will be abstracted electronically. The abstracted data of the primary literature will no longer be written on paper but entered directly in a structured manner into microcomputers and stored on magnetic diskettes. After several automatic plausibility and redundancy checks, these data will be copied onto the mainframe computer and loaded into the database. The researching scientist has immediate access to the factual data of the current literature.

The general public online access to the Fifth Supplementary Series abstracts will be limited to compound classes which are not published in the Handbook in the near future, i.e., the acyclic and isocyclic compounds. Subscribers to the Fifth Supplementary Series Handbook will most likely be given additional access to the heterocyclic abstracts.

THE DATABASE CONCEPT

In respect of the different sources of available information, the database can be divided into a *Short File* and a *Full File*. The *Full File* will consist of all the data which have been published in the Handbook series and which have been checked for errors and redundancies by Beilstein information specialists. The *Short File* is built directly from the abstracts of the primary literature without any further checking. Factual data of the current, primary literature will be added to the *Short File* on a regular basis. The *Full File* will be continuously extended by processing the raw material of the *Short File* *via* an error checking and removal of redundancies.

The structure of the database can be divided into two parts: the numeric factual file and the structure file (Figure 2). These two parts are subsequently described in more detail.

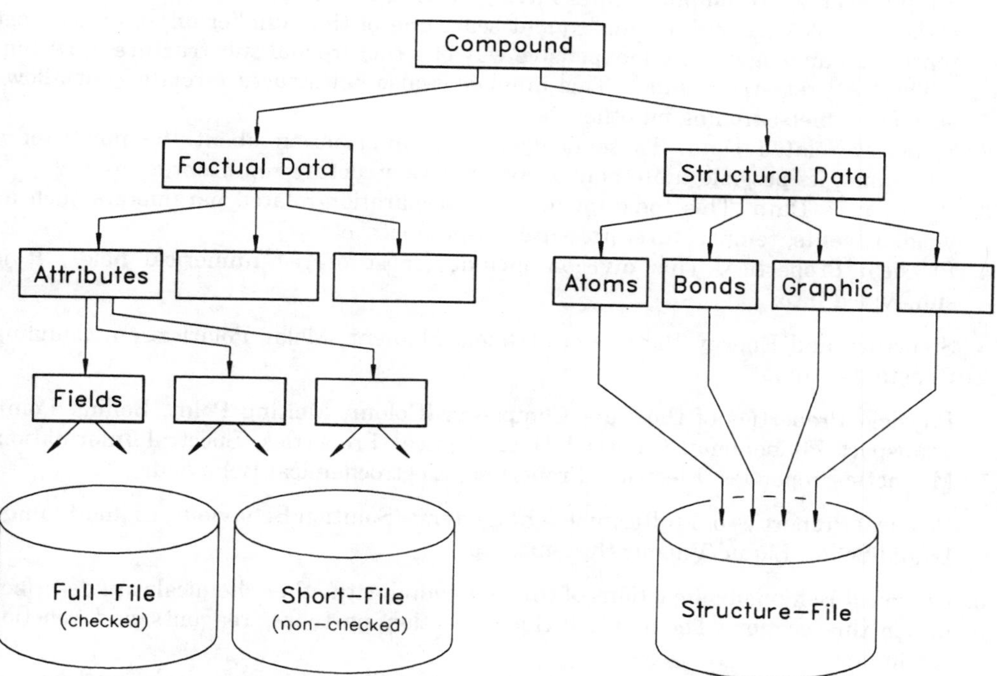

Figure 2

The Factual File

The Factual File has a relational structure and contains three types of fields:

1. Numerical Fields. Most numerical parameters can be stored as 2-byte integers, but some physical parameters require a 4-byte floating point format.
2. Boolean Fields. These fields store the presence (or absence) of a keyword or a parameter.
3. Alphanumeric Fields. Literature citations, comments for a preparation, etc. are stored as character strings.

All Boolean fields and most of the numerical fields are searchable separately or in combination.

Chemically, the data structure can be divided into 7 parts:

1. Identifiers. These parameters contain the molecular formulae and registry numbers for identification and search of the compounds. Three registry numbersare stored for each compound:

   A structure-independent compound identifier for the internal organisation of the database.

   A structure-dependent non-unique hash code, the Lawson Number. Since this number is structure-related, several structurally closely related compounds can have the same Lawson number. The structure-ordering according to this number is very similar to the Beilstein Handbook Ordering System. The number can be computed on microcomputer (after having entered the structure graphically) using a rather complex algorithm. Subsequent searching of this number on an online host represents an elegant and inexpensive way of structure and substructure browsing.

   The CAS registry number. This number is also not structure-related but allows easy cross-file searching on other databanks.
2. Structure-related Data. These fields include information about the purity of a compound, its possible tautomers or alternative structure representations.
3. Preparative Data. This topic includes all preparation-related parameters such as yield, solvents, temperature, pressure, by-products, etc.
4. Physical Properties. This division includes most of the numerical fields. It is subdivided into:

   Structure and Energy Parameters (Dipole Moment, Molar Polarisation, Coupling Constants, etc.).

   Physical Properties of the Pure Compound (Colour, Melting Point, Boiling Point, Transport Phenomena, Calorical Data, Optical Properties, Spectral Information, Magnetic Properties, Electrical Properties, Electrochemical Behaviour).

   Physical Properties of Multicomponent Systems (Solution Behaviour, Liquid/Liquid-, Liquid/Solid-, Liquid/Vapour-Systems, etc.).

5. Chemical Behaviour. Reactions of this compound with other chemicals are described under this section. The fields include reaction partners, reagents and reaction conditions.
6. Physiological Behaviour and Applications. Use, Toxicity, Biological Function and Ecological Data Parameters are described under this topic.
7. Characterisation of Derivatives and Salts.

There are 510 fields in total. Of these, 305 fields are searchable separately or in combination (208 Boolean fields and 97 numerical fields).

The Structure File

The structures of the Beilstein compounds are stored in connection tables (CT's) to allow a very flexible structure and substructure search. Since most commercially available structure/substructure handling programs such as MACCS (MDL) or DARC (Télésystèmes/Questel) work on the basis of CT's, the Beilstein Registry Connection Table (BRCT) can be easily adapted for in-house systems.

The following criteria governed the design of the BRCT format.

Completeness and Unambiguity. The Registry CT must contain all of the structural information which is necessary to describe completely the chemical structure of a compound for the purpose of its storage and retrieval.

By virtue of its completeness, the Registry CT must provide an unambiguous representation of each compound in the Registry file, so that a Registry CT describes one and only one compound in the Registry file.

Uniqueness. The Registry CT must provide a unique representation of the structure of each compound in the Registry file, so that each compound has one and only one Registry CT.

Compactness. The Registry CT must be compact, in order to minimise the storage requirements of the Registry File, and to minimise the quantity of data which may be required to be transferred between different applications programs.

Flexibility. The Registry CT must be flexible, although precisely defined, in order to enable upgrading or other modification of applications programs without requiring the Registry CT to be redefined.

Treatment of Tautomerism and Resonance. The Registry CT must provide for a satisfactory treatment of the phenomena of tautomerism and resonance, so that different 'valence bond' descriptions of the same molecule can be recognised as equivalent and can be represented in the Registry file by the same Registry CT.

In addition, the Registry CT must provide for different tautomeric forms of a molecule to be represented by different Registry CT's in cases where this is desirable.

The Beilstein Registry CT consists of a Header vector and a number of 'lists'. The Header vector is obligatory, and must be present in the CT. Certain of the lists are also obligatory, while others are optional and are present in the CT only when necessary.

The Header vector contains information which controls the length of the obligatory lists, and the number of optional lists which are present in the CT. The lists themselves describe the structure graph, modifications of the structure graph and any supplementary information which is necessary to describe completely the structure of the chemical compound.

THE INTERMEDIATE FILE

After input, the data are loaded from the floppy disks into an intermediate file on the mainframe. The data are run through several plausibility checks and when necessary are corrected. In this intermediate file all data from the various input sources are converted to the same file and data structure according to our data structure definition.

THE DATABASE

After having been converted to the same file structure the data wi11 be loaded into an ADABAS-managed database. More than 10 different commercial database management systems have been evaluated during our systems analysis phase in 1984/85. The two highest scoring systems have been bench-marked with an artificial Beilstein-structured database with 200,000 compounds. ADABAS scored best in practically all tests (loading, updating,retrieval etc.).

An ADABAS-based update and retrieval system is currently being developed in our Institute together with a German software house. This system will be used to make final corrections to the database, to append compounds and retrieve compounds for checking, and for writing Handbook articles. It will also be licensed to in-house customers (see below) of the Beilstein database. There will be a micro-and a mainframe-based version of this software.

DATABASE ACCESS

The data structure of the Beilstein database will enable many different accessing methods to the information dependent only on the retrieval software used. We are planning the following access methods with our inhouse retrieval software:

1. *Via* a graphical or alphanumerical input, depending on the type of computer terminal (structure or substructure search).

   By searching for numerical terms (physical data, keywords or Boolean terms, such as the existence of spectra, etc.). These terms can be searched separately or in Boolean combinations ('AND', 'OR', 'NOT').
2. By combination of the two search methods described above.
3. By searching other key fields, such as the molecular formula, the CAS registry number or the molecular, structure-related Beilstein Registry Number (Beilstein Prime Key).

The Beilstein factual database can be accessed through various Institutions and computers respectively:

1. *Via* public online hosts.Contracts have been signed with Dialog Information Systems and STN-Karlsruhe. Negotiations are currently under way with various other major online hosts.
2. Large chemical companies can acquire a licence for the Beilstein database to run it on their own computer centre. Employees of these companies can have in-house access to the Beilstein data and are able to add their own data to the database. Subscribers to the Beilstein database will receive update continuously. Several

contracts have been signed already. Delivery of the data will start in the third quarter of 1987.

3. Individual customers can access selected parts of the Beilstein database with their own microcomputers when attached to a Winchester or Optical Storage Device (CD-ROM).

POSTER SESSION: ECDIN, ENVIRONMENTAL CHEMICALS DATA AND INFORMATION NETWORK

Ole Norager
Chemistry Division, The Ispra Establishment of the Joint Research Centre of the Commission of the European Communities, I-21020 Ispra(VA), Italy.

ABSTRACT

The ECDIN databank contains data related to chemicals which actually or potentially occur in the environment as pollutants. The data content in ECDIN is illustrated in a series of examples. The organisation of the databank is discussed together with a presentation of the software used to access and to maintain the databank.

INTRODUCTION

The present paper is written based on a poster presented at the Chemical Structures Conference in Noordwijkerhout in June 1987. The scope of the poster was to give an overall presentation of the ECDIN databank, and not to discuss any particular aspects of the databank in detail. The poster was presented to a forum of experts in chemical information, and the poster and the paper include discussions of features of the databank which are not visible to the normal user.

THE ECDIN DATABANK

ECDIN is a factual databank, created at the Ispra Establishment of the Joint Research Centre (JRC) of the Commission of the European Communities as a part of the Commission's Environmental Research Programme. The databank enables people engaged in environmental management and research to obtain reliable information on chemical compounds. The databank brings together a wide variety of data on chemicals which actually or potentially occur in the environment as a result of human activities. The broad spectrum of parameters and properties contained in ECDIN provides the user with general information on chemical substances and supports the evaluation of the risks connected to the use of these substances. The ECDIN database establishes links between various types of data is therefore a powerful tool for correlation analyses.

The data for the ECDIN databank are collected by specialised research institutions working under contract to the Commission. The contractors scan the published and unpublished literature and extract data for ECDIN. Information about the source for the individual data items is stored in the databank together with the extracted data. The data collection is supervised by a group of the Commission's scientific staff working at the Ispra Establishment of JRC. The ECDIN group in Ispra also takes care of the loading of the data into the database and of the general maintenance of the databank.

W. A. Warr (Ed.)
Chemical Structures

The data stored in ECDIN are facts: typically numerical results of experiments. Facts can also be stored in the form of text, such as a summary of a legal text concerning the use of a chemical, an indication of experimental conditions, or a description of the effects of a chemical on test animals.

The data are organised in files. Table 1 contains the titles of database files as the user sees them.

Table 1. ECDIN, Data Categories and Data Files

A IDENTIFICATION A Chemical Synonyms B Chemical Structure Diagram C Identifiers and Definitions	F TOXICITY A Classical Toxicity B Aquatic Toxicity C Carcinogenicity D Effects on Soil Microorganisms E Mutagenicity
B PHYSICAL-CHEMICAL PROPERTIES A Physicochemical Properties	
C PRODUCTION AND USE A Chemical Producers B Chemical Processes C Production and Consumption Statistics D Export Statistics E Import Statistics F Uses	G CONCENTRATION AND FATE IN THE ENVIRONMENT A Concentration in Environmental Matrices (Cost 64b) B Concentration in Human Media C Concentration in Animal Media D Metabolism in Soil E Bioaccumulation in the Aquatic Environment
D LEGISLATION AND RULES A Legislation (IRPTC-legal) B Directive 67/548/EEC C Waste Treatment (IRPTC-Waste)	H DETECTION METHODS A Analytical Methods B Odour and Taste Threshold Limits
E OCCUPATIONAL HEALTH AND SAFETY A Human Health Effects B Occupational Exposure Limits C Occupational Poisoning Reports D Occupational Diseases Prevention E Symptoms and Therapeutic Treatment	I HAZARD INFORMATION (in collaboration with Brandweerinformatie Centrum voor Gevarlijke Stoffen)

The two files D(A) and D(C) are in the ECDIN databank as a result of a collaboration between the ECDIN project and the International Register of Potential Toxic Chemicals. This collaboration implements an agreement between the Commission of the EC and the UN Environment Programme. The category I is not fully implemented at the time this paper is written, but these data should be made available to users during the summer of 1987.

SOME EXAMPLES OF DATA

The data are stored in the ECDIN database in a structured way. Different data items are put into different fields, and data read directly from the databank are only useful if you know from which field in which file they have been read. The user sees the data after they have passed a formatting program, which inserts the retrieved data in an explanatory text. The following examples are shown as the user sees the data.

Chemical names and synonyms are the most important keys to retrieval of data from the ECDIN databank. The synonyms stored in the ECDIN files can be divided into three groups, synonyms from the files of Chemical Abstracts Service, names used by the European Customs Union, and names from other sources. The first group of names is mainly in the English language, whereas the two other groups contain names in more European languages. Figure 1 shows different names for butanoic acid.

Chemical structure information is an important part of the ECDIN database. From the start of the project the Commission has supported collaboration between the ECDIN

Figure 1. Chemical Synonyms

```
Substance 2339X  Butanoic acid

   Preferred & 9th. Coll. Index
       Butanoic acid

   8th. Coll. Index
       Butyric acid

   Italian
       acido butirrico

   German
       Buttersaeure

   French
       acide butyrique

   Dutch
       boterzuur

   Danish
       smørsyre

   CAS Synonyms (Copyright 1987 by the American Chemical Society)
       n-Butanoic acid
       n-Butyric acid
       Butyric acid
       Ethylacetic acid
       1-Propanecarboxylic acid
       Propylformic acid
       Butyrate
```

project and other European research activities on structure information. An early version of the DARC system was installed on the computer in Ispra as a result of the collaboration with Professor Dubois' group in Paris, and structure-activity relationships have been analysed in Ispra using ECDIN data and programs based on software developed by Professor Ugi's group in Munich.

Despite the availability of systems to graphically display structure diagrams it was decided first to implement ECDIN with software which could be used by teletype terminals without facilities for display of graphics. The structure representation in the publicly accessible part of the databank is based on the CROSSBOW system. The records in the structure file contain a structure diagram which can be reproduced using single characters. It further contains the Wiswesser Line Notation, a CROSSBOW connection table, and a fragment screen. Parts of the CROSSBOW system have been modified to allow the structure retrieval programs to work directly on the ECDIN database under the database management system ADABAS.

Figure 2 shows how the structure diagrams are displayed.

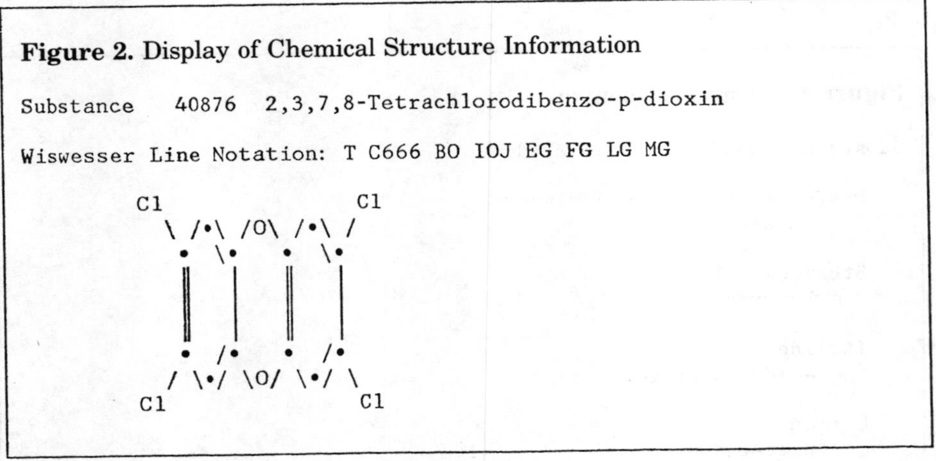

Figure 2. Display of Chemical Structure Information

The file A(C) Identifiers and Definitions (see Table 1) is a new file, which at present only a few users outside the JRC can access. The file will be made available to users in general during the summer of 1987. The file will contain data related to the identification of chemical substances for legislative purposes. The first data to be accessible in the file will come from the lists of chemicals maintained by the Commission, but, later, data originating from national files and from other international organisations will also be entered. The current file contains only data related to the EINECS Inventory.

The EINECS Inventory, the European Inventory of Existing Chemical Substances, is an inventory of the chemical substances which were on the European Community market between January 1971 and September 1981. Substances which are not in the EINECS inventory are considered new and must pass a notification procedure before they can be marketed in the EC member states. The inventory has been compiled by the Commission of the European Communities (Directorate General IX, Environment, Consumer Protection and Nuclear Safety) in collaboration with the member states. The Commission first created a core inventory, ECOIN, with 33,000 substances based on existing files. Industry was then invited to report substances which were on the market, but not in the core inventory.

All the substances in ECOIN and all the substances reported by industry have been registered by Chemical Abstracts Service. This registration was finished in the autumn of 1986. The official publication of the EINECS Inventory is awaiting a translation of the inventory into all the official languages of the EC, but a provisional version is being published in the English language only.

Staff from the ECDIN project participated in the creation of the core inventory, took care of the technical management of the reporting phase, and are still involved in support of translation and printing. The data flow between the member states, the Commission and CAS was controlled by an automatic system built in parallel to the proper ECDIN databank. The EINECS inventory has been implemented as a part of the ECDIN databank and ECDIN is the authoritative online version of EINECS.

Information on whether a compound is in the EINECS inventory is found in the file (A)C, Identifiers and Definitions. Figure 3 shows two examples of display of data from this file.

Other displays are shown in Figures 4–9.

Figure 3. Display of data from the file Identifiers and Definitions

```
Substance       16 Chloropyriphos

EINECS Number:  2208644             CAS Registry Number:  2921-88-2

Compound: Phosphorothioic acid, O,O-diethyl O-(3,5,6-trichloro-2-pyridinyl)
          ester
Molform : C9H11Cl3NO3PS



Substance 2152838 Zeolites

EINECS Number:  2152838             CAS Registry Number:  1318-02-1

Compound : Zeolites
Definition:
         Crystalline aluminosilicates, composed of silica (SiO2) and alumina
         (Al2O3), in various proportions plus metallic oxides. Produced by
         hydrothermal treatment of a solid aluminosilicate or of a gel
         obtained by the reaction of sodium hydroxide, alumina hydrate and
         sodium silicate. The initially obtained product, or a naturally
         occurring analog, may be partially ion-exchanged to introduce other
         cations. Specific zeolites are identified by notations indicating
         crystal structure and predominant cation, e.g., KA, CaX, NaY.
```

PUBLIC ACCESS TO THE ECDIN DATABANK

The ECDIN databank is currently accessible from the DC host centre in Copenhagen through the public data transmission networks. The DC host centre is operating the

Figure 4. Physical-Chemical Properties

```
Substance   194 2-Propanone

MOLFORM: CH3COCH3
MELTPT :  -94.6 deg C                     BOILPT :  56.0-56.5 deg C
       :  -95.35 deg C at 760 mm Hg              :  56.2 deg C at 760 mm Hg
CRITTMP:  234.95 deg C <01>               CRITPRS:  47 atm
DENSITY:  0.797 at 15 deg C
       :  0.7899 GM/CC at 20 deg C
DENSITY (gas rel to air):  2 <02>

 vapour pressure   temperature          comments          reference
____________________________________________________________________
400 mm Hg          40 deg C
____________________________________________________________________

                 Physical-Chemical Properties (Solubility)

Infinitely soluble in WATER
Infinitely soluble in ETHYL ALCOHOL
Infinitely soluble in ETHER

                 Bibliographic References


<01>     ENGINEERING SCIENCES DATA UNIT-INDEX LONDON

<02>     "SICHERHEITSTECHNISCHE KENNZAHLEN BRENNBARER GASE UND DAMPFE"
         BY K. NARBERT AND G. SCHON PTB.
```

databank on behalf of the Commission. Access takes place through a user interface called the ECDIN Display System, which has been developed at the JRC as a part of the ECDIN project. The major steps in the dialogue between the user and the Display System are outlined below.

Selection of Access Point

A Chemical Substances
B Directive 67/548/EEC

Selection of Search Type

(here for access point A):

A Substance name
B ECDIN number
C Chemical Abstracts number

D GUD number (Gestion de l'Union Douanière)
E RTECS number (Registry of Toxic Effects of Chemical Substances).

Input of Search Parameter

(here a Substance Name):
Phenol

Figure 5. Display of Production and Consumption Data

Substance 828 Chlorine

		Total (KT)		
		Capacity (A)	Production (B)	Consumption (C)
BELGIUM	1983	730.0		
EEC	1979			8100.0
	1980	8900.0	7000.0 *	
	1981		6800.0	
	1982		6630.0	6600.0
	1984	8600.0	8000.0	8000.0
FRANCE	1977		1700.0	
	1982		1250.0	
	1983	1500.0		
GREAT BRITAIN	1977		1050.0	
	1979		1000.0	
	1981		870.0	
	1983	1300.0		
GREECE	1983	40.0		
ITALY	1979		900.0	
	1981		825.0	
	1983	1300.0		
NETHERLANDS	1983	350.0		
UNITED STATES	1983		9000.0	9200.0
WEST GERMANY	1977		2800.0	
	1979		3200.0	
	1982		2900.0	
	1983	3300.0		
WORLD	1973		24600.0	
	1982		25000.0 **	

Figure 6. Display of data concerning Uses

```
Substance    828  Chlorine

    Direct Use
        DISINFECTANT FOR WATER TREATMENT. AND IN PULP AND PAPER
        USED IN PULP AND PAPER INDUSTRY.
        CATALYST.

    Indirect Use
        INTERMEDIATE FOR CHLORINATED SOLVENTS, VINYL CHLORIDE,
        FLUOROCARBONS AND OTHER CHEMICALS.

    Consumption Summary
        25% USED FOR SOLVENTS
        30% FOR VCM
        12% FOR PROPYLENE OXIDE     ( EEC, 1985 ).
```

Figure 7. Display of Occupational Safety and Health Data

```
Substance   40876   2,3,7,8-Tetrachlorodibenzo-p-dioxin
    Warning

        THE ONLY INFORMATION ON THE HEALTH EFFECTS IN HUMANS FROM
        EXPOSURE TO TCDD IS FROM CLINICAL OR EPIDEMIOLOGICAL STUDIES
        OF POPULATIONS WHO WERE OCCUPATIONALLY AND NON-OCCUPATIONALLY
        EXPOSED TO 2,4,5-T AND TCP CONTAMINATED WITH TCDD.
        TO DATE, NO STUDIES OF HUMANS INCLUDE A QUANTIFICATION OF
        EXPOSURE TO TCDD.
        HUMANS EXPOSED TO MATERIALS REPORTED TO BE CONTAMINATED WITH
        TCDD HAVE DEVELOPED CHLORACNE AND OTHER SIGNS OF SYSTEMIC
        POISONING.  SOFT TISSUE SARCOMA HAS BEEN OBSERVED IN EXCESS
        AMONG WORKERS EXPOSED TO PHENOXY HERBICIDES.  THESE DATA ARE
        INCONCLUSIVE REGARDING TCDD TOXICITY IN HUMANS BECAUSE THE
        POPULATIONS STUDIED HAD MIXED EXPOSURES MAKING CAUSAL
        RELATIONSHIPS BETWEEN EXPOSURE AND EFFECT UNCLEAR.  THE DATA
        ARE, HOWEVER, SUGGESTIVE OF AN ASSOCIATION BETWEEN EXPOSURE TO
        PHENOXYACETIC HERBICIDES CONTAMINATED WITH TCDD AND EXCESS
        LYMPHOMA AND STOMACH CANCER.  ATTEMPTS TO ASSOCIATE REPRODUC-
        TIVE EFFECTS WITH TCDD EXPOSURE ARE INCONCLUSIVE BECAUSE OF
        THE INADEQUATELY DEFINED POPULATIONS STUDIED AND THE DIFFICUL-
        TIES OF DEFINING EXPOSURE.
        ON THE BASIS OF ANIMAL EXPERIMENTS THERE ARE SEVERAL
        CLASSIFICATIONS FOR IDENTIFYING A SUBSTANCE AS A CARCINOGEN.

    Bibliographic References

  -NIOSH:
   CURRENT INTELLIGENCE BULLETIN, 40.
   2,3,7,8-TETRACHLORODIBENZO-P-DIOXIN (TCDD).
   U.S. DEPARTMENT OF HEALTH AND HUMAN SERVICES, PUBLIC HEALTH
   SERVICES, CENTERS FOR DISEASE CONTROL,
   NATIONAL INSTITUTE FOR OCCUPATIONAL HEALTH AND SAFETY (USA),
   1984.
```

Figure 8. Display of Aquatic Toxicity data

```
Substance        9X  Dieldrin

   Latin Name          : BUFO ARENARUM
                         (AMPHIBIAN)
   Life Stage          : EMBRYO(11 DAY)

   Test Parameter      : ACETYLCHOLINESTERASE ACTIVITY IN VIVO
   Effect              : 22.7-26.7 % DECREASE

   Concentration       : 2.0 MG/L
   Exposure Time       : 4.0 HR

               ---- T E S T   C O N D I T I O N S ----

   Waterflow: Static, Medium: Growth Medium, Place: Laboratory,
   Temperature: 20.0-22.0 °C, Purity: 100.0 %,
   Solvent: ACETONE


        Bibliographic References

   DE LLAMAS M.C. ET AL
   ARCH.ENVIRON.CONTAM.TOXICOL. 14, 161-166  1985
```

Selection of Data Files for Report Generation

From the list of Data Categories and Data Files in Table 1.

Selection of Menu Type for Record Selection

(here shown for Classical Toxicity):

A By type
B By species
C By route

Selection of Records

(for choice A above)

type of experiment	number of records present
A Unspecified	21
B Acute	27
C Chronic	21

Figure 9. Display of Mutagenicity data

Substance 112992 **Methanesulfonic acid, methyl ester**

Type of Study : METAPHASE ANALYSIS
Genetic end Point : CHROMOSOME ABERRATIONS
Species and Strain : HAMSTER CHINESE
Cell : CHL FIBROBLAST

EXPOSURE CONCENTRATION

Concentration : 0.04 MG/ML   Duration : 24.0 HR

METHOLOGICAL DETAILS

Activating System : None
Solvent : PHYSIOLOGICAL SALINE

Method Details : CELL DIVISION WAS BLOCKED BY COLCEMID 0.2 UG/L
STAINED WITH 1.3 % GIEMSA

RESULTS

Effect : POSITIVE

Exposed	Control
17.0	1.0
GAPS/100 METAPHASES ANALYSED	
90.0	1.0
BREAKS/100 METAPHASES ANALYSED	
89.0	0.0
QUADRI- AND TRIRADIAL EXCHANGE BETWEEN CHROMOSOMES/100 METAPHASES	
4.0	0.0
EXCHANGE BETWEEN ARMS IN CHROMOSOMES/100 METAPHASES ANALYSED	
1.0	1.0
% OF POLYPLOID CELLS	
97.0	2.0
% OF CELLS WITH STRUCTURAL CHROMOSOME ABERRATIONS	

Comment : DOSE AT WHICH ABERRATIONS WERE DETECTED IN 20% OF METAPHASES
D20=0.013 MG/ML

Bibliographic References
M. ISHIDATE, JR. AND STAFFS: A NEW DATA BOOK ON CHROMOSOMAL ABERRATION TESTS USING A CHINESE HAMSTER LUNG FIBROBLAST CELL LINE, CHL. NATIONAL INST. OF HYGIEN. SCI TOKYO 158 JAPAN

D LDL_o	17
E LD_{100}	3
F LD_{50}	28
G Local	9
H Subacute	10

ECDIN AND ADABAS

The ECDIN databank is operated under the ADABAS database management system. Also, the version available on the host is stored in an ADABAS database. Public access to ECDIN takes place through a program written in the COBOL programming language. Maintenance of the database and research activities using the database in Ispra use other pieces of software mainly written in the NATURAL programming language. Batch updates of the database are performed through an ECDIN Update Program, which in addition to operating on the database carries out different controls on the data before they are loaded. This software configuration is outlined in Figure 10.

An ECDIN ADABAS file is described as a set of fields. Table 2 contains a part of the layout for the file with data concerning physicochemical properties.

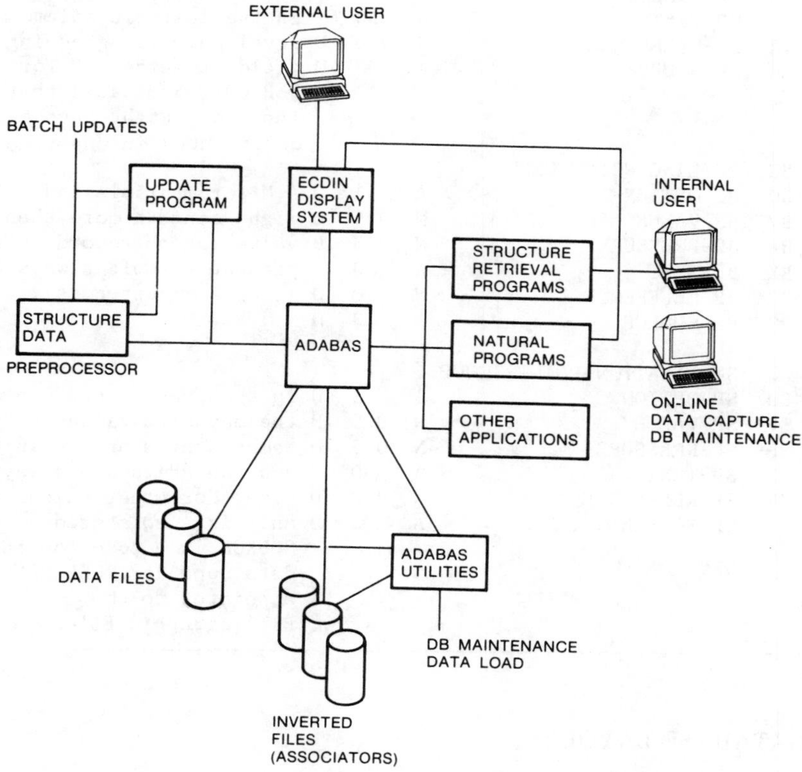

Figure 10. Outline of the ECDIN Software Configuration

Table 2. Part of the File Description for the file with Physical-Chemical properties data

D indicates that the field is a descriptor and can be used as serach-key

TYL	DB	NAME	F	LENG	D	REMARKS
1	EN	ECDIN-NO	A	7	D	The ECDIN number links
						the files together
1	MF	MOLECULAR-FORMULA	A	64		The Molecular Formula
1	M1	MF-SOURCE	A	3		is not searchable in
1	M2	MF-REFERENCE	N	9.0		this file
1	MO	MW-OPTION	A	1	D	Option can be =, <, >, or
M 1	MV	MOLECULAR-WEIGTH	N	10.5	D	R for Range. In the latter
1	MR	MW-REFERENCE	N	9.0	D	case two values are stored
1	MS	MW-SOURCE	A	3	D	in the value field, MV
P 1	LG	MELTING-POINT-GROUP				
2	LO	MP-OPTION	A	1	D	The field TEXT is used to
M 2	LV	MP-VALUE	N	10.5	D	add any comment to the
2	LP	MP-PRESSURE	N	10.5	D	stored value. The numbers
2	LX	MP-TEXT	A	30		in the leftmost column are
2	LR	MP-REFERENCE	N	9.0	D	level numbers, grouping
2	LS	MP-SOURCE	A	3	D	fields together. P for
						PERIODIC indicates that
						the group might have more
						occurrences in one record
P 1	BG	BOILING-POINT-GROUP				
2	BO	BP-OPTION	A	1	D	A M for multiple indicates
M 2	BV	BP-VALUE	N	10.5	D	might contain more than one
2	BP	BP-PRESSURE	N	10.5	D	value in one record. (In
2	BX	BP-TEXT	A	30		present example always 0,
2	BR	BP-REFERENCE	N	9.0	D	1, or 2 occurrences).
2	BS	BP-SOURCE	A	3	D	
P 1	SG	SUBLIMATION-POINT-GROUP				
2	SO	SP-OPTION	A	1	D	In the present example are
M 2	SV	SP-VALUE	N	10.5	D	the measured values either
2	SP	SP-PRESSURE	N	10.5	D	temperatures or pressures
2	SX	SP-TEXT	A	30		and the units are always
2	SR	SP-REFERENCE	N	9.0	D	same. For other values the
2	SS	SP-SOURCE	A	3	D	unit is also stored.
						SOURCE is a code for the
						data supplier, REFERENCE
						a pointer to the
						Bibliographic file.

ECDIN DATABASE LAYOUT

The ECDIN database consists of a set of ADABAS files. There are more than 30 production files and a similar number of files used for new developments. Each of the logical files seen by the user accessing the database through the ECDIN Display System (they are the files in Table 1) corresponds to an ADABAS file, but the database contains other files in addition to these.

Most of the ECDIN ADABAS files contain data related to a chemical compound, and in these files, each record contains an ECDIN number, which links the record to one record in the central Compound File. The ECDIN number can also be used to relate data in two files directly to each other without involving the Compound File. Figure 11 illustrates a part of the database design including files which are not linked directly to the Compound File.

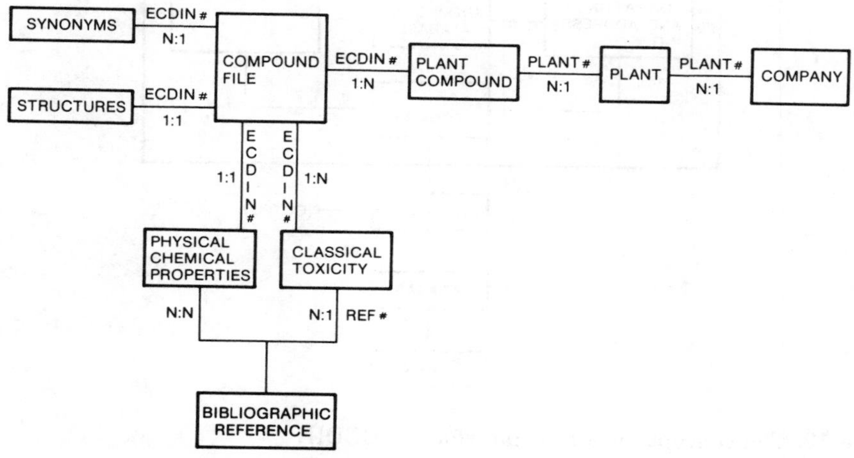

Figure 11. ECDIN Database Design (part)

THE ECDIN-ACCESS PROJECT

In the framework of the INSIS programme of the Commission's Directorate General XIII (Telecommunication, Information Industry and Innovation) which aims to improve the access to databases of the Community institutions, a project has been started at the beginning of 1987 for the development of a new user interface to the ECDIN databank. The project has been named ECDIN-Access. Although the interface should first be implemented for ECDIN it is the intention that the concept of the interface should also be applicable for other databanks and other Database Management Systems. The project is in an initial phase, analysing user needs and technical feasibility. The current ideas involve an expert system with knowledge of the database and of the Common Command Language (CCL). The expert system helps the user to build CCL commands, and an interface converts the Common Command Language commands to ADABAS calls. Figure 12 illustrates this concept.

ECDIN AND NATURAL

NATURAL is a high-level language particularly designed to access ADABAS databases, but also applicable for data processing in other environments. The

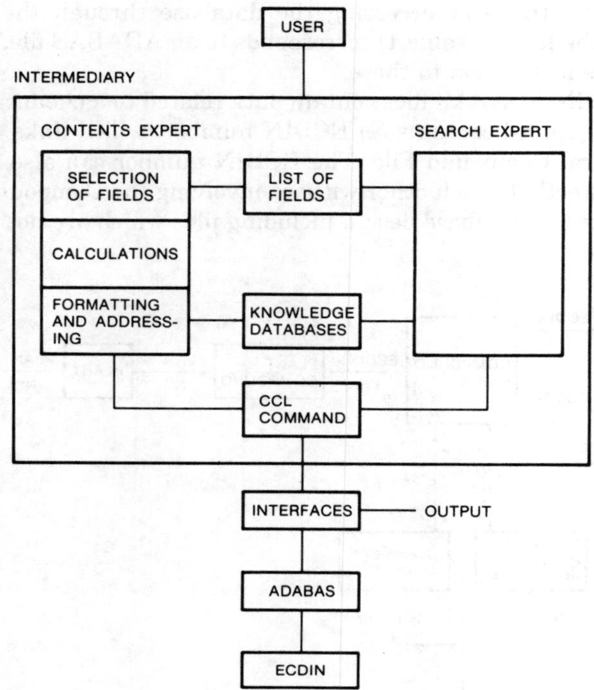

Figure 12. The concept for a new interface to ECDIN

programming language is supplied by Software AG, which also produces the ADABAS DBMS. The program in Figure 13 shows a simple structure retrieval using fragment codes together with a search for a truncated name. The fragments describe the ring system of Benomyl, a pesticide, and the name adds the further condition that the compounds should be derived from carbamic acid. After the structures have been retrieved the Compound Name is retrieved from the Compound File and toxicity data are retrieved from the Classical Toxicity File. The data are only displayed if the desired toxicity data are found. Some examples of the output produced by the program are included in the Figure. The fragmentation used is based on a modified version of the CROSSBOW system.

Figure 13. A NATURAL program

```
0010 * POSTER
0020 FIND STRUCTURE WITH FRAGMENT =  101 AND FRAGMENT = 105
0040                 AND FRAGMENT =  134 AND FRAGMENT = 118
0060                 AND FRAGMENT =   78 AND FRAGMENT =  97
0080 FIND FIRST NAMES WITH ECDIN-NUMBER = ECDIN-NUMBER (A20)
0090            WHERE SEARCH-NAME = 'CARBAMIC ACID,' THRU 'CARBAMIC ACID,9'
0100 FIND COMPOUND WITH ECDIN-NUMBER = ECDIN-NUMBER(0080)
0151 FIND UNIQUE NAMES WITH EN = EN(0080) AND NAMETYPE = 'PSN'
0160 FIND CL-TOX WITH ECDIN-NUMBER = ECDIN-NUMBER(0080)
0170      WHERE ROUTE = 'ORAL'
0180       AND (    SPECIES = 'RAT'
0190             OR SPECIES = 'MOUSE'
0200             OR SPECIES = 'WILD BIRD')
0210 AT START OF DATA (0160) DISPLAY / EN COMPOUND-NAME(0100) /
0211                                 CHEMICAL-NAME (1-10) (ES=T)
0220 DISPLAY 10X SPECIES(A10) ROUTE(A5) TYPE(A08) CV(N10.5) CU
0240 END
```

```
ECDIN                            COMPOUND-NAME
NUMBER                           CHEMICAL-NAME
------- ------------------------------------------------------------

   627x Benomyl
        Carbamic acid, <1-<(butylamino)carbonyl>-1H-benzimidazol-2-yl>-,
        methyl ester
        WILD BIRD  ORAL  LD50                 100.0 MG/KG

  29297 1H-Benzimidazole
        RAT        ORAL  LDLO                 500.0 MG/KG
        MOUSE      ORAL  LD50                2910.0 MG/KG

  29480 2-Methylbenzimidazole
        RAT        ORAL  LDLO                 500.0 MG/KG
```

POSTER SESSION: POTENTIAL ENHANCEMENTS TO THE CAS CHEMICAL REGISTRY SYSTEM

Gerald G. Vander Stouw
Chemical Abstracts Service, P.O. Box 3012, Columbus, Ohio 43210, U.S.A.

ABSTRACT

The Chemical Abstracts Service (CAS) Chemical Registry System began operation in 1965 as a tool for CAS staff to manage information about substances indexed from *Chemical Abstracts*. Today the Registry, which contains records for more than eight million chemical substances, is the core of an online search service which is used worldwide. As the use of Registry data has evolved, CAS has identified some areas in which Registry structuring conventions might be altered to make the Registry more useful to online searchers and other users as well as to CAS staff. The CAS Registry System Enhancements Feasibility Study project is examining these areas and developing Proposals for new or modified structuring conventions, as well as other enhancements and modifications to the system. These proposals will be based on input provided by a broad spectrum of online searchers and other users of chemical information. Implementation of these recommendations over the next few years will result in the Registry File becoming an improved base for online structure searching.

HISTORY OF THE CAS REGISTRY SYSTEM

Chemical Abstracts Service (CAS) began operation of the Chemical Registry System at the beginning of 1965. Initially, the primary purpose of this connection table-based system was to detect the re-occurrence of substances indexed from *Chemical Abstracts* (CA), and thus provide a means of managing bibliographic references and other information about substances. The key to the Registry System, the CAS Registry Number, is a computer-assigned identification number which serves to identify uniquely substances on the basis of molecular structure.[1] In 1968 the scope of the system was expanded, principally through the addition of conventions for representing polymers and co-ordination compounds. The modified system became known as Registry II. During the next several years, Registry II became incorporated into an extensive computer-based system for the production of CA Indexes. Fundamental to this production system was the Registry's ability to determine when a substance being indexed was identical to one already on file. In such a case, the CA Index Name already assigned was retrieved and used in the printed Index, and duplicate naming was avoided. The result was a significant saving in production costs. This use of the Registry as the basis for Index production has continued to the present day.

In 1974 CAS made another major change to the Registry System, resulting in the system known as Registry III.[2] The changes made in 1974 were primarily in the internal representation of the connection table stored in the system, and were designed

W. A. Warr (Ed.)
Chemical Structures

primarily to support the CAS nomenclature generation process and the algorithmic generation of structure diagrams from connection tables. The principal change was in the explicit identification of ring systems present in a molecule; another major change was a systematisation of the Text Descriptors used to describe stereochemistry in the Registry record.[3] The change from Registry II to Registry III affected the entire file (then approximately three million substances) and involved a massive, time-consuming conversion effort. The resulting system has remained in use until the present time.

CHANGES IN ENVIRONMENT FOR USE OF REGISTRY DATA

Since the implementation of Registry III in 1974, many changes have occurred in the ways in which chemical information is used, and these have created new and different requirements for the system. Some of these changes simply result from continued growth, as the file of three million substances has now grown to more than eight million. Computer technology has seen great advances, particularly in the development of online search techniques, high speed value-added communications networks, and the proliferation of personal computers. The greatest change for the Registry System, however, has been the evolution of what was once primarily an in-house tool into a resource which is widely used by people outside CAS. The Registry System has become recognised world-wide as an authority for substance identification. The use of the Registry Number as an identifier has become wide-spread and is frequently mandated by law. Several major industrial firms have chosen to use the CAS Registry technology to manage their own chemical substance files; other firms have selected alternative approaches that are also connection table-based. The most important development is that structure and nomenclature files derived from Registry data have been searched online since 1980 by users around the world, through the CAS Registry File[4] available through the STN International scientific and technical information network and through the DARC file available from Télésystèmes Questel.[5]

The development and acceptance of online structure and substructure search have constituted a fundamental change in the way in which chemical information is accessed and used. This change has created new expectations for the contents and quality of the structure files being searched. Users of the online files have had to develop some understanding of the structuring conventions used by CAS for the substances in these files. As indicated previously, these conventions were developed many years ago and were intended primarily for in-house use. For the great majority of chemical substances, especially organic compounds, these structures are adequate for today's needs. Some structuring practices cause difficulties, however, and some of these have been the subject of complaints by users during the last several years. In response, CAS has begun an effort which has been titled 'Registry Enhancements'. The overall purpose of this effort is to determine how the Registry System might be modified to be more responsive to the needs of today's users. The end result will be a modified Registry System that will provide an effective basis for chemical informationhandling and searching into the next century. The initial research project in this effort is already under way. The project is charged with identifying the areas of Registry that should be considered for possible changes, and developing recommendations for these. The recommendations will be implemented in a series of projects over the next several years.

A very important aspect of the current project is an extensive effort to obtain suggestions from users and potential users of Registry-based services. Users will help CAS identify the areas needing attention, clarify the questions in these areas, and determine the changes to be made. Tools being used by CAS to obtain comments and suggestions from users include: written and telephone surveys; contacts by account representatives and by staff at exhibit booths at national and international professional meetings; visits with persons who have expertise in specific technical areas; focus groups of knowledgeable persons who meet at CAS or other locations; and conversations among CAS staff and scientists at technical meetings such as this.

AREAS OF REGISTRY BEING CONSIDERED FOR MODIFICATIONS

Although other areas may well be considered, initial discussions of possible modifications have identified the following areas as a basis for soliciting and classifying suggestions from users: polymers, biological macromolecules, stereochemistry, coordination compounds and inorganic compounds, materials, and general issues.

Polymers

Current CAS registration of polymers emphasises records based on the structures of their monomers.[6] About 18 per cent of Registry Numbers for polymers are assigned to structure repeating units (SRU), which are used only when the basic structure of the polymer is known or can be defined. One suggestion being evaluated is that SRU structures be used wherever possible, even when this means formulating variable SRU's. These SRU records would be linked to the monomer-base records, so that a searcher could retrieve a polymer either through SRU-based or monomer-based searching. In addition to this fundamental question about extending polymer handling, a number of other issues, affecting smaller numbers of polymers, have been identified and are being evaluated.

Biological Macromolecules

The current growth of biotechnology creates potential new opportunities for the delivery of structure-based services. CAS currently abstracts and indexes a very substantial portion of the literature related to the preparation and use of such biochemical entities as nucleic acid and protein sequences. However, the current Registry System enforces size limits which prevent most of these biological macromolecules from being represented structurally; instead, they receive so-called 'manual' registration and can be searched only by their names or molecular formulae. Databases that are based on sequences of symbols for amino acid or nucleotide residues, such as GenBank and those in the Protein Identification Resource, are coming into increasing use. An important issue being investigated by this project is whether CAS should make such databases available, and what capabilities should be offered for searching such databases.

Stereochemistry

The Registry III system is based on a two-dimensional atom-bond connection table. Molecules that differ only in their stereochemistry are assigned different Registry Numbers; however, the stereochemical detail known about them is recorded in a linear character string known as a Text Descriptor.[3] Users have frequently expressed a need for explicit three-dimensional information stored as part of connection tables, and for display of stereospecific structures. In recent years, there has been increasing use of computer programs for such purposes as molecular modelling and prediction of organic synthesis; these applications also need stereochemical data. Techniques for storage of atom-by-atom atom stereochemical data were described by Petrarca *et al.*[7] and adapted for use by Wipke[8] as part of a system for organic synthesis prediction. The current project will evaluate the possibility of using techniques such as those, or other methods for storing three-dimensional data as part of the Registry record. A particular concern here is whether and how to upgrade the many structure records that already have stereochemistry recorded as Text Descriptors.

Co-ordination Compounds

CAS structuring conventions for coordination compounds[6] do not distinguish sigma and pi bonds; metal to ligand bonds are treated as ordinary covalent bonds. The project will consider possible changes in conventions for these structures so that their records more nearly reflect the ways in which chemists think of these structures.

Materials

This heading inludes entities such as alloys, which are currently machine-registered, and other materials for which automatic registration techniques have not been formulated and which, therefore, receive 'manual' registration.[9] The general question to be addressed in this area is how to provide users with better access, especially through online searching, to the references and data that they need regarding these substances. Considerations in this area will also have to take into account the development of files of numeric data for materials, and how best to interface with these files.

General Issues

This heading covers questions which have implications throughout the Registry System, and not only within specific areas. Stereochemistry is one such area, but it has been judged to be of sufficient importance to be discussed separately here. Other issues which will be examined within these projects include questions such as whether modifications are needed to the existing conventions for handling tautomerism and alternating bonds,[10] and whether new or different kinds of linking are needed between, for example, acids and their salts, ketones and their derivatives, etc.

SUMMARY

CAS is currently beginning a broad review of its Chemical Registry System leading to changes which will improve its usefulness for a wide variety of users. It is vital that these changes be based on a thorough understanding of user needs and desires. Meetings such as this conference on Chemical Structures are one important source of the information which will help in this effort.

REFERENCES

1. Morgan, H.L. 'The Generation of a Unique Machine Description for Chemical Structures'. *J. Chem. Doc.* 1965, *5*, 107–113.
2. Dittmar, P.G., Stobaugh, R.E., Watson, C.E. 'The Chemical Abstracts Service Chemical Registry System. I. General Design'. *J. Chem. Inf. Comput. Sci.* 1976, *16*, 111–121.
3. Blackwood, J.E., Elliott, P.M., Stobaugh, R.E., Watson, C.E. 'The Chemical Abstracts Service Chemical Registry System. III. Stereochemistry'. *J. Chem. Inf. Comput. Sci.* 1977, *17* 3–8.
4. Dittmar, P.G., Farmer, N.A., Fisanick, W., Haines, R.C., Mockus, J. 'The CAS ONLINE Search System. I. General Design and Selection, Generation, and Use of Search Screens'. *J. Chem. Inf. Comput. Sci.* 1983, *23*, 93–102.
5. Attias, R. 'DARC Substructure Search System: A New Approach to Chemical Information'. *J. Chem. Inf. Comput. Sci.* 1983, *23*, 102–108.
6. Ryan, A.W., Stobaugh, R.E. 'The Chemical Abstracts Service Chemical Registry System. IX. Input Structure Conventions'. *J. Chem. Inf. Comput. Sci.* 1982, *22*, 22–28.
7. Petrarca, A.E., Rush, J.E., Lynch, M.F. 'A Method for Generating Unique Computer Structural Representations of Stereoisomers'. *J. Chem. Doc.* 1967, *7*, 154–165.
8. Wipke, W.T. 'Computer-Assisted Three-Dimensional Synthetic Analysis'. In *Computer Representation and Manipulation of Chemical Information*; Wipke, W.T., Heller, S.R., Feldmann, R.J., Hyde, E., Eds; John Wiley and Sons:New York, 1974, pp. 147–174.
9. Moosemiller, J.P., Ryan, A.W., Stobaugh, R.E. 'The Chemical Abstracts Service Chemical Registry System. VIII. Manual Registration'. *J. Chem. Inf. Comput. Sci.* 1980, *20*, 83–88.
10. Mockus, J., Stobaugh, R.E. 'The Chemical Abstracts Service Chemical Registry System. VII. Tautomerism and Alternating Bonds'. *J. Chem. Inf. Comput. Sci.* 1980, *20*, 18–22.

SUBSTRUCTURE ANALYSIS AS BASIS FOR INTELLIGENT INTERPRETATION OF SPECTRA

Wolfgang Bremser
Central Research, BASF Aktiengesellschaft, 6700 Ludwigshafen, West Germany

ABSTRACT

Intelligent computer assisted interpretation of spectroscopic data should be based on the knowledge from large structure oriented data collections. Both the inspection of spectral features and the statistical evaluation of similar structures (from library searches) can provide a set of probability ranked substructures which are readily assembled to target structures. The idea of substructure analysis allows the chemist to combine the results of different interpretation strategies, different databases and different spectroscopic methods to yield the structural information desired. Thus in a multidimensional data system like SPECINFO 'structural noise' can be effectively suppressed, if all information available in the spectroscopic laboratory is combined in a central intelligent computer system.

Equally important is the prediction of spectral parameters in order to rank the candidates of the structure generator. Chemical shifts and coupling constants may serve as an example where substructure oriented strategies lead to an acceptable estimate of the spectral curve.

REFERENCES

1. Bremser, W. 'Structure Elucidation and Artificial Intelligence'. *Angew. Chem. Int. Ed. Engl.* 1988, *27*, 247–60.
2. Bremser, W., Neudert, R. 'Automation in the Spectroscopic Laboratory – Solutions and Perspectives'. *Eur. Spectrosc. News* 1987, *75*, 10–27.
3. Neudert, R., Bremser, W., Wagner, H. 'Multidimensional Computer Evaluation of Mass Spectra'. *Org. Mass Spectrom.* 1987, *22*, 321–329.
4. Passlack, M., Bremser, W. 'IDIOTS – Structure-oriented Databank System for the Identification and Interpretation of Infra-red Spectra'. In *Computer-supported Spectroscopic Databases*; Zupan, J. Ed., Ellis Horwood: Chichester, 1986.

W. A. Warr (Ed.)
Chemical Structures

COMPUTER-AIDED SPECTROSCOPIC STRUCTURE ANALYSIS OF ORGANIC MOLECULES USING LIBRARY SEARCH AND ARTIFICIAL INTELLIGENCE

Henk A van't Klooster
National Institute of Public Health and Environmental Protection (RIVM), P.O. Box 1, 3720 BA Bilthoven, The Netherlands

Peter Cleij and Hendrik-Jan Luinge
Faculty of Chemistry, University of Utrecht, Croesestraat 77A, 3522 AD Utrecht, The Netherlands

Gerard J Kleywegt
Faculty of Chemistry, University of Utrecht, Padualaan 8, 3584CH Utrecht, The Netherlands

ABSTRACT

The success of computer-aided methods for the identification or structure elucidation of organic molecules is determined by the reliability of the analytical results, which in many cases can be assessed through the statistical significance of the conclusions. A key factor determining the usefulness of computer-aided library search systems is the extent to which the reproducibility of the relevant data is accounted for in the design of the similarity measure. Based on mathematical statistical models of the reproducibility of the data involved, library search systems for mass, ^{13}C-NMR and high resolution ^{1}H-NMR spectra, as well as for ultra-violet spectra combined with HPLC retention data have been developed (in FORTRAN and Pascal). These systems are accessible (provisionally on an experimental basis) to Dutch universities and other laboratories through the national CASSAM Centre (Computer-Aided Spectroscopic Structure Analysis of Molecules) located at Utrecht. For the analysis of combined infra-red and mass spectral data a pilot version of an expert system has been developed (in PROLOG) using artificial intelligence techniques and information theory. Here too, the significance of results is indicated by numerical (relative) probabilities. General concepts and results are discussed.

INTRODUCTION

For computer-aided extraction of information concerning the identity or structure of organic molecules from spectral and/or chromatographic data two main approaches apply:

1. comparison of analytical results with reference data, using library search or pattern recognition techniques;
2. application of artificial intelligence (AI) techniques using empirical rules (whether

W. A. Warr (Ed.)
Chemical Structures

or not automatically generated) for the interpretation of spectral and/or chromatographic features of the unknown compound.

The main aims of library search systems are:

1. Straightforward identification by retrieval of candidate compounds/molecular structures. Examples of such systems are: Biemann's MS search,[1] McLafferty's MS/PBM system,[2] and our MS, ^{13}C-NMR, high resolution ^{1}H-NMR and LC-UV library search systems.[3–7]
2. Classification of the unknown molecule or (sub)structure analysis by retrieval of compounds with (sub)structures similar to the unknown. Examples: Henneberg's SISCOM (MS),[8] McLafferty's STIRS (MS),[9] ^{13}C-NMR search systems of Bremser[10] and Clerg[11] and our ^{1}H-NMR retrieval system.[6]
3. Identification of substructures of the unknown molecule by (statistical) interpretation of (multiple) search results (STIRS).

Artificial intelligence techniques are applied for the identification of substructures of unknown molecules and for the generation of complete candidate structures based on a substructure analysis and verification of possible structures.[12–20]

Library search is mainly used for the identification of more or less frequently occurring compounds of which reference data are likely to be available. If no reference data are available, i.e., when dealing with unknown or completely new compounds, an interpretative search, or a STIRS-like method, or the AI-approach can be employed. This situation might occur in research laboratories where new compounds are being developed and in environmental research, for example when new chemical products (e.g., pesticides) are released.

The fact that mass, infra-red, NMR and ultra-violet spectra and, to a certain extent, also chromatographic retention data can be considered as 'molecular fingerprints', forms the basis of most computerised library search systems. Retrieval methods for characteristic chemical data and techniques for the comparison of human fingerprints have similar elements: the first step is to clean up the raw data, then in many cases a data reduction is carried out by selection of prominent features. Finally, there is the comparison of unknown and reference data patterns, which, for a useful result, requires a statistical correlation to be established. In this paper no attention will be paid to feature selection.

In library search four main factors can be distinguished:

1. the specificity of the fingerprint data;
2. the reproducibility of the fingerprint data;
3. the method of feature selection;
4. the design of the similarity measure.

Factors 1 and 2 are correlated: usually, a decrease/increase of the reproducibility of the fingerprint data will cause a decrease/increase of the specificity of that data. Factors 2 and 4 are also correlated. As a matter of fact, the reproducibility of the data involved in a search system is a crucial element of the design of the similarity index. Especially the interlaboratory reproducibility plays an important role, whenever multisource databases are being used. This reproducibility is determined by differences in samples, instruments, experimental conditions, performance of analysts and operators, introduction of coding errors, etc. As a consequence, a major factor determining the usefulness of computer-aided library search systems is the extent to which the reproducibility of the relevant data is accounted for in the design of the similarity measure.

The following questions are imperative:

1. What is considered to be 'similar', what is 'different'?
2. What difference is 'acceptable'?
3. What difference is 'significant'?
4. Which rational (formal) criterion is to be used?

A REPRODUCIBILITY-BASED SIMILARITY INDEX

The concept of a reproducibility-based similarity index will be introduced for a simple one-dimensional (fictitious) example. Suppose we want to identify a compound which we already know is a methylpentanol isomer, by library search based on gas-chromatographic data. We have measured a retention value of 740 units with a standard deviation of 5 units. The following small library of reference data is available.

Compound		*Retention value* (arbitrary units)
2-methyl-1-pentanol		804
3-	-1-	816
4-	-1-	803
2-	-2-	717
3-	-2-	778
4-	-2-	736
2-	-3-	762
3-	-3-	747

Evidently, the retention values of the 4,2-isomer and the 3,3-isomer are closest to the measured value of 740. But should we reject the 2,2-isomer or the 2,3-isomer? In other words: which reference values are significantly different from the measured value (e.g., on a 1 per cent level of significance) and which are not? Our library search system should provide a clear answer to these questions.

In comparing measured and reference data the statistical theory of hypothesis testing applies. The null hypothesis H_0 for every comparison is that the unknown compound is identical to the reference compound. The test parameter is the 'difference quantity' Δq, representing in our example, the difference between two retention values:

$$\Delta q = Ret_{measured} - Ret_{reference}. \quad (1)$$

In this case the reproducibility model of Δq is simply a normal probability distribution function $p(\Delta q)$ under the null hypothesis, with mean = 0 and standard deviation = 7. The parameter for testing H_0 is the P-value, or, as we call it here, the similarity index (SI), for our simple library search system defined as the integral of the reproducibility function, in this case a symmetrical Gaussian curve:

$$SI = 2 \cdot \int_{\Delta q = \Delta Q}^{\infty} p_0(\Delta q)\, d\Delta q \quad (2)$$

with ΔQ being the actual measured value of Δq.

The integration is done from the point of the actually measured value of the difference (ΔQ) to infinity (2-sided) (see Figure 1). For example: the similarity index

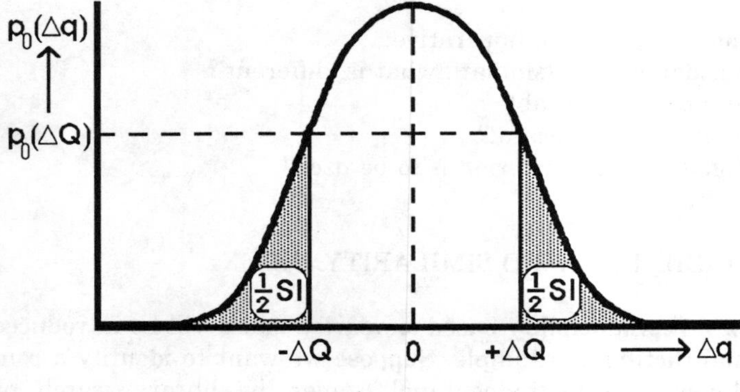

Figure 1

value for an observed difference of 7 units is equal to 2 x 16% = 32%. If the observed difference is zero, SI equals the whole area under the curve: 100%.

Our library search program, based on this similarity index, allows the specification of a minimum value for the similarity index. If, for example, a threshold value of 1 per cent is specified, references with an SI less than 1 per cent are rejected as possible candidates. For these candidates the null hypothesis is rejected at a significance level (α) of 1 per cent. The output of the retrieval system is a list of compounds (if any) that have SI values above the preset threshold, i.e., compounds for which the null hypothesis is accepted. For the pentanol problem our library search system gives the following output:

IN SEARCH FOR: UNKNOWN PENTANOL SI (THRESHOLD) = 1.0%
SEARCH RESULTS:

NO	*SI(%)*	*COMPOUND NAME*
1	57.0	4-METHYL-2-PENTANOL
2	32.4	3-METHYL-3-PENTANOL

END OF HIT LIST

Thus, applying a significance level of α = 1%, the null hypothesis is accepted for the 4,2-and the 3,3-isomer and is rejected for all other isomers. The conclusion is that the unknown methylpentanol is probably the 4,2-or the 3,3-isomer, or some compound not represented in the reference file!

LIBRARY SEARCH OF 'MOLECULAR FINGERPRINTS'

For the comparison of analytical data, e.g. molecular spectra and/or chromatographic retention data, a similarity index in the form of a P-value, as used in hypothesis testing, has been developed.[3] This similarity index requires that unknown and reference compounds be characterised by a set of continuous 'feature quantities' $q_1 \ldots q_i \ldots q_m$. In the case of mass spectra these may be the peak intensities at a certain number of selected masses, whereas for ^{13}C-NMR spectra the chemical shifts could be used.

The actual comparison is made on the basis of the values of a set of difference quantities $\Delta q_1 \ldots \Delta q_i \ldots \Delta q_m$, representing the differences in value of the feature quantities for the unknown and reference data, by calculating the value of the similarity index SI, given by:

$$SI = \underset{R\,[\Delta Q_1 \ldots \Delta Q_i \ldots \Delta Q_m]}{\int_{\Delta q_1} \ldots \int_{\Delta q_i} \ldots \int_{\Delta q_m}} p_o(\Delta q_1 \ldots \Delta q_i \ldots \Delta q_m)\; d\Delta q_1 \ldots d\Delta q_i \ldots d\Delta q_m \quad (3)$$

where p_o is a probability density function called the reproducibility function of the set of difference quantities Δq_i, in case the reference compound considered is identical to the unknown compound (the null hypothesis H_o). ΔQ_i represents the actually observed value of the i'th difference quantity, and R is in the region of the multidimensional space of difference quantities defined by the condition:

$$p_o(\Delta q_1 \ldots \Delta q_i \ldots \Delta q_m) < p_o(\Delta Q_1 \ldots \Delta Q_i \ldots \Delta Q_m) \quad (4)$$

The model of reproducibility of the search data, as expressed by the reproducibility function, forms the basis of the similarity index.

An important property of the SI is that for correct matches the SI values are uniformly distributed in the range of 0–100%, which implies that the average SI value for correct matches is 50 per cent. This property is independent of the type of data compared.

Application of the similarity index provides a classification of the references by separating these in two classes: compounds that may be and compounds that cannot be the same as the unknown. In terms of hypothesis testing, this is equivalent to acceptance or rejection of the null hypothesis that the unknown and the reference compound are identical. A library search system based on this principle should retrieve all references of the 'may be' class, i.e., all references with a similarity index exceeding a predefined threshold value, rather than the 5 or 10 'best matches' (which may also be very bad matches).

Based on this general concept, library search systems for mass spectra,[3,4] ^{13}C-NMR spectra,[5] high resolution ^{1}H-NMR spectra[6] and ultra-violet spectra combined with HPLC retention data[7] were developed. Reproducibility models for molecular spectra were elaborated from some hundreds (sometimes more than thousand) of pairs of replicate spectra: different reference spectra of a compound, recorded under various experimental conditions.[3,7]

THE CASSAM CENTRE

The systems described above have been implemented at the national CASSAM Centre in the Netherlands (CASSAM being an acronym for: Computer-Assisted Spectroscopic Structure Analysis of Molecules). The CASSAM Centre is (provisionally on an experimental basis) accessible through national networks to workers at Dutch universities and other research institutes. Participants in the CASSAM project are the University of Utrecht, the Netherlands Organisation for Applied Scientific Research (TNO) and the Netherlands National Institute of Public Health and Environmental Protection (RIVM).

CASSAM data bases include:

- mass spectra (Wiley/McLafferty collection)
- ^{13}C-NMR spectra (TNO collection)
- 500 MHz ^{1}H-NMR spectra of carbohydrates (University of Utrecht collection)
- UV spectra and LC retention data of organophosphorus pesticides (RIVM collection).

Extensions and updates are planned with new releases of commercial databases (also of infra-red spectra) and reference files made available by Dutch research institutes (CASSAM users).

CASSAM software consists of various programs for the comparison or interpretation of molecular spectra. A brief description of some of these programs and examples of results are given in the following sections.

LIBRARY SEARCH OF MASS SPECTRA

The Mass Spectral Reproducibility-based Retrieval (MSRR) system of the CASSAM Centre uses the Wiley-McLafferty database of electron impact mass spectra. The 1987 release of this database contains 120,000 mass spectra of 100,000 organic compounds. A reproducibility model of the mass spectra was developed from 1400 pairs of replicate spectra, originating from an earlier version of the database (containing 39,000 spectra). This model is a mathematical statistical description of the observed systematic and random differences in peak intensities. For the search upto 24 intensities at selected masses are used.[3,4] The criterion for the selection of the peaks is based on the (empirical) fact that peaks with relatively high intensities and peaks at relatively high masses contain more information than small and low-mass peaks. The systematic differences include the mass discrimination effect of certain types of mass spectrometers and chromatographic effects. For the similarity index of the MSRR system a value can be specified, analogous to the one-dimensional chromatographic example discussed above. The system can be used in two main modes: one for mass spectra of compounds believed to be pure, the other mode for spectra of contaminated samples or mixtures.

An example of output from the MSRR system is given in Figure 2.

For an 'unknown' mass spectrum (A), in this case of a component (known to be 2-hydroxyxanthone) extracted from a plant, the system retrieves from a library of 39,000 reference spectra five references having a similarity index (SI) value of at least 5 per cent (as compared with spectrum A). First on the 'hit list' is 2-hydroxyxanthone,of which the reference spectrum (B) has an SI-value of 69.2 per cent. References 2, 3 and 4 are isomers, of which the mass spectra are expected (on mass spectrometric grounds) to show much similarity with that of 2-hydroxyxanthone. Reference 5 is a different compound, which, however, has some structural features in common with hydroxy-xanthones. As to the SI-values: only an exact copy of a reference spectrum yields an SI value of 100 per cent. Two different spectra of a compound, recorded under different experimental conditions, however, will always exhibit systematic differences, which in some cases may be quite substantial. The similarity index used in this library search system takes account of such differences.[3,4]

Apart from a library search, CASSAM also allows the plotting of specified unknown and/or reference spectra, as well as of molecular structures.

```
**************************************************************
                 CASSAM CENTRE - UTRECHT
              MASS SPECTRAL RETRIEVAL SYSTEM
           DATABASE: WILEY/MCLAFFERTY COLLECTION
**************************************************************
IN SEARCH FOR:    test sample, known to be:  2-hydroxy-xanthone
RETRIEVAL RESULTS:
```

NO	SI(%)	SER NR	MOLW	FORMULA	COMPOUND NAME
1	69.2	14606	212	C13H8O3	2-hydroxy-xanthone
2	47.1	14604	212	C13H8O3	3-hydroxy-xanthone
3	46.9	14605	212	C13H8O3	1-hydroxy-xanthone
4	40.1	14603	212	C13H8O3	4-hydroxy-xanthone
5	8.1	14601	212	C13H8O3	1,4-dihydroxy-fluorenone

```
END OF HIT LIST
**************************************************************
```

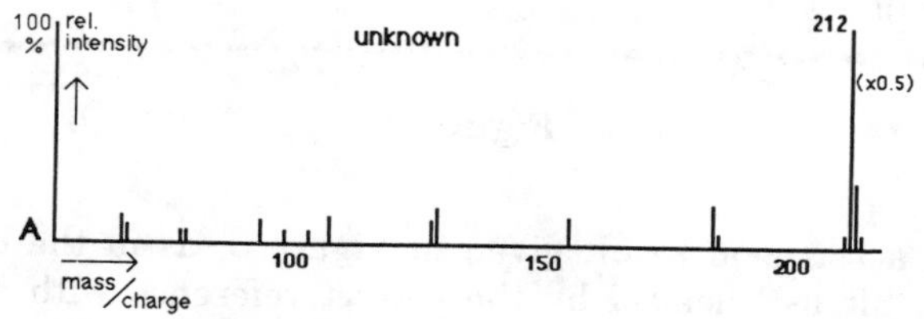

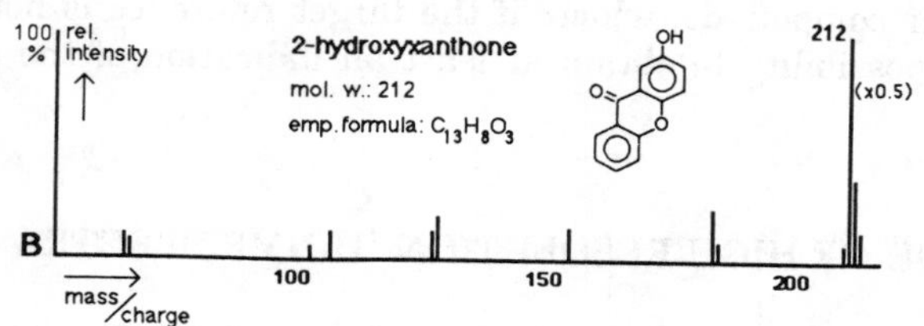

Figure 2

LIBRARY SEARCH OF ^{13}C-NMR SPECTRA

The ^{13}C-NMR Reproducibility-based Retrieval ('C13RR')system uses only chemical shifts as features as these appear to contain sufficient characteristic information. Peak intensities are not useful since they exhibit a very poor reproducibility, and multiplicities cannot be used because the general concept requires that the features be of a continuous nature. The similarity index was developed on the basis of a reproducibility model of chemical shifts using 200 pairs of replicate ^{13}C-NMR spectra.[5] The database contained 6000 spectra originating from the Netherlands Organisation for Applied Scientific Research (TNO).

In a pre-selection, reference spectra having more than one peak more or less than the unknown and references showing chemical shift differences of corresponding peaks of more than 15 ppm are discarded.

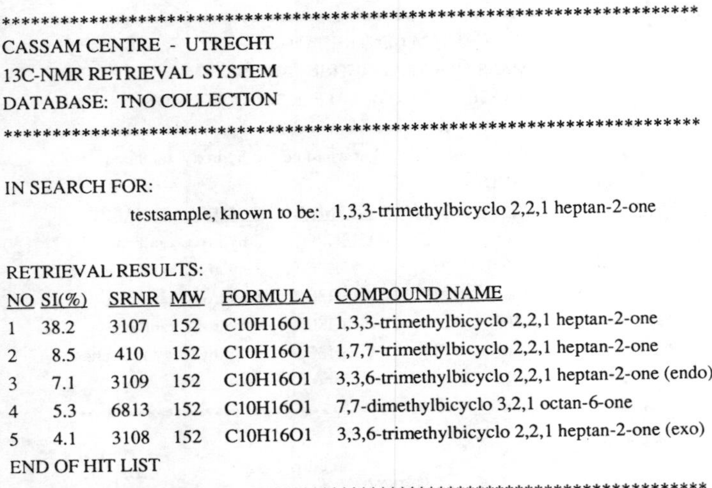

****

CASSAM CENTRE - UTRECHT

13C-NMR RETRIEVAL SYSTEM

DATABASE: TNO COLLECTION

****

IN SEARCH FOR:

testsample, known to be: 1,3,3-trimethylbicyclo 2,2,1 heptan-2-one

RETRIEVAL RESULTS:

NO	SI(%)	SRNR	MW	FORMULA	COMPOUND NAME
1	38.2	3107	152	C10H16O1	1,3,3-trimethylbicyclo 2,2,1 heptan-2-one
2	8.5	410	152	C10H16O1	1,7,7-trimethylbicyclo 2,2,1 heptan-2-one
3	7.1	3109	152	C10H16O1	3,3,6-trimethylbicyclo 2,2,1 heptan-2-one (endo)
4	5.3	6813	152	C10H16O1	7,7-dimethylbicyclo 3,2,1 octan-6-one
5	4.1	3108	152	C10H16O1	3,3,6-trimethylbicyclo 2,2,1 heptan-2-one (exo)

END OF HIT LIST

****

Figure 3

The example of a retrieval result given in Figure 3, as in the case of the MSRR example, shows a 'hit list' headed by the correct reference with an SI value being significantly higher than the SI values of the other candidates. These, however, are all structurally similar compounds, which, if the target reference is not in the database, often provide the possibility to obtain at least an indication of the type of compound being analysed.

LIBRARY SEARCH OF HIGH-RESOLUTION ^{1}H-NMR SPECTRA

A system for computerised retrieval of high resolution (500 MHz) ^{1}H-NMR spectra was developed for identification of carbohydrates derived from glycoproteins. The system also allows interpretation of search results in terms of substructures (probably) being present in the unknown compound.[7] The '^{1}H-NMR Reproducibility-based Retrieval' ('1HRR') system uses chemical shifts (and no intensities or multiplicities, for the same reasons as with the 13CRR system) as features to characterise the spectrum. The (experimental) database consisted of some 60 reference spectra. As the criterion of matching, a combined similarity index computed from two primary similarity indices is used. One of these similarity indices is based on a statistical model describing the reproducibility of chemical shifts and is applied to the matching peaks in spectra of the unknown and the reference compound. The other similarity index ($SI_{mismatch}$) is based on the probability distribution of the percentage of mismatches between two spectra.

A pre-selection is applied, checking the presence or absence of peaks in three pre-specified regions (5.60–4.77; 1.95–1.60; 1.30–1.00 ppm), each indicative of the presence of a particular structural element. Furthermore, reference spectra for which an $SI_{mismatch}$ value less than 2 per cent is calculated are not processed further. In the library of carbohydrate reference ^{1}H-NMR spectra, only the chemical shift values of signals within certain ranges (1.0–3.5 and 4.0–5.6 ppm) are stored as output of the NMR computer, that is, having 4 decimal digits with a precision of 0.0001 ppm.

The result of the search again is a list of references having a similarity to the unknown above a specified threshold value. In the example shown in Figure 4 for a spectrum of a known test compound consisting of two chains attached to an N-acetylgalactosaminol (Ga1NAc-o1) unit, the system has found three target reference spectra in the database ('correct positives', ranked as numbers 1, 2 and 3). The retrieved reference compounds with numbers 4 and 5 have lower SI-values, but are all structurally related to the target reference compound.

Declaration of codes used:

Galβ(1→3) = D-galactose-β(1→3)
Galβ(1→4) = D-galactose-β(1→4)
GalNAc-ol = N-acetyl-galactosaminol

```
*******************************************************************
CASSAM CENTRE - UTRECHT
HR-1H-NMR RETRIEVAL SYSTEM
DATABASE: CARBOHYDRATES (RUU COLLECTION)
*******************************************************************

UNKNOWN :
test sample, known to be:  NeuAcα(2->3)Galβ(1->4)GlcNAcβ(1->6)\
                                                                GalNAc-ol
                                          NeuAcα(2->3)Galβ(1->3)/

SI THRESHOLD: 13%
-------------------------------------------------------------------
RETRIEVAL RESULTS:

NR   IDENTIF   SI (%)   STRUCTURE
-------------------------------------------------------------------
1    AP.001    54       NeuAcα(2->3)Galβ(1->4)GlcNAcβ(1->6)\
                                                          GalNAc-ol
                                    NeuAcα(2->3)Galβ(1->3)/

-------------------------------------------------------------------
2    AP.002    53       NeuAcα(2->3)Galβ(1->4)GlcNAcβ(1->6)\
                                                          GalNAc-ol
                                    NeuAcα(2->3)Galβ(1->3)/

-------------------------------------------------------------------
3    AP.003    51       NeuAcα(2->3)Galβ(1->4)GlcNAcβ(1->6)\
                                                          GalNAc-ol
                                    NeuAcα(2->3)Galβ(1->3)/

-------------------------------------------------------------------
4    AP.004    23                   Galβ(1->4)GlcNAcβ(1->6)\
                                                          GalNAc-ol
                                    NeuAcα(2->3)Galβ(1->3)/

-------------------------------------------------------------------
5    BH.001    14                                         GalNAc-ol
                                    NeuAcα(2->3)Galβ(1->3)/

-------------------------------------------------------------------
END OF HIT LIST

*******************************************************************
```

Figure 4

GlcNAcβ(1→6) = N-acetyl-galactose-β(1→6)
NeuAc-α(2→3) = N-acetyl-neuraminic acid-α(2→3)
NeuAcα(2→6) = N-acetyl-neuraminic acid-α(2→6)

LIBRARY SEARCH OF COMBINED HPLC AND UV DATA

The library search system for identification of pesticides based on diode-array UV spectra combined with HPLC retention data uses an (experimental) reference database consisting of some 200 LC-UV data sets of organophosphorus pesticides, recorded (at the RIVM) from standard solutions under various experimental conditions.[7] To enable comparison of the UV data (normalised) absorbance values at 107 wavelengths were used as the feature quantities. HPLC data are characterised by a feature quantity in the form of a capacity factor. Similarity indices for the UV, HPLC and combined UV + HPLC data were developed based on the reproducibility models of the UV and HPLC data. The search is based on a comparison of the combined data if the same eluent was used for both the unknown and the reference (which is usually recorded and, hence, can be checked), otherwise only the UV data was used. The sample output in Figure 5 shows the identification of an organophosphorus pesticide (a known compound as a test case). The hist list consists of only one identity: the correct one. For 95 per cent of some 100 'unknowns' the target reference (correct positive) was first on the 'hit list', with similarity index values being significantly higher than those found for false positives (if any). When the same eluent was used both for the unknown and the reference optimal results are obtained with combined UV and LC data.[7]

```
****************************************************
CASSAM CENTRE - UTRECHT
HPLC/UV  RETRIEVAL  SYSTEM
DATABASE:  RIVM COLLECTION/PESTICIDES
****************************************************
IN SEARCH FOR:
            testcase, known to be:     carbophenothion /eluent 2
RETRIEVAL RESULTS:
NO   SI(%)   DATA    ELUENT   NAME
 1   98.0    UV+LC     2      CARBOPHENOTHION
END OF HIT LIST
****************************************************
```

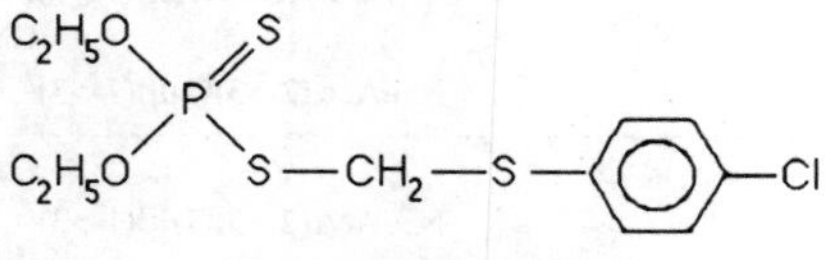

$(C_2H_5O)_2P(=S)-S-CH_2-S-C_6H_4-Cl$

carbophenothion

Figure 5

EXSPEC: AN EXPERT SYSTEM FOR STRUCTURE ANALYSIS OF MOLECULES

The purpose of a chemical expert system is to provide fast, easy, efficient and effective access to chemical information and knowledge in a specific domain of expertise, *via* computer-representation of integrated reference data, theoretical and empirical knowledge (e.g., in the form of 'rules'), system models and reasoning mechanisms. An expert system should not only contain (integrated) relevant 'hard facts' (e.g., numerical correlations and statistics) but also the expertise ('soft knowledge') of experienced specialists in the field. For structure elucidation of organic molecules various spectrometric methods are available. In many cases different methods such as mass, infra-red, NMR and ultra-violet spectroscopy provide complementary structural information. This is one of the basic elements of EXSPEC, an expert system for computer-aided interpretation of combined spectral data.[12–16]

EXSPEC is written in PROLOG and runs on Apple Macintosh (II) computers. PROLOG is is a fifth generation computer language, especially designed for the application of artificial intelligence.[21] Interpretation rules describe the relationship between certain molecular substructures and spectral features. On the basis of a set of reference compounds, containing these substructures of which infrared and mass spectra are available, an automated 'rule generator' was developed.[15] Using Bayes' theorem, the conditional probability p(S | m/e) or p(S | w) of the presence of a structural unit S, given a certain special feature (i.e., for MS a peak representing a molecular or fragment ion at mass m/e, for IR an absorption in wavelength interval w) can be calculated.

The available spectra can be read from disk or typed in on the keyboard. Apart from options *input* and *interpret* there is also an option *explain*, which provides the possibility of requesting the reasoning process used to reach the conclusion. In the example shown in Figure 6, a secondary alcohol has been identified as a functional group with a probability of 99 per cent. Examples of dialogues with EXSPEC options *input*, *interpret* and *explain* are given. Under *explain* the rules specifying the relevant correlations between spectral and structural features and the conditional probabilities are listed.

THE EXSPEC STRUCTURE GENERATOR

In EXSPEC the generation of molecular structures is carried out in three steps.[16] As a first step (not shown here) possible elemental compositions being compatible with a specified molecular weight andpossible elements are calculated. In the second step the user selects a formula (here C_4H_9ClO) and the program determines which molecular fragments (small or larger groups such as methyl, hydroxyl and phenyl) are compatible with the selected formula. The user then makesa choice. In the example (see Figure 7) the fragment '-OH' has been selected, whereas the fragment '-O' has not been chosen. By this the user indicates that he is interested in alcohols and not in ethers having the formula C_4H_9ClO. The program then determines which combinations of fragments are plausible. In the third step the user selects a set of fragments, for which the program then generates all possible unique (acyclic) structures in the form of a 'structure list'. As an illustration of the interpretation of this list the corresponding structures are drawn. Eventually, by processing the other selected fragment sets the isomeric structures can be found (exhaustively and without duplications). As shown above the

```
:start
*** EXSPEC IR/MS INTERPRETATION ***
Option: Input
Do you want to enter new spectra (n) or
update old ones (u) ? n
Do you want any old spectra to be deleted ? n
What spectra are available ? Infrared Mass
Do you want to read an Infrared spectrum from disk ? y
Infrared spectrum number ? 178
Do you want to read a Mass spectrum from disk ? y
Mass spectrum number ? 178
Please, enter the molweight of the compound: 116
```

```
Option: Interpret
Possible molecular formulas are:
  C5H8O3
  C6H12O2
  C7H16O1
Alcohol has probability 0.9
  Primary_alcohol has probability 0.01
  Secondary_alcohol has probability 0.99
  Tertiary_alcohol has probability 0.01
  Phenol has probability 0.01
```

```
Option: Explain
What functionality do you want to have explained ?
Secondary_alcohol
Secondary_alcohol was found to have probability 0.99
because:
  fragment (45 any) --> P(S | m/e) = 0.64
  fragment (44 any) --> P(S | m/e) = 0.66
  absorption (1003 950 moderate) --> P(S | w) = 0.65
  fragment (42 any) --> P(S | m/e) = 0.61
  absorption (2977 2923 very_strong) --> P(S | w) = 0.62
  fragment (19 any) --> P(S | m/e) = 0.80
```

Figure 6

EXSPEC module *interpret* 'translates' spectral information into structural information, e.g., in the form of statements concerning the presence or absence of a secondary alcohol group. Integration of these modules provides the possibility to consider larger fragments, which strongly reduces the number of possible structures. If, for example, it has been deduced that the unknown compound contains a tertiary alcohol group, the total number of structures with formula C_4H_9ClO is reduced to only two.

```
:start
*** EXSPEC STRUCTURE GENERATOR Ver 4 ***

Molecular formula: (C 4 O 1 Cl 1 H 9)                    Molweight is 108.5
List of possible fragments :
1  = "-CH3"        2 = "-CH2-"        3 = "-CH<"
4  = ">C<"         5 = "-OH"          6 = "-O-"        7 = "-Cl"
Please enter numbers : (1 2 3 4 5 7)
Selected fragments :
  ("-CH3" "-CH2-" "-CH<" ">C<" "-OH" "-Cl")
*** I am thinking. . .

*** Suitable combinations of fragments :
1 = ("-Cl" (1) "-OH" (1) ">C<" (1) "-CH2-" (1) "-CH3" (2))
2 = ("-Cl" (1) "-OH" (1) "-CH<" (2) "-CH3" (2))
3 = ("-Cl" (1) "-OH" (1) "-CH<" (1) "-CH2-" (2) "-CH3" (1))
4 = ("-Cl" (1) "-OH" (1) "-CH2-" (4))
Please enter number : (3)
Number of bonds : 10
Number of fragments : 6
Connectivity OK !
*** I am thinking. . .

*** 6 unique structures found
*** UNIQUE SOLUTION # 1 > Structure list :
  ("-CH<" ("-Cl") ("-CH2-" ("-OH")) ("-CH2-" ("-CH3")))
*** UNIQUE SOLUTION # 2 > Structure list:
  ("-CH<" ("-OH") ("-CH2-" ("-Cl")) ("-CH2-" ("-CH3")))
*** UNIQUE SOLUTION # 3 > Structure list :
  ("-CH<" ("-CH3") ("-CH2-" ("-Cl")) ("-CH2-" ("-OH")))
*** UNIQUE SOLUTION # 4 > Structure list :
  ("-CH2-" ("-CH2-" ("-Cl")) ("-CH<" ("-CH3") ("-OH")))
*** UNIQUE SOLUTION # 5 > Structure list :
  ("-CH2-" ("-CH2-" ("-OH")) ("-CH<" ("-CH3") ("-Cl")))
*** UNIQUE SOLUTION # 6 > Structure list :
  ("-CH2-" ("-CH2-" ("-CH3")) ("-CH<" ("-OH") ("-Cl")))
```

```
    Cl               OH               C
    |                |                |
C-C-C-C-OH       C-C-C-C-Cl       Cl-C-C-C-OH

  OH               Cl                   OH
  |                |                    |
C-C-C-C-Cl       C-C-C-C-OH       C-C-C-C-Cl
```

Figure 7

CONCLUSIONS

It has been shown that computer-aided library search and artificial intelligence techniques applied to the comparison and/or interpretation of molecular fingerprints, such as mass, infra-red, NMR and UV spectra, whether or not combined with chromatographic data, are useful aids in the identification and structure elucidation of organic molecules. The utility of library search systems is to a great extent determined by the quality of the relevant data, the main quality factors being the specificity and the reproducibility. A similarity index, formulated in general terms and based on a mathematical-statistical model of the reproducibility of the relevant fingerprint data, has formed the keystone in the development of several library search systems. The results of these systems are quite promising, both for identification and for substructure analysis. Furthermore, it has been demonstrated that artificial intelligence can usefully be applied for the elucidation of molecular structures. PROLOG, a fifth generation computer language, has proved to be a powerful tool in the development of expert system modules for the interpretation of spectral data, explanation of conclusions and generation of molecular structures from identified substructural units.

ACKNOWLEDGMENTS

The authors gratefully acknowledge valuable discussions with Prof J H van der Maas, Dr M J A de Bie, Dr A Sicherer-Roetman and Mr Ch Schalkwijk (Univ. Utrecht), Dr Ch L Citroen and Dr W H Dekker (TNO) and Dr Ch E Goewie and Dr G van de Werken (RIVM).

REFERENCES

1. Hertz, H. S., Hites, R. A., Biemann, K. 'Identification of Mass Spectra by Computer-searching a File of Known Spectra'. *Anal. Chem.* 1971, *43*, 681-691.
2. Pesyna, G. M., Venkataraghavan, R., Dayringer, H. E., McLafferty, F.W. 'Probability-based Matching System Using a Large Collection of Reference Mass Spectra'. *Anal. Chem.* 1976, *52*, 1362–1368.
3. Cleij, P., van't Klooster, H. A., van Houwelingen, J. C. 'Reproducibility as the Basis of a Similarity Index for Continuous Variables in Straightforward Library Search Methods'. *Anal. Chim. Acta* 1984, *150*, 23–36.
4. Cleij, P. 'Reproducibility of Mass Spectral Peak Intensities as the Basis of an Automated Library Search Method for Identification of Organic Compounds'. D.Sc. Dissertation, State University of Utrecht, 1984.
5. Bally, R. W., van Krimpen, D., Cleij, P., van't Klooster, H. A. 'An Automated Library Search System for ^{13}C-NMR Spectra Based on the Reproducibility of Chemical Shifts'*Anal. Chim. Acta* 1984, *157*, 227–243.
6. Bot, D. S. M., Cleij, P., van't Klooster, H. A., van Halbeek, H., Veldink, G. A., Vliegenthart, J. F. G. 'Identification and Substructure Analysis of Oligosaccharide Chains Derived from Glycoproteins by Computer-retrieval of High-resolution 1H-NMR Spectra'. *J. Chemom.* 1988, *2*, 11–27.
7. Boessenkool, H. J., Cleij, P., van't Klooster, H. A., Goewie, C. E., van den Broek, H. H. 'Computer-aided Library Search of Combined LC Retention and Diode-array UV Spectral Data'. *Mikrochim. Acta* 1987, *2*, 75–92.
8. Henneberg, D. In *Advances in Mass Spectrometry*, Vol. 8B; Quale, A. Ed.; Heyden: London, 1980, pp. 1511–1531.

9. McLafferty, F. W., Venkataraghavan, R. 'Computer Techniques for Mass Spectral Identification'. *J. Chromatogr. Sci.* 1979, *17*, 24--29.
10. Bremser, W., Klier, M., Meyer, E. 'Mutual Assignment of Subspectra and Substructures. Structure Elucidation by Carbon-13 NMR Spectroscopy'. *Org. Magn. Reson.* 1975, *7*, 97–105.
11. Schwarzenbach, R. S., Meili, J., Knitzer, H., Clerc, J.T. 'A Computer System for Structural Identification of Organic Compounds from Carbon-13 NMR Data'. *Org. Magn. Reson.* 1976, *8*, 11–16.
12. Luinge, H. J., van't Klooster, H. A. 'Artificial Intelligence Used for the Interpretation of Combined Spectral Data'. *Trends Anal. Chem.* 1985, *4*, 242–243.
13. Kleywegt, G. J., van't Klooster, H. A. 'Chemical Applications of PROLOG. Interpretation of Mass Spectral Peaks'. *Trends Anal. Chem.* 1987, *6*, 55–57.
14. Kleywegt, G. J. 'Artificial Intelligence in Chemistry'. *Lab. Microcomput.* 1987, *6*, 74–81.
15. Luinge, H. J., Kleywegt, G. J., van't Klooster, H. A., van der Maas, J. H. 'Artificial Intelligence Used for the Interpretation of Combined Spectral Data.Part 3. Automated Generation of Interpretation Rules for Infrared Spectral Data'. *J. Chem. Inf. Comput. Sci.* 1987, *27*, 95–99.
16. Kleywegt, G. J., Luinge, H. J. van't Klooster, H. A. 'Artificial Intelligence Used for the Interpretation of Combined Spectral Data. Part 2. PEGASUS: a PROLOG Program for the Generation of Acyclic Molecular Structures'. *Chemom. Intell. Lab. Syst.* 1987, *2*, 291–302.
17. *Applications of Artificial Intelligence for Organic Chemistry; The DENDRAL Project*; Lindsay, R. K., Buchanan, B. G., Feigenbaum, E. A., Lederberg, J. Eds.; McGraw-Hill: New York, 1980.
18. Wipke, W. T., Rogers, D. 'Rapid Subgraph Search Using Parallelism'. *J. Chem. Inf. Comput. Sci.* 1984, *24*, 255–262.
19. Trulson, M. O., Munk, M. E. 'Table-driven Procedure for Infra-red Spectrum Interpretation'. *Anal. Chem.* 1983, *55*, 2137–2142.
20. Tomellini, S. A., Hartwick, R. A., Woodruff, H. B. 'Automatic Tracing and Presentation of Interpretation Rules Used by PAIRS: Program for the Analysis of IR Spectra'. *Appl. Spectrosc.* 1985, *39*, 331–333.
21. *Programming in PROLOG*; Clocksin, W. F., Mellish, C. S., Eds.; Springer Verlag: Berlin, 1984.

SYSTEMATIC DRUG STRUCTURE-ACTIVITY EVALUATION/ CORRELATION

Ernst Meyer and Ehrhard Sens
ZN/D -C 6, BASF Aktiengesellschaft, D-6700 Ludwigshafen, West Germany

ABSTRACT

A computer-assisted approach called KOWIST (KOrrelation von WIrkungen mit STrukturmerkmalen) is presented which gives a better survey of well-known drug classes. A dialogue system was created for the construction of genealogical trees of pharmacophore families in order to reveal promising branches hitherto unexamined. The program also allows the detection of other classes which have escaped attention because by chance no strikingly active representatives had been examined. The method has proved to be useful for the systematic evaluation of a collection of computer-stored structural formulae and their biological test results for the most promising ones of 120,000 actually occurring substructures.

INTRODUCTION

In many chemical and pharmaceutical companies and institutions huge amounts of biological test data have been stored in computer-readable form but evaluated only rather superficially for active drugs and compound classes. These 'data cemeteries' could be examined far more systematically and thoroughly with the assistance of computer methods.

Drug research tries to find new active compound classes as well as more useful representatives of drug classes already noticed. For the latter purpose certain systematical methods such as Quantitative Structure-Activity Relationships (QSAR) and Molecular Design have been created in order to save synthesis and test effort and achieve the optimum of activity faster; nevertheless, an improved survey on a drug family might be very helpful in getting further hints on promising areas of the 'chemical space' around a typical representative (a 'lead') of a drug class. The former goal of drug research, i.e., the detection of new drug classes, was achieved in most cases less systematically and more by chance when a strikingly active compound was tested which did not belong to a well-known drug class. We imagine that many hidden drug classes have escaped attention only because their representatives hitherto tested showed too little activity. Such classes, however, might be revealed by systematic evaluation of available test data.

Our work is based on the usual assumption that many compounds containing a common substructure (a so-called pharmacophore) show similar biological activities and therefore form a drug class. The pharmacophore can be stepwise augmented by substituents which might increase or decrease its activity, and we wanted to get a survey of the effects of the different augmentations. So we had to construct a family tree

W. A. Warr (Ed.)
Chemical Structures

of fragments (of increasing sizes) for each test direction and for each drug class. Each family member had to be characterised by an activity score and perhaps also by a confidence indicator for this score value (Figure1).

The construction of many such family trees is a laborious task, and so we wanted computer assistance for it, e.g. fast access to the fragment structures as well as to the lists of all related smaller or bigger fragments, respectively. And becausethe chemists doing this job would be curious to know more about the background, we also gave them immediate access to the sources, i.e., to the structural formulae of the compounds tested containing the current fragment as well as to the corresponding test results.

From our collection tested we tried to evaluate as many significant substructures as possible which occurred there frequently enough for statistical evaluation. The primary goal was to assess what active pharmacophores existed in our collection. Considering the different families of substructures, however, we found a problem deciding which of the members actually could be called a pharmacophore. We wanted to survey the relations among the various substructures of different sizes, and to arrange them in an order useful for speculations in which direction further syntheses and tests should go.

APPROACH

First we needed a mean score of activity for a set of compounds containing the same substructure, and we defined a so-called 'Höfigkeitsquotient H' as the ratio of 'active' to 'inactive' compounds containing the corresponding substructure, related to the same ratio for all compounds tested. The discrimination of these two sets was carried out after setting a suitable threshold value for the activity score in the test under consideration.

From our collection of compounds tested we generated all occurring (i.e., 2.5 million) different substructures of 5 through 15 non-hydrogen atoms (where rings were not cleaved).Then we chose a specific biological test and removed all those substructures which occurred less than 16 times in the corresponding tested compounds; fewer occurrences did not seem to us to be helpful for statistical evaluation. After counting the 'active' and 'inactive' compounds we computed the 'Höfigkeitsquotient H' for each remaining substructure ('fragment').

In order to assist the construction of a drug family tree (a mockup of which is shown in Figure 1) we wrote a dialogue program which immediately gives access to the fragment structures and to their H-values and occurrences as well as to the structural formulae of the compounds including these fragments and to their test results (Figure 2).

The commands for this dialogueare shown in Table 1. They allow browsing in all directions through fragment hierarchies as well as through the structural formulae containing the current fragment of compounds tested together with their activities. The bench chemist interested in a drug family draws the substructures in levels according to their size and enters bars for the genealogical relations. Then he adds the fragment numbers, the calculated H-values, and the occurrences. By comparing the H-values and the confidence measure he can judge in what direction the best chance for the synthesis and test of active compounds might be given.

Looking at the genealogical substructure tree the chemist can consider what extensions of the pharmacophore *in*creased the mean activity of the corresponding drug compounds and which ones *de*creased it. For getting more detailed information during his reflections he is able (by command F) to call the first/next structural formula containing the current fragment onto the screen, and by command W he also can call

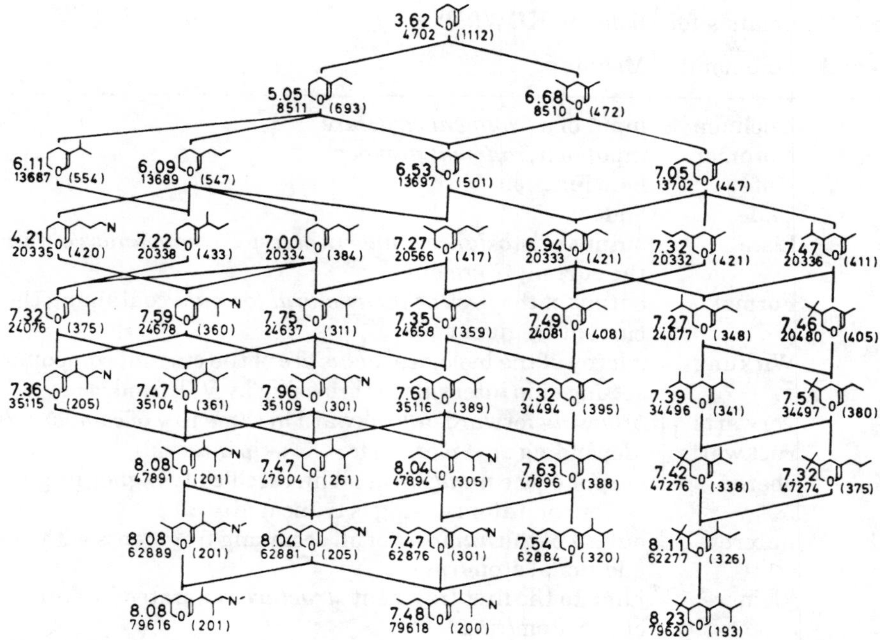

Figure 1. Mockup of a fragment hierarchy arranged by fragment size, such as that composed manually with assistance of the dialogue program. The bars show the genealogical relationships; the big figures represent the Höffigkeitsquotient H, and the smaller ones the fragment numbers and the occurrences.

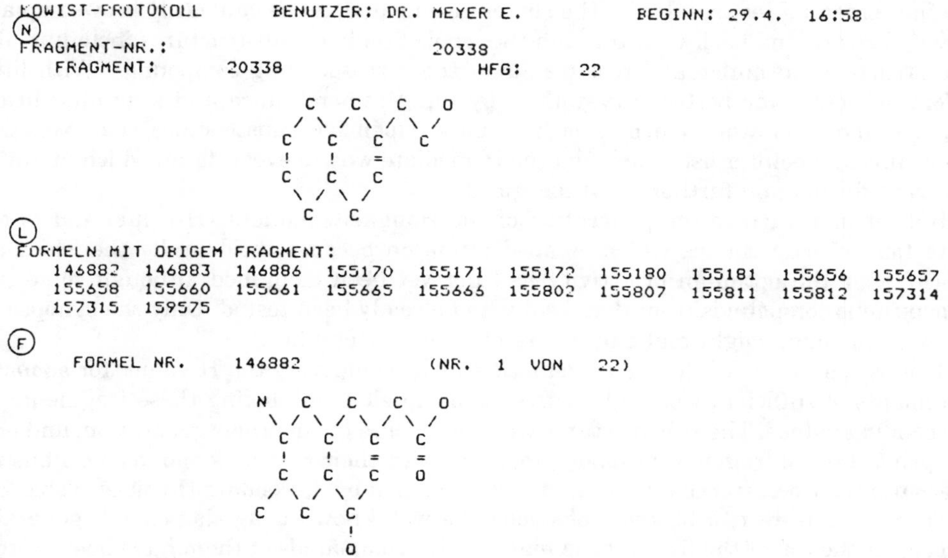

```
KOWIST-PROTOKOLL        BENUTZER: DR. MEYER E.          BEGINN: 29.4.   16:58
(N)
FRAGMENT-NR.:                            20338
   FRAGMENT:        20338                      HFG:      22

                              C    C    C    O
                             / \  / \  / \  /
                            C    C    C    C
                            !    !    =
                            C    C    C
                             \  / \  /
                              C    C
(L)
FORMELN MIT OBIGEM FRAGMENT:
   146882  146883  146886  155170  155171  155172  155180  155181  155656  155657
   155658  155660  155661  155665  155666  155802  155807  155811  155812  157314
   157315  159575
(F)
      FORMEL NR.      146882               (NR.  1  VON   22)

                         N    C    C    C    O
                          \  / \  / \  / \  /
                           C    C    C    C
                           !    !    =    =
                           C    C    C    O
                          / \  / \  /
                         C    C    C
                                   !
                                   O
```

Figure 2. Part of a dialogue procedure (O = commands).

Table 1. Commands for Dialogue-KOWIST

Command	Mnemonic	Meaning
Z	Zeichnen	input of a *fragment structure*
N	Nummer	input of a *fragment number*
H	Hilfe	help function
E	Ende	end
L	Liste	output of *substance numbers of tested compounds* containing the current fragment
F	Formel	output of the first/next *structural formula* containing the current fragment
W	Wirkung	survey of the biological *activities* of the current compound
± n		setting skip intervals for browsing by V, R, and =
V	vorwärts	*browsing* forward or backward in same row of *compounds*
R	rückwärts	(depending on +n or −n for the skip interval)
O	obere	output of the *numbers* of all hierarchically higher fragments (where one atom or ring is truncated)
U	untere	output of the *numbers* of all occurring fragments with one additional atom/ring
K	kleinere	jump to the first fragment *structure* generated by truncation of one atom/ring
G	größre	jump to the first fragment *structure* augmented by one atom or ring
=	gleich	*browsing* to the next *fragment* (of equalsize) in the row (possibly skipping some after setting interval defaults to ±n)

up the screening test results for the current compound. So he can easily browse up and down, forward and backward through the whole family of substructures including also the structural formulae and test results of the corresponding compounds. With little effort he gets a far better survey than by fuzzy remembrance and searching in old reports and laboratory journals or by many expensive substructure searches. Our programs and pointer lists show him the immediate way to every detail which might be relevant for judging further synthesis goals.

But let us return to the generation of the Höfigkeitsquotient (H-value) and to the detection of drug classes which escaped attention because incidentally the 'chemical space' near the optimum of activity had not yet been examined although some less conspicuous compounds from that family had already been tested. Only the synopsis of these compounds might make us aware of such new families.

This synopsis was made in a systematic way by computing the H-values for as many fragments of sufficient size and occurrence as possible and listing these fragments by descending values. The substructures were numbered during their generation, and now we printed these fragment numbers, together with their H-values and (as a confidence measure) their occurrence numbers, in a list arranged by descending H-values (Table 2).

There is, however, a blurring obstacle: the well-known drug classes will generally appear at the top of the list because many active compounds of them have been tested, and because precisely the most active part of their 'chemical space' will have been preferred for examination. This fact obscures somewhat the drug classes not yet

Table 2. Clipping of a fragment list arranged by H-values

H-Value	Occurrence	Fragment Number	H-Value	Occurrence	Fragment Number	...
6.192	52	1094987	6.192	52	1096136	...
6.171	89	29465	6.143	37	6706	...
6.110	31	112228	6.110	31	869166	...
6.088	56	134059	6.088	28	137039	...
6.088	28	604474	6.088	28	800551	...
6.026	22	866254	6.026	22	867516	...
6.026	22	871209	6.005	41	159156	...
5.981	19	17429	5.981	19	62667	...
5.981	19	146866	5.981	19	148677	...
5.967	73	13702	5.953	35	159577	...
5.919	16	13702	5.919	16	90955	...
.	.	.	.	.	.	.
.	.	.	.	.	.	.
.	.	.	.	.	.	.

noticed, and therefore it is advisable to remove these well-known classes from the collection before searching for new leads. Thus the fragment structures corresponding to the upper part of the list are printed, and an expert in that field marks those pharmacophores or substructures which should be used for the exclusion of the corresponding tested compounds before the procedure of selecting the fragments and computing the H-values is repeated.

The list then constructed should contain in its top part the most promising fragments which escaped attention. Nevertheless, in judging these fragments, not only the H-values of the fragments but also their occurrence numbers should be considered.

Family trees can then also be constructed for those promising fragments, and the test results of the corresponding compounds tested should be inspected in more detail, too, before synthesising new compounds. The family trees can also be extended this way beyond fragment sizes of 15 non-hydrogen atoms if desired.

CONCLUSIONS

A survey of the ten different steps of our approach is given in Table 3. The most expensive part of our approach was the generation of all suitable 2.5 million substructures from our collection. This could be done with reasonable effort only after finding a special algorithm and optimising the program. From these substructures of 5 through 15 non-hydrogen atoms we evaluated only those which occurred in more than 15 compounds tested, i.e., about 120,000 fragments. Considering a specific biological test and setting a suitable threshold limit of activity, we subdivided the compounds tested into 'active' and 'inactive' sets and calculated for each fragment the quotient of its occurrences in these two sets as a score for its 'quality'. Then we listed the fragment numbers according to decreasing H-values together with these values and the occurrences (Table 2).

After removing the already known drug classes from our collection in order to make the result more distinct we repeated the calculation of the quotient H and sorted the fragments according to their descending H-values. At the top of this list some promising pharmacophore families showed up which had escaped attention because by chance no strikingly active representatives of these families had been tested.

Table 3. Working Scheme for KOWIST

structural formulae of compounds tested

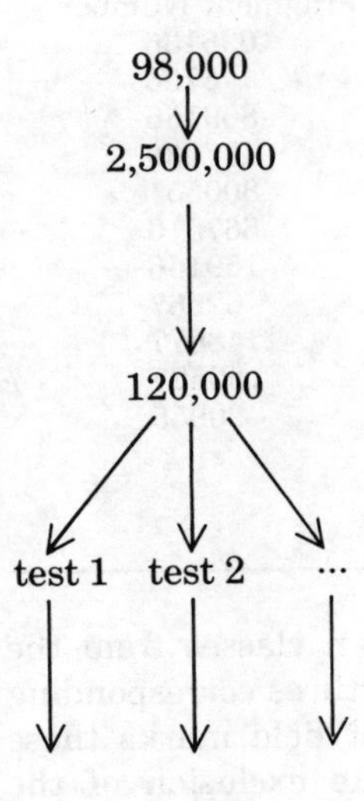

(1) generating the occurring substructures

substructures of 5 . . . 15 non-hydrogen atoms

(2) eliminating those substructures which occurred in less than 16 compounds tested

fragments for correlation

(3) choice of a suitable test, setting a threshold value of activity, and for each interesting test:

(4) computing the H-value for each fragment; listing the fragments in descending order of their H-values

(5) printing (or plotting) the structural formulae containing the fragments with the greatest H-values

(6) scanning these formulae and designating the well-known drug classes; substructure search for these classes and elimination of their members from the collection

(7) repeating steps (4) and (5)

(8) using the dialogue program and looking for drug families not yet noticed; drawing family trees and browsing through the formulae and test results of the compounds containing the new pharmacophores

(9) comparing the H-values and activities and looking for promising holes, i.e., areas not yet examined but possibly bearing maxima of activity

(10) synthesising new compounds from these holes and testing them in order to verify or disprove the guesses.

Using a dialogue program for accessing the fragment structures, their quotients H, and their occurrence numbers in compounds tested , our bench chemists could easily construct genealogical trees of pharmacophore families in order to get a better survey of the influence of different substituents or other structural features on the mean activity of the corresponding test compounds. For still deeper evaluation, the program easily also gives access to the structural formulae of the corresponding compounds as well as to their test results.

ACKNOWLEDGMENTS

We thank Mr. G. Herr for the machine selection of tested material and for the evaluation of the threshold values, as well as Dr. U. Schirmer for designating the well-known pesticide classes.

THE IMPACT OF MICROCOMPUTERS ON CHEMICAL INFORMATION SYSTEMS

William G Town
Hampden Data Services Ltd, Abingdon Road, Clifton Hampden, Oxfordshire, OX14 3EG, U.K.

ABSTRACT

Chemists need to be able to combine chemical structure diagrams with text to produce reports, scientific papers, and safety and marketing material. They also need to be able to conduct searches which combine structural concepts with text and/or data concepts and they require microcomputer systems which would replace the traditional index card file. Ideally, the personal chemical information manager would provide offline query negotiation and uploading and would also allow downloading of both chemical and textual information from public and company files into the desktop system as well as providing access to CD-ROM based products. The extent to which these concepts are realised by existing software is analysed and projections for the future are made.

HISTORICAL BACKGROUND

Developments of hardware affect operating environments, application software, and in turn the way in which we work with computers. The first generation of computers, between 1938 and 1953, used relays and valves and virtually no operating software. The second generation (1953–1963) utilised transistors and batch processing operating systems evolved. The third generation (approximately 1963–1973) employed small and medium scale integration, making possible the development of interactive computing systems. Fourth generation computing (1973–1983) used large, and very large scale integration and the resulting increase in computing power led tomore complex operating environments involving networking and workstations. We are now in the era of fifth generation computers (1983–1993?) with very large scale integration and parallel processors and the current emphasis is on development of intelligent knowledge-based systems.

The effect of this evolution of hardware and operating systems may be observed in the area of chemical structure information systems. A typical software application in a batch processing environment was CROSSBOW in-house, which appeared in about 1967.[1] Software of the interactive era, firstly using teletype-compatible terminals, is typified by the development of the NIH/EPA Chemical Information System, CIS,[2] which through its use of character matrix graphics on teletype terminals made chemical information systems available to the scientist in his laboratory. In the early 1980s interactive, graphics in-house systems such as DARC[3] and MACCS[4] and the publicly available CAS ONLINE[5] and Questel-DARC services began to appear. Now, in the distributed processing environment of the later 1980s, new types of systems are being made possible by the advent of microcomputer-based software. Some examples of this

W. A. Warr (Ed.)
Chemical Structures

new generation are the appearance in 1984 of PSIDOM[6] and in 1985 of CPSS[7] for use in-house and SuperStructure[8] in 1985 as a front-end to the online system CIS.

DISTRIBUTED INFORMATION SYSTEMS

Figure 1 shows a schema for an ideal distributed chemical information system. Several authors in this book refer to the need for standard interfaces. Ultimately, the personal computer will provide the graphics interface not only to personal computer databases but also to company databases running on the company mainframe, and possibly also through the same network to public hosts, so that the chemist using a personal workstation will be able to create queries which can be addressed to local files, company files and public files. Soon, chemical databases will be available on Compact Disk Read Only Memory (CD-ROM) searchable by both substructure and text. These too fit into the scheme of Figure 1. Databases such as infra-red spectra libraries will have structure-searchable components either on the personal computer or on the laboratory instrument and will also be used through the same graphical interface.

NEW TECHNOLOGIES

Progress in the development of processors, data storage and graphics coprocessors is very rapid and itmay be helpful to consider briefly these technologies before considering the applications on which they impact.

Processors

New 32-bit machines are appearing on the market with much higher processor speeds (16 MHz clock) and larger memory address capacities (4 gigabytes). Typical chips are the Intel 80386, Motorola MC68020 and National Semiconductor 32032. It is interesting to compare the computing power of the top end of IBM's new PS/2 range with that of the last generation of mainframe computers. The Model 80 of the PS/2 range is roughly equivalent to an IBM 370/168 which as recently as the late 1970s could have been at the centre of a computer service for a whole research organisation. We are therefore in a situation where unprecedented computer power is becoming available on the bench top, and this, in turn, will have a major impact on the power and size of problems which can be tackled at the workstation.

Peter Willett's paper elsewhere in this book refers to some developments in novel architectures: transputers and clusters of processors such as the Intel iPSC.

Data Storage

Optical data storage, and, in particular, CD-ROM and its variant Compact Disk Interactive (CD-I) are considered in more detail in Patrick Gibbins' chapter. Write-Once-Read-Many (WORM) technology, another form of optical memory, is already a standard option offered by IBM in its PS/2 series. CD-ROM and WORM offer very large amounts of storage at the workstation, up to 200 megabytes in the case of WORM, which is largely an archival type of storage facility, and over 500 megabytes in the case of CD-ROM.

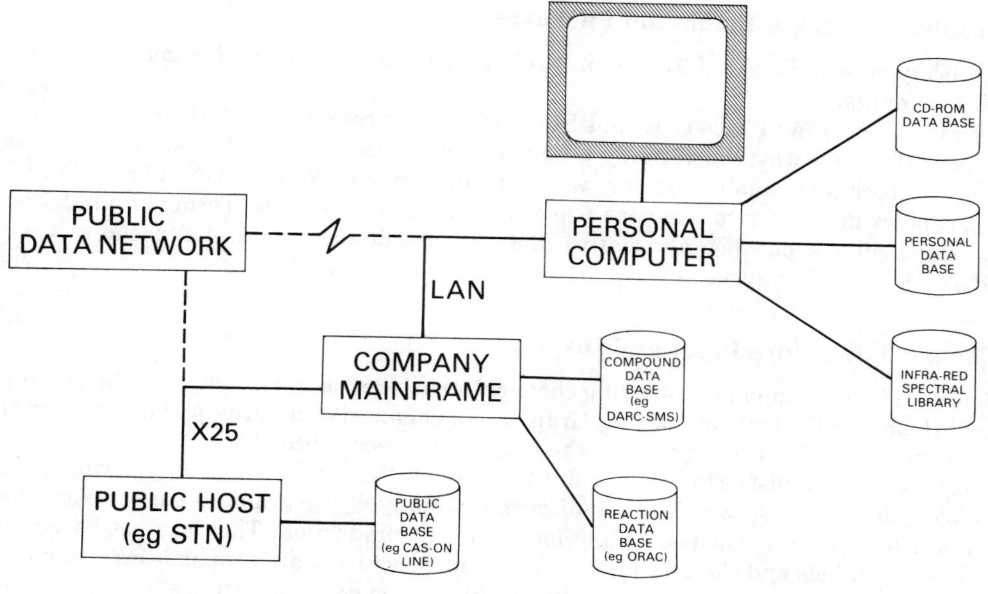

Figure 1

Newer storage technologies, for example, magneto-optical and molecular/biological storage, are also being developed.

Graphics Coprocessors

Several new chips have been announced recently (e.g. the Intel 82786 and the TMS 34010 from Texas Instruments) and these are beginning to appear in graphics boards which offer much higher resolution, more colours and faster interfaces. Somewhat surprisingly, the new IBM series machines do not actually contain graphics coprocessors but some of the boards which are being developed to mimic those machines are being based on these chips. Ultimately we are likely to see IBM using these coprocessor chips in their products too and through these types of facilities, the IBM PCs and PS/2's are going to have the graphics capability of the Macintosh, and at last we are going to see in IBM computers the user-friendly and fast interfaces which the Macintosh has already had for some years.

IMPACT OF MICROCOMPUTERS ON CHEMICAL STRUCTURE INFORMATION SYSTEMS

The five major areas where microcomputers have made an impact relate to graphics terminal emulation (which was the first area where the influence was felt), structure input and display, personal chemical structure databases, offline structure query negotiation and 'downloading' of chemical structures.

Graphics Terminal Emulation Packages

A variety of terminals (e.g. Tektronix 4010, VT640) can be emulated on more than one microcomputer.

Products such as EMU-TEK and PC-PLOT have probably revolutionised chemical information in a way which many people do not realise. Until emulation packages of this sort became available the only way of using services like CAS ONLINE or DARC in graphics mode was to use very expensive graphics terminals. Terminal emulation packages offer a cheaper alternative and as a result have increased usage of CAS ONLINE and other services.

Chemical Structure Input and Display

Software for drawing and displaying chemical structures on microcomputers has been a growth area in the last year or two. Graphics for chemical structures, and integration of the structures with text or data, is the subject of a recent book.[9]

A fundamental distinction can be drawn between those software packages which are image-only and those which are multipurpose. Many of the graphics packages on the market have been developed simply for document production. They are not based on connection tables and therefore cannot have substructure search capabilities. They use a command language or graphic input techniques, or both. Sometimes, editing of the displayed structure is allowed and in some cases file creation capabilities are present. Examples of packages of this type are ChemDraw,[10] WIMP[11] and MPG.[12] However, only the multipurpose packages such as PSIDOM and CPSS construct a connection table as the basis for the diagram. With these packages the structures created on the PC can be used for purposes other than document production or database creation: local substructure searching becomes possible, connection tables can be 'uploaded' into a company's mainframe database, and structures can be 'downloaded' onto the PC.

Personal Chemical Structure Databases

The earliest chemical databases which became available on PC's were not structure-based. In the UK one of the first chemical databases distributed on personal computers was the hazards database CHEMDATA[13] which was developed at the Chemical Emergency Centre at Harwell. This was a database created primarily for the emergency services to help deal with chemical emergencies. Initially, the database was offered as an online service, but obviously the need to keep this service operational 24 hours a day, 7 days a week, put tremendous strains on the operators of the service and as soon as the first microcomputers were introduced, the Chemical Emergency Centre recognised their potential for distributing the data in another form.

Four or five software systems now offer structure searching on PC's and databases are available for use with them. Some of these systems have facilities for data handling or interfaces to database management packages (for example, PSIDOM can be used in conjunction with dBASE III).

Personal databases are at present distributed on floppy disks or cartridges. Very soon the first chemical structure CD-ROM product will appear, from a collaborative venture between KnowledgeSet, Hampden Data Services and Pergamon Orbit Infoline. New distribution media such as softstrip also have potential in this area.

Products on the market to date tend to be a repackaging of existing products, for example ISI's Index Chemicus subsets and the BIOSIS Information Transfer Subsystems (B-I-T-S). They are aimed at the end user, not the information specialist, and are priced at a level that the end-user can afford. They are packaged with retrieval software and the vendors are aiming for high volume production.

Offline Structural Query Negotiation

To understand the advantages which this facility would give to users we need first of all to look at how online graphics systems are currently used. When a chemist draws a query in the CAS ONLINE or DARC systems, he is interacting with a large mainframe computer, often many thousands of miles distant from his desk and to create his query diagram he is sending information commands across this large distance, through the telecommunications network to the central computer, which in turn is interpreting the commands and creating the query image for display at the user's terminal. It is clear that there are tremendous advantages to be gained if the diagram could be created at the PC. This would reduce the load on the central computer, reduce telecommunications traffic and reduce the cost to the user. The stress caused by trying to remember all the commands that are required to be able to create that diagram while the costs are ticking up (the so-called taxi-meter syndrome) would also be reduced. Using such an offline query negotiation the user can think about the query, and ensure that it is right before logging onto the online host and transmitting the query down the line. Before this can happen, it does require that the host defines standard formats for queries and also provides the appropriate hooks in the host's retrieval system. As yet, the only system of this type which is publicly available is offered by the Chemical Information System (CIS) in which the front end is a package called SuperStructure running on the PC and interacting with the CIS host machine. Before very long it is certain we will see offline query building facilities developing in the more commonly used hosts like STN, or Télésystèmes DARC.

Downloading of Chemical Structure Information

The converse of offline query negotiation and uploading is downloading of information from online hosts. At present there are many packages available for assisting users in the interaction with text-based systems. Some, such as SCIMATE, INTERCEPTOR, CSIN and IT, also allow the user to capture text from the public system into the microcomputer. We will ultimately see corresponding developments in the area of chemical structure information. It will then be possible to download structure information which has been retrieved from the public file into the microcomputer so that it can be further processed at the microcomputer level. Molecular Design Limited, MDL, has two products, ChemTalk and ChemHost[7] which allow PCs to interact with MDL's in-house software but as yet there is no facility for downloading connection tables from public structure databases. One facility which is already widely used, is the ability to capture pictures from the online hosts. Although these pictures can be captured, stored, re-displayed and printed, there is no commercial software for taking the vectors which are captured from the online hosts, (i.e. the diagram) and converting them into a connection table. Until connection tables are available on the PC from the online host, local searching of the downloaded chemical structure information will not be available.

IMPACT OF MICROCOMPUTERS ON THE CORPORATE APPROACH TO INFORMATION HANDLING

It is necessary to view chemical information handling in the context of a company's needs for information handling in general. In most companies microcomputers are used primarily for word processing, spreadsheet packages and database management systems. If we were to analyse the word processing requirements of a chemical company we would find that a large proportion of the word processing that is done in the company is not chemical in the sense that it does not require chemical diagrams to be embedded in the document. Thus it is important that whatever solution is proposed to meet the needs of chemical word processing, also fits in with the general philosophy of the company and does not dictate the main features of the word processing system. Chemical word processing should be an add-on to an existing standard word processing package rather than a specialised system that has to be imposed on a whole company, when 90% of the users may not need that facility. Similarly, when we consider database management, it is important that chemical diagrams can be pulled into standard environments, such as dBASE III or SMART, rather than the database management requirements' being forced into chemical packages.

Most large companies are developing office automation systems, as an adjunct to their word processing capability, using local area networks for electronic mail. If they have a need for chemical structure handling within those facilities, in other words if they have chemical documents or reports which they want to send through an electronic mail system, the standard facility should be used and the chemical application software should fit within the standard facility.

CONCLUSION

The evolution of chemical information systems has been influenced strongly by the computer hardware available. Chemists communicate structure-related concepts by means of the conventional two-dimensional diagram. Early computer systems were ill-adapted to meeting their needs. Although the eagerness with which chemists and chemical information system developers have embraced new computer technology can lead to fears that the developments are technology-led, in reality, the chemist is only now within grasp of the ideal chemical information system, namely, one in which he can communicate his requirements and review the answers to those requirements, through one universal graphical user inteface to all sources of chemical information relevant to his needs.

REFERENCES

1. Eakin, D.R. 'The ICI CROSSBOW System'. In *Chemical Information Systems*; Ash, J.E.; Hyde, E., Eds; Ellis Horwood: Chichester, 1974; pp. 227–242.
2. Feldmann, R.J. 'Interactive Graphic Chemical Structure Searching'. In *Computer Represent-*

ation and Manipulation of Chemical Information; Wipke, W.T.; Heller, S.R.; Feldmann, R.J.; Hyde, E., Eds; John Wiley and Sons: New York, 1974; pp. 55–81.

3. Attias, R. 'DARC Substructure Search System: A New Approach to Chemical Information'. *J. Chem. Inf. Comput. Sci.* 1983, **23** (3), 102–108.
4. Dill, J.D.; Hounshell, W.D.; Marson, S.; Peacock, S.; Wipke, W.T. 'Search and Retrieval Using an Automated Molecular Access System', paper presented at 182nd National Meeting of the American Chemical Society, New York, August 1981.
5. Farmer, N.A.; O'Hara, M.P. 'CAS ONLINE: A New Source of Substance Information from Chemical Abstracts Service'. *Database* 1980, *3* (4), 10–25.
6. Meyer, D.E. 'Microcomputer-Based Software for Chemical Structure Management: A Comparison'. In *Graphics for Chemical Structures Integration with Text and Data*; Warr, W.A. Ed.; ACS Symposium Series 341; American Chemical Society: Washington, 1987; pp. 29–36.
7. de Rey, D. 'Applications of Personal Computer Products for Chemical Data Management in the Chemist's Workstation'. In *Graphics for Chemical Structures Integration with Text and Data*; Warr, W.A. Ed.; ACS Symposium Series 341; American Chemical Society: Washington, 1987; pp. 48–61.
8. McDaniel, J.R.; Fein, A.E. 'Design and Development of an Interactive Chemical Structure Editor'. In *Graphics for Chemical Structures Integration with Text and Data*; Warr, W.A. Ed.; ACS Symposium Series 341; American Chemical Society: Washington, 1987; pp. 62–79.
9. *Graphics for Chemical Structures Integration with Text and Data*; Warr, W.A. Ed.; ACS Symposium Series 341; American Chemical Society: Washington, 1987.
10. Johns, T.M. 'Chemical Graphics : Bringing Chemists into the Picture'. In *Graphics for Chemical Structures Integration with Text and Data*; Warr, W.A. Ed.; ACS Symposium Series 341; American Chemical Society: Washington, 1987; pp. 18–28.
11. The Wisconsin Interactive Molecule Processor WIMP, was written by H.W. Whitlock, University of Wisconsin, Madison, Wisconsin, USA and is marketed by the Aldrich Chemical Company, Milwaukee, Wisconsin, USA.
12. Molecular Presentation Graphics, MPG, is marketed by Hawk Scientific Systems, Kinnelon, New Jersey, USA.
13. Cumberland, R.F. 'CHEMDATA – a Microcomputerised Databank for the Rapid Identification of Chemicals and their Hazards in an Emergency', paper presented at Emergency Response '86, Vancouver, Canada, September 1986.

USE OF MICROCOMPUTER SOFTWARE TO ACCESS AND HANDLE CHEMICAL DATA

Daniel E Meyer
Chemical Information Division, Institute for Scientific Information, 3501 Market Street, Philadelphia, PA 19104, U.S.A.

ABSTRACT

This paper focuses on the rapidly developing area of microcomputer-based chemical graphics software and provides an overview of the five major categories of software which handle chemical structure representations. One of these categories, Structure Management Software, is discussed in greater detail. Four newly available packages in this category will be compared for hardware requirements and data handling capabilities.

INTRODUCTION

In the past five years, more than 40 microcomputer-based software packages have been introduced to display and manipulate chemical structure data. These programs can be divided into five categories based on their primary function:

1. Structure Drawing.
2. Emulation/Communication.
3. Structure Management.
4. Modelling.
5. Special Application.

STRUCTURE DRAWING SOFTWARE

These programs allow for the graphic input of structural diagrams and a high-quality output (print/plot) for use in reports and publications. A listing of some of programs in this category would include:

ChemText (Molecular Design Limited).
Spellbinder Scientific (Lexisoft).
PsiGen (Hampden Data Services).
Molmouse (Beilstein Institute/Springer Verlag).
Molstruc (Compudrug).
Wimp (Aldrich Chemical Company).
ChemDraw (Cambridge Scientific Computing).
Egg (Peregrine Falcon Company).
T^3 (TCI Software Research).

W. A. Warr (Ed.)
Chemical Structures

GIOS (B. Thieme Verlag).
Vuwriter (Vuman Computer Systems).
Superstructure (Fein Marquart).
MPG (Hawk Systems).
Chemintosh (Softshell Company).

EMULATION/COMMUNICATION SOFTWARE

Emulation packages allow PC users to access graphic-based chemical systems such as CAS ONLINE and DARC and to conduct structure searches. Most programs in this group can capture downloaded vector files from the host systems and display the information at a later time for viewing and/or printing. A listing of some of the packages in this category would include:

ChemTalk (Molecular Design Limited).
TGraf 05 and TGraf 07 (Grafpoint).
Emutek (FTG Data Systems).
PC-Plot (Microplot).

MODELLING SOFTWARE

PC-based modelling programs are becoming more numerous as the hardware capabilities of microcomputers expand to provide sufficient memory and processing speed. These programs utilise graphic input to display 3-D images and/or calculate physical properties. Some of the programs in this category include:

Xiris Molecular Modelling System (Xiris Corporation).
The Molecular Animator (American Chemical Society).
Molecular Graphics (COMpress).
Modeler (Appillion Software Associates).
Molec (COMpress).
Molecular Animator (COMpress).
CamSeq/M (Weintraub Software Design Associates).

SPECIAL APPLICATION SOFTWARE

Programs which allow graphic input and display but are focused to a specific application, product, or online system are grouped together in this category. To show an example, SANDRA (Beilstein) provides the capability to input a structure and then the program generates a code indicating in which sections of the Beilstein Handbook a researcher can locate information on the compound. Programs in this group include:

SANDRA (Beilstein Institute/Springer Verlag).
TopFrag (Derwent Publications).
TopKat (Health Designs).

STRUCTURE MANAGEMENT SOFTWARE

Structure Management programs allow for the input, storage, and retrieval of chemical diagrams and structure. This third feature, retrieval, differentiates these programs from Structure Drawing packages and the vector files in Emulation Programs. Thus in using these programs, researchers can create small structure databases (1,000–4,000 compounds) and then conduct exact match and substructure searches to retrieve specific groups of compounds. The six programs currently available in this category include:

ChemBase (Molecular Design Limited).
ChemFile/ChemLit (COMpress).
ChemSmart (ISI Press).
PSIDOM (Hampden Data Services).
HTSS (Tree) Dr Peter Bruck/(TDS, Inc.).
TopDog (Health Designs).

Four of these newly available packages, ChemBase, ChemFile, ChemSmart, and PSIDOM, were reviewed and the basic systems features compared. The other two packages, HTSS and TopDog, are not reviewed in this study but are both available for IBM-compatible microcomputers.

SOFTWARE DESCRIPTIONS

ChemBase

ChemBase, from Molecular Design Limited, is a very powerful structure and reaction management program which is part of a series of programs called the Chemist's Personal Software Series (CPSS). As part of this group, ChemBase interfaces with ChemText, a text editor program which allows for graphics to be easily inserted into text, and ChemTalk which is a communication/graphic emulation program to interface with Molecular Design Limited's mainframe software systems. As can be seen in Table 1, ChemBase requires a minimum of 640K memory and a mouse input device. An initial database of structures and reactions is provided with the software as well as a large library of structure templates. ChemBase provides considerable system expertise by checking for proper valence, automatically generating molecular formula and molecular weight, and highlighting atom overlap. These features facilitate the input of structure records or queries. A sample structure display from ChemBase is shown in Figure 1a. (All figures were produced on a dot matrix printer rather than a laser printer to show the output quality from these programs on standard office/laboratory equipment).

ChemFile

ChemFile, from COMPress, is a very easy to understand structure management package which is marketed with a similar package called ChemLit. The main difference between these two packages is that ChemFile allows for text to be entered in

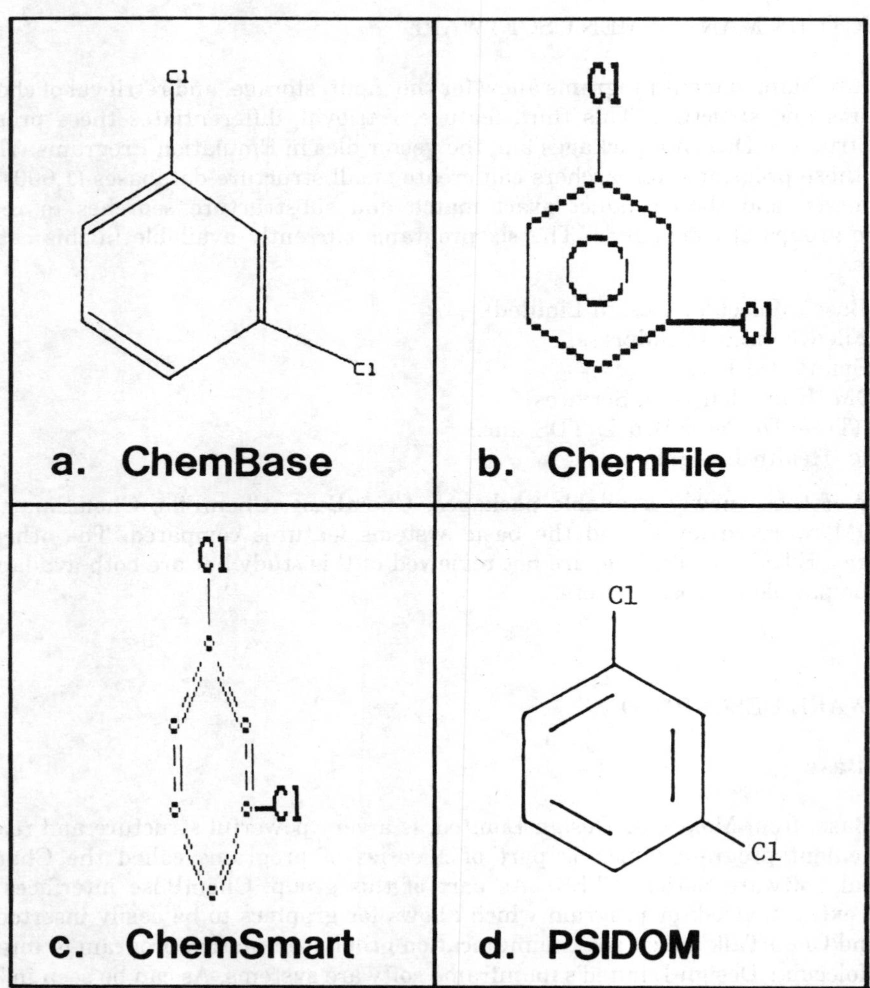

Figure 1

specific fields while ChemLit has a single note field where free-text notes and abstracts can be entered and then retrieved *via* string searching. A feature of ChemFile's field format is the ability to search for ranges of numeric data such as molecular weight, boiling point, etc. However, each data field is restricted in size to less than 20 characters. ChemFile works easily on a two-floppy, 256K system and when searching a file, asks if the user wants to continue the search on another diskette. This feature is a great help in allowing user files to grow beyond the storage limits of a single data diskette. The software comes with two versions of the programs: one for using the cursor keys and the other version for using a mouse. A sample display from ChemFile is shown in Figure 1b.

ChemSmart

ChemSmart, from ISI Software, is an easy-to-use program which is the first product from a series of microcomputer-based products designed to handle chemical data. ChemSmart allows for structure input and has a corresponding note card for each entry to store data such as compound name, molecular formula, molecular weight, physical/chemical properties, etc. Of the text, only the molecular formula, name and registry number fields are searchable, ChemSmart works easily on a two-floppy, 256K system and the single program diskette handles either cursor or mouse controls. A database of 250 structures and a sample reaction areprovided with the software as well as a number of standard templates. ChemSmart provides limited system expertise by checking for proper valence when structures are drawn. A sample structure display is shown in Figure 1c.

PSIDOM

PSIDOM, from Hampden Data Services, is a series of modules for handling chemical structure and text data. The basic module for structure input and display is called PsiGen. A second module, PsiBase, is now available which permits structure/substructure searching against PSIDOM-created files. Both of these programs will be treated as one unit so as to be directly comparable with the other packages reviewed. Other PSIDOM modules are available to integrate text files with structures (PsiText) and generate reports containing structures from databases (PsiRep). PSIDOM requires a hard disk and a minimum of 512K memory. A mouse or cursor controls can be used for input. PSIDOM has a broad range of input techniques ranging from standard free-hand drawing to convenient ring and chain command notations. Feldmann notation input (e.g., 66U6D5 for a steroid skeleton) is also available which allows for the rapid building of large ring systems. The system automatically generates molecular formulae and has limited valence checking. A sample structure display is shown in Figure 1d.

SYSTEM DESCRIPTIONS

Hardware Requirements

Table 1 lists the basic hardware requirements for the four programs. ChemBase and PSIDOM require a hard disk drive and a minimum of 512K memory (640K is recommended for ChemBase). ChemFile and ChemSmart work well with a standard two-floppy machine with 256K memory. Only ChemBase requires a mouse input device but all four programs can utilise the mouse.

Data Input Capabilities

All of the programs permit the standard input features of different atom and bond types. Table 2 compares a few of the input features and shows that all but ChemFile provide the additional stereochemical bond representations. All of the programs allow

Table 1. Hardware requirements

	MACHINE TYPE	MEMORY	DRIVES	MOUSE	GRAPHICS BOARD
CHEMBASE	IBM	640K	TWO FLOPPY DISK DRIVES (360 KB EACH) (HARD DISK DRIVE RECOMMENDED)	REQUIRED	IBM & COMPATIBLE, HERCULES
CHEMFILE	IBM	256K	TWO FLOPPY DISK DRIVES (HARD DISK OPTIONAL)	OPTIONAL	IBM & COMPATIBLE
CHEMSMART	IBM	256K	TWO FLOPPY DISK DRIVES (HARD DISK OPTIONAL)	OPTIONAL	IBM & COMPATIBLE
PSIDOM	IBM	512K	HARD DISK DRIVE	OPTIONAL	IBM & COMPATIBLE, HERCULES

Table 2. Data input capabilities

	TEMPLATES	DRAW MODE	STEREOCHEMISTRY
CHEMBASE	Y	NORMAL CONTINUOUS RUBBER BAND	UP DOWN EITHER
CHEMFILE	NONE PROVIDED (FROM USER STORED STRUCTURES)	CONTINUOUS RUBBER BAND	(DOTTED LINE ONLY)
CHEMSMART	Y	BOND COMPASS	UP DOWN
PSIDOM	Y	CONTINUOUS RING AND CHAIN COMMANDS FELDMANN NOTATION COMMANDS	UP DOWN EITHER

for templates to be used for input and all but ChemFile provide templates with the package.

Manipulating Structures

Table 3 shows that all of the programs offer the standard features necessary to arrange structures on the screen. ChemBase, however, is the only program which offers a very

Table 3. Manipulating structures

	ENLARGE	SHRINK	CLEAN	DELETE	ROTATE	MOVE
CHEMBASE	Y	Y	Y	Y	Y	Y
CHEMFILE	Y	Y	N	Y	N	Y
CHEMSMART	Y	Y	N	Y	Y	Y
PSIDOM	Y	Y	N	Y	Y	Y

powerful 'clean' command which takes a poorly drawn structure and generates a 'clean', chemically correct version.

Searching Capabilities

The searching capabilities, which make these programs different from previously available drawing programs, are shown in Table 4. ChemBase has the most powerful search capability which includes all text fields and the ability to retrieve a structure

Table 4. Searching Capabilities

	EXACT STRUCTURE	SUBSTRUCTURE	TEXT	REACTIONS
CHEMBASE	Y	Y	Y	Y
CHEMFILE	Y	Y	Y	N (AVAILABLE ON CHEMLIT)
CHEMSMART	Y	Y	LIMITED TO MF AND NAME	Y
PSIDOM	Y	Y	LIMITED TO NAME	REACTION SCHEME DISPLAYABLE ONLY

from a reaction scheme specifically as either a reactant or product. ChemFile has a range feature for each specific numeric field but is limited in the amount of data that can be stored (less than 20 characters per field).

Data Provided with Software

Table 5 shows the amount and types of data provided with each product. ChemBase and ChemSmart both provide sufficient databases and templates to give initial search examples and increased utility to the user. PSIDOM provides a large supply of templates and all of the programs allow stored structures to be recalled as templates.

Table 5. Data provided with software

	STRUCTURES	TEMPLATES	REACTIONS
CHEMBASE	100	57	100
CHEMFILE	NONE	NONE	NONE
CHEMSMART	250	9	1
PSIDOM	NONE	80	NONE

SUMMARY

All of these structure management packages offer the capability to build personal databases and then recall structures via registration number, text, and structure. This last feature, structure search capability, makes these packages different and more powerful than the structure drawing or data capturing programs which have been available for the past few years. Some of the existing drawing programs will most likely be modified to incorporate search capability in the near future. PSIDOM was first released as a personal structure file manager with individual record recall and now has been expanded to permit structure and substructure retrieval against its stored records. More entries into the market seem likely in the next 12–18 months.

The advent of the microcomputer-based systems such as those described here provides a new opportunity for individuals to store and manipulate personal and proprietary data. Small data collections of up to a few thousand compounds can be stored to keep track of new research, inventories, and laboratory data. Such information has traditionally been kept in card files or unindexed file folders. When stored as a PC-based structure file, these data can be rapidly searched and specific information easily retrieved and evaluated.

Several papers have appeared recently which describe how personal structure management systems can be used for structure retrieval[1] or utilised to search for structure-activity relationships.[2] The lower cost of the microcomputer environment makes this technology affordable to a larger audience, including individual researchers and students.

In addition, these programs can serve as the central link in a chemical information workstation which allows researchers to input a structure and manipulate the structure to gather information, prepare a camera-ready copy, create a personal file, and a host of other activities. Researchers will be able to upload structures to online and on-site systems and capture retrieved answers, and move relevant compounds into modelling programs, and other applications. Several of the programs included in this paper have this type of utility now and more interfacing of programs can be expected.

REFERENCES

1. Contreras, M.L.; Deliz, M.; Galaz, A.; Rozas, R.; Sepulveda, N. 'A Microcomputer-based system for Chemical Information and Molecular Structure Search'. *J. Chem. Inf. Comput. Sci.* 1986, *26*, 105–108.
2. Meyer, D.; Cohan, P. 'Designing New Compounds with PC Databases'. *Am. Clin. Prod. Rev.* 1986, *5* (12), 16–19.

CD-ROM: A NEW ELECTRONIC PUBLISHING MEDIUM

Patrick Gibbins
Archetype Systems Ltd., Boundary House, 91-93 Charterhouse Street, London EC1M 6LN, U.K.

ABSTRACT

An outlineof the benefits and limitations of publishing on CD-ROM is presented. CD-ROM is positioned differently from other optical storage media. Typical applications in fields other than chemistry are identified. Chemistry and chemical structure information is particularly suited to exploit CD-ROM as a distribution medium. The value of the information is high and the information itself is frequently durable. CD-ROM products require special software considerations which are described.

CD-ROM: A NEW ELECTRONIC PUBLISHING MEDIUM

In the late 1970s the first 12″ video disks were launched by Philips in competition with video tape. These were analogue devices; this paper is not concerned with analogue storage technology. Digital optical disks emerged in the early 1980s. Digital optical storage breaks down into two broad areas. The earliest digital video disk devices based on 12″ format disks, were designed for archiving application; this technology continues to be available and is growing in the range of applications it finds. The compact disk format emerged from quite a different route. It emerged out of the CD audio development which started to take off in 1985 and in the last 2 to 3 years has achieved phenomenal growth; growth far beyond the expectations of the original designers and manufacturers.

It is necessary to draw a basic distinction between CD-ROM technology, which is based on the same technology as CD audio, and WORM technology. WORM stands for Write Once Read Many Times. It is important to understand the positioning of these two technologies. CD-ROM is positioned as a publishing medium. WORM devices are positioned as essentially an alternative to magnetic storage for data processing applications. A WORM device has the ability to write onto a disk as well as read from it. The WORM drive is connected to the computer in just the same way as a hard disk drive or a floppy disk drive. WORM devices are not standardised. If data is written on to a WORM device and sent to a recipient, the recipient will need exactly the same device in order to be able to read the disk back. The data storage medium(the disk) is itself costly, between £50 and £100 per blank disc.

WORM technology is positioned as an alternative to magnetic storage and will find a large number of applications in the data processing environment, in the corporate computer centre. It will succeed primarily where there is an important archiving element. There are some benefits which it offers. The storage medium is cheaper than magnetic disks, it has a longer shelf live, and is less sensitive to environmental change.

W. A. Warr (Ed.)
Chemical Structures

Building societies, banks, and other large administrations are adopting optical storage technology to solve their large storage problems. WORM technology is not suitable for publishing applications because there is no cheap method for rapid replication and because formats arenot standardised.

CD-ROM on the other hand is a replication technology. A master disk is made in virtually the same way that a CD audio master disk is made, and in the same way that a black plastic 12″ audio disk is made. A master is produced and copies are stamped out. CD-ROM has a number of basic characteristics. The technology is standardised. It is physically standardised using the same standards as CD audio. The unit prices of the disks fall with the copy volumes that are made. In other words to make afew disks is expensive; with a run of 1000 disks each copy is very cheap. The cost of the CD-ROM players is also falling with increased volume, starting with the advantage that the manufacturing of CD audio players underlies it. The engineering for a CD-ROM player is 90 per cent the same as the engineering for a CD audio player. The only difference is an additional piece of electronics which sends out a digital stream rather than an analogue stream to an amplifier.

CD-I is a variant on the theme of CD-ROM. The basic difference is that it is a specification for a device rather than just a specification for the physical format of a disk. The CD-I device will include a player plus a processor. It has a UNIX-like processor incorporated; CD-I will not need to plug into an existing computer. Philips believe that CD in a computer-based form has wide applications for the domestic market. They also felt that it would take too long for the total standardisation of CD-ROM to take place, and that the availability of PC's in the domestic consumer market was so low that it was necessary to offer a device which has a computer built in. CD-I has confused potential publishers and indeed potential consumers. Many CD-ROM proponents are unhappy that CD-I was announced at such an early stage. It is also difficult to assess the potential of the domestic market which in the past has shown resistance to 'edutainment' products. CD-ROM is primarily aimed at the professional user of information. What makes CD-ROM attractive for electronic publishing applications? Firstly we have a standardised disk of sufficient capacity that we can put together several related works on a single disk. Publishers can obviously achieve much better retrieval than can be obtained with paper-based products. Retrieval can work across all publications on a disk or selectively. Publication integrity is an important point; by making a master disk and stamping out copies, publishers are following a process similar to hard copy publishing. The process of publishing is about fixing information in time. It is an important distinction to make to people who propose the use of erasable optical disks or WORM devices for electronic publishing. The large capacity of CD-ROM makes it an ideal distribution medium for large collections of data. Between 500 and 600 megabytes can be stored depending upon how the data are formatted. This is equivalent to 200,000 A4 pages, about 30–50,000 typeset pages, 15 magnetic tapes or 1500 floppy disks. The example that is popularly quoted is 2 complete copies of the Encyclopaedia Britannica.

CD-ROM gives users direct access to large databases using a personal computer. Compared with online there is no dial-up access needed and no time-related charge applied. CD-ROM imposes no penalties on the occasional user. The clock is not ticking while the users try to find their way around the system. CD-ROM also opens the possibility of delivering databases of graphics to the PC. Although chemical structures are delivered over dial-up lines they are a very special class of graphics. They require relatively small amounts of data to be transmitted. For engineering drawings or more complex material such as photographs the telephone line bandwidth is insufficient to handle the required volumes of data.

But because CD-ROM is a replication technology it is not suitable for fast changing information. It costs something of the order of £3,000–£4,000 to have a master CD-ROM made. If the product requires updating every month then mastering becomes a very significant fixed cost in the publishing equation. Publishing an updated CD-ROM every week, as Lotus are with their 'One Source' Financial Information Service, creates not only high fixed costs but also the logistical headache of having to get the data to a mastering plant in time for the week's production cycle. It is not viable to consider CD-ROM for distributing information updated more frequently than once a week.

What are the types of applications which appropriately use CD-ROM? Directories, catalogues, parts lists and maintenance procedures manuals are already available and in production. DEC have already distributed a directory of software to their main distribution centres. They are also working towards distributing software on CD-ROM. Readers will be familiar with the use of CD-ROM to distribute popular encyclopaedias. However what we are seeing move fastest in North America is bibliographic databases for the library market and high-value collections of information aimed at selected professional sectors. Several well-designed products have been launched aimed at the finance industry, collections of corporate financial information,and data on historical movements of stocks and shares. These products typically sell at $10-20,000 for a year's subscription. Products are also available which provide marketing information based on geodemographic data to help the professional marketeer in targeting his sales effort. Again, this is high-value information aimed at niche markets.

Corporate publishing and technical documentation are different from commercial publishing. Products are aimed at a captive audience. The justification for the cost and the effort of using a new medium is made on cost savings or improved functionality. The producer does not have to rely on acceptance by the consumer for the product to be a success. Technical documentation will be one of the main areas of application for CD-ROM. There is work going on using CD-ROM to deliver maintenance manuals and interactive parts catalogues. Trials are being carried out by the aerospace industry, and by the automotive industry (Honda, Chrysler, General Motors and Ford of America all have projects underway). CD-ROM is an attractive alternative replacement technology for microform in a world which is increasingly data processing dependent. Instead of finding the appropriate part on a reel of microfilm and then having to transfer data manually to a stock control system, there can be an automatic link from the part number to the stock control system, the accounting system and so on.

During the early development of CD-ROM there has been emphasis on distributing bibliographic databases. There are limitations in the application of CD-ROM to this area because of the frequency of updating needed and costs associated with updating. Those who have worked in the online industry know that the higher value bibliographic databases tend to be very large in size. There will be considerable customer resistance to taking one disk out and trading in another in order to look at a particular section of the database. In time CD-ROM will have the effect of forcing online systems to concentrate on what they are really good at, which is dealing with volatile information and with very large data collections.CD-ROM will ultimately be positioned as an alternative to print and microform rather than directly challenging online. We will also see hybrid products, which allow back files of data to be stored locally on CD-ROM but at the same time provide telecommunications software so that the online host can be dialled up and more up-to-date data accessed.

Software is an essential component of every CD-ROM-based information product. CD-ROM engineering is based on constant linear velocity. This means that the disk is rotating over the laser at the same speed irrespective of location on the disk. The motor has to slow down as data near the centre are accessed and speed up at theouter tracks. This makes access to CD-ROM intrinsically slow by data processing standards. Access times are somewhere between a half a second and one second. Data transfer rates, when the need gets to the required section of the disk, are good. They are better than is achieved with floppy disk and approach what is obtained with a hard disk. When software is written to work to an acceptable level with CD-ROM it has to be written with the engineering limitations in mind. The basic principle has been established that designers have to minimise the number of I/O's needed to access a point on the disc. If more than 2 accesses are needed then the chances are that the retrieval software will give a slow response time. These characteristics have not been considered in the development of magnetic-storage-based retrieval systems and a number of organisations which have tried to convert from one to the other have found it difficult. There are however very successful and effective software packages designed specifically to work with CD-ROM.

A further basic change has taken place as a result of the introduction of CD-ROM into the electronic publishing arena. An online service operation has to assume that the device to which it is delivering its source is a dumb terminal. The operation cannot make any assumptions about what equipment is on the end of the telecommunications line. This means that little can be done to control the appearance of the data on the screen. Although more and more PC's are being used as online terminals the online host keeps sending data in character mode formats. The problem is that nothing can be done to control the screen, make the data more attractive or find new ways to organise and present information. Windows and GEM give similar functionality on the PC. It is very important for electronic publishing that it should catch up with the standards set by the publishers of PC-based software, such as Lotus and Microsoft. Those products are attractive. Most of the information that is delivered from online hosts is not presented attractively. Developers of online services are limited in the repertoire of character set used and the layout of the screen. These limitations disappear with CD-ROM but they also bring a new set of technical challenges to the product designer.

CD-ROM will prove in the future to be an important medium for distributing chemical information. To date relatively few products have been launched, however we can envisage three broad categories within which to group types of chemical information suitable for delivery in this way. Firstly there are bibliographic databases. A number of bibliographic databases in the field of chemistry have already appeared on CD-ROM. The benefits to the consumer are potentiallyto reduce costs, if the consumer is a heavy user of that database, and a more rational charging scheme. The benefit to the information provider is to provide additional sources of income and to reach new markets resistant to using online. The author has already stated some reservations aboutusing CD-ROM in direct competition with online. There remains the problem of the size of some of the more valuable databases. For many online databases value lies in their comprehensiveness, so publishing subsets on CD-ROM in many cases will diminish that value. The database publisher also has a significant problem with pricing. Because a disk can be replicated for £10 or £15, readers should not expect to be able to buy Chemical Abstracts for £15 a copy on a CD-ROM. The publisher of the bibliographic database, receiving much of his revenue from online, will have serious difficulties adjusting to the idea of selling the database on CD-ROM. Database producers typically sell copies of their products on tape at between £10,000 and £50,000. It is not realistic to expect to sell subscriptions at this price for a product which is designed to be used by a single user. Therefore the publisher has a difficult decision in deciding the price of his database on CD-ROM.

The second category of chemical information which is potentially more interesting, is chemical directories on CD-ROM. There is a project which the Beilstein Institute has underway, to investigate the delivery of subsets from their large database on CD-ROM, incorporating a substructure search capability. Pergamon are working on transferring one of their comprehensive chemistry series to CD-ROM. The publishers of the Heilbron Dictionary of Organic Compounds are also thinking about delivery on CD-ROM and Wiley's Kirk-Othmer Directory is already available in the U.S.A.

The third, and potentially the most interesting application of CD-ROM in chemistry, is the delivery of collections of raw data for processing locally on the PC. For example, physical chemistry databanks, spectra databanks, crystallography databanks and so on. There is a tendency, conditioned by the attitudes of print publishers, to think of an electronically published product as being used just for something with no onward processing. This is a very blinkered approach. CD-ROM is an effective alternative to magnetic tape. It is more stable, it is cheaper, it is more standardised and where the objective is to distribute the data to a diffuse group, there is considerable potential for delivering files of processable data in this way. Some organisations already have projects in hand. Some of the databanks that are offered on the CIS system would be good target application. Companies like Molecular Design are well positioned, with their established customer base, to exploit this approach. Software for manipulating the data collections can be delivered with the CD-ROM but a neutral database format would allow the consumer to use his own software for manipulating the data.

In conclusion, CD-ROM is a very important new medium for electronic publishing. It is unlikely to displace online, but rather to complement it. CD-ROM also opens up substantial new opportunities to distribute other types of chemical information in electronic form.

ANALOGY AND INTELLIGENCE IN MODEL BUILDING (AIMB)

W Todd Wipke and Mathew A Hahn
Department of Chemistry, University of California, Santa Cruz, California 95064, U.S.A.

ANALOGY IN MODEL BUILDING

Analogical reasoning concludes that if two things are alike in some respects, they may be alike in others. In model building it follows that if two compounds have similar structure they may have similar geometry. As the difference between the two compounds approaches zero, their geometries approach identity.

Analogy is a non-numerical process. It is the key to the creation of new ideas and the generation of hypotheses and is fundamental to science.

GOALS OF AIMB

The objectives of the AIMB project were to use analogy to solve problems previously thought to be solvable only by intensive numerical calculation.[1]

The goals were to use knowledge and analogy as a chemist does and to build 3-D models rapidly and symbolically, avoiding molecular mechanics. An objective was to take from the field of expert systems the concept that programs that make inferences should justify their results. Molecular mechanics does not do that. AIMB should be extendible to conformational search.

COMPONENTS OF AIMB

AIMB is a type of expert system and so it has a knowledge base. This is a file of known 3-D structures (a library of X-ray crystal data and abstracted knowledge of X-ray data). There are also a 3-D graphical interface; a problem analyser and decomposer; an analogy finder; an analogy (similarity) evaluator; a model constructor; and an explanation and next best model module.

SYSTEM CONFIGURATION

The library of known 3-D structures is on a VAX11/750 which did not even have a floating point accelerator at first. The dynamic graphical system uses a PS300 with hard copy facilities (laser printer and Tektronix plotter).

AIMB PROCEDURE

This is based on the conception of what a chemist does.

A library of models is created. The structural diagram of a target molecule is entered.

W. A. Warr (Ed.)
Chemical Structures

The program perceives the target and analyses the ring structure. It looks to see if the target molecule or an analogy is present in the library. If such a molecule is not present, the program divides the problem into subproblems and solves each. The solved subproblem parts are then assembled and the degree of fit of the subparts is computed. The explanation and supporting data are prepared and the model is displayed.

Various examples were given.

Assembling a model at Depth 1 takes between 18 and 33 seconds. Using Depth 10 only doubles the time. A human being would take much longer.

CONCLUSIONS

AIMB is fast, even without the use of a floating point accelerator. It would be reasonable to use the program to convert thousands of structures into three-dimensional ones. It is fast because it is symbolic.

AIMB is also accurate.

Using prior knowledge is efficient. If the knowledge base gets *bigger* the problem is solved *faster*.[2] This is contrary to most expert systems and database management systems. It is also true that the solution is better if the knowledge base is bigger because the bigger the knowledge base the more likely the program is to find a close analogy and the closer the analogy the better the result.

AIMB is applicable to complex functionality and to libraries of computed data, even though experimental data has been preferred in the current knowledge base.

The quality of the model increases with experience.[2]

The process is easy to understand and commonsense suggests that chemists are more likely to use a system that is easy to comprehend.

AIMB can enumerate conformational space. Many other programs generate non-ring models in 99 per cent of cases but AIMB generates only rings and every one is energy minimised (or almost minimised). This is an exciting approach to the conformation problem.[3]

The program is still a prototype and there are many remaining questions to explore.

ACKNOWLEDGMENTS

The graphics work was supported by the National Science Foundation (grant number DMB-8521802).

This summary of Todd Wipke's paper has been prepared by the Editor from tape recordings and notes made at the Conference.

REFERENCES

1. Wipke, W. T.; Hahn, M. A. 'Analogy and Intelligence in Molecular Model Building'. In *Applications Artificial Intelligence in Chemistry*, Pierce, T.; Hohne, B. American Chemical Society, Symposium Series No. 306, 1986, pp. 136–146.
2. Hahn, M. A.; Wipke, W. T. 'Poster Session: Analogy and Intelligence in Model Building'. In these Proceedings.
3. Wipke, W. T.; Hahn, M. A. 'AIMB: Analogy and Intelligence in Model Building. The Benefits of Experience'. *Tetrahedron Computer Methodology*, 1988, *1*, in press.

POSTER SESSION: ANALOGY AND INTELLIGENCE IN MODEL BUILDING (AIMB)

Mathew A Hahn and W Todd Wipke
Department of Chemistry, University of California, Santa Cruz, California 95064, U.S.A.

ABSTRACT

The AIMB program is a symbolic, non-numerical approach to molecular model building and conformational analysis. The purpose of this project is to develop a rapid, accurate and automatic model builder that can be applied to a wide range of chemical structures. A further aim is to create a program that will facilitate the exploration of the conformational space of molecules and molecular fragments. The program utilises knowledge about the shape and structure of known molecules (taken from crystallographic data) for model construction and conformational search. Results are presented on the ability of AIMB to model a cyclooctane ring compound and its utility in exploring low energy conformations of the compound.

INTRODUCTION

Currently, molecular conformation of organic molecules may be determined through experimental techniques such as X-ray crystallography, manual construction of models, or computer calculation using molecular mechanic mechanics programs like MM2. X-ray analysis is precise but is time-consuming and in general gives only one energy minumum structure with no indication of alternative conformations. Manual construction of models is advantageous in that chemists usually build good models quickly by hand, and they can readily analyse alternative conformations. It is difficult, however, to enter the co-ordinates of a hand-held model into a computer for further computational analysis. Molecular mechanics (MM) programs[1] give good results but require an initial starting geometry before minimisation can begin. The process of generating one or more starting geometries for MM calculations is often a very tedious process.

We saw a need for a modelling program that operated with the positive characteristics of a chemist (flexibility, speed) but gave results comparable to X-ray analysis or MM calculations. This paper reports on a new modelling program called AIMB in which we have approached the problem of model building from a new direction.

Analogy in Problem Solving

Humans are flexible problem solvers. Our problem solving ability is based on:

W. A. Warr (Ed.)
Chemical Structures

1. Our ability to reason within a problem domain. We use logic and make inferences based on the knowledge we have at hand.
2. Our ability to utilise knowledge from different problem domains. This is the concept of reasoning by analogy.
3. The total amount of knowledge we can apply to the problem. We acquire knowledgethrough specific experiences.

In solving an unfamiliar or unexpected problem, we find the most closely related knowledge we have to apply to the problem. Drawing conclusions from related experience is an example of analogical reasoning. In planning a synthesis a chemist will read related journal articles with the hope that reaction schemes and mechanisms which were successful in different systems may be applicable to his own. In conformational analysis, an experienced analyst can predict the geometries of low-energy conformers of a flexible molecule based on his knowledge of such things as preferred ring and chain conformations, steric interactions, and experience with similar molecules of known geometry.

As humans we are adept at using experiences to reason by analogy. We do this very efficiently and seem to know which analogies to consider first and which analogies are most closely related to a problem. Much work has been done recently in the field of Artificial Intelligence to get computers to reason and learn by analogy.[2–5] Our program, AIMB, applies analogy to the domain of molecular modelling.

Goals

The foremost goal of this project is to have the computer use knowledge and apply analogy like a chemist. Chemists have over the years acquired a great deal of knowledge about the shape and structure of molecules. AIMB will have the advantage of access to such knowledge.

A second objective is to build accurate models rapidly given only a description of connectivity (connection table[6] or user drawn structural diagram) and do this symbolically, avoiding any use of molecular mechanics. After all, chemists build very good models manually and mentally without minimisation.

A significant goal is to have our model builder explain and justify its answer. We would like to provide indications of experimental and literature precedent, showing why our program can or cannot build a particular model well. The program should make clear what the causes for uncertainty are as well as probable magnitude. Further, we would like to provide an assessment of the model and some indication of which parts of the model are most strongly supported and which are most tenuous.

A final goal is to apply the same methods to generate alternative reasonable conformers of a compound; in other words, to develop a symbolic conformational exploration capability.

Components

Figure 1 shows the components of AIMB. Reasoning by analogy can be accomplished only when knowledge or experience is available. AIMB, therefore, needs an initial source of knowledge. This knowledge should contain information about the three-

Library of X-Ray Crystal Compounds
Abstracted Knowledge of X-Ray Data
3-D Graphical Interface
Problem Analyser and Decomposer
Analogy Finder
Analogy (Similarity) Evaluator
Model Assembler
Explanation Facility
Conformational Analysis Module

Figure 1

dimensional conformation of molecules i.e., bond lengths and angles, torsional angles, preferred ring and chain conformations etc. We chose to use X-ray crystal structures as source of knowledge because X-ray data contains a vast amount of information about molecular conformation.[7] Information about bond angles, preferred conformations, steric interactions etc. is implicitly contained in X-ray data.

Currently we have a library of about 10,000 compounds taken from the Cambridge Crystallographic Database.[8–11] Each structure in the library represents a specific experimental result and is not an 'averaged structure'.[12] We could, of course, have used a library of computed structures, but no such library existed, and we favoured having an experimental basis for our inferences.

After choosing X-ray data as a source of knowledge, we had to address how to represent this knowledge. Analogy is based on the assumption that two situations share common attributes. In AIMB, analogy is performed by attempting to match subcomponents of the target molecule with subcomponents of the compounds in the crystal file. The program looks for examples of substructures in the crystal file which are similar to substructures of the target.

To make the search for analogies efficient, AIMB abstracts from the crystal library classes of substructural components. The classes of substructural pieces are then organised hierarchically. The broadest classification scheme places components into one of three classes: ring assemblies, rings, and chains. The components of each class are then differentiated on the basis of size. The components are further differentiated on the basis of such things as atom and bond type.

The next component of AIMB is a 3-dimensional graphical interface. Input by the chemist is performed by sketching 2-dimensional structural diagrams. 3-D output is used to observe results and operation of the program.

After input, the problem analyser perceives the target structure to identify rings,[13] chains, aromaticity, and stereochemistry.[14] The analyser first determines if the target or close analogues are contained in the crystal library. If a match is found it selects the most closely related and presents it as a possible geometry for the target. If there are no analogues to the target the decomposer uses a 'divide and conquer' strategy to create substructures of the target, each of which is treated as a new problem. The best subdivisions are those that minimise connections between subsystems. In our case, we select ring assemblies[13] and chains as subdivisions. Ring assemblies are defined as a group of one or more rings fused by at least two mutual atoms.

The analogy finder performs the search for each subproblem looking for analogies to each subproblem in the knowledge base. Because of the hierarchical structure of the knowledge base AIMB will first look for components with identical structure and will relax matching constraints until one or a prespecified number of analogues are found. This process is similar to human analogy: the best analogies are found first and increasingly distant analogies are drawn upon if the problem has not been solved.

The analogy evaluator is responsible for scoring each match to determine the best analogy. In evaluating a match, AIMB bases closeness of analogy on a number of atom and bond attributes. These attributes include atomtype, charge, valence, hybridisation, stereochemistry, and bondtype.

As subcomponents are found the model constructor assembles the solved pieces. How well the subcomponents fit together is computed and the pieces being annealed are displayed on the screen.

The explanation portion of the program prepares supporting data for the model, describes analogies used and matching constraints that were or were not met. One may obtain literature references for the parent compounds of the substructural pieces used. Because our program operates in a 3-dimensional domain much of the program's explanation is not text but 3-D pictures. Examples of such explanation include displaying the parent compound of the substructural piece used, displaying all analogies being considered as candidates for the current subproblem, and displaying actual pieces being fitted together in the annealing process.

An important feature of this program is its ability to provide conformation search capability and to generate next best models. Because the program finds and evaluates more than one analogy for each subcomponent, the user can ask to see what the model would look like if a specified ring assembly, ring or chain is replaced with the closest substructure analogy.

RESULTS

In this section we present the results and describe the operation of the AIMB program applied to a cyclooctane based ring compound.

Figure 2 shows the screen of the AIMB during program execution. The compound in

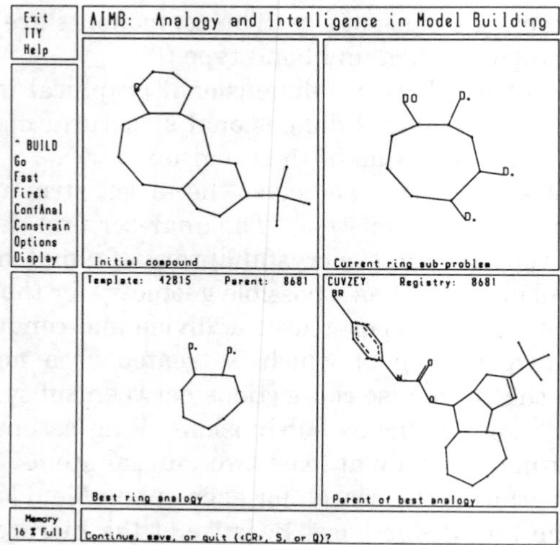

Figure 2

the upper-left box is the initial cyclooctane target structure as it has been drawn in (in 2-D) by the user. After input, the program perceives the structure and determines that it is comprised of three rings and one ring assembly. It then looks for any compound analogues in its library. No matches are found and so the problem decomposer is invoked. The first decomposition divides the problem into two subproblems: the entire 6-8-6 ring assembly and the t-butyl side chain. The program prioritises the subproblems and solves the ring assembly first. No matches for the ring assembly are found and the ring assembly problem is then divided into three subproblems: the cyclooctane ring, the cyclohexane ring and the tetrahydropyran ring. The upper-right box of Figure 2 shows the cyclooctane ring subproblem. The program finds and evaluates ten examples of cyclooctane rings.

The lower-left box of Figure 2 displays the best ring analogy found. The parent compound of this ring is displayed to the right. The analogy is fairly poor, but represents the only example in the library of a saturated eight-membered ring fused to another saturated ring. In a similar manner, AIMB evaluates analogies for the two six-membered rings and the t-butyl group. After each subproblem is solved the model constructor is invoked to anneal the pieces together. The entire process up to this point is performed symbolically and does not involve any geometrical co-ordinates. The co-ordinates for each substructure are now retrieved for the annealing process. Figure 3 shows the three sequential annealing steps necessary to construct the model.

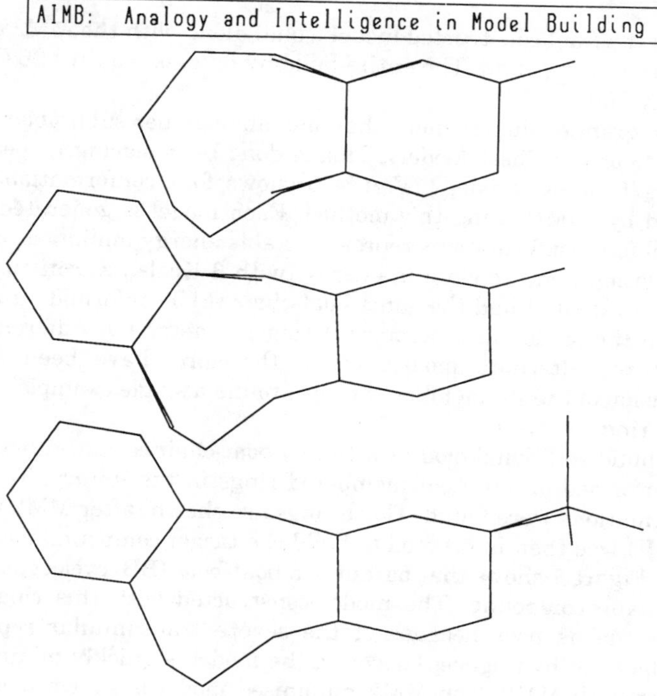

Figure 3

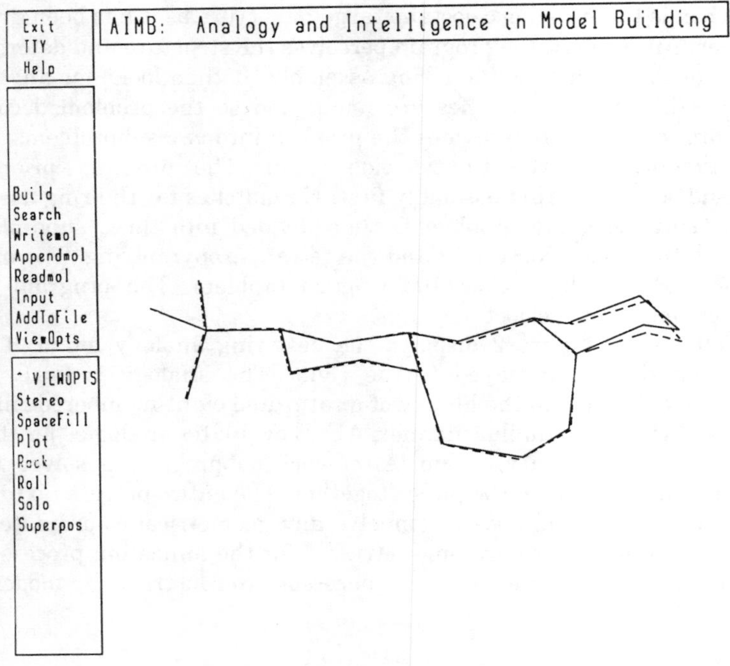

Figure 4

Figure 4 shows the final constructed model (solid) along with the MM2 conformation (dashed) of the target compound. The entire building process requires 26 CPU (35 real) seconds on a VAX 750.

Because the program evaluates more than one analogy per subproblem it is able to generate alternate or next best models. This is done by replacing a specified ring or chain with its next closest analogy. Figure 5 shows four conformationally different models generated by AIMB using this method. Each model is generated in less than three seconds. All four conformations represent stable energy minima as demonstrated by MM2 minimisation and are close in energy (with 3 Kcals). Surprisingly, all of the cyclooctane rings evaluated had the same boat-chair (BC) conformation and, thus, all four models have the same eight-membered ring geometry but different ring fusion positions. The three alternate models could, therefore, have been generated by permuting the fusion of the six-membered rings around a single example of a boat-chair eight-membered ring.

In an effort to build additional models with non-boat-chair central ring conformations a search was performed on 50 eight-membered rings in the library. Three non-boat-chair ring conformations were found. These rings are shown (after MM2 minimisation) in Figure 6. AIMB was then instructed to build the target compound using these non-boat-chair rings. Figure 7 shows that parent of a boat-boat (BB) cyclooctane taken from a bicyclo3.3.1nonane compound. The model constructed with this ring is shown in Figure 8. The model is poor because of the severe transannular repulsion of the overlapping bridgehead hydrogens. However, the model is quickly minimised to a low energy conformer with MM2. The MM2 minimised model is shown below the AIMB model in Figure 8.

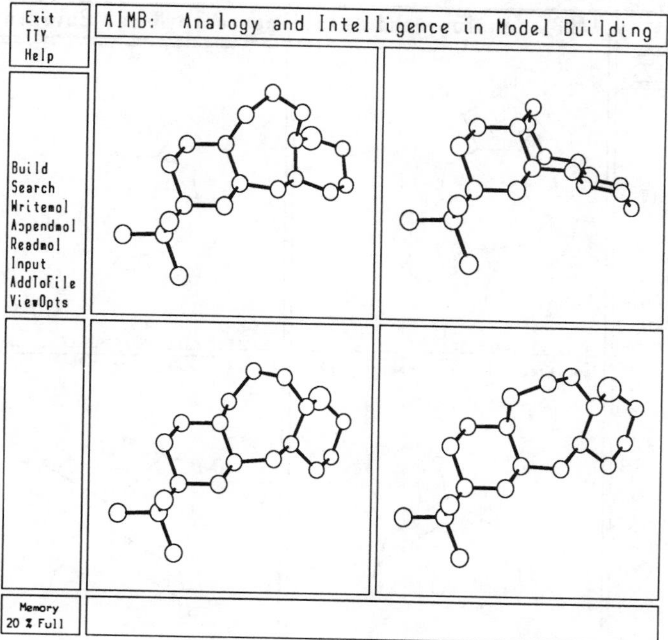

Figure 5

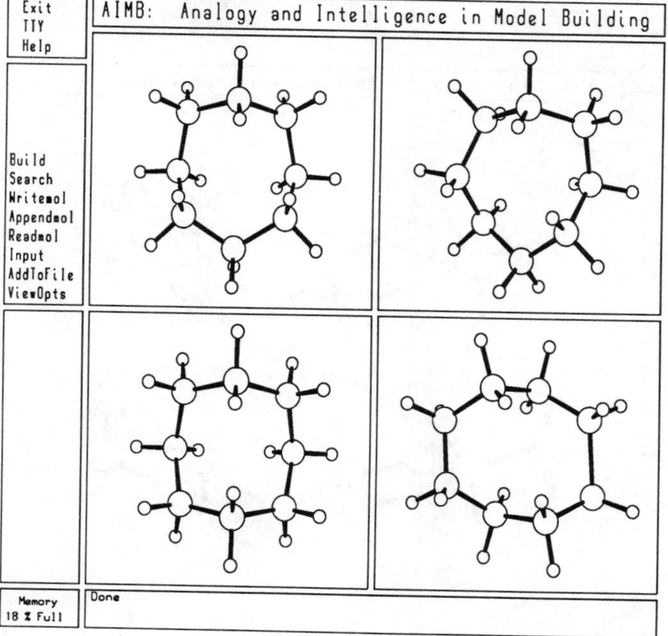

Figure 6

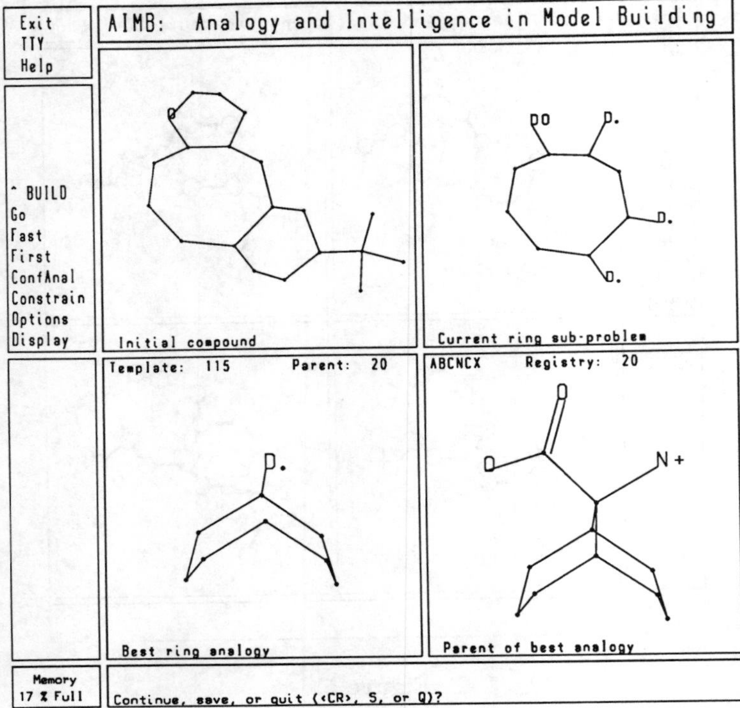

Figure 7

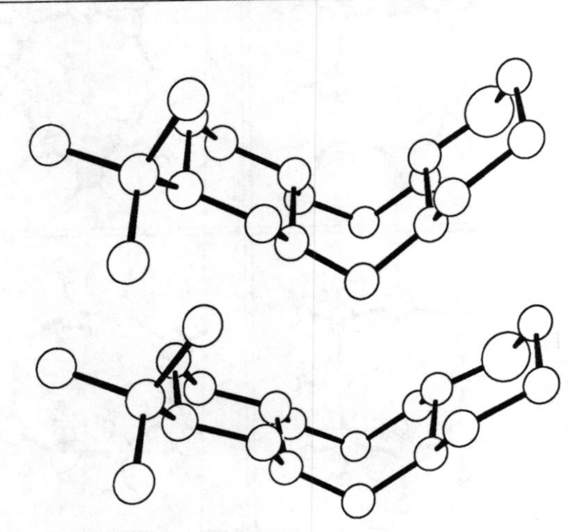

Figure 8

CONCLUSION

The use of analogy in model building leads to a fast, accurate process. Relying on previous experience is an efficient way to build models. While we based our knowledge on crystal data, one could use computed structures separately or in conjunction with crystal data. We currently have a small subset of the Cambridge Crystal database in the library. As the size of the knowledge base increases so does the quality of model construction. Because of the knowledge representational scheme, there is no increase in the time taken for model building with increase in knowledge base size. This is in contrast to most current expert systems. Further, there is actually a decrease in build speed as more knowledge is added to the database. Better analogies are found sooner because less matching constraints need to be relaxed before analogies are found. This is similar to human problem solving, where experience with a problem domain increases both the quality of problem solution as well as solution speed.

The program builds models rapidly, and construction time increases linearly with the structural complexity of the target. This is in contrast to energy minimisation methods which increase exponentially with the number of atoms in the target.[15]

The process that the program uses is simple to understand and is like the process chemists use in constructing models. The program does not use force fields or complicated mathematics. The models AIMB generates are supported by experimental data and justification for or against a model can be easily evaluated. The program can be used to enumerate multiple conformations and in doing so selects local energy minima conformations. Traditional enumeration programs spend much of their time searching over non-energy minima space. Currently, however, the program cannot automatically explore the conformational space.

AIMB is still a prototype and there are still many questions and problems to answer. The program cannot build rings or chains if there are no examples in the crystal library. This has been a problem for very large rings and chains. We are currently working on this problem. Long range non-bonded interactions such as hydrogen bonding, steric and charge interactions are currently not explicitly handled. However, such interactions are implicit in crystal data and we would like to apply the same methods to account for these interactions.

The rules of analogy need to be explored and optimised. We would like to see the effect changing these rules has on model construction. And lastly, we would like to explore the application of the program in areas where conventional programs have not been used, areas involving complex functionality where the forces and interactions may not be well understood (inorganics and organometallics), but where there are many known crystal structures.

REFERENCES

1. *Molecular Mechanics*; Burkert, U., Allinger, N. L., Eds; ACS Monograph *177*; American Chemical Society: Washington, 1982.
2. Lenat, D. 'Software for Intelligent Systems'. *Sci. Am.* 1984, *251*, 204.
3. Lenat, D., Mayank, P., Shepherd, M. 'CYC: Using Common Sense Knowledge to Overcome Brittleness and Knowledge Acquisition Bottlenecks'. *AI Mag.* 1986, *6* (4), 65–85.
4. Winston, P. 'Learning and Reasoning By Analogy'. *Commun. ACM* 1986, *23* (12), 689–703.

5. Carbonell, J., Minton, S. 'Metaphor and Common-Sense Reasoning'. Technical report CMU-CS-83-110, Computer Science Department, Carnegie-Mellon University, 1983.
6. Corey, E. J., Wipke, W. T. 'Computer-Assisted Design of Complex Molecular Synthesis'. *Science* 1969, *166*, 178–192.
7. Wilson, S., Huffman, J. 'Cambridge Data File in Organic Chemistry. Applications to Transition-State Structure, Conformational Analysis, and Structure/Activity Studies'. *J. Org. Chem.* 1980, *45*, 560–566.
8. Allen, F., Brice, M., Cartwright, B., Doubleday, A., Hummelink, T., Hummelink, B., Kennard, O., Motherwell, W., Rodgers, J., Watson, D. 'The Cambridge Crystallographic Data Centre: Computer-based Search, Retrieval, Analysis and Display of Information. *Acta Crystallogr., Sect. B* 1979, *35*, 2331–2339.
9. Kennard, O., Allen, F., Brice, M., Hummelink, T., Motherwell, W., Rodgers, J., Watson, D. 'Computer-based Systems for the Retrieval of Data: Crystallography'. *Pure Appl. Chem.* 1977, *49*, 1807–1816.
10. Murray-Rust, P., Motherwell, W. 'Computer Retrieval and Analysis of Molecular Geometry. I. General Principles and Methods'. *Acta Crystallogr., Sect. B* 1978, *34*, 2518–2526.
11. Murray-Rust, P., Bland, R. 'Computer Retrieval and Analysis of Molecular Geometry. II. Variance and Its Interpretation'. *Acta Crystallogr., Sect. B* 1978, *34*, 2527–2533.
12. Taylor, R., Kennard, O. 'The Estimation of Average Molecular Dimensions for Crystallographic Data'. *Acta Crystallogr., Sect. B*1983, *39*, 517–525.
13. Wipke, W. T., Dyott, T.M. 'Use of Ring Assemblies in a Ring Perception Algorithm'. *J. Chem. Inf. Comput. Sci.* 1975, *15*, 140–147.
14. Wipke, W.T., Dyott, T.M. 'Simulation and Evaluation of Chemical Synthesis. Computer Representation of Stereochemistry'. *J. Am. Chem. Soc.* 1974, *96*, 4825–4834.
15. *A Handbook of Computational Chemistry: A Practical Guide to Chemical Structure and Energy Calculations*; Clark, T. Ed.; John Wiley and Sons: New York, 1985.

A MULTIPROCESSOR ARCHITECTURE FOR SUBSTRUCTURE SEARCH

Peter Jochum and Thomas Worbs
Softron GmbH, Wuermstr. 55, D-8032 Graefelfing, West Germany

ABSTRACT

The most time-consuming step in substructure search is the atom-by-atom match (ABAM). This is the reason why all commercially available search systems try to reduce the size of the search set by a preceding 'pre-screening step'. The high CPU load imposed by ABAM is especially true when stereo information (configuration and conformation) is included in the matching algorithm. The need for an unusually large number of operations compared to the amount of data (i.e., 2 connection tables) on which it is operating is typical for an ABAM algorithm. This is in strong contrast to common numerical problems. This unusual requirement led to the construction of a search machine (TOPFIT) consisting of many independent parallel processing units and an administrating master processor which distributes the search data to the individual CPU's and collects their answers consisting of one bit: does or does not fit.

INTRODUCTION

The multiprocessing system presented in this paper has been the result of a joint development effort of the Beilstein Institute of organic chemistry and Softron. While Beilstein provided its chemical know-how, hand-selected test data, and, last but not least, the funding, Softron defined the system architecture, built the hardware, wrote the operating system, and developed the search algorithms.

The system has been designed for both an optimised vectorial coprocessing unit connected to a mainframe to free the main CPU from the very special burden of substructure search, and a stand alone microprocessor-based substructure search system for selected parts of the Beilstein structure file (Beilstein Selects).

The prototype which has been in test since summer 1986 has been given the name TOPFIT as an acronym for TOPological FITter. It exceeded our expectations: already the prototype has achieved about 20 MIPS (Mega Instructions Per Second) and this can be upgraded to more than 50 MIPS.

MOTIVATION

It is well known that substructure search is not so much a chemical but rather a graph theoretical problem. Compared to 'common' numerical problems, graph theoretical problems show an unpleasant behaviour. Consider for example a finite element problem. Here the number of operations to solve the problem is linearly proportional to

W. A. Warr (Ed.)
Chemical Structures

the number of data points being the element nodes in three-dimensional space. This is in strong contrast to the problem of finding a subgraph in a given graph. It can be proven that the number of operations (comparisons) necessary to solve the problem grows faster than any power of the number n of nodes in the given graph. This unfortunate property can be experienced when performing an atom-by-atom match (ABAM) in any of the commercially available substructure search systems.

But there is also good news. Substructure search consists of a large number of very similar tasks which could easily be performed *simultaneously*. Hence a parallel processing machine sometimes called a vector processor would be ideally suited for the problem. Moreover, the average number of atoms per molecule is presently 25. Beilstein does not describe compounds with more than 250 atoms (there are no polymers). Thus 1 byte is enough to store atom numbers in a connection table and 8-bit operations will do when comparing two nodes. There is no need to 'abuse' a 32-bit-wide mainframe opcode when a cheap microprocessor's opcode renders the same performance. The essential power of a mainframe in performing floating point operations very fast is not needed for this application. Summing up, the problem has the following characteristics:

1. Few data per molecule (100 Bytes).
2. Many operations (and even more when stereo information is included).
3. 8-bit operations.
4. No floating point needed.
5. High parallelism.

These are the optimum conditions for a micro-based vector processor.

ARCHITECTURE

Figure 1 shows the hardware architecture of TOPFIT. There is one master processor which communicates with N slave processors through dual-ported RAM. The dual-ported RAM can be addressed by both the master bus and the slave bus. Each slave is an independent microcomputer with its own processor, memory, bus and operating system. Its computing power is close to 1 MIPS.

The dual port design guarantees a very high communication bandwidth and, therefore, avoids a traffic bottleneck for the data exchange between master and slave. The number N of slaves to be controlled and served by the master processor is variable and can be installed by the operating system. It is limited only by the electrical characteristics of the master bus.

To obtain optimum performance a new operating system (TOPOS) has been developed which could be tuned for the special purpose TOPFIT was designed for. TOPOS has the following main functions:

1. Installation and configuration: number of slaves, protocols, error handler, communication.
2. Downloading the slave operating system and booting the slaves.
3. Host communication: blocks of connection tables must be received from and answers must be transferred back to the host.
4. Task scheduling and administration: connection tables must be distributed to the slaves at the right times. To obtain optimum performance the slaves must kept busy as far as possible. Hence task administration is completely interrupt driven.

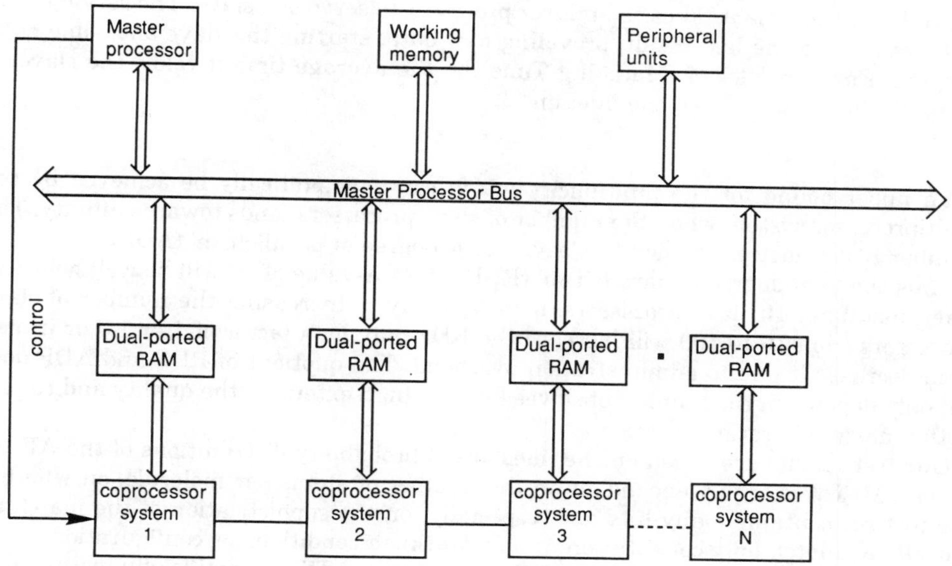

Figure 1. Architecture of TOPFIT

5. Slave supervision and control: proper function of the slaves must be supervised. If for example a slave does not deliver the result within a certain time-frame a time-out condition is generated.
6. Error handling and protocol: communication errors must be recognised and erroneous data retransmitted. Slaves which are not working properly or which have broken down must be detected and deactivated to keep the mean time between failures as low as possible. These are only two examples from a long list of possible errors which are detected and, if possible, corrected during normal operation. Therefore, a slave break down does not cause any fatal situation. There is only one consequence: the performance decreases by a certain percentage, e.g., for a machine equipped with 15 slaves about 7 per cent.

The slave microcomputer consists of a CPU, random access memory, a memory management unit and a master bus interface. It has no ROM (read only memory). The whole slave operating system and the retrieval software are both downloaded from the master. Thus software updates and upgrades are easily implemented even on machines equipped with a large number of slaves. The slave operating system includes the following functions: selfcheck, a simple command interpreter and master communication.

RESULTS

It is well-known that due to administration overhead multiprocessor systems never reach their theoretical performance bound of 'N times' compared to a monoprocessor machine. This can easily be understood by the following consideration.

Let TS be the time it takes the master processor to serve one slave. The serving time includes picking the last result, providing new data, starting the slave and some book keeping. Further, let TM (Matching Time) be the average time it takes the slave to compute the answer. Then the quotient

$$A = TM/TS$$

is an upper bound for the multiplicity which can asymptotically be achieved by the multiprocessor system when the number of slave processors tends towards infinity. The number A is sometimes called the 'asymptotic degree of parallelism' (ADP).

Thus the 'real degree of parallelism' (RDP) at an A-value of 10 will be well below 10 when installing 10 slave processors, it may be say 8. Increasing the number of slave processors from 10 to 100 will increase the RDP only by a factor of 1 or 1.5 or it may even decrease it due to administration overhead. The quotient of RDP and ADP does not only depend on the number of slaves but to a high extent on the quality and tuning of the master operating system.

Our test results are based on the (measured) probability distributions of the ABAM times TM. TM is highly dependent on the number of atoms per molecule, on whether the test set contains many hits and, especially, on the sophistication of the matching model, i.e., match-only constitution or also Markush conditions or configuration.

For generating the plot in Figure 2 the serving time TS was artificially increased to about 2 msec. The matching time TM is used as a parameter. Thus the asymptotic behaviour could be demonstrated with a limited number of slaves. Our benchmarks show that a RDP value of 30 is easily achievable even for small molecules and that a value between 50 and 100 is realistic when not only the constitution but also the configuration is matched.

An RDP value of 30 corresponds to a computing power of nearly 30 MIPS which exceeds that of an IBM 3090-150 mainframe by a factor of at least 3 for a price which is only a small fraction of an IBM mainframe.

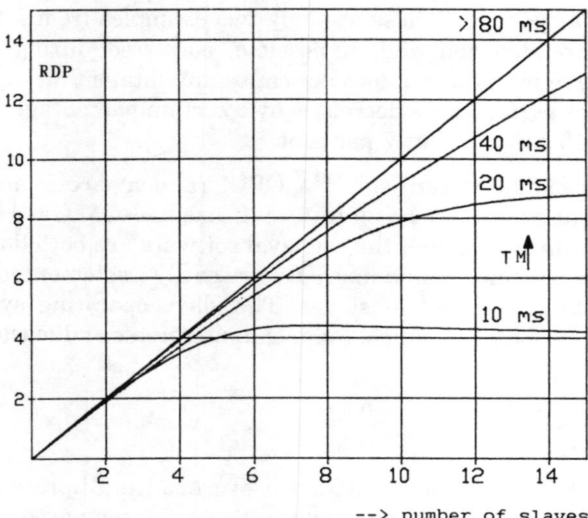

Figure 2. RDP as a function of the number N of slaves and time TM in milliseconds.

THE EVOLUTION OF THE CAS PARALLEL STRUCTURE SEARCHING ARCHITECTURE

Nick Farmer, John Amoss, William Farel, Jerry Fehribach, Christian Zeidner
Chemical Abstracts Service, 2540 Olentangy River Road, Columbus, Ohio 43210, U.S.A.

ABSTRACT

CAS began work on designing a structure search system based on a parallel computer architecture in the late 1970s. The original design concepts shaped the implementation of a 1980 prototype system which permitted screen-only searching for about 500,000 structures. The prototype was scaled up to support structure searching for the full CAS Registry Structure file in 1982 – then about 5 million substances. Since then, the system has grown according to the original design to keep pace with the growth in search usage and the growth in size of the CAS Registry File (currently more than 8 million structures).

Over the past several years work has been underway on the second generation parallel structure search architecture. This architecture is a completely new hardware and software implementation which builds on the already proven architectural principles of the first generation system as well as taking advantage of inverted file searching rather than sequential searching to implement the screen search algorithm.

OVERVIEW

In the late 1970s, Chemical Abstracts Service (CAS) began requirements for and design of a chemical structure searching system for the CAS Registry structure file. (Note that in this paper, structure searching means exact structure searching as well as substructure searching.) The system was to provide a rich complement of chemical structure search capabilities such as Markush queries, use of shortcuts and query modelling from file structures so that a wide variety of structure queries could easily be asked. Query structures were to be input using graphics input and display. In addition to the importance of the functionality the users of the system would see, it was critical that the performance and capacity of the system should be incrementally and readily extendible as the use of the system increased, as the CAS Registry structure search file grew and as new functions requiring computing resources were added.

In response to these requirements, a structure search system was designed and by mid-1980 a pilot system was built and being tested. The pilot system had only a subset of the features that users would require and it searched only about 5 per cent of the entire CAS Registry structure file. However, it did demonstrate the feasibility of using serial sequential screen searching which was a critical design concept of the system.

From this early 'proof of concept' pilot, the capabilities for query framing input were enhanced to support both graphics and text structure input. Query framing was also expanded so that a user could specify variable atoms or bonds or entire substructural

W. A. Warr (Ed.)
Chemical Structures

THE SEARCH QUESTION

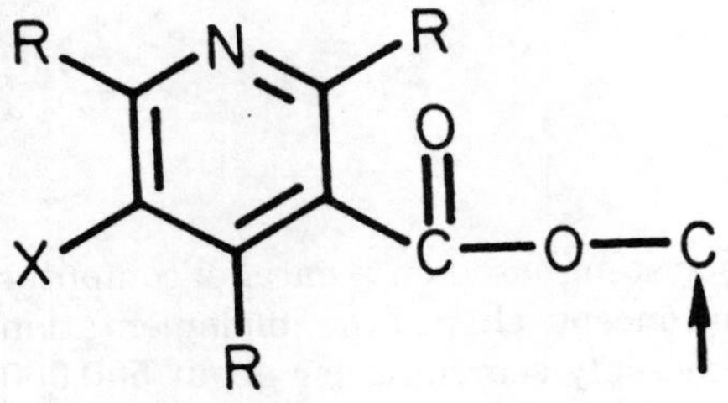

- X = any halogen
- R = anything including halogen
- Isolated ring system
- C (″) may be in a ring or a chain

Figure 1. Typical Substructure Search

units as part of the query (see Figure 1 for such a sample Markush query). Underlying these search capabilities was the CAS structure search machine complex which performed the actual searches against the CAS Registry structure file. The search machine complex had been expanded so that by 1982 eight pairs of minicomputers were performing parallel screen and iterative (atom-by-atom) searching on the entire file of 5 million substances.

The search machine complex, which grows at the rate of two computers for every 700,000 substances, has proven to be a solid corner-stone for structure searching the CAS databases over the past seven years. However, the following factors have convinced us to replace it with a new system:

1. Additional structure-based databases like the CAS Markush file are being built which require new features, including the ability to more closely couple text and structure searching.
2. The search machine hardware and foundation software is becoming obsolete.
3. CAS has developed a state-of-the-art inverted file search capability in support of text searching and numeric file searching that can readily beapplied to structure screen searching.

This next generation structure search system is being called the CAS search engine complex. This system will build on many of the principles used by the search machines and will remedy areas of weakness in the first implementation. The second generation search engines implementation will also be built on more general foundation software, thus making it more portable, more robust and easier to modify. The search engine complex will be able to support the new databases and features that are planned for STN International (the Scientific and Technical Network operated by Chemical Abstracts Service in Columbus, Ohio; Fachinformationszentrum Energie, Physik, Mathematik GmbH in Karlsruhe, Federal Republic of Germany; and the Japan Information Centre of Science and Technology in Tokyo, Japan). Finally, the hardware being used for the search engine implementation will be easier to maintain and upgrade than the now outdated search machine hardware.

This paper summarises the principal features of the search machine architecture. It then describes the search engines and how they address the new needs and environment that have emerged over the past several years.

Retrieve all substances which contain the above substructure where:

SEARCH MACHINES – THE FIRST GENERATION

The CAS search machine architecture was shaped by several factors:

1. The user requirements for effective structure searching.
2. The need to build a system that could grow incrementally using price performance effective hardware to accommodate growing usage and file size.
3. The existence of already proven algorithms, foundation software and a computing environment that could be used as building blocks.

Early in the design process it was decided to try out novel, not yet widely implemented searching techniques including parallel searching, serial screen searching using a fixed size screen set, and overlapped processing which returned results as they became available. These techniques were particularly well-suited to the problem of searching such a large file with consistent response time over a wide variance in query types and query loads.

The structure query input and query compilation functions were separated from the actual searching function. Query compilation converts the structure input representation as specified by a user into the screen and connection table forms needed by the screen and iterative search processes for the actual searching function. Query negotiation and query compilation were performed on the IBM mainframe computer to take advantage of already existing software. Also, these functions are not compute-intensive and so were not an appreciable factor when considering the relative cost of computing resources.

Figure 2 gives an overview of the major functions and where they are performed.

Parallel Searching

In contrast to query compilation, the searching function requires considerable computing and I/O resources. As a result, it naturally lends itself to implementation on a multi-minicomputer search complex.

The file was divided into separate segments so that it could be searched in parallel on multiple minicomputers. One computer and its associated I/O paths were assigned to search every file segment. As the file grew, more minicomputers were added, keeping the search time constant for a fixed number of users. As the number of users grew, individual file segments could be further divided, thus keeping the search time constant for each user and balancing overall system load. This ability to add searching capacity incrementally and the inherently better price-performance ratio of minicomputers made for a more cost-effective implementation when compared to a mainframe implementation.

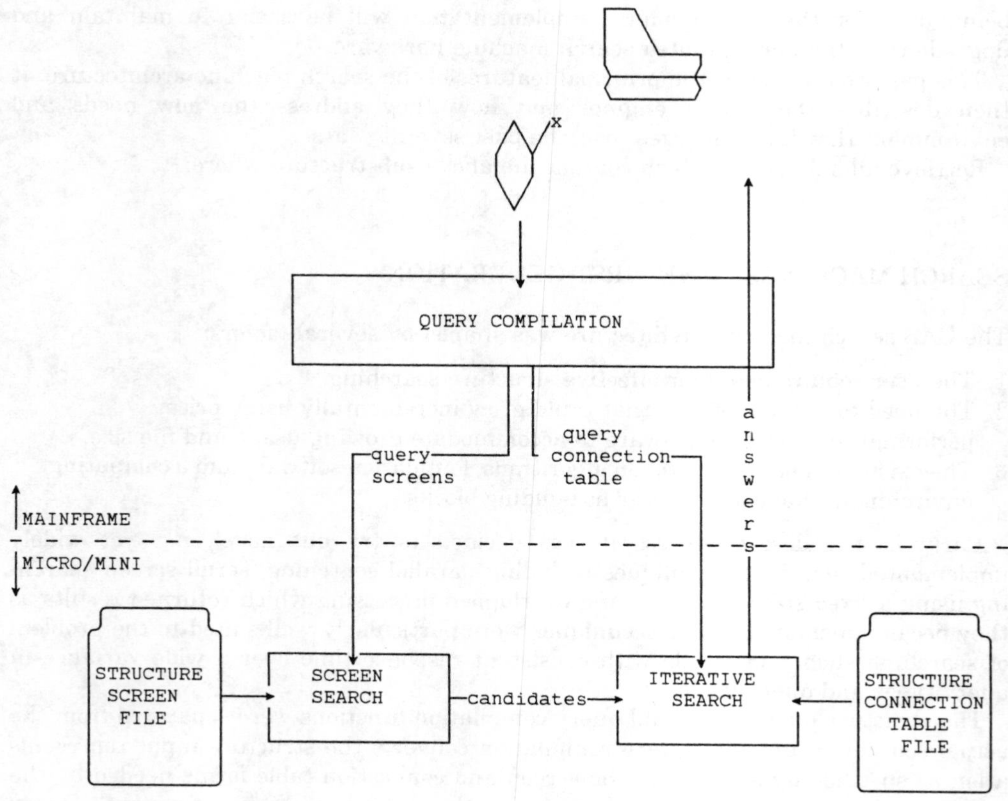

Figure 2.

Separating Screen and Iterative Searching

Performing structure searching that gives only those answers specified in a query with reasonable response time on a large file requires a two-step search process. The first step, called screening, narrows the file to a manageable number of candidate answers. During screening, no valid answers may be missed. Screening is based on identifying certain structural features (like sequences of certain atoms and bonds, the presence of certain ring systems or the total number of atoms of a certain type) in the file structures. In the CAS structure search system the presence or absence of each of about 2000 structural features makes up the structure screens. The query structure is analysed for this set of structural features. In the screening step, any file structure that has the features required by the query is identified as a candidate.

The second step, called iteration, eliminates false drops from the candidate answers. Iteration is done by attempting to match exactly the query structure with the file structure. The query structure is repeatedly overlaid on the file structure until a match is found or until all possible overlays have been attempted without success.

Figure 2 shows the separation of these two steps.

Screen searching and iterative searching are done on separate minicomputers for each Registry structure file segment. The 2000 structure screens for each file substance can be represented in 250 bytes. The representation for the actual structure record, called the structure connection table, also occupies an average of about 250 bytes. Including overheads, the average size per screen and connection table record is about 330 bytes permitting about 700,000 structures to be stored on one pair of 250 MB drives (one 250 MB drive each for the screen and iterative files). In recent years the average structure size has increased somewhat reducing the number of structures that can be stored on a pair of disks.

Sequential Screen Searching

Screen searching is done serially rather than using a more traditional inverted file approach. That is, the entire screen file for a particular file segment is read sequentially from disk into memory. All sets of query screens on the current list of queries to be searched are compared against each file structure screen set. If a query screen set matches a file screen set, then the candidate file structure is passed on to iterative search.

Incoming queries are simply added to the query list and the current location being read from the screen file is noted. After queries have been on the list for an entire read cycle of the disk, they are complete and can be removed.

This sequential screen search had the following advantages:

1. It was straightforward to implement, particularly the file maintenance functions.
2. It permitted screen processing to be overlapped with iterative search processing since candidate answers were available for iterative searching assoon as they were successfully screen matched.
3. The response time for the screen search component was not sensitive to disk contention as usage increased. Consequently it remained consistent until CPU contention to perform the screen matching becomes a factor.

It also had some disadvantages:

1. New screen types (e.g., increasing from 2000 to 3000 screens) could not be added easily without rebuilding the entire file.
2. The minimum response time to perform screen searching was the time it took to pass the entire disk. For the Registry structure file implementation, this was about five and a half minutes.

Overlapped Processing

Once a user query has been compiled, it is sent to the screen and iterative processors for each file segment. The screen processor immediately adds the query to its query list and begins comparing it to the structure screens being read from the file. When a query screen set matches a file candidate, its file key is passed on to the iterative processor, which in turn looks that structure up on its file and begins iteration processing. If the candidate matches during the iteration phase then it is an answer to the query and can be returned to the user for display. In this way, screen searching, iterative searching and answer display are all overlapped in time.

Foundation Software Environment

The software environment which was used for developing the search machines was one that CAS had developed starting with the single user DEC RT-11 operating system. Two previous CAS project efforts had built a multitasking and memory management capability on top of RT-11 which was called MERT-11 (Memory Extended RT-11). Application code developed under MERT-11 is written in the DEC Macro-11 Assembler language.

MERT-11 did not have enough networking and file management capabilities to meet the search machine requirements. The networking functions were developed in a general way by further extending MERT-11. This included process-to-process communication between MERT-11 and the IBM mainframe (running under the MVS operating system), a broadcasting capability, a route-through capability to transmit messages through one computer to another, flow control, virtual circuit management, packeting capabilities and an overall network addressing scheme.

Finally, since there were no general data management and file management facilities the application programs had to define and implement the necessary file structures as well as carrying out any utility functions. These included file print utilities for debugging, file copy utilities for back-up and file load utilities for restoring data.

System Management and Control

Figure 3 provides an overview of the search machine configuration and is an aid in understanding the discussion in the next several sections.

There were significant challenges in designing and implementing control and error recovery mechanisms for the many concurrently operating processes. There were also challenges in providing adequate system and database management flexibilities as well as software controls and installation tools with only the minimal underlying foundation software support.

The control for an individual query is handled by an IBM mainframe process. This process broadcasts the query to all iterative processors and then waits until all of the iterative processors acknowledge that the query has been started. To provide that acknowledgement, the iterative process has to propagate the query on to the screen processor and receive an acknowledgment from it.

Overall control processing was simplified by assuming that the iterative processors had to be paired 1 to n with screen processors. This means that an iterative processor controls all of its screen processors and screen processors only need to know about a single iterative processor. This limits configuration flexibility somewhat in that screen CPU capacity and file space have to be multiples of iterative CPU capacity and file space.

The database management facilities are limited because a general database management system was not available and consequently files cannot be readily moved around and re-configured across disks. This makes dividing a file segment in half (e.g., to increase user capacity) possible but cumbersome.

System management functions are handled from two physical consoles that can communicate with all of the minicomputers. System commands can be sent to a single computer or multicast to multiple computers. Diagnostic output from each of the individual computers is also collected at these consoles.

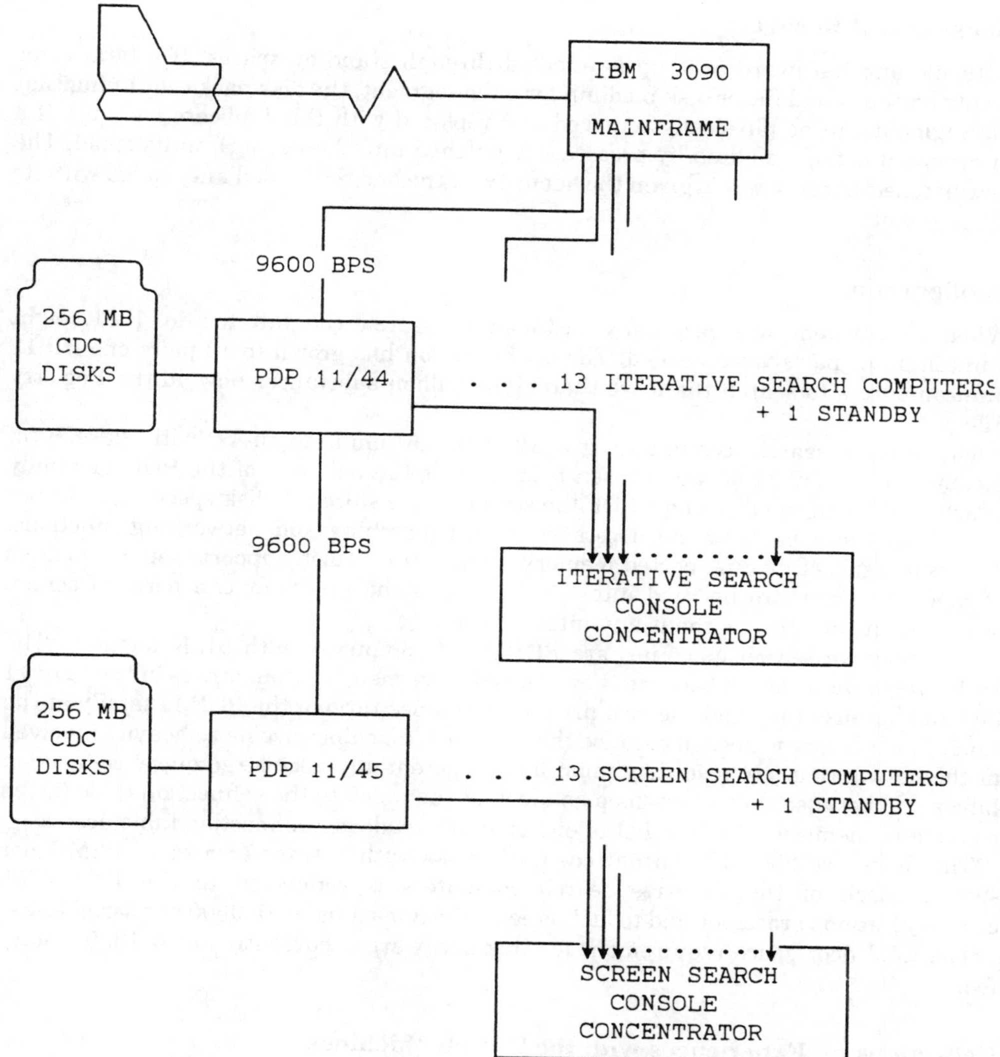

Figure 3. Search Machine Hardware Configuration

Software control is managed by keeping three versions of the software on each machine. One version is the 'current version,' one version is the 'previous version' and one version is the 'new version' that is being installed. When new software is installed, by downloading it from the IBM mainframe, each machine is manually booted up under the new version. If an abort occurs, the machines reboot themselves under the current version, not the new version. In this way, new software can be phased in without significantly affecting operation. When the new software has 'aged' sufficiently, the machines are notified to mark it as the current software and the cycle starts allover again.

Back-up and Recovery

Both file and hardware back-up is handled through stand-by spares. If a fatal error occurs on the actual file or disk medium for a file segment, the disk pack containing that file segment can be physically removed and replaced with the duplicate back-up. If a minicomputer fails, a stand-by processor is patched into the network in its stead. The new patched-in processor is given the 'identity' of the one that failed and loaded with its file segment.

Configuration

When the system was originally installed to search the full file in 1981, eight minicomputer pairs were needed. The configuration has grown to 13 pairs of PDP 11 minicomputers to search the more than eight million substances now on the Registry File.

The screen search computers are PDP 11/45 minicomputers with 256KB of memory. The PDP 11/45 was chosen because it is the only one of the PDP 11 family which could read the disks on which the screens were stored at disk speed (1.2 MB per second) and still perform the necessary screen matching and networking functions. This is a result of its dual-ported memory architecture which supports separate banks of memory. Data can be read into one bank and the processor can perform screen matching in the other without any interference.

The iterative search machines are PDP 11/44 computers with 512K memory. The PDP 11/44s were chosen because they allowed more memory than others in the PDP 11 line and because they had the best price-performance ratio in the PDP 11 family at the time. The memory is needed because the iteractive machines are more heavily involved in the networking, because they must handle potentially very large query connection tables and because they must keep as much of the index to the connection table file as possible in memory to reduce I/O to one or fewer reads per connection table access.

The disks are 256 MB (formatted) CDC disks with a transfer rate of 1.2MB per second. Each of the iterative search computers is connected to the IBM 3705 communications processor and to 'its' screen search machine with 9600 bps serial links.

The total floor space requirement for this hardware is currently about 1350 square feet.

Reflections on Experiences with the Search Machines

Over the lifetime of the system the response time has remained at a constant five to six minutes per search although usage has continued to increase steadily. The cost of the computing equipment has generally dropped, especially as it has become available on the used equipment market. The architecture has proven to be readily operable. Availability and reliability have been generally comparable to the CAS IBM mainframe environment – currently availability is in the area of 99 per cent for the search machines. Some tuning and operational enhancements were implemented in the early years, but such activity has subsided considerably in recent years.

SEARCH ENGINES – THE SECOND GENERATION

A number of things have changed since the search machine was designed and implemented in the late 1970s and early 1980s.

1. Several additional requirements must be addressed in this implementation including the capability to perform effective combined structural and textual searches and the ability to support Markush file searches ofgeneric substances.
2. Since the implementation of the search machines, CAS has developed software for searching and maintaining large inverted text files. Inverted file searching was first implemented on the IBM mainframe computer to support Messenger text searching. Recently a parallel processing version of this software has been designed and will be implemented during 1988 as part of the US Patent Office Automation project to support searching the US Patent File. The parallel processing version of the software has been written in the C language for the UNIX operating system making it operable on avariety of different machines.
3. The networking software capabilities have also been upgraded considerably to support complete process-to-process communication among IBM mainframe processes and UNIX processes using a local area network. Use of a high-speed local area network permits different design options.
4. The hardware has continued to become more price-performance effective as well as physically smaller. Non-removable disks have taken the place of removable media disks leading to different back-up and recovery considerations.
5. Finally, a number of things have been learned in operating the searchmachines. Good system management capabilities are important in providing operational flexibility and in maintaining high reliability and availability. Performance management features were not essential on the search machines, but it would have been desirable to have them under some circumstances. Configuration flexibility is important as system loads change, as new data is added, as improved response times are required or as new features are needed.

As a result of all of these factors, the second generation architecture, which is being called the 'search engine complex', is considerably different from the first generation 'search machine'.

Configuration

The CAS search engines are being implemented using 5 UNISYS-Sperry 5000/90 computers. The 5000/90 computers can be configured with 1–4 Motorola 68020 CPU's. The disk drives that will be used are 550 MB Fujitsu drives. Each computer can be expanded to a total of about 8 GB of disk storage. The search engine complex computers will be connected *via* an Ethernet network. They will be connected to the Messenger software at the IBM mainframe with an IBM channel to Ethernet interface. An overview of the hardware configuration is shown in Figure 4. The total floor space required for the search engine implementation is about 500 square feet.

Parallel Searching

The concept of parallel searching will be retained with the search engines. For applications with large files and heavy use, this is a critical factor. It permits a close match of computing power with usage and response time requirements thus optimising cost/performance. Because of the heavy use the CAS Registry File receives and the response times users demand, it is cost-effective to separate the file into multiple segments with one computer searching each segment.

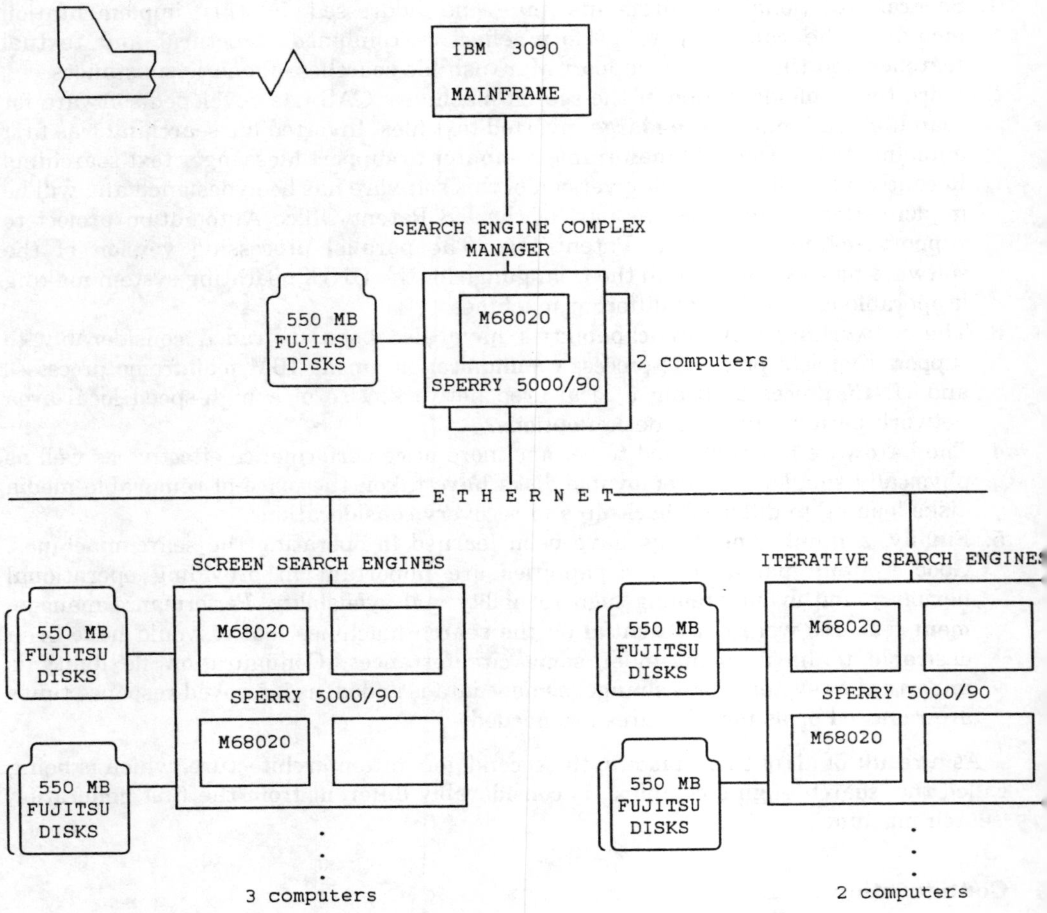

Figure 4. Search Engine Complex Hardware Configuration

In 1986 a cost study was conducted to compare the cost of searching the CAS Registry File on a parallel bank of Motorola 68020-based minicomputers *versus* the cost of searching the file on an IBM mainframe. The cumulative costs over a five year period were noticeably higher on the mainframe if response times comparable to the five minute search machine response times were desired. If response times were reduced from about five minutes to about one minute for the average search, then cumulative mainframe costs were more than twice as great as the minicomputer costs.

For smaller files, the complexity and overheads for distributed parallel searching increase in proportion to the computing power required for the actual search processing and thus the attractiveness of this approach diminishes.

Separating Screen and Iterative Searching

The logical separation of screen and iterative searching remains in the search engines. However, with the multitasking capabilities of the UNIX operating system and the increased processing power (including multiprocessor configurations) and memory capacity of current UNIX computers, it is possible to run multiple instances of either of the search processes or even both search processes in the same computer. This increases the configuration flexibility and at the same time improves the operability of the system because of the reduced number of operating system images.

Inverted Screen Searching

The search engine architecture will substitute inverted file searching for sequential screen searching. This will permit the structure screen set size to be increased because new screens simply become new terms in the inverted dictionary. It will also permit other numeric or textual information associated with a structure to be added to the inverted screen file and processed as part of the inverted screen searching process.

The inverted file screen searching can considerably improve upon the current five to six minute response time, given sufficient search engine processing capacity. Most searches are expected to be completed in about one to two minutes. However, some searches will take considerably longer and under some circumstances could take longer than with the search machines. The search engine search times will be more sensitive to user load as well. This is analogous to current inverted file text search systems but is further exaggerated because of the additional iterative search processing that must be done – a step not necessary when text searching.

Iterative Searching

The iterative search algorithm is being completely rewritten for the search engine implementation to support the use of generic search queries against a file of Markush structure representations. That is, both the queries and the structures on a file may have nodes represented as variable substructures or certain generic nodes (e.g., heterocycle).

Overlapped Processing

A consequence of the inverted file screen search is that one aspect of overlapped processing will no longer be possible. With a list intersection algorithm, which supports the inverted file searching, the entire list intersection process must be completed before any of the candidates are identified. Thus no candidates are available for iterative searching until the entire screen searching step has been completed. When the iterative searching begins, the entire list of candidates is waiting to be processed. As valid answers are found they can be sent back to the user for display while the iterative search is processing the remaining canditates. Current plans are not to implement the 'display while searching' feature in order to maintain compatibility with the current Messenger text searching interaction. The plans to improve considerably response times for most queries are expected to make this feature unnecessary.

Foundation Software Environment

The foundation software environment available for the development of the search engines is considerably more complete than the one that was available for the search machines. Starting in the early 1980s CAS adopted the UNIX operating system and the C language. CAS programming and office automation are done under UNIX as are many application development efforts. In 1984, CAS began developing foundation software that would permit the construction of sophisticated distributed applications using both the UNIX and IBM environments. (Some of this software has been developed in support of the US Patent and Trademark Office Automated Patent System. CAS is a software subcontractor to Planning Research Corporation for this effort.)

One of these foundation software components is the UNIX version of the FIDO (Facility for Integrated Data Operations) database management facility. CAS developed FIDO to operate in the IBM MVS environment in the early 1970s and has since used it extensively for application development. FIDO provides a standard way of representing, cataloguing, dumping, moving and in general, managing large application files with large variable length records.

Other foundation software facilities that will be heavily used by the search engines are UNI (UNIX Network Interface) and MNI (Messenger Network Interface). UNI and MNI permit UNIX and MVS application processes to communicate freely with one another over a variety of underlying physical and logical network interfaces and configurations without the application knowing anything about them. (The UNI software was also developed in support of the Automated Patent System Contract.)

System Management and Control

The control structure will also be changed considerably with the implementation of the search engines. Rather than locating the control of the multiple parallel search processes in an IBM mainframe query specific process, control will be centralised in a single search engine complex resident process called the Search Engine Complex Manager (SECM). On the mainframe side, where query negotiation and query compilation are performed, the SECM will interface to the rest of the Messenger search and retrieval system in exactly the same way as the mainframe process that performs inverted file text searches does.

On the other side, the SECM will be attached to and control the screen search and iterative search processes. It will be responsible for broadcasting new queries.As screen searching is completed, the candidate answer file keys will be sent to the SECM. The SECM will combine the candidate answer lists from all of the screen search engines and then separate them appropriately by the iterative search file segment boundaries. These realigned candidate answer lists will then be sent to the proper iterative search process. By performing this combination and separation process, it will no longer be necessary to pair the screen file segments and processors with iterative file segments and processors. This allows greater flexibility in matching available computer resources with resources actually needed for each process.

File management functions will be greatly enhanced with the search engines implementation because FIDO will be used as the underlying database management system for the search and retrieval files. This will permit much greater freedom in moving, reorganising, backing up and extending the files and the information in them.

The system management capabilities are also being improved with the search engines. The use of the SECM as the control and co-ordination point for the entire search engine complex will permit many of the system management functions to be carried out in a single place. This includes such activities as querying the well-being of individual components, monitoring current activity, enforcing security controls and collecting diagnostics and performance statistics.

The software distribution, management and control will generally be built using standard UNIX features and capabilities. UNIX files will be used for the software entities;control files and most temporary work files.

Back-up and Recovery

The search engine design and implementation allows for a greater variety of options for back-up and recovery than the search machines do. The SECM itself may be replicated so that a 'hot' back-up is available. One or more SECMs can be configured to take over automatically for a SECM that fails, losing only searches that are currently in progress. If multiple SECMs are configured redundantly for a search engine complex, they can share the load when they are all operational.

The entire search file can be replicated, so that the loss of a file segment will only interrupt in-progress search sessions. Finally, by replicating the search file, and by using dual-ported disk controllers, the search engine complex can be configured in such a way that only searches currently in progress are lost if a single component (e.g., a computer or a disk drive) fails. The search engine complex software will support all of these options by specifying the appropriate configuration information.

RELATED AND POTENTIAL FUTURE ACTIVITIES

An additional effort currently underway is an IBM mainframe implementation of structure searching. This implementation is intended for small to moderately sized files that have moderate usage. It will interface to the rest of the Messenger software in the same way that the search engines do. It will use the same inverted file search software currently used for Messenger text searching and it will use the same C language implementation of the iterative search algorithm being developed for the search engines. The IBM version will not implement parallel searching of the file although it could, given the multiprocessormulti I/O channel IBM configurations that are available.

Interesting future parallel searching work would involve investigations into control structures that permit parallel searching at levels lower than at the file segment level. While there are no direct functional benefits to the user, it is possible, depending on usage levels and file sizes, that there would be benefits in terms of price and performance, particularly given the I/O considerations and the price-performance of individual computer hardware components. However, limitations in the speed of processing and data transfer that can be achieved in the hardware will indeed be reached at some point. The cost per unit of processing power, I/O bandwidth and storage capacity is likely tocontinue to drop even though performance ceilings will be reached. These two factors in combination with the user's need to search and retrieve from rapidly growing databases with more sophisticated queries (augmented by intelligent query generation aids) leads one to the conclusion that parallel searching will play an increasingly important role in modern search and retrieval systems.

REACTION INDEXING: AN OVERVIEW OF CURRENT APPROACHES

A Peter Johnson
Wolfson CADOS Unit, Department of Chemistry, University of Leeds,
Leeds, LS2 9JT, U.K.

ABSTRACT

Early attempts to use computers to aid reaction indexing and retrieval were based on the use of keywords or phrases to retrieve textual information, but lacked any facility for handling chemical structure diagrams. Once modern systems for structure handling and graphical display had been developed, it was relatively easy to modify these systems to handle the storage and manipulation of the multiple structures involved in reaction representation. At this stage the main search methods were either exact structure search or variants on a simple substructure search. The latter had the disadvantage of being relatively slow and prone to false drops, particularly for systems where a product substructure search was the only option available.

The second generation of reaction retrieval systems were more specifically tailored to the problem of efficient reaction retrieval, and specifically stored information relating to the atom-to-atom correspondences in going from reactants to products. Searches which made use of these correspondences were fast and essentially free from false drops. In addition these systems permitted extensive use of searches based on keywords or phrases, which are particularly useful when associated with a hierarchically structured thesaurus.

Despite the sophistication of the best of the currently available systems, there remain a number of problems still to be solved, of which perhaps the most interesting would be the ability to search across a number of reaction steps in a multi-step sequence.

INTRODUCTION

The major commercially available reaction indexing programs are REACCS (Molecular Design Limited), SYNLIB (started by Clark Still, developed by Smith, Kline and French and now sold by Dan Chodosh), ORAC (developed at the University of Leeds) and DARC-RMS (Télélsystèmes). Pfizer's in-house system CONTRAST is noteworthy but is not available outside Pfizer. The release of CASREACT by Chemical Abstracts is awaited with interest.

Major theoretical contributions to the field have been made by Mike Lynch, Peter Willett, George Vladutz and Robert Fugmann.

W. A. Warr (Ed.)
Chemical Structures

SEARCH KEYS

The user may want to search on author, or authors, on the source of the information (journal title, patent etc.) or on year (or range of years).

In chemical structure terms, he may wish to search for an exact structure of a reactant or product (although this is probably more useful with an in-house than with a literature database) or for a substructure in a reactant or product. However, what he would most like to do is generic reaction search.

He may also want to search for reaction keyword descriptors, for reagent keyword descriptors, for actual reagents or for the name of a reaction.

He is not likely to want to search on yield, reaction temperature and pressure, and so on, but this information may produce useful constraints on the search.

AUTHOR SEARCH

Author records must be indexed for rapid retrieval. Truncation is necessary because the user may be doubtful about the author's initials or the spelling of his name.

JOURNAL SEARCH

It is less likely that a user will wish to search on this but in case he does, the system must allow for a variety of common abbreviations (e.g. both *JACS* and *J. Amer. Chem. Soc.*) for ease of input and search.

EXACT STRUCTURE SEARCH

The same canonical name matching used for chemical structure retrieval systems works very well also in reaction indexing systems and provides rapid retrieval.

SIMPLE SUBSTRUCTURE SEARCH

The substructure search feature is similar to that of MACCS, DARC and OSAC, but has the additional possibility of specifying that the substructure is to be in the product or in the reactant or in both or in one but not in the other. (There is a dangerous pitfall here. A search for O=C-O-C in the product but not in the reactant would miss the reaction of $CH_3COC_6H_4OH$ going to $CH_3COC_6H_4COCH_3$).

It is essential that the system must handle stereochemistry.

REACTION SEARCH

Three methods have been used.

Method 1

This consists of a product substructure search supplemented by the specification that certain atoms or bonds in the substructure are part of the reaction site. The method usually produces a large number of hits but some people feel that this gives 'browsability'.

Definitions of Reaction Site. There are three ways of looking at this. The graphical reaction site consists of all the bonds which have undergone a change and the atoms attached to those bonds. The chemical reaction site (as used in ORAC) extends the graphical reaction site by the addition of those functional groups which are absolutely essential for the reaction to proceed. The subjective reaction site (as used in SYNLIB) consists of those atoms and bonds the inputter considers significant.

Method 2

Here the user does a substructure search for a fragment of the reactant followed by a substructure search for part of the product and he then intersects the two hit lists using AND logic.

This is likely to lead to false drops. For example a search for C=C being reduced to CH-CH would also find the olefin being converted to CH(OH)-CH(OH), even if the carbon atoms were flagged as being in the reaction site.

Method 3

The use of atom-atom mapping or bond mapping in the substructure search completely overcomes the problem of false drops. For example there are no problems in searching for the conversion of a nitrile to a primary amine if the nitrogen atom is flagged with a number 1 in both reactant and product and the carbon atom is flagged 2 in both. This technique also permits questions about chemoselective reactions, for example the conversion of a ketone to a secondary alcohol in the presence of an aldehyde.

This atom-atom mapping feature has been present in ORAC since the Spring of 1985. The feature is also available in CONTRAST.

KEYWORDS

Are they necessary? The author feels that they are since they can be used to describe concepts which are difficult to convey simply through structural formulae, in particular terms relating to reaction mechanism such as SN2 or sigmatropic rearrangement.

In searching for the reduction of an α,β-unsaturated ketone to CH-CH-C=O, it may help to use the keyword 'reduction' if your system does not cope with the two C-H bonds formed in this reaction. (ORAC, it should be noted, would not have a problem here. It would not be necessary to use the keyword 'reduction'.)

ORAC Version 7 has a heirarchical thesaurus of keywords. Thus both a reagent, and the families it belongs to, can be handled.

PUBLIC REACTION DATABASES

A well known abstracted database is Theilheimer which is searchable with REACCS.
More than one of the vendors of reaction indexing software use consortia of abstractors in the Universities to construct reaction databases.
ISI do direct abstracting from the primary literature to produce the Current Chemical Reactions (CCR) database.
A development expected in the future is the production of reaction database items by an independent vendor in 'master' format i.e. *any* system can use the data. An example is CHEMINFORM, from FIZ, described by Alex Parlow elsewhere in these Proceedings.

Database Sizes

REACCS can currently offer about 100,000 reactions, SYNLIB 35,000 and ORAC 30,000. The author confidently expects ORAC to offer 110,000 by mid-1988.
CASREACT is in a different league – 170,000 reactions are being abstracted every year. It is an interesting question how many reactions a user needs in a selective database. The critical question is whether there are *redundant* examples. The only way of pruning a database is to remove similar extraneous examples.
There may be a need for specialist datasets e.g. β-lactam chemistry, peptide chemistry.

INTEGRATION

As far as ORAC is concerned integration is viewed as follows. (Other systems have similar links). The reaction retrieval system can be integrated with PC software, in this case Hampden Data Services' PsiGen. Integration with the structure handling system OSAC is in place. OSAC itself is integrated with the database management system 1032 and will soon be integrated with ORACLE. (Other database management systems could easily be accommodated.) It is also planned to integrate ORAC with synthesis planning programs such as LHASA and CASP. A start has been made on this – the Host Language Interface will be used. ORAC will also have links to CAMEO (for product prediction).

HOST LANGUAGE INTERFACE

This operates at three levels: import and export of reaction data in SMD (Standard Molecular Data) format, interactive communication with other devices and processes *via* an ASCII datastream (not yet implemented), and a callable level.

THE FUTURE

The following problems remain.
The problem of tautomers and mesomers is exacerbated in reaction systems.

Metal pi complexes are not easily handled.

Multi-step reactions (A goes to B goes to C goes to D) pose problems. One would like to be able to input the separate reactions but to search also across steps e.g. A to C, B to D and A to D. Chemical Abstracts Service has made some progress in this area (see the poster described by Blower *et al.* elsewhere in these Proceedings).

Reaction keyword perception should become more and more automatic.

It would also be interesting to work on the automatic creation of a synthesis planning knowledge base using the factual information from a reaction database.

ACKNOWLEDGMENTS

The author acknowledges the efforts of his co-workers Tony Cooke, Kevin Higgins and Paul Hoyle.

This chapter has been written by the Editor using tape recordings and notes made at the Conference, with the approval of the author.

THE IMPLEMENTATION OF ATOM-ATOM MAPPING AND RELATED FEATURES IN THE REACTION ACCESS SYSTEM (REACCS)

Tom E Moock, Jim G Nourse, David Grier and W Douglas Hounshell
Molecular Design Limited, 2132 Farallon Drive, San Leandro, Ca 94577, U.S.A.

ABSTRACT

Computer-aided reaction indexing has become an increasingly important tool for synthetic chemists in the past few years. New products, as well as newer versions of established products, continue to advance the state-of-the-art in areas of structural search specificity and efficiency.

The addition of atom-atom mapping to REACCS' reaction substructure search represents another step in the efforts of our group to improve its searching capability. Related searching features, such as stereochemical transformations (inversion/retention of configuration) and exact-transformation search properties are supplied to allow the user to take better advantage of atom-atom mappings. These features also set the groundwork for the implementation of searching reaction schemes.

The REACCS program is supported by a number of large databases of reactions. To assign mappings to such a huge number of reactions with minimum time/labour costs, an automatic method of assigning mappings and reacting centres was developed. This 'automapping' module has the unique ability to perceive non-stoichiometric reactions, in addition to well-behaved transformations.

INTRODUCTION

A synthetic chemist, either an industrial or an academic one, often finds himself searching the literature for specific chemical reactions. In a typical literature search in the library, there are no good methods for finding the type of chemical transformation he is searching for, especially in a specific structural environment. If he is lucky, he may stumble upon a review of relevant chemistry. If not, he has a great deal of work to do. He must wade through a tremendous amount of information and, in a sense, prepare his own review. After he is done, he may as well write up his review and save someone else the trouble in the future.

A graphical reaction indexing system can be thought of as a method of creating a specialised review. As each search over a database can be different, each review can be focused on a different subject. Such a computerised search over the literature is much faster and more versatile than any manual method. A search conducted over large databases of reactions gives the user the ability to gather together chemical knowledge which could be scattered over the library, or separated by several feet on a bookshelf.[1]

With a full-featured reaction indexing system such as the Reaction Access System (REACCS),[2,3] there exists a wide variety of criteria for creating such a review. Table 1

W. A. Warr (Ed.)
Chemical Structures

contains a list of methods for searching for reaction information, of structural, textual and numeric formats. These are not necessarily functions unique to REACCS; most commercial reaction indexing systems contain at least some of these features.

Of the graphical search types listed, one of the most important is Reaction Substructure Search, or RSS. The reaction indexing group at Molecular Design initiated a new project to increase the flexibility and specificity of RSS searching. That project is the subject of this paper.

Table 1. A List of Reaction-based Search Types in REACCS.

Reaction-based Graphical Searches	**Molecule-based Graphical Searches**
Exact Reaction Match	Exact Structure Match on Reactants, Products, Solvents, Reagents (all reaction components)
Reaction Substructure Search	
With/without Reacting Centre Bonds	Substructure Match on all components
With/without Atom-atom Mappings	Formula Search on all components
With/without Stereochemical Transformations	Tautomer Search on all components
Reacting Centre Substructure Search	**Reaction-based Data Searches**
	Literature References
	Reaction conditions
	Solvent and Reagent Names
	Keywords
	(Etc.)

The requirements of the project were:

1. To use atom-atom mappings (defined in the next section) in queries and database entries.
2. To create a method for the automatic assignment of atom-atom mappings and reactingcentre bonds.
3. To allow for more flexible, specific and convenient searching of transformations involving hydrogens, such as oxidations and reductions.
4. To allow for better handling of stereochemical transformations, such as epimerisations.

ATOM-ATOM MAPPING OF REACTANTS AND PRODUCTS[4]

In REACCS an RSS search query is composed of one or more substructures, which represent reactants and/or products. Reacting centre bonds (those which change as a result of the reaction) are often marked on these structures to indicate where the changes are occurring. An example is shown in reaction A of Figure 1. A chemist viewing this reaction might interpret the reaction verbally as 'find all the reactions in the database which convert a carbonyl to a carbon-oxygen single bond.' The query would indeed find all such transformations, including the ketone reduction in reaction B.

(A) Query

(B) Hit

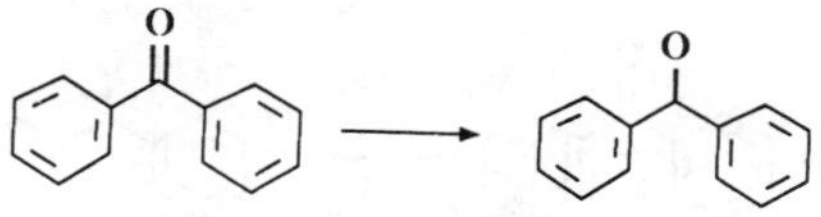

(C) Hit?

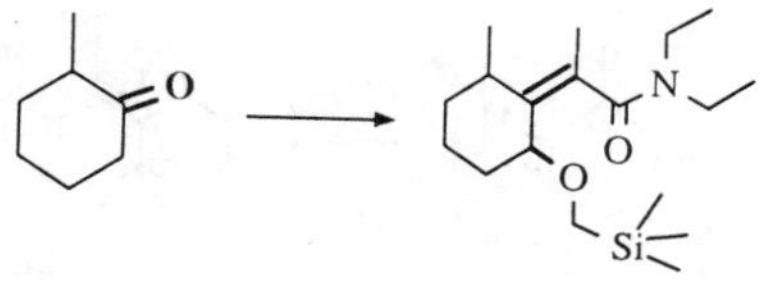

Figure 1. The specificity of the query is insufficient to differentiate between the correct hit (reaction B) and the noise hit (reaction C).

There is enough vagueness in the query, however, to bring up other unrelated reactions, such as reaction C. This reaction satisfies the query because there does exist a carbonyl in the reactant, an alcohol in the product, and the reacting centre bonds are all consistent with the query. Nowhere in the query is there an indication that the same carbon and oxygen atoms should be used in both the reactants and products.

To remove unwanted 'noise hits', such as reaction C, more information in the query is needed: the user must specify in the query that the carbon and oxygen atoms represent the same atoms in the reactant as well as the product. In a sense, the atoms in the reactant are 'mapped' to the atoms in the product. Either the user or an automated method must be used to establish the atom correspondences.

The advantages of atom-atom mapping are several:

1. It increases the specificity of searches, and is the only method of searching for certain types of transformations.
2. It can increase the efficiency of RSS searches by closely tying together reactants and products.
3. It is a requirement for the searching of reaction schemes.

DISPLAY AND ASSIGNMENT OF ATOM-ATOM MAPPINGS

For the chemist to assign the mappings manually, an editor was created within REACCS which allows him to mark the atoms and display the mappings. The display of the mappings is shown in Figure 2. The query of Figure 1 is shown in reaction A marked and with mappings displayed. With this notation, the user is made aware that the central carbon atom in both the reactant and product are the same atom, because it has the same 'mapping number' (number 1). Similarly, the oxygens are indicated as being the same, since they both have the same mapping number of 2. Both the query and databases entries must have atom correspondences for RSS.

Reaction B is found in RSS. Its atom mappings are consistent with the query reaction A. Reaction C, however, does not show mappings consistent with the query, and therefore would not be hit by a search using atom-atom mappings.

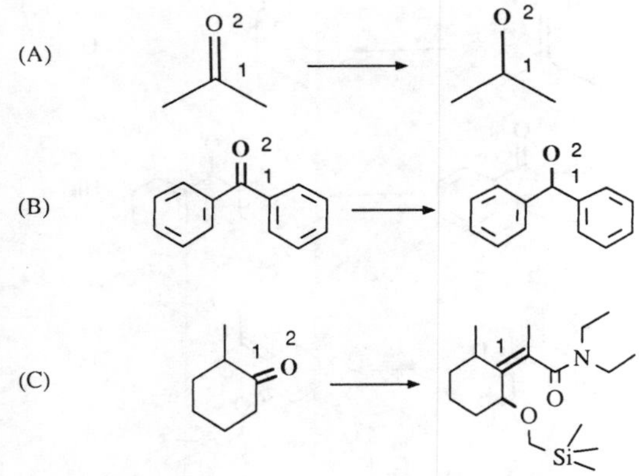

(A)

(B)

(C)

Figure 2. The ambiguity of the query in Figure 1 has been resolved by using atom-atom mapping. The mapping in reaction B is consistent with the query, but the mapping in reaction C is not.

AUTOMATIC ASSIGNMENT OF ATOM-ATOM MAPPINGS

REACCS has associated with it a number of public databases, comprising a total of nearly 90,000 reactions.[5] Moreover, there exist a large number of reactions in private databases held by many of REACCS' 70+ installations worldwide. These reactions contain reacting centre bonds, but do not have atom-atom mappings assigned. It would be prohibitively expensive to assign the mappings manually, so a procedure was developed for assigning atom-atom mappings automatically.

A complicating factor is that there are many non-stoichiometric reactions in REACCS databases. Reactions are represented in REACCS in a manner similar to the way they appear in the literature: grossly unbalanced, some structures left out, large fragments have disappeared, and so forth. Many reactions represent summaries of several steps, where reacting centres are poorly localised. These reactions are characterised by the following attributes:

1. The presence of alternative products, such as mixtures of isomers or the creation of side-products.
2. Multiply used reagents.
3. Missing pieces from either side, to focus attention on other structures.
4. Deceptively simple transformations due to symmetry or similarity. Diels-Alder reactions using simple dienes (such as cyclopentadiene) are a common example of this in REACCS databases.

All of these factors place severe demands on the automatic assignment of reacting centres and atom-atom mapping.

Reaction A in Figure 3 is an example of a well-behaved reaction.[6] It is reasonably balanced, and the reacting centres are well localised. Reaction B, on the other hand, is a reaction with unbalanced stoichiometry.[7] The triphenylphosphorous fragment is apparently lost, while both the *cis* and *trans* isomers of the heterocycle are produced. Reactions such as this are extremely difficult problems for automatic reacting centre perception (ARCP) programs.

The requirements of the ARCP project were to produce a program that:

1. Has high reliability over large databases of realistic, complex reactions.
2. Handles unbalanced and other ill-behaved reactions.
3. Allows human intervention when necessary. For difficult reactions, the program must serve as an assistant to the user, who can guide the program with a few, strategically placed markings.
4. During batch processing of databases, the program recognises potentially poor solutions. With nearly 90,000 reactions, it would be prohibitively expensive to examine all of them while searching for the 1–5 per cent which failed.

The problem of automatic reacting centre perception is similar to the problem of finding common subgraphs. In the identification of reacting centres and atom-atom mappings, the most important subgraphs are the largest, or nearly the largest. These are often discovered computationally using some variant of the Morgan algorithm.[8] Once these subgraphs are found, the classical approach[9,10] is to remove them, focus on the subgraphs which are left, and assign them as reacting centre bonds or atoms.

The ARCP module in REACCS works by iteratively discovering many common subgraphs, not simply the largest, and evaluating each possible solution using various criteria to select the best one. An example of this method[11] is described schematically in Figure 4. In this Diels-Alder reaction, using the largest common subgraph (LCS) leads to the wrong solution. REACCS does indeed generate a solution based on the LCS, but it also finds another, correct solution. Using an extensively tuned evaluation function, REACCS chooses the correct solution as the best one.

REACCS uses this iterative scheme, as well as other methods, to solve non-stoichiometric reactions. The ARCP module looks for conditions such as the existence of alternate subgraphs (alternate products), and many-to-one mappings of identical

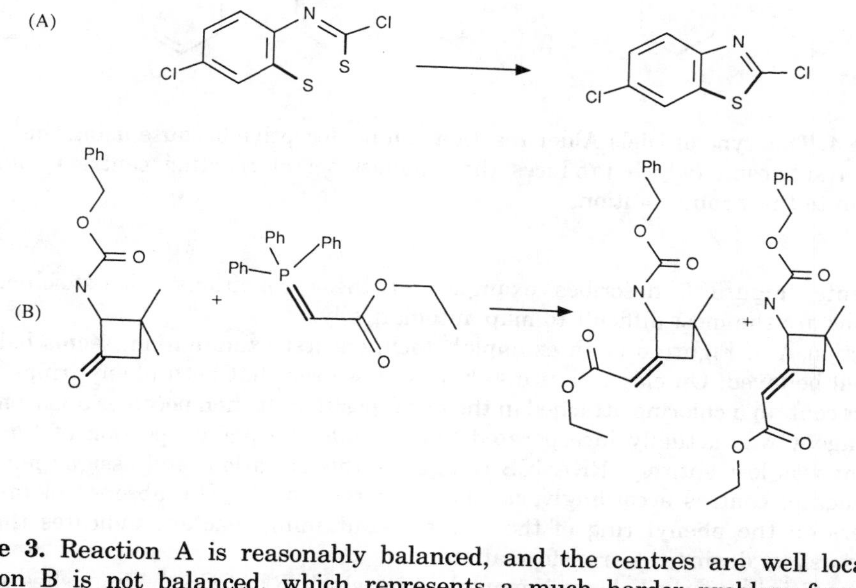

(A)

(B)

Figure 3. Reaction A is reasonably balanced, and the centres are well localised. Reaction B is not balanced, which represents a much harder problem for ARCP programs.

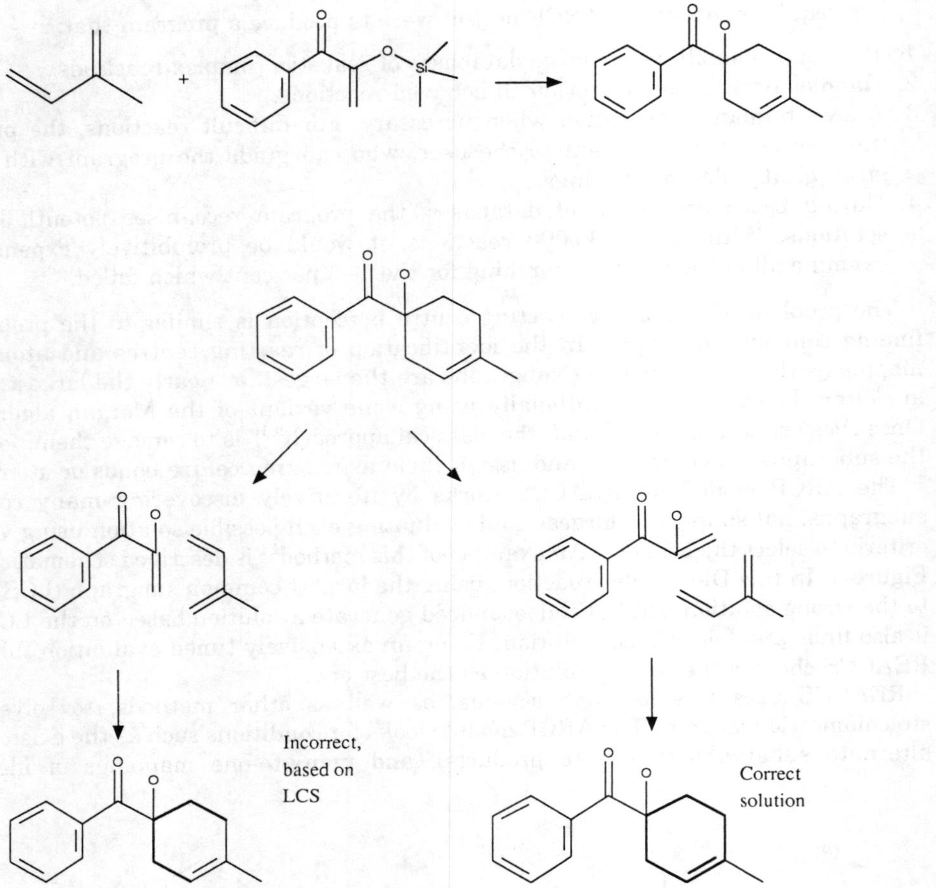

Figure 4. This type of Diels-Alder reaction can be deceptive because using the largest common subgraph (which produces the smallest set of reacting centres) leads the program to the wrong solution.

fragments. Figure 5 describes examples of these conditions. Non-stoichiometric reactions are the most difficult to map automatically.

Reaction A of Figure 6 is an example[12] hich, on first examination, seems balanced and well behaved. On closer scrutiny, however, we see that both phenyl rings in the product contain a chlorine attached in the *para*- position. It then becomes apparent that one reagent was actually incorporated twice, while the phenyl portion of the other reactant was lost entirely. REACCS recognises this situation, and assigns mappings and reacting centres accordingly, as shown in reaction B. The absence of mapping numbers on the phenyl ring of the nitrogen-containing reactant indicates that the group was lost during the transformation.

Figure 7 is a nearly balanced example[13] in which REACCS distinguishes similar reagents. One might get the impression that the newly formed heterocyclic ring in the

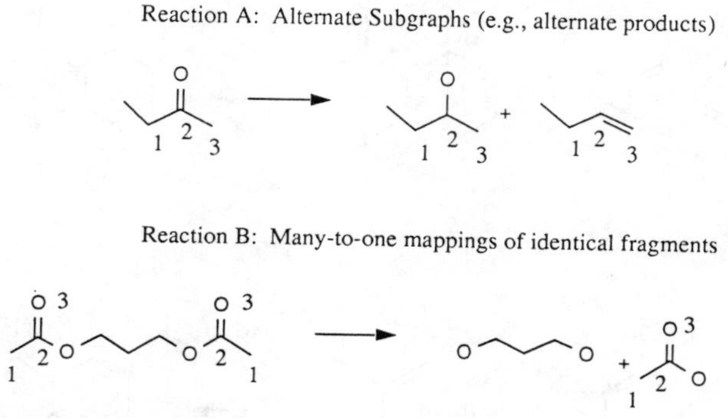

Reaction A: Alternate Subgraphs (e.g., alternate products)

Reaction B: Many-to-one mappings of identical fragments

Figure 5. Types of conditions used to analyse and solve unbalanced reactions.

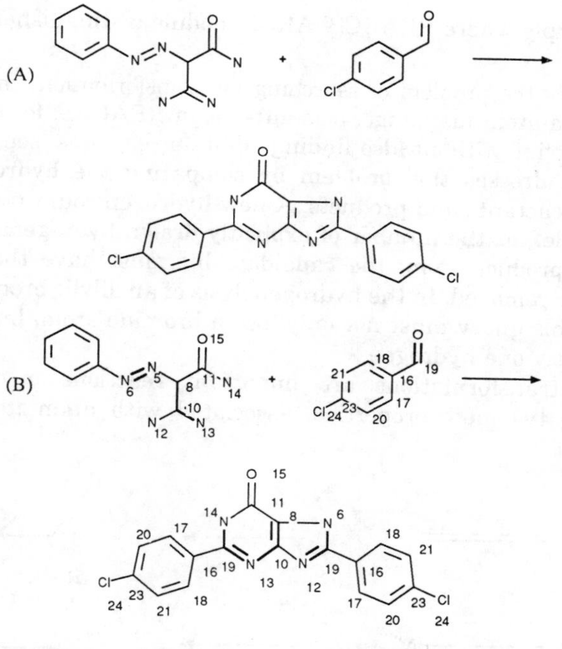

Figure 6. Analysis of an unbalanced reaction which looks deceptively well-balanced.

product comes from the most similar reagent (the first one). In fact this is not true, as the mapping numbers found by REACCS indicates in reaction B.

RELATED PROPERTIES

As mentioned in the introduction, the RSS project is more broad than atom-atom mapping and ARCP alone. Other methods have been developed based on our procedures for atom-atom mapping.

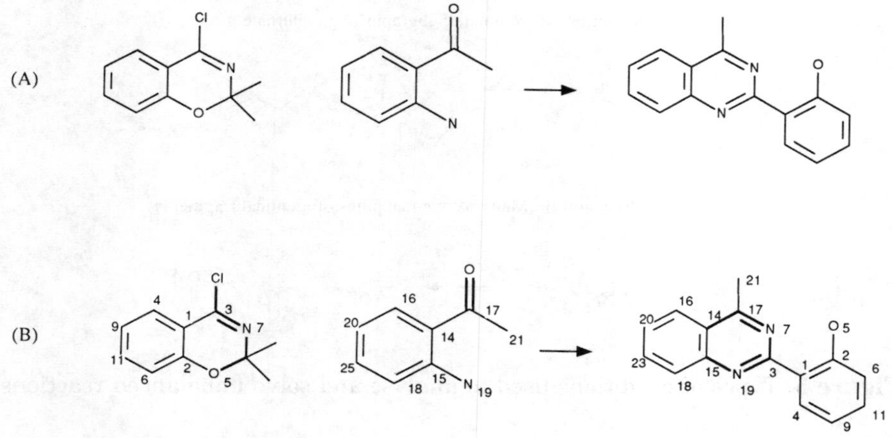

Figure 7. An example where REACCS' ARCP module distinguishes similar reagents.

One method solves the problem of searching for transformations involving hydrogens. Without using atom-atom mappings, it is difficult in REACCS, for example, to search for reductions of olefins without also finding addition reactions (see Figure 8).

REACCS now addresses this problem by comparing the hydrogens on similarly mapped atoms in reactants and products. A new hydrogen count property (on the SSS menu in Build mode), or the number of explicitly drawn hydrogens, is compared with the reactant and product atom; the candidate hit must have the same number of hydrogens added or removed. In the hydrogenolysis of an allylic bromine (Figure 9), the candidate hit for this query must not only lose a bromine atom, but must replace the bromine with exactly one hydrogen.

Stereochemical transformations are important reactions in modern chemistry databases, and so two new properties, associated with atom-atom mapping, were

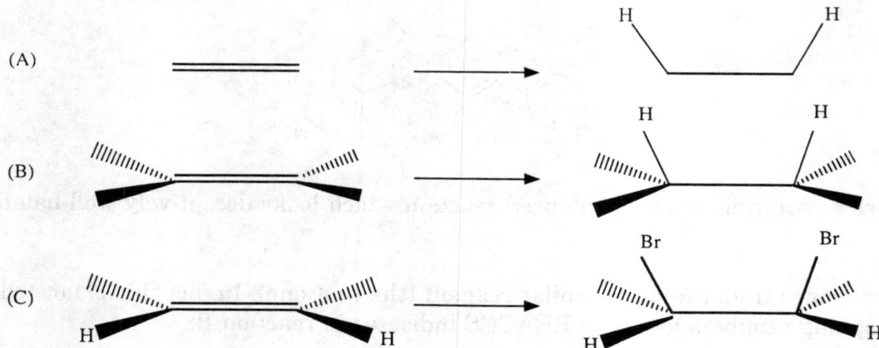

Figure 8. Without atom-atom mapping, searching for reactions involving hydrogen is difficult. The query is unable to distinguish between the reduction and the addition reactions. The use of atom-atom mapping allows REACCS to compare changes in hydrogens between mapped atoms in the reactants and products, and resolve this ambiguity.

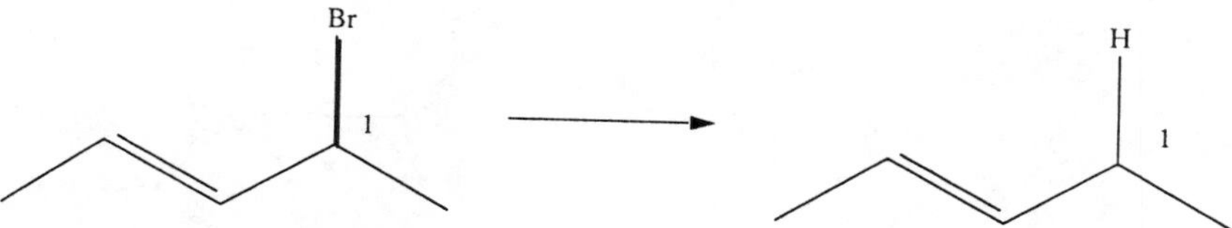

Figure 9. REACCS searches for reactions involving hydrogen by comparing explicit hydrogens on mapped atoms between reactants and products.

created. Users can apply the inversion or retention properties to queries or database entries, indicating that the transformation must proceed with the correct stereochemistry. These properties are also automatically assigned by REACCS' ARCP module. Reaction A of Figure 10 is an example where the inversion property was applied to an Sn2 displacement reaction. This property is especially useful in searching for epimerisation reactions (reaction B), where no other reaction exists and there are no bonds to mark as reacting centres.

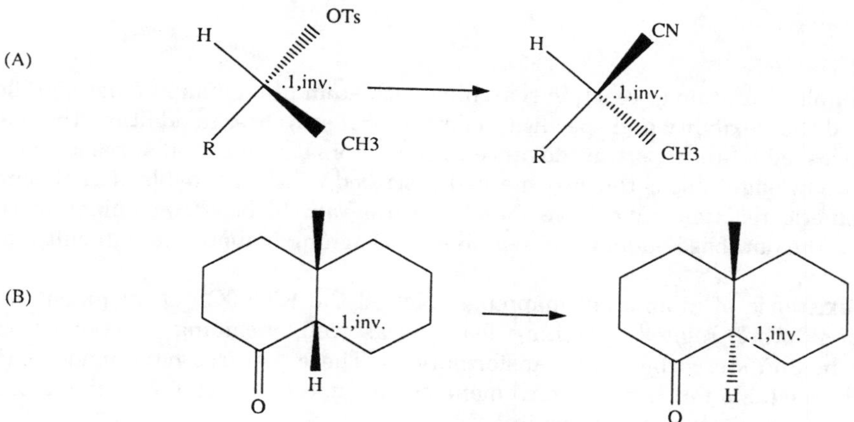

Figure 10. Use of the Inversion/Retention properties to search for stereochemical reactions.

REACCS also supports an atom property for searching for very specific reactions. This property, called Exact Change, is used to define completely all changes occurring at the atom it is applied to. If the property is assigned, no other changes may occur at that atom in the candidate hit. For example, reaction A in Figure 11 is a query designed to search for the creation of disulphides from thioalcohols. The query will find such reactions, as in Reaction B. However, it will exclude reaction C because, at the site where the exact change property was applied, another change also occurred: a bond to oxygen was formed. Had the exact change property been omitted, reaction C would also have been found.

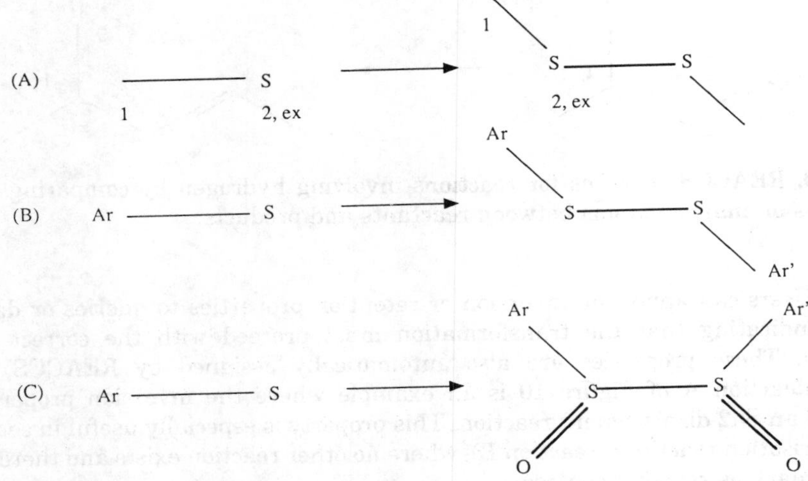

Figure 11. The use of the Exact Change property completely describes the changes allowed at an atom in an RSS query. In this case, reaction B is hit but not reaction C.

CONCLUSION

The implementation of reaction searching with atom-atom mapping has significantly increased the flexibility and specificity of structural searches. In addition, the mapping procedures led to an important advance in the automatic detection of reacting centres. To our knowledge, this is the only method described which is capable of analysing non-stoichiometic reactions. We have found the program to be a convenient method of assisting the database builder and searcher in assigning mappings to difficult reactions and queries.

The existence of atom-atom mappings allowed the REACCS development team to implement other related searching features as well, including reactions involving hydrogens and stereochemical transformations. These features have made searching for such reactions much more convenient or, in some cases, made available searches which were previously impossible in REACCS.

REFERENCES

1. Wipke, W. T., 'Exploring Reactions with REACCS'. Presented at the 188th National Meeting of the American Chemical Society, Philadelphia, PA, August, 1984.
2. Wipke, W. T., Dill, J. D., Peacock, S., Hounshell, D. 'Searchand Retrieval Using an Automated Molecular Access System'. Presented at the 182nd National Meeting of the American Chemical Society, New York, NY, August 1981.
3. Wipke, W. T., Dill, J. D., Hounshell, D., Mock, T., Grier, D. 'Exploring Reactions with REACCS'. In *Modern Approaches to Chemical Reaction Searching*: Proceedings of a Conference Organised by the Chemical Structure Association at the University of York, England, July 1985; Willet, P. Ed.; Gower: Aldershot, 1986.
4. Other commercially available (or announced) graphical reaction indexing systems which search databases containing atom-atom mappings include ORAC, RMS-DARC and the CAS ONLINE Reaction Database. The SYNLIB program uses another approach. These programs

were described in talks at the Modern Approaches to Reaction Indexing meeting, York, England, July, 1985. See reference (3).

5. At the time of this writing, the public databases available from Molecular Design, and their numbers of reactions are:

Theilheimer	46,000
Journal of Synthetic Methods	14,000
Organic Synthesis	5,000
Current Literature File	21,000

6. *J. Synth. Methods*, 8(2), 75386X; Cassella AG, Europe, 39905.
7. *J. Synth. Methods*, 8(9), 77215X; Al-Zaudi, S., Stoodly, R. J. *J. Chem. Soc. Chem. Comm.* 1982, *17*, 995–996.
8. Morgan, H. L. 'The Generation of a Unique Machine Description for Chemical Structures – a Technique Developed at Chemical Abstracts Service'.*J. Chem. Doc.* 1965, *5*, 107–113.
9. Vledutz, G. E., 'Development of a Combined WLN/CTR Multilevel Approach to the Algorithmic Analysis of Chemical Reactions in View of Their Automatic Indexing', British Library, Researchand Development Department Report No. 5399, 1977.
10. Lynch, M. F., Willet, P. 'The Automatic Detection of Chemical Reaction Sites', *J. Chem. Inf. Comput. Sci.* 1978, *18*, 154–159.
11. *J. Synth. Methods*, 8(5), 76177X; Sasaki, T., Ishibashi, Y., Ohno, M. 'Catalysed Cycloaddition Reactions of α-Silyloxy-α,β-Unsaturated Ketone and Aldehyde'. *Tetrahedron Lett.* 1982, *23* (16), 1693–1696.
12. *J. Synth. Methods*, 8(4), 75842X; Yoneda, F, Koga, R., Higuchi, M. 'A One-Step Synthesis of Purine Derivatives by the Reaction of Phenylazomalonamidamidine with Aryl Aldehydes'. *Chem. Letters* 1982, *3*, 365–368.
13. *J. Synth. Methods*, 8(4), 75857X; Tochikawa, R., Wachi, K., Terada, A. 'Studies on 1,3-Benzoxazines. VI. Formation of Quinazolines and 4H-3,1-Benzoxazines by the Reaction of 4-Chloro-2H-1,3-benzoxazines with Aminoacetophenone, Aminobenzophenone and Aminobenzyl Alcohol Derivatives'. *Chem. Pharm. Bull.* 1982, *30* (2), 559–563.

POSTER SESSION: REACTION INDEXING IN AN INTEGRATED ENVIRONMENT

Guenter Grethe, Donna del Rey, Judy G. Jacobson, Melisande VanDuyne
Molecular Design Limited, 2132 Farallon Drive, San Leandro, California 94577, U.S.A.

ABSTRACT

Management of chemical information through computerised programs based on graphical I/O has become a powerful tool for chemists. Synthetic chemists increasingly depend on the availability of effective interactive reaction indexing systems in their daily work. These programs should satisfy the requirements of chemists in different environments and should be able to handle different synthetic or preparative information.

REACCS, the reaction management system developed by Molecular Design Limited, effectively handles information on reactions as well as on molecules. The use of this program in environments as different as laboratories in basic research or process

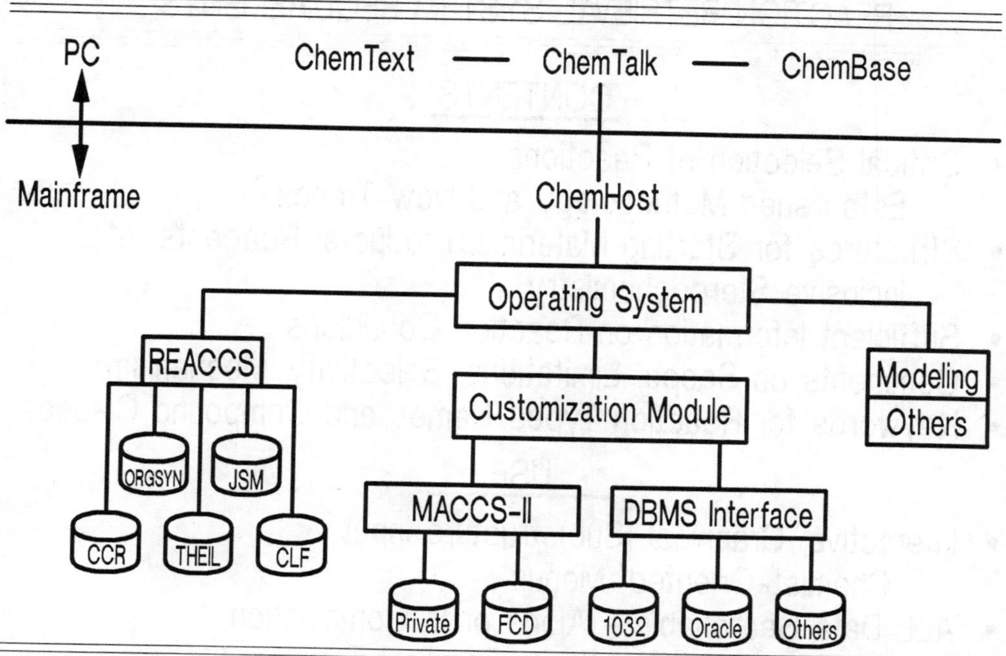

Figure 1. Scheme for an Integrated Information System.

W. A. Warr (Ed.)
Chemical Structures

development will be described. The flexibility and power of the program will be shown through search examples utilising the various features of the program and the generation of specialised, private databases will be demonstrated. Finally, we will discuss our efforts towards integration of this program with other mainframe or microcomputer programs towards the generation of an integrated chemical information system.

DESCRIPTION OF POSTERS

Increasingly, synthetic chemists are using computerised chemical and biological information in their daily routine directly at their workplace. The growth of information and the increasing number of available programs to manage these data makes integration imperative to guarantee efficient use by chemists in the future.

In the last few years structure-based, interactive reaction retrieval has joined other programs as an important tool for the synthetic chemist. In this paper we shall discuss several applications of Molecular Design Limited's reaction information management system REACCS in the laboratory. Special emphasis will be placed on integration with other programs, not only at the mini/mainframe level but also with PC-based programs. Figure 1 shows this integration schematically based on MDL's MACCS-II system and the Customisation Module. Any efficient reaction retrieval system has to show the features outlined in Figure 2. These will be amply illustrated in the subsequent

REACTION RETRIEVAL SYSTEM REQUIREMENTS

CONTENTS

- Critical Selection of Reactions
  Established Methodology and New Trends
- Structures for Starting Materials, Products, Reagents, etc.
  Inclusive Stereochemistry
- Sufficient Information on Reaction Conditions
- Comments on Scope, Limitations, Selectivity, Mechanism
- Keywords for Reaction Types/Names and Compound Classes

USE

- Interactive, Graphical (Sub)structure Input
  "Chemist-Oriented" Menus
- ALL Data Searchable - Alone or in Combination
- Flexible Display of Results in "Familiar" Form

Figure 2. Requirements for an Efficient Reaction Retrieval System.

applications. In combination with integration these features provide a powerful tool for the synthetic chemist as a regular end-user.

To access REACCS or any other program available through the integrated information system the chemist calls up a site-specific menu (Figure 3) and graphically selects the program and/or the selected database; if REACCS is the program of choice only one of the database buttons has to be activated. The menus can be generated readily using a sequence of commands in a high-level fourth generation language (see below) as part of the MACCS-II system; they can be tailored to meet the needs of different institutions and applications. After calling up REACCS, the chemist first chooses BUILD mode to formulate his search query graphically. Information is needed for the preparation of dihydropyrans substituted in the 2,5-positions by a hetero Diels-Alder reaction. After drawing on the screen the minimally required query fragments for this reaction, the user makes refinements by indicating the relative stereochemistry of the substituents. In order to eliminate noise, the automapping feature of the program is invoked to map corresponding atoms in reactant and product and to indicate reacting centres (Figure 4). A reaction substructure search (RSS) is performed with this query automatically over all databases accessible by the program, or only over selected ones, which generates a number of hit lists. Figure 5 shows an example from the Current Literature File. The reagent in this Lewis acid mediated reaction is $ZnCl_2$. To search for the availability of this reagent and at the same time to prepare a summary the chemist again utilises the MACCS-II system. By selecting appropriate menu buttons on the menu shown in Figure 6 the chemist can search for the reagent over the Fine Chemicals

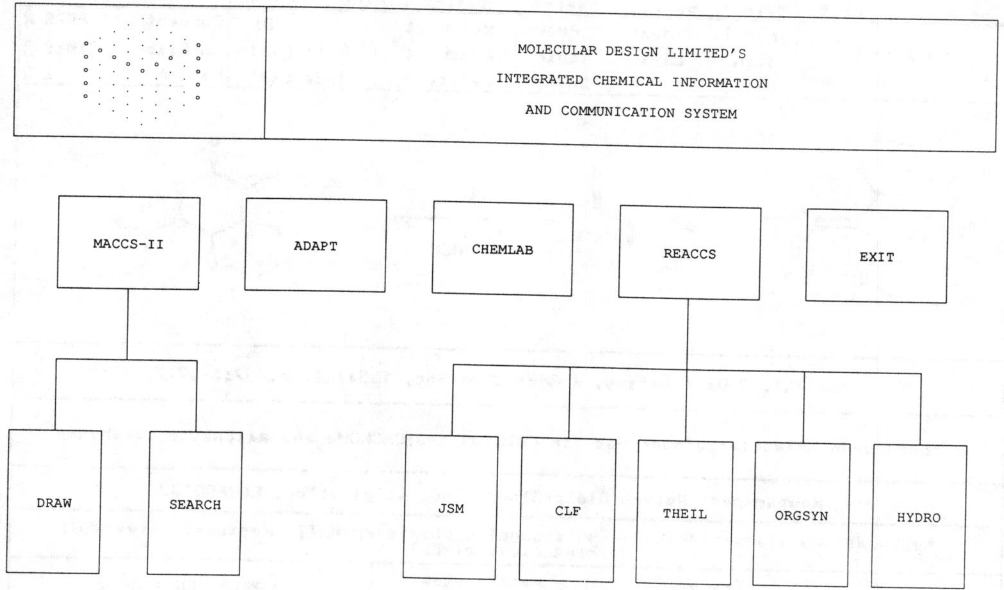

Figure 3. Customised Entry Menu.

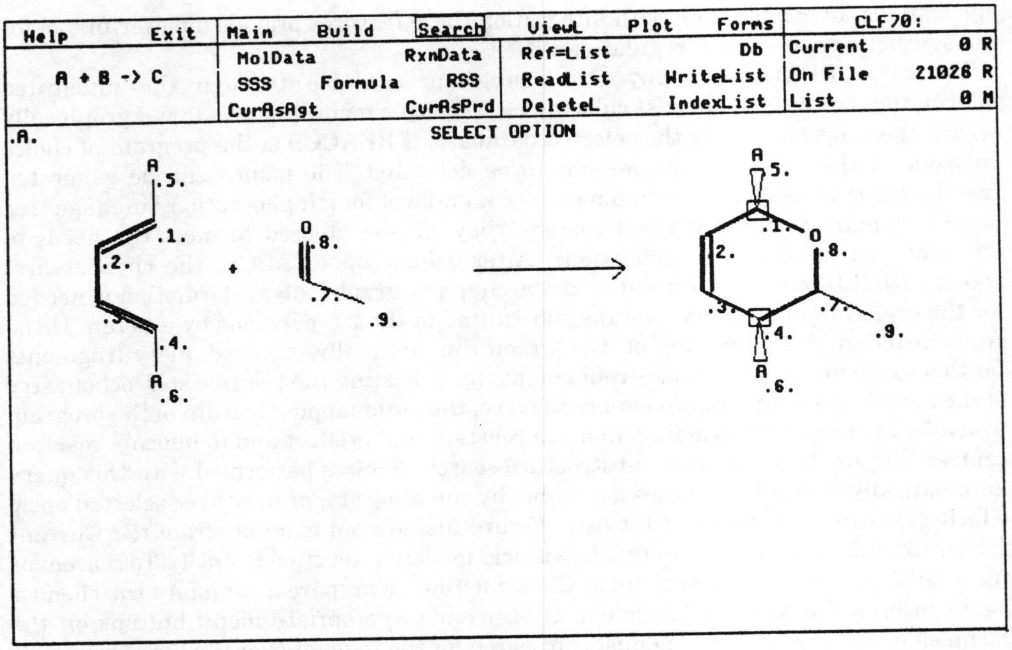

Help Exit	Main Build Search	ViewL Plot Forms	CLF70:
A + B -> C	MolData RxnData SSS Formula RSS CurAsAgt CurAsPrd	Ref=List Db ReadList WriteList DeleteL IndexList	Current 0 R On File 21026 R List 0 M

.A. SELECT OPTION

A .5. .1. .2. .3. .4. A .6. + O .8. .7. .9. → A .5. .1. O .8. .2. .3. .7. .9. .4. A .6.

Figure 4. RSS Query with Atom-Atom Mapping.

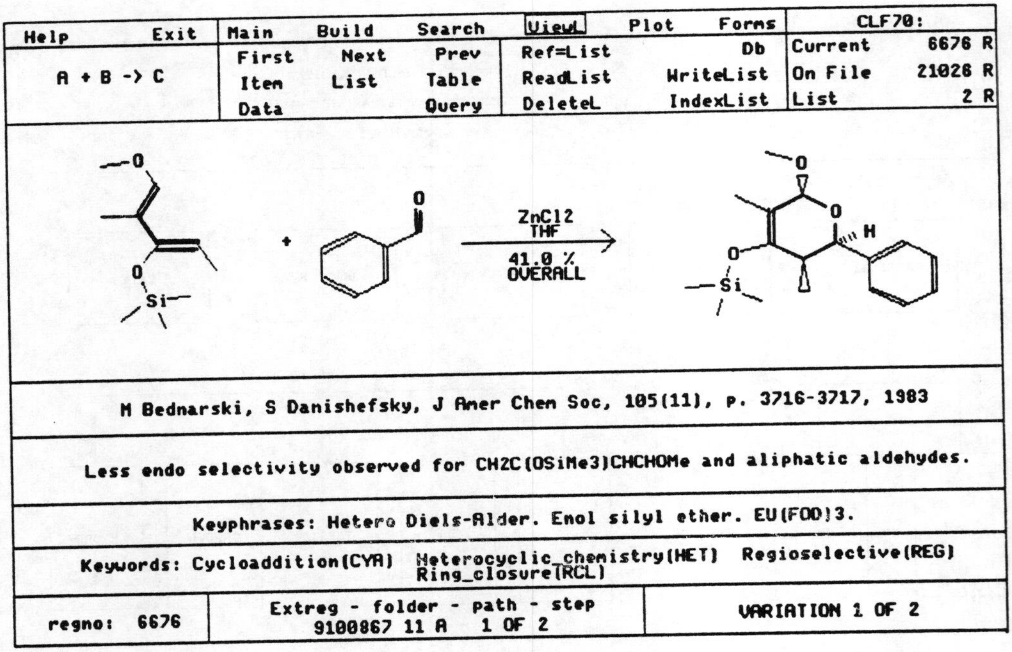

Help Exit	Main Build Search	ViewL Plot Forms	CLF70:
A + B -> C	First Next Prev Item List Table Data Query	Ref=List Db ReadList WriteList DeleteL IndexList	Current 6676 R On File 21026 R List 2 R

ZnCl2
THF
41.0 %
OVERALL

M Bednarski, S Danishefsky, J Amer Chem Soc, 105(11), p. 3716-3717, 1983

Less endo selectivity observed for CH2C(OSiMe3)CHCHOMe and aliphatic aldehydes.

Keyphrases: Hetero Diels-Alder. Enol silyl ether. EU(FOD)3.

Keywords: Cycloaddition(CYA) Heterocyclic_chemistry(HET) Regioselective(REG) Ring_closure(RCL)

regno: 6676	Extreg - folder - path - step 9100867 11 A 1 OF 2	VARIATION 1 OF 2

Figure 5. REACCS Display of Hetero Diels-Alder Reaction.

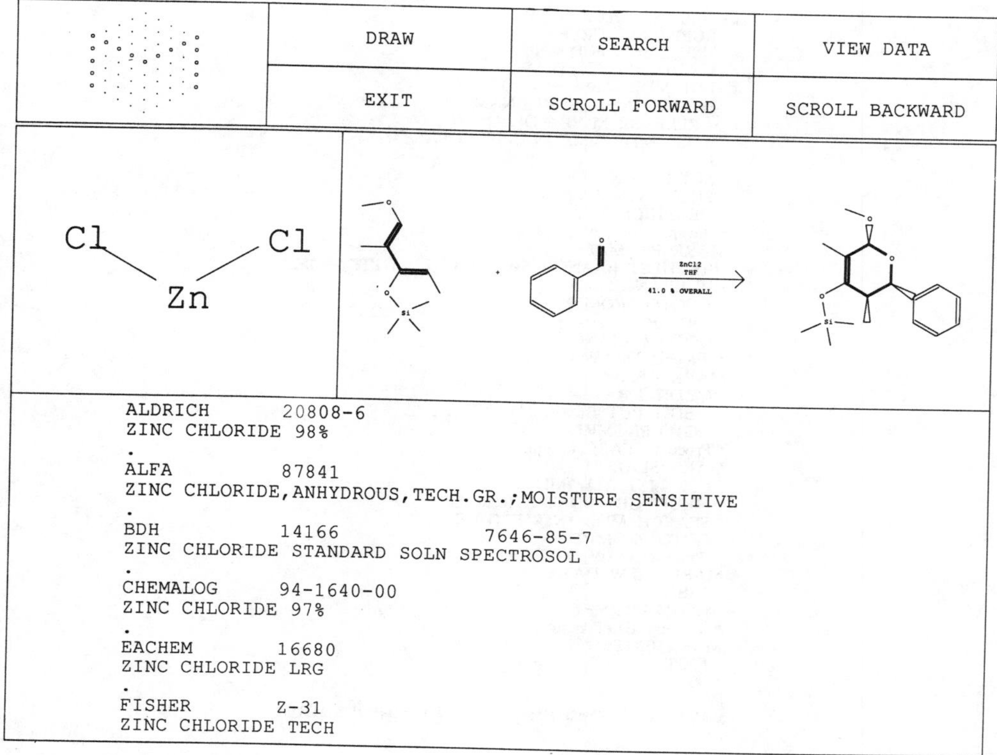

Figure 6. Customised MACCS-II Form for Data Display.

Directory database and display data and metafiles. The series of commands required to automate this procedure is written in sequence language and is shown in Figure 7. Scrolling buttons are also incorporated in the menu in order to display data that may not fit into the box.

While this example is typically encountered in basic research laboratories, researchers in process or technical development frequently have quite different needs. High among these are methods to store and search for information associated with the work carried out to improve an individual reaction by changing the conditions and/or reagents. In REACCS this information can be stored under one reaction entry in the form of variations. Furthermore, the ability of REACCS to handle molecular information – structures and associated data – allows the formulation of unique queries as illustrated in Figure 8. The search in this case is for all reactions in which a cyclopentene thiolester is formed. These conversion queries change the result of a search for molecules into reactions or *vice versa*.

One of the hits generated by the search displays in the lower right corner of the display form the information 'Variation 1 of 6' (Figure 9). This notice indicates that the reaction was carried out six times under different conditions. Again, the chemist uses the menu generating facility of the MACCS-II system to prepare a summary of the data as shown in Figure 10.

```
* Initialize screen
  SUPPRESS SCREEN
  USE MENU FCD.MNU
  USE FORM FCD.FRM
  LET MDBNAME = 'FCD:'
  IF $MDNAMI <> MDBNAME
    RELEASE MDB /NOPROMPT
    USE MDB &MDBNAME /NOPASS
  END IF
  SET EDGES OFF
  SHOW SCREEN
  GRAPHICS
* Prompt for button hit
MARK PROMPT
  REQUEST BRANCH /NOPROMPT /BUTTON=OFLD
  IF BRANCH = ' '
    GOTO PROMPT
  END IF
  GOTO &BRANCH
* Process DRAW request
MARK DRAW
  MEDIT DRAW
  MEDIT PUT BDAT STRUCTURE
  GOTO PROMPT
* Process SEARCH request
MARK SEARCH
  USE LIST ALL /MD
  MEDIT GET BDAT STRUCTURE
  SEARCH MDB *STRUCTURE
  GOTO PROMPT
* Process VIEW DATA request
MARK VIEW DATA
  VIEW
  GOTO PROMPT
* Process EXIT request
MARK EXIT
  EXIT
```

Figure 7. Sequence Language Example.

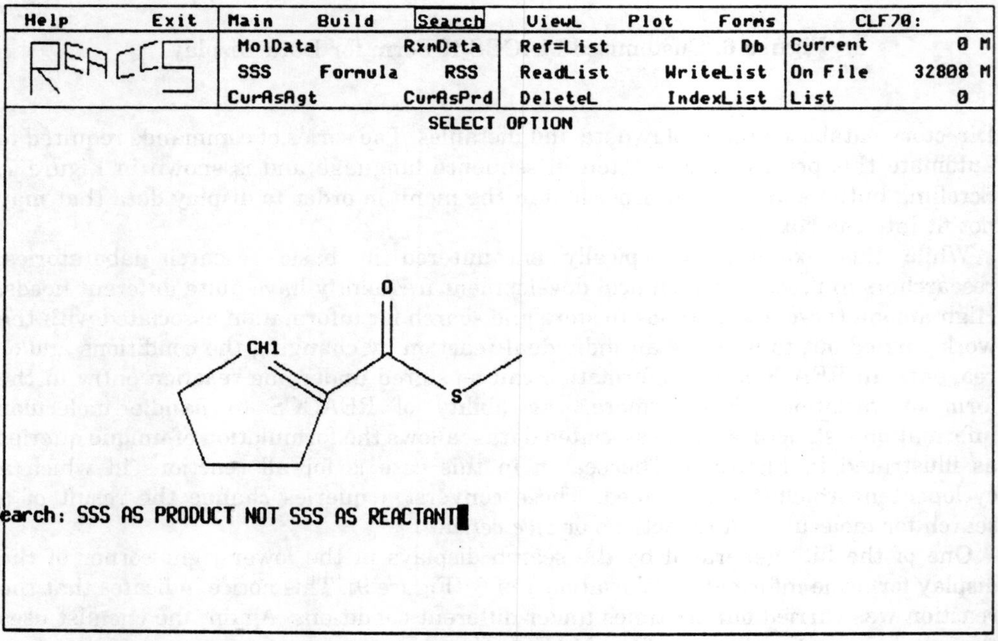

Help Exit	Main Build Search ViewL Plot Forms		CLF70:	
REACCS	MolData RxnData	Ref=List Db	Current	0 M
	SSS Formula RSS	ReadList WriteList	On File	32808 M
	CurAsAgt CurAsPrd	DeleteL IndexList	List	0

SELECT OPTION

O

CH1

S

Search: SSS AS PRODUCT NOT SSS AS REACTANT

Figure 8. Conversion Search Query.

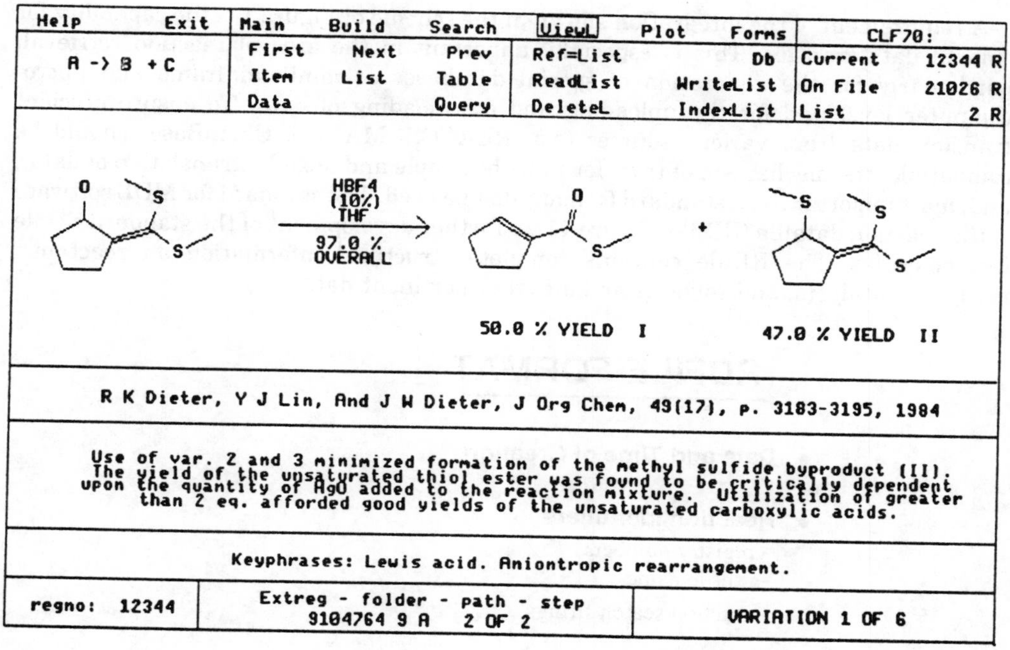

Help Exit	Main Build Search ViewL Plot Forms	CLF70:
A -> B + C	First Next Prev Ref=List Db	Current 12344 R
	Item List Table ReadList WriteList	On File 21026 R
	Data Query DeleteL IndexList	List 2 R

HBF4 (10%) THF

97.0 % OVERALL

50.0 % YIELD I

47.0 % YIELD II

R K Dieter, Y J Lin, And J W Dieter, J Org Chem, 49(17), p. 3183-3195, 1984

Use of var. 2 and 3 minimized formation of the methyl sulfide byproduct (II). The yield of the unsaturated thiol ester was found to be critically dependent upon the quantity of HgO added to the reaction mixture. Utilization of greater than 2 eq. afforded good yields of the unsaturated carboxylic acids.

Keyphrases: Lewis acid. Aniontropic rearrangement.

regno: 12344	Extreg - folder - path - step 9104764 9 A 2 OF 2	VARIATION 1 OF 6

Figure 9. Result of Conversion Search Query.

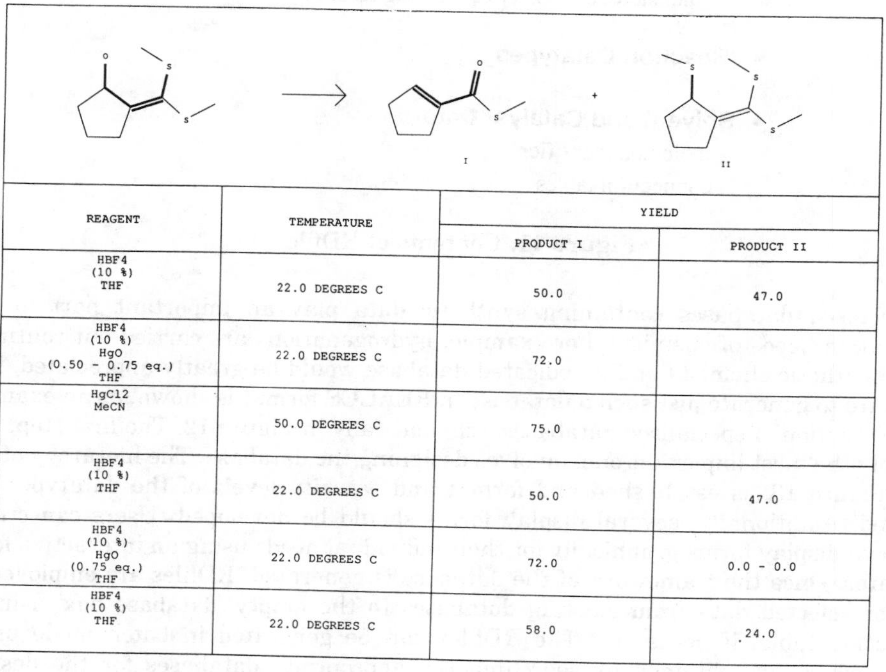

I

II

REAGENT	TEMPERATURE	YIELD	
		PRODUCT I	PRODUCT II
HBF4 (10 %) THF	22.0 DEGREES C	50.0	47.0
HBF4 (10 %) HgO (0.50 - 0.75 eq.) THF	22.0 DEGREES C	72.0	
HgCl2 MeCN	50.0 DEGREES C	75.0	
HBF4 (10 %) THF	22.0 DEGREES C	50.0	47.0
HBF4 (10 %) HgO (0.75 eq.) THF	22.0 DEGREES C	72.0	0.0 - 0.0
HBF4 (10 %) THF	22.0 DEGREES C	58.0	24.0

Figure 10. Customised Menu for Display of Variations.

A requirement of the integration shown in the scheme of Figure 1 is the capability for smooth data transfer. This is especially important in the area of reaction retrieval, particularly for the generation of special databases at mini/mainframe and microcomputer level and for the uploading and downloading of data. To ensure efficient transfer, data from various sources (e.g. REACCS, MACCS, ChemBase) should be compatible, the mechanism of transfer must be simple and flexible (translation of data), and, most importantly, a standard format must be used. The standard for MDL software is the reaction datafile (RDfile, Figure 11), a further development of the standard SDfile for molecules. The RDfile contains complete structural information on reactants, products, catalysts, and solvents and all other pertinent data.

RDFILE FORMAT

- **Date and Time of Creation**
- **Reaction Identifiers**
  - registry numbers
  - rxnfile name
  - reaction search strings
- **Connection Tables of Reactants and Products**
  - atom and bond descriptors
    (incl.stereochemistry & reacting centers)
- **Reaction Datatypes**
- **Solvent and Catalyst Data**
  - molecule identifiers
  - connection tables

Figure 11. Contents of RDfile.

Specialised databases containing synthetic data play an important part in the information needs of chemists. For example, hydrogenations are carried out routinely by all synthetic chemists and a dedicated database would be greatly appreciated. The procedure to generate just such a database in REACCS format is shown, as an example for the creation of specialised databases, schematically in Figure 12. The first step, and probably the most important one, involves designing the database. The hierarchy of the data (Figure 13) is established and format and security levels of the datatypes are installed. Additionally, several display forms should be developed. Users can create their own display forms graphically for their individual needs using an interactive form generator. Once the framework of the database is generated, RDfiles are employed to transfer selected data from existing databases to the empty 'database box' using a translation table, if necessary. The RDfiles can be generated in batch mode using hitlists which are obtained by searching the appropriate databases for the desired reactions. In our case, we searched for reactions in which the datatype 'atmosphere'

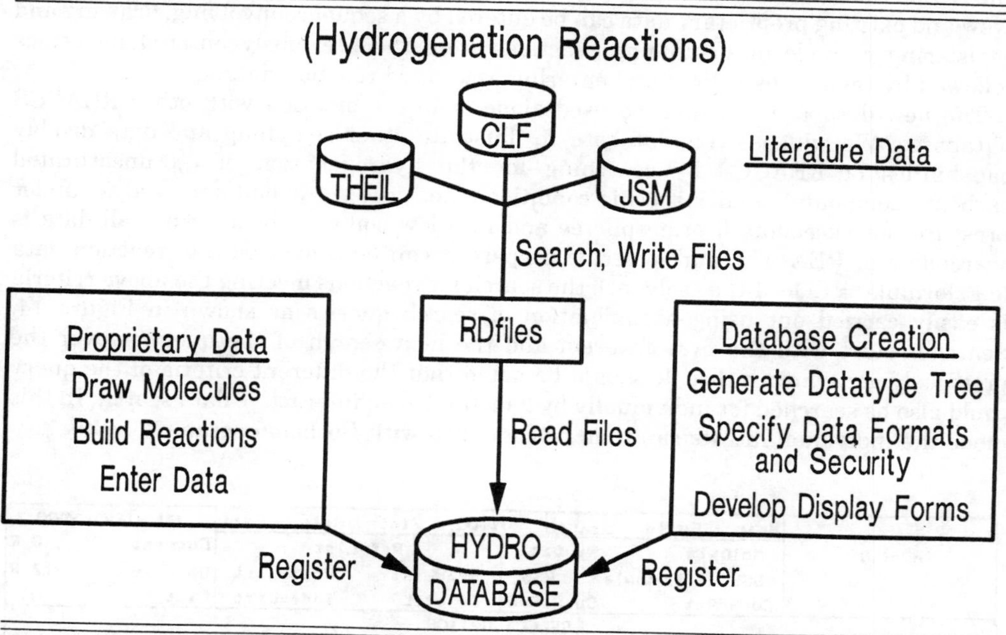

Figure 12. Schematic View for Generation of a Specialised Database.

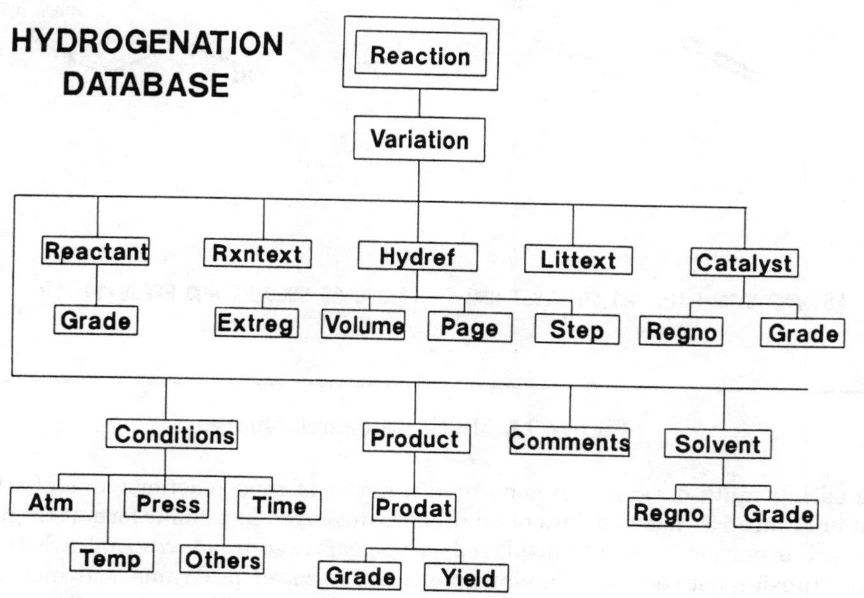

Figure 13. REACCS Database Hierarchy.

equals 'H2'. In addition to the structures of all molecules participating in the reactions the RDfile contains all the reaction data which should be passed into the new database. New and existing proprietary data can be entered by a sequence involving, drawing and registering of molecules; building reactions from the previously entered molecules followed by their registration; and entering associated reaction data.

The new database can then be used alone or in combination with other REACCS databases. We will use this database to illustrate other searching and data display capabilities of REACCS by searching for the hydrogenation of α,β-unsaturated carbonyl compounds under specific conditions, i.e., using a palladium catalyst under pressure not exceeding 5 atmospheres and in a low-boiling solvent. Since all data is searchable in REACCS, and molecule properties can be converted into reaction data (e.g. formula = C(1−3)O as solvent), the search for reactions meeting the above criteria is easily carried out using acombination of search queries as shown in Figure 14. Searches start with the first criterion and the next search of the query is over the previously generated hitlist. It should be noted that the different criteria of the query could also be searched for individually by a menu-driven interactive data search. In this mode the individual hitlists can then be combined with Boolean operators.

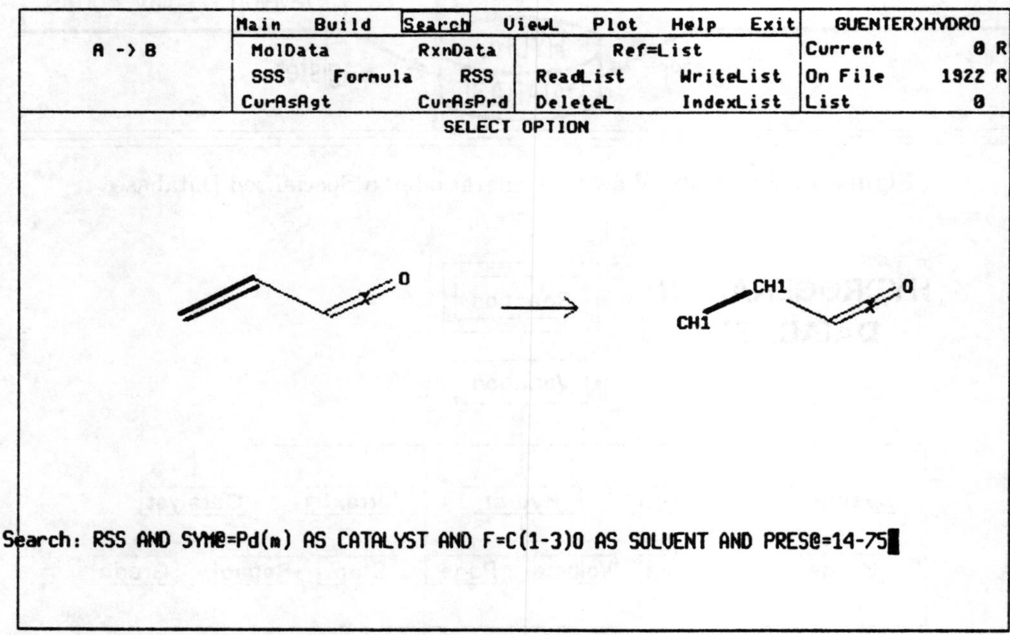

Figure 14. Multidata Search Query.

The earlier multi-data search generated a hitlist of nine reactions, one of which is shown in Figure 15. Selected data of all hits are displayed in tabular format (Figure 16) for a quick overview. The data displayed can be captured in an external ASCII file.

Data transfer between mini/mainframe and PC-based programs will increasingly play an important role as more PC's find their way into the laboratory. The ease and speed with which programs like ChemBase draw molecules make these programs ideal

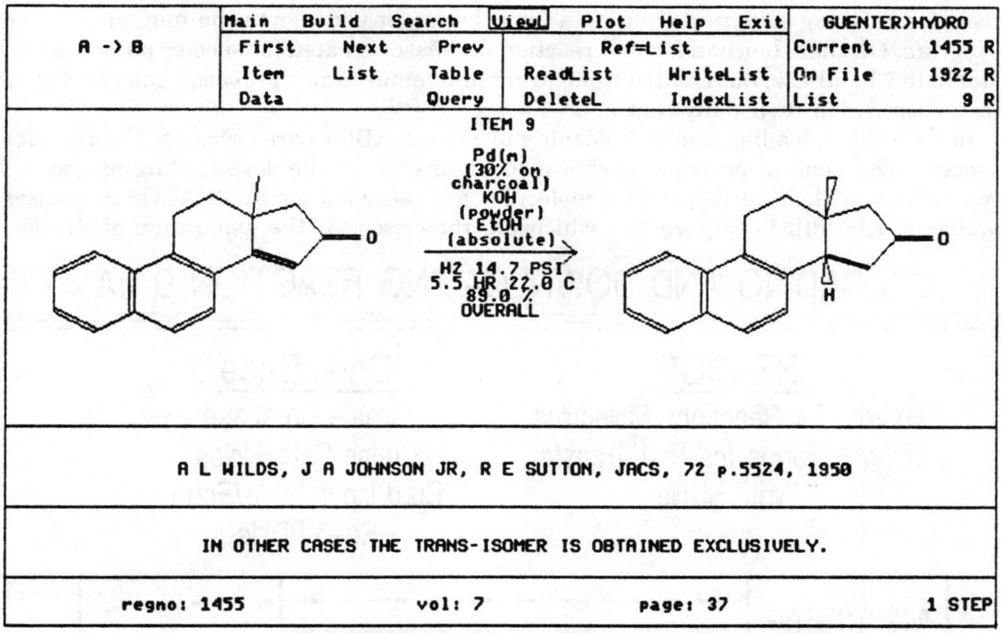

A -> B	Main Build Search View Plot Help Exit	GUENTER>HYDRO
	First Next Prev / Ref=List	Current 1455 R
	Item List Table / ReadList WriteList	On File 1922 R
	Data Query / DeleteL IndexList	List 9 R

ITEM 9

Pd(m) (30% on charcoal) KOH (powder) EtOH (absolute)

H2 14.7 PSI 5.5 HR 22.0 C 89.0 % OVERALL

A L WILDS, J A JOHNSON JR, R E SUTTON, JACS, 72 p.5524, 1950

IN OTHER CASES THE TRANS-ISOMER IS OBTAINED EXCLUSIVELY.

regno: 1455 vol: 7 page: 37 1 STEP

Figure 15. Result of Multidata Search.

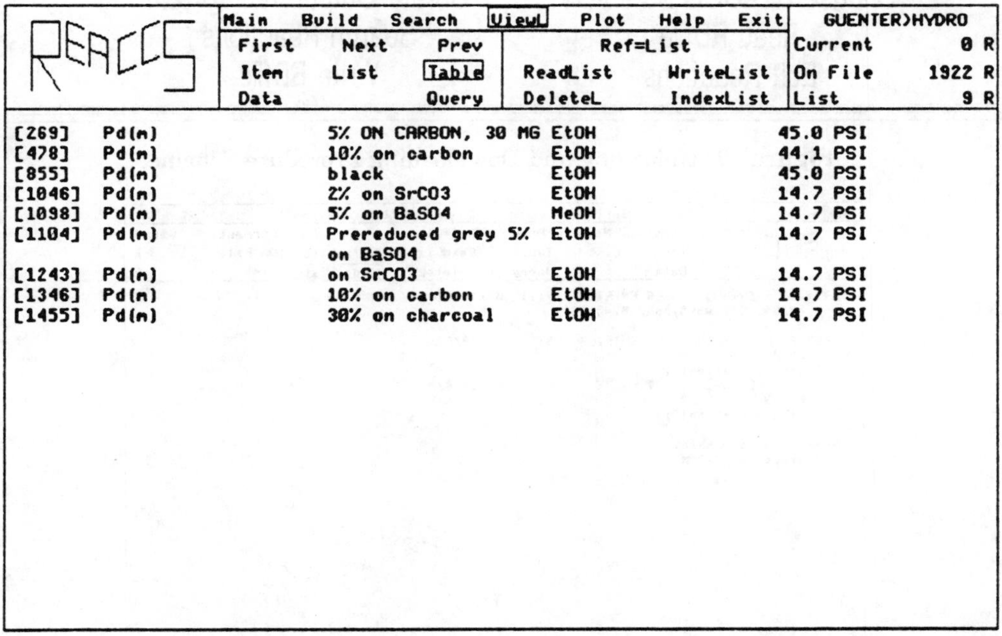

REACCS	Main Build Search View Plot Help Exit	GUENTER>HYDRO
	First Next Prev / Ref=List	Current 0 R
	Item List Table / ReadList WriteList	On File 1922 R
	Data Query / DeleteL IndexList	List 9 R

[269]	Pd(m)	5% ON CARBON, 30 MG	EtOH	45.0 PSI
[478]	Pd(m)	10% on carbon	EtOH	44.1 PSI
[855]	Pd(m)	black	EtOH	45.0 PSI
[1046]	Pd(m)	2% on SrCO3	EtOH	14.7 PSI
[1098]	Pd(m)	5% on BaSO4	MeOH	14.7 PSI
[1104]	Pd(m)	Prereduced grey 5% on BaSO4	EtOH	14.7 PSI
[1243]	Pd(m)	on SrCO3	EtOH	14.7 PSI
[1346]	Pd(m)	10% on carbon	EtOH	14.7 PSI
[1455]	Pd(m)	30% on charcoal	EtOH	14.7 PSI

Figure 16. REACCS Data Display in Table Format.

tools for entering structures and data for subsequent transfer to the mini/mainframe program. On the other hand, small reaction databases created for specific purposes and stored in ChemBase, but created from larger mini/mainframe databases, can effectively help chemists in their daily routine.

In both the uploading and downloading processes, RDfiles are crucial to the transfer process. The general principle is shown in Figure 17. In the downloading procedure, reactions and their corresponding molecules are searched for in REACCS databases and stored in hitlists (Figure 18), which are then used for the generation of RDfiles.

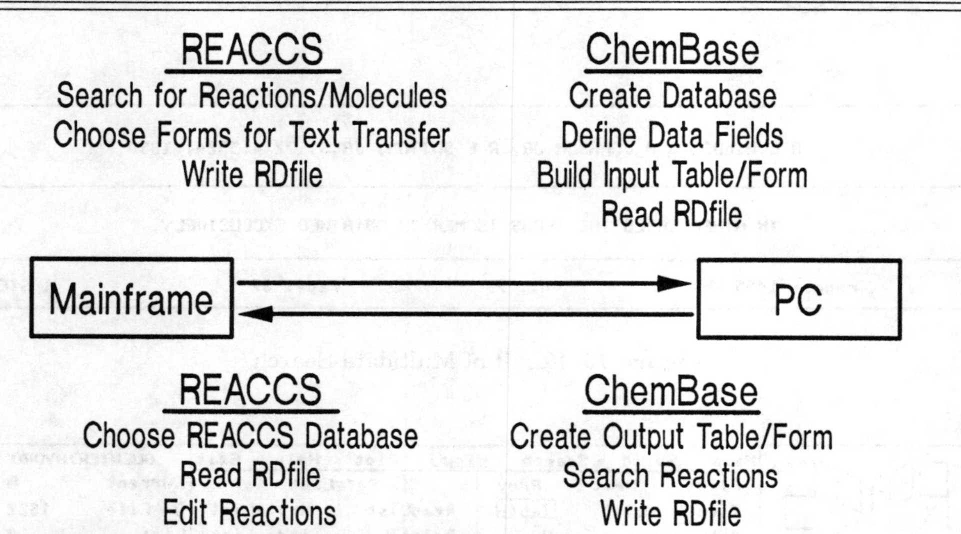

Figure 17. Uploading and Downloading Procedure Scheme.

```
Help      Exit   | Main   Build   Search  | ViewL   Plot   Forms   |  *>JSM
                 | First  Next    Prev    | Ref=List        Db     | Current   0 R
 REACCS          | Item   List    Table   | ReadList  WriteList    | On File   0 R
                 | Data           Query   | DeleteL   IndexList    | List      1 R

LIST 1    [RXNS]      6 ENTRIES  17:11  02/18/87
    r2104,3051,4369,7849,8740,9611
    rss search
LIST 2    [RXNS]      2 ENTRIES  09:54  02/19/87
    r2104,7849
    with variation(2)
LIST 3    [RXNS]      1 ENTRY    10:18  02/19/87
    r7849
    with variation(3)

Summary by database:
  3 lists for *>JSM
```

Figure 18. Lists of Reactions for Transfer to ChemBase Database.

Additionally, the choice of an appropriate REACCS display form is important. Information shown on or below the reaction arrow and below the reactant(s) and/or the product(s) is passed into ChemBase as screentext and appears in the ChemBase form as it did in the REACCS reaction display.

Using ChemTalk, the communication program from Molecular Design Limited, the newly created RDfile is then transferred from the mini/mainframe to the PC. From ChemTalk the user can easily enter ChemBase to create a new reaction database with data fields corresponding to the REACCS data which is being transferred. With the aid of an input table, designed to accept designated data, the RDfile is now read into the ChemBase database. This input process is carried out on individual reactions and allows the user to accept or reject information on an individual basis. The process can also be automated and all reactions are entered, without qualification, into the database. Figure 19 displays a reaction obtained from the new PC database using a Transform Substructure Search (TSS). This new type of reaction substructure search, developed especially for ChemBase, looks for the transformation indicated in the query.

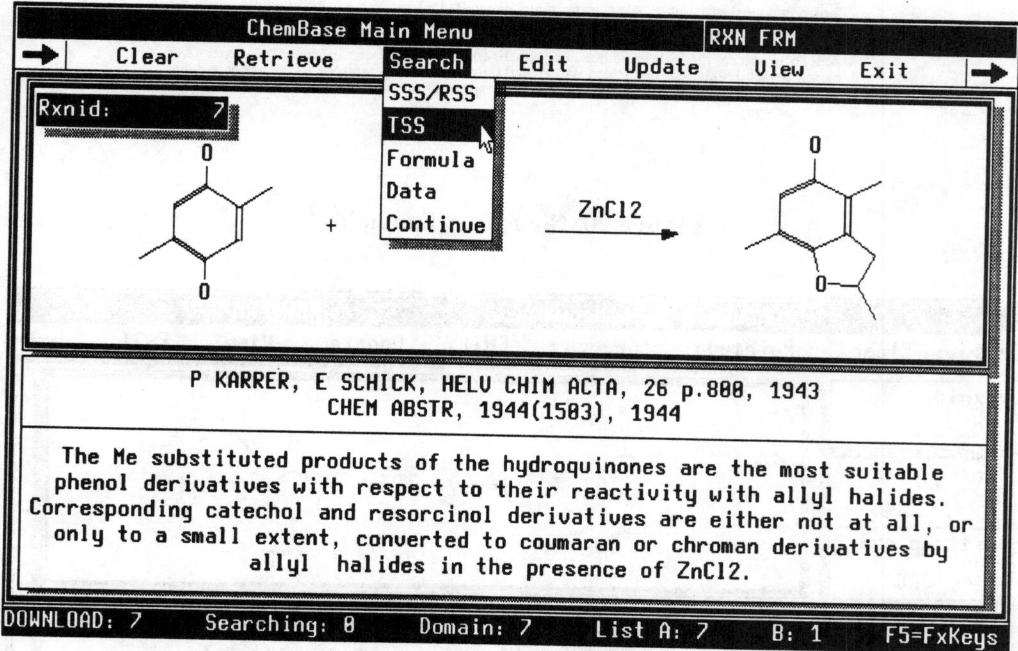

Figure 19. Result of a TSS Search in ChemBase.

The versatility of the integrated system discussed thus far is further enhanced with the capability of creating a document combining data (transferred from the new database) with text. Figure 20 illustrates how ChemText, MDL's chemical wordprocessing system, could have been used to prepare this manuscript for publication (the text was taken from the original literature).

The uploading process of a ChemBase database into REACCS differs from the previously discussed downloading procedure primarily in the order of events. In the first step a ChemBase database is created with data fields that reflect the datatypes in the corresponding REACCS database. A form is generated containing the data fields

Es schien uns nun von Interesse, auch C-Methylderivate des Hydrochinons und Resorcins auf ihre Fähigkeit, mit Allylhalogeniden zu Chroman- oder Cumaranderivaten zusammenzutreten, zu prüfen. Dabei sollten im wesentlichen die Bedingungen der Tocopherolsynthese eingehalten werden, d. h. wir arbeiteten mit den freien Phenolen in Benzol oder Ligroin als Lösungsmittel und Zinkchlorid als Katalysator.

Experimenteller Teil

Versuche zur Kondensation von 4,5-Dimethyl-resorcin und 2,5-Hydrochinon mit Allylbromid

Diese Versuche haben wir in ähnlicher Weise wie die Kondensation von Trimethyl-hydrochinon mit Phytylbromid zu *d,l*-α-Tocopherol[1] und mit Allylbromid zu 2,4,6,7-Tetramethyl-5-oxycumaran[2] ausgeführt. Als Lösungsmittel diente Benzol, als Katalysator Zinkchlorid. In der Wärme reagierten die Mischungen unter starker Bromwasserstoffentwicklung. Nach der Zersetzung der Reaktionsmasse mit Wasser zog man mit Äther aus, trennte die ätherlöslichen Anteile in solche, die sich in verdünnter wässeriger Lauge lösten und in Neutralprodukte. Aus diesen beiden Fraktionen versuchten wir sowohl durch chromatographische Reinigung wie durch Herstellung von Allophanaten einheitliche Reaktionsprodukte abzutrennen.

1) Helv. **21**, 520 (1938).
2) Helv. **21**, 939 (1938).

Figure 20. ChemText Example.

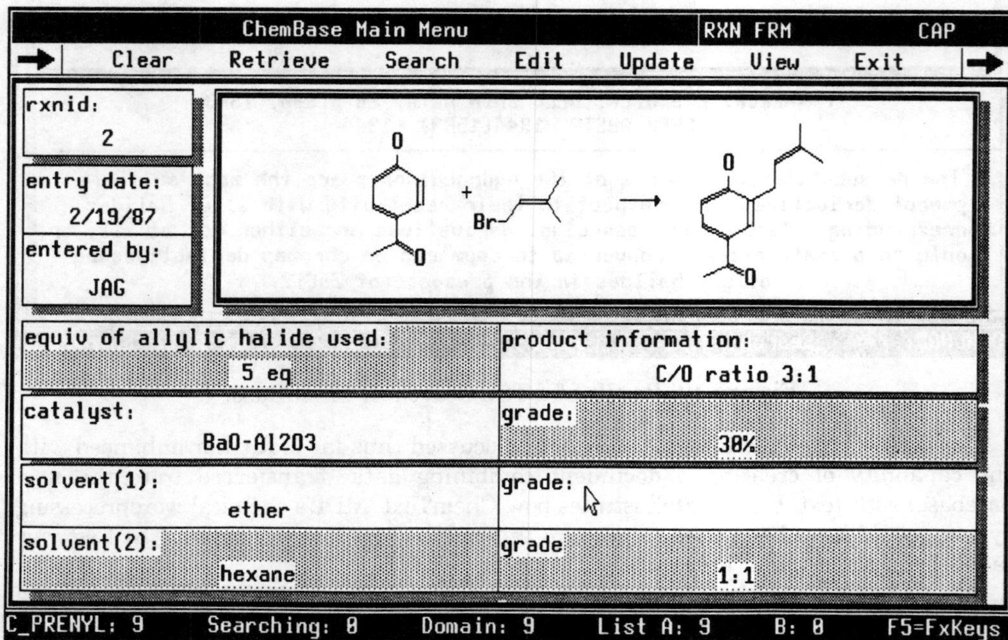

Figure 21. ChemBase Output Form.

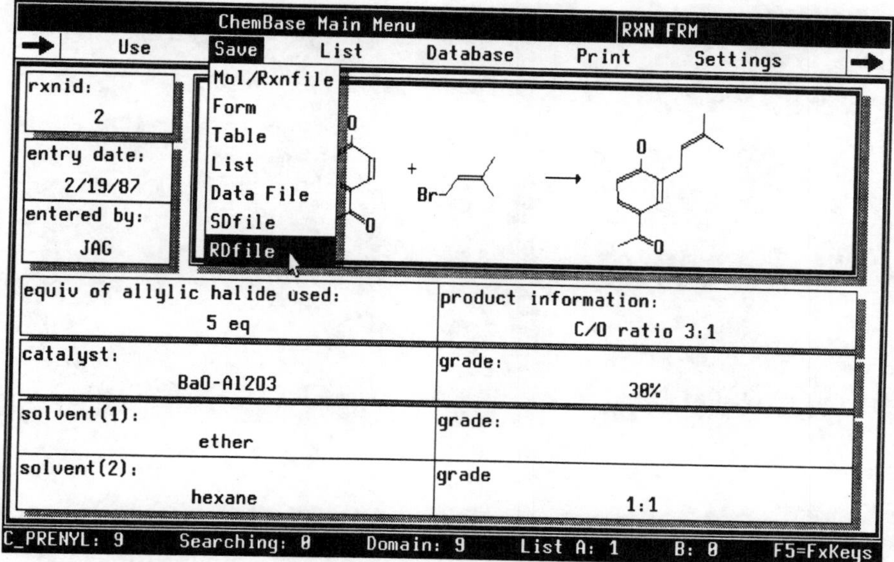

Figure 22. Data for RDfile Generation in ChemBase.

Help Exit | Main Build Search ViewL Plot Forms | *>DB

A + B -> C | Clear Db ReadFile WriteFile IndexData | Current 1 R

Find Register Molecule Reaction | On File 4 R

Delete Data Current | List 0

SELECT OPTION

Regno: 1 | Entry_date: 2/19/87 | Entered_by: JAG

OH + Br 5 eq → (BaO-Al2O3, 30%, Hexane, Ether, 1:1, 24.0 HR, 25.0 C) OH, O, C/O ratio 3:1, 39.0 % OVERALL

Glusenkamp, K.; and Buchi, G.; J. Org. Chem, 51, 4481-4483 (1986) Bajwa, B.S.; Khanna, P.L.; Seshadri, T.R.; Indian J. Chem., 9, 1322 (1971)

Figure 23. Result of Transfer in REACCS Format.

and is filled with reaction information (shaded areas in Figure 21). An RDfile is then written generated from a search in the ChemBase database (Figure 22). A resulting REACCS reaction is shown in Figure 23.

The versatility and flexibility of these data transfer processes demonstrate the power and usefulness of an integrated system for the chemist.

POSTER SESSION: COMPUTER ASSISTED SYNTHESIS DESIGN USING CHIRON AND REACCS

David W Elrod
Computational Chemistry, Research Laboratories, The Upjohn Company, Kalamazoo, Michigan 49001, U.S.A.

ABSTRACT

Aranciamycinone, an anthracycline antitumour antibiotic, has three adjacent chiral centres on its A ring. Most of the published syntheses for anthracyclines have resulted in racemic products because of the difficulties inherent in asymmetric synthesis. CHIRON, developed by Steve Hanessian at the University of Montreal, uses a pattern recognition algorithm to relate the stereochemistry of the target to a library of chiral starting materials. CHIRON allows the precursors to be saved in a format compatible with MDL's REACCS program. REACCS was used to work out a possible pathway for the synthesis of optically active Aranciamycinone starting from chiral precursors suggested by CHIRON.

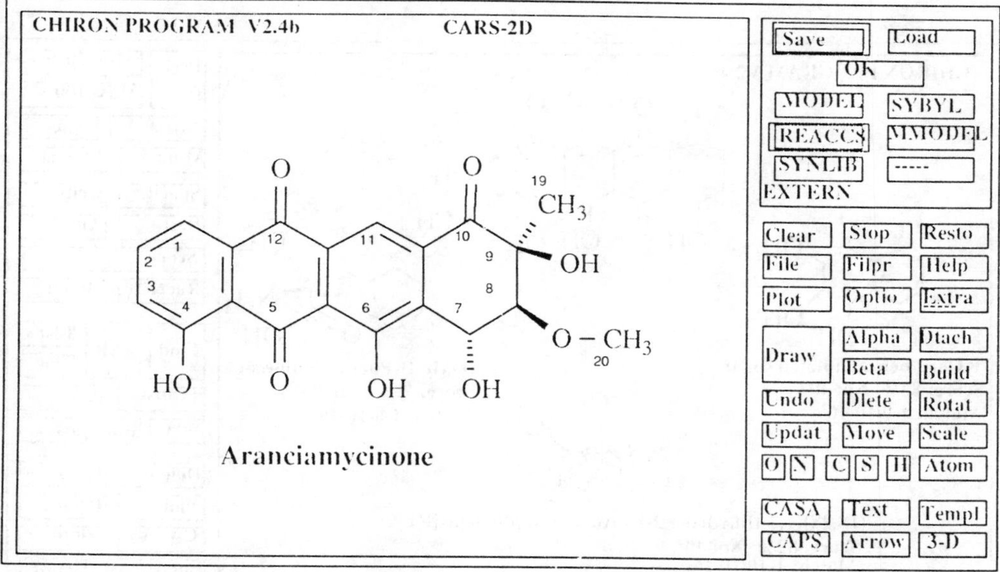

Figure 1

W. A. Warr (Ed.)
Chemical Structures

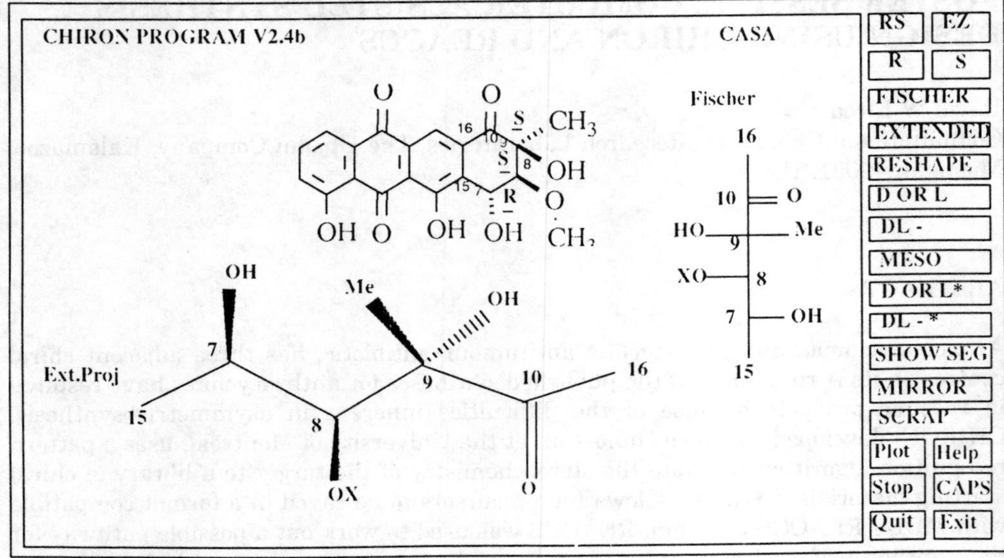

Figure 2

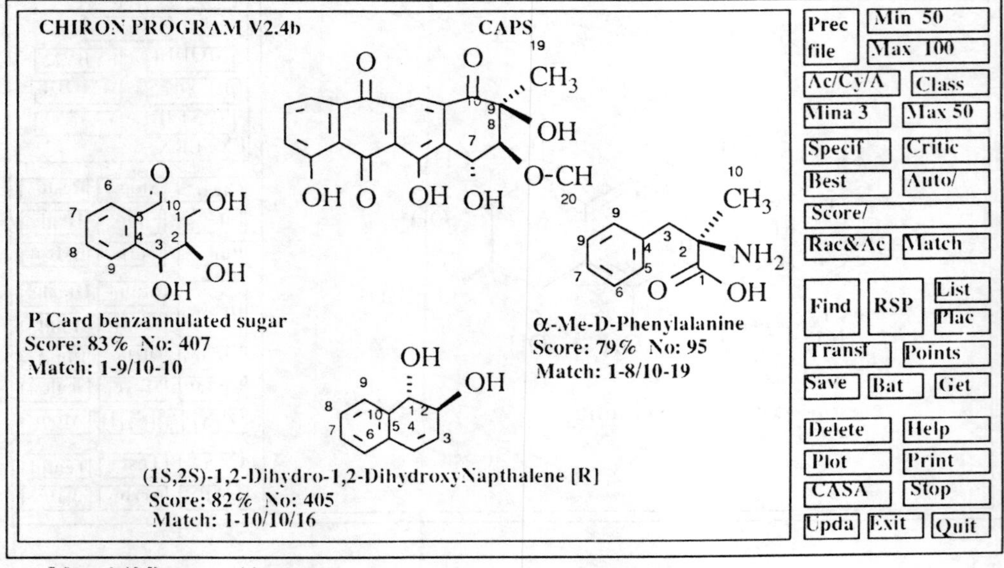

Figure 3

```
CHIRON PROGRAM

PRECURSORS FOUND    =  43
----------------

-----------------------------------
PRECURSOR #1 : P. Card benzannelated sugar (acyclic form)
   Score :  83    No : 407
   Reference : P. Card, JOC,47,2169(1982)
   Match P/M   1:9* 2:8* 3:7* 4:15 5:16 6:11 7:17 8:14 9:6 10:10
-----------------------------------
PRECURSOR #2 : (1S,2S)-1,2-Dihydro-1,2-Dihydroxy NAPHTHALENE [R]
   Score :  82    No : 405
   Reference : R. Miura et al, T.L.,5271(1968)
   Match P/M   1:10 2:9* 3:8* 4:7* 5:15 6:6 7:14 8:17 9:11 10:16
-----------------------------------
PRECURSOR #3 : (R)-MANDELIC ACID
   Score :  80    No : 491
   Reference : Comm.
   Match P/M   1:8* 2:7* 3:15 4:16 5:11 6:17 7:14 8:6
-----------------------------------
```

Figure 4

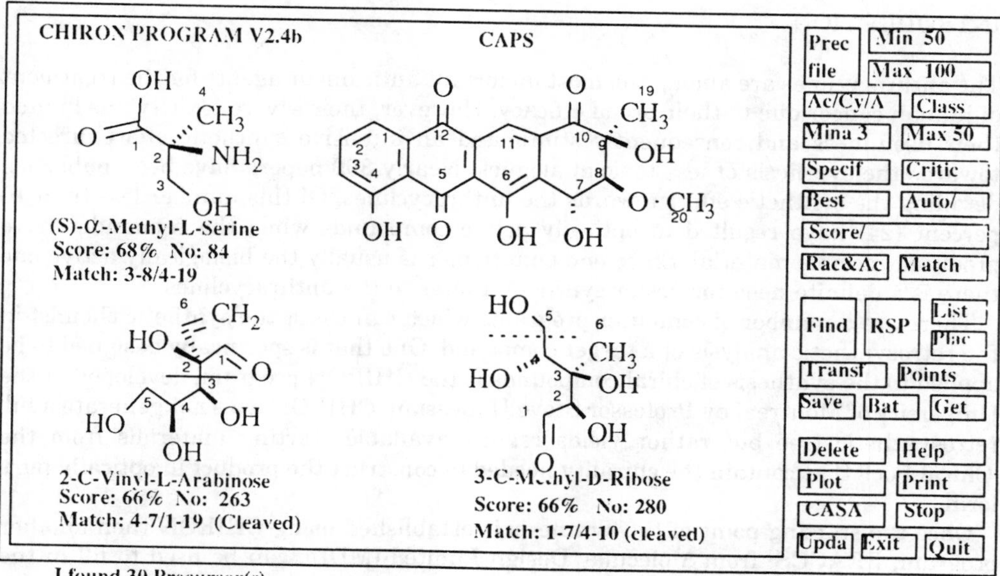

Figure 5

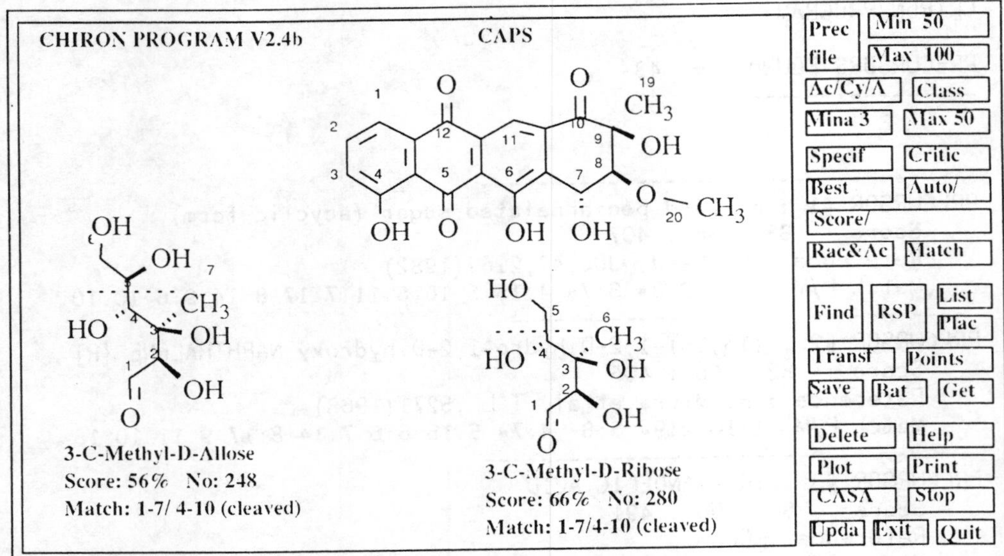

Figure 6

INTRODUCTION

The anthracyclines are among the most important antitumour agents for the treatment of human cancer due to their broad efficacy. However, their severe toxicity has limited their usefulness and consequently stimulated an intensive synthetic effort directed towards the synthesis of less toxic analogues. Nearly 300 papers have been published detailing the synthetic effort towards the anthracyclines. Of this number less than 10 percent (24) have resulted in optically active compounds while the remainder have produced racemic material. Since one enantiomer is usually the biologically active one there is a definite need for better synthetic routes to the anthracyclines.

There are a number of computer programs which can assist the synthetic chemist in the retrosynthetic analysis of a target compound. One that is specifically designed to be applied to the synthesis of chiral compounds is the CHIRON program, developed at the University of Montreal by Professor Steve Hanessian. CHIRON does not generate a full retrosynthetic tree but rather yields readily available starting materials from the 'Chiral Pool' that contain the chirality needed to construct the product in optically pure form.

Once the starting point of the synthesis is established using CHIRON then another program, REACCS, from Molecular Design Limited (MDL), can be used to fill in the details of the synthetic pathway that leads to the chiral product.

The synthetic problem used to illustrate this process is the generation of a proposed route for the chiral synthesis of the antitumour antibiotic aglycone Aranciamycinone. Aranciamycinone has been synthesised in racemic form by Kende *et al.*[4]

```
CHIRON PROGRAM

PRECURSORS FOUND  = 35
----------------

-----------------------------------
PRECURSOR #1 : (S)-alpha-Methyl-L-SERINE
    Score :  68    No :  84
    Reference : D. Seebach et al,
                T.L.,25,2545(1984)
    Match P/M  1:10 2:9* 3:8* 4:19
-----------------------------------
PRECURSOR #2 : 2-C-Vinyl-L-ARABINOSE
    Score :  66    No : 263 (Cleave)
    Reference : W.G. Overend et al,
                JCS,3433(1965)
    Match P/M  1:19 2:9* 3:8* 4:7* 5:10
-----------------------------------
PRECURSOR #3 : 3-C-Methyl-D-RIBOSE
    Score :  66    No : 280 (Cleave)
    Reference : A. Rosenthal et al,
                Carb.Res.,16,337(1971)
    Match P/M  1:7* 2:8* 3:9* 4:10
-----------------------------------
PRECURSOR #4 : 2-Hydroxymethyl-D-ERYTHROSE
    Score :  64    No : 268 (Cleave)
    Reference : P.T. Ho, T.L.,1623(1978)
    Match P/M  1:8* 2:9* 3:10 4:19
-----------------------------------
PRECURSOR #6 : 3-C-Aminomethyl-D-GLUCOSE
    Score :  58    No : 274 (Cleave)
    Reference : A. Rosenthal et al,
                Carb.Res.,13,113(1970)
    Match P/M  1:7* 2:8* 3:9* 4:19 5:10
-----------------------------------
PRECURSOR #12 : 3-C-Methyl-D-ALLOSE
    Score :  56    No : 248 (Cleave)
    Reference : F.A. Carey et al,
                Carb.Res.,12,463(1970)
    Match P/M  1:7* 2:8* 3:9* 4:10
```

Figure 7

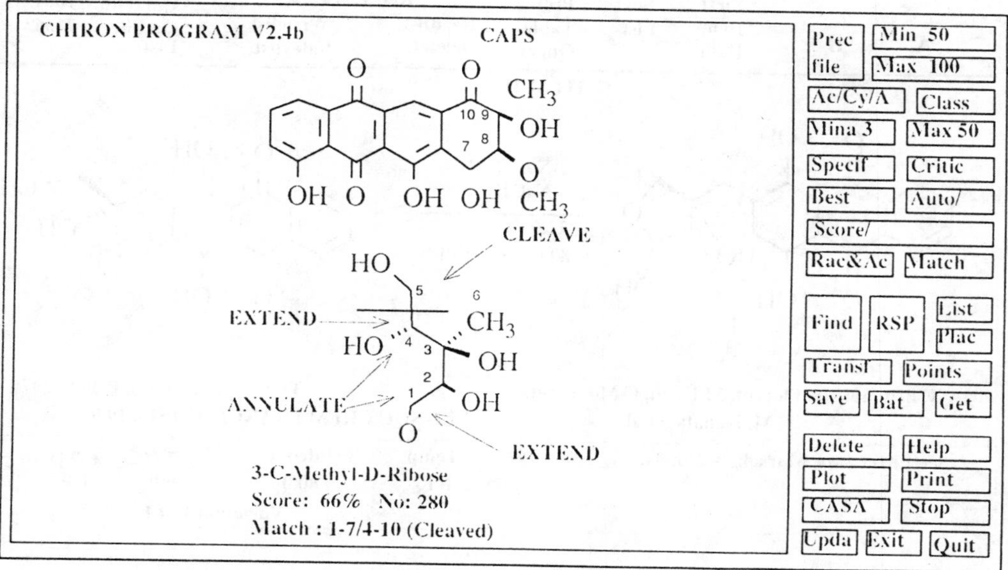

Figure 8

REACCS	Main	Build	Search	Viewl	Plot	Help	Exit	CLF:
A → B	First Item Data	Next List	Prev Table Query	Ref=List Readlist DeleteL		Writelist Indexlist		Current 19965 R On File 21363 R List 21 R

ITEM 21

NaOEt (4 mmol)
MeNO2 (3 ml)
EtOH (100 ml)
92% Overall
N2 24-48 hr

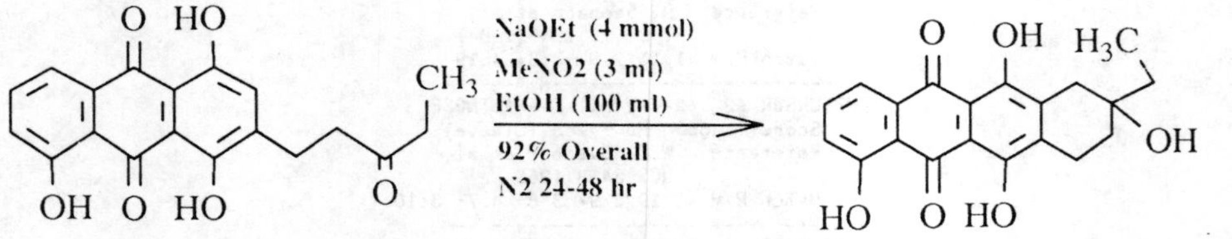

K Krohn, W Priyono, ANG CHEM INTERN ED, 25 p.339, 1986

Variation 1 of 1

Figure 9

REACCS	Main	Build	Search	Viewl	Plot	Help	Exit	CLF:
A → B	First Item Data	Next List	Prev Table Query	Ref=List Readlist DeleteL		Writelist Indexlist		Current 10647 R On File 21363 R List 2 R

ITEM 21

Na2S2O4
NaOH
22.0 C
80.0% Overall

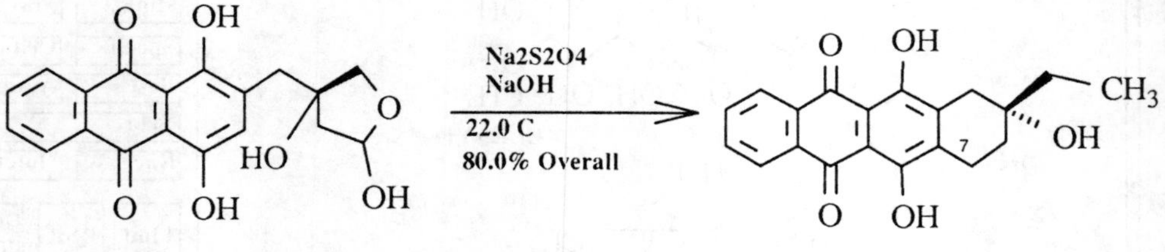

F Bennani, JC Florent, M Koch, C Monneret, TET LETT, 25(36), P.3975-3976
C Marschalk, et al, BULL SOC CHEM FRANCE, p. 1545, 1936

See Litref 2 for Marschalk Conds.

Temp	% 7-deoxy	% 7-α	% 7-β
RT	80.0	0.0	0.0

Variation 1 of 1

Figure 10

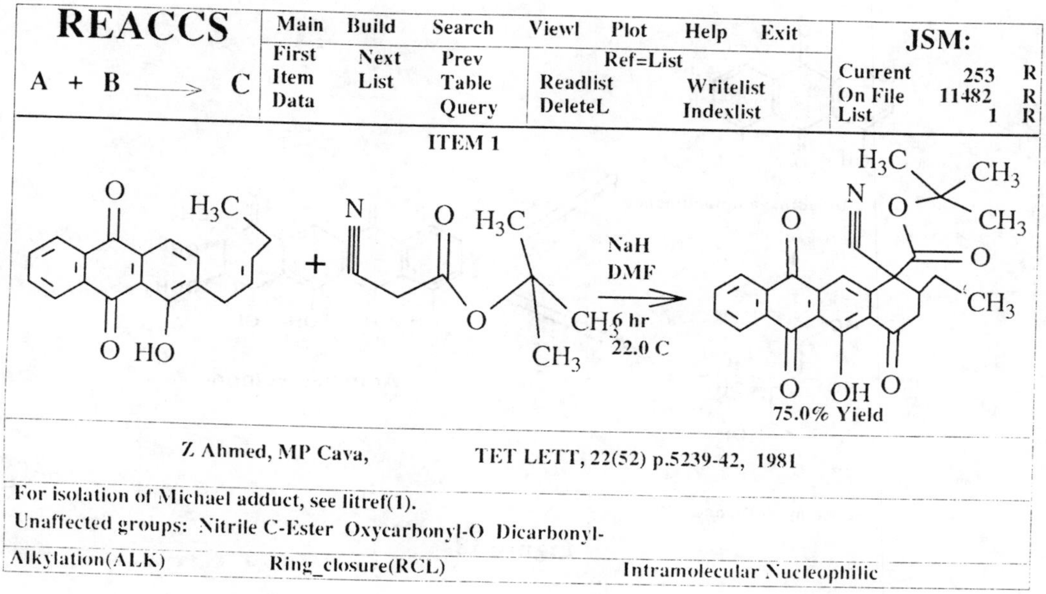

REACCS

A + B ⟶ C

Main Build Search Viewl Plot Help Exit

First Item Data | Next List | Prev Table Query | Ref=List Readlist DeleteL Writelist Indexlist

JSM:
Current 253 R
On File 11482 R
List 1 R

ITEM 1

NaH
DMF
6 hr
22.0 C

75.0% Yield

Z Ahmed, MP Cava, TET LETT, 22(52) p.5239-42, 1981

For isolation of Michael adduct, see litref(1).

Unaffected groups: Nitrile C-Ester Oxycarbonyl-O Dicarbonyl-

Alkylation(ALK) Ring_closure(RCL) Intramolecular Nucleophilic

Figure 11

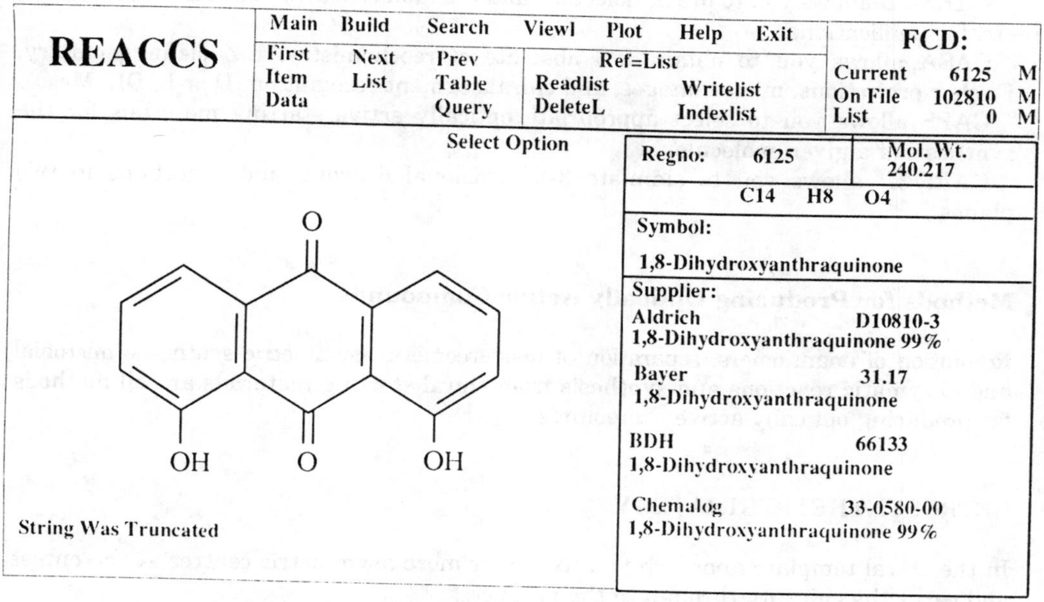

REACCS

Main Build Search Viewl Plot Help Exit

First Item Data | Next List | Prev Table Query | Ref=List Readlist DeleteL Writelist Indexlist

FCD:
Current 6125 M
On File 102810 M
List 0 M

Select Option

Regno: 6125 | Mol. Wt. 240.217

C14 H8 O4

Symbol:

1,8-Dihydroxyanthraquinone

Supplier:

Aldrich D10810-3
1,8-Dihydroxyanthraquinone 99%

Bayer 31.17
1,8-Dihydroxyanthraquinone

BDH 66133
1,8-Dihydroxyanthraquinone

Chemalog 33-0580-00
1,8-Dihydroxyanthraquinone 99%

String Was Truncated

Figure 12

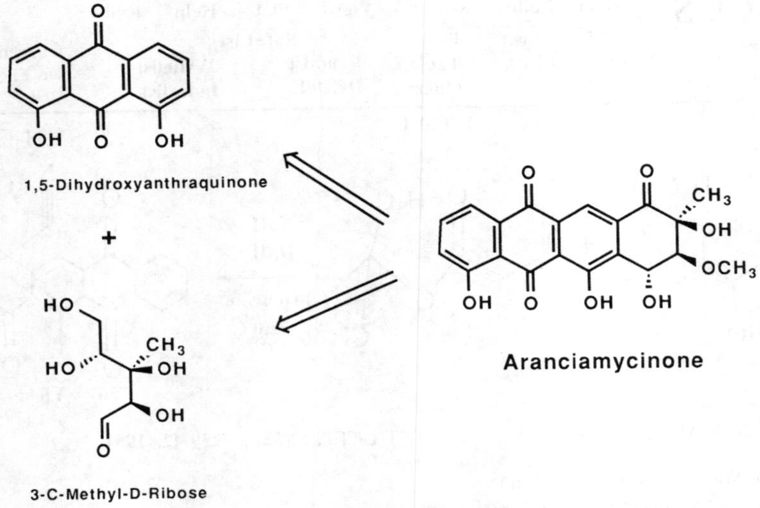

1,5-Dihydroxyanthraquinone

+

3-C-Methyl-D-Ribose

Aranciamycinone

Figure 13

METHODS

CHIRON is an interactive computer program for the analysis and perception of stereochemical features in molecules and for the selection of precursors in organic synthesis.

CARS-2D allows you to draw molecules and reaction schemes with arrows, reagents, etc. for publications.

CASA allows you to obtain R/S absolute stereochemistry, E/Z olefin geometry, Fischer projections, mirror images, and chiral segment recognition (D or L, DL, Meso).

CAPS allows you to select appropriate optically active starting materials for the synthesis of a given molecule.

CARS-3D allows you to simulate 3-Dimensional drawing and reflections in two planes.

Methods for Producing Optically Active Compounds

Resolution of enantiomers, separation of diastereomers, asymmetric synthesis, microbial and enzymatic reactions and synthesis from chiral starting materials are all methods for producing optically active compounds.

RETROSYNTHETIC STRATEGY

In the chiral template approach you fix one or more asymmetric centres at the outset and carry the chirality through to the product.

CHIRON is used to do retrosynthetic analysis of Aranciamycinone to give chiral starting materials.

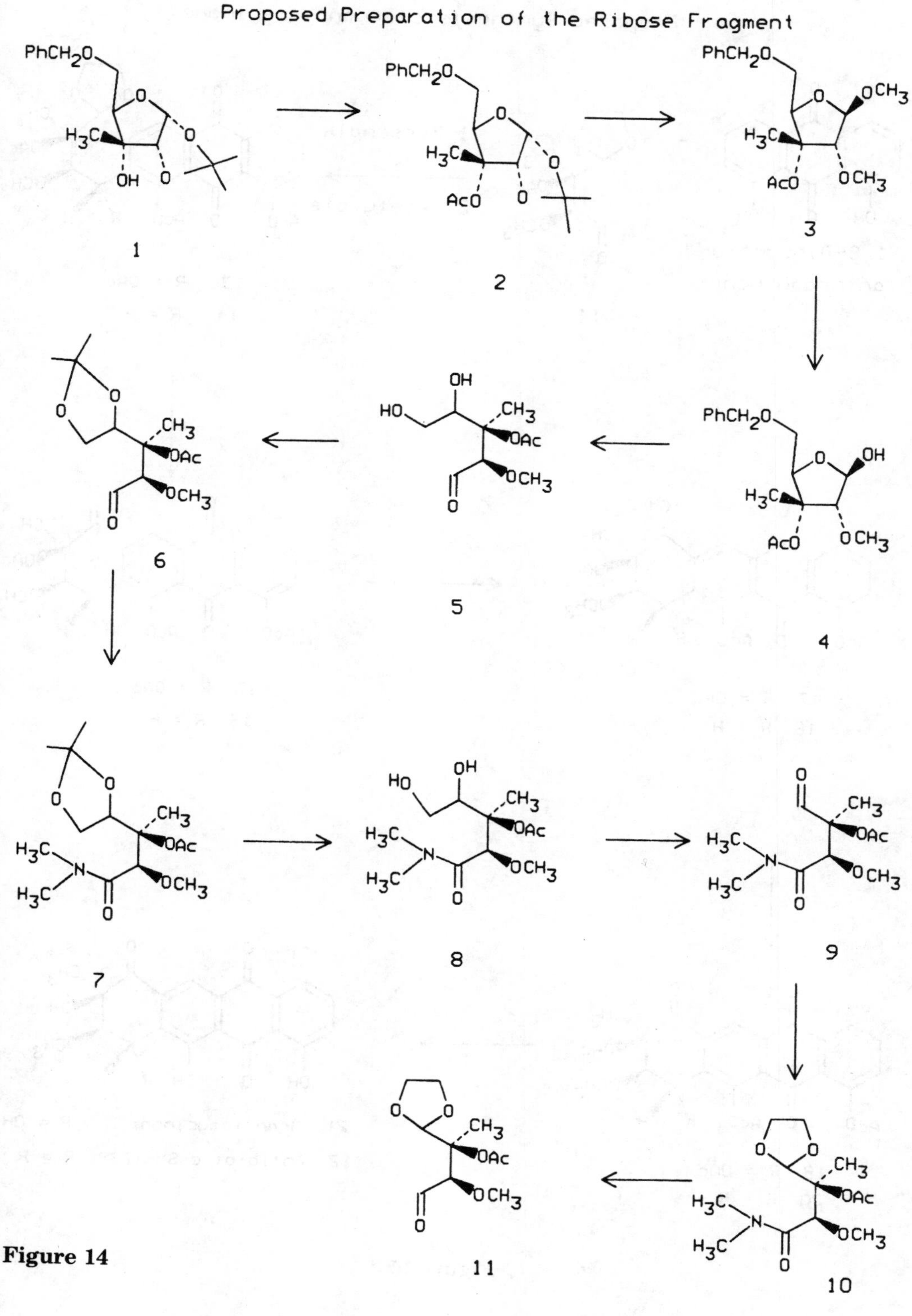

Proposed Preparation of the Ribose Fragment

Figure 14

Proposed Synthesis of Aranciamycinone

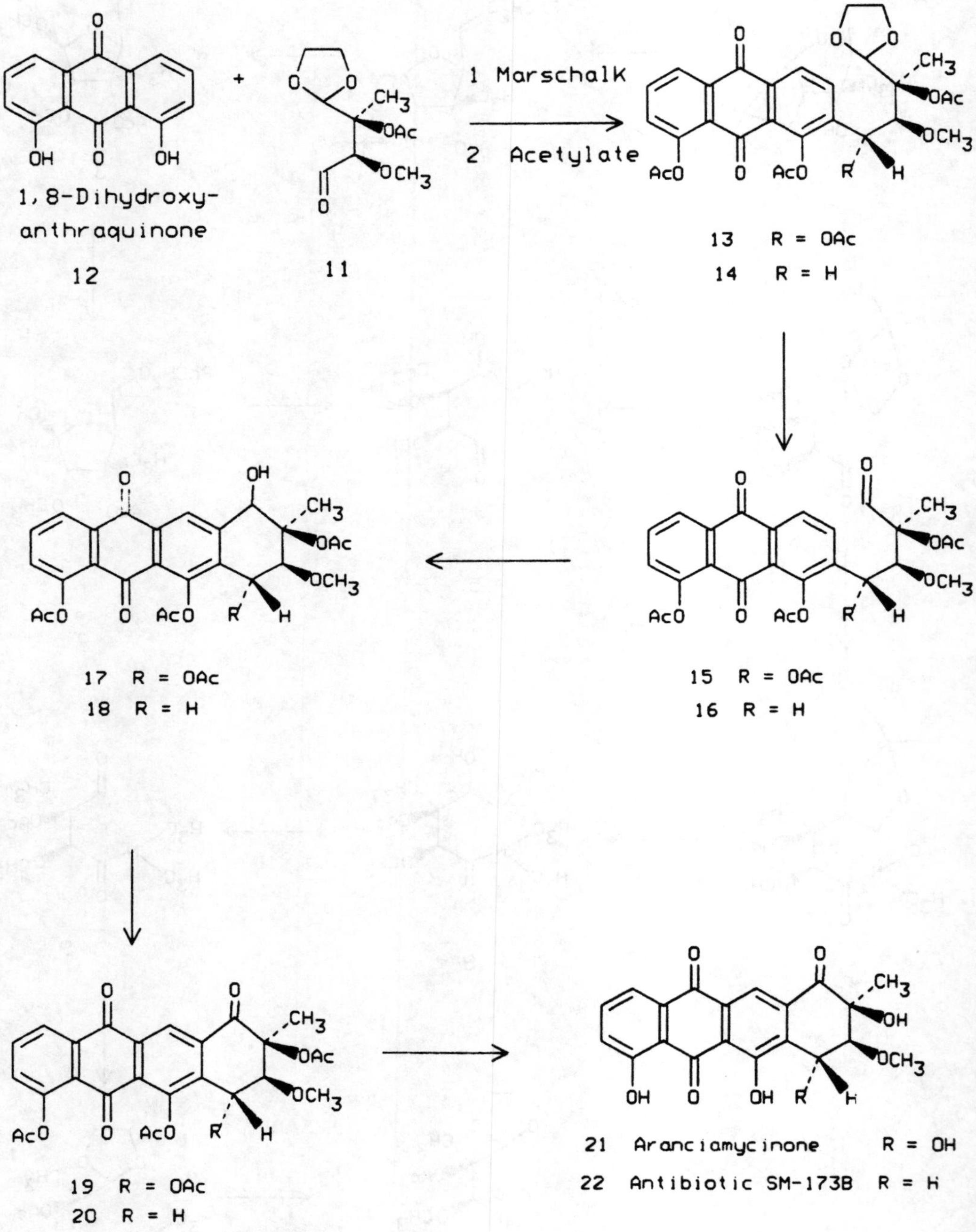

1,8-Dihydroxy-anthraquinone

12

11

1 Marschalk

2 Acetylate

13 R = OAc

14 R = H

15 R = OAc

16 R = H

17 R = OAc

18 R = H

19 R = OAc

20 R = H

21 Aranciamycinone R = OH

22 Antibiotic SM-173B R = H

Figure 15

You manually generate the synthetic pathway from chiral precursors to Aranciamycinone and then use REACCS to find reaction conditions and references for each step of the synthetic route.

CHIRONCOMPUTER ASSISTED PRECURSOR SELECTION (CAPS)

This performs retrosynthetic recognition of symmetry elements, functional group interrelations, and carbon framework patterns.

The following rules are applied:

Rule 1: Do not break a bond between 2 adjacent chiral centres.

Rule 2: Bond breaking in a retrosynthetic sense avoids disturbing chiral centres asmuch as possible to give the longest carbon chain that contains the maximum overlap of chirality and functionality with the precursors.

Rule 3: Match all possible disconnections generated by rules 1 and 2 against all possible orientations of the precursors.

CONCLUSIONS

CHIRON has generated chiral precursors which contain the necessary chirality for the construction of optically active Aranciamycinone.

The chemist must choose the most likely precursors based upon the chemist's knowledge, experience, and creativity.

CHIRON and REACCS can transfer structures back and forth between the programs.

REACCS can be used to fill in the details of the synthetic pathway.

REFERENCES

1. Bennani, F., Florent, J.C., Koch, M., Monneret, C. 'An Efficient Synthesis of Optically Active 4-Demethoxyanthracyclinones'. *Tetrahedron Lett.* 1984, *25* (36), 3975-3978.
2. Cava, M.P., Ahmed, Z., Benfaremo, N., Murphy, R.A., Jr., O'Malley, G.J. 'Anthraquinone Dye Intermediates as Precursors of Aklavinone-type Anthracyclinones'. *Tetrahedron* 1984, *40* (22), 4767–4776.
3. Hanessian, S., Major, F., Leger, S. 'Synthetic Design of Enantiomerically Pure Medicinal Agents – Aspects of Precursor Recognition by Visual and Computer-assisted Perception'. *New Methods Drug Res.* 1985, *1*, 201–224.
4. Kende, A.S., Johnson, S. 'Anthracyclinones by Oxidative Dearomatisation: Total Synthesis of SM-173B and Aranciamycinone'. *J. Org. Chem.* 1985, *50* (5), 727–729.
5. Krohn, K., Priyono, W. 'Synthetic Anthracyclinones: Cyclisation of an Intermediate Hydroxynitronate'. *Angew. Chem. Int. Ed. Engl.* 1986, *25*, 339–340.

PREDICTION OF CHEMICAL REACTIVITY AND DESIGN OF ORGANIC SYNTHESIS

Johann Gasteiger
Institute of Organic Chemistry, Technical University of Munich, D-8046 Garching, West Germany

Michael G Hutchings
Imperial Chemical Industries plc, Organics Division, Hexagon House, Blackley, Manchester M9 3DA, U.K.

Heinz Saller and Peter Löw
CHEMODATA Computer-Chemie GmbH & Co Kg, D-8038 Gröbenzell, West Germany

ABSTRACT

The representation of organic reactions in the computer should reflect our level of insight into chemical reactions. A detailed discussion of a reaction rests on the reaction mechanism. Empirical methods have been developed that allow the quantification of various chemical effects like heat of reaction, bond dissociation energies, charge distribution, inductive, resonance, and polarisability effects. The values obtained with these procedures are used in a general approach for the calculation of the reactivity of each individual bond in a molecule or reaction intermediate. Based on these reactivity values, the evolution of reaction mechanisms is achieved. This general protocol is used in EROS, a system for the prediction of the course of chemical reactions and the design of organic syntheses.

INTRODUCTION

An understanding of chemical reactivity is of paramount importance for the design of organic syntheses. It is essential to find good chemical reactions and to have a knowledge about their feasibility and limitations. Therefore computer-assisted synthesis design has to face the problem of predicting chemical reactions and reactivity. This is true for synthesis design systems based on a database of known reactions and also for those based on methods of formal reaction generation. And it is necessary to consider the reactivity of synthesis precursors when applying short-range and long-range strategies.

Thus, a short-range strategy for the synthesis of 4-hydroxy-1-phenyl-pentanone-2 (I) will lead to the two carbonyl compounds methyl benzyl ketone and acetaldehyde as synthesis precursors (Figure 1). Having reached this conclusion it is absolutely necessary to consider whether those two precursors will react by the anticipated aldol condensation to the target compound (I). A closer inspection based on insight into chemical reactivity has to decide that this reaction will not proceed unequivocally to the desired product (I). Rather, a mixture of compounds will be obtained including the

W. A. Warr (Ed.)
Chemical Structures

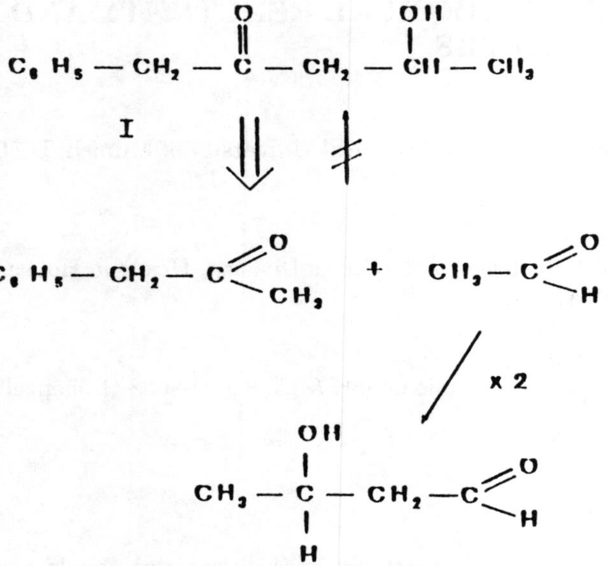

$$C_6H_5-CH_2-\overset{O}{\overset{\|}{C}}-CH_2-\overset{OH}{\overset{|}{CH}}-CH_3$$

I

$$C_6H_5-CH_2-C(=O)-CH_3 \quad + \quad CH_3-C(=O)H$$

x 2

$$CH_3-\underset{H}{\overset{OH}{C}}-CH_2-C(=O)H$$

Figure 1. Short range strategy for the synthesis of the β-hydroxyketone I.

product of the self-condensation of acetaldehyde. This leads to the conclusion that the initial strategy has to be discarded or that a protection of functional groups or different reaction conditions have to be applied.

Long-range strategies, too, need a knowledge about the feasibility and course of chemical reactions to fill up the gap between starting materials and the target molecule. In an interesting 14-step synthesis (1) of the antitumour agent AT125 the five carbon atoms of the target molecule were derived from cyclopentadiene (Figure 2). Let us assume that this central idea of the synthesis was the result of a sudden flash of insight. Then, to come up with the individual steps of the pathway from the starting material cyclopentadiene to the target molecule required careful consideration of the reactivity of each intermediate in the synthesis.

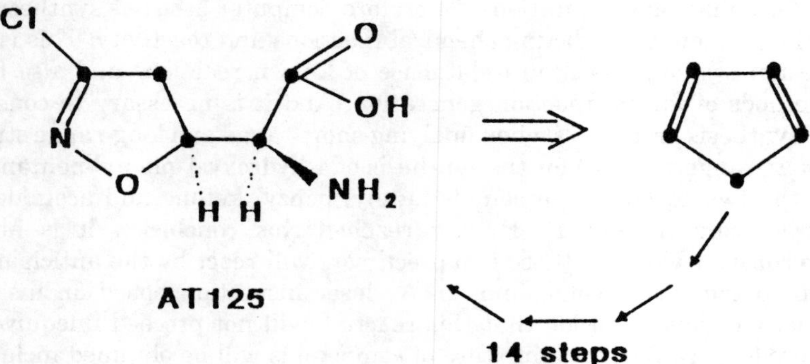

Figure 2. Long range strategy for the synthesis of AT125.

Several problems that show up and have to be clarified in synthesis design require an understanding of chemical reactivity: multi-site reactivity, alternative reactions, side reactions, protection of functional groups, and reaction conditions.

Organic chemists are used to discussing chemical reactivity in terms of reaction mechanisms. Predictions on the course of chemical reactions are made by a detailed development of individual mechanistic steps. It is our belief that the computer-modelling of chemical reactions should follow such a procedure and make predictions on reaction mechanisms.

FOUR LEVELS OF SOPHISTICATION FOR REACTION REPRESENTATION

With progress in the understanding of chemical reactions more and more information about a reaction has become available. This increase in chemical information can be structured into four levels and is illustrated by the representation of the hydrolysis of an ester (Figure 3).

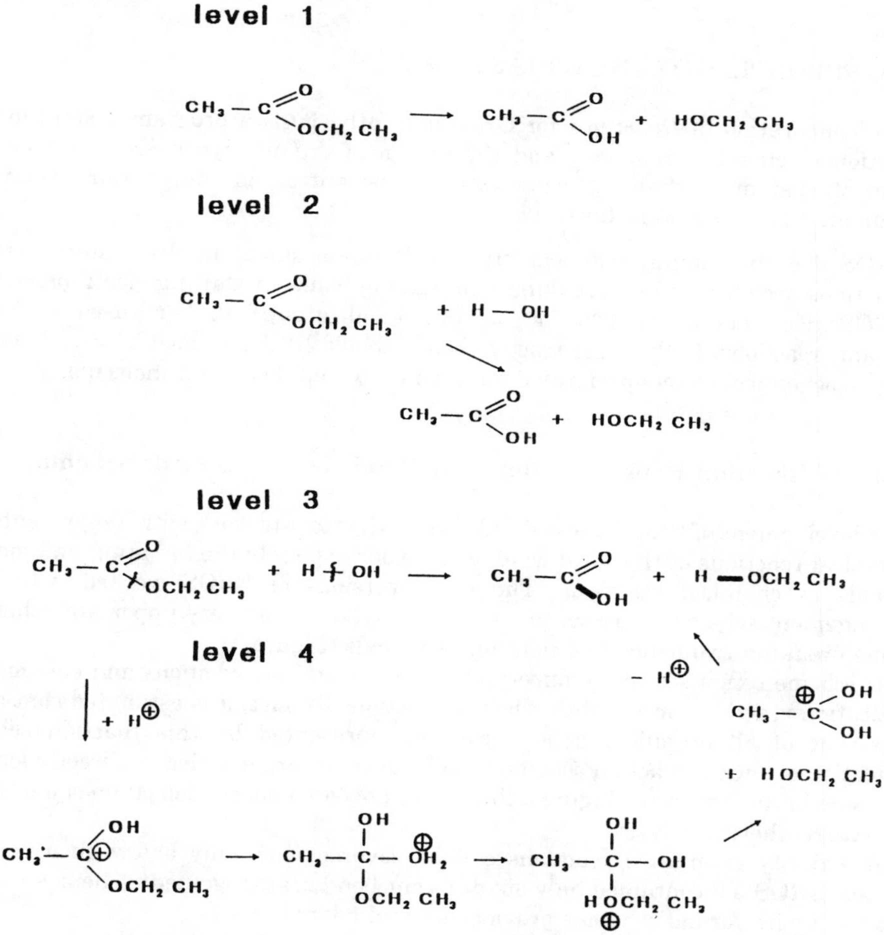

level 1

$$CH_3-C(=O)OCH_2CH_3 \longrightarrow CH_3-C(=O)OH + HOCH_2CH_3$$

level 2

$$CH_3-C(=O)OCH_2CH_3 + H-OH \longrightarrow CH_3-C(=O)OH + HOCH_2CH_3$$

level 3

$$CH_3-C(=O)\text{⫽}OCH_2CH_3 + H\text{⫽}OH \longrightarrow CH_3-C(=O)OH + H-OCH_2CH_3$$

level 4

$$CH_3-C(=O)OCH_2CH_3 \xrightarrow{+H^{\oplus}} CH_3-C^{\oplus}(OH)OCH_2CH_3 \longrightarrow CH_3-C(OH)(\overset{\oplus}{O}H_2)OCH_2CH_3 \longrightarrow CH_3-C(OH)(OH)H\overset{\oplus}{O}CH_2CH_3 \longrightarrow CH_3-\overset{\oplus}{C}(OH)OH + HOCH_2CH_3 \xrightarrow{-H^{\oplus}} CH_3-C(=O)OH + H-OCH_2CH_3$$

Figure 3. Four levels of representing the hydrolysis of an ester.

At the most basic level a reaction is specified as the conversion of starting materials to products. Quite often, only compounds that are considered to be important are given. In the example, water is omitted from the first reaction equation in Figure 3.

A more detailed description of a chemical reaction has to take account of all starting materials and products, and has to take account of all atoms in a chemical reaction (level 2 in Figure 3). It is realised that there are quite a few chemical reactions where this is not possible as not all products are known.

Even more information is given when the bonds that are broken and made are specified (level 3). This already needs some knowledge of the reaction mechanism. In the example given a decision has to be made between an O-alkyl or O-acyl cleavage of the ester.

Finally, the structures of the intermediates of a chemical reaction have to be given in order to specify the detailed reaction mechanism (level 4). In the example given, it is the one observed under acidic conditions.

The deeper the level of reaction representation the more information can be accessed, the more insight can be gained.

BASIC PRINCIPLES OF THE EROS SYSTEM

EROS (*E*laboration of *R*eactions for *O*rganic *S*ynthesis) is a program system for the prediction of chemical reactions and the design of organic synthesis. Work on this system started more than 12 years ago. At the outset, in the overall design two fundamental decisions were taken (2):

1. EROS does not obtain reactions from a database stored in the system. Rather, reactions are obtained by breaking and making bonds or shifting electrons.
2. EROS does not contain a list of functional groups that is used to access organic reactions. Rather, the reactive bonds should be determined by calculations of physicochemical effects performed for each individual bond of a molecule.

Level 3 of Reaction Representation: The Bond Rearrangement Scheme

In the development of EROS it was decided from the start to work with a representation of chemical reactions of the third level, to consider explicitly the breaking and making of bonds in chemical reactions. The early versions of EROS worked with bond rearrangement schemes representing overall reactions. One very important scheme is the one breaking two bonds and making two bonds (Figure 4).

This scheme comprises many important organic reactions: additions and eliminations, substitution reactions or certain pericyclic reactions. In fact, it is estimated that about 50 percent of all organic reactions can be represented by this reaction scheme. Application of such a reaction scheme onto bonds of a target molecule directly leads to precursors in the synthesis. Figure 5 illustrates how an aldol condensation can be found in a retrosynthetic analysis.

The majority of organic reactions can be covered with only a few such reaction schemes. EROS 3.2 contained only six different bond rearrangement schemes.

Working with formal schemes provides several advantages:

1. *No database of reactions* is necessary. Thus, one has no problems of building up and

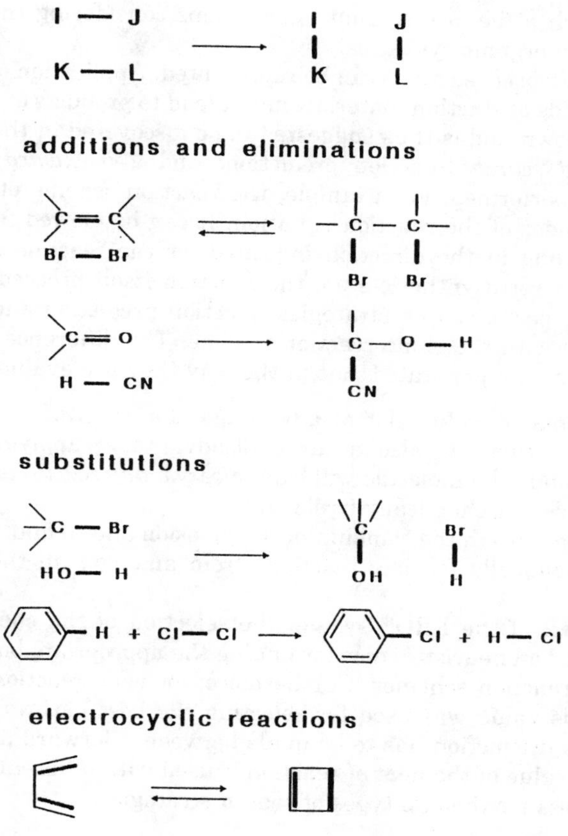

I — J ⟶ I J / K — L K L

additions and eliminations

substitutions

electrocyclic reactions

Figure 4. Examples of reactions that break two bonds and make two bonds.

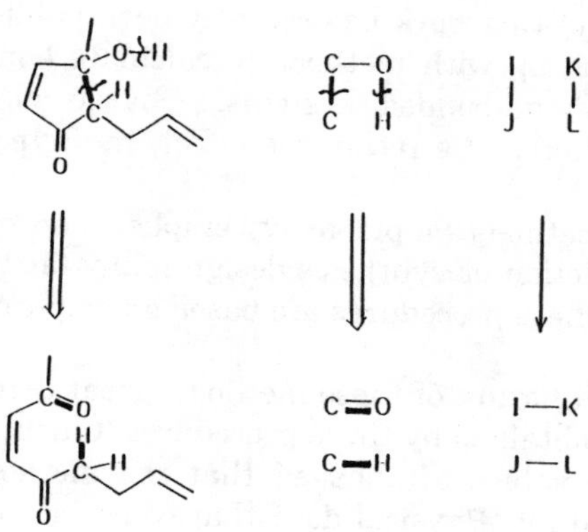

Figure 5. Obtaining an aldol condensation by the reaction scheme of Figure 4.

maintaining such a file, a task that is enormous considering the large number of reactions used in organic synthesis.

2. Both *known and novel* reactions can be represented. Application of a formal reaction scheme onto bonds of starting materials might lead to products of a reaction that has not yet been known and is thus suggested to be discovered in the laboratory.
3. Both a *forward search* (reaction prediction) and a *backward search* (synthesis design) can be performed. For example, the reaction scheme of Figure 4 has two bonds on both sides of the reaction equation. It can be viewed from both sides, the reaction proceeding in the direction indicated, or the reaction as obtained in the program to be a retrosynthetic step, the reaction itself proceeding in the reverse manner. Both types of search strategies, reaction prediction and synthesis design, can be obtained with the same reaction scheme. The difference does not lie in the way the reactions are generated, but in the way they are evaluated (*vide infra*).

These three points provide a lot of advantages for working with formal reaction schemes. However, there is also a large disadvantage: applying formal reaction schemes onto all bonds of a molecule will lead to *very many reactions*, reactions that are formally conceivable but chemically irrelevant.

The big task is to contain the combinatorial explosion and to find ways to generate or select only the chemically feasible reactions from amongst all the formally possible ones.

Up to version 3.2 of the EROS system the selection of the chemically interesting reactions was based on heuristic rules for finding the appropriate bonds that were to be submitted to the reaction schemes. Furthermore, for each reaction the enthalpy was calculated and this value was used for selecting the best pathways.[3] Clearly, in the selection process a distinction has to be made between a forward and a retrosynthetic search. Thus, the value of the heat of reaction is used with different sign and weight in the selection process for the two types of search strategies.

Level 4 of Reaction Representation: The Reaction Mechanism

For a more sophisticated evaluation and selection scheme we decided to model reaction mechanisms. Ten years ago we embarked on a program to quantify the factors and effects a chemist uses to explain the course of chemical reactions, to derive reaction mechanisms. A summary of this work has recently been published.[4]

Over the years we came up with methods to calculate bond dissociation energies, charge distribution in sigma-bonded systems,[5] charge distribution in pi-bonded systems,[6] the inductive effect,[7] the resonance effect, the hyperconjugation effect and the polarisability effect.[8]

In the design of these methods we put heavy emphasis on speed of calculation as in problems of reaction prediction or synthesis design many fairly large chemical species have to be evaluated. All these procedures are based on empirical methods developed to model these effects.

Because of the empirical nature of these methods, great care was taken to show the significance of the values obtained by these procedures. Calculations and correlation of physical data with these values established that the methods indeed modelled the effects they were designed for. Physical data that could be calculated included dipole moments, ^{1}H- and ^{13}C-NMR shifts, $^{1}J_{C\text{-}H}$ coupling constants, ESCA (C-1s, F-1s and N-1s) chemical shifts, Auger shifts, and IR stretching frequencies.

Having established the importance of the calculated parameters in the studies of physical data we turned our attention to the problem of chemical reactivity. It could be shown that the values for the inductive and polarisability effect can be used to calculate reactivity data of some fundamental polar reactions in the gas phase (Figure 6). Linear equations could be developed to calculate proton affinities of amines,[7,8] of alcohols and ethers,[9] and of thiols and thioethers,[9] as well as gas phase acidity data of alcohols.[9]

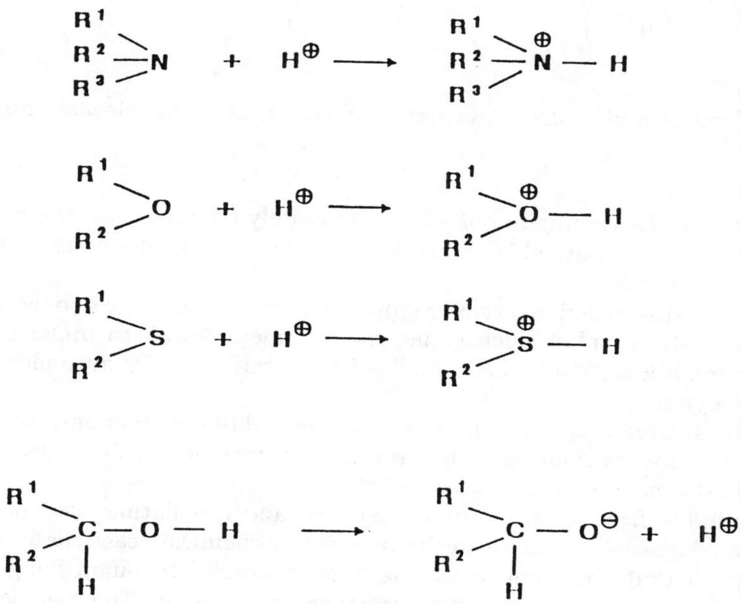

$$R^1R^2R^3N + H^{\oplus} \longrightarrow R^1R^2R^3\overset{\oplus}{N}-H$$

$$R^1R^2O + H^{\oplus} \longrightarrow R^1R^2\overset{\oplus}{O}-H$$

$$R^1R^2S + H^{\oplus} \longrightarrow R^1R^2\overset{\oplus}{S}-H$$

$$R^1R^2CH-O-H \longrightarrow R^1R^2CH-O^{\ominus} + H^{\oplus}$$

Figure 6. Fundamental polar reactions for which data have been calculated from values on the inductive and polarisability effect.

As an example, Equation 1 gives the results for the proton affinities (PA) of ethers and alcohols. The proton affinities, the heats of reaction released on protonation of alcohols and ethers, can be calculated from the electronegativity parameters, X_{12}, being a measure of the inductive effect, and α_d, representing the polarisability effect.[9]

$$PA = 272.8 - 19.29\,X_{12} + 5.22\,\alpha_d \qquad (1)$$

(in kcal/mol).

The different signs of the coefficients of the two parameters express the fact that the inductive effect destabilises the positively charged protonated species, whereas the polarisability effect is a source of stabilisation for those ions.

The successful extension of these studies to proton affinities of aldehydes and ketones opened the way to calculate enthalpies for all four reactions of the cycle connecting alcohols and carbonyl compounds (Figure 7).[10]

For all four reactants of this cycle, linear equations of the form of Equation 1 could be developed based on parameters of the inductive, the polarisability, and, in certain cases, the hyperconjugation effect. These results are significant as this cycle contains all conceivable polar reaction types: the dissociation of a neutral species into a positive

$$R^1R^2C{=}O + H^{\oplus} \longrightarrow R^1R^2C{=}O^{\oplus}{-}H$$

$$\downarrow + :H^{\ominus} \qquad\qquad \downarrow + :H^{\ominus}$$

$$R^1R^2HC{-}O^{\ominus} + H^{\oplus} \longleftarrow R^1R^2HC{-}O{-}H$$

Figure 7. Proton- and hydride-transfer reactions connecting alcohols and carbonyl compounds.

and negative ion; the recombination of two oppositely charged ions; the reaction of a neutral species with a positively charged ion; and the reaction of a neutral species with a negatively charged ion.

Furthermore, this reaction cycle (Figure 7) contains the reaction of a carbonyl compound with the simplest nucleophile, the hydride anion, both under neutral and under acidic conditions. Thus, these studies bear significance for an understanding of acid-base catalysis.

Figure 8 illustrates this reaction cycle again in a different manner. The equations developed allow the calculation of the relative heights of the four species involved depending on the substituents at the free sites.

Having found a firm basis for understanding and calculating data of gas phase reactions we addressed ourselves to the problem of chemical reactions in solution. It could be shown that the same parameters so successful in quantifying gas phase reactivity data are also applicable to reactions in solution. The reactions studied

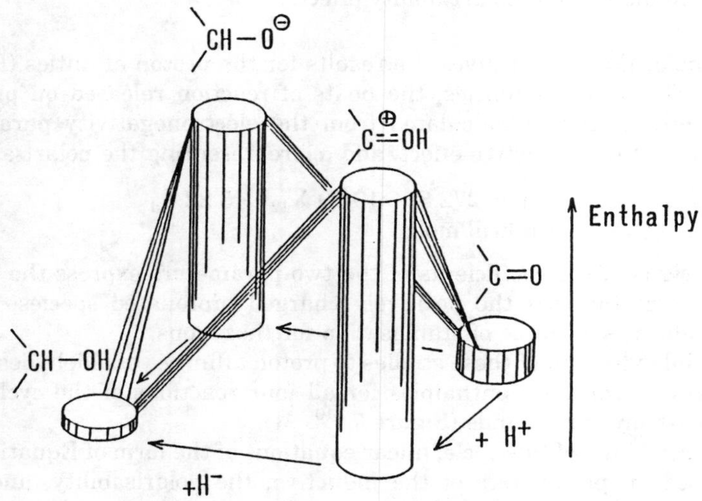

Figure 8. Energy diagram for the four reactions of the cycle in Figure 7.

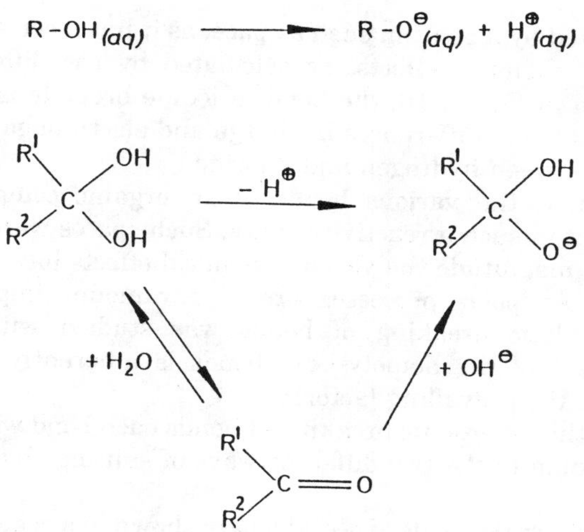

$$R\text{-}OH_{(aq)} \longrightarrow R\text{-}O^{\ominus}_{(aq)} + H^{\oplus}_{(aq)}$$

Figure 9. Aqueous phase reactions for which data have been calculated.

included aqueous phase acidity of alcohols and gem-diols, and hydration of carbonyl compounds (Figure 9).[11]

These studies with simple albeit important gas phase and solution reactions demonstrated that the methods developed for the quantification of the various chemical effects provide parameters useful for the calculation of reactivity data. This encouraged the development of a general approach to quantification of chemical reactivity.

The complicating situation with chemical reactions is that their reactivity is simultaneously influenced by many factors and it is so to various extents. To account for that dependence on many variables, the multidimensionality of chemical reactivity, we have taken the various chemical effects as co-ordinates of a space, the reactivity space. (Figure 10).

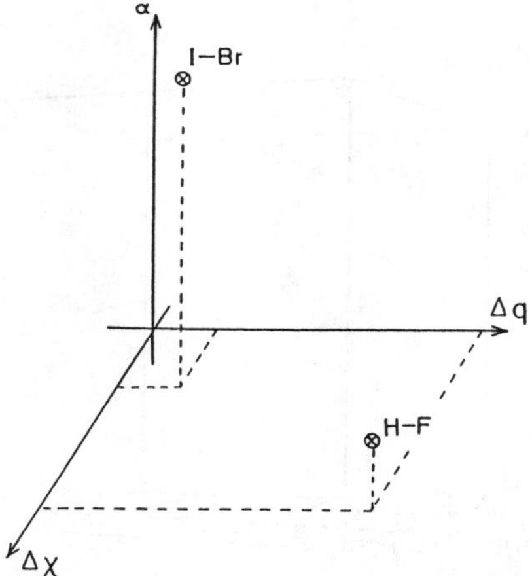

Figure 10. A three-dimensioned reactivity space spanned by polarisability, α, and the differences in charge, Δq, and electronegativity, ΔX.

A bond is represented by a point in such a space, as it has a certain value for each of the co-ordinates, the chemical effects, as calculated by the different methods. For example, as indicated in Figure 10, the bond in iodine bromide is distinguished by a high polarisability but small differences in charge and electronegativity. The converse is true for the bond between hydrogen and fluorine.

In a similar manner, the various bonds of an organic compound will each be represented by a point in such a reactivity space. Such spaces were used to further an understanding of the magnitude the various chemical effects have on the reactivity of organic compounds. As polar processes are of paramount importance in organic chemistry the heterolytic breaking of bonds was studied with preference. (The quantitative treatment of the homolysis of bonds is inherently easier as the bond dissociation energy is the prevailing factor).

When considering the heterolytic breaking of bonds each bond will be represented by *two* points corresponding to the two different ways of shifting charges to the atoms of the bond.

In Figure 11 various single bonds of acetaldol are shown in a reactivity space spanned by the polarity in the sigma electron distribution, Q^σ, the bond dissociation energy, BDE, and the resonance effect, R.

For each bond of aldol, calculations were made by the methods developed in our group for the bond dissociation energy, the sigma charge distribution and the amount of resonance stabilisation of the charges generated by polar breaking of that bond. The two ways of heterolysis of a bond are shown with the points 4 and 7. Point 4 corresponds to the dissociation of OH^-, whereas point 7 represents the loss of OH^+.

As an additional feature, the bonds of aldol represented in the three-dimensional reactivity space of Figure 11 were marked according to their reactivity. Bonds that

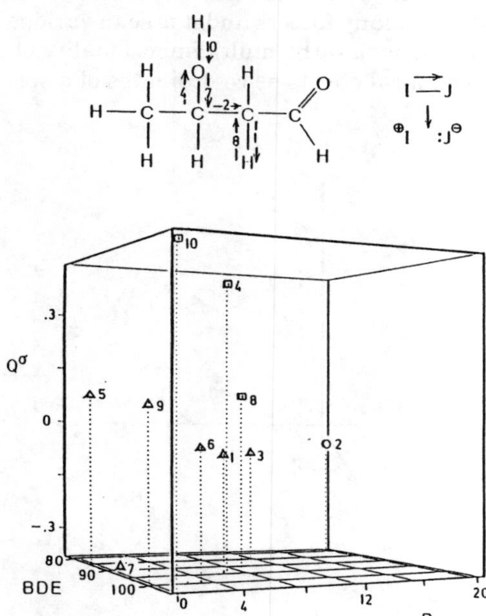

Figure 11. Reactivity space representing the polar breaking of various bonds in acetaldol.

were considered by the chemist as being reactive (breakable) are distinguished by small cubes, non-reactive bonds are indicated by little pyramids. No classification was made for the central C–C bond of aldol represented by point 2. (The carbon-oxygen double bond is not represented in this space as the analysis concentrated on understanding the reactivity of single bonds).

It can be seen that reactive and non-reactive bonds clearly separate. Non-reactive bonds are found in the lower and left parts and more to the front of this perspective of the 3-D reactivity space. Reactive bonds are moved more to upper regions, to the right, and further away from the observer. The clear separation of breakable and non-breakable bonds shows that the property that is under investigation, chemical reactivity, is well represented in this space.

It is tempting to assume that the more reactive a bond, the further the point that represents this bond should be away from the plane of separation. Thus, from the observation point taken in Figure 11, the more reactive bonds should be found more into the direction of the upper right hand rear corner. This is true and can be illustrated with Figure 12. This Figure is a duplication of Figure 11, but some additional points are contained in this 3-D space corresponding to bonds of two ions obtainable from aldol. Calculations of bond dissociation energies, sigma charge distribution and charge stabilisation through resonance were made for all bonds of the anion of aldol and for the

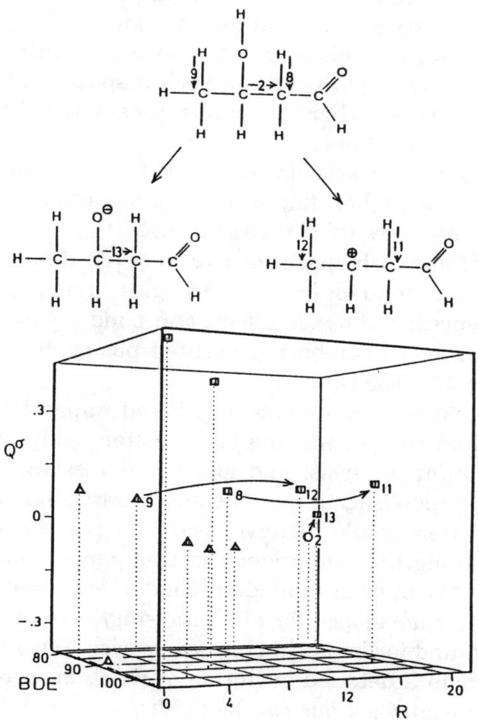

Figure 12. The reactivity space of Figure 11 containing additional points for bonds of the two ions derived from aldol and showing the representation of increases in reactivity.

carbocation obtained from aldol after removing an OH^- group. From all those bonds only three are entered into the reactivity space so as not to overcrowd the picture.

It will be generally agreed that the central C–C bond of aldol becomes more reactive after deprotonation giving way to the possibility of a retro-aldol condensation. This increase in reactivity is reflected in the shift of point 2 to point 13, away from the observer, in a direction of higher reactivity as was anticipated in the discussion above. The removal of a proton from carbon atom 2 of aldol leading to the enolate anion and represented by point 8 was considered to be a feasible event. In the carbocation, removal of this proton is represented by point 11. This must be facilitated since the incipient carbanion will be stabilised by the adjacent positive centre leading to an α,β-unsaturated carbonyl compound. This anticipated increase in reactivity is reflected by a shift of the point 8 to point 11, to the right into a direction of higher reactivity.

Removal of a proton from carbon atom 4 of aldol was considered not easy (point).[9] In the carbocation, however, this proton should be much more acidic as again the carbanion can be stabilised by the adjacent positive charge. Thus, this bond becomes reactive and the point representing this heterolysis is again shifted more to the right of the reactivity space. However, the product is a β,γ-unsaturated carbonyl compound, which is not as stabilised as the α,β-isomer. This is reflected in the fact that point 12 is not as far to the right as point 11.

To summarise, the three variables used in constructing the reactivity space of Figures 11 and 12 and quantified by our methods are very well-suited to the representation of the reactivity of aldol and the ions derived from it. In this space, bonds that are classified by a chemist as either reactive or non-reactive are clearly separated. Furthermore, the distance between points in such a space is of chemical significance. The more reactive a bond, the further the point representing the breaking of this bond will be from the plane of separation.

Figures 10–12 have shown reactivity spaces with three dimensions. However, in general, reactivity spaces of higher dimensionality have to be considered as chemical reactivity might depend on more than three factors.

We are routinely working with spaces of five to seven dimensions, the co-ordinates being given by the values of bond dissociation energies, sigmaand pi charge differences, electronegativity differences, resonance effect, and bond polarisability. Spaces of such high dimensionality can no longer be represented pictorially. However, they can be investigated by statistical methods.

We haveemployed a variety of unsupervised and supervised pattern recognition methods such as principal component analysis, cluster analysis, k-nearest neighbour method, linear discriminant analysis, and logistic regression analysis, to study such reactivity spaces. We have published a more detailed description of these investigations.[12] As a result of this, functions could be developed that use the values of the chemical effects calculated by the methods mentioned in this paper. These functions allow the calculation of the reactivity of each individual bond of a molecule.

Different functions were developed for aliphatic single bonds, for multiple bonds, for bonds to aromatic rings, and for bonds in charged species. With these functions and the values obtained from them a detailed evolution of reaction mechanisms can be made. Thus, the course of chemical reactions can be predicted. For the starting materials the most reactive bonds are determined by calculating the various chemical effects and by weighting them with the reactivity functions. Breaking the most reactive bonds leads to charged intermediates that are subjected in turn to functions that determine whether additional bonds are to be broken or new bonds are to be made. This continues until stable reaction products are obtained. In addition, the heat of reaction is calculated for the overall process.

Figure 13 gives the results for the reaction of 3-chloro-3-methyl-butanol with water, obtained with the present version 5.1 of EROS.

The first reaction is an exchange of OH and C1, the second, fourth and fifth reactions are fragmentation reactions leading to an olefin and either carbonyl compound and HC1 or an addition product from the latter two compounds. The product of the third reaction is obtained by a step involving the protonation of ethylene which occurs as an intermediate. This step is unrealistic under basic conditions.

For reactions 4 and 5 the details of the reaction mechanism as developed by EROS 5.1 are shown in Figure 14.

The calculations of the chemical effects and the use of these values for the estimation of the reactivity of bonds automatically reveal the functional groups in a molecule, since the bonds in the functional groups are in many cases the most reactive ones. However, the detailed modelling of reaction mechanisms also overcomes limitations of the concept of functional groups. Based on a search for functional groups a carbon-carbon single bond will not be considered as reactive. However, the example of Figure

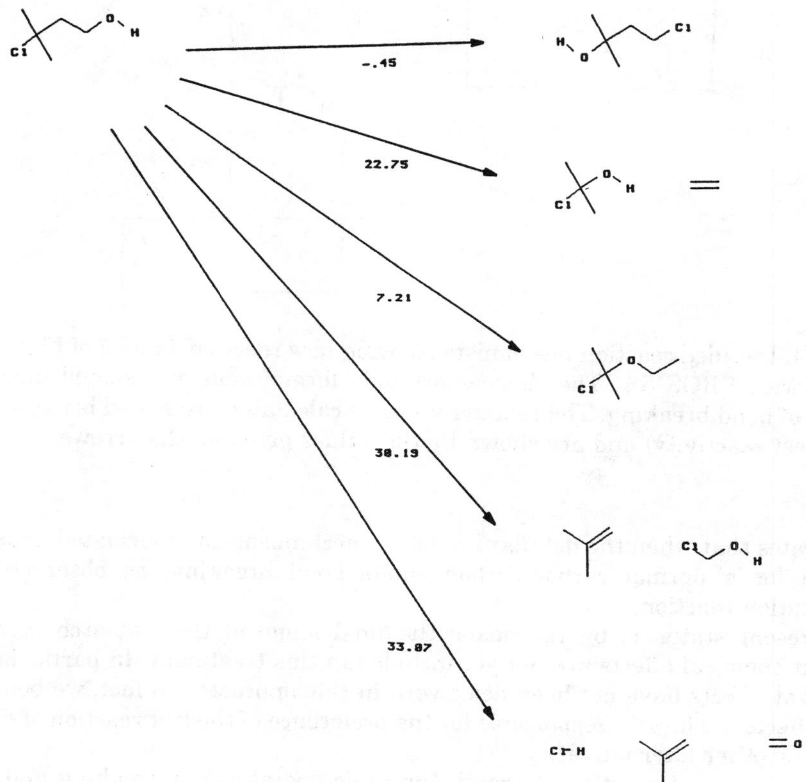

Figure 13. The reactions predicted for the treatment of 3-chloro-3-methyl-butanol with water (reaction enthalpies in kcal/mol).

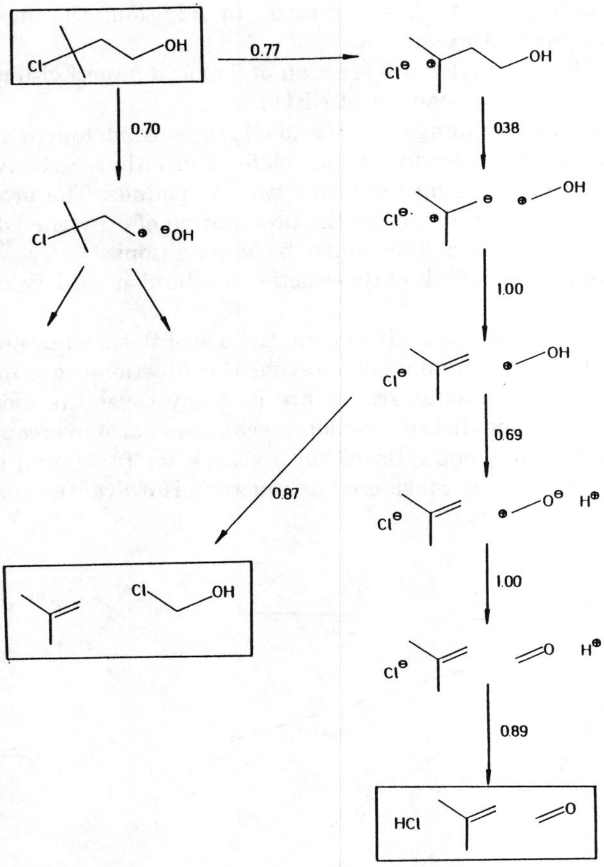

Figure 14. Detailed reaction mechanism showing how reaction 4 and 5 of Figure 13 are obtained with EROS 5.1. The charges are only formal ones and should indicate the direction of bond breaking. The reactivity values calculated are scaled between 0.0 and 1.0 (highest reactivity) and are shown by the values given on the arrows.

14 illustrates that when the details of reaction mechanisms are elucidated, reasons can be found for a normal carbon-carbon single bond breaking, as observed in this fragmentation reaction.

The present status is by no means the final stage of this research. Some very important chemical effects are not yet included in this treatment. In particular, steric and solvent effects have not been dealt with in this approach. In fact, we believe that solvent effects are largely responsible for the preference of the last reaction of Figure 13 over all the other alternatives.

The generation of reactions through the basic operations of breaking and making bonds allows generation of all conceivable overall reaction schemes. For example, EROS 3.2 worked, among others, with the two reaction schemes (reaction generators) RG21 and RG22 (Figure 15). This Figure also illustrates how both schemes can now be

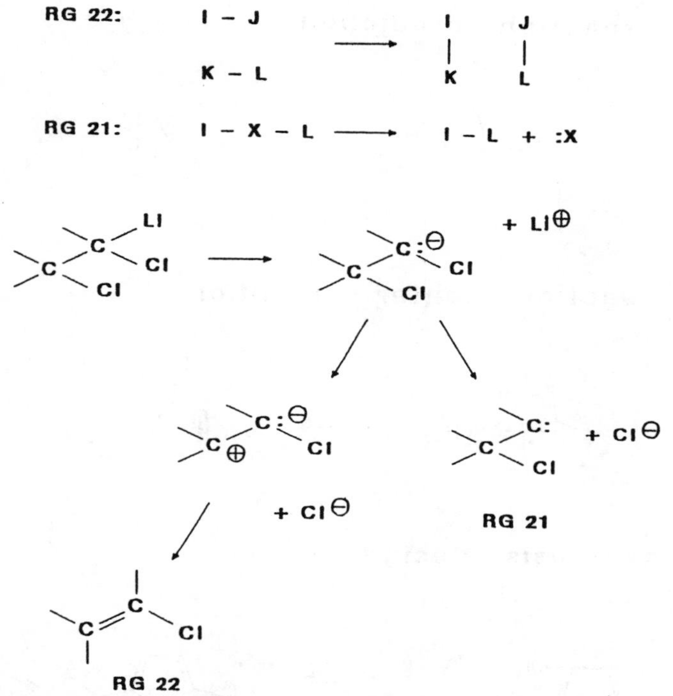

Figure 15. The treatment of reactions as sequences of individual mechanistic steps gives access to different types of net reactions (overall reaction schemes).

automatically obtained through the step-wise generation of reaction mechanisms in EROS 5.1.

The master-plan for a general treatment of chemical reactivity presented here opens up approaches to a wide variety of chemical problems. EROS 5.1 can be used to predict the course of chemical reactions, leads to identification of reaction centres, and is being further developed for support of the planning of organic syntheses. Figure 16 gives one representative each. Some more detailed examples have been published.[4]

PROGRAM IMPLEMENTATIONS

The early versions of EROS were coded in PL/1, a programming language that was used up to version EROS 4.1.[4] Recently a complete redesign of the system was performed and the coding was made in standard FORTRAN77. More than 30 man-years of work have gone into the development of the various versions of EROS. EROS 5.1 consists of about 250 subroutines and comprises about 30,000 lines of code. EROS is highly portable and has been implemented on various machines (CDC Cyber 995 under NOS2.0, VAX under VMS, PCS Cadmus under MUNIX, IBM 6150 under AIX, BULL SPS7 under UNIX, SUN3/160C under UNIX).

It has been shown here how methods for calculation of various chemical effects[3,5–8] can be used for the construction of reaction mechanisms. In the course of the development of these procedures it was shown that a variety of physical and chemical

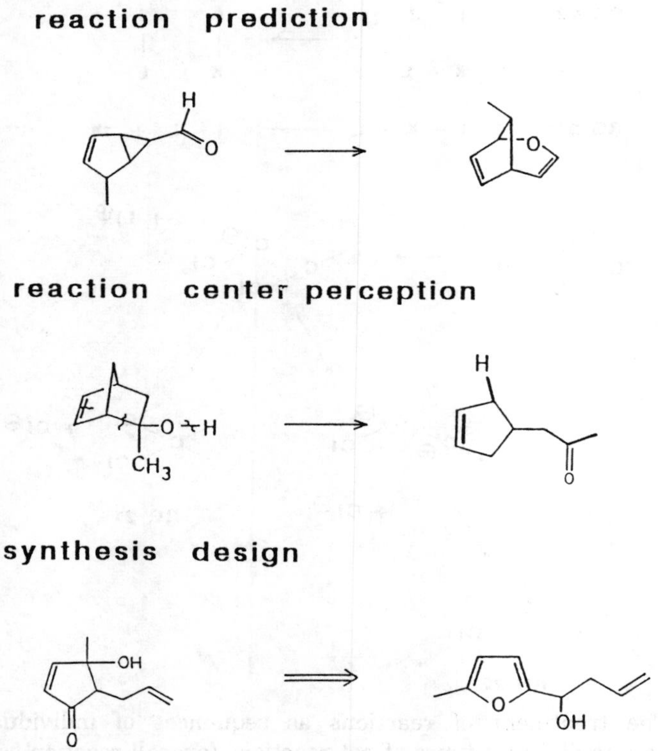

reaction prediction

reaction center perception

synthesis design

Figure 16. Fields of application for EROS 5.1.

data can be calculated with them. The values obtained are also useful for the correlation of biological data in quantitative structure activity (QSAR) studies as they provide a detailed description of energetic and electronic effects. Realising the importance of the methods for quantifying various chemical effects over and above reaction prediction has led us to assemble these methods in in a separate program package. This system PETRA (*P*arameter *E*stimation for the *T*reatment of *R*eactivity *A*pplications) can be used to calculate parameters for correlating and calculating physical, chemical, and biological data.

ACKNOWLEDGMENTS

We want to gratefully acknowledge the support of this work by Deutsche Forschungsgemeinschaft, Sumitomo Chemical Co., Japan, Tecnofarmaci S.p.A., Italy, and, particularly, by a generous grant through a Joint Research Scheme from Imperial Chemical Industries plc, England. The redesign and coding of the system in FORTRAN77 rested on many shoulders: our thanks are extended to everyone involved in this endeavour.

REFERENCES

1. Kelly, R. C., Schletter, I., Stein, S. J., Wierenga, W. 'Total Synthesis of alpha-amino-3-chloro-4,5-dihydro-5-isoxazoleacetic acid (AT/125), An Antitumour Antibiotic'. *J. Amer. Chem. Soc.* 1979, *101*, 1054–1056.
2. Gasteiger, J., Jochum, C. 'EROS, a Computer Program for Generating Sequences of Reactions'. *Top. Curr. Chem.* 1978, *74*, 93–126.
3. Gasteiger, J. 'Automatic Estimation of Heats of Atomisation and Heats of Reaction'. *Tetrahedron* 1979, *35*, 1419–1426.
4. Gasteiger, J., Hutchings, M. G., Christoph, B., Gann, L., Hiller, C., Löw, P., Marsili, M., Saller, H., Yuki, K. 'A New Treatment of Chemical Reactivity: Development of EROS (Elaboration of Reactions for Organic Synthesis), an Expert System for Reaction Prediction and Synthesis Design'. *Top. Curr. Chem.* 1987, *137*, 19–73.
5. Gasteiger, J., Marsili, M. 'Iterative Partial Equalisation of Orbital Electronegativity: A Rapid Access to Atomic Charges'. *Tetrahedron* 1980, *36*, 3219–3228.
6. Gasteiger, J., Saller, H. 'Calculation of the Charge Distribution in Conjugated Systems by Quantification of the Mesomerism Concept'. *Angew. Chem.* 1985, *97*, 699–701 and *Angew. Chem. Int. Ed. Engl.* 1985, *24*, 687–689.
7. Hutchings, M.G., Gasteiger, J. 'Residual Electronegactivity – An Empirical Quantification of Polar Influences and Its Application to the Proton Affinity of Amines'. *Tetrahedron Lett.* 1983, *24*, 2541–2544.
8. Gasteiger, J., Hutchings, M. G. 'New Empirical Models of Substituent Polarisability and Their Application to Stablisation Effects in Positively Charged Species'. *Tetrahedron Lett.* 1983, *24*, 2537–2540. Gasteiger, J., Hutchings, M. G. 'Quantification of Effective Polarisability. Applications to Studies of X-ray Photoelectron Spectroscopy and Alkylamine Protonation'. *J. Chem. Soc. Perkin Trans. 2*, 1984, 559–564.
9. Gasteiger, J., Hutchings, M. G. 'Quantitative Models of Gas-phase Proton-transfer Reactions Involving Alcohols, Ethers, and Their Thio Analogues. Correlation Analysis Based on Residual Electronegativity and Effective Polarisability'. *J. Amer. Chem. Soc.* 1984, *106*, 6489–6495.
10. Hutchings, M. G., Gasteiger, J. 'A Quantitative Description of Fundamental Polar Reaction Types. Proton- and Hydride-Transfer Reactions Connecting Alcohols and Carbonyl Compounds in the Gas Phase'. *J. Chem. Soc. Perkin Trans. 2*,1986, 447–454.
11. Hutchings, M. G., Gasteiger, J. 'Correlation Analyses of the Aqueous-phase Acidities of Alcohols and Gem-diols, and of Carbonyl Hydration Equilibriums Using Electronic and Structural Parameters'. *J. Chem. Soc. Perkin Trans. 2*, 1986, 455–462.
12. Gasteiger, J., Saller, H., Löw, P. 'Elucidating Chemical Reactivity by Pattern Recognition Methods'. *Anal. Chim. Acta* 1986, *191*, 111–123.

REACTION RETRIEVAL AND SYNTHESIS PLANNING

Willi Sieber
Pharma Research, Sandoz AG, Basel, Switzerland

ABSTRACT

The search for new biologically active compounds requires a very large number of different syntheses. Reaction retrieval and computer-assisted synthesis planning have therefore a great potential for finding strategically important keysteps and suggesting the application of new synthetic methodology. The two approaches are in a way complementary to each other. In combination they can produce better results than each one separately. A concept for their integration and preliminary results obtained are discussed. Our research chemists have access to a variety of additional programs providing information about reactions, structures and experimental data, where similar aspects are important.

REACTION RETRIEVAL AND SYNTHESIS PLANNING

When synthesis planning was introduced in our company several years ago it was the first computer program with graphical structure manipulation. With its rapidly growing transform library the Computer-Assisted Synthesis Planning (CASP) was capable of finding new synthetic sequences or key steps and with its bond breaking strategy it was also capable of finding alternative reactions.

Together with graphical structure searching it triggered the introduction of graphics terminals into every synthesis laboratory. A big educational shift took place from the information specialist to the laboratory chemist. They learned to draw structures with light pen or mouse, to create a command file, to write scientific reports and send them by electronic mail.

As more and more scientific software became available we faced the problem of permanent education for our users and had to look for an alternative solution.

After careful examination of the user's needs we determined that there were about a dozen different tasks most frequently performed and simple enough to be standardised.

Among the structure-oriented tasks the most important is to determine if a structure is known or unknown. If known, the next important task is to get all related information from public or proprietary databases.

Our full structure search system determines the possible Chemical Abstracts Service (CAS) registry numbers and internal compound numbers for a given structural drawing. Additionally the user can select from a set of predefined keyword profiles and submit a job for CAS ONLINE. A minicomputer will set up the query, call the host system, perform the search, capture the answers and inform the user that the job was successfully completed. The user can now play back the answers at his terminal with structural formulae, bibliography and abstracts. The only activities are structure drawing, menu selection, waiting and scanning the results.

W. A. Warr (Ed.)
Chemical Structures

Among the synthesis-oriented tasks the most frequent is to get references with experimental procedures for a given reaction from public or proprietary databases. The next most frequent task is to find a potential synthetic sequence based on similar syntheses that were successfully done.

The first task can be done very conveniently with the available reaction retrieval systems. The second task is either a synthesis retrieval or a synthesis proposal task. The reaction retrieval systems cannot yet handle the synthesis retrieval job, but the synthesis proposal task can be done with one of the synthesis planning programs.

In our specific environment the chemist has the following synthesis tools available: ORAC, REACCS, SYNLIB, CASP, CAMEO and CHIRON. Although all these programs work with graphical (sub)structure drawing, each one works with slightly different commands and with strongly different substructure conventions, not to mention the confusing world of keywords. Additionally all those programs are dynamically upgraded with new versions and additional databases.

How can the ordinary user keep track of the possibilities? How can he find the right approach for his problem?

In this situation we decided to try an integrated approach. Integration was necessary both at the front-end and behind the scenes. The chemist should have a single program to draw and submit his synthesis problem. He should be able to do it in a manner independent of the query languages of the different programs. All databases should be equally available for searching.

The first step was to create a program for structure and reaction drawing independent of the retrieval programs. We called it reaction box or RXNBOX (Figure 1).

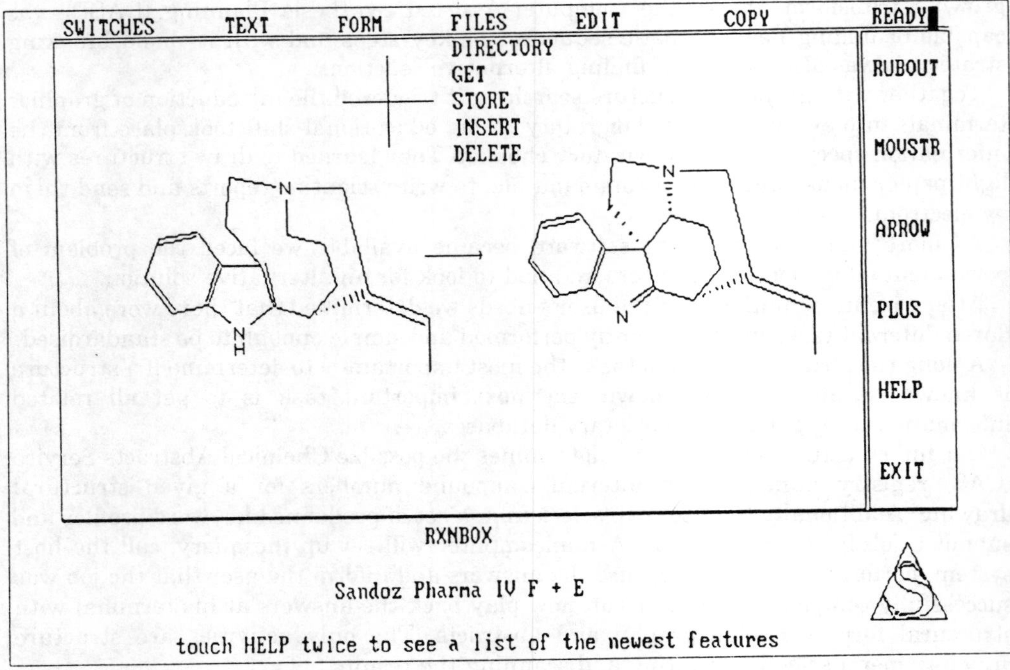

Figure 1

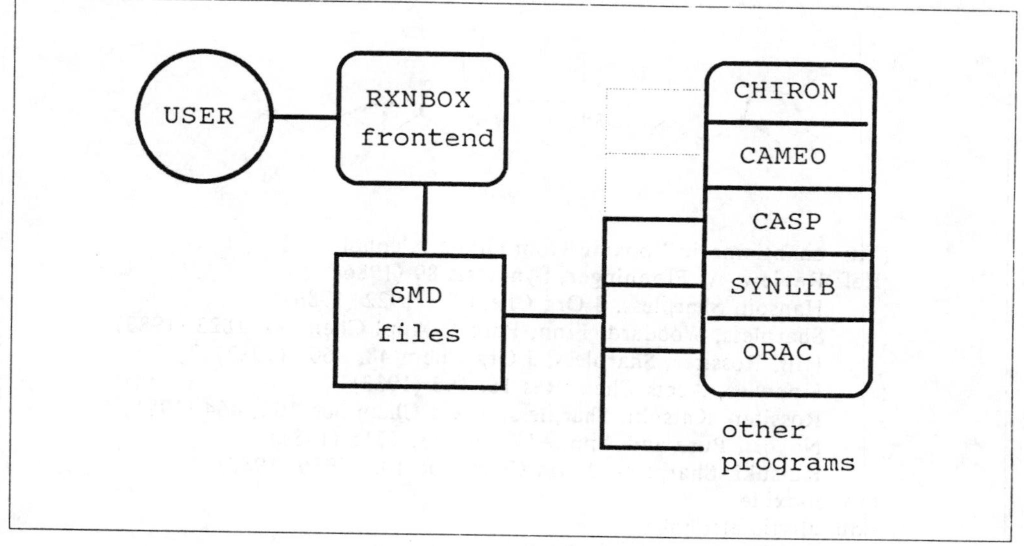

Figure 2

The first version is used for input of reference reactions from publications and for the proprietary database of syntheses. Later versions will beable to create queries for reaction and synthesis searches (Figure 2).

The program creates files in our standard molecular data format (SMD) in order to transfer molecular data between all our programs. They consist of various blocks of information such as connection table, coordinates, bibliography, reagents, yield. The format allows new types of data to be added without problems for existing files. A program reading such files will only interpret the blocks known to it.

The second step was to make our existing CASP files of reference reactions available not only for synthesis planning but also for reaction retrieval. We currently have two files, some 25,000 so-called 'certain' reactions and some 1000 complete reference reactions. The 'certain' reactions were generated from product molecules and the corresponding CASP-transforms. They are partially simplified literature examples and have only collective bibliography that can be displayed separately (Figure 3). The reference reactions consist of reactants, products and specific bibliography input with the RXNBOX program (Figure 4).

The two databases were imported into ORAC and can be searched together with the other databases available.

The next step is to find all reactions in the other ORAC databases that correspond to a CASP synthesis proposal or to a CASP transform in general. For this purpose CASP can deliver the reactant and product substructures with corresponding differences and put them into a query file for ORAC to search. The detected cross references can be stored in the CASP transform library for quick retrieval. Doing this manually so far for our ongoing transform work we found that many transforms have no corresponding ORAC (or SYNLIB) reactions. It seems that some classical chemistry is not yet represented therein.

What are the further steps to take? We will import the proprietary databases of total

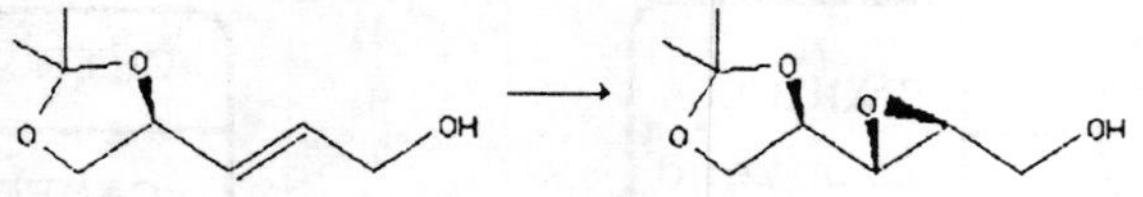

;Tit: enantiomeric Epoxide from allylic Alcohol
;Ref: Review: A. Pfenninger, Synthesis 89 (1986)
; Hanson, Sharpless, J Org Chem 51, 1922 (1986)
; Sharpless, Woodard, Finn, Pure & Appl Chem 55, 1823 (1983)
; Hill, Rossiter, Sharpless, J Org Chem 48, 3607 (1983)
; Knowles, Accts Chem Res 16, 106 (1983)
; Rossiter, Katsuki, Sharpless, J Am Chem Soc 103, 464 (1981)
; Noyori, Pure and Applied Chem 53, 2315 (1981)
; Katsuki, Sharpless, J Am Chem Soc 102, 5976 (1980)
;Pro: epoxide
;Edu: allylic alcohol
;Rea: TBHP = t-butylhydroperoxide and
; Ti-tartrate in CH2Cl2 at -20 C (Sharpless reagent)
; (from D(-) or L-(+)-diethyltartrate)
;Met: enantioselective epoxidation with 70 - 90 % yield and >90 % ee
TBHP-Epoxidation 33D

Figure 3

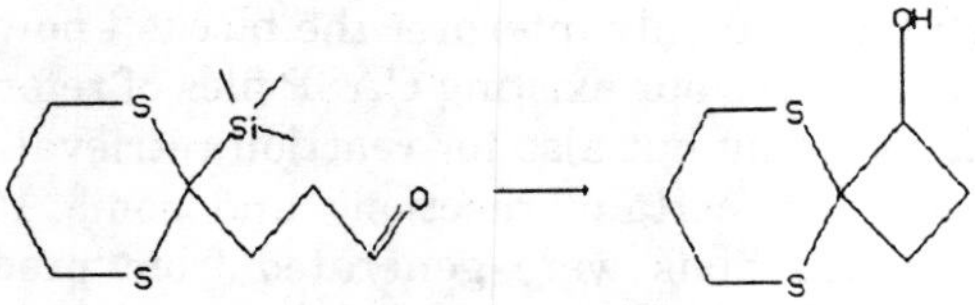

Ref: J Chem Soc Chem Comm 306 (1981)
Aut: Grotjahn D.B. Andersen N.H.
Met: Carbanion Addition to Aldehyde and Ketone
Rea: Bu4NF
Sol: THF
Tmp: rt
Yld: 56%
Trf: 226A (Gourraud Sandoz)

Figure 4

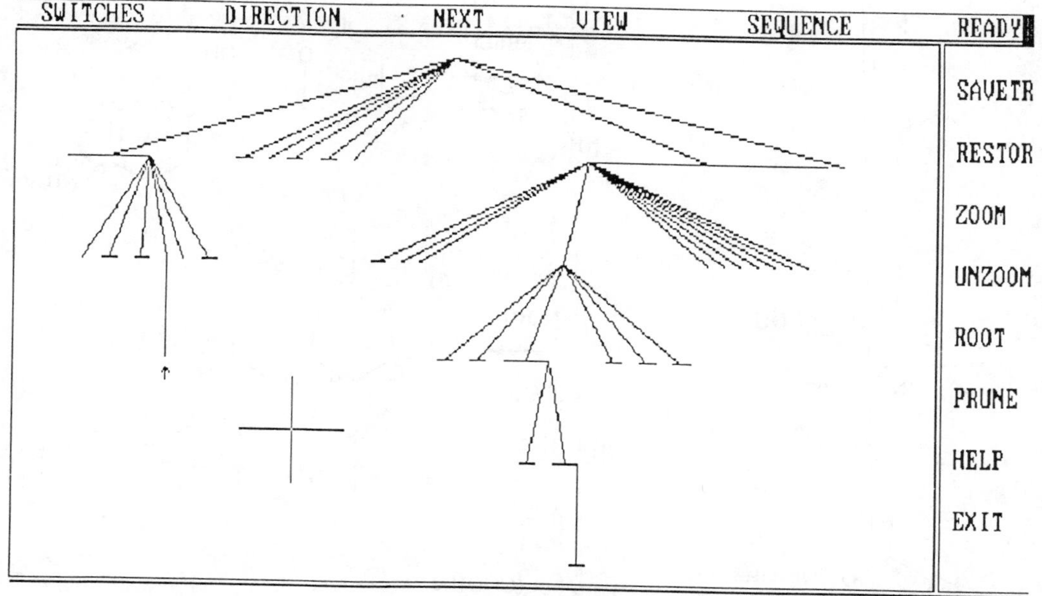

Figure 5

synthesis into ORAC. If a reaction search hits a single step of such a sequence it is desirable to examine quickly the rest of the total synthesis. This should be possible by calling the TREE module of CASP (Figure 5).

Interfacing between reaction retrieval and synthesis planning has so far been used for exchange of database information. But why go further into an integrated system capable of query formulation and query translation?

Reaction retrieval deals with very precise full structure information but you have to be aware of different possible ways to define the reaction centres. Another problem is the various notations for certain functional groups (e.g., sulphur and phosphorus groups). The most restricting factor yet is the inability to search for reaction sequences. Because of this you have to guess how a one-pot reaction or a multistep sequence was stored in the database. As a consequence you miss the information (Figure 6).

Synthesis planning on the other hand is based on derived and interpreted information only but is not affected by reaction centre or keyword definitions. You simply draw your product molecule and make use of strategic constraints if desired. The problem for the user is to get keysteps and trivial reactions all together and he usually gets too many proposals anyway. With the help of the reference reactions he can easily figure out if a proposal is closely related to reactions published.

How would our integrated system eventually serve the user? Let us distinguish between reaction searching and synthesis searching tasks.

For reaction searching the chemist will draw a substructure reaction with our RXNBOX front-end and submit it for searching in ORAC. If he gets the desired information, the job is done. If not, the system will formulate a corresponding query for CASP and submit it for processing. If he finds an interesting reaction proposal the system will search ORAC again for specific literature references.

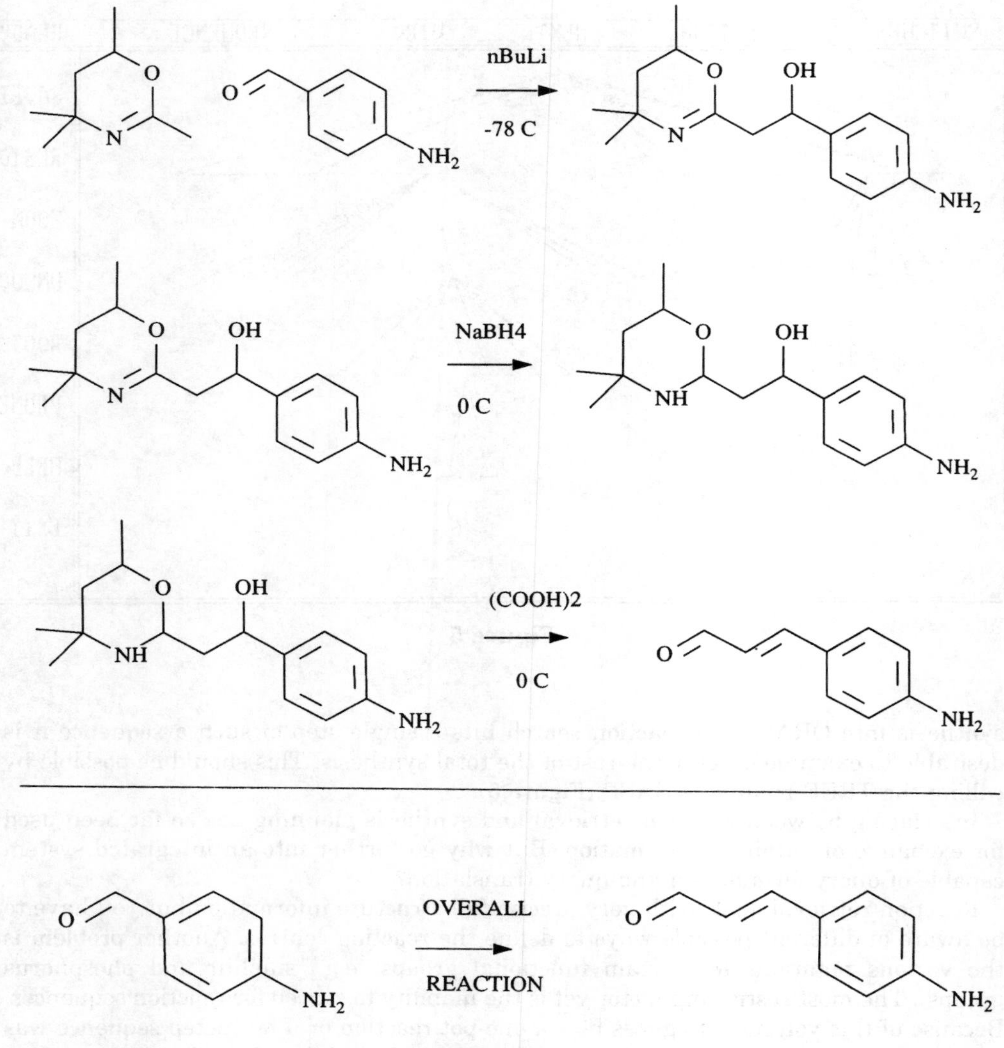

nBuLi

-78 C

NaBH4

0 C

$(COOH)2$

0 C

OVERALL

REACTION

Figure 6

For synthesis searching the chemist will draw a target molecule with our RXNBOX front-end and submit it to CASP. From the generated synthesis tree he can select an interesting sequence. CASP will formulate corresponding reaction search queries and submit them to ORAC. The answers can be single steps, sequences or total synthesis. The user will decide if they confirm the plausibility of the selected CASP sequence or not.

With a database of total synthesis and a few enhancements to ORAC we can go a step further. We will be able to search for multistep sequences as long as they are marked with a common tree identifier. We can even link subtrees of formal total syntheses together.

Our future integrated system of retrieval and proposal for reactions and synthesis can do much more for the chemist than the two approaches separately.

CAOS/CAMM SERVICES: SYNTHESIS PLANNING AND MOLECULAR MODELLING TOOLS

Jan H Noordik
CAOS/CAMM Centre, Faculty of Science, University of Nijmegen, Toernooiveld, 6525 ED Nijmegen, The Netherlands

ABSTRACT

The advanced and sophisticated computer tools for molecular design which have become available over the last few years, require considerable resources in terms of hardware and expertise when they are to be used as an integrated system. However, the bench chemist, and in particular the academic bench chemist, generally lacks both. To provide easy access to the new tools, The Dutch National Facility for Computer Assisted Organic Synthesis and Computer Assisted Molecular Modelling has been established in 1985. The CAOS/CAMM Centre provides all the software tools, databases and computing facilities needed for molecular design, by making them accessible online at all Dutch Universities. A user-friendly, graphics menu interface allows even the novice user direct access to the module(s) of his choice. Training sessions and workshops given at the Centre are another means to introduce the new techniques in the chemical laboratory.

MOLECULAR DESIGN METHODS

Computer-assisted molecular design methods use a variety of techniques, ranging from relatively simple search methods on remote host computers to complicated computational chemistry calculations on supercomputers.

The following techniques are the basic ingredients of any integrated molecular design environment: compound-oriented information storage and retrieval, reaction-oriented information storage and retrieval, synthesis planning and molecular modelling.

COMPOUND-ORIENTED LITERATURE SEARCHING

This technique includes both searching on remote host systems and consulting in-house systems. Examples of the first are CAS ONLINE (Chemical Abstracts), Derwent (patent literature) and ChemQuest (Fine Chemicals Directory). Generally these searches are not an activity of the research scientists but are performed by specialised (library) personnel.

Examples of the second technique are Beilstein (currently with SANDRA on PC), MACCS (for storage and retrieval of structural data) and DARC-SMS (for storage and retrieval of structural data). Both MACCS and DARC-SMS can be coupled with general Database Management Systems (ORACLE, 1032) to store descriptive data. Other examples are CCSD, the Cambridge Crystallographic Database, PDB, the Brookhaven

W. A. Warr (Ed.)
Chemical Structures

Protein Database, ICSD, the Inorganic Crystallographic Database and METALS, the Metals Database. These last four systems can be viewed as a source of building blocks for molecular modelling. Finally there is MMSA, Macro Molecular Sequence Analysis, data files and software tools for analysis of amino-acid and nucleotide sequences in proteins and nucleic acids. These systems (unlike those of the first group) are usually consulted by research scientists.

REACTION-ORIENTED LITERATURE SEARCHING

This technique can be used to search the public literature, to retrieve the 'best' literature precedents for a desired reaction. The search question can be generally formulated as 'Which possibilities are available to convert starting material P into target material T?' Systems to answer this question are SYNLIB, ORAC and REACCS.

The first two systems have their databases drawn from the recent chemical literature. The last one mainly uses Theilheimer and Organic Syntheses as its sources of data.

Reaction-oriented searching can also be used as an in-house tool to store and retrieve company reactions. The DARC-RMS and MACCS/REACCS systems are typical examples of this usage, although SYNLIB and ORAC could alsobe used for this purpose.

A general characteristic of reaction searching is the use of graphics input and output.

SYNTHESIS PLANNING

In synthesis planning the general search query is 'How can one prepare target material T and starting with what?' There are different approaches to answer this question. From a chemical point of view one can work in a synthetic and a retrosynthetic direction. From a computer science point of view one can distinguish between logic-based and information-based approaches.

A typical example of an information-based retrosynthetic system is LHASA. This system generates a synthesis tree utilising a system of 'transforms' which describe individual steps to construct a target.

Integration of a synthesis planning system and a reaction retrieval system to explore individual steps for literature precedents would result in an extremely powerful design tool.

MOLECULAR MODELLING

In the molecular modelling techniques, one has to differentiate between molecular construction tools and molecular modelling tools. The first try to establish a physically allowable and logically correct 3-D geometry for a molecule. The second take the logically correct 3-D structure and create a chemically correct 3-D structure.

The tools used in molecular construction are hand-sketching in 2-D or 3-D (where use is made of graphics devices and graphics programs) and a library of graphics blocks or structure fragments, basically retrieved from one of the databases mentioned above.

The tools used in molecular modelling are computational chemistry programs such as MM (Molecular Mechanics), QM (Quantum Mechanics) and MD (Molecular Dynamics).

The general characteristics of molecular modelling techniques are use of (advanced) graphics devices and availability of considerable computer resources.

NECESSARY RESOURCES FOR AN INTEGRATED CHEMICAL DESIGN SYSTEM

Hardware

Dedicated computer (*de facto* standard VAX)
Fast Local Area Network
Access to host-systems with library files
Low performance graphics systems (PC's)
High performance graphics system (*de facto* standard Evans and Sutherland).

Software

Reaction retrieval system	SYNLIB; ORAC; REACCS
Synthesis planning system	LHASA; CAMEO; . . .
Spectral identification	PBM/STIRS; . . .
Molecular modelling system	CHEM-X; SYBYL; COGS; MACROMODEL; . . .
Interfaces to MM, QM and MD MD	
MD programs	AMBER; CHARMM
Macromolecular sequencing	

CAOS/CAMM CENTRE IMPLEMENTATION

All of the above tools are available under a monitor program or a menu (see Figure 1). This (graphics) menu offers submenu control of all the chemical utilities, commands which are independent of machine or operating system, format conversions and elaborate online help facilities.

FACILITIES OF THE DUTCH NATIONAL CENTRE FOR COMPUTER-ASSISTED CHEMISTRY

In January 1985, the CAOS/CAMM Centre was founded with financial support from the Ministry of Education, the Dutch National Science Foundations SON and STW, and the Faculty of Science of the University of Nijmegen.

Facilities

The Centre is situated at the University of Nijmegen and operates a VAX 11/785 computer with 16 Mb core and approximately 1.3 Gb external memory. Advanced colour display facilities are available at the Centre for use by all academic scientists. A large number of medium cost/medium performance graphics workstations are put up by the Centre at all Dutch academic chemical laboratories.

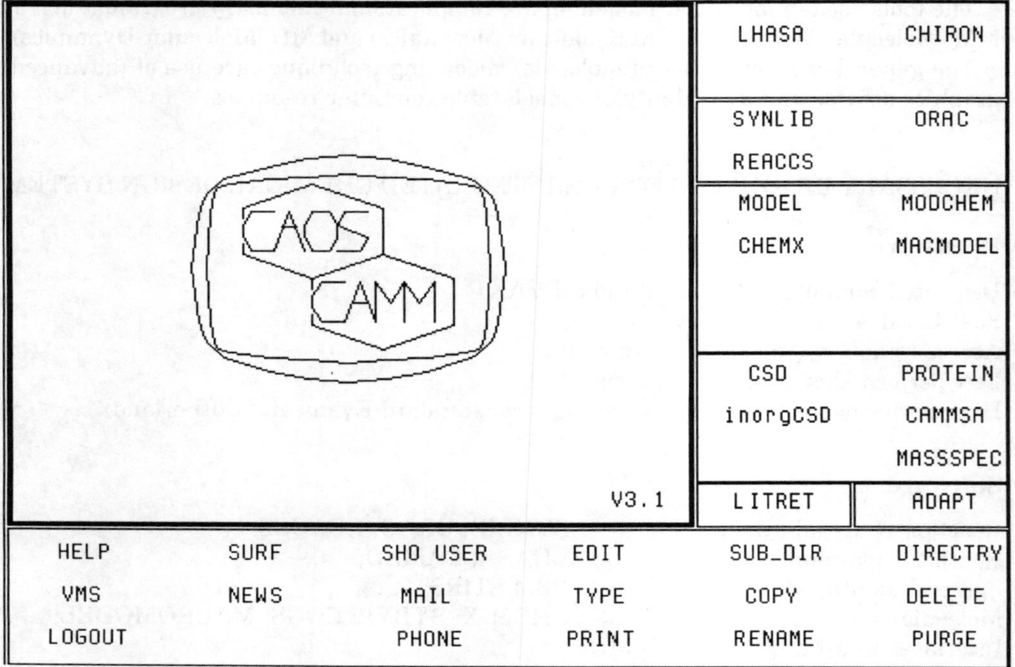

Figure 1. Selects a tool by pointing to a screen button. All buttons have elaborate help files.

Software Tools

Under the acronym CAOS, the Centre provides facilities to aid the chemist in designing organic synthesis. Structural formulae, the natural language of chemists, are used to interact with the computer. Promising synthesis paths can be selected and evaluated using 'knowledge bases'. Individual steps of these paths, the chemical reactions which convert starting material to product *via* intermediates, can be tested for precedences in the literature, by searching databases containing many thousands of reactions.

Under the acronym CAMM, a large number of programs are provided to aid the scientist in studying molecular properties. Simple as well as complex molecules can be studied and visualised using modern colour display terminals. Various databases containing information on crystal structures, proteins and nucleotides are available online.

Non-academic Users

For-profit organisations can gain access to the services of the CAOS/CAMM Centre provided that appropriate licences on commercial software have been obtained.

JOINT COMPOUND/REACTION STORAGE AND RETRIEVAL AND POSSIBILITIES OF A HYPERSTRUCTURE-BASED SOLUTION

George Vladutz
Institute for Scientific Information, 3501 Market Street, University City Science Center, Philadelphia PA 19104, U.S.A.

Scott R Gould
Crozer Chester Medical Center, Chester, Pennsylvania, U.S.A.

ABSTRACT

Reaction retrieval systems are usually built without due connection with systems built earlier for the retrieval of compounds. This paper discusses options for a file organisation for a unified structural database of compounds and their reactions. Such a database has a potential for improved retrieval capabilities and high browsability.

MICROCOMPUTER-BASED RETRIEVAL SYSTEMS

One of the few examples of commercially available retrieval systems encompassing both compounds and reactions is Molecular Design's microcomputer-based ChemBase system.[1] It saves molfiles (molecular structure files) separately from its reaction file without any elaborate linkages between the files. This makes it rather difficult to move information between the files and requires the conversion of a molfile to move it in or out of a reaction file. Additionally, searching a compound structure for where it appears in a reaction is not a simple operation. Lastly, reaction sites are neither identified nor searchable. These limitations could be overcome if the reaction site could be associated with the molfiles or reaction files stored for searching.

Another microcomputer-based retrieval system, aimed at compounds as well as reactions is ChemSmart.[2–5] This system saves a compound as a separate file associated with its own two-page notebook. Within the design mode there is a mechanism for labelling one or more atoms in the structure, but these are not searchable as of yet. ChemSmart uses compound files directly to define a reaction's reactants and products, and then associates the reaction with its own reaction notebook. The only way to label a reaction site presently in ChemSmart, is to label the reaction site fragments in the compounds used in the reaction. The labelled atom positions will show the label even in the reaction display. However, the reaction cannot be searched for the compounds present in it, or the presence of this labelled reaction site. The reaction site can be used as a query for substructural search on the compounds, if the compounds can be saved in the search file. The reaction display module can take compounds from either the searchable or non-searchable files, so this is not a final solution. Even if the reaction site could be found quickly by using the reaction site as a substructure for compound

W. A. Warr (Ed.)
Chemical Structures

searching, the compounds cannot be searched to identify if they are part of a specific reaction.

The optimal solution would be to have the reaction site saved in a searchable format. A reaction site is smaller than a molecule and, therefore, more space efficient. Also, smaller files will search faster and provide for greater data compression. One reaction site can be associated with a number of compounds and reactions.

ChemSmart's future development plans include the utilisation of an algorithm based on the above approach which will solve the task by an efficient linkage between its compound and reaction files. ChemSmart's general design makes the implementation of such analgorithm possible.

INTEGRATION OF DATABASES OF COMPOUNDS AND THEIR REACTIONS

Below we will examine in more detail different methods of integration of databases of compounds and their reactions, having in mind mainly the case of large and very large databases.

The usage of the registration numbers of compounds in an existing compound database in lieu of the actual structural diagrams of reactant and product molecules is the simplest way of creating a useful link between a compound and a reaction database. A reaction equation recorded using three-byte binary compound registration numbers takes on the average not more than 15 bytes. Such records will enable the user to search for reactions with reactant and/or product molecules having specific structural features. Such searches will be reduced to substructure searches aimed firstly at finding the structures with the given features and secondly at identifying all the specific reactions in which the found structures are either reactant or product molecules. For the latter purpose each compound of the database has to carry cross-references to the ID numbers of the corresponding reactions. In the case of a large collection of data the number of reactions will exceed the number of compounds and each such cross-reference can require a five-byte binary reaction ID number, with an average of four cross- references per reaction. This brings the additional memory expenditure to 15+20=35 bytes/ specific reaction.

Another way of exploiting the links between the reaction database (recorded using compound registry numbers) and the compound database is to process the reaction records serially, calling for each reaction the corresponding reactant and/or product structures and performing on them the structure searches specified in the query or queries. Using this technique the cross-referencing from compounds to reactions is unnecessary. This type of organisation of a joint compounds and reactions database limits the reaction retrieval possibilities to the above types of queries, dealing with structural features of the participating compounds. Queries dealing with the peculiarities of the proper structural changes brought about by the reaction, cannot be answered.

SUPERIMPOSED REACTION SKELETON GRAPHS

The simplest expression for describing the essence of structural changes occurring in a reaction is the record of the reaction skeleton graph (RSG) including all the bonds altered by a specific reaction together with the atoms involved. The concept of RSG roughly equivalent to the traditional more or less fuzzy concept of reaction site', or 'reaction centre' has been formally defined and exemplified.[6–8] For illustration purposes below are two examples:

②O H ④ ②'O↔H ③'

①C C ③ ①'C↔C ④' (1)

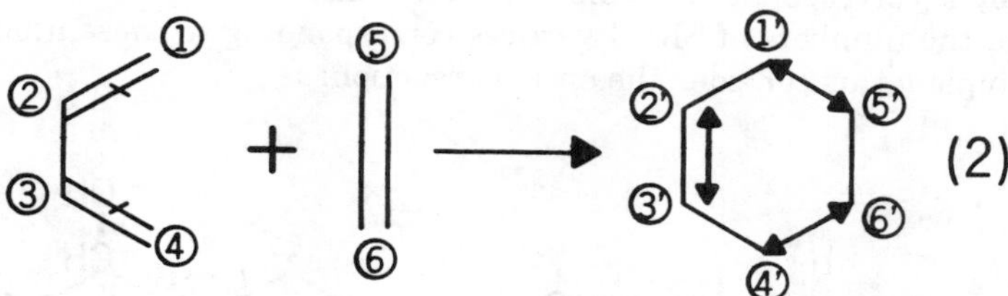

In these examples the symbols '$\not-$' and '$\leftrightarrow$' are used correspondingly for the bonds broken and made. In this form of presentation a table describing the mapping between the atoms of the left- and right-hand sides of the RSG is a necessary part of the RSG's record. A more compact form of presentation of RSG's is the 'superimposed' form (9,7,8); these superimposed reaction skeleton graphs (SRSG) make the recording of the above mapping unnecessary since the atoms of the reaction site are represented only once. The superimposed forms of the above two examples are given below:

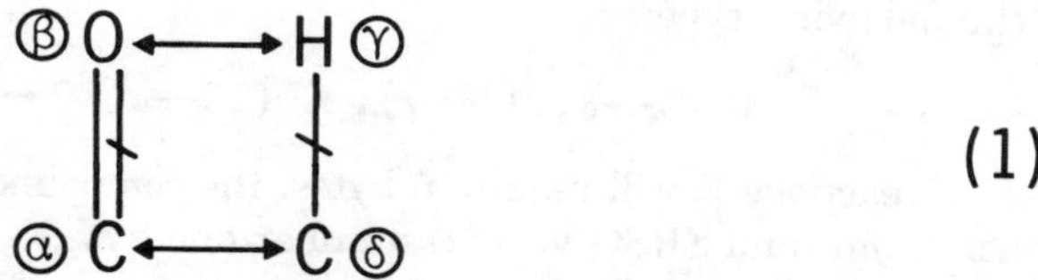

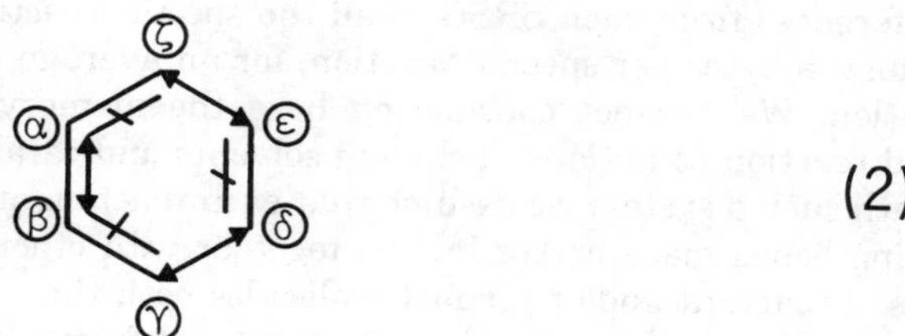

Such SRSG's are labelled graphs which can be easily recorded using connectivity tables based on some arbitrary (or canonical) numbering of their atoms (vertices). Such numbers are represented in the above examples by encircled Greek letters. Allowing one byte for an atom symbol, six bits for an SRSG atom number and four bits for an

SRSG bond symbol the connectivity table type record of an average SRSG such as (2) will take 36 bytes.

The total number of known (or even theoretically possible) SRSG's can be estimated as not exceeding 10,000. That means that the memory needed for recording all SRSG's will be of very reasonable size, 360-400 kbyte. Each SRSG refers to a very significant number of specific reactions. In order to record a collection of specific reactions using a database of SRSG's and a pre-existing compound database, references have to be made from each SRSG vertex to the corresponding atom numbers in both the reactant and product molecules. This means that the records of specific reactions will consist of a serial ID number of the given reaction, the ID number of its SRSG and the registration numbers of the reactant and product molecules. Each of those registration numbers will be followed by a correspondence table between some of the numbers of atoms of the structure and the numbers of SRSG vertices corresponding to those atoms.

As an example let us consider the specific reaction:

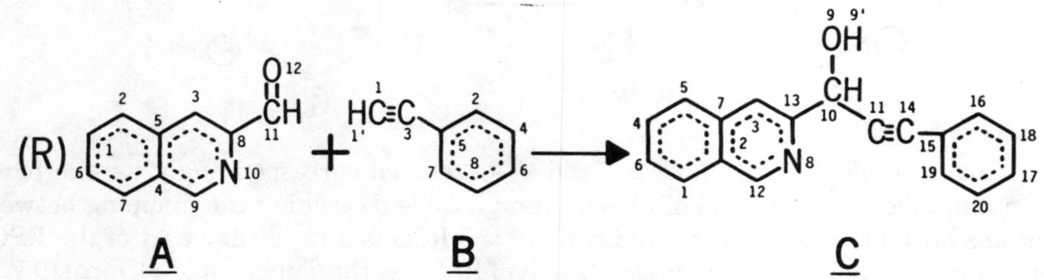

A B C

If R is the ID number of this reaction and **A**,**B**,**C** are the registry numbers of the corresponding compounds, their (arbitrary) atom numbering is as shown, and the above (1) is the ID number of the SRSG corresponding to this reaction, then the record of this reaction will be of the following type:

R: (1), A ($11 \leftrightarrow \alpha, 12 \leftrightarrow \beta$), B ($1 \leftrightarrow \delta$, $1' \leftrightarrow \gamma$), C ($10 \leftrightarrow \alpha$, $9 \leftrightarrow \beta, 9' \leftrightarrow \gamma, 11 \leftrightarrow \delta$)

For a large collection of reactions R will require 5 bytes, the compound numbers (**A**,**B**,**C**) 3 bytes each and each atom (and SRSG vertex) number one byte.

The average length of a specific reaction recorded this way can be estimated as taking 42 bytes. This includes the full, although implicit, recording of the mapping of reactant and compound atoms. To this memory expense one has to add the space taken by necessary cross-references from each SRSG to all the specific reactions it covers. This requires an additional 5 bytes per specific reaction, for an average of around 50 bytes per reaction equation. We are not considering here the memory requirements for recording yield and reaction conditions, including solvents and catalysts.

When dealing with such a system we assume that each reaction query contains some element(s) concerning bonds made and/or broken together with other details concerning structural features of reactant and/or product molecules with the

indication of the position or distance of those structural features in relation to the altered bonds. For instance we may be looking for reactions in which C–C bonds are made and there is an C≡C group in α-position to the newly formed bond. Such queries are then handled by a substructure search performed on the SRSG database, which will produce the SRSG's containing the bonds made and/or broken required by the query. For each found SRSG the corresponding specific reaction records will be called consecutively, and for each specific reaction, using the compound numbers it includes, the corresponding compound structures will be called from the compound database. In each such compound the position of the altered bond required by the query will be determined using the atom number correspondence tables contained in the reaction records. In order to find out whether the specific reactant (or product) compound contains the structural feature indicated in the query a directed substructure search is made starting with the identified altered bond and seeking the required structural elements in the requested position relative to the altered bond(s). For instance if the making of a C–C bond is specified in the query during the initial search in the SRSG database the above SRSG (1) will be found and thereafter the specific reaction (R) processed as one of the reactions corresponding to this SRSG. In this reaction the position of the made C–C bond (α↔β bond in SRSG (1)) will be inferred as bond 10–11 in the product molecule **C** and therefore if the query asks for a C≡C group in the vicinity of this bond the substructure search performed in **C** will look for this group only in the immediate vicinity of atoms 10 and 11 of structure **C**.

STORAGE REQUIREMENTS

In a few years the number of known chemical compounds can be expected to reach 10,000,000 or 10^7. The average size of compounds can be estimated to be of 25 non-hydrogen atoms, engendering on average not more than 5 rings, which corresponds to about 30 bonds/compound. The compact recording of such structures, using non-redundant connectivity tables, will take about 120 bytes/compound. In any standard system each compound record will be associated with some screening information. We will assume that the screens are organised in inverted lists, on average 20 screens are associated with each individual structure, and the screens are recorded in inverted lists using 3 byte binary compound numbers. This will add an average of 60 bytes per compound, bringing the total memory size to an average of 120+60=180 bytes/compound. For the collection of all known compounds the total memory size can be estimated as $10^7 \times 1.8 \times 10^2 = 1.8 \times 10^3$ megabytes. That would fit on just four standard read-only compact (CD-ROM) disks of 550 megabytes each.

An estimate of the total number of specific reactions described in the literature can be based on the observation that the majority of compounds are synthesised just by one specific reaction, whereas a relatively small fraction of compounds are obtained *via* many different reactions. Assuming that each tenth compound is a product in an average of 40 reactions and each of the remaining compounds is obtained by just one reaction the total number of reactions will be of the order of 50,000,000 or 5×10^7. This estimate is possibly an upper limit since the number of new specific reactions being described in current literature does not exceed annually 500,000, which would correspond to a period of 100 years for reaching the above mark. Although organic chemistry has been active in describing reactions for more than a century, the scale of research activities was much more limited in that past so that a 20–30 million guess for the total number of known specific reactions may be more realistic. Taking nevertheless the 5×10^7 figure and the estimated average size of 50 bytes/reaction we obtained a total memory size of the order of 2.5×10^3 megabytes, the equivalent of 5 CD-ROM disks. The size of the necessary SRSG database component becomes negligible in this context. It is likely that this size will be doubled if yields, solvent, catalyst and other reaction conditions are also recorded.

One can see that for a globally performing joint compound/reaction storage and retrieval operation the memory requirements for its reaction storage component significantly exceed the requirements of its compound component. Nevertheless one can easily see also that the performance of the joint system will depend in a significant degree on the performance of the compound substructure search module since the processing of both compound-oriented and reaction-oriented queries will be reduced ultimately to substructure searches.

HYPERSTRUCTURES

In this connection it is appropriate to examine here some possible non-standard organisation methods for a large scale structural diagram database which have potential for enhancing its performance, especially in respect of atom-by-atom matches, which are the ultimate and most time-consuming stage in substructure searches. We will examine the possibilities of an approach based on creating 'hyperstructures' by superimposing more or less similar individual structural diagrams.

The idea is illustrated below using the example of the following four compounds **A–D**:

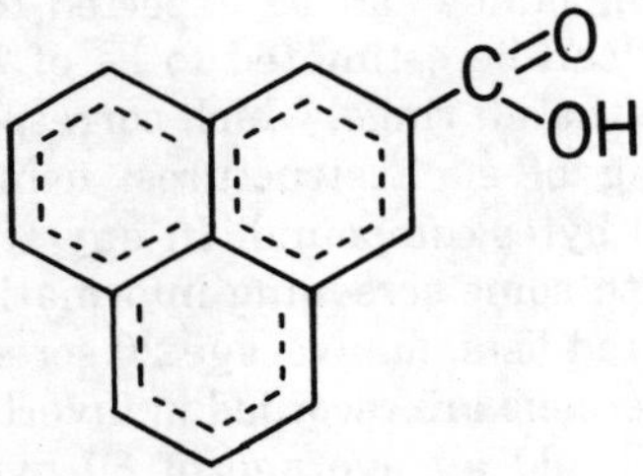

A

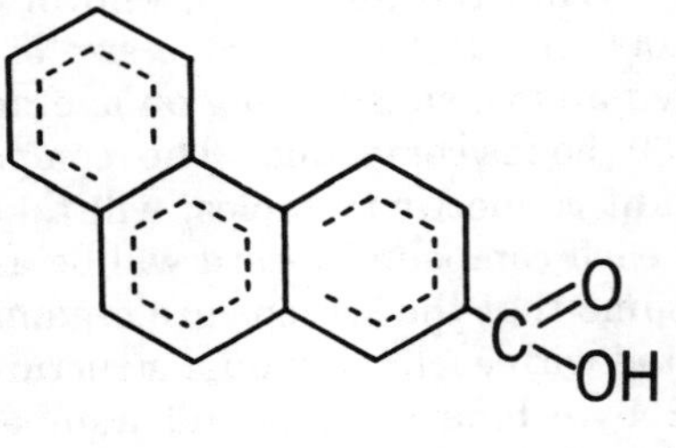

B

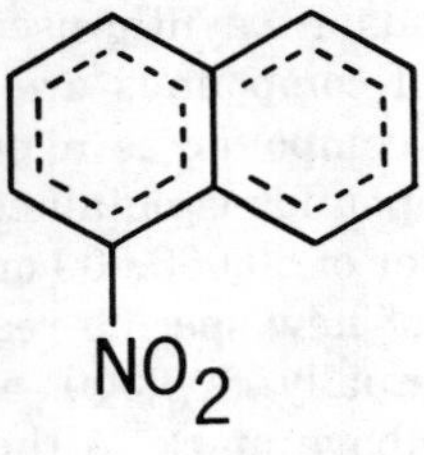

C

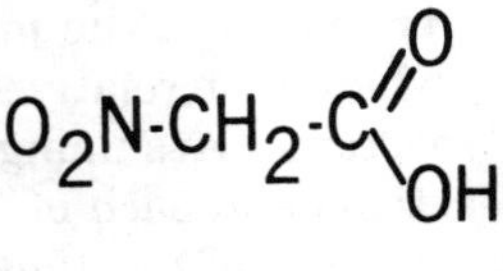

$O_2N\text{-}CH_2\text{-}C(=O)OH$

D

These compounds have some common fragments and they can be 'compacted' by representing them by one 'hyperstructure' which fully incorporates all of them and at the same time indicates the individuality of each of the superimposed structures.

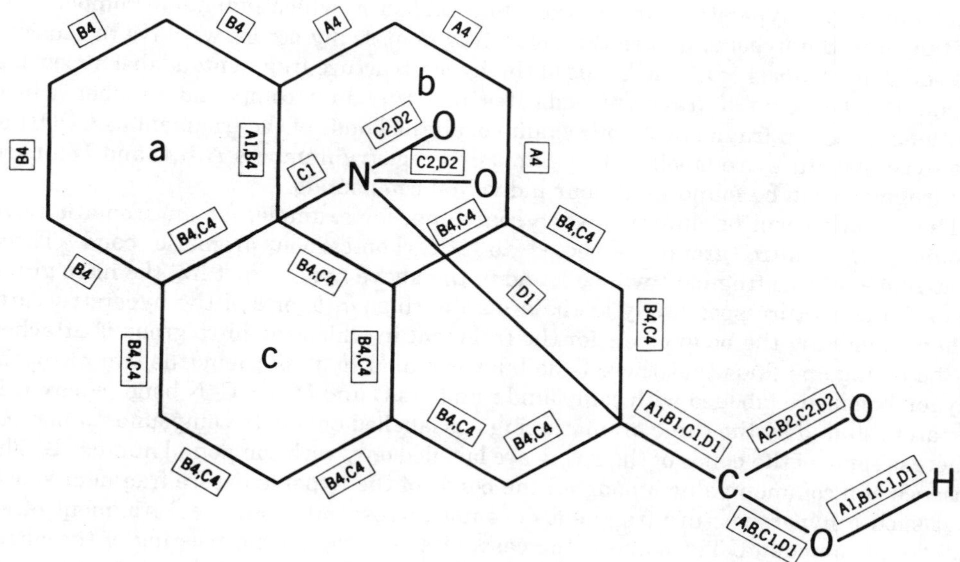

In this hyperstructure the nature (single, double, aromatic etc.) of bonds is not represented but each bond has a set of labels indicating the registry numbers of the individual diagrams whose bonds are represented by the given bond; each such label is followed by the indication of the nature of the given bond in the respective compound. Therefore by selecting only the bonds labelled by some individual compound registry number and using the bond nature indications attached to those registry numbers one can reproduce each individual diagram contained in the hyperstructure.

The 4 individual compounds **A** – **D** contain a total of 64 bonds, whereas the hyperstructure has only 27 bonds. This corresponds to a compaction ratio of 1:2.4 of the number of bonds, which reflects at least an equal size reduction of the corresponding records of the connectivity table type. Counting 4 bytes needed for recording one compound bond and three bytes for recording a hyperstructure bond we arrive at a total of 252 bytes for the individual diagrams and 91 bytes for the hyperstructure. There are 64 bond labels on the hyperstructure (one label for each bond of the initial structures) and, depending on the size of the compound database, these labels can occupy either 128 bytes (2-bytebinary numbers) or 192 bytes (3-byte binary numbers), yielding correspondingly either some summary memory savings or a memory overspending compared to the summary memory size needed for recording the initial compounds.

ATOM-BY-ATOM SEARCH OF HYPERSTRUCTURES

The justification for creating a hyperstructure may be the possibility for improving the performance of an atom-by-atom substructure searching procedure. For instance if we are looking for compounds containing a C–COOH fragment, instead of performing such a search separately on the four above individual compounds, we could perform one single search on the hyperstructure using the same atom-by-atom matching procedure we would use consecutively on the individual compounds. After identifying this fragment in the hyperstructure in order to establish in which individual compound(s) contained in the hyperstructure the found fragment really occurs we have to consider the set of bond labels of all the bonds of the hyperstructure fragment: a label appearing on all the bonds of a fragment indicates an individual compound number which contains the given fragment. In our example, all the bonds of the fragment C–COOH of the hyperstructure are labelled with compound registry numbers **A**,**B**,**C** and **D** and so the fragment can be found in all four individual compounds.

The situation will be different when searching, for example, for an aromatic nitro group, i.e., a nitro group connected to a carbon-carbon aromatic bond. Three occurrences of this fragment will be found in the above hyperstructure, the nitro group being attached correspondingly to the aromatic rings *a*, *b*, or *c* of the hyperstructure. When examining the bond labels for the fragment in which the nitro group is attached to the *a* ring one finds that there is no label common to all of them: the two nitrogen-oxygen bonds are labelled with compound numbers **C** and **D**, the C–N bond (where **C** is the atom shared by the three aromatic rings) is labelled only with compound number **C**, whereas some of the bonds of the *a* ring are labelled only with compound number **B**. The absence of a common value among all the bonds of the hyperstructure fragment shows that such a hyperstructure fragment does not correspond to any real fragment of an individual compound. The same is the case with the fragment consisting of the nitro-group attached to ring *b*, some of the bonds of this ring being labelled with **A**, a number not occurring as a label of the nitro group's bonds. In the case of the ring *c*, whose bond labels are **C** and **B**, the **C** is the common value appearing on all its bonds as well as on all the bonds of the nitro group. Therefore one infers that the fragment consisting of the nitro group attached to ring *c* does correspond to a real fragment of compound **C**.

Since inusing an atom-by-atom matching procedure the identification of a query fragment in the hyperstructure happens step-by-step, it is reasonable to check at each such step whether the already matched part of a fragment corresponds to some real occurrence in an individual structure. For this purpose, after successfully matching a first and a second bond of a query substructure against two bonds of a hyperstructure one has to perform the operation consisting of the intersection of the two sets of corresponding bond labels of the hyperstructure in order to find the common label(s) shared by these two hyperstructure bonds. The set of common labels thus found indicates the individual compounds containing the already matched part of the query fragment. This set has to be saved and intersected with the set of labels of every additional successfully matched hyperstructure bond. If at any point of the substructure search the result of the intersection of the labels set of the currently matched bond with the saved result of the previous intersections happens to be empty, one infers that the last bond matching is an unacceptable one since the given substructure does not correspond to any real substructure of some individual compound. When all the bonds of the query substructure have been successfully matched with hyperstructure bonds the result of the label set intersections is a set of individual compound numbers containing the given substructure.

In the case of our hyperstructure example when the query structure is the aromatic nitro group let us assume that the very first bond of the query fragment is an O=N bond which is successfully matched with one of such hyperstructure bonds labelled with compound numbers **C** and **D**. Let the second query fragment bond be an N–C bond which matches with the hyperstructure N–C bond labelled with **C**; the intersection of the sets {**C**,**D**} and {**C**} yields {**C**}. If the third query fragment bond is an aromatic C–C bond which is, for example, matched with a hyperstructure aromatic C–C bond labelled {**A**,**B**} the intersection with the saved {**C**} will result in an empty set indicating an unacceptable matching. By 'backtracking' the adjacent aromatic hyperstructure bond C–C labelled with the set {**B**,**C**} can be selected and such a fragment consisting of three already matched bonds will correspond to a real fragment of the individual compounds whose numbers are given by {**C**} ∪ {**B**,**C**} = {**C**}.

In our example of a hyperstructure covering 4 individual structures, the question of the justification of its introduction and use boils down to comparing the computational effort required for performing the four separate substructure searches in the four individual structures, with the summary computational effort required for performing one such search in the hyperstructure, combined with the set intersection operations following each separate bond matching step in the atom-by-atom substructure procedure.

Although the size of our hyperstructure is very small the essence of the problems one encounters dealing with hyperstructures encompassing large and very large compound databases remains the same. Below we will consider the case of such a large database and we will attempt to analyse the problems involved in creating and using hyperstructures in this context.

CONSTRUCTION OF HYPERSTRUCTURES

Formally speaking, a 'hyperstructure covering a set of individual structures' can be defined as a labelled graph having the following properties. Each 'covered structure' is contained in the hyperstructure as a subgraph and each edge (bond) of the hyperstructure belongs to at least one such subgraph. Thereby each hyperstructure bond corresponds to a bond of at least one of the structures covered. Each hyperstructure is labelled by a set of 2-tuples. The first element of such a 2-tuple is a pointer, or number, indicating a covered structure, one of whose bonds corresponds to the given hyperstructure bond. The 2-tuple's second element is the bond label (single, double, triple or aromatic) which the corresponding bond has in the structure indicated by the first element of the 2-tuple.

In so far as the construction of a hyperstructure from a set of individual structures is concerned two approaches are possible. The first of these approaches is aimed at achieving a maximal degree of 'superimposition' of the covered individual structures, in other words the maximal degree of overlap between the corresponding substructures, representing in the hyperstructure the individual structures. This approach seems to ensure the minimal size of the resulting hyperstructure. One way of realisation of the above approach is to use in the construction of the hyperstructure a stepwise procedure of superimposition whereby the consequently superimposed structures have significant overlapping fragments (maximal common subgraphs). This approach was recently experimentally explored by Wipke and Rogers;[10] working with a collection of 1000 compounds, they have found that the so-constructed hyperstructure (called by them 'unique storage tree') achieved a 21 per cent storage space saving and at the same time 'reduces search efforts by 70 per cent to 95 per cent relative to conventional unstructured sequential systems'. A serious problem this approach can be expected to face, when applied to very large collections of compounds, is the size of effort necessary to build the hyperstructure: determining the 'best' sequence of processing (ensuring high degrees of superimposition) of the individual structures may require computationally intensive pre-processing of the collection.

A seemingly different approach, especially attractive for very large compound collections, involves the construction of the hyperstructure by random superimposition of the covered structures. We will explore theoretically this approach having in mind a quasi-global collection of 10^7 compounds assuming for simplification purposes some non-essential limitations concerning the maximal size of compounds and the number of different heteroatoms they can contain. We will assume in particular that we are dealing with compounds consisting of not more than 100 non-hydrogen atoms, which can include the atoms of following elements: C, N, O, P, S, Si, B, F, Cl, Br, Li, Mg, Fe. We will also allow up to 100 carbon atoms in any individual structure, but not more than 20 atoms of any of the rest of elements. The great majority of known compounds fits these limitations, but it is more important to stress that such limitations are used here only to simplify the illustration of our ideas and could be easily eliminated in any real-life implementation. The idea is to use as hyperstructure a 'complete graph' which, by definition, will contain any smaller individual structure as its substructure. This means that instead of gradually building the hyperstructure by stepwise superimposition of individual covered structures we will use a complete graph as a pre-constructed hyperstructure and we will fit mechanically, in a random manner, the covered individual structures into this pre-constructed hyperstructure. This way the effort needed for hyperstructure construction will be minimised; it remains to be seen what the storage size savings and search effort savings, if any, may be.

In the context of our example, under a 'complete graph' we will assume a structure consisting of 100 C atoms numbered from 1 to 100, 20 nitrogen atoms numbered from 101 to 120, 20 oxygen atoms numbered from 121 to 140, and so on with the rest of the elements enumerated above, ending with 20 iron atoms numbered 341–360. We will assume also that each of the 360 atoms of the hyperstructure is connected with the rest of its 350 atoms. This will result in a hyperstructure consisting of a total of 360 atoms and $(360\times359)\div2 = 64{,}620$ bonds. The non-redundant connectivity table type record of this structure (in which no bond types are specified), using 5 bytes per bond will require 323,100 bytes, i.e., ⅓ megabyte.

The 'fitting' of an individual structure into this hyperstructure will involve a renumbering of the original (maybe arbitrary) numbering of atoms of the individual structure in such a way that all the carbon atoms get consecutive numbers from 1 to the highest possible value (necessarily $<$ 100), all the nitrogen atomsget consecutive numbers from 100 but not exceeding 120 and so on for all the rest of the elements, with the atoms of each of the elements getting numbers allowed by the corresponding permissible atom number range in the hyperstructure. Thereafter the fitting of an $\langle$I, l, J$\rangle$ bond of the individual structure number M into the hyperstructure will consist in the inclusion of the 2-tuple $\langle$M,l$\rangle$ inclusion into the set of labels associated with the $\langle$I,J$\rangle$ bond of the hyperstructure, where I,J are respectively the numbers of individual structure atoms linked by a bond, and l is the label of such a bond in the individual structure. As a technical detail one should notice that C–H bonds are omitted in both hyperstructure and individual structure whereas the heteroatom-hydrogen bonds of individual structures are fitted by creating the corresponding additional bonds.

When all the bonds of all the structures of the compound database have been fitted into the hyperstructure, those of its bonds who have empty label sets will be deleted. One can estimate that this way a significant fraction of the 64,620 bonds of the hyperstructure will be eliminated, but this proves not to be an important consideration since the total size of the connectivity table type record of the graph part of the hyperstructure (0.3 megabytes) is quite negligible in our quasi-global context. Another necessary part of the record of the hyperstructure is the list of labels associated with each used bond. Such a list of labels will consist of 2-tuples, the first element of which (the individual compound number) will require in the case of a 10^7 compounds collection a 3-byte binary number, while the second element, the label of the individual compound's bond, takes 1 byte for a total of 4 bytes per hyperstructure bond label. The total number of label set elements for the hyperstructure's bonds is equal to the total number of individual compound bonds; assuming, on the average, 30 bonds per compound we get $30 \times 10^7 \times 4 = 1200$ megabytes. This shows that the hyperstructure bond label lists make up practically the entire storage space necessary for recording the hyperstructure.

The comparison of this figure with the total memory size estimated above for the sequential standard storage of the same collection of compounds shows no dramatic differences. Given the approximate nature of our estimates we can conclude that in a real life implementation a hyperstructure of compounds may or may not save memory space depending on the specific implementation details.

EFFICIENCY OF THE SEARCH PROCESS

The important question remains whether the compound hyperstructure can offer substantial advantages in processing time for substructure searches which are the ultimate step in answering various query types dealing with both compounds and reactions.

An atom-by-atom search using the hyperstructure will be performed in a way similar to searches performed on an individual structure, by a stepwise matching and assignment of the hyperstructure's bonds to the query fragment's bonds. Each time a query fragment bond has been successfully matched to a hyperstructure bond an additional operation has to be performed in order to find out the set of individual compounds which contain the already matched portion of the query fragment. When a hyperstructure bond is assigned to the first bond of a query fragment the set of individual compounds for which such an assignment is valid will consist of the set of compound numbers associated with the hyperstructure's given bond. After another hyperstructure bond has been assigned to the next query fragment bond the above set, saved as the set of compounds potentially containing the query fragment, has to be intersected with the set of compound numbers associated with this second hyperstructure bond and the resulting smaller set has to be saved as the new set of compounds potentially fitting the query. Each time when a new hyperstructure bond is being matched with and tentatively assigned to a next query fragment bond, the saved set of potential individual compound hits has to be intersected with the set of compounds associated with the new hyperstructure bond. If the set resulting from any of such intersection operations proves to be empty, the tentative bond matching has to be rejected as not corresponding to any real specific compound covered by the hyperstructure. The final compound number set, resulting from the intersection of the compound set associated with the hyperstructure bond matched with the last bond of the query fragment, is a partial list of exact hits for the given substructure search.

After each bond matching step the intermediary sets of hits, corresponding to an already matched portion of the query fragment are saved with the last matched fragment bond. When backtracking is performed, and a number of previously matched query fragment bonds are freed for reassignment, the hit set corresponding to the fragment portion preserving its assignments can be picked up as the intermediary hit set saved with the last fragment bond preserving its assignment. This saved intermediary hit list is then used for intersection with the compound set of the first newly matched hyperstructure bond, and the resulting new intermediary hit set is saved with the fragment bond participating in this new match; thereby the old intermediary hit set (preceding backtracking) is replaced by a new intermediary hit set, corresponding to the new match. The union of all hit sets sequentially associated with the last fragment bond during the different successful matching is the final and full list of hits for the given substructure search.

The total number of bonds in our global collection of 10^7 compounds can be estimated as 3×10^7; dividing this figure by the total number of bonds in our hyperstructure (about 6.4×10^4) we get a figure of 4,687 for the average number of individual compound bonds superimposed onto one hyperstructure bond. A more realistic estimate is obtained by considering the carbon-carbon bonds which can be judged as making up about 70 per cent of the total bonds of specific compounds, i.e., a total number of about 2×10^8 bonds. On dividing this number by the total number of carbon-carbon bonds in our hyperstructure (4,950) we get on average about 40,000 specific compound carbon-carbon bonds superimposed on each hyperstructure C–C bond. This means at the same time an equal average compound number in the sets associated with each hyperstructure C–C bond; it also means that one of the sets to be intersected after each carbon-carbon bond matching step will be on the average of this size. The average size of the saved intermediary hit list set, which is the second participant of the set intersection operations can be estimated as 1,000 assuming that an average size query fragment includes about 10 bonds and that the average total number of hits for such a fragment is around 100.

A single bond matching operation involves many steps, in particular, firstly, positioning on the next fragment bond, secondly, selecting in the compound a yet non-assigned bond to be matched with the given fragment bond, thirdly, checking the correspondence of the atoms and bond labels of the bonds being matched, and fourthly, recording the assignment of the successfully matched bonds. Estimating that each of the above steps takes on the average not less than 25 elementary computer operations, one can estimate that each single bond matching operation involves about 100 elementary computer operations. Therefore the total computational effort needed to perform, during substructure searches sequentially performed on individual compounds, the matching of the 40,000 individual compound bonds superimposed in one hyperstructure bond, can be estimated as involving 4×10^6 elementary operations. The procedure of intersecting a set of 40,000 3-byte numbers with a similar set of 1,000 numbers can be envisaged as performed by 1,000 consequent binary searches of each of those 1,000 numbers in the longer list of 40,000 numbers. The number of steps of a binary search in a list of 40,000 items is about $\log_2 40{,}000 = 16$, therefore 10^3 binary searches will involve 16,000 steps. Each of those steps can be estimated as involving around 10 elementary operations and that brings the total number of elementary operations required by a single set intersection procedure to 160,000. Taking in account the efforts needed to perform the one time operation of matching the query fragment bond with the hyperstructure bond, including the steps described above for saving the intermediary hit lists, as well as the steps required thereafter to actually 'extract' from the hyperstructure the individual compound diagrams corresponding to the compound numbers from the final hit list of a given substructure search, we will more than double the estimated number of elementary operations, up to 400,000 elementary operations for each procedure of matching a query fragment bond with a hyperstructure bond. This still leaves us with a tenfold advantage in terms of the summary computational efforts needed for an atom-to-atom substructure search performed on a global hyperstructure compared with the same type of searches performed sequentially on individual compound structures. This projected advantage can become even more substantial if more efficient computational techniques can be found and implemented for performing set intersection operations. We have envisaged a specical bit/byte encoding based procedure, which could enhance the effectiveness of set intersection operations by one ortwo orders of magnitude.

The above considerations concern atom-by-atom substructure searches, which in practice are only the final steps of a substructure query handling cycle since, usually, some screening steps are utilised to substantially narrow down the number of potential hits. Only the relatively few hit candidates are then subject to atom-by-atom searches. It is worth considering the effects of such screening on the expected comparative performance of sequentially performed *versus* hyperstructure based (i.e., quasi-parallel) procedures.

For this purpose we will hypothesise that for an average 10-bond size query fragment, as a result of screening, the number of potential hits for such a substructure search has been narrowed down to 1,000, which is only 10 times more than the average final number of hits for such a query. The 1,000 compounds, containing an estimated number of 20,000 carbon-carbon bonds, represented by the approximately 5,000 C–C bonds of the hyperstructure would produce on average the superimposition of only 4 individual compounds carbon-carbon bonds on each carbon-carbon bond of the hyperstructure. At such alow level of superimposition, the sequential atom-by-atom substructure search can be expected to have advantages over the hyperstructure based search. Under such circumstances the computational effort needed to perform the screening can become of decisive importance.

Before concluding this study of a compound hyperstructure-based solution to the joint compound/reaction storage and retrieval problem we want to note that the collection of SRSG's, an important part of the reaction retrieval component of the proposed system, can be also organised in a hyperstructure. Although the average size of an SRSG is much less than the average size of a compound and the total number of SRSG's in a global collection (expected not to exceed 10,000) is also small, it can be shown that a hyperstructure of SRSG's can enhance the speed of SRSG substructure searches, which are the first step of handling reaction site (and reaction type) oriented queries.

CONCLUSION

The examination of different possible types of linkages between compound and reaction databases shows that the design of a global integrated compound/reaction database has clear advantages in comparison with systems separately designed for compounds and reactions. The study of the possibilities of a system including a compound hyperstructure component indicates that, although somehow ambiguous in the context of screening systems, the usage of hyperstructures may present significant advantages.

REFERENCES

1. Meyer, D., Cohan, P. 'Designing New Compounds with PC Databases'. *Am. Clin. Prod. Rev.* 1986, *5* (12), 16–19.
2. Gould, S. R., Meyer, D. E. 'A Chemical Information System for Microcomputers'. *Am. Lab.* 1987, *19* (3), 126–127.
3. Meyer, D. E., Gould, S. R. 'Software for Accessing Chemical Information'. *Am. Lab.* 1987, *19* (6), 124–126.
4. Gould, S. R. 'ChemSmart – A Chemical Compound/Reaction Database'. Paper presented at the 192nd Meeting of the American Chemical Society, Anaheim, California, 1986, CHED abstract number 32.
5. Gould, S. R. 'Chemical Structure Searches on an IBM PC from a 5¼ "Floppy Disk"'. Papers presented at the 192nd Meeting of the American Chemical Society, Anaheim, California, 1986, COMP abstract number 14.
6. Vladutz, G. 'Concerning One System of Classification and Codification of Organic Reactions'. *Inf. Storage Retr.* 1963, (1), 117–146.
7. Vladutz, G. (Vleduts, G. E.), Geyvandov, E. A. In *Avtomatizirovannye Informatsionnye Sistemy dlya Khimii* (Automated InformationSystems for Chemistry). Nauka: Moscow, 1974.
8. Vladutz, S. G., 'Do We Still Need a Classification of Reactions?'. In *Modern Approaches to Chemical Reaction Searching*; Willett, P. Ed.; Gower: Aldershot, 1986, pp. 202–220.
9. Kiho, Y. K. 'Formal'noe Opredelenie Nekotorykh Ponyatiy Kolichestvennoy Organicheskoi Khimii'. (The formal definition of some concepts of quantitative organic chemistry). *Org. React. (Tartu)*, 1972, *8* (2).
10. Wipke, W. J., Rogers, D. 'Tree Structured Maximal Common Subgraph Searching. An Example of Parallel Computation with a Single Sequential Processor'. In the press.

REACTION PLANNING (COMPUTER AIDED REACTION DESIGN)

Rainer Herges
Institut für Organische Chemie, Universität Nürnberg Henkestr. 42, 8520 Erlangen, West Germany

ABSTRACT

A computer aided systematic search for new organic reactions with the program IGOR[1] is presented. The algorithms of the program are briefly described. Three organic reactions of different types, which have been predicted by the program, have been realised in the laboratory.[2]

REACTION PLANNING

The subject is related in a very broad sense to synthesis planning. In synthesis planning the target is a molecule, which is to be synthesised by a sequence of known reactions. The target in reaction planning is the reaction itself. A striking difference to synthesis planning is the fact that data (reactions stored in a database) are not manipulated but new information (novel reactions) generated. In the sense of artificial intelligence (AI) the program could be called creative.

The name of the program used is IGOR (Interactive Generation of Organic Reactions).[1] The program was developed in Prof. I. Ugi's group (Technical University Munich), written by a computer scientist (J. Bauer) and improved in collaboration with the author to a level of applicability for chemical purposes.

The first programs with the ability of generating information in chemistry were those which constructed all isomers of a given molecular formula.[3] There are for instance 3 isomeric pentanes, 5 hexanes, 9 heptanesand 366319 decanes. A more sophisticated example is shown in Figure 1. The 20 valence isomers of cyclooctatetraene have been calculated with IGOR using only a few seconds CPU time.

The algorithms of this type of computer program are based on graph theory. A. T. Balaban, a scientist working in the field of application of graph theory for more than 20 years[4] wrote in 1985 in *J. Chem. Inf. Comput. Sci.*:[5]

> . . . The field is continuously expanding and may even reach the heuristic stage, when new potential reactions will be uncovered by these computer programs; . . .

The first reaction predicted by an artificial intelligence type program which could successfully be performed in the laboratory is shown below and was published in the same year in *Chimia*:[1]

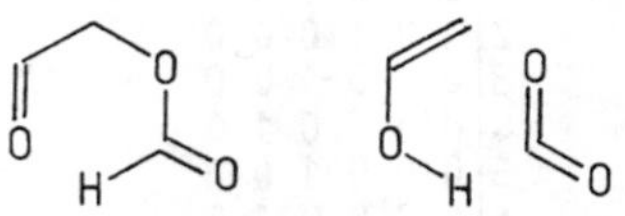

W. A. Warr (Ed.)
Chemical Structures

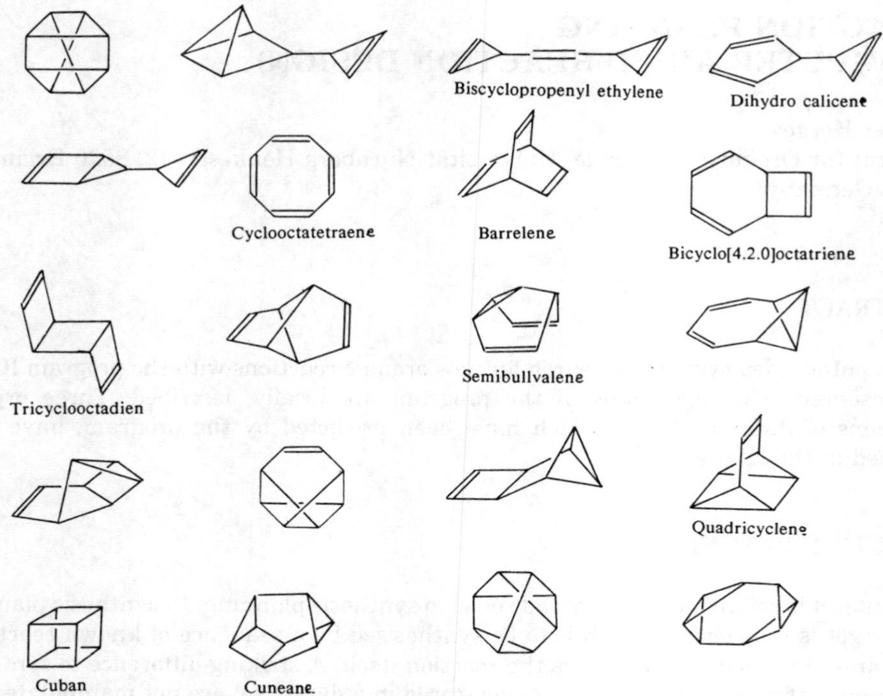

Figure 1. The 20 Valence Isomers of $(CH)_8$ computed with IGOR.

To use graph theory in order to search for new chemical reactions, one needs a graph theoretical representation of chemical reactions. More importantly one needs a representation of a reaction as a single graph. Most of the reaction indexing algorithms use two graphs, the graph of the reactants and the graph of the products (e.g., ChemBase from MDL). For our purposes this is impractible and this kind of coding of reactions is probably impracticable for any 'intelligent' reaction database.

A model which complies with our precondition is the Ugi-Dugundji model.[6] Reactants and products are represented by their connectivity matrix and the reaction by simply subtracting the two matrices mathematically. The R-matrix represents the changes in bond order during the reaction and consists of a single graph.

	1	2	3	4	5	6
1	0	2	0	0	0	0
2	2	0	1	0	0	0
3	0	1	0	1	0	0
4	0	0	1	0	1	0
5	0	0	0	1	0	2
6	0	0	0	0	2	0

Edukt

+

	1	2	3	4	5	6
1	0	-1	0	0	0	1
2	-1	0	1	0	0	0
3	0	1	0	-1	0	0
4	0	0	-1	0	1	0
5	0	0	0	1	0	-1
6	1	0	0	0	-1	0

R-Matrix

=

	1	2	3	4	5	6
1	0	1	0		0	1
2	1	0	2	0	0	0
3	0	2	0	0	0	0
4	0	0	0	0	2	0
5	0	0	0	2	0	1
6	1	0	0	0	1	0

Produkt

Thus a reaction can be separated into an R-matrix representing the changes during the reaction and an invariant 'sigma-frame'.

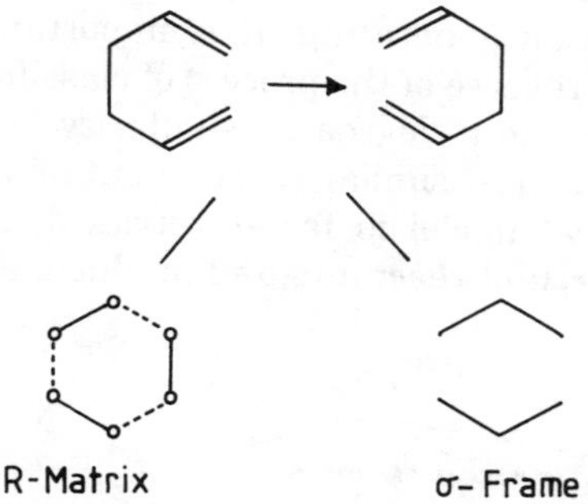

By varying the sigma-frame (without changing the R-matrix) reactions can be generated. The basic principle is exemplified in Figure 2. In this example we start with a reaction, which is known as a 'group transfer reaction'. If we add on both sides an additional bond (indicated by arrows) we come to the 'ene reaction'. Since we perform the operation on the reactant and the product side, the R-matrix remains constant. The 1,5 sigmatropic shift, the Diels Alder reaction and the Cope rearrangement can be derived from the ene reaction by the same procedure. The algorithm implemented in this computer program is more sophisticated since one has to consider symmetry to avoid multiply generated reactions.

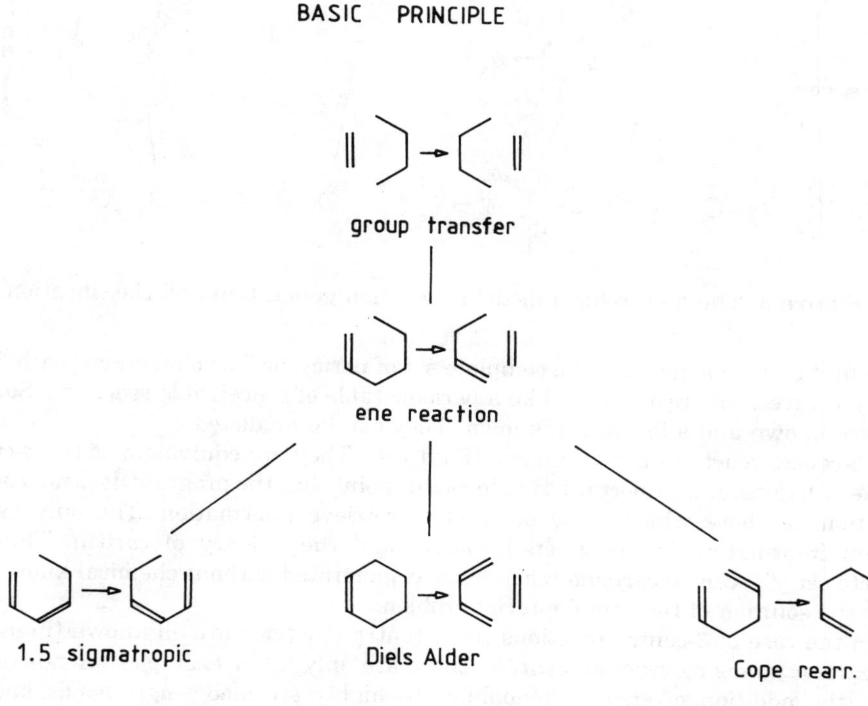

Figure 2. The basic principle of the generation of 'basic reactions'.

The next step in our reaction generating procedure is to introduce heteroatoms into the reaction centres. We call this level the level of 'hetero reactions'. The last step, generating the level of specific reactions is introducing substituents. Figure 3 shows the hierachical scheme of reaction generating. It is important to note that the reaction generation procedure is the reverse of the process of classification of chemical reactions. The model allows one to define a degree of 'similarity' between reactions. Thus one would be able to retrieve the most similar reaction out of a reaction database. Another advantage of the hierachical model is faster retrieval, since one does not have to perform a substructure search of reactants and products in the database.

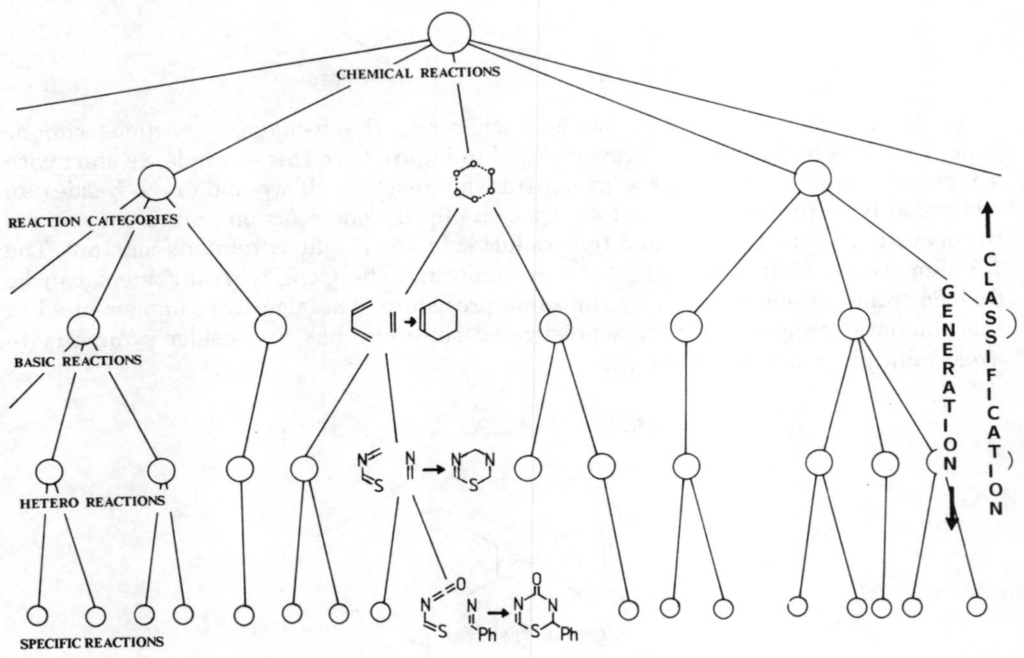

Figure 3. The hierarchical model of reaction generation and classification.

As a first test we generated the complete set of pericyclic 'basic reactions' with 3 to 6 reaction centres. The result looks like a periodic table of conceivable reactions. Some of them are known and a few are unknown. They can be predicted.

The 3-centre reactions are all known (Figure 4). They are equivalent to the 3-centre carbene reactions. It is important to note at this point that the program does not contain a reaction database which could be used to retrieve information. The only system inherent information is the given R-matrix and the valency of carbon. Thus the complete set of 3-centre carbene reactions was generated without chemical knowledge, but by the solution of the combinatorial problem.

As in the case of 3-centre reactions the 4-centre reactions are all known (Figure 5). With carbon atoms as reaction centres, there are only a few examples for reactions 5 and 6 (the addition of strong dienophiles to highly strained single bonds and the rearrangement of cubane to cuneane).

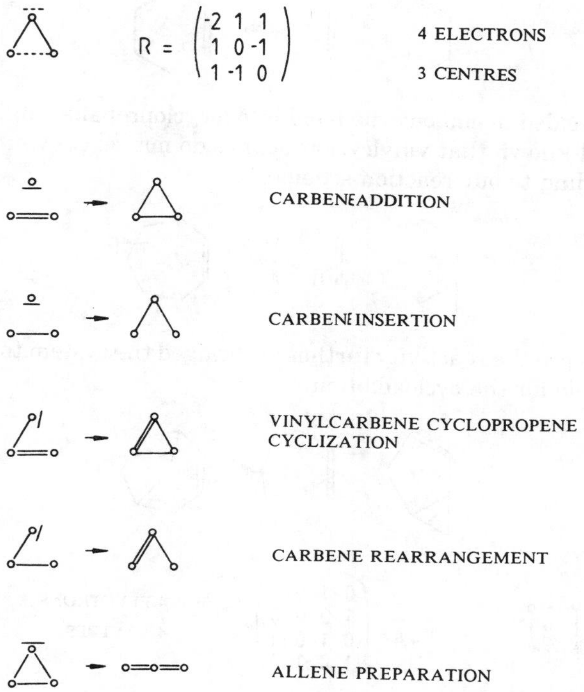

$$R = \begin{pmatrix} -2 & 1 & 1 \\ 1 & 0 & -1 \\ 1 & -1 & 0 \end{pmatrix}$$

4 ELECTRONS

3 CENTRES

CARBENEADDITION

CARBENEINSERTION

VINYLCARBENE CYCLOPROPENE CYCLIZATION

CARBENE REARRANGEMENT

ALLENE PREPARATION

Figure 4. The 3-centre pericyclic basic reactions.

But among the 5-centre reactions there are some unknown species. Reactions 1, 2 and 3 are well-known and synthetically useful reactions (see Figure 6).

We chose reaction 3, the 1,3 dipolar cycloaddition, to test out hetero reaction generation procedure. By introducing the elements carbon, nitrogen and oxygen into the reaction centres of the 1,3 dipole we generated all conceivable allyl type compounds. Unfortunately the structures (shown in Figure 7) are known and they had all been used in 1,3 dipolar additions. Thus our hope to find a new 1,3 dipole was lost.

There are 12 conceivable 6-centre pericyclic basic reactions (see Figure 8). Among the known reactions there are classical pericyclic reactions like the electrocyclic reaction, the Cope rearrangement and the 1,5 sigmatropic shift. Reactions 6, 8, 10, 11 and 12 with carbon atoms as reaction centres are unknown. Reactions 8 and 12 are trimolecular and therefore not favoured on entropy grounds. Reactions 10, 11, and 12 involve two or more sp_3-hybridised centres reacting with each other, a fact which is not favourable on energy grounds.

There is only one reaction left which looks quite likely: reaction 6. With hydrogen in the 4-membered component the reaction is known as the 'ene reaction'. We chose this basic reaction with carbon atoms at all reaction centres for a preparative attempt, since this type of reaction should be synthetically useful for C-C connections.

We considered the breaking of the C-C single bond in the 4-membered component to be the crucial step of the reaction.

Therefore we decided to embody the bond into a cyclopropane ring to put strain on it. However it is well-known that vinylcyclopropanes do not react even with the strongest dienophiles according to our reaction scheme.[8]

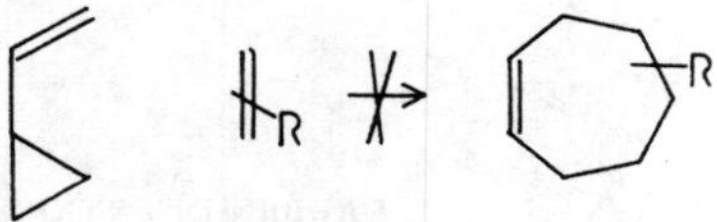

In order to enhance the reactivity further we bridged the system to fit it in a geometry which is favourable for the cycloaddition.

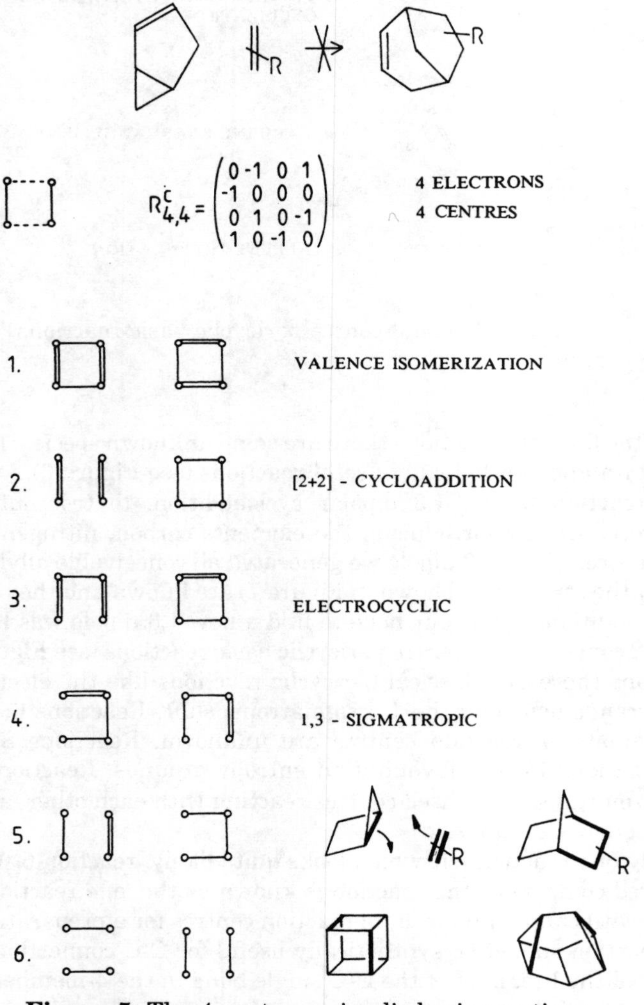

Figure 5. The 4-centre pericyclic basic reactions.

6 ELECTRONS
6 CENTRES

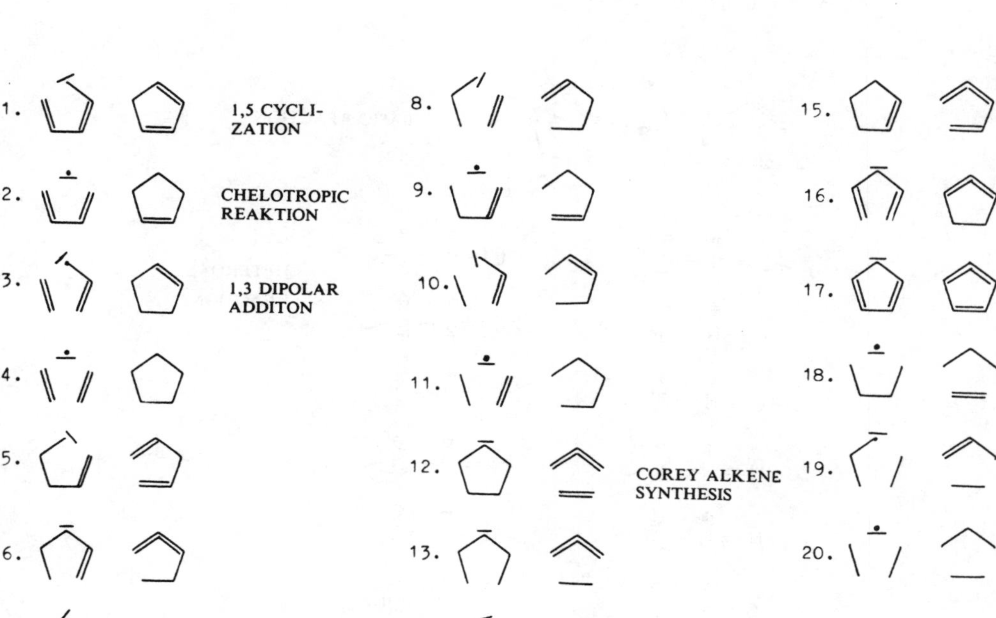

Figure 6. The 5-centre pericyclic basic reactions.

But even the bicyclohexene is not reactive enough to react according to our mechanism. The C-C bond obviously needs further activation by additional strain or by activating substituents. In order to find the best substituent and the right position we carried out a number of quantum mechanical (MNDO) calculations on various substituted vinylcyclopropanes. It turned out that the most effective activation was achieved with an electron donating substituent in the 1-position.

X = O, N, S

The most simple way to combine the two ideas, fixing of the geometry in *cis* conformation and activation by an electron donating substituent, is bridging the system with the substituent. Our first trial was homofurane and tetracyanoethylene.

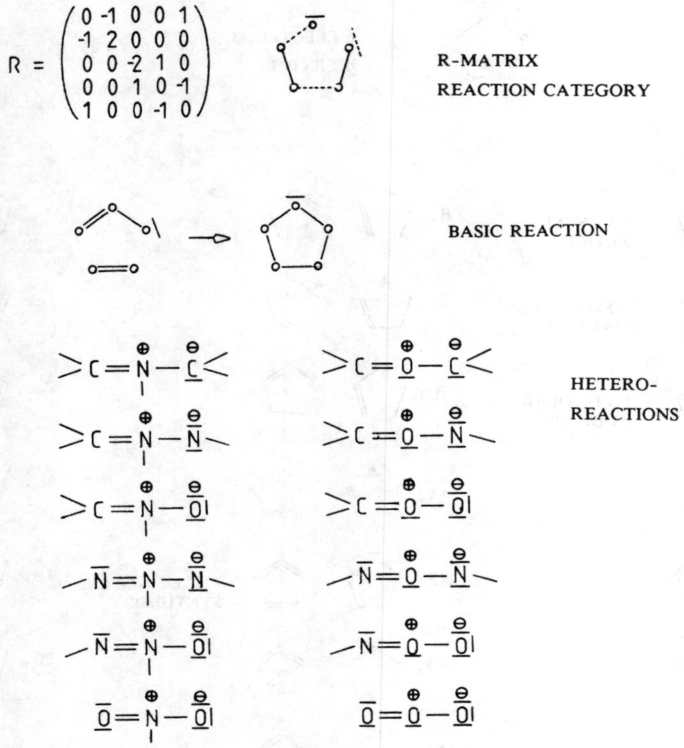

$$R = \begin{pmatrix} 0 & -1 & 0 & 0 & 1 \\ -1 & 2 & 0 & 0 & 0 \\ 0 & 0 & -2 & 1 & 0 \\ 0 & 0 & 1 & 0 & -1 \\ 1 & 0 & 0 & -1 & 0 \end{pmatrix}$$

Figure 7. The allyl type 1,3 dipoles with C, N and O as reaction centres.

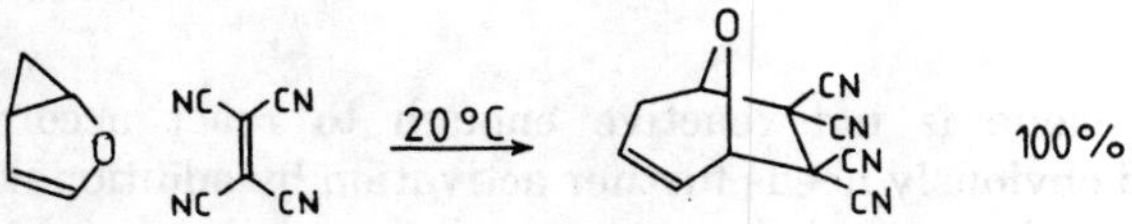

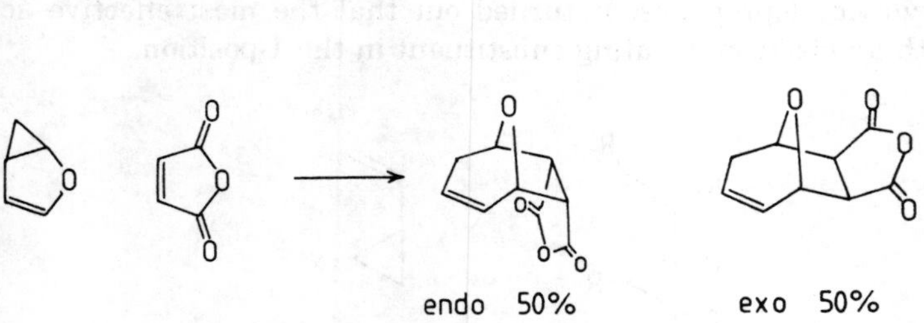

The yield was 100 per cent, the reaction time 20 minutes and the reaction temperature 20°C.[9] The corresponding sulphur substituted compound homothiophene was even more reactive. With less symmetrical dienophiles, like maleic anhydride, the two possible stereoisomers *exo* and *endo* are formed in a 1:1 ratio. The reaction is of potential synthetic use, since there are only a few methods for 7-membered ring synthesis. With nitrogen as heteroatom the skeleton of the atropine alkaloids is formed within one step. There are numerous alkaloids and drugs within this class of compounds (e.g., scopolamine). During a careful study of the literature we found that the reaction was not as 'virginal' as we thought. There are 3 examples of this $[\pi_2\sigma_2 + \pi_2]$ addition in the literature.[8–10]

For our second example we again chose the 6-centre pericyclic reactions. From the 12 conceivable basic reactions of this reaction category we elected theretro ene reaction scheme for further study (Figure 9). It is the reverse scheme of the homofurane cycloaddition, our first example. The reaction is not very likely to proceed with carbon atoms as reaction centres, since two C-C single bonds have to be broken within one step. We decided to prefer hydrogen to carbon as the group which is to be transferred during the reaction and oxygen as the sp_2-centre to which the hydrogen is transferred. All other centres can be either carbon or oxygen.

With these preliminary conditions IGOR generated 8 hetero reactions. Some of them are known (e.g. the ester pyrolysis).

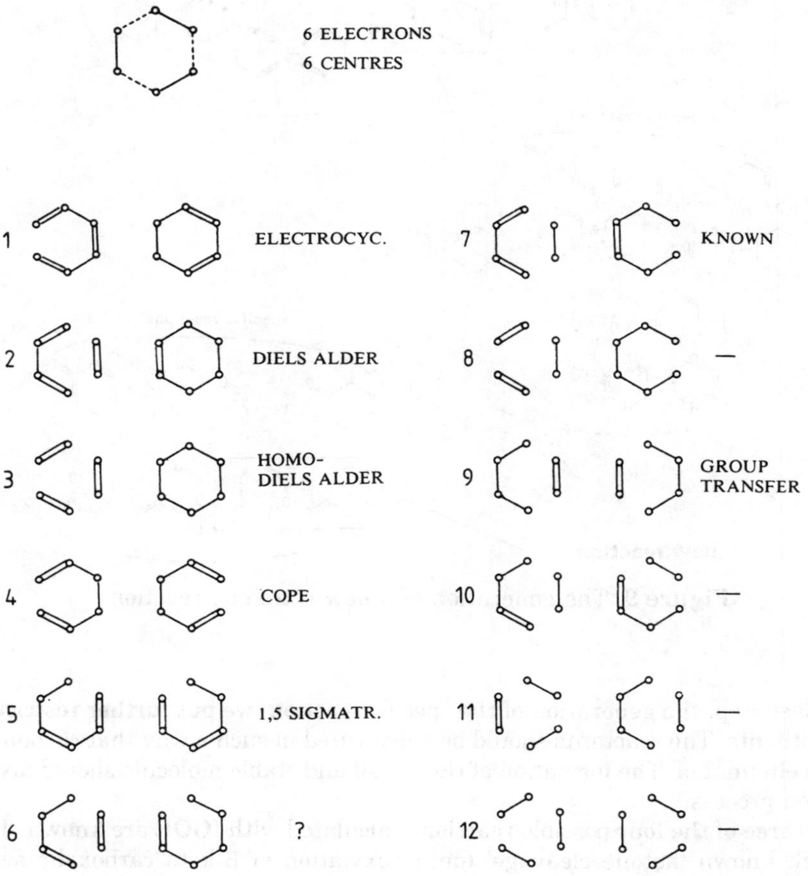

Figure 8. The 6 centre pericyclic basic reactions.

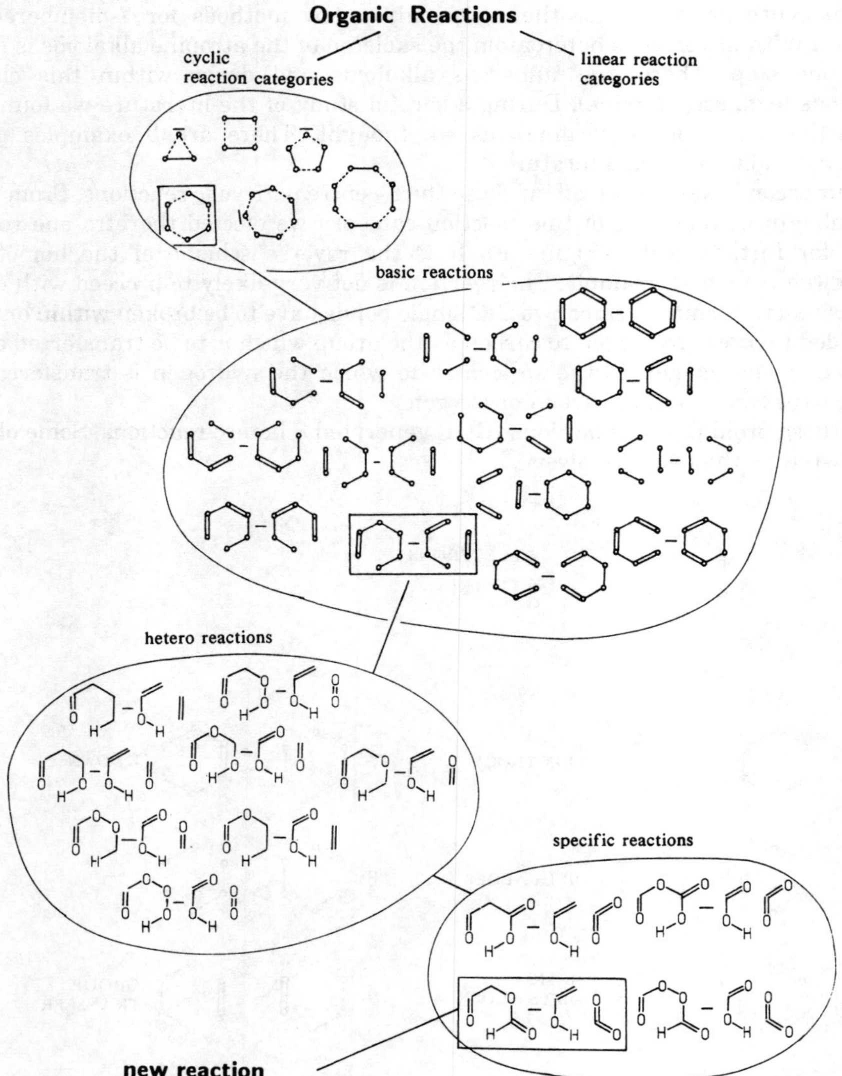

Figure 9. The generation of a new retro ene reaction.

In the last step, the generation of the specific reaction, we put further restrictions on the substituents. The reactants should be substituted in such a way that carbon dioxide should be eliminated. The formation of this small and stable molecule should favour the elimination process.

Indeed three of the four possible reactions calculated with IGOR are known. The first is the well known 'ketone cleavage' (decarboxylation of β-keto carboxylic acids and β-dicarboxylic acids) at about 100°C. The second is the decarboxylation of the mixed anhydride of a carboxylic acid with carbonic acid. The third, the decarboxylation of peroxyester is reported in the literature,[11] but the fourth is unprecedented according to a careful search of the literature ('new reaction' in Figure 9).

The reactant can be easily prepared by esterification of an acyloin with formic acid. The pyrolysis was carried out in cyclohexane solution at 290°C. The yields are quite low in the liquid phase but almost quantitative in the gas phase.

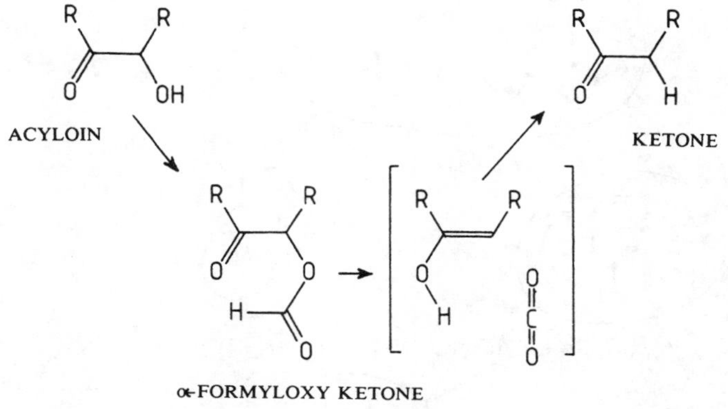

ACYLOIN

KETONE

α-FORMYLOXY KETONE

R = Ph 30 %
R = CH_3 48 %
R = $(CH_2)_4$ 15 %

The versatility of our reaction planning algorithms is demonstrated in our third example. The design of an isonitrile preparation method exemplifies the fact that reactions can be designed for specific purposes.

In the past all attempts to synthesise isonitriles with strong electron withdrawing groups have failed. Strongly electrophilic isonitriles isomerised during preparation to the corresponding nitriles or reacted with the nucleophilic solvent which is needed for their formation. Therefore we tried to develop a new method.

In order to avoid solvent problems we again chose the pericyclic reactions as the source of our new reaction, since these reactions are usually not solvent dependent. In consideration of the known thermal lability of this class of compounds we only took the photochemically allowed reaction categories into account (those with 4n electrons in the transition state, Figure 10). Photolysis at low temperatures might be the suitable condition for the formation.

In the course of the reaction a formal lone pair must be formed on carbon. There are only two reaction categories satisfying this restriction, the 3-membered and the 7-membered pericyclic reactions. We considered the 7-membered pericyclic reaction to be too large a system to provide an elegant synthesis and concentrated on the 3-membered reaction category.

This category consists of 5 conceivable basic reactions (see Figure 4). However only two of them comply with an isonitrile substructure. Introducing the suitable elements and substituents into these two basic reactions yields three possible specific reactions for an isonitrile synthesis.

We chose the elimination reaction of the cyclopropenone imine for our first trial in the laboratory, since the reactant can be easily prepared.

SYNTHESIS OF STRONG ELECTROPHILIC ISONITRILES Z–N=C|

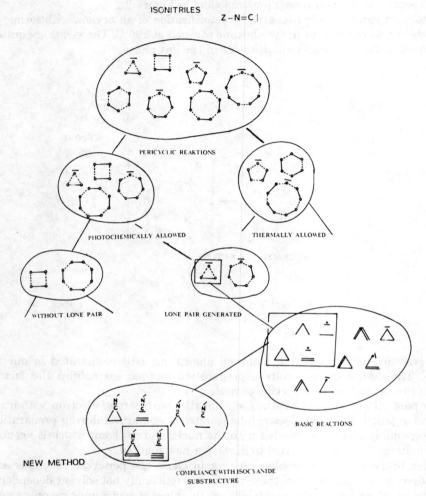

PERICYCLIC REAKTIONS

PHOTOCHEMICALLY ALLOWED

THERMALLY ALLOWED

WITHOUT LONE PAIR

LONE PAIR GENERATED

BASIC REACTIONS

NEW METHOD

COMPLIANCE WITH ISOCYANIDE SUBSTRUCTURE

Figure 10. The design of an isonitrile synthesis.

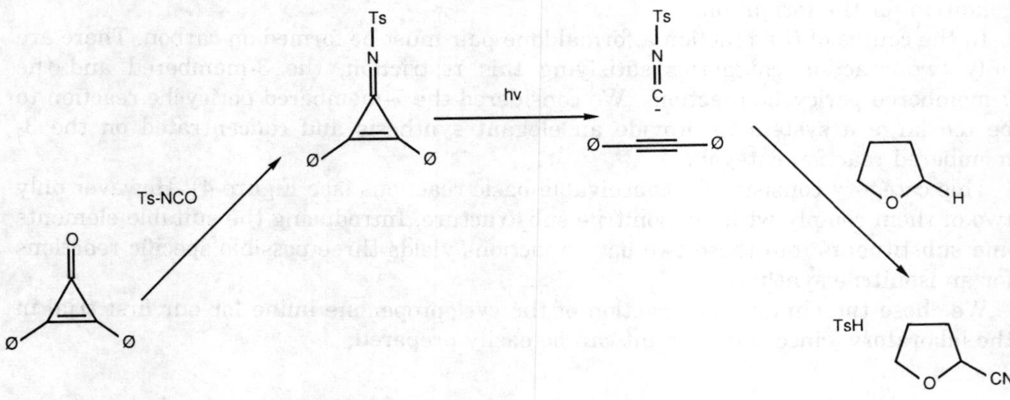

The photolysis was carried out at −80°C. As predicted by quantum mechanical calculations the strong electrophilic isonitrile reacted like a carbene and inserted into the C-H bond of the α-hydrogen of the solvent (tetrahydrofuran).

How 'new' are these three computer predicted organic reactions? One could raise the objection, that there are 'similar' known organic reactions (the last example is a chelotropic reaction). The answer is easy to give within our hierachical model of reaction classification. We can define a 'degree of novelty' or innovation of chemical reactions:

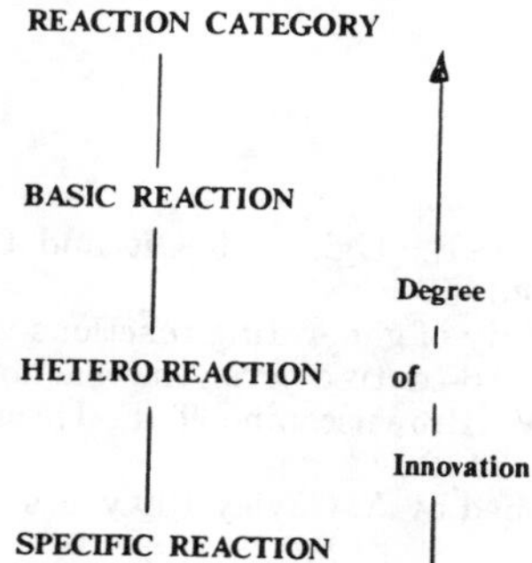

The lowest degree of innovation would be to find a new specific reaction, a known reaction with different substituents. I would call this the level of undergraduate research. The next level of novelty is the level of hetero reactions, a new variation of hetero atoms in a known reaction scheme (e.g., a new hetero Diels Alder reaction).

The level of basic reactions is a rather high level of innovation. There have been some chemists awarded the Nobel prize for discovering new basic reactions (e.g., Wittig, Diels, Alder, Friedel, Crafts etc.).

The highest level of innovation would be to find a new reaction category. This would be very difficult since they are probably all known.

Since this presentation started with a citation of a theoretical chemist I would like to close with a citation of a preparative chemist. D. H. R. Barton is one of the most distinguished chemists in natural product synthesis and wrote in 1973 in *Chemistry in Britain*:[12]

'. . . I would not like to neglect to mention that we are still very poor in synthetic methods in organic chemistry. Most of the synthetic work is done with organic reactions of the type which have been known for a long time. If you know 20 organic reactions you probably know most of the steps used in synthetic work, particularly in industry, but I am quite sure there must be hundreds of other organic reactions to be discovered. We have not in the past thought about these problems in the right way. When we have been faced with a problem of effecting a chemical synthesis we have sought *known* methods. We have not paused to think why we do not invent a *new* method every time. If we adopt this philosophy we are going to be extremely busy till the end of the century, a) trying to equal the enzymes and b) thinking up new ways of synthesis.'

I believe that 'reaction planning' could contribute to the solution of this problem. It is an effective tool for the exploration of so far unknown areas of organic chemistry. But besides the discovery of new organic reactions, the research in 'computer aided chemistry' has a nice didactical side effect. If one is forced to squeeze one's chemical knowledge into algorithms which can be programmed on a computer, one has to be aware of the heuristic processes of chemical thinking. I personally learned more about organic chemistry doing this kind of 'computer chemistry' than out of any textbook.

ACKNOWLEDGMENTS

Prof. I. Ugi, Dr. J. Bauer and the Fonds der Chemischen Industrie.

REFERENCES

1. Bauer, J., Herges, R., Fontain, E., Ugi, I. 'IGOR and Computer Assisted Innovation in Chemistry'. *Chimia* 1985, *39*, 43–53.
2. The first program with the ability of generating reactions was written by T. F. Brownscombe in his thesis. He listed the 4- and 6-centred pericyclic reactions and used enthalpy estimates as selection critera. Stevens, R. V., Brownscombe, T. F. Thesis, Rice Unviversity, Houston, TX 1972; *Diss. Abstr. Int. B* 1973, *34* (3).
3. The first algorithm was published by A. Cayley 130 years ago: Cayley, A. *Philos. Mag.* 1857, *13* (1), 172.
4. Balaban, A. T. 'Valence Isomerism of Cyclopolyenes'. *Rev. Roum. Chim.* 1966, *11*, 1097–1116.
5. Balaban, A. T. 'Applications of Graph Theory in Chemistry'. *J. Chem. Inf. Comput. Sci.* 1985, *25*, 334–343.
6. Dugundji, J., Ugi, I. 'An Algebraic Model of Constitutional Chemistry as a Basis for Chemical Computer Programs'. *Top. Curr. Chem.* 1973, *39*, 19–64. Ugi, I., Bauer, J., Brandt, J., Friedrich, J., Gasteiger, J., Jochum, C., u.W. Schubert. 'New Applications of Computers in Chemistry'. *Angew. Chem.* 1979, *91*, 99–111.
7. See also Hendrickson, J. B. 'The Variety of Thermal Pericyclic Reactions'. *Angew. Chem.* 1974, *86*, 71–100.
8. Volkenburgh, R. V., Greenlee, K. W., Derfer, J. M., Boord, C. E. 'A Synthesis of Vinylcyclopropane'. *J. Am. Chem. Soc.* 1949, *71*, 3595–3597. Nishida, S., Moritani, I., Teraji, u. T. 'Thermal 2+2 Cycloaddition of Cyclopropylethylene with Tetracyanoethylene'. *J. Org. Chem.* 1973, *38*, 1878–1881.
9. Herges, R., Ugi, I. 'Synthesis of Seven-membered Rings by $[(\pi 2+\sigma 2)+\pi 2]$ Cycloaddition to Homodienes'. *Angew. Chem.* 1985, *97*, 596–598. Herges, R., Ugi, I. 'Cycloadditionen von Homodienen'. *Chem. Ber.* 1986, *119*, 829–836.
10. Fowler, F. W. 'Cycloadditions of N-methoxycarbonyl-2,3-homopyrrole'. *Angew. Chem.* 1971, *83*, 148–149. *Angew. Chem. Int. Ed. Engl.* 1971, *10*, 135–136. Baldwin, J., Pinschmidt, R. K. jr. 'The Cycloaddition of Bicyclo2.1.Opent-2-ene with Tetracyanoethylene'. *Tetrahedron Lett.* 1971, 935–938. Christl, M., Brunn und E., Lanzendoerfer, F. 'Reactions of Benzvalene with Tetracyanoethylene,2,3-Dichloro-5,6-dicyano-p-benzoquinone, Chlorosulphonyl Isocyanate, and Sulphur Dioxide. Evidence for Concerted 1,4-Cycloadditions to a Vinylcyclopropane System'. *J. Am. Chem. Soc.* 1984, *106*, 373–382.
11. Kharasch, M. S., Kuderna, J., Nudenberg, W. 'Reactions of Atoms and Free Radicals in Solution. XXXVI. Formation of Optically Active Esters in the Decomposition of Optically Active Diacyl Peroxides in Solution'. *J. Org. Chem.* 1954, *19*, 1283–1289.
12. Barton, D. H. R. 'To Jump the Gap: An Interview with Professor Sir Derek Barton'. *Chem. Br.* 1973, *9*, 149–153.

POSTER SESSION: MACHINE GENERATION OF MULTI-STEP REACTIONS IN A DOCUMENT FROM SINGLE-STEP INPUT REACTIONS

Paul E Blower, Jr., Steve W Chapman, Robert C Dana, Howard J Erisman and Dale E Hartzler
Chemical Abstracts Service, 2540 Olentangy River Road, P O Box 3012, Columbus, Ohio 43210, U.S.A.

ABSTRACT

Chemical Abstracts Service (CAS) has created a protocol for machine generation of multi-step reactions within a document from the single-step reactions which are input. This protocol is being used in the CASREACT database for both explicit and 'implied' multi-step reactions. Both the input data and the output are described. Special features of this tool include:

1. ability to handle multi-product input reactions, discarding for later steps all those not involved in the multi-step sequence;
2. elimination of duplicates, but distinguishing reactions with identical reactants and products when other reaction participants and/or yield are different; and
3. accurately tracking all side- and branch-reactions converging into the main reaction sequence.

MACHINE GENERATION OF MULTI-STEP REACTIONS

Chemical Abstracts Service has been preparing the database for its online reaction retrieval service, called CASREACT, since October 1984. The initial database for the service will include reactions of organic substances (including organometallics and biomolecules) reported in 104 journals since January 1, 1985 and subsequently abstracted in *Chemical Abstracts* Organic Chemistry Sections (CA Sections 21–34). Those who use CASREACT will be able to search for reactions by specifying a structure, a substructure, or a Registry number and indicating whether the substance's role in the reaction is as reactant, product, reagent, catalyst, or solvent. Answers will involve a reaction diagram containing structures for reactants and products and the CA reference for the reaction.

Although most useful chemical preparations involve between one and three individual chemical reactions, there are important syntheses that require many steps. Most of the 'classical' natural product syntheses require at least five steps; many require between ten and twenty reactions, with additional resolution of optical isomers usually necessary. A further characteristic of these 'classical' syntheses is that two or more synthetic pathways may converge to yield the final product.

For example, in 1978 Nakata and co-workers prepared the ionophore antibiotic Lasalocid A by a convergent protocol in which the critical intermediate A was prepared

W. A. Warr (Ed.)
Chemical Structures

in 11 steps from racemic 7-ethyl-3-methylene-7-nonen-4-ol, then coupled with intermediate B, which had been prepared in 3 steps from benzyl 2-hydroxy-3-methylbenzoate, and finally converted to the desired natural product by an additional reaction after the coupling.[1] A schematic diagram of this synthesis is shown in Appendix B.

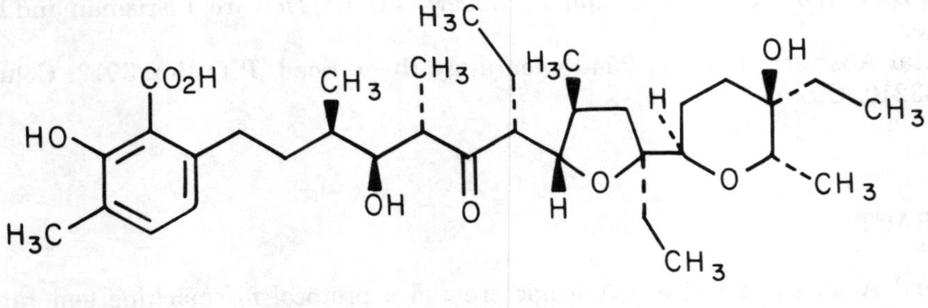

Lasalocid A

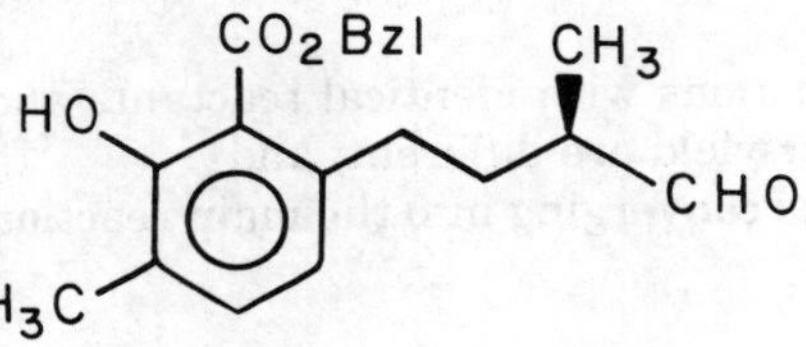

Intermediate A

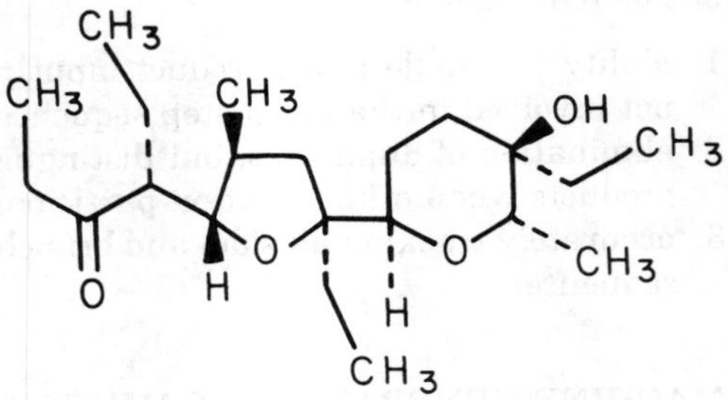

Intermediate B

Alternative syntheses of Lasalocid A by Ireland and co-workers have also been reported.[2]

Those who use CASREACT should be able to find such synthetic protocols, regardless of their length or complexity. Thus, if the synthesis mentioned above had been reported after 1984 and thus included in the CASREACT database, the user should be able to search for substructures of the nonenol and/or the benzoate as reactants, and either the Registry number or a substructure for Lasalocid A as a product, and find this complex synthetic protocol. It should not matter that there are 17 separate chemical reactions required for this transformation, or that the two possible reactants are from the two distinct 'branches' of this convergent synthesis. The reactants could have been specified by Registry number also, as queries in CASREACT can involve either substructures (by creating and crossing an answer set from the Registry File) or exact Registry numbers.

At the same time, CAS recognised that good use of resources required that the document analysts should not try to create the reaction links required to tie all the single-step reactions together. Although the typical natural product synthesis usually shows the overall strategy clearly enough for the analyst to create the links, there are more modest, but multi-step, pathways that are not obvious in the literature. Hence, document analysts only specify single-step reactions, with the multi-step connections being created by machine procedures. This paper describes our approach to this problem.

Linkage reactions are machine-generated by the following general method.

When the product of a reaction is used as the reactant of another reaction, a new linkage reaction is generated that consists of all the reactants of the first reaction, all of the products of the second reaction, any additional reactants that were introduced in the second reaction, the combined catalysts, solvents, and reagents of both reactions, and the combined reaction notes of both reactions.[3]

Additional programming rules support the following special features of the linkage reaction generation, so that the system:

1. identifies linear sequences and their combinations to form linkage reactions;
2. handles multi-product reactions and eliminates additional reactants that are not part of the linkage reaction;
3. identifies product-reactant loops and discards;
4. eliminates duplicate linkage reactions;
5. does not consider duplicate reactants and/or products within a single-step reaction for generating linkage reactions.

The hypothetical reaction map shown in Figure 1 leads to single-step reactions that could be input. The example is intended to be very complex and to include possibilities not found in most chemical syntheses, e.g., two paths to A.

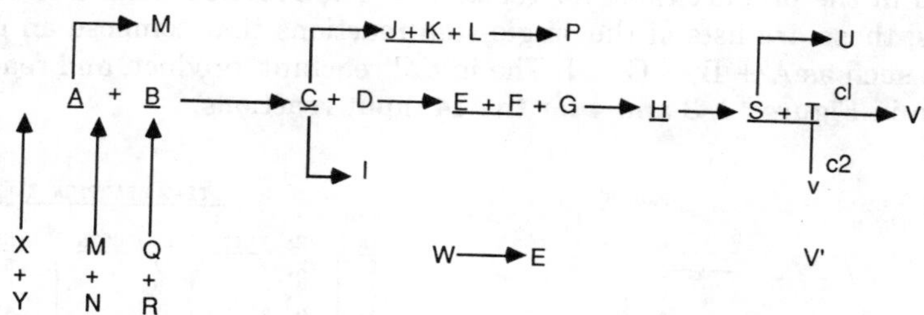

Figure 1

* In this diagram, the underlined letters represent the products of a reaction and the non-underlined letters represent additional reactants added.
** V and V′ are the same product, but produced using different catalysts (cl and c2).

The document analyst specifies the single-step reactions to be entered in the database:

1. X + Y → A
2. M + N → A
3. Q + R → B
4. A + B → C
5. W → E
6. C + D → E + F
7. E + F + G → H
8. C → I

9. C → J + K
10. A → M
11. J + K + L → P
12. H → S + T
13. S → U
14. T → V
15. T → V′

The document analyst specifies both reactions 14 and 15, but reaction 15 will be ignored when constructing the linkage reactions. A total of 96 linkage reactions is generated for the above reaction scheme; they are shown in Appendix A.

The system numbers the added linkage reactions, starting with numbers following those of the input reactions. This example does not show our handling of the 'implied' multi-step reactions – those in which there are several successive steps without indication of intermediate products. Such reactions are treated as if they were explicit multi-step reactions, but with unknown reactants and products.

The overall process involves the following steps:

1. Build tables for the reactants, products and reactions from the specified single-step reactions. The reactant and product tables list the reference for each unique compound and all reaction numbers for the reactions in which they appear with the corresponding role. Duplicates are eliminated as the tables are being built. Thus, in the above example, compound C would appear in the reactant table for reactions 6, 8 and 9 and in the product table for reaction 4. The reaction table also shows linear sequences; these are lists of the single-step reactions that compose an unbranched sequence, such as A + B → C → I. The initial reactant, product, and reaction tables are shown in Figures 2, 3 and 4 for the 14 input reactions.

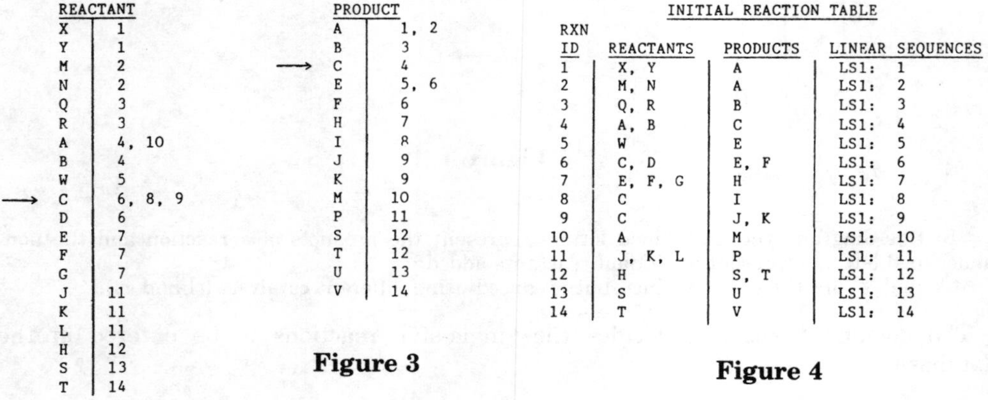

REACTANT	
X	1
Y	1
M	2
N	2
Q	3
R	3
A	4, 10
B	4
W	5
→ C	6, 8, 9
D	6
E	7
F	7
G	7
J	11
K	11
L	11
H	12
S	13
T	14

Figure 2

PRODUCT	
A	1, 2
B	3
→ C	4
E	5, 6
F	6
H	7
I	8
J	9
K	9
M	10
P	11
S	12
T	12
U	13
V	14

Figure 3

INITIAL REACTION TABLE

RXN ID	REACTANTS	PRODUCTS	LINEAR SEQUENCES
1	X, Y	A	LS1: 1
2	M, N	A	LS1: 2
3	Q, R	B	LS1: 3
4	A, B	C	LS1: 4
5	W	E	LS1: 5
6	C, D	E, F	LS1: 6
7	E, F, G	H	LS1: 7
8	C	I	LS1: 8
9	C	J, K	LS1: 9
10	A	M	LS1: 10
11	J, K, L	P	LS1: 11
12	H	S, T	LS1: 12
13	S	U	LS1: 13
14	T	V	LS1: 14

Figure 4

2. The products are matched against the reactants in these tables to create linkage reactions that can be added to the reaction table. A match means that the reactant can be substituted in the reaction which gave the product. Thus, compound C is involved in input reactions as follows:

(4) A + B → C
(6) C + D → E + F
(8) C → I
(9) C → J + K

(Reaction numbers are those of the input reactions).

Matching of the reactant and product tables for C shows that A and B as reactants can be substituted for C when it appears as a reactant. Thus:

(25) A + B + D → E + F
(46) A + B → I
(52) A + B → J + K

(Reaction numbers are those of the created linkage reactions – see Appendix A).

3. Linear sequences are detected and recorded using the following method.

If a linkage reaction is being generated in which the reactant that was matched with a product is part of an existing linkage reaction, but is not part of the first single-step reaction in the existing linkage reaction, then the new generated linkage reaction has more than one linear sequence. The reaction with the matched products is contained in the new linear sequence.

For example, linkage reaction number 44 (X + Y + Q + R → I) reflects two linear sequences. Linear sequence one contains input reactions 1, 4 and 8; while linear sequence two contains input reactions 3, 4 and 8.

4. A 'phase' of the multi-step generation is complete when all products have been processed. Overall, the reaction table resulting from the initial matching process (first phase) of the multi-step generation would be as shown in Figure 5 (the reaction ID numbers are not those of the final set of linkage reactions shown in Appendix A).

REACTION TABLE

RXN ID	REACTANTS	PRODUCTS	LINEAR SEQUENCES
1	X, Y	A	LS1: 1
2	M, N	A	LS1: 2
3	Q, R	B	LS1: 3
4	A, B	C	LS1: 4
5	W	E	LS1: 5
6	C, D	E, F	LS1: 6
7	E, F, G	H	LS1: 7
8	C	I	LS1: 8
9	C	J, K	LS1: 9
10	A	M	LS1: 10
11	J, K, L	P	LS1: 11
12	H	S, T	LS1: 12
13	S	U	LS1: 13
14	T	V	LS1: 14
15	X, Y, B	C	LS1: 1,4
16	X, Y	M	LS1: 1,10
17	M, N, B	C	LS1: 2,4
18	Q, R, A	C	LS1: 3,4
19	A, B, D	E, F	LS1: 4,6
20	A, B	I	LS1: 4,8
21	A, B	J, K	LS1: 4,9
22	W, F, G	H	LS1: 5,7
23	C, D, G	H	LS1: 6,7
24	E, F, G	S, T	LS1: 7,12
25	C, L	P	LS1: 9,11
26	H	U	LS1: 12,13
27	H	V	LS1: 12,14

Figure 5

5. New reactant and product tables are then created to start the next phase. The new reactant table drops any reactants that were not involved in linkage reactions. The new product table contains both input and new linkage reaction numbers. The new reaction table includes both input and linkage reactions.

In all subsequent phases, the new product and reactant tables are again matched, resulting in additional linkage reactions. This process continues until no new linkage reactions result from the matching. There can be no more than (x-1) phases, where x is the number of input reactions.

6. When reactions lead to multiple products which are then used as reactants in a single-step reaction, the substitution leading to a linkage reaction must involve all of them. Thus:

Input reactions:
(9) C → J + K
(10) J + K + L → P

Linkage reactions created:
Valid (60) C + L → P
Invalid C + K + L → P (substitution only for J, not J + K)

APPENDIX A

Generated Linkage Reactions for Model Reaction Scheme

15. X + Y + B → C
16. M + N + B → C
17. A + Q + R → C
18. X + Y + Q + R → C
19. M + N + Q + R → C
20. X + Y + B + D → E + F
21. M + N + B + D → E + F
22. A + Q + R + D → E + F
23. X + Y + Q + R + D → E + F
24. M + N + Q + R + D → E + F
25. A + B + D → E + F
26. X + Y + B + D + G → H
27. M + N + B + D + G → H
28. A + Q + R + D + G → H
29. X + Y + Q + R + D +G → H
30. M + N + Q + R + D +G → H
31. W + F + G → H
32. A + B + D + G → H
33. C + D + G → H
34. C + D + W + G → H
35. C + B + D + W + G → H
36. X + Y + B + D + W + G → H
37. M + N + B + D + W + G → H
38. Q + R + A + D + W + G → H

39. X + Y + Q + R + D + W + GH
40. M + N + Q + R + D + W + G→ H
41. X + Y + B → I
42. M + N + B → I
43. A + Q + R → I
44. X + Y + Q + R → I
45. M + N + Q + R → I
46. A + B → I
47. X + Y + B → J + K
48. M + N + B → J + K
49. A + Q + R → J + K
50. X + Y + Q + R → J + K
51. M + N + Q + R → J + K
52. A + B → J + K
53. X + Y → M
54. X + Y + B + L → P
55. M + N + B + L → P
56. A + Q + R + L → P
57. X + Y + Q + R + L → P
58. M + N + Q + R + L → P
59. A + B + L → P
60. C + L → P
61. X + Y + B + D + G S + T
62. M + N + B + D + G S + T
63. A + Q + R + D + G → S + T
64. X + Y + Q + R + D + G → S + T
65. M + N + Q + R + D + G → S + T
66. W + F + G → S + T
67. A + B + D + G → S + T
68. C + D + G → S + T
69. E + F + G → S + T
70. C + D + W + G → S + T
71. A + B + D + W + G → S + T
72. X + Y + B + D + W + G → S + T
73. M + N + B + D + W + G →S + T
74. Q + R + A + D + W + G → S + T
75. X + Y + Q + R + D + W + GS + T
76. M + N + Q + R + D + W + G→ S + T
77. X + Y + B + D + G U
78. M + N + B + D + G U
79. A + Q + R + D + G →U
80. X + Y + Q + R + D + G → U
81. M + N + Q + R + D + G U
82. W + F + G → U
83. A + B + D + G → U
84. C + D + G → U
85. E + F + G → U
86. H → U
87. C + D + W + G → U

88. A + B + D + W +G → U
89. X + Y + B + D + W + G → U
90. M + N + B + D + W + G → U
91. Q + R + A + D + W + G → U
92. X + Y + Q + R + D + W + G→ U
93. M + N + Q + R + D + W + G→ U
94. X + Y + B + D + G → V
95. M + N + B + D + G → V
96. A + Q + R + D + G → V
97. X + Y + Q + R + D + G → V
98. M + N + Q + R + D + G → V
99. W + F + G → V
100. A + B + D + G → V
101. C + D + G → V
102. E + F + G → V
103. H → V
104. C+ D + W + G → V
105. A + B + D + W + G → V
106. X + Y + B + D + W + G → V
107. M + N + B + D + W + G → V
108. Q + R + A + D + W + G → V
109. X + Y + Q + R + D + W + G→ V
110. M + N + Q + R + D + W + G→ V

APPENDIX B

Flow Chart for the Lasalocid A Synthesis

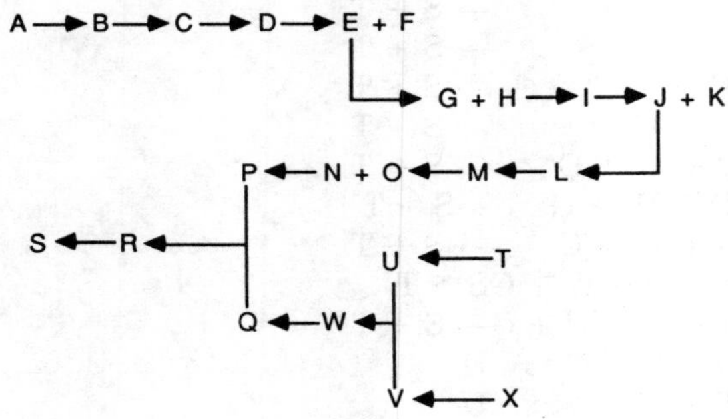

P and Q are Intermediates A and B, respectively, and S represents Lasalocid A. This 'reaction map' leads to the following input reactions:

1. A → B
2. B → C
3. C → D
4. D → E + F
5. E → G + H
6. G + H → I

7. I → J + K
8. J → L
9. L → M
10. M → N + O
11. N + O → P
12. P + Q → R
13. R → S
14. T → U
15. U + V → W
16. W → Q
17. X → V

Use of the multi-step generator creates a total of 205 linkage reactions, therefore requiring a total of 222 reactions to track all the single and multi-step possibilities in the above reaction map.

REFERENCES

1. Nakata, T., Schmid, G., Vranesic, B., Okigawa, M., Smith-Palmer, T., Kishi, Y. 'A Total Synthesis of Lasalocid A'. *J. Amer. Chem. Soc.* 1978, *100*, 2933–2935. Nakata, T., Kishi, Y. 'Synthetic Studies on Polyether Antibiotics. III. A Stereo controlled Synthesis of Isolasalocid Ketone from Acyclic Precursors'. *Tetrahedron Lett.* 1978, 2745–2748.
2. Ireland, R. E., Anderson, R. C., Badoud, R., Fitzsimmons, B. J., McGarvey, G. J., Thaisrivongs, S., Wilcox, C. S. 'A Total Synthesis of Ionophore Antibiotics. A Convergent Synthesis of Lasalocid A (X537A)'. *J. Amer. Chem. Soc.* 1983, *105*, 1988–2006. Ireland, R. E., Thaisrivongs, S., Wilcox, C. S. 'Total Synthesis of Lasalocid A (X537A)'. *J. Amer. Chem. Soc.* 1980, *102*, 1155–1157. Ireland, R. E., McGarvey, G. J., Anderson, R. C., Badoud, R., Fitzsimmons, B. J., Thaisrivongs, S. 'First Successful ENDOR Studies of Organic Radical Ions in Liquid Crystals'. *J. Amer. Chem. Soc.* 1980, *102*, 6180–6181.
3. 'Reaction notes' are input by the document analyst to report reaction conditions that are important and cannot be seen from the reaction participants; safety information associated with the reaction; or other reaction-related information. Reaction conditions are described by terms such as 'photochem or 'high pressure'. The reaction note is searchable.

POSTER SESSION: CHEMINFORM – PRINTED ISSUE AND DATABASE

Axel Parlow
Fachinformationszentrum Chemie GmbH (FIZ CHEMIE), Steinplatz 2,
D-1000 Berlin 12, West Germany.

ABSTRACT

This paper describes publishing an abstracting journal and creating a reaction database in parallel. ChemInform (formerly Chemischer Informationsdienst) has been produced since 1970 by FIZ CHEMIE and Bayer AG. It is an abstracting journal covering mostly publications on new reactions and synthetic methods from about 250 major journals. It is published weekly, since 1987 fully in English. An experienced staff of scientists is making the greatest possible effort to produce a journal that is well-known for its reliability, its critical evaluation, its current awareness and its attractive layout. Special emphasis is laid on the construction of a graphical reaction scheme that makes it possible to grasp even complicated reaction sequences 'at a glance'. The structures of the reacting compounds are drawn together with the most important reaction data (catalysts, solvents, yields, etc.). Multi-step reaction sequences are combined into one scheme to show the reaction flow. At present, this publication is mostly used as an current awareness service, for browsing through the new literature or selecting abstracts for a personal card file. For retrospective searches only an author index is available. These restricted possibilities will be overcome by the developments presented here, in order to enhance the usefulness of this product.

DESCRIPTION OF PLANS

The main goal for the software system under development is to make available in machine-readable form all the information contained in ChemInform. In this way different kinds of indexes may be produced as well as data dealing with selected parts of chemistry for specialised information services, either printed or on computer-readable media. But the most important fact is that the information in ChemInform contains all the data needed for a reaction database.

Since the production process (400 abstracts weekly, mostly with at least one reaction scheme, totalling about 50,000 single reactions per year) may not be delayed or increased in price, we decided to do the input of text and data only once. This means that the material for the printed issue, for subfiles, indexes and for the database has to be derived from one data source.

The input of data is done at two different locations: at FIZ CHEMIE in Berlin and at Bayer in Leverkusen. For textual input, which has been computer-based since 1986, high-quality typesetting systems are used which allow for a variety of special characters (boldface, Greek symbols, subscripts and superscripts at the same column,

W. A. Warr (Ed.)
Chemical Structures

etc.). The text from Leverkusen is transferred on diskette to Berlin, where it is combined with the text from the Berlin staff and where the production of the offset films for the printed journal is done.

Until 1986 the reaction schemes have been drawn by hand and combined with the text, also by hand. This 'glue-and-scissors' technique has now been overcome as follows.

For graphical input, we use our own enhancements of a CAD system based on the GKS (Graphics Kernel System) standards. This system (Formula Design System, FDS, by GTSGral), which is running on individual PC's, maintains comprehensive internal data structures which allow for storage of facts that have more than just a graphical meaning. For example, a double bond is not only two lines close together, but the meaning 'special sort of connection between two defined centres' is known to the program; or text over an arrow has its meaning within the reaction and has to be interpreted as reaction data, and so on.

Text and reaction schemes are then combined within a typesetting system which is able to use GKS 'metafiles' as graphical inserts. These 'metafiles' contain information about the graphical segments (vectors, text strings and their co-ordinates). In this manner it is possible to produce the film for offset printing without intermediate paperwork. The printing of the journal itself continues as usual.

The implementation of the new technique is done step by step. At present, almost 80 per cent of the graphical schemes 'drawn' in Berlin have already been input into the computer.

On the other hand, software development is underway to use the information stored by the textual and graphical inputs for different purposes rather than just printing them. We decided to build a base file which will keep all the information that is contained in the different input sources. From this base file several sorts of output may be derived later.

When building the base file, some problems arise which have not yet been solved by any available software system. The information in the reaction scheme has been compressed for better readability, and this compression has to be undone.

For instance, multi-step sequences are drawn in one scheme, the product of one step being the reactant of the next. In order to get information on individual reactions, these sequences have to be split up by the program, and the intermediate compounds are assigned to either reaction. Thus, several 'single' reactions are derived from one ChemInform scheme.

Since the structures of all reaction participants will be stored with their complete topology, including stereochemistry, shortcuts like -Ph or -COOH or -CH_3, which are heavily used in the reaction schemes, have to be expanded. In a similar way, generic groups are replaced by defined substructures. (Generic groups within ChemInform apply only to defined substructures, not to terms such as 'alkyl', so all compounds are in principle fully defined). When this is finished, the connection tables of all compounds can be calculated.

Furthermore, the other data in the schemes have to be identified and stored according to their meaning. These are normally text strings which denote reagents, catalysts, solvents, reaction conditions like temperature or pressure, yields, stereoselectivity, and others. The identification is done by positional information, by look-up tables and key characters.

A special survey is being carried out to implement the coding of reacting centres and the mapping of atoms over several reaction steps. This feature will be useful for highly sophisticated reaction management systems which will allow for retrieval of reactant-product pathways over several steps whereby the user will not need to know the intermediates. Such systems will be available in the future.

Besides the data derived from the schemes, the metafiles (the 'pure' graphic) and, of course, the data from the textual input (title, abstract, bibliographic data) will also be stored in the base file. This will be the source for indexes or subfiles, and for reaction databases for existing retrieval systems. As far as the used retrieval system will allow, all of these data except the display-only graphic metafile may be searchable.

It is planned to produce an online version of the database, which will be implemented on STN International. For this purpose, a data file in standard format that can be used under Messenger software will be derived from the base file. (Stereochemical information will be kept in this file, even though Messenger cannot handle this information at present). The enhancements to Messenger developed at CAS for CASREACT, the online reaction retrieval system developed by CAS, will also be used for this database. The display capability of Messenger will, in this configuration, also allow for the display of the whole (multi-step) reaction schemes.

On the other hand, the data in the base file may be converted into an SMD file (Standardised Molecular Data) which is used by the companies of the CASP pool in Europe for interchanging substance and reaction data. Interfaces to commercially available reaction retrieval systems, using the SMD data structure, exist or are being developed. In this way, in-house versions of the database to be used with REACCS, ORAC, SYNLIB, etc., will be made available.

POSTER SESSION: CONVENTIONS, PRACTICES AND PITFALLS IN DRAWING CHEMICAL STRUCTURES

Kurt L Loening
Chemical Abstracts Service, Columbus, Ohio 43210, U.S.A.

ABSTRACT

Chemical structures are thought to be the international language of chemistry as illustrated by the theme of this meeting. To be truly unambiguous, however, great care must be taken in their representation, and in some cases further standardisation is needed. Pertinent IUPAC recommendations and common practices with regard to drawing selected chemical structures are reviewed. Some areas where recommendations do not exist or are inadequate are identified. A few examples from the literature showing ambiguous or misleading structures are presented.

IUPAC RECOMMENDATIONS

Stereoformulae

Conventions: *Thickened, broken, dotted, wavy lines, dots and wedges.*

Reference: IUPAC, 'Rules for the Nomenclature of Organic Chemistry. Sections E: Stereochemistry'.[1]

Rule E-6.3. Formulae that display stereochemistry should be prepared with extra care so as to be unambiguous and,whenever possible, self explanatory. It is inadvisable to try to lay down rules that will cover every case, but the following points should be borne in mind.

(**———**) A thickened line denotes a bond projecting from the plane of the paper towards an observer.

(------------) A broken (or dashed) line denotes a bond projecting away from an observer, and, when this convention is used,

(————) a full line of normal thickness denotes a bond lying in the plane of the paper.

(∿∿∿∿) A wavy line may be used to denote a bond whose direction cannot be specified or, if it is explained in the text, a bond whose direction it is not desired to specify in the formula. (If a racemate is to be designated, this is done by reference to its optical activity, (±)- IUPAC Rule [3AA-4.5]).

(................) Dotted lines should preferably not be used to denote stereochemistry, and never when they are used in the same paper to denote mesomerism, intermediate states, etc.

W. A. Warr (Ed.)
Chemical Structures
©Springer-Verlag Berlin Heidelberg 1988

(▶ ◀) [Some] difficulties can be overcome by using wedges in place of lines, the broader end of the wedge being considered nearer to the observer. Wedges should not be used as a complement to broken lines.

● Single large dots have sometimes been used to denote atoms or groups attached at bridgehead positions and lying above the plane of the paper, but this practice is strongly deprecated.

H Hydrogen or other atoms or groups attached at sterically designated positions should never be omitted.

Comment: The broader end of the wedge is considered nearer to the observer.
Example: Norbornane or Bicyclo[2.2.1]heptane

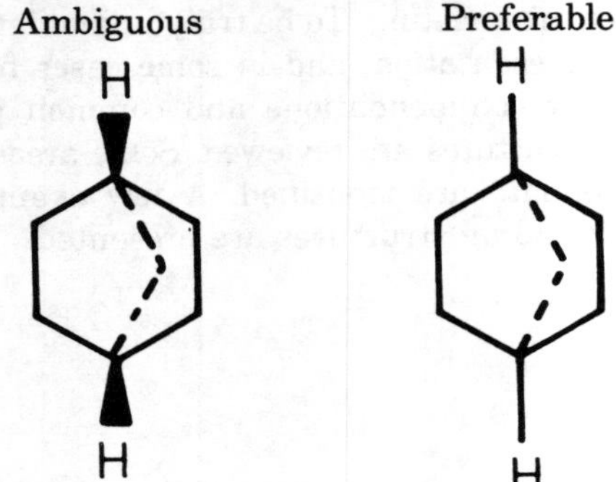

Comment: A racemic mixture is designated by placing (±) in close association with the structure.
Example: Racepinephrine

HO
HO
$CHCH_2NHCH_3$
OH

(±)

Comment: Although, not specifically provided for in the IUPAC rules a stereospecific structure for which only the relative configuration is known can be designated by placing (*rel*)- in close association with the structure.
Example:

$CH_2N(CH_3)_2$
HO
H

(*rel*)

Convention: The Reentrant Angle

Rule E-2.3. . . . When the *cis-trans*-designation of substituents is applied, rings are considered in their most extended form; re-entrant angles are not permitted: for example:

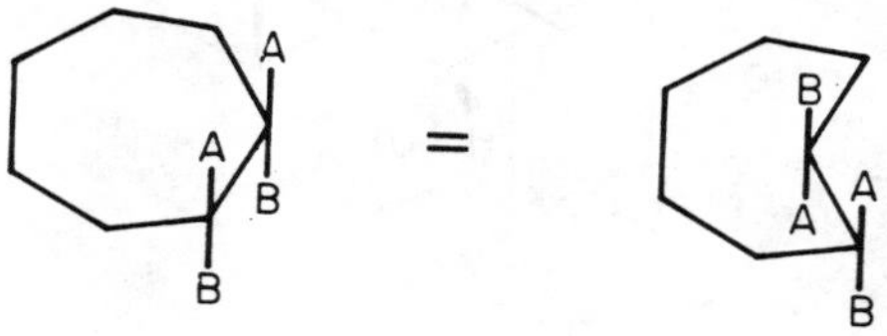

cis apparently *trans*

Comment: The re-entrant angle is also a problem in assigning the correct configurational descriptors, R and S or R^* and S^*.[2]

Example: 2-Methylcyclodecanol

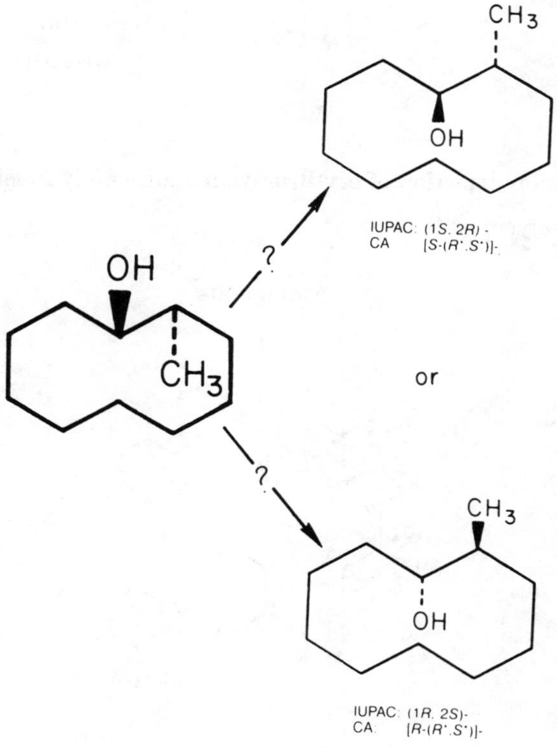

Comment: The problem is not limited to ring systems, but is also encountered in chains drawn in zig zag fashion, especially those with epoxy bridges.

Example:2,3-Anhydro-D-mannose

Ambiguous

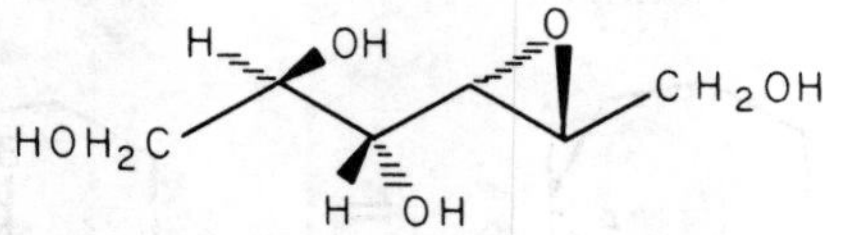

Preferable

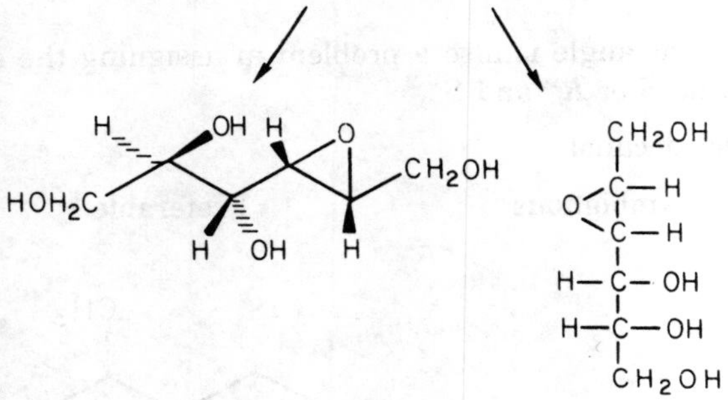

Comment: A common depiction of erythromycin additionally illustrates the problem.

Example: Erythromycin

Ambiguous[3]

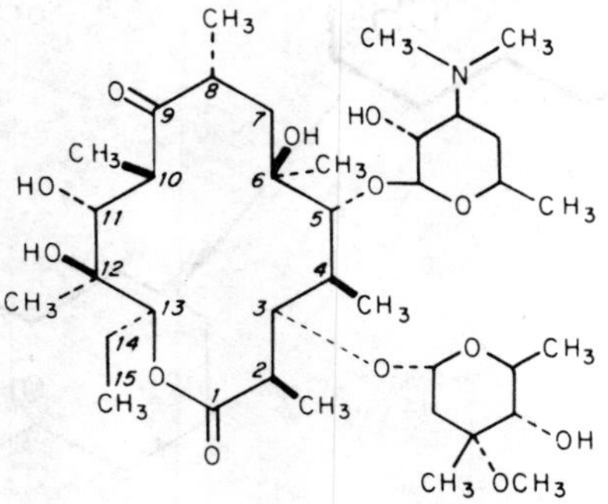

Preferable[4,5]

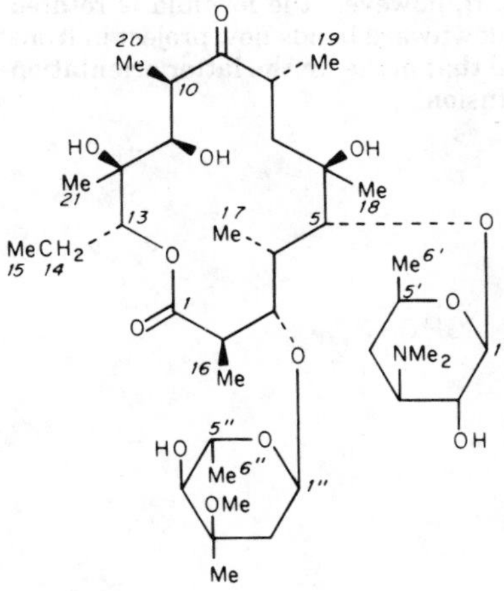

or

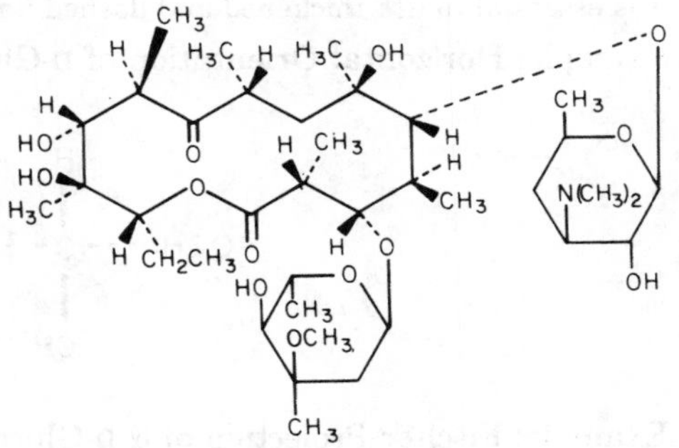

Carbohydrates

Convention: The Fischer Projection

Rule E-6.1. In a Fischer projection the atoms or groups attached to a tetrahedral centre are projected on to the plane of the paper from such an orientation that atoms or groups appearing above or below the central atom lie behind the plane of the paper, and those appearing to left and right of the central atom lie in front of the plane of the paper, and that the principal chain appears vertical with the lowest-numbered chain member at top. (See also IUPAC-IUB, 'Tentative Rules for Carbohydrate Nomenclature' Section 2(a)) (6).

Example: D-Glucose

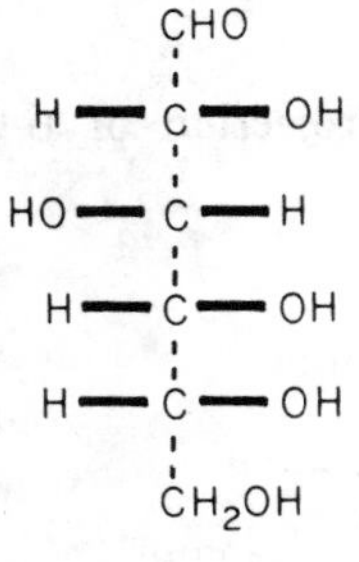

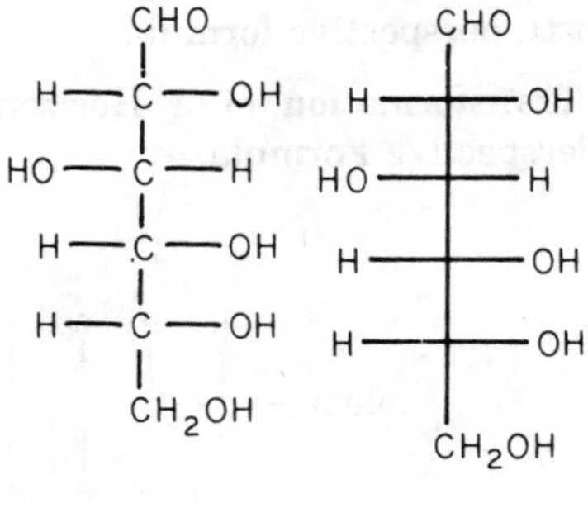

Vertical Orientation

Fischer Projections

Note: The first of the two types of Fischer projection should be used whenever convenient.

Note: If a formula in the Fischer projection is rotated through 180 degrees in the plane of the paper, the upward and downward bonds from the central atom still project behind the plane of the paper, and the sideways bonds project in front of that plane. If, however, the formula is rotated through 90 degrees in the plane of the paper, the upward and downward bonds now project in front of the plane of the paper and the sideways bonds project behind that plane. In the latter orientation it is essential to use thickened and dashed lines to avoid confusion.

Example: Horizontal Orientation of D-Glucose

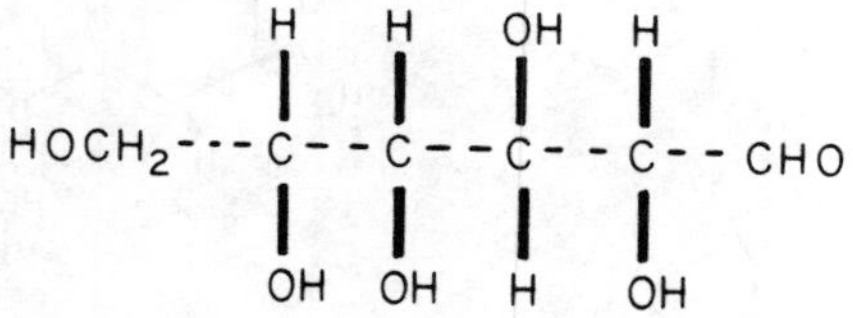

```
          H     H     OH    H
          |     |     |     |
HOCH2---- C --- C --- C --- C --- CHO
          |     |     |     |
          OH    OH    H     OH
```

Example: Fischer Projection of α-D-Glucopyranose

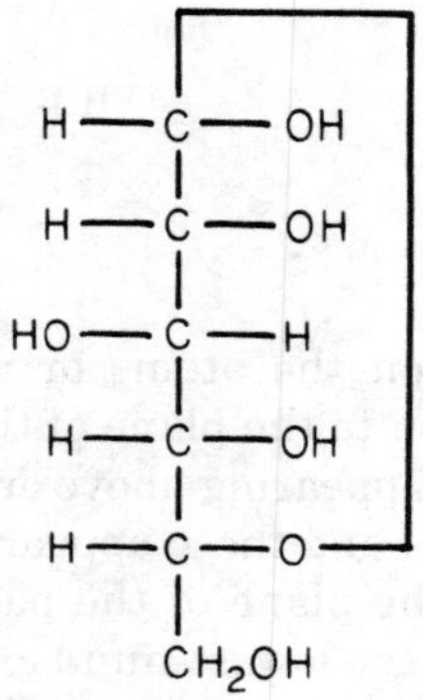

```
         ____________
        |            |
   H ── C ── OH      |
        |            |
   H ── C ── OH      |
        |            |
  HO ── C ── H       |
        |            |
   H ── C ── OH      |
        |            |
   H ── C ── O ──────
        |
       CH2OH
```

Convention: The Perspective Formulae of Haworth

Reference: IUPAC-IUB, 'Tentative Rules for Carbohydrate Nomenclature'.

Comment: Often it is more convenient as well as more accurate to represent the cyclic forms of the monosaccharides by a ring diagram. Such a representation is the widely used Haworth perspective formula.

Example: Transformation of a Horizontal Fischer Projection of D-Glucose into a Haworth Perspective Formula.

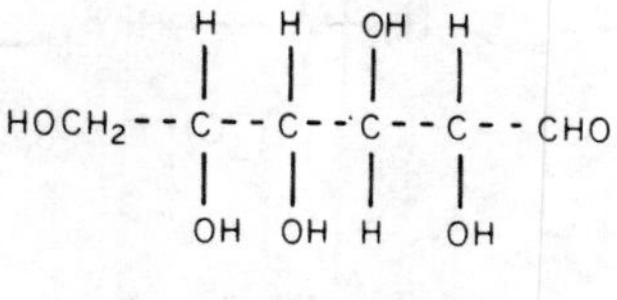

```
          H    H    OH   H
          |    |    |    |
HOCH2 --- C -- C -- C -- C --- CHO
          |    |    |    |
          OH   OH   H    OH
```

D-Glucose

```
          CH2OH
          |
     H    |
     |  / C ──── OH
     | /  |
     C    H
          OH     H  CHO
     |\   |      | /
     | \  |      |/
   OH    C ───── C
         |       |
         H       OH
```

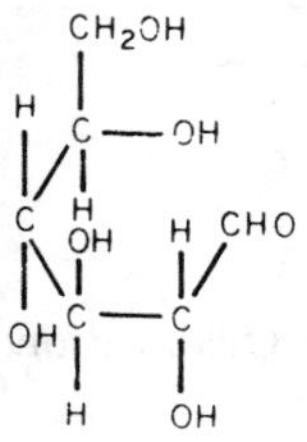

D-Glucose

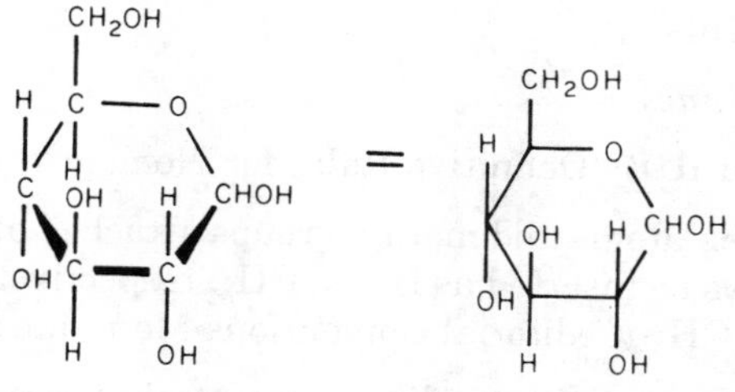

D-Glucopyranose = Haworth Projection

Haworth Representation of D-Glucopyranose

Comment: It must be emphasised that, since Haworth representations are not projection formulae as is the Fischer formula, they *cannot* be rotated in the plane of the paper. A Haworth-type formula can be rotated without change about a central vertical axis. It is always necessary to keep the same groups above and below the plane of the ring.

Example: α-D-Glucopyranose Does Not Equal-L-Glucopyranose

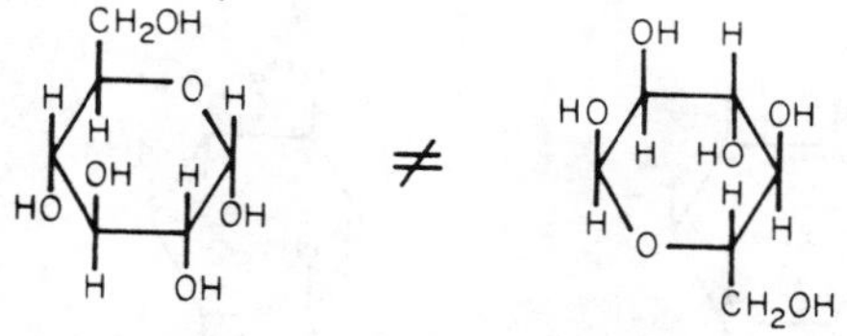

α-D-glucopyranose ≠ α-L-glucopyranose

Example: Equivalent Representations of -D-Glucopyranose

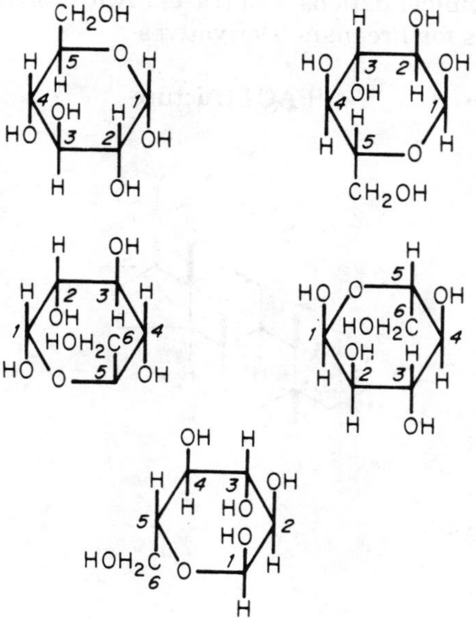

Steroids Versus Cyclitols

Convention: Unlettered Lines

Reference: IUPAC and IUB, 'Definitive Rules for Nomenclature of Steroids'.[7]

Rule 2S-1.4. All hydrogen atoms and methyl groups attached at ring-junction positions
CH_3 must always be inserted as H and CH_3, respectively CH_3 (Me may be used
Me in place of CH_3 if editorial conventions Me require it).

The practice, sometimes followed in steroid formulae, of denoting methyl groups by bonds without lettering is liable to cause confusion and should be abandoned. This isessential in view of customs in other fields (for example, cyclitols), and applies also to other groups of compounds such as cyclic terpenes and alkaloids for which steroid conventions are commonly used.

Reference: IUPAC and IUB, 'Nomenclature of Cyclitols'.[8]

Comment: In the Fischer-Tollens projections for cyclitols, unlettered lines denote hydroxy groups.

Examples: *myo*-Inositol;1,3,5-Triamino-1,3,5-trideoxy-*scyllo*-inositol

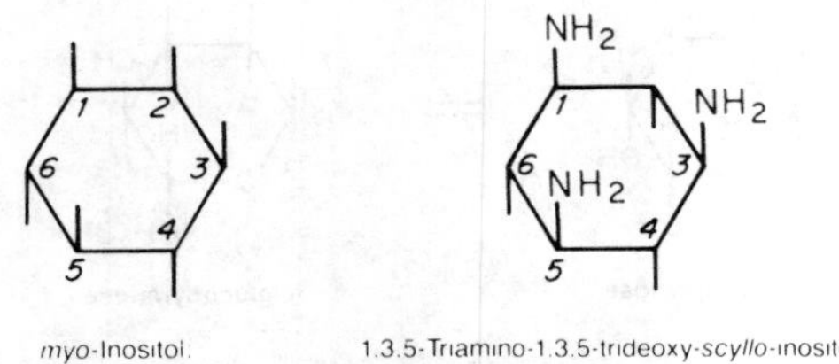

myo-Inositol. 1,3,5-Triamino-1,3,5-trideoxy-*scyllo*-inositol

Examples: IUPAC Recommendations Contrasted with Structures Extracted from Various Reference Works for Pregnane Derivatives

IUPAC Structure

Non-IUPAC Structures[9–12]

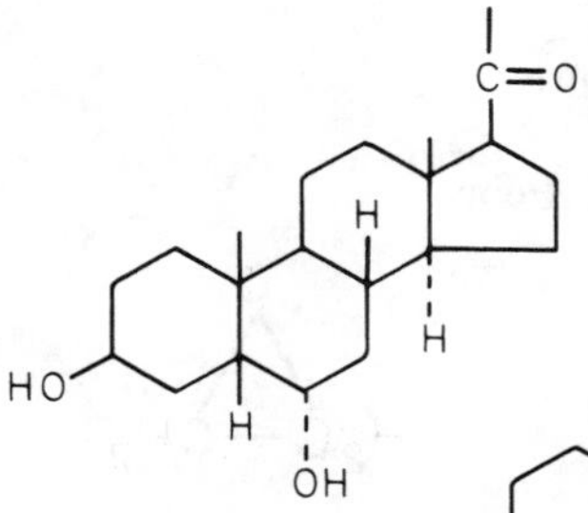

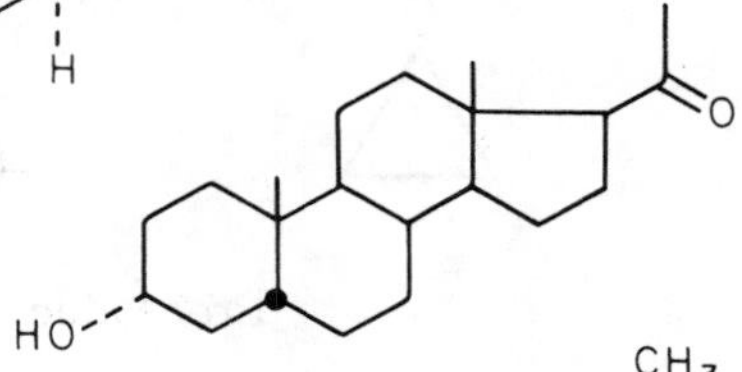

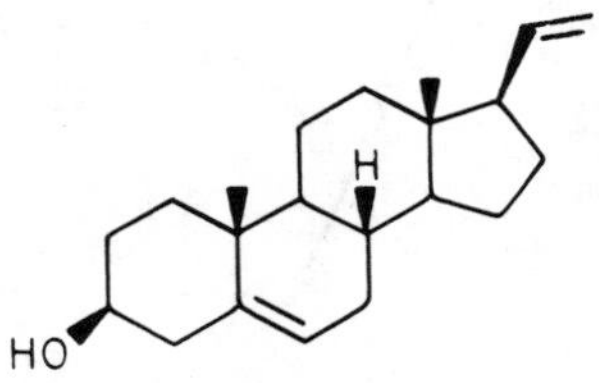

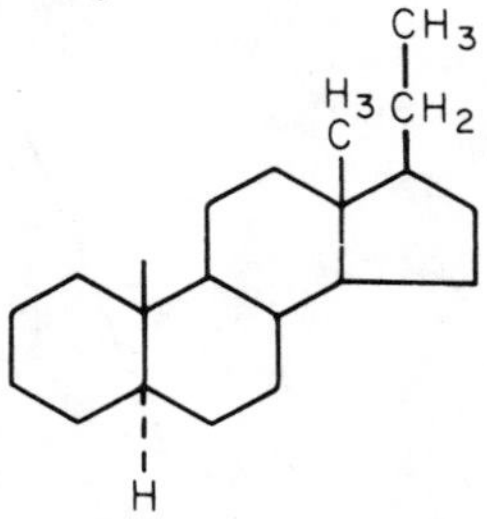

General Organic Structures

Convention: 'R' and 'X' Groups
Reference: IUPAC, 'Nomenclature of Organic Chemistry'.[1]

R Except where specifically stated to the contrary, symbols R, *etc.*, are used to denote univalent radicals attached by means of carbon and derived from aliphatic, carbocyclic, or heterocyclic compounds, which may be saturated or unsaturated, and unsubstituted or substituted, but they are not used for -CN, -CNO, -CNS, or -CNSe groups or for groups attached directly through >C=X where X is O, S, Se, Te, NH, or substituted NH. Up to three such groups may be designated R, R′, and R″, or R^1, R^2, and R^3; for more than three different groups the sequence R^1, R^2, R^3, R^4 . . . is recommended. R_2, R_3, and R_4 should be used only to denote two, three, or four, repectively, of the same group denoted by simple R.

Example:

Ambiguous[13] Preferable

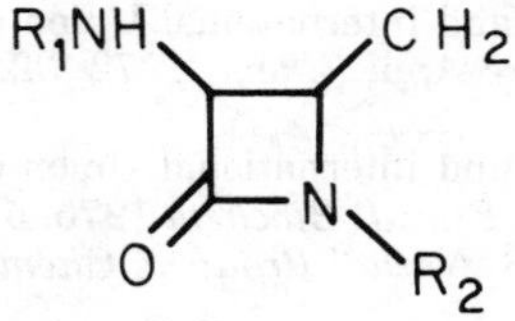

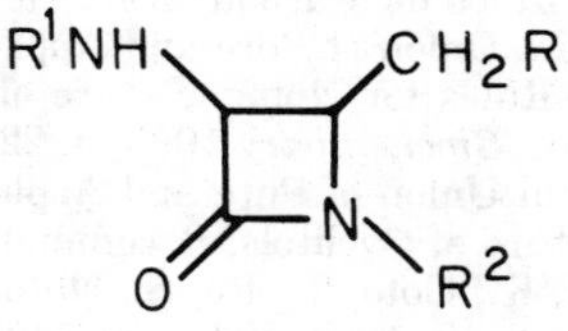

Comment: Though not specifically stated in the IUPAC rules, every atom or group should be attached to a ring by a bond.

Example: Oxirane Versus Cyclopropanone

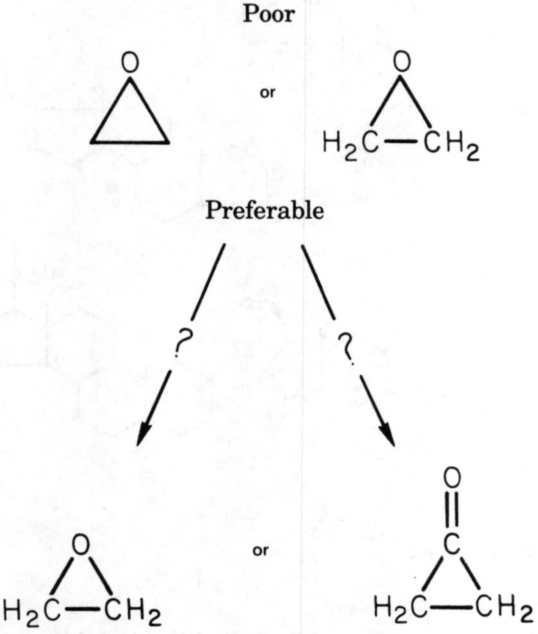

REFERENCES

1. International Union of Pure and Applied Chemistry. *Rules for the Nomenclature of Organic Chemistry, Sections A,B,C,D,E,F and H*; Pergamon Press: Oxford, 1979.
2. Kupchan, S. M., Maruyama, M., Hemingway, R. J., Hemingway, J. C., Shibuya, S., Fujita, T., Cradwick, P. D., Hardy, A. D. U., Sim, G. A. 'Tumour Inhibitors. LXVII. Eupacunin, a Novel Antileukemic Sesquiterpene Lactone from Eupatorium Cuneifolium'. *J. Amer. Chem. Soc.* 1971, *93*, 4914–4916.
3. Nakanishi, K., Goto, T., Ito, S., Natori, S., Nozoe, S. *Natural Products Chemistry*; Academic Press: New York, 1975; Vol. 2, p. 75.
4. Chemical Abstracts Service. *Index Guide*, 'Erythromycin'; American Chemical Society: Washington, D.C., 1987, p. 659G.
5. Chemical Abstracts Service. *Index Guide*, 'Erythromycin'; American Chemical Society: Washington, D.C., 1972, p. 871G.
6. International Union of Pure and Applied Chemistry, and International Union of Biochemistry. 'Tentative Rules for Carbohydrate Nomenclature'. *Eur. J. Biochem.* 1971, *21*, 455–477.
7. International Union of Pure and Applied Chemistry, and International Union of Biochemistry. 'Definitive Rules for Nomenclature of Steroids'. *Pure Appl. Chem.* 1972, *32*, 285–322. Also published in: *Biochemistry* 1969, *8*, 2227–2242.
8. International Union of Pure and Applied Chemistry, and International Union of Biochemistry. 'Nomenclature of Cyclitols, Recommendations 1973'. *Eur. J. Biochem.*1975, *57*, 1–7.
9. Nakanishi, K., Goto, T., Ito, S., Natori, S., Nozoe, S. *Natual Products Chemistry*; Academic Press: New York, 1974; Vol. 1, p. 540.

10. Devon, T. K., Scott, A. I. *Handbook of Naturally Occurring Compounds*; Academic Press: New York, 1972; Vol. 2, p. 435.
11. Dorfman, R. I., Ungar, F. *Metabolism of Steroid Hormones*; Academic Press: New York, 1965; p. 16.
12. *Dictionary of Organic Compounds*; 5th ed. Chapman and Hall: New York, 1982; Vol. 5, p. 4759.
13. Sendai, M., Hashiguchi, S., Tomimoto, M., Kishimoto, S., Matsuo, T., Kondo, M., Ochiai, M. 'Chemical Modification of Sulfazecin. Synthesis of 4-(substituted Methyl)-2-azetidinone-1-sulfonic acid Derivatives'. *J. Antibiotics* 1985, *38*, 347.

GRAMMAR IN CHEMICAL INDEXING LANGUAGES

Robert Fugmann
Alte Poststrasse 13, 6270 Idstein, West Germany

ABSTRACT

Any natural language comprises vocabulary and grammar as its essential constituents, but in indexing languages grammar of the more than merely rudimentary kind is only rarely encountered. High performance and reliability is imparted to any indexing language in which a more sophisticated kind of grammar is realised. Presently, topological methods of structural formula documentation and role indicators constitute the most common variation of grammar in an indexing language.

Grammatical-syntactical methods can be used to large advantage for the representation of other kinds of concept relations, too, if the principles of the analytico-synthetic classification are obeyed. This is demonstrated for segmentation, relation indicators, syntactical terms, concept number syntax and for the TOSAR-system.

In IDC a copious armamentarium of syntactical tools of the kind discussed has been employed, some of them having been in use for more than 25 years. In this way the advantages of the topological method are imparted to the handling of concept relations different from the purely structural ones: the vocabularies are and remain small and easy to overview, and indexing is performed with a correspondingly high reliability, which renders retrieval correspondingly accurate.

Presently, renunciation of indexing language grammar can only historically be justified and originates from an era in which the necessary equipment in the form of computers had not yet been available. Grammarless indexing languages and the appertaining information systems cannot meet the requirements that are increasingly made on them through the continual growth of the search files and of search frequency and through the ever-continuing influx of new concepts.

INTRODUCTION

In the field of chemistry many information systems have already been placed in operation, but in the long run only very few have proven equal to the demands that have developed in the course of time. Under purely empirical observation, nearly all of these many systems in their initial stages raised the highest hopes, and their deficiencies have been revealed only in a later development stage.

The life history of such information systems is summarised schematically in Figure 1. Plotted above the time axis is the *noise* that is observed when such systems are used, and below it is plotted the *information loss* that afflicts searches in such information systems.

No searches or at most only a few test searches are made during the starting-up period of such a system. When the system is fully operational, the end-user or

W. A. Warr (Ed.)
Chemical Structures

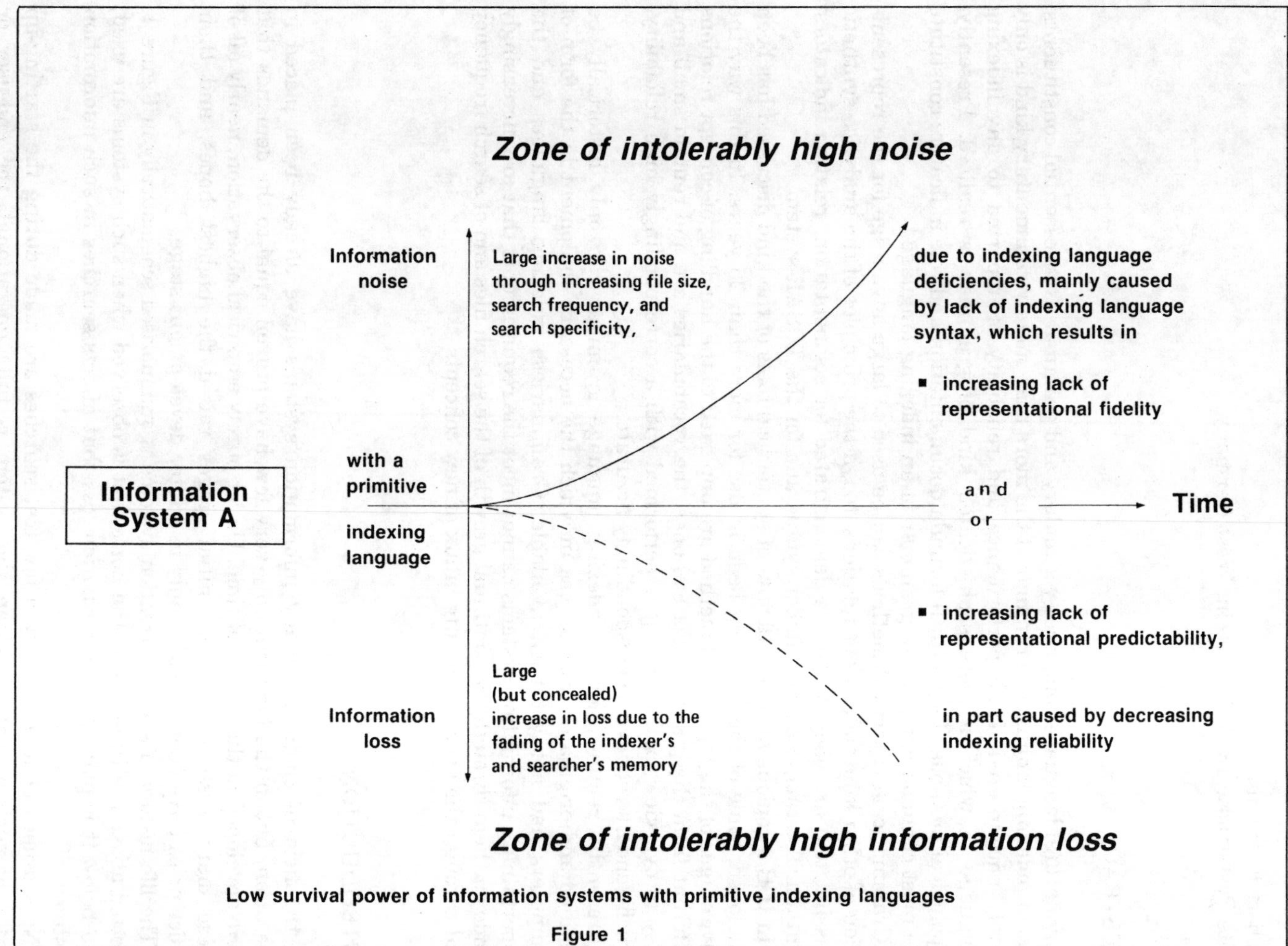

Low survival power of information systems with primitive indexing languages

Figure 1

information scientist is faced with the task of recognising and weeding out the *noise* among the search responses. This work increases at a faster rate than can be accounted for by the growth of the file. This work *of sifting through search responses* can in the course of time easily become intolerably expensive. The use of an information system has often been discontinued or at least restricted for these reasons.

The situation is rendered even more critical by the fact that more and more frequently a *loss of information* is found to occur, and to an increasing extent. This is largely due to the increase in the work that the indexer must (even if mostly unconsciously) perform in searching through the indexing language vocabulary, and it is also due to the fact that searching through the file is, though only in a concealed way, based on the indexer's memory. In the course of time this basis diminishes gradually, or in the case of a change in personnel, it declines in big jumps.

These two circumstances together, the increase in noise and the increasing loss of information, have frequently forced the abandonment of information systems, after which the work of indexing had to be commenced again with a presumably (and hopefully!) better information system.

What is so insidious about such a development is that under purely empirical observation it does not reveal itself until very late, and that initially such systems are even highly valued by the less experienced user.

But such a development is by no means a necessary occurrence, for another type of information system is well known that is not subject to these faults, such as the registry system of Chemical Abstracts Service and the GREMAS System (Generic Retrieval by Magnetic Tape Storage) of IDC. The degeneration of information systems is due to the indexing language with which they work. The causes are, as we now know, deficient *representational fidelity* and inadequate *representational predictability*. Both these are in turn caused by a lack of grammar in the indexing language, specifically the lack of *syntax* in such a grammar.

Figure 2 uses the same scheme of representation as in Figure 1 to depict the life history of an information system whose fate has taken a more auspicious course. These systems must, however, go through a critical initial phase. They must necessarily work with more expensive and arduous indexing and storage procedures, and they are not yet able to demonstrate their superiority, which lies in their survival power. They even seem to display, on purely empirical criteria, an inferiority with respect to other, more primitive systems, especially when an economic yardstick is applied, even though these seemingly superior systems already carry the germ of the fatal disease that will soon bring about their decline.

All this has been recorded in greater detail[1–3] and within the scope of this paper I must restrict myself to reference to these publications.

Now, however, grammar in an information system can exert its survival insurance effect on information systems in other areas than that of pure structural chemistry, if this grammar is applied in accordance with the rules of the analytico-synthetic classification, as developed by the Indian School of Ranganathan.

In the following, I would like to report on the efforts and applications within IDC (International Documentation in Chemistry, Otto Volger Strasse 19, 6231 Sulzbach, Federal Republic of Germany) outside the area of chemical structural formulae. This covers the following syntactic tools:

1. Semantic categories.
2. 'Subject-oriented' segmentation and linkage.

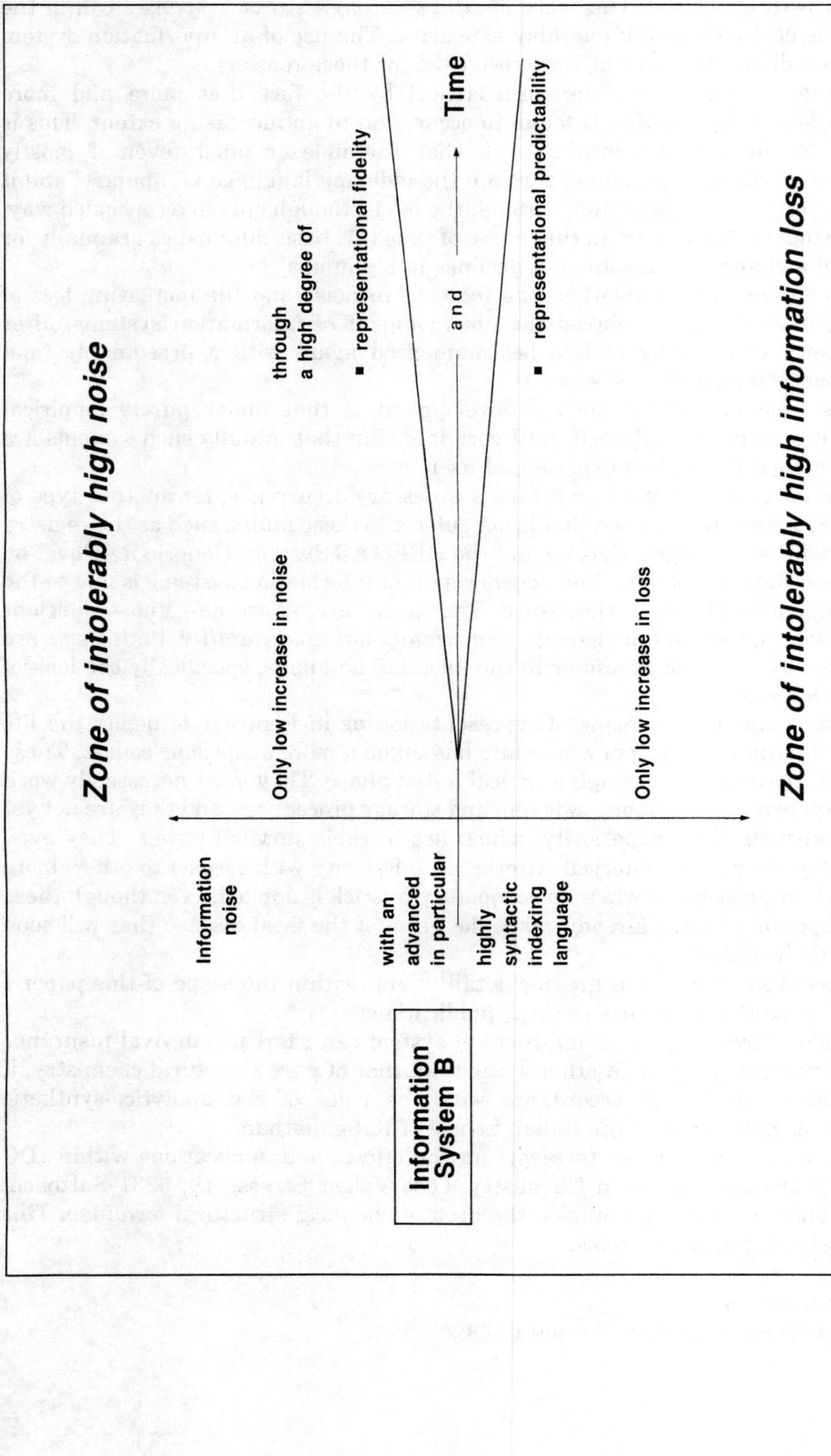

Zone of intolerably high noise

Information noise

Only low increase in noise

through a high degree of

- representational fidelity

Information System B

with an advanced in particular highly syntactic indexing language

and → Time

- representational predictability

Only low increase in loss

Zone of intolerably high information loss

High survival power of information systems with advanced, syntactical indexing languages

Figure 2

3. Syntactic terms.
4. Relation indicators.
5. The TOSAR System.
6. The 'Concept Number Syntax'.

We will base our consideration on the following text from the field of applied chemistry:

'Improved adhesion of protective coatings on sheet iron through two-stage etching with phosphoric acid

As a first step sheet iron is etched in a dispersion of chlorinated, liquid hydrocarbons in aqueous phosphoric acid (20%) at 20°C for 15 minutes, during which the layer of grease normally adhering to the surface is also removed. In the second stage the metal is etched again with 5% phosphoric acid at 50°C for only 5 minutes. A surface that has been thus prepared is especially well suited to accept a rolled-on layer of nickel, such a layer of iron-manganese alloy, or one of iron-chromium-nickel alloy in the ratio 74:18:8 or for dip-coating with zinc. Trichloroethylene is preferred for degreasing. Sheet iron that is coated in this manner possesses outstanding resistance to attack by seawater.'

We first represent this text in a more lucid, graphic form (see Figure 3). The time sequence of the individual steps proceeds from top to bottom. The relation between concepts by the Boolean AND and OR operator is represented by a dot and a circle respectively, as is common in the TOSAR system (*To*pological representation of *S*ynthetic and *A*nalytical *R*elations of Concepts).[4]

SEMANTIC CATEGORIES

In natural language there are many ways of expressing the relations between concepts, for example by the order of words in a sentence, the use of prepositions and cases, by relative pronouns, etc. Since an author is free to choose between all these possibilities, at the time of formulating a query one cannot know which special syntactic tools were used in the texts being sought. Thus, if one wants to use relations between concepts as search parameters, special rules for representing them in the file must be established and consistently applied, independently of how these relations had been expressed in the original text.

The first step in this direction is for indexing to agree on a small set of *semantic categories*, in accordance with the rules of analytico-synthetic classification. In our documentation work in the field of pure and applied chemistry the set of semantic categories shown in Figure 4 has proven successful.

In many respects these semantic categories form the conceptual backbone of an information system, but in particular they can serve to define the relations between concepts on more than the merely superficial criteria of their casual arrangement in an original text. Then one can specify the syntactic device that is to be used to represent a particular kind of relation in the search file. For example, it can be agreed that segmentation or linkage is always to be used to represent the especially important relation between a *substance and its properties*.

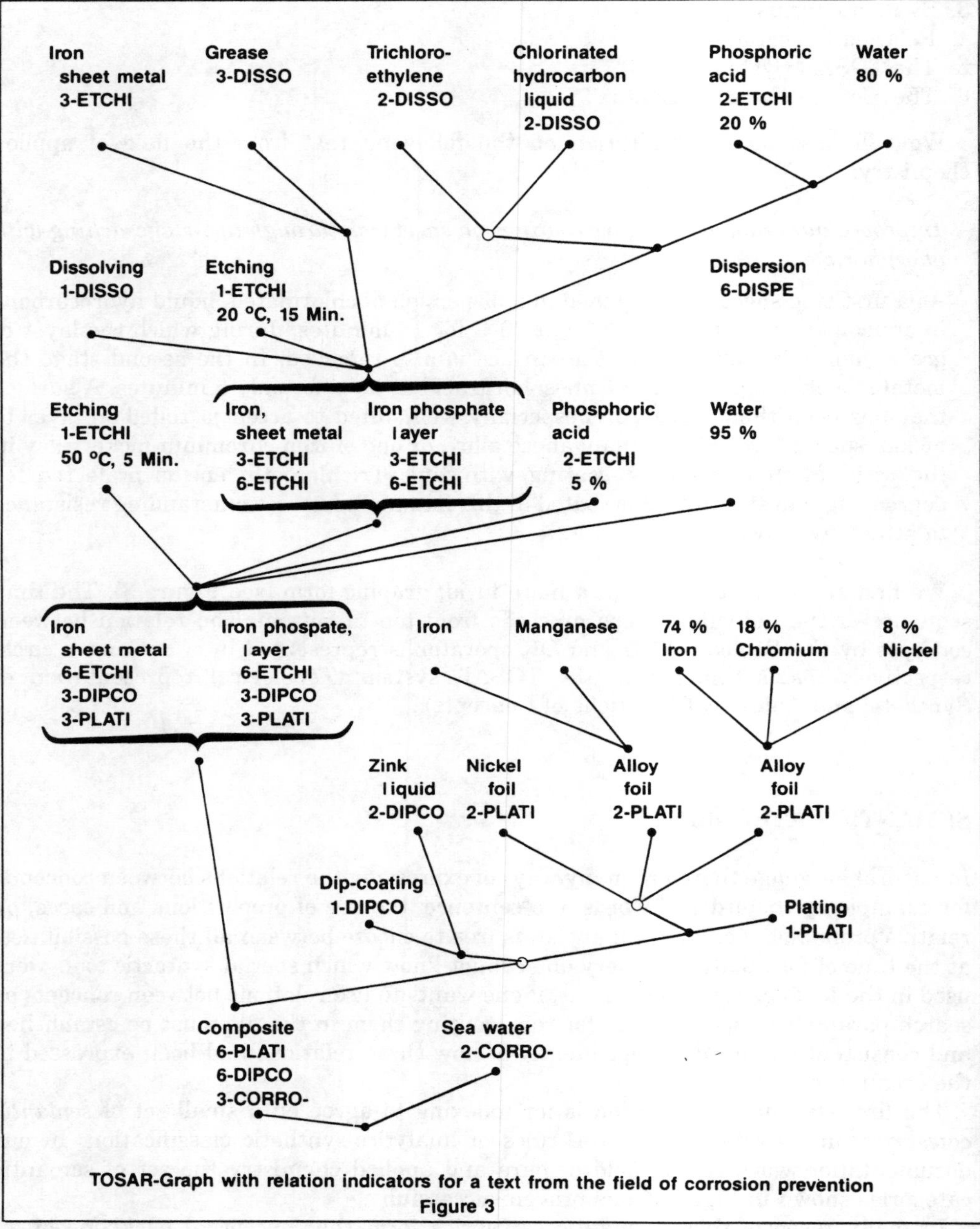

TOSAR-Graph with relation indicators for a text from the field of corrosion prevention

Figure 3

Substance	Attributes
Apparatus	property
Living being	function
Process	state
physical	form
chemical	use

Example of a set of semantic categories in an information system

Figure 4

SEGMENTATION

In segmentation a document is divided into separate subdocuments, which are often called 'segments'. This is carried out in such a way that within each segment an especially close relation exists between the concepts that co-occur there. In IDC extensive use is made of segmentation, particularly to represent the especially important relation between a substance and all its properties recorded or implied in the document under consideration, or, more generally, between a substance and its attributes. Thus, this relation is represented from a subject-oriented view and is independent of the casualities of the natural language mode of expression encountered in a document.

Then, one no longer has to content oneself with requiring the co-occurrence of substance and substance attributes merely within a common document. Rather one can more specifically require the co-occurrence within a *common segment* with hardly any danger of information loss. If, for example, the vocabulary already contains the descriptor 'iron' and also 'sheet metal', then no further descriptor is needed to express 'iron sheet', because this concept can be expressed justas precisely by a segment comprising the concepts iron and sheet metal.

Similarly well founded would be a subject-oriented use of links (i.e., independent of natural language).

In Figure 3 each group of concepts at one point of the graph means one such segment in the file.

THE SYNTACTIC TERM

As the next syntactic device let us consider three-position descriptors, each position of which provides information of a particular kind. These pieces of information are firmly associated in one term, which creates an especially close connection between them. Such terms have proved successful in the fragment code GREMAS for chemical structures and also for the documentation of patent literature in the fields of plant protection and agriculture (see Figure 5).

With regard specifically to the latter field, the first position of such a term is used to indicate whether a living being is itself actively involved in a process (which is expressed by '7'), whether it is negatively ('8') or positively ('9') affected or involved only in an indifferent manner in a process ('0').

The second position indicates the type of living being concerned; for example, 'E' stands for fungi and 'G' for seed plants.

The third position is used to designate the type of process in which the living being is involved, for example the supply of nutrients ('J') or direct exposure to chemical action ('G'). Thus, a fertiliser for seed plants is '9GJ' and a preparation for the control of fungi is '8EG'.

Again the indexer is forced to provide complete information, which is very much in the interest of complete and precise searches. At the same time the number of specific descriptors required in the vocabulary has been drastically reduced, since these syntactic terms can be formed by the indexer and the enquirer as required.

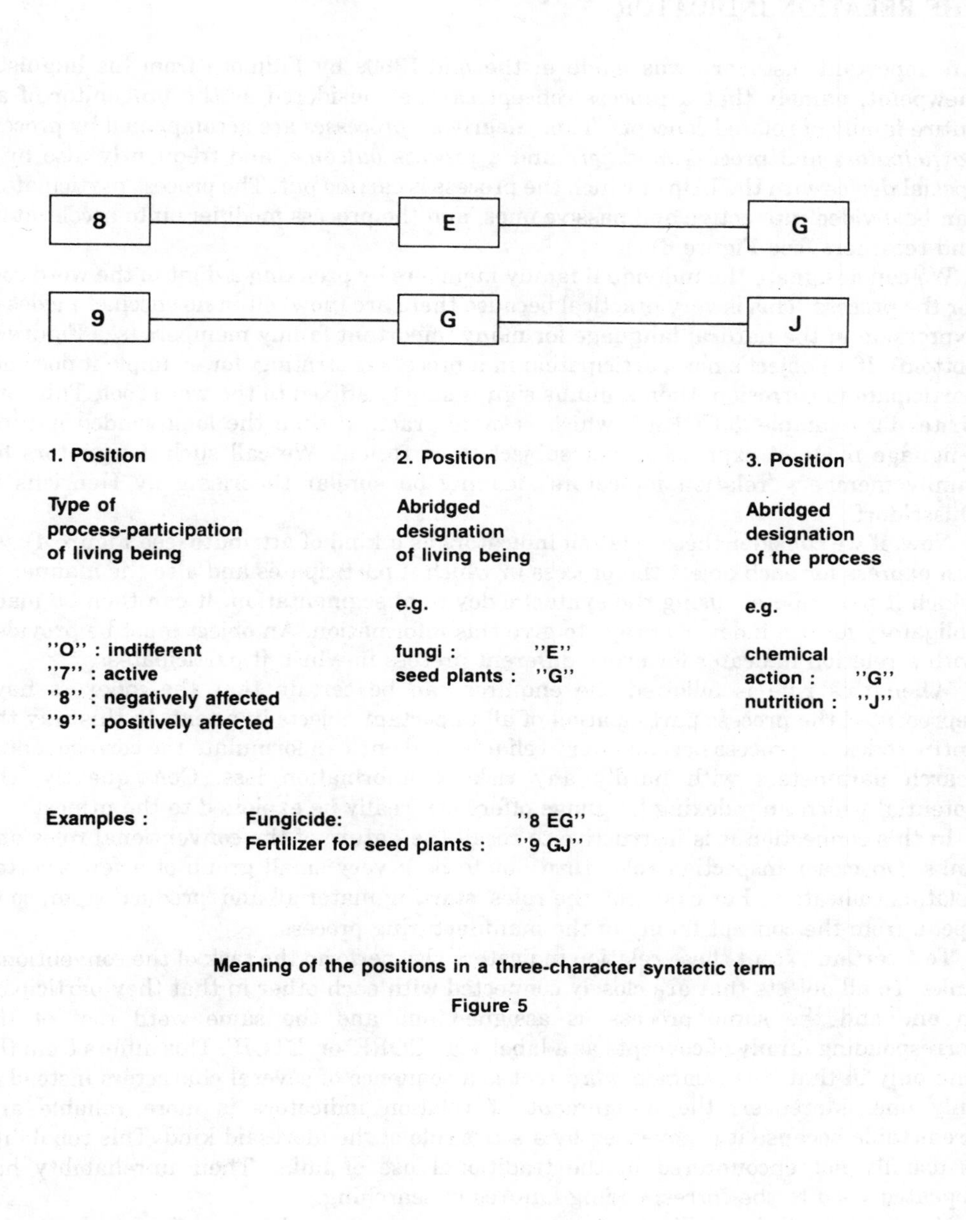

Meaning of the positions in a three-character syntactic term

Figure 5

THE RELATION INDICATOR

An important discovery was made in the mid-1960s by Fillmore from his linguistic viewpoint, namely that a process concept can be considered as the progenitor of an entire family of related concepts. Thus, nearly all *processes* are accompanied by process *participators* and process *modifiers* and a process *outcome*, and frequently also by a special *device* with the help of which the process is carried out. The process participators can be divided into active and passive ones, and the process modifiers into accelerators and retarders (see Figure 6).

We can designate the individual family members by prefixing a digit to the word root for the process. This is very practical because there are (now) often no succinct modes of expression in the natural language for many important family members (see Figure 6, bottom). If an object's non-participation in a process is striking, for example it does not participate in corrosion, then a minus sign is simply affixed to the word root. Thus one writes for example '3-CORR-', which is more practical than the long-winded natural language mode of expression 'not subject to corrosion'. We call such designators for family members 'relation indicators', leaning on similar theorising by Henrichs in Düsseldorf.

Now, if we consider these relation indicators as a kind of attribute (see Figure 4), we can express for each object the process in which it participates and also the manner in which it participates, using the syntactic device of segmentation. It can then be made obligatory for the indexer always to give this information. An object must be provided with a relation indicator for every different process in which it participates.

When this rule is followed the enquirer can be certain that the indexers have represented the process participation of all important objects in a text. In this way the entire indexing process becomes very reliable, and one can formulate the corresponding search parameters with hardly any risk of information loss. Consequently, the potential which an indexing language offers can really be exploited to the utmost.

In this connection it is instructive to recall the nature of the conventional roles and links. On closer inspection roles turn out to be a very small group of a few selected relation indicators. For example, the roles 'starting material' and 'product' stem, so to speak from the concept family of the manufacturing process.

To a certain extent these relation indicators also perform the task of the conventional links. To all objects that are closely connected with each other in that they participate in one and the same process, is assigned one and the same word root of the corresponding family of concepts as a label, e.g. 'CORR' or 'ETCH'. This differs from the link only in that this common word root is a sequence of several characters instead of only one. Moreover, the assignment of relation indicators is more reliable and predictable because it is governed by a strict rule of the aforesaid kind. This reliability is usually not encountered in the traditional use of links. Their unreliability has repeatedly led to the corresponding failures in searching.

Hence the relation indicators prove to be a simple and very effective device for indexing with a high degree of representational fidelity and predictability without requiring special features of the hardware or software with which they are used.

Process	Process-related object properties (relation indicators)					
	Process participants		Process modifiers		outcome	apparatus
	active	passive	accelerating	retarding		
1	2	3	4	5	6	7
Etching	etching substance	subject to etching			etched material	
Dissolving	solvent	subject to dissolution			solution	
Corrosion	corrosive	subject to corrosion	corrosion accelerating	corrosion inhibiting	corroded material	
	not corrosive	not subject to corrosion				
Plating	plating material	subject to plating			plated material	plating unit
Dip-coating	dip-coating material	subject to dip-coating			dip-coated material	dip-coating unit
Filtration		subject to filtration	filtration aid	filtration inhibiting	filtrate residue	filter

encoded relation indicators:

1 - ETCHI	2 - ETCHI	3 - ETCHI			6 - ETCHI	
1 - DISSO	2 - DISSO	3 - DISSO			6 - DISSO	
1 - CORRO	2 - CORRO	3 - CORRO	4 - CORRO	5 - CORRO	6 - CORRO	
1 - PLATI	2 - PLATI	3 - PLATI			6 - PLATI	7 - PLATI
1 - DIPCO	2 - DIPCO	3 - DIPCO			6 - DIPCO	7 - DIPCO
1 - FILTR		3 - FILTR	4 - FILTR	5 - FILTR	6 - FILTR 6 - FILTR	7 - FILTR

Relation indicators for the text of Figure 3

Figure 6

THE TOSAR SYSTEM

Let us now turn to Figure 3 in which we have represented the content of a text by a readily comprehensible graph. Originally TOSAR was conceived for the purpose of retrieval. In particular the program utilised the logical relations prevailing between the stored concepts. It would be undesirable, for instance, if this text were retrieved in response to an enquiry about chromium-manganese alloys, since in this text the two metals belong to different alloys. Furthermore, these metals even exclude each other because they belong to alloys which are connected with one and the same OR node (c.f. last paragraph of Introduction).

In the TOSAR system the time sequence of the various processes recorded there can also be used as a search parameter, and one can require that certain substances should participate in a reaction and others should not.

Unfortunately, input with the TOSAR system is relatively expensive, and it has not yet become established in practice, at least not for retrieval. But for merely rendering transparent the occasionally very confusing connections of concepts described in text form the TOSAR graphs have made a good start among the companies belonging to IDC.

THE CONCEPT NUMBER SYNTAX

In the long run in some areas of chemistry one cannot dispense with the indexing of logical relations that are described as existing between the various concepts in a text. We have already mentioned the importance of distinguishing between the relation 'chromium AND manganese' on the one hand and 'chromium OR manganese' on the other.

Similarly, in the polymer field one would not wish as a response to an enquiry about the copolymerisation of acrylonitrile AND butadiene a text in which the use of EITHER acrylonitrile OR butadiene is described. If in such areas the logical concept relations are not available as restrictive search parameters, then the noise in the retrieval responses can easily become intolerably large in the course of the years.

For use in IDC a technique was therefore developed, which, similar to the TOSAR system, expresses these logical relations between concepts, but without offering the many other refinements of the TOSAR system that are so costly to use.

During indexing '*concept numbers*' are assigned to all concepts of which the mutual logical relations are to be expressed and made available for retrieval. These concept numbers are arbitrarily selected by the indexers and need only be unambiguous within a single document. After that, the relation between the components of a composite (e.g., of a mixture of monomers or of the various layers of a coated material) is recorded in a *logical formula* which is entered into the record of the composite under consideration, i.e., by using the syntactic device of segmentation. For example, the logical formula for concept number '14' is added to the segment of the resistant composite at the root of the graph in Figures 3 and 7. The numbers appearing in quotation marks are concept numbers.

The machine program processes this logical formula during storage and develops all alternative *combinations* comprised by this formula and finally replaces the logical formula by the entire set of these combinations.

A search in such a file proceeds in two steps. In the first, 'semantic' part of the search all documents are retrieved in which the components of a composite being sought are recorded. Only these documents can possibly satisfy the strict search parameters of the subsequent part of the search. In this first step the concept numbers are also determined that have previously been assigned by the indexers to the individual components recorded in the document under consideration.

In the second, 'syntactic' step the search program replaces the original search parameters for the components of the composite in question with the concept numbers that were encountered for these components in the document under consideration. For example, the search parameter 'chromium' will be converted into '11' and that for manganese into '9'. In case of a search for an alloy comprising chromium *and* manganese the program checks whether these concept numbers co-occur in the logical AND relation in at least one of the (alternative) combinations developed from the logical formula. If this is the case, the syntactic search parameters are satisfied, too, and the document under consideration can be issued as a relevant response. In the case of the example of Figure 7, however, the combination 'chromium AND manganese' (intermediately to be expressed as '9 AND 11') will not be found among the alternative combinations and this document will, consequently, not be found as a noise response.

In this relatively simple way most of the important relations of concepts in the polymer field can be accurately represented in a predictable form, i.e., in a form that is useful for retrieval.

This model of concept number syntax can be applied to the representation of many other relations between concepts. To do this one needs only assign distinct concept numbers to the concepts in a document and to represent the interesting relation between the concepts in a correspondingly defined logical formula. In this way, for example, one can express the relations between the components of a pharmaceutical preparation or the type of relation that prevails among the compounds that belong to the same reaction sequence. The exclusion of a specific component in a composite can also be phrased as a search parameter.

One will choose this technique whenever retrieval in a field is (or will in future become!) impaired by an excessively high noise ratio because of a lack of syntax.

CONCLUSION

In the area of structural formula documentation very reliably indexed and very precisely and completely searchable files have been created through the introduction of the topological approach. Its success is attributable to the distinctly syntactic character of this approach. The vocabulary is extremely small (comprising only the elements of the periodic system and a few distinct types of chemical bond), and the major part of expressiveness is contributed by the highly advanced topological syntax. If this 'analytico-synthetic principle', having more or less subconsciously been applied here in a specific variation, is applied to other concepts and areas, similarly great progress can also be achieved in these areas.

Highly syntactic indexing languages have become possible only through the advent of high-performance computers. This must be seen as one of the most important benefits of the use of computers. The use of computers in information systems with indexing languages which are restricted to a vocabulary and hardly employ any grammatical-syntactic devices, means that the potential of computer technology is being poorly

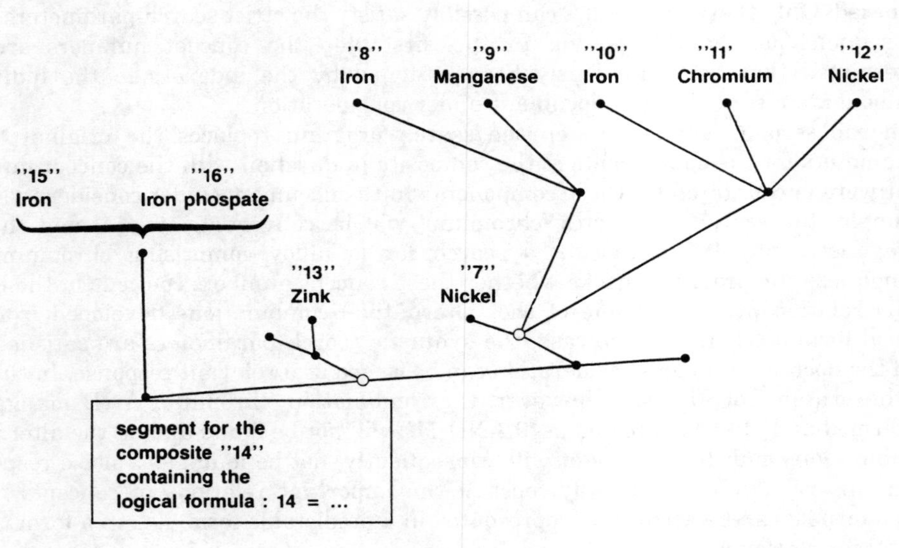

14 = 15 AND 16 AND [13 OR 7 OR (8 AND 9) OR (10 AND 11 AND 12)]

= 15 AND 16 AND 13 or

15 AND 16 AND 7 or

15 AND 16 AND 8 AND 9 or

15 AND 16 AND 10 AND 11 AND 12

Logical formula for the composite of Figure 3 and its resolution into the individual combinations comprised by the logical formula

Figure 7

exploited. The use of computers must not be allowed to disguise the fact that in reality these systems are obsolete.

ACKNOWLEDGMENTS

Substantial contributions to the development of the syntactic devices reported here have been made by Dr. Maria Isenberg, Ass. Günter Kusemann, Drs. Ingeborg and Herbert Nickelsen, Gottfried Ploß, and Jakob H. Winter at Hoechst AG.

REFERENCES

1. Fugmann, R. 'The Five-axiom Theory of Indexing and Information Supply'. *J. Amer. Soc. Inf. Sci.* 1985, *36*, 116-129.
2. Fugmann, R. 'Pecularities of Chemical Information from a Theoretical Viewpoint'. *J. Chem. Inf. Comput. Sci.* 1985, *25*, 174–180.
3. Fugmann, R. 'Role of Theory in Chemical Information Systems'. *J. Chem. Inf. Comput. Sci.* 1982, *22*, 118–125.
4. Fugmann, R., Nickelsen, H., Nickelsen, I., Winter, J. H. 'Representation of Concept Relations Using the TOSAR System of the IDC: Treatise III on Information Retrieval Theory'. *J. Amer. Soc. Inf. Sci.* 1974, *25*, 287–307.

INDEX

As this is a Proceedings, containing relatively concise individual articles, the index entries are intended to direct the reader to pertinent sections of pertinent articles—rather than to document all page occurrences of a given concept. Cross references are limited to *see* and *see also* notes, and initialisms are avoided as index entries.

The indexing method used was the Nested Phrase Indexing System (NEPHIS) developed by Professor Timothy Craven of the University of Western Ontario. It was implemented using the IOTA-2 (Information Organisation based on Textual Analysis, Two-Disk Version) indexing system developed by Professor James D. Anderson of Rutgers University. The assistance of Professor Anderson in preparing this index is gratefully acknowledged.

The Muscovite

Books by Alison Macleod

THE HERETIC

THE HIRELING

CITY OF LIGHT

THE MUSCOVITE

ALISON MACLEOD

The Muscovite

HOUGHTON MIFFLIN COMPANY BOSTON 1971

First Printing c

First American Edition

ISBN: 0-395-12714-9
Library of Congress Catalog Card Number: 73-124356

Printed in the United States of America

To Dora-Scarlett

The discovery of Russia by the northern ocean, made first, of any nation that we know, by Englishmen, might have seemed an enterprise almost heroic; if any higher end than the excessive love of gain and traffic had animated the design.

John Milton, *A Brief History of Moscovia*

This pece of service was verie acceptable, wherof I much repent me.

Sir Jerome Horsey, *Travels*

FOREWORD

Jerome Horsey was a real person. In 1573 the Muscovy Company of London, also known as the Russia Company, sent him to its trading post in the White Sea. During the next eighteen years he became dangerously involved with Russia's rulers, and made a fortune. The other Englishmen in Russia accused him of intolerable pride, of embezzlement, of informing against his own countrymen, and even of high treason. The Russians accused him of spying. An English ambassador to Russia, Giles Fletcher, warmly defended him.

Jerome Horsey's own account of these events, together with Giles Fletcher's book, *Of the Russe Commonwealth,* and other documents, were printed by the Hakluyt Society in 1856 under the title *Russia at the Close of the Sixteenth Century.* Another Hakluyt Society volume, *Early Voyages to Russia and Persia,* contains the travels of Sir Anthony Jenkinson. Recently J. L. I. Fennell, M.A., Ph.D., has issued Prince A. M. Kurbsky's *History of Ivan IV* and *The Correspondence between Prince Kurbsky and Ivan IV,* in two scholarly volumes with Russian on one side of the page and English on the other. (Both are published by the Cambridge University Press.) This is very nearly all the authentic information available in English, and it is full of mysterious gaps.

To find out why Horsey passed over so many of the charges against him in silence, why he omitted all mention of one of his journeys to Russia, why he seemed such a villain to his countrymen, and why he nevertheless won the hearts of some powerful friends, I have had to pursue my researches in Russian, German, Italian and Danish. I have not found all the answers, but there are clues to what the answers may have been. I have not altered any established facts, except the date of one murder, and the year of Horsey's trial.

The London Library has been a tireless guide and friend.

1

I assured the men from London that my husband would be back by nightfall. He had been at the Red Lion in Wendover, dining with other Justices of the Peace, but he had sent his groom to tell me that he would be home for supper. There was no need, I said, to go in search of him.

Since I spoke so calmly, the two officers took my word. I invited them to sit in the parlour, and brought them cakes and wine. I saw to it that their men were made comfortable in the kitchen.

Then I went upstairs and put on my diamond necklace, my husband's wedding gift. I put on every ring I had. I wrapped myself in my old cloak, making sure that it hid the necklace, and put on my gloves to hide the rings. Then I left the house by the garden door.

There was an hour of daylight yet. I went by the plank bridge over the stream, and then behind our home farm (whose hedge would screen me) over the stile and into the lane. A few steps along the lane, and then, to avoid Great Kimble village, I slipped off to the left, by the path our cowherds use. And here, beyond the gate, was a sea of mud; I held up my skirts, almost to my knees, as I plashed through. Then I came to clear green fields again. The ground was wet, but I could run, or almost run, for half a mile. I had to jump the ditches; they were full of water. But I came at last to the stile on the Upper Icknield Way. As I leaned over the stile the branches tore at my headdress. I felt hair fall over one eye, and did not push it away, because I was already late. I had missed my husband. He was there, but riding away, with his back to me.

I screamed out, 'Jeremy! Oh, Jeremy!'

He reined his horse and look round at me, startled. Angry, too, I thought; but for once I did not care.

He wheeled the horse hard round, and came to me shouting, 'What is it, Bess? For God's sake—what's the matter?'

I jumped over the stile. And then I found I had no breath. I could only lean against the stile and cough, while Jeremy still shouted, 'What's the matter with you, Bess?' He slid off the horse and seized me by the shoulders, hurting me. 'For God's sake—is it Felicity?'

I managed to say, 'No, Felicity's well enough.'

His grip on my shoulders grew gentler. 'You silly girl! Felicity's well, and you're well, or you couldn't have run so hard. Then what else matters? What news is there that won't wait until I come home?'

I said, 'You may not want to come home. There are men waiting for you.'

'What sort of men?'

'Officers of the Queen. They have six armed soldiers with them, Jeremy. They're to bring you before the Privy Council, to answer a charge of high treason.'

He did not seem surprised. With a heavy sigh he asked, 'Who accused me?'

'They didn't say.'

'Hm! I can guess.'

He stood looking to right and left, though there was nothing to look at, the month being January. I thought he was casting about for a way of escape. I pulled off the necklace and the rings.

'Take these!' I said. 'With money you can go anywhere. You know all the lanes; you can find your way in the dark; and by tomorrow night you'll be on the high seas.'

He stared at the jewels, and at me, and said, 'Why, Bess, do you think I'm guilty?'

I looked into the stern dark face, which I had once thought ugly. He went on, 'Haven't you the sense to see that, if I run away, I make myself out guilty? I suppose you think I am.'

'Oh, Jeremy, no! But your friends at court are dead. How can you be sure of a fair trial?'

He looked round, once more, at the daddy-man's-beard in the hedges, at the bare brown woods of Chequers, and at Little Kimble church, which lies all by itself in this lost place. The people of Little Kimble, half a mile away, will not come near the churchyard towards nightfall. That was why I had gone to meet Jeremy there; nobody would hear us talk.

But it seemed he did not want to talk. He looked a long time at the church, and then again at me. At last he said, 'I have not broken any English law. If that's not enough to save me here, I shall not be safe anywhere else.'

Then he held me in his comfortable arms, and kissed me. He put my necklace round my neck again, and smoothed my hair back under my headdress. He hoisted me on to the saddle-bow, and remounted.

'Let's ride home quietly,' he said, 'like honest people.'

We rode where everyone could see us, through Great Kimble village, and past the gate of the home farm, and by the front way to our house. Felicity came running out, with old Meg puffing behind her. Jeremy called, 'Your mother has taken your place!'

The child set up a wail, because what she liked best in the world was to ride on Jeremy's saddle-bow. We dismounted, and he picked her up. The wailing stopped as always, always when she found the protection of his arms.

Carrying the child, he went into the parlour to confront the men from London. I kicked off my muddy shoes and ran upstairs, and called my maid. She soon had me brushed and redded, in clean shoes and stockings, with nothing but the flush in my cheeks to show I had run so hard. Looking calm and stately, I came into the parlour.

The officers were treating Jeremy with respect, calling him by his title, Sir Jerome Horsey. It was no pleasure to them, they said, to come on such an errand to a gentleman of his reputation, Lord of the Manor of Kimble, a Justice of the Peace.... Seeing them so friendly, Jeremy asked if he might spend the night under his own roof.

They agreed, asking only whether we could lodge them for the night. 'I shall have the best room made ready for you,' I told them. 'And the servants' rooms leading off it, for your men.'

'Quite right, Bess!' my husband said. 'Make these gentlemen welcome! They are doing their duty to the Queen. Once this misunderstanding is put right, we shall meet again as friends.'

When you have to set an example to a great many people, it is wonderful how calm you can be. We took our guests to supper as if they had been any other guests. As on any other evening, our upper servants were waiting behind their chairs. They bowed to us and to our guests, and we all stood, waiting for Jeremy to say grace. Beside me, Felicity joined her little hands.

In a steady voice, Jeremy thanked God for letting us live amid such abundance, in a country which, even in the dead of winter, gave us fresh milk and eggs.

We did indeed have abundance; it was my pride that I would never be put out by unexpected guests. There was a ham, and chickens from our home farm, and pheasants from our woods, and a great cheese from our dairy, and bread from our bakehouse, and apples from our orchards, besides the cinnamon custards Jeremy loved. We had no galleons or castles of pastry-cook's work. It was honest food, for eating, not for display; and, though a good deal was cleared away untasted, I knew that it would be thriftily finished up in the kitchen.

In honour of our guests, we drank French wine, instead of ale. The wine made us talk cheerfully; most of all Jeremy, though he drank the least. He told the officers how rich our pastures were, and how in spring the horses, if they were not carefully watched, would make themselves ill by overeating. He told them what good hunting there was hereabouts, especially when Sir William Hawtrey, 'a good cousin and a good neighbour', invited us to Chequers. 'One part of Sir William's estate,' he said, 'is called the Happy Valley. It must be the rabbits and hares who are happy; they live in such entangled woods that nobody can catch them.'

When supper was done, and it was Felicity's bedtime—then I saw how long he held the child, and how hard he found it to give her up to old Meg. And still he said nothing but what was cheerful. Our guests remarked on the beauty of our damask table cloth, and on the napkins made to match it.

'A present from the merchants of Hamburg,' Jeremy said.

When they admired our great silver loving-cup, he told them, 'A present from the merchants of Lübeck.' Then he added, 'If there were a moon, I would take you out and show you why my neighbours think me crazed. I have a bath-house, built Russian fashion, where I bathe in hot water and steam once a week.'

'What—even in the winter?' one of the officers asked.

'Especially in the winter,' Jeremy said. 'Though I don't run out naked and roll myself in the snow, as the Russians do.'

We took our guests back to the parlour, and Jeremy called for more wine. I guessed that he wanted to make the officers talk. Both, indeed, grew tipsy, but they swore they did not know the name of Jeremy's accuser.

'It can only be my namesake,' Jeremy said. 'The other Sir Jerome—Sir Jerome Bowes, once Her Majesty's ambassador in Russia. He has never let me be since the last time I came back from Russia—since 1591. Almost six years! He always works through someone else. He tried to persuade the Russia Company that I had robbed them. I have had a world of trouble going through my old account books, finding forgotten receipts and bills of lading.'

I had been hostess to guests from the Russia Company, from the time I was married, five years before. But, though Jeremy would often tell me stories about his life in Russia, he was apt to grow suddenly angry when I asked him questions. I had never understood until that moment what the trouble was. Sir William Hawtrey should have told me; besides being our neighbour, and my father's cousin, he was the oldest member of the Russia Company. But he had always answered my enquiries by reminding me I was a woman.

One thing even a woman could understand. I said, 'The other day the Russia Company invited us to dinner.'

Jeremy smiled and said, 'Yes, we are good friends now. When the accounts had been sifted, the Company discovered that *they* owed *me* money. Sir William Hawtrey had told them that, all along. So, last week, they asked if they might give me and my wife a ceremonial dinner at the Guildhall next April. We accepted, but now . . .'

'Why, sir, you may be free in good time for your dinner,' the younger officer said. He added, 'Your journey with us . . . You may bring your own manservant.' The older one agreed with too much heartiness. 'Yes, that's the custom—for anyone born a gentleman.'

'Oh, I was born that,' said Jeremy. 'Born a beggar, as well.'

The officers laughed and said that they too had been younger sons.

'I was the youngest of six,' Jeremy said. 'My father was the squire of Melcomb Horsey in Dorset, but he was so poor that he could hardly borrow the premium to have me apprenticed to the Russia Company. Which was a rough apprenticeship.' He smiled. 'Whatever hardships I may find in the Tower will scarcely be new to me.'

'Don't speak of hardship!' said the younger officer. 'You'll have a room to yourself, a proper bed. . . . And your wife may lodge nearby, and visit you every day.'

I cried out, 'Oh! Then may I come to London with you?'

Jeremy shook his head. 'That's impossible, sweetheart. You must be here, to look after Felicity.'

'But old Meg looks after her!' I said. 'She brought me up safe enough. Why, when the Russia Company's invitation came we agreed that we could both go to London for a few days.'

'This will be more than a few days,' Jeremy said. 'Leave

servants to themselves a month or more, and you come back to a thieving, brawling rabble.'

'Our servants wouldn't brawl or thieve!' I protested. 'They're old friends, most of them; they came with me from my father's house. I scarcely tell them what to do; they know their work.'

'My lady,' said the junior officer to me, 'you are right. Quite right.' He swayed a little. 'There's a wonderful harmony in your house. We could feel it from the moment we crossed your threshold.'

The older man wagged an unsteady finger at Jeremy. 'Take my advice!' he said. 'Take the advice of a man who's known a good many in your situation. Let Lady Horsey come! Now I wouldn't say this to a man with a *brabbling* wife. But a quiet, wise, discreet lady, a lady who keeps her own house in such good order.... Believe me, a good wife is the best friend a prisoner can have. Who else can run to and fro to seek out your friends at court?'

Jeremy said, 'My friends at court are dead.'

'Then you need your wife all the more,' the younger officer told him.

The older man leaned over his glass and winked. 'You have one friend at court,' he said. 'The Queen herself. When who-ever-it-was made his accusation, and Lord Burleigh brought it before the Queen, she had no choice but to sign a warrant for your arrest. But even while she signed it, she said, "I still believe Jeremy Horsey will prove himself honest." '

I was greatly comforted by this, but I could see, from Jeremy's face, that he was in two minds whether to believe it.

Just before we went to bed, the junior officer asked the question which I had been longing to ask. 'Sir, this enemy of yours, this Bowes—what makes him so bitter against you?'

Jeremy smiled. 'I saved his life in Russia. He is one of those people who cannot forgive a good turn.'

The officers were sober enough to station two of their armed men outside our bedroom door. Finding them there when I came to bed, I did not cry out at this evidence that my husband was indeed a prisoner. I smiled at the men and asked them whether they had been well fed in the kitchen. They said, 'Yes, thank you, my lady.'

Inside the bedroom I found Jeremy shaking with laughter. 'Thank you, my lady!' he whispered, holding me in his arms. 'Oh, my lady, my lady, my quiet, wise, discreet lady! None of them know you! My wild girl, jumping over the stile like a gipsy! And giving up her jewels! Wanting to defy the Queen and all her laws for me!' He pulled off my headdress. 'Let me see you again with your hair falling over your face!'

Because of those men outside the door, we made love quietly, stopping every time the bed creaked. At last these very pauses made the pleasure past all bearing, and we had to let the bed creak as it would.

The sleepy after-time was always dear to me, but never so dear as that night, when I wanted to comfort him in his danger. I murmured that I would come to London with him, find a lawyer, appeal to the Queen. . . . He interrupted me.

'I have to tell you, Bess, about this house. You don't believe now that I married you to get it?'

I had thought this matter long ago settled between us. 'You know I don't think that!'

'But you did, Bess. You did!'

'Never! I knew my father was so deep in debt to you that you could have seized the house. I was only . . . well, at first I thought . . .'

'What, Bess? What did you think?'

Why did he force this out of me? 'I thought you did not want to seize the house, because then everyone would hate you as a moneylender. There would be nobody to visit you, or hunt with you. I though you married me because that was your only way to get the good opinion of the neighbours, and the house as well.'

'Then why did you consent?'

'I wouldn't at first. When my father told me what was arranged, I cried out against it. I asked him how he dared think of marrying me to such an ugly old man. Well, I knew you were thirty-seven.'

I thought he would laugh at that, but he did not. He only stroked my hair.

'You *are* too young. Too young for the trouble I've brought you. But you yourself agreed to it. I wouldn't have taken your father's consent for yours; I've seen enough of those marriages . . . Bess, why did you consent?'

'Father told me that nobody else would want me. Not with my long nose, and my flat chest, and my bad temper. . . .'

At this he did laugh. 'If you ever had a temper, you must have had one then.'

'Yes; I screamed and cried. Until I looked in the glass and saw my long nose red with weeping. Then I thought father was right; no man on earth could want a girl who looked like me. When you came courting me I didn't believe a word you said.'

'Not a single word?'

'You said my hair was cloudy sunshine. I liked that. But then father told me that the things men say to women they don't reckon as lies.'

'Your dad's not very fond of me, is he? He thinks a money-lender is mother and father to the Devil. My own dad thought the same. These country squires! It never enters their heads that they ought not to get into debt.'

'Now that's not fair! Father never was in debt, until the Queen stayed with us. You don't know what that's like—feeding a hundred courtiers, putting up tents for the servants to sleep in, handing out bread and cheese and beer to all the country people. . . . You can't think how far people will walk for free beer! They came from the other side of Aylesbury. There were three thousand watching, when the Queen made father a knight. It was all a great honour, of course, but it ruined us.'

Jeremy said, 'I ought to thank God for that. If your dad hadn't wanted money, I should never have seen you. Do you remember, Bess, when he brought you to the Russia Company? You were a child—about fourteen, I think.'

'Thirteen. My mother had just died.'

'That's right; you were all in black. The black hood and the pale hair—that's what caught my eye. Yes, and your nose.' He kissed it. 'I liked your nose. A long straight nose is a rare thing in Russia. But you looked so impossibly sad!'

'Why impossibly? I was sad. I knew father had come to the

Russia Company for another loan, and I knew they'd never give him one, because he hadn't repaid the first.'

'What did you think, Bess, when I spoke to you?'

'I thought—what a strange man, all in furs, like a great bear! And this jutting beard of yours, black and fierce! And when you said you were just back from Russia, it was like a fairy tale. Do you remember, Jeremy—I asked you if you had seen the great whales drawn into the Maelstrom, and if it were true that they uttered piteous cries.'

'I remember,' Jeremy said. 'I told you that the ships I travelled in had always given a wide berth to the Maelstrom, but that the story of the whales must be true, since it was reported by Sir Anthony Jenkinson. You lost your sad look then, Bess, asking me if I knew Sir Anthony. You told me that you'd read his book of his travels again and again. You asked me—had I sailed where he sailed, down the Volga, and hoisted the English flag on the Caspian Sea, and ridden on a camel to Bokhara? And had I sat talking to the Shah of Persia, as Anthony Jenkinson did? I had to tell you no; my furthest reach was to sit talking to the Tsar in Moscow.'

'But that was wonderful enough! You made me feel so envious! I thought—if only I'd been born a boy....'

'So you told me then, Bess. I'll never forget that. You were looking out of the window, down Seething Lane, to the river. You said that when you saw the Thames at high tide you could bear it, because it looked sluggish and still. But now, at ebb tide, it was running, running to the sea. It was going to the Maelstrom, and the whales, and Russia, and leaving you behind. I thought of myself as a boy, in that very office, looking out of that very window. I'd been homesick, black and blue from the cane, eating broken victuals and sleeping on straw, but always, always with one consolation—that I was going to sail down that river. And you could never do that, being a girl. I was well accustomed to the sight of women grieving; I'd seen some with much more to grieve over than you. And yet you made me feel what I had never felt. I might have had that bad luck. I might have been born a woman.'

'Was that why you came running after father and me?'

'I knew what the trouble was, when I saw your dad come out of Mr. Burrough's office, with Mr. Burrough telling him, "It's not that we doubt your word, Sir Griffith; it's just that we don't do that sort of business." And then your dad saying, "But my cousin, Sir William Hawtrey. . . ." And Mr. Burrough, "That arrangement last year was a special favour to Sir William. But the other directors. . . . Well, this year things are tighter. Much tighter." He pursed his lips to show how tight they were. I couldn't speak to your dad then and there, so I followed him down to the street. I told him that as agent for the Company I couldn't help him, but on my own account I might. He was all gratitude—do you remember? And glad enough to borrow from me again, the next time I came back from Russia. But you see how he thought of me! A moneylender, with horns and a tail. Bess, I swear I never threatened to foreclose on Kimble Manor.'

'He didn't say you'd threatened him. Only that you might. And if you did, he said, I shouldn't have six feet of earth to dig myself a grave in. Hampden House is entailed on my brother. There'd be nothing for me. Who'd want me without a dowry? I looked in the glass again. I tell you, Jeremy, in those days I envied our silly kitchen maid, who bruised her bottom jumping down the stairs. She was married hugger-mugger to the groom, her baby's father. The other servants laughed at her, but I thought—at least she was desired for her body, not her estate! Jeremy, I could not believe that you liked me.'

'Not . . . Not when we walked about your father's estate together, and you showed me how you kept house? Do you remember taking me to the still room where the ale was brewing, and down to the cellar where the cheeses were ripening, and round the home farm to see the new litters of pigs? Didn't you know that I took some pleasure in being with you?'

I tried to remember. 'Oh yes! I did believe you, when you praised me for my thrifty housekeeping. I could see it would save you a great deal of money, to have a wife who understood the ways of a country estate.'

'So whatever I said, it proved I was thinking of nothing but money! I still want to know, Bess—why did you consent to marry this grubbing moneylender?'

'Well, I was a girl, so I couldn't be a traveller. And I was plain, so I couldn't be beloved. All I could do was keep house, and rather than do that for my cantankerous dad, I chose to do it for a stranger.'

'Then—did you never think you would have to go to bed with this stranger?'

'I asked a girl I knew who was lately married, and she said it could be endured, if you lay quite still and thought about something else.'

'So that's what you were trying to do, that first night?'

'No. I found I couldn't even try. Because it wasn't what my friend said. She told me, "All it is—he throws himself on top of you, and provided you don't struggle, it's over in two minutes." '

'Who is this friend of yours? Do I know her husband?'

'Jeremy! You mustn't ask me that.'

'Well! I still don't understand why you cried so much, when you found out I wasn't like him.'

'That was why. Because you were so gentle, and wooed me so sweetly. I began to understand that this thing could be done for delight. I thought—if only I had done it with a man I loved—'

'And you hated me. You must have done, to say . . . what you did say.'

'Oh, Jeremy! Don't bring that up again. I was young; I had no sense.'

'If you'd said it on that first night. . . . But no, it was the second night, when you'd had time to think. You were deliberately trying to hurt me. Why?'

'Because I didn't know! I never dreamed you'd want me again the second night. My friend never said. . . . And I was tired. I'd had no sleep. . . .'

'I had no sleep either, once you'd said it. I got out and walked out of the room, so as not to hit you.'

'I thought you were never going to come back. . . .'

My voice failed me, as I remembered that night. Alone, in silence, hearing nothing but the echo of my own appalling words. *Why can't you leave me be? You've got what you married me for; you're living in Kimble Manor.* And the fear that he would never again give me what I had thought I did not want. And then the hot and beating wave of shame; the knowledge that, since I was thirteen years old, my tempers and my sulks had been because I wanted nothing else.

'Perhaps I never would have come back,' Jeremy said, 'if the servants hadn't begun to creep out, and peer, and giggle. I thought I looked a fool, walking up and down the landing. I thought I *was* a fool, to marry a girl of eighteen, and a hard, proud, cruel girl at that. So I went back, determined not to touch you, but when I saw you all swollen and blubbered with tears. . . .'

Now I was weeping at the memory. He kissed me and explained, 'I didn't bring this up to make you cry. It's because of the house. Bess, what made me so angry then was that you knew the terms of the marriage settlement. You knew that I had only a life interest in the house. It belongs to trustees, who hold it for you and your heirs. I could have taken it in payment of your father's debts; I could simply have taken it as your dowry. But I went out of my way—I spent hours with lawyers—to see that so far as the law permits a married woman to own anything, you should own Kimble Manor.'

'I know, sweetheart, I know.'

'But what you don't know is that I've been settling other property on you—property I have in London. I thought, you see, that something like this might happen. Thank God I saw to it in time! Now the Queen can't seize it.'

'Why should she seize it?'

'Because that's the law, once I'm found guilty of treason. You'll have to go to the lawyer, Bess—my lawyer, not your dad's. Go to Mr. Oulton and ask him for the documents. You'll find that you and Felicity are quite rich.'

'What use is that to me—to be rich without you? Jeremy, I'll come to London; I'll plead for you—'

'That won't make any difference.'

'It will! Those men said I could help you if I came.

'Bess, don't you understand yet? Nothing you do or say can help me in my trial. That will all depend on things that happened years ago, which . . . which you can't alter.'

'But they said—'

'I know what they said. But what they meant I understood well enough, when they began hinting their questions about what rank I was born to. I'm a gentleman, Bess, not a nobleman. I have no right to a quick death. I can be dragged to execution at the horse's tail, hung till I'm not quite dead, cut down, my belly ripped open. . . .'

I was winding myself round him as if my body could protect his body from the knife.

'Castrated, into the bargain,' Jeremy said. 'I admit I don't fancy that. So you might help me there, Bess. You might persuade the Queen to let me be beheaded. You could remind her that she once had some reason to like me.'

'She does like you. Those men said so.'

'If that's true. . . .'

'It must be true. Else why were they so friendly?'

'Out of pity. If not for me, then for you. You were looking so young . . . Bess, go on coiling yourself round me. I like that . . . How could anyone call you flat-chested?'

'They don't now—not since I fed Felicity.'

'But even before that, you had sweet little breasts. Like lemons. . . . The thought of death enlivens me! This must be why soldiers are always at it.'

This time, we were past caring how the bed might creak.

Flooded with happiness from this man's overflowing life, I could not believe in the nearness of his death. Why did he feel so sure of it? I murmured, 'My darling, since you're innocent—'

'Did I say I was innocent?'

'You said you had not broken any English law.'

'That's true. If the Privy Councillors find me guilty, it will be their injustice. But it may be the justice of God.'

I could understand that. I had myself often been in fear of God's justice, for all the sins which do not break man's laws.

For losing my temper, often and often, for carelessly dropping my best ring and slapping my maid when she could not find it.... For being impatient of my mother's complaining, when she was mortally ill.... That alone might account for our present trouble.

We slept, Jeremy soundly, myself by fits and starts. I heard the trampling as fresh guards relieved those men outside our door. I heard the clatter as the kitchen boy got up to light the fires. At that I jumped out of bed. It was before my usual time, but none too soon, if I were to set out with Jeremy.

So I hurried my clothes on before I could feel the cold. I woke my maid Rosemary and told her to pack. She would have to come with me; I could not go to court without a servant. Next I woke Jeremy's man Paul. I told the kitchen boy to rouse the cooks; breakfast must be early. I woke the grooms to tell them which horses we should need. Then I wrote to my father, telling him our trouble, and sent a man on horseback to deliver this letter to Hampden House.

I was about to write to my father's cousin, Sir William Hawtrey. He was a director of the Russia Company, so who could help us more? But then I remembered that he was in London, and I myself might reach him faster than a letter.

On my way back to the bedroom, I heard a quiet singing. Felicity was awake. I went into her room and picked her up. Old Meg (though she had slept through the singing) felt that the child was gone from the bed beside her; she moaned and raised herself on one elbow.

'Sleep, Meg!' I said. 'You'll have work enough soon.'

I set Felicity on her pot. 'I can go by myself, Mama,' she said. 'I'm not a baby now.'

I washed the child and wrapped her in a shawl. Then I picked up her clothes, and took her into the bedroom, where Jeremy was lying fast asleep still. My maid had lit the fire. I sat down by it and began a task which as a rule I left to Meg. I dressed my Felicity. As I covered, one by one, her soft perfections, I wondered how I ever came to bear a creature of such beauty. My hair might be cloudy sunshine; hers was unclouded; and since her third birthday it had grown thick and wavy and long.

I talked softly to the child, explaining that her father and I must go away for a while. She did not cry, but looked at me anxiously, saying two or three times, 'But will Meg be here? Meg won't go away?'

Our talking soon woke Jeremy. He put his head through the bed-curtains, and cried out, 'Oh, let me have her!'

'In a moment!' I said. I picked up a comb and began to take the tangles out of that beautiful hair.

Jeremy cried out like a beast under the butcher's knife. The chid and I both turned to stare at him. I still held the comb upraised.

'Put that thing down!' Jeremy shouted. He jumped out of bed, as if to snatch the comb from my hand. Then Felicity began to wail with fright, because he looked so strange and haggard in his night-shirt. He recollected himself, and took the child in his arms, and crooned to her.

To me he said, 'It's only that . . . I have so little time. You shouldn't waste it with your fussing. You can comb her hair when I'm gone.'

I said, 'I'm coming with you.'

'No, Bess. Later, when the trial's over.'

'I'm coming now.'

'Bess, you don't understand. This trial is not going to be straightforward. There are things I can't explain to the Privy Council. I don't know that I can tell them even to a lawyer.'

'Then tell them to me.'

'Don't meddle with matters you can't possibly understand!'

I did not answer. I looked at him. He came to me gently, and put the child in my lap.

'Look after her!' he said. 'Look after my darling for me. That's how you can help me best.'

I said, 'She needs both of us alive, and safe.'

He went into the next room, where his manservant was, to dress. I sat enjoying the warmth of the fire, and the sweeter warmth of the child, whose arms were about my neck. I wished with all my heart that Jeremy were right, that I had no need to leave her.

When he came back he stood looking down at me and Felicity. He asked, 'Are you sure she'll be safe with Meg?'

'Of course.'

'Then come to London if you like, Bess. God knows I want you to. Only—no silly questions!'

'No questions,' I agreed. I knew that it was useless to ask, even, why it hurt him to see me comb the child's hair. If his life were to be saved, he would have to tell me this. But he must think he was telling me of his own accord.

It was pitch dark, still, as we mounted our horses. In the confusion, with Felicity crying, and Meg shushing her, and myself declaring loudly (for my voice had to reach the servants as well as the child) that we should both be back from London soon, I did not at once know that my father and brother were beside me. It was my father's hand I felt first, reaching out to seize my arm. Leaning over from the saddle, he whispered in my ear, 'Oh, Bess, forgive me!'

'What for?' I asked.

'For bringing you into this marriage.'

'That was the best thing you ever did for me,' I told him.

By the lantern light I saw him shake his head. He led my horse a little aside, so that we could talk privately.

'For God's sake, Bess!' he said. 'Don't go to London!'

'Why not?'

'Believe me, Bess, there's no need. Nobody expects you to follow where this man's going.'

'He's my husband.'

'Oh yes, well, of course, your duty. . . . But your duty's here, looking after your house and your child. Why should you go to London? What can you do there?'

'I'll go and see Cousin Hawtrey. And the Queen. I'm sure she won't refuse to see me.'

'Oh, she might see you. When she stayed with us, I remember, she recited some Virgil, and asked if you knew what came next, and when you did she was delighted with you. But, Bess, the Queen herself can't interfere with justice. There is far more against this man than you know.'

'What is there against him?'

'We've all tried to keep it from you, Bess. He came back from Russia with more money than any honest servant of the Company could make.'

'I know all about that. It's been settled. The Russia Company went through his accounts and found that they owed him something. Why, if Cousin Hawtrey were here you could ask him.'

'What Cousin Hawtrey doesn't know,' said my father, 'what nobody knows, is where the money did come from. Have you asked Jeremy that?'

'No, but—'

'Do you know how rich he is? He's given those brothers of his in Dorset enough to live like princes—and they were in a beggarly state before.'

'Are you reckoning it against him that he's generous?'

'He still keeps plenty for himself. I tell you, Bess, when any merchant venturers need capital for a new enterprise, they go first to Jeremy Horsey. Provided that they can convince him it'll pay ten per cent.'

'But that's your answer. If he makes ten per cent on everything he lends, he's bound to be rich.'

'At any rate we could all be rich. The difficult thing is to find the money to start with. Where did he find it, Bess?'

'What's all this to do with a charge of high treason?'

'Men have been paid before now for betraying their country.'

'That's impossible! I mean, if it were Spain.... But we're not at war with Russia; we can't be; it's too far away.'

'Do you know what they call him in the City? The Muscovite!'

A father is a father; it was my duty to kneel down and ask his blessing. I did that directly, slipping off my horse and plumping down on the cold cobblestones. It was the only way to stop him talking.

My brother helped me to remount. As he did so he whispered, 'What a comfort it is, Bess, that this house is settled on you!'

I could scarcely bear to hear my father talk to the Queen's officers. 'I'm Sir Griffith Hampden, of Hampden House. I entertained Her Majesty there once. Lady Horsey is my daughter....

Oh no, Sir Jerome is no relation to anyone in these parts....'

Just as we were setting out, we had another visitor. Our neighbour from Monks Risborough, Sir Henry Ewer, came riding up in a hurry. Even through the murk I could see that his broad face was covered in sweat.

'My dear, dear fellow,' he cried out to Jeremy. 'I have only just heard.... Is there anything I can do?'

Jeremy thanked him, and said that he might indeed help, if he would ride round our estate sometimes, and see that our people were not neglecting their work.

'No, I mean—can I give evidence on your behalf?'

Jeremy said, 'I believe my accuser is a man I knew some years ago, in Russia. I shall have to find witnesses who were there at the time.'

'But witnesses to your character!' said Sir Henry. 'They'll surely admit those. I could give evidence—a dozen of us could—about what you said at the Justices' dinner yesterday.'

'Why—what did I say?'

'You remember! When we were all talking about poachers, and old Rowland said that it seemed a great waste of time for us to sit on the bench and hear witnesses, and pleadings, and the clerk showing off his points of law, when the accused man was only some dirty poacher. And then you said something... something like... well, Jeremy, you put it better than I could.'

'Oh, I remember,' Jeremy said. 'I told Rowland that he would not like to live in a country where nobody troubled about points of law. I said that in Russia the noblemen flogged and imprisoned their own peasants without regard for law.'

'Yes, and old Rowland said, "A good thing, too!" Then you told him that the king of that country has the same rights over the noblemen, and he might not like that so much. There was quite a bit of chat, you know, after you'd gone. Some of us were saying you should be in Parliament. But old Rowland was grumbling away, making out you were on the side of the poachers. I told him not to talk such nonsense. "You find me a better rider than Jeremy Horsey," I told him. "You find me a better companion for a day's hunting or hawking." Never you fear, Jeremy—when you come to trial, I'll tell them you can't be a traitor.'

Such evidence might not convince the Queen, but I was glad my father and brother could hear it. I took the hand of Sir Henry between my hands, and thanked him with all my heart. He kissed me courteously, as a gentleman should kiss a neighbour's wife. Then he whispered, 'Bless you, my dear; you've proved your husband's innocence. Running to warn him, giving him your jewels—and still he wouldn't run away!'

I did not ask how he knew. Even if I had slipped out unseen, my maid had found my jewel-case empty, and innumerable eyes had watched me come home riding on Jeremy's saddle-bow. All our servants had brothers and sisters and cousins in the household of Sir Henry. In country places nothing is a secret. That thought comforted me now, as in the growing light our tenants and villagers came to wish us well. If the country people respected Jeremy, it was because they knew him.

My father was telling the officers that the lane through Chequers and past Hampden was nothing like so muddy as the Princes Risborough road. My brother agreed. Nobody said—at least, not in my hearing—that it was unfit for a Hampden girl to ride through populous places in the company of a prisoner. But the officers evidently understood. As if apologising to my father, they told him that they could not avoid High Wycombe. But they would, as he advised, approach it by the most secluded lanes.

Sir Henry rode with us as far as my father and brother did, to the very gates of Hampden House. My brother's boy, who like himself is called John Hampden, came running to meet us. He called out, 'Uncle Jeremy! Tell me a story!'

'Some other people want to hear my stories,' Jeremy said.

The child hugged him, and so did Sir Henry. That shamed my father and brother into embracing Jeremy too.

At parting, my father whispered, 'Don't expect any comfort from old Cousin Hawtrey! A man who brings lawsuits against his own grandchildren.... No family feeling! Never had. If he hadn't been such a miser, I need never have gone cap in hand to the Russia Company. I should never have been in debt to this man, and you need never have. . . .'

I said, raising my voice, 'Thank you, father, for finding me the best husband in the world.' Then I waved him goodbye.

As Jeremy and I rode on, our servants and our guards fell behind us, out of earshot. We were free to talk, with little interruption from the horses, who plashed quietly through the mud.

Jeremy asked me, 'Is old Sir Henry the man who never troubles his wife more than two minutes?'

I shook my head. I could have my secrets too.

'Well! A good fellow, at any rate. Your dad and your brother, too—that was good of them, to show that they still dared be seen with me.'

'Why shouldn't they? They ought to know the charge against you isn't true.'

'What a child you are!' Jeremy said.

I asked indignantly, 'How do you mean—a child?'

He shook his head, smiling. That sort of smile made me feel so lost, so helpless, that I found myself stopping and looking round, as if for somebody to take my part. But of course there was only our escort. Jeremy followed my gaze.

'Look at them, Bess! What innocence! It's not that they're giving me a chance to escape. I couldn't ride fast, in this mud. But look how they straggle along! They've never heard of ambush or surprise. And these are men of the Queen's bodyguard—the only regular soldiers we have in this country, God help us! If they had to face a horde of Tartars, all shooting as they gallop—even turning to shoot backwards as they gallop, so that their retreat is more deadly than another army's advance. . . . The Tartars could conquer us in the twinkling of an eye, if they had ships to bring them here.'

'God forbid!'

'Oh, they won't come. But the Spaniards nearly did . . . what is it? . . . nine years ago.'

We were now taking a gentle way downhill. When we looked leftwards the low sun dazzled us, and so, whenever we took our eyes off the path, it was to look right, into the woods. Our noble beech trees, when they lose their colours, take a new colour from the winter sun. They glowed so redly, you might fancy autumn back again. Then, looking at the thick trunks green with moss, you might hope soon to see the green of spring. Beyond the woods the hillside was already ploughed in tidy furrows.

Jeremy said, 'If the Spaniards had landed on the south coast, they would have been here within a week.'

'But our men were all in arms, to fight them.'

'Bess, if you had seen your father murdered, and had been raped by one soldier after another, and had then gladly consented to live with an officer—'

'Why gladly?'

'Because he'd keep the common soldiers off you. You'd have to play and sing for him, of course. You'd have to turn to his religion, and betray your fellow-countrymen to his revenge. And if after all this he still abandoned you, and you came crawling back to the ruins of Hampden House, to give birth to a dead child.... Oh, God forbid, as you say! But if this had happened to you, Bess, or something like it, you might understand what it was for the Russians to be ruled by the Tartars for two hundred years, and to be in continual danger from them still. It's made them... different from us....' He frowned. 'I should have sent someone with a letter to Oxford, to Giles Fletcher.'

'Is he one of your witnesses?'

'He was in Russia.... Not at the same time as Bowes. I might be able to explain this business to him. When I talk, he knows what I'm talking about.'

'Don't I know what you're talking about? Why, Jeremy, do you think I have no imagination?'

'Oh, you may have. But imagination.... When I was a boy, in that Office in Seething Lane, I thought I had imagination. We all had to go into mourning because of the Tartars. That was 1571, the year they reached Moscow. They burned every house outside the Kremlin wall. Twenty-five of the Company's men died in the flames of the Company's warehouse. Three Englishmen escaped by sheltering in a cellar built of stone—a rare thing in Moscow. They brought the news back to Seething

Lane. When I heard them talk I thought I could picture it all—the flames racing through the little wooden houses, licking up to melt the churches' golden domes; then the people running away, the flames running after them, the few left alive coming back to bury their dead....

'Our masters doubted whether we apprentices would ever go to Russia. Could the Company go on trading there? But soon we heard that the Tartars had not attacked the north, where our other warehouses were. Our men were safe in Vologda, in Yaroslavl, and—most important of all—on Rose Island in the White Sea, where ships from England anchor. It's these northern parts that are the richest in pearls and furs and wax.

'The Tartars did not take the Kremlin, or beseige it. All they wanted with Moscow was to see it burn. While the ashes were still hot they went back to the far south, to the Crimea, their horses laden with loot, their prisoners tied to the horses' tails. The question I heard my elders debating was what could stop them coming again.

'Till now the answer had always been—Tsar Ivan. I thought I could imagine him too, this great warrior. He had fought the Tartars all his life. It was his father and grandfather who drove their Tartar overlords out of central Russia. But it was Ivan who took Kazan and Astrakhan, and made the Volga from end to end a Russian river. It was because of Ivan's conquests that Sir Anthony Jenkinson was able to sail down the Volga to the Caspian Sea and Persia.

'But (by the talk I heard) Ivan was losing heart. He was not even present when the Tartars burned Moscow; he'd run away to the north. But soon after the Tartars retreated we heard that Ivan was back in the Kremlin, and active, in his own way. That is, he was torturing and executing those generals who had not fought for him. And he had hired some German mercenaries, to keep the Tartars from attacking again.

'That was reassuring, but soon a letter came from Mr. Rowley, the Company's chief agent on Rose Island, to say that Ivan had taken a great dislike to the English. He had sworn to have the head of Sir Anthony Jenkinson, if ever he set foot again on

Russian soil. Mr. Rowley did not know, when he sent this warning, that Sir Anthony was already on his way to Russia.

'I think Sir Anthony was the bravest man who ever lived in the world. Not for the reasons you know, Bess. Not because he made a rampart from his merchandise and his camels, and fought the robbers who tried to bar his way to Bokhara. Not even because he sat calmly in the Persian court, when he knew that the courtiers were urging the Shah to kill him. Sir Anthony did these things when he was young, and perhaps foolhardy. But when he sailed on that last voyage to Russia, his hair was already streaked with grey. He had a comfortable home in England; he had a wife and children whom he loved. He landed at Rose Island, and Mr. Rowley told him that Tsar Ivan had sworn to have his head. Sir Anthony knew that the Tsar was not accustomed to utter idle threats. Besides, Ivan had shown his displeasure against the English already. Our men at Rose Island and the other trading centres were alive, but that was all. They had been forbidden to trade, or collect money due to them, or even to stir from their own houses.

'Mr. Rowley thought it possible that, if Sir Anthony went on to Moscow, he might improve matters for his fellow-countrymen. It was also possible that he might make the Tsar still angrier. The choice was for Sir Anthony alone. The ship which took him to Rose Island could as quickly have brought him back again. Instead, the ship brought a letter from him to the Company, to say he was going on.

'The rest of the story we did not know that year. Sir Anthony did go on, but only for two days. When he reached Kholmogory, near the mouth of the Dvina, he was told that he could not go further. For five months every way to Moscow was guarded and barricaded, because of an outbreak of plague.

'At first the plague stopped even letters, but at last, in the middle of January, came the Tsar's decree that Sir Anthony should come to his court. He came safely to Pereslavl, and the Tsar's officers led him to a house they had made ready for him there. But they would not let him leave it. Sir Anthony knew some Englishmen were stranded in the town, but he was not allowed to see them.

'A month later, the Tsar decided to receive him. And still Sir Anthony did not know whether Ivan meant to make friends, or to watch his execution. He was led unarmed past the ranks of the palace guard, every man of whom had his sword unsheathed and upright. Sir Anthony prayed God not to let him tremble, or, at least, not to let anyone see that an Englishman was trembling. When he had made the proper obeisances, he begged the Tsar to accept some gifts from the Queen of England. Ivan examined the great candlesticks which our Queen had chosen, not for the richness of the gilding, but for its intricacy—a reminder that our workmen far surpassed theirs. Sir Anthony presented his personal gifts, a silver basin and ewer, a bunch of ostrich feathers, and a hand mirror. Ivan gazed into the glass with every appearance of satisfaction. Then he sent the courtiers away, all but his chief secretary, Shelkalov, and his favourite executioner, Malyuta. (This man's real name was Grigory Lukyanovich Skuratov-Byelsky; Malyuta, which means "baby", was the Tsar's own joking nickname for him.) Sir Anthony had to dismiss his own servants, keeping only his interpreter. He could not rely on his own sketchy knowledge of Russian. But he understood well enough when Ivan explained what had made him so angry with the English. It was that they had not sent him soldiers. He bitterly accused Sir Anthony of not passing on to the Queen his previous appeals for military aid.

'Sir Anthony assured the Tsar that he had faithfully conveyed his messages to the Queen. He reminded the Tsar that our ships brought him gunpowder, shot, muskets and skilled men for his cannon foundries. And that they returned full of Russian furs, hemp, wax and seal oil. This trade had continued for nearly twenty years, ever since our seamen found the way to the White Sea. Had their coming not been to Russia's advantage? England, which would not sell arms to any other country, had sold them freely to Russia. (Which was true; this was the only country too far away to attack us.)

'While Sir Anthony paused, and the interpreter translated, the chief secretary, Shelkalov, looked very sour. Malyuta kept his eyes fixed on the Tsar's face, which was thoughtful.

‘Suddenly the Tsar rose, advanced on Sir Anthony, threw his arms around him, and kissed him three times.

‘Every one of his requests was granted. The Company’s men were free again. Our trade began to prosper; all the more so as, that year, the Tartars’ usual summer attack had been routed. The German mercenaries had helped in this victory, and so had the cannons made under English direction.

‘Sir Anthony recounted all this on the day he appeared, safe and well, in the Russia Company’s office. It was then September, 1572. He had been away fifteen months. The great men of the Company, when they had heard his story, made him repeat it in the general office, for the clerks and apprentices to hear. Sir Anthony spoke to us quietly, without one vainglorious word, and then asked our pardon for talking so long. When he left us he said he hoped soon to sail north again; he meant to find the north-east passage round Siberia, to Cathay.

‘I longed then for one thing only—to be like him. And though some things happened afterwards, to make me change my mind, I wish to this day I had shown one tenth of his courage!’

My duty as a wife was to tell Jeremy that he had plenty of courage. His own journeys had surely been dangerous enough? But I held my peace. He was now silent, wrestling with some remorse which was dear to him, and secret; and I must not intrude. So we rode in silence through the fertile valley of Hughenden.

Jeremy began to look about him, his face growing more cheerful, as we approached High Wycombe. I had always thought it a dull place, like any other town with nothing to do between market days. The low thatched houses faced one another across a wide High Street which I knew to be cobbled, but which might have been under deep snow for all the noise we could hear, until we ourselves entered the town.

We stopped by the town pump and waited for our escort. Jeremy smiled, so happily that I asked him if he were day-dreaming. He said quickly, ‘No!’ And after a little while, ‘Bess, I’m ashamed to tell you—but you’re right. I was dreaming about what Sir Henry told me. To be in Parliament.... To represent a town like this.... To have that butcher over there come to me with some little grievance—which is no more little to him than the nail in the shoe of a poor traveller.... To go to Westminster and have the matter put right.... To be welcomed back by these people.... To make them a speech from the upper window of...what's this inn? Oh yes, a Red Lion, like the one in Wendover. Do you think that's a petty ambition, Bess?'

'I think it's very reasonable. You could do all this.'

'I could do it easily, and it is petty. Not the sort of power I had once within my grasp. In this country all ambitions must be petty; we are bridled by the laws, and praise be to God for that! I tell you, Bess, if I should come alive out of this trouble, I should like more than anything else to be a Member of Parliament.'

I was so glad to find him no longer fixed in his resolution of dying, that I felt as if I held the Queen's reprieve already in my hand. With a light heart I led the way into the Red Lion, to order dinner.

I had not understood that Jeremy would have to treat his captors. The officers sat with us in the inn's upper room, ordering the best of everything. They also asked the landlord to make sure that their six men (who sat with our two servants, at the common table below) were not stinted of good beef and beer.

Jeremy only smiled and said, 'Serve them some sweetmeats, too.' Then he asked the landlord what state the road was in, between us and Gerrard's Cross. The landlord said that it was 'diabolical'; the roads in these parts were getting worse and worse. I saw the little smile on Jeremy's lips, and knew that he was busy with his daydream. As Member of Parliament, he could see to the mending of the roads....

My mind was on the present. I said to Jeremy, 'We're still only twenty-five miles from Oxford. You could send Paul with a letter to your friend Giles Fletcher.'

He shook his head. 'Fletcher's not a young man. His health ...I don't want to call him as a witness unless I have to.'

The landlord brought the reckoning. Jeremy paid. I whispered, 'If this goes on all the way to London, you'll be fined before you're tried.'

He whispered back, 'So long as it keeps them in a good temper! I don't want to ride with my feet tied under the horse's belly.'

In fact we rode on as before, our guards leaving us free to talk. Jeremy said, 'Don't fret about the money, Bess! I have enough—and what a blessing that is! What irked me most when I was young—worse than the hard beds and the beatings—was having nothing, beyond the odd shilling my father could send me. When I used to look down Seething Lane to the Thames, and dream of sailing away, I also dreamed I would come back a rich man. How, I did not know. The Company's agents were forbidden to trade on their own account.

'But when Sir Anthony inspired us by his courage, he inspired us also by his wealth. He had not made much by acting for the Company. The secret was that he had been acting for the Queen.'

I cried out, 'But everyone says the Queen won't part with a penny!'

Jeremy shushed me, with a glance at our guards. But they were well behind. He said, 'It's true she seems reluctant, sometimes, to pay for board and lodging. Your father's not the only one to find that out. And yet the men who serve her well retire to very comfortable houses, on good estates. Besides, the present-giving in Moscow was not all one way. Sir Anthony showed us a diamond Ivan had given him—flawless, and so large that it would make a dowry for a princess.

'All I wanted, then, was the Queen's favour. But that's all any man in England wants, and some of them have powerful friends. The Company had a good friend at court, Robert Dudley, Earl of Leicester. But he was not likely to notice one poor apprentice.

'During the time when we had no news out of Russia, and it was thought our trade there might be at an end, the Company made use of its idle ships to trade with Rouen. I was allowed to go there once, to do the clerking work. Then the Protestant towns in the Low Countries shut their gates against the Spaniards, and raised the standard of the Prince of Orange. Our Queen took no part in the war, but she looked the other way when all the London companies sent ships to supply the Protestants. The Company sent me on one of these journeys—and then on another, since I had proved myself indifferent to storms and cannon fire.'

I said, 'Then why do you think you're lacking in courage?'

'More silly questions, Bess! Be quiet a minute; let me put my thoughts in order.'

2

It's not courage, Bess, to think that the cannon ball is going to hit someone else. Or that the ship won't founder. Or that, if it does, you'll be able to swim where nobody else can swim, in a raging sea. To believe these things, all you need is to be eighteen. And vain. God, who lets others be killed, would never consent to the death of such an interesting and important person. What! When I had not yet made my fortune!

Casting about for a means to make it, I saw that there were very few good interpreters. On my journeys to and fro, I tried to learn Dutch. I was making some progress, when Sir Anthony's return set us thinking of Russia once more. From the middle of September, 1572, until the following spring, we apprentices took up our former studies, grappling with the Russian way of writing.

I cannot say 'of speaking'. Where in London would you find a real Russian? An Englishman who had spent some time in the country taught us useful words, and even useful sentences, but on what principle the words fitted into the sentences he could never explain. Before I was bound apprentice I had studied at home with my tutor, a man so kind that he made learning a delight. He had grounded me well in Latin. This was encouraged by the Company. If you're shipwrecked in any place that has a schoolmaster, you may make use of Latin. But my tutor had also taught me Greek, and for this I was ridiculed. It was a language for the universities, not for practical men. Sometimes the resemblance between a Greek word and a Russian one would so delight me that I would begin to explain it aloud, interrupting the lesson. The teacher would rap my knuckles, while my companions jeered. What we had to learn was to make out bills and receipts, and never to conduct the smallest transaction without them. It was impressed on us that in any lawsuit between an Englishman and a Russian, if there were no documents, the court would settle the matter by making them draw lots. The Russians called this 'the judgment of Heaven'. The Company preferred the judgment of paper and ink.

It was important, too, that we should learn the Tsar's full titles. If ever we came into his presence, we must recite every word. So, at the start of each lesson, we would chant, 'Great Sovereign Ivan Vasilyevich, Tsar and Grand Prince of all Russia, of Vladimir, Moscow and Novgorod, Tsar of Kazan, Tsar of Astrakhan, Sovereign of Pskov . . .' and so on, while I twisted on my hard bench, trying to catch a glimpse of the river which would one day bring me to these places.

We had also to learn Danish and German, because it is difficult to reach Russia without speaking one or the other. Here we had better teachers. There are plenty of Danes and Germans in London, and the Company found educated men, who knew the Latin grammar and could explain their own.

My studies were interrupted. In the first months of 1573, we had word that a Dutch town, Haarlem, was hard pressed by the Spaniards. As I knew some Dutch already, I was chosen to go with a supply ship. We slipped past the Spanish patrols in the fog, and sailed by the Haarlem Lake to the beleaguered city. Out of that fog things came in unfamiliar shapes. The soldiers who stood with muskets at the ready, while we unloaded, were women. The great stones used to mend the breaches in the city wall also looked like women. They were statues of saints, torn out of the cathedral by these fierce Protestants. When the fog lifted, one of the girl soldiers took me up the cathedral tower, to show me the beseiging army. As we stood there the tower swayed under our feet; the Spaniards were bombarding it with cannon. Yet this did not frighten me so much as the sight of the Spanish army. Thousands of men were moving to their tasks; from a distance thousands more were marching to reinforce them. Every man who stood still was utterly still; every man who was moving moved in perfect harmony. I was glad to sail out of Haarlem with my life.

My journey to Russia began in May 1573. My father and mother came to see me off. They were like strangers to me; since I was fourteen I had seen them only once a year. But my mother was in tears at this final parting. Like you, she had been reading Sir Anthony Jenkinson's book of his travels. She was afraid that I might fall among the cannibals, the Samoyedy of Siberia. Was it true, she asked one of the sailors, that these heathens would cook one of their own children to feed a traveller? Or was it not more likely that they would cook the traveller to feed their children? The sailor told her that the Samoyedy were far east of where we were going.

'You wouldn't fret so, madam, if you could see Rose Island. It's a perfect little England. Your boy will find everything as pleasant as can be. Why, he'll scarcely need to see a foreigner.'

Our six tall ships went eastwards, and then northwards, to the whales. We found anchorage in a Norweyan inlet, and went ashore in small boats to exchange our grain for fresh fish. The people were glad to see us; until the first English traders came they had lived on fish from one year's end to another, because corn will not grow in that country. Grass was growing there, though; and a white flower with heart-shaped leaves, which the sailors called 'scurvy grass'. They picked it and ate it, advising us apprentices to do the same. They said it had such virtue that not only would it prevent the scurvy; it would make a man already ill begin to mend within the hour.

We sailed on, past the wild Lapps living in tents, with garments of reindeer skin wrapped so close about them (though it was the summer) that we could see nothing of them but their eyes. Our sailors muttered prayers whenever they caught sight of a Lapp. It was, they said, a 'Well-known fact' that these barbarians used witchcraft to call up storms.

Beyond the cape which is the very northern point of Europe, a storm caught us. Our ships were scattered. But, in that savage place, the ship I was on found shelter at a harbour built of stone, guarded with a strong fortress, and garrisoned by soldiers dressed much like ourselves. This was Vardö—the Wardhouse, our sailors called it—an island belonging to the King of Denmark.

Though they could not refuse us a refuge from the storm, the Danes were unfriendly, and asked for our harbour dues in advance. The Governor came aboard, bowed coldly to our captain, and made a speech. The wind suddenly dropped, and I heard every word.

The Governor hoped, he said, that we would see in this thunderstorm the anger of God. No doubt we were carrying gunpowder for Ivan, the scourge of all his neighbours. Every country in Europe had refused to supply him with arms. Even the free German cities, which had profited most from the Russian trade, now enforced the ban. If our Queen were a real Protestant she would send gunpowder only to her fellow-Protestants in Denmark.

Our captain looked round at us, and asked, 'Does anybody know what this fellow's talking about?'

I waited for someone older than myself to translate. There were sailors and agents of the Company who had passed that way before. They could haggle over harbour dues, and say, 'Good morning' in Danish—our captain had said that much. But now they only stared. I looked at my six fellow-apprentices. We had all studied the language together. I could not believe their minds were as blank as their faces.

The captain asked again, impatiently, for an interpreter. So I came forward, and told him what the Governor had said. He frowned.

'Tell him we're not carrying any gunpowder.'

'What,' I whispered 'if he orders his men to search the ship?'

'Then tell him we'll use the gunpowder to blow his men up. Yes, and tell him not to insult our Queen.'

This last part I took first. I told the Governor that our Queen was a true Protestant. She had not prevented our Company from sending supplies to Haarlem. I had myself travelled with these supplies, and had seen the gratitude of Protestants for English help.

This, though I delivered it with some hesitation, was clear enough to the Danes. The Governor began eagerly asking me what I had seen of the fighting in the Low Countries. Would Haarlem hold out? His officers gathered round me. They were all starved for news; their own ships came rarely, and their only neighbours were the wild Lapps.

My fellow-apprentices watched me open-mouthed. For the first time I understood that, in their minds, these foreign words had been only racks and thumbscrews, devised on purpose to torment them. They had never thought of using them as human speech.

In exchange for my news the Danes told me some of their own. The reason they were bitter against the Russians was that, without any declaration of war, the Russians had raided Oesel, a Danish island in the Baltic. This was eighteen months ago, and there had been no other open fighting, but they saw preparations for it everywhere. Russian monks 'worse than Papists' were infesting these northern parts, pretending they had come to preach to the Lapps. The monks' leader, one Feodorite, was a known emissary of the Tsar. The Governor had constantly to send out patrols to the frontier. Hence he was very short of men, and nobody had been on leave for more than a year.

Having told me all this, the Governor grew more friendly. One of the officers remarked: 'The English will never help us; they're too greedy for profit.'

The Governor checked him. 'We have our own greedy men. Prince Magnus is taking the Tsar's money.'

At the name 'Magnus' they all spat.

I do not say it was my skill in their language which prevented the Danes from searching our ship. It delayed them a little, and meanwhile the five other ships of our storm-parted company came sailing into Vardö. This meant that the Danes were outnumbered. They could do nothing but allow us to repair our ships, and sail away unsearched. The Governor, at parting, gave me a gold chain. That was the first money of my own I ever made. I think even your dad, Bess, would admit it was honestly earned.

Now that I knew how to value my own skill in languages, I looked forward to hearing Russians talk. Would I recognise the words I had learned? How quickly could I learn more?

I looked eagerly at the first Russian place we passed. It was a sprawl of low buildings round a church, built solidly enough, but all of wood. 'Pechenga,' someone said. We could see monks hoeing, monks digging, monks hauling in fishing nets.... Was every single soul in these parts a monk? They must all be spies, as the Danes had said; why else would they live in such a lonely place? We did not anchor there, but sailed on, day after day, past an unpeopled country, through seas as wild, until we came to the calm waters of the White Sea.

When we anchored at Rose Island, it was our own tongue that welcomed us ashore. This was indeed a little England. The island was in circuit some seven or eight miles, and at our first approach it seemed quite wild, a mass of rosemary and roses. But near the neat stone harbour wide acres had been cleared to make pasture for sheep, and even for a few cows. There were meadows almost ready for mowing, and fields of rye, and vegetable gardens of cabbages and beans. Pigs were being fattened, and hens went scratching about, as in an English farmyard. True, our warehouses and our living quarters were built in the Russian fashion. The walls and roofs were logs laid side by side; the cracks between the logs were stuffed with tow. The windows were small, and made of mica, not glass. Each room had a stove built of stone, which took up nearly a quarter of the space. Our beds, which we had to reach by ladder, were on a sort of shelf above the stove. But in our dining hall there was an English Bible and a *Foxe's Book of Martyrs,* besides English histories, poems, and plays. A cheer went up from our fellow-countrymen, whose exile we were now to share, when their servants brought ashore a crate full of new books. Mr.

Rowley, the Company's chief agent there, grumbled that there were too many romances translated from the Italian, and not enough sermons. The man who was to instruct us newcomers, Mr. Taylor, had a different complaint. There were no jest books. He said that in the long winter he grew tired of hearing the same jokes over and over.

Mr. Rowley, by his look, was not a man for jokes. When he said grace for supper, he thanked God for our safe coming. He added, 'We pray Thee, oh Lord, so to prosper our affairs as may be, first, for Thy glory, secondly, for the honour of England, and, next, for the Company's profit.'

This first meal ashore was a feast, in its way. There was a noble choice of wildfowl and game, and abundance of excellent fish, but no beef. The cattle were too precious to be killed. We had a little mutton, but it did not compare with English mutton. There was not much butter, and the bread was made of rye. The cranberries (the only fruit of those parts) were delicious as we ate them, cooked with honey. They made me thirsty, though. Soon after supper I went to the pump in the courtyard. As I was about to drink, Mr. Taylor, that lover of jokes, handed me a cup. 'This'll quench your thirst,' he said. It looked like water. I took a great gulp, and began to choke. My throat was on fire; tears came pouring down my cheeks. The other men gathered round me, laughing, and saying, 'What, never tasted vodka? You'll come to like it.' (They were wrong; I never did.) As I clung to the pump, still gasping and weeping, someone began patting my back. A woman's voice was saying in broken English, 'Tears—not! Tears—oh no!' Then the woman gave me a cup of real water, murmuring, 'At least you should not let them see you cry.' These last words were in Danish.

I turned and looked at her. She was a tall, gaunt creature, wearing a greasy apron—a kitchen maid. I asked her in Danish, 'How do you come to be here?'

Her smile, on hearing her own language, made me see what her face must have looked like years before. I showed her my gold piece, with her king's picture on it, and she burst into tears.

I felt a tap on my shoulder. Someone said, 'Mr. Rowley would like a word with you.'

I hastily washed the tears off my face, and ran to Mr. Rowley's office. The other five apprentices were already standing before him.

'Now that we are assembled,' said Mr. Rowley, looking coldly at me, 'there are certain things I have to tell you. It has cost the Company a great deal of money to send you here. I have to see that none of this is wasted. There is to be no idleness, no gaming, no dicing, no falling into bad habits of any kind. When a young man is going to the bad, the first sign is usually his dress. You all have your new Company clothes you were given in London? Two good coats of Suffolk cloth, one doublet, three sets of underlinen, three good hard-wearing shirts, three pairs of hose, a workday coat and a workday cap, a cap for Sundays, and a pair of knitted gloves. And all alike, in this plain russet colour—that's what I like to see. Your winter coats, fur hats, fur gloves and so on will be provided here. And next summer, when the year's first shipment of cloth comes from England, our tailor will make you new outfits. You see, you are not short of any necessities. What I have to warn you against—' I thought, by the look on his face, he was going to say *sodomy* '—is embroidery. You'll see a good deal of it, when you go to market. If you must buy the stuff, don't let me see it! Send it home to your mothers and sisters. If I catch any English lad in a coat encrusted with gold and silver, I rip it straight off his back. And give him something to remember me by! Pride in dress is a most abominable thing, in a young lad still apprenticed. The only pride we should show is a proper pride in being English. Which means that we know how to handle a boat, how to swim, how to trade honestly, stay awake in the afternoon, and keep sober.' Here he looked at me again. 'We drink ale or mead here, not vodka. That word means "little water"—a Russian kind of joke. It's no joke to me. I've seen a man who took to it kill himself in a month. Another thing I won't endure is cruelty. If I find one of you lads being cruel, I'll beat him into pulp. Our servants here are mostly Danes and Germans. The Danes look after our farm; they have a talent that way. The Germans are good at carpentry, blacksmithing and so forth. When you go to our rope factory at Kholmogory, you'll see that we have some very good German workmen. By Russian law these people are slaves, because they were captured in war. That doesn't make them cattle, and certanly not *your* cattle. In particular the women—' once more a glance at me '—are honest Protestants, and not to be corrupted. So don't let me catch you offering them money, shaming them, making them cry! Now, as to the Russians. When you go into Kholmogory, and see how they live, and what sort of religion they follow, you may think ... what you like. Think it, but don't say it, even in English. They've learned some of our words—the worst ones—and they take offence. One of our lads was murdered by the peasants, a few years ago. When we asked for justice, we were told he had been making game of their religion, and they always punish blasphemy by death. Well, this is their country!'

Mr. Rowley's clerk came in to say that the abbot of St. Nicholas was in our dining hall.

'Tell the servants to bring cakes and wine,' said Mr. Rowley. He turned to us. 'You lads had better come with me. You'll be meeting these monks of St. Nicholas often; they're on the mainland, less than a mile from here. Even a Russian boat can manage the journey. As soon as you see the abbot, bow from the waist. When he raises his hand to bless you, kneel down at once. Your hose will stand it; they were made in England. And remember I shall beat the stuffing out of anyone who laughs!'

I did not feel inclined to laugh, but rather to gape, when I saw the Russian abbot. His white beard cascaded over his golden robes; his long white hair flowed from under a hat so thickly jewelled that I could not see what material was beneath. Behind him were his attendants, each one dressed in richly embroidered robes, and each one different. Because their garments were heavy with pearls, these men stood rigid, showing not a sign that they had human limbs. The door was open to let in the endless brightness of the summer evening; as it struck their gold and silver it seemed the brightness of Paradise. We knelt to these men as to beings from another world. And in the language of another world, in deep mournful vowels and consonants which carried the swish of angels' wings, the abbot pronounced his blessing. I could not understand one word.

Speaking the sort of Russian I had learned in London, Mr. Rowley begged the abbot and his followers to take some refreshment. I understood that, and the abbot's hypocritical refusals. The stiff robes could bend well enough, when the men inside them were eager to sit down and sample the French wine our ships had brought. Mr. Rowley waved us apprentices out of the room, but not before I had seen the slavering of the abbot's old lips, the greedy smiles that broke across those other holy faces.

The next Russians I saw were the Governor of Kholmogory and his attendants, who came to us four days later. In the Tsar's name they bought the guns and gunpowder which had come with us on the ships. Then, still in the Tsar's name, they picked out our best bales of cloth. All the rest we might sell wholesale to Russian traders, but must not sell retail in the market place. What I heard of this I understood; these were the words we had been taught. Amid the grim, suspicious Russian faces Mr. Rowley showed a dignity which I was coming to admire. When everything was happily settled, he presented a gold goblet, finely worked, to the Governor. As he took it, the Russian seemed to become another man, his smile was so sudden and so sweet. I said as much to Mr. Taylor.

'Sweet!' said Mr. Taylor. 'That bird of prey! Yes, I suppose his life is sweet to him. All the more, as he knows it's likely to be short. The last governor they had here was boiled.'

I laughed, but Mr. Taylor meant it. 'It's one of old Ivan's favourite methods,' he explained. 'They have two great cauldrons of water side by side; one with a fire under it, the other freezing cold. The executioners dip the man into each one by turns till he dies. Which takes quite a time, I believe.'

I asked what the man had done to earn such a punishment.

'Oh, he was governor while Sir Anthony Jenkinson was here. Six English vessels full of grain came when Sir Anthony did, and the old governor wouldn't let us sell them. He asked us how we dared insult his country; there was no shortage of grain in

Russia. This was when we knew that the harvest had failed. We could see with our own eyes that the peasants were eating the barks of trees. When Sir Anthony told the Tsar about it, we were given leave to sell the grain—a year too late. A lot of the peasants had starved to death in the meantime.'

'Then it's understandable,' I said, 'that the Tsar should put that governor to death.'

'Oh yes! It would be. If he hadn't been carrying out the Tsar's orders.'

We apprentices went with Mr. Taylor to sell the English cloth. For two days the flat-bottomed barges travelled up the River Dvina to Kholmogory. It was eighty miles, and for most of those miles we did not see one human being. There was a small monastery, called the Archangel. Round it the forest had been cleared, and sheep were grazing. After another long stretch of scrub and forest, we saw some pretty wooden houses, painted in bright colours, and a wooden church. But then once more everything was wild, until we neared Kholmogory, and found naked men searching the river for pearls. They greeted us, and shouted to each other, incomprehensibly.

At the quayside we apprentices were supposed to think only of checking the bales of cloth as they were brought ashore. But how could we keep from looking at the people who came to look at us? The women took my breath away. I could not see what shape they were; their dresses hung too loose; but everything they wore seemed to blaze with white and scarlet. And their lips were painted scarlet, their faces powdered white, so thickly that they looked as if they had been beaten about the face with a bag of flour. Strangest of all, these brazen creatures rode their horses astride, like men.

Mr. Taylor nudged me and muttered, 'Don't stare at them! They're not whores. A Russian may beat his wife black and blue if he likes, but he must let her paint her face.'

I knew by now that pearls were common in these parts, but still I thought it strange to see a peasant girl, walking to market on bare and filthy feet, carrying on her head a very halo of pearls. Some of these headdreses were simple curves, and some like the rays of the sun, or the petals of a flower.

All the women wore (besides the pearls) fine embroidery. The men's kaftans too were richly embroidered, though the hems might be ragged and splashed with mud.

At the English rope factory everyone was plainly dressed and working, Germans on the rope-walk, Englishmen in the office. Here I changed my piece of Danish money for roubles, and set out to look at the market.

The market place was deep in dust, and there were no cobblestones underneath. I thought at first there was only bottomless dust, but here and there I could see that the place was paved—if that's the word—with logs laid side by side. Someone had swept the horse-dung into heaps, and on these heaps, or near them, people threw their old rubbish. Considering that the heaps must be worth something as manure, it surprised me that the peasants were so slow to cart them away.

Dirty as the ground was, all the people suddenly cast themselves down on it, not kneeling merely, but beating their foreheads against the earth. The Governor of Kholmogory was riding through. The people seemed ready to lick up the very dust at his horses' feet. Being English, I bowed only from the waist, and he saluted me graciously. When he and his followers had gone past, the marketing could begin again.

In the midst of the crowd, in the midst of the stench, a man stood playing a stringed instrument, and chanting. The peasants listened in rapture, and so did I, though I understood nothing yet. I thought the song had rivers in it, and forests, and the cry of men lost looking for Cathay. *'Lyooblyoo'* I heard—and is there in the world a better way to say, 'I love'? For the sake of this one word I knew, I gave the singer a coin.

There was not much else to spend my money on, except the things Mr. Rowley had warned us against. In a stinking hut the men were swilling vodka, and on trestle tables the women were showing embroidery. These market women chanted a sort of slow blank verse; their customers made brief retorts, which often seemed to rhyme. Since we apprentices were all dressed alike, the people knew us as Englishmen. 'Eeng-leesh, Eeng-leesh!' one woman cried to me. Her stall was covered with white linen towels, each with a border of strange birds and beasts and flowers, in drawn-thread work. Proudly, she held a towel up, and said one Russian sentence I understood. 'I made this one myself.'

There, all in drawn thread, were our English ships, just as they sailed the White Sea. And a row of little dancing men—ourselves, in our English clothes. I looked at this thing, which any woman, anywhere in the world, would have been proud to make. Then I looked at its maker, at her garish headdress, at her mask of paint and powder, and at the black eye which the powder could not hide. Picking out words I remembered, I asked her if she could read and write. She laughed at the very thought, and her laugh showed that somebody had knocked out her front teeth.

I bought the towel for my mother. It would make her think that the Russians were civilised.

In my stroll through the market, I had lost Mr. Taylor. He reappeared suddenly at my elbow and began to hustle me along. 'This way, Jeremy. No time to lose. . . . What's the German for "stonemason"?'

'Steinmetz,' I said. This was a lucky chance. Our German teacher in London had impressed on us the words for the trades.

'There!' said Mr. Taylor. 'The other lads told me you'd be sure to know. Mr. Rowley particularly asked me to get him one, and I never thought to ask him the word. He wants a carpenter, too. I know that's a simmer-something.'

'Zimmermann.'

'Ssh! Don't say it loud enough for the Germans to hear, or they'll tell us that's what they all are. And take a good look at their hands. We don't want to be landed with some gentleman who's never worked in his life.'

We had now reached the most crowded part of the market place. The peasants were gaping at some fifty men and women, whose hands were bound behind their backs. Two Russians with whips and muskets guarded them, while a third Russian chanted the praises of these, his goods. The slaves looked with contempt at their masters. Most of us Englishmen stood head and shoulders above the Russians, but these Germans were taller yet. The men had been stripped of all but their underlinen; the women were in ragged shifts, not reaching much below the knee. It was all the more clear how beautiful they were. What fine high breasts the women had, what broad shoulders the men! Disaster had ravaged their faces; but these faces were beautiful to the bone.

The Russian peasants, who had licked the dust before their Governor, now shouted insults at the slaves they meant to buy. I heard the second whole sentence I understood. (We had been taught it with a solemn injunction not to use it.) 'Oh, go and rape your mother!'

Mr. Taylor pushed the Russians aside, and began to question the slaves. In a mixture of languages, he established that one was a stonemason and another a carpenter. He made the slave-dealer untie these two pairs of hands, which he examined carefully. Then he felt the muscles, and held the mouths open to look at the teeth. Through all this, and the bargaining that followed, the Germans gave no sign of any feeling, except a calm contempt. But when the sale was concluded, and Mr. Taylor made ready to lead the two men away, there was an outcry from the others. The women went down on their knees. They were begging us to buy them, to set them to work in our rope factory, to let them scrub our floors—anything rather than leave them among savages.

'What are you snivelling for?' Mr. Taylor said to me. 'This happens every market day. We can't buy them all. It's not our fault that the Tsar captured so many in his wars.'

The slaves thought it was our fault. The two Germans told me so, as we went back to Rose Island on the barge. They said they came from Livonia, that part about the eastern Baltic, where the Germans rule the ports and the market towns. These towns were strongly fortified; they had for centuries resisted all Russian attacks. Two years before this the Russians had come in greater strength than ever, and with them were their new allies, the Tartars.

I cried out: 'But the Tartars are the Russians' chief enemies!' I appealed to Mr. Taylor, who was dozing, having lost track of all this German talk. He said with a yawn, 'Oh, Tsar Ivan's got some Tartars to fight on his side. Shows you what a clever fellow he is.'

The Germans declared that they could have defeated Russians, Tartars and all, if only these barbarians had not been armed with English weapons.

'Even with *your* cannon,' the carpenter told me, 'they could not take Reval. They had to retreat from Reval. But as they retreated they worked off their spite on the rest of us.'

'They used *your* gunpowder to blow up our houses,' added the stonemason. 'They stuck *your* muskets in our backs, and tied us to their horses' tails. Then they whipped up the horses, and made us trot behind. People fell, and we could not reach out a hand to help them, because our hands were bound. We had to see them dragged along the ground until they died.'

'We are so much taller than those Russians!' exclaimed the carpenter. 'And stronger! We could have cut them into shreds, if it had not been for *your* weapons. Didn't you see that the slave-drivers in the market were carrying English muskets?'

I thought it strange that Sir Anthony Jenkinson had never spoken of this. He had spent five months in Kholmogory; he must have seen the slave market. He must have known how many of the Company's workmen were slaves. He could not help but know how the Tsar had used English weapons. Was there nothing in this to offend that pattern of manly virtue?

And Mr. Rowley. . . . A religious man, who never let a Sunday pass without two hours of prayers and Bible readings (which he conducted himself, like a captain at sea, since we had no clergy among us). How did Mr. Rowley like this trade?

I did not dare to approach him. But a few days later he called me into his office to ask me about a missing bill of lading. I found it at once, which put him into a good mood. So I ventured to ask if he did not think it wrong for the Company to sell arms to Tsar Ivan.

'You fret about those things, do you?' he said. 'That's unusual here. . . . You see, Jeremy, it's not simple. The Tsar needs our guns, to defeat the Crimean Tartars. And that's to our advantage. We can't trade with Tartars. They don't want to buy anything. Their greatest leaders wear patches and rags, to show that they despise the comforts of life. Whereas Russians . . . . well, yes, we sell arms to the Tsar. I wonder the Germans dare to complain of that. If anyone taught old Ivan the art of war, it was the Germans. Why, he learned their language for that very reason. He never could have conquered Kazan and Astrakhan without them.'

I said I had heard in London that Ivan used German mercenaries against the Tartars. What was new to me was that he used Tartar mercenaries against the Germans.

'You mean he uses Cherkasses,' Mr. Rowley said. 'They are Tartars of a sort. The Tsar married one of their princesses, to get the help of her brothers. And then he was too clever to use them against his main enemies, the Crimean Tartars. They're too much alike; they might have made common cause. No, he had to use the Cherkass Tartars against the West. You can't call them a mercenary army; like all Tartars, the Cherkasses won't raise a finger for money. What they fight for is the satisfaction of seeing cities burn. And, of course, the prisoners.... All right, they do make them slaves. But the people we buy—the fact is, I never think of them as slaves. They lie as warm as we do in the winter, wear the same furs.... Why, Jeremy, there's a great many servants in England worse fed and worse treated than ours are. We never even beat them.'

'No, sir. There's no need, is there? All we have to do, if they don't work, is threaten to sell them back to the Russians.'

'Well? Is it my fault we're in Russia? What can I do?'

'Sir, couldn't our slaves be bound to serve the Company for a certain term of years, as we are, and then go home?'

'Home? The Baltic shore's almost a thousand miles from here, and it's always being fought over. They'd only be captured again.'

I persisted. 'The Danes at least, sir, could travel on one of our ships to the Danish fortress at Vardö.'

'Oh, there has been some talk of that,' Mr. Rowley said. 'The trouble is, we're not on good terms with Denmark. Their ships attack ours, and then their King makes out that these weren't his men, only some wandering pirates. Our Queen tells him—if they're pirates, punish them! That's put the Danes in a temper. Now they're asking that these people we've bought and paid for should be handed over to them free of charge. Where's the profit to the Company in that?'

The following week, when Mr. Taylor went again to Kholmogory, he asked Mr. Rowley if I might go with him.

'Why Jeremy?' said Mr. Rowley sourly. 'Why not one of the other lads?'

'Jeremy's my German interpreter,' Mr. Taylor explained.

'Well, don't you teach him any of your evil ways! None of those painted Russian whores! And see if you can find me another stonemason. This fellow you bought me is just what I wanted—showing our own lazy lot how to work. One more like him, and we'll have that stone wall finished before the winter. Just as well. The Tsar seems friendly to us now, but . . . we all know what happened at Novgorod.'

On the barge, going up the Dvina, I asked Mr. Taylor, 'What happened at Novgorod?'

'Oh, didn't Sir Anthony Jenkinson tell you? Perhaps he didn't know. It was kept secret a long time. You'd think that impossible, but the fact is, while he was dealing with Novgorod, the Tsar made his Tartars kill all the travellers they found on the roads, so that the news wouldn't spread. People *did* escape, of course, but what they said wasn't believed. I wouldn't believe it myself. Novgorod, the richest place in Russia, the only place in Russia with traders who knew how to trade—finished, gone!'

'You mean the Tsar wiped out one of his own cities?'

'When you say wiped out. . . . The town's there still. There are people. But nobody to trade with. The Company lost a lot of money over that. Because, a couple of years before it happened, two of our men spent the whole summer finding out short cuts from Rose Island to Novgorod. Then, in January 1570, the Tsar turned up at Novgorod with his Tartars. It wasn't only that he killed people. It was the way he picked them out. The first he went for were the priests, the nobles, the merchants —all the people with a bit of education. His Tartars burned the warehouses; there was melted wax and tallow flowing down the gutters, mixed with blood. They say in six weeks there were sixty thousand dead. I don't know who could have counted.'

We were sailing past some peasant women. They were in the woods, gathering cranberries. As it was a working day, they did not wear their paint or pearls. They looked up to see us pass, and their faces, framed in simple kerchiefs, were healthy and full of goodwill. Some of them, with shy smiles, waved to us. I asked Mr. Taylor, 'Do people like these know what happened at Novgorod?'

'Oh, they know everything. The ballad singers tell them. That's why the Russian army wouldn't fight. It was after Novgorod that the Crimean Tartars invaded. The soldiers let the Tartars go past and burn Moscow.'

'Then do these people hate their own Tsar, worse than they hate the Tartars?'

'Sometimes. They have moods.... They won't say a lot in front of us English. Ask them what they think of old Ivan, and they'll tell you it's to be expected he should lose his temper, when he found his cousin was a traitor.'

'His cousin?'

'Don't you know about that either? That was the reason for Novgorod—or the excuse. The Novgorod people were supposed to have been plotting with Ivan's cousin, Prince Vladimir. And this Vladimir, so they say, had been bribing the Tsar's cook to poison him. So the Tsar sent for Vladimir and told him to drink poison. You might say that was justice, as justice goes in Russia. But why did Vladimir's wife and sons have to drink the poison too?'

I asked, 'How old were the sons?'

'Ten... twelve... thereabouts. Vladimir had a daughter as well—Maria. The Tsar put her in a convent. He's just fetched her out, from what I hear, to marry his friend Magnus. You surely must have heard of Magnus?'

'Is that the one the Danes can't mention without spitting?'

'That's the lad! A few years ago he was their hero. He's the brother of the King of Denmark. He governed the Isle of Oesel, in the Baltic, and he used to help the Baltic Germans in their wars against Ivan. Did quite well, too. Then Ivan bought him.

Promised he should be a king, like his brother—"King Magnus of Livonia". Magnus was on his way to meet Ivan, when he heard what had happened to Novgorod. He nearly turned back. But then, they say, he thought how much he wanted to be a king. So he went on. More than I should have done! I don't care how good the trade is in this country; when my seven years are up you won't see me for dust. Just so long as Ivan doesn't kill us all first!'

'You think he wants to?'

'Does he know what he wants? Did he know what he was going to do to Novgorod? We'll be on the safe side; we'll fortify Rose Island. Not so that it looks fortified; only with strong stone walls, and a stone granary for our food, and a few guns hidden away. He can't attack us in the summer; he hasn't any ships. He'll wait until the sea is frozen hard enough to hold his troops. We can beat them off. We have gunpowder enough to break the ice and drown the Russians. Then in spring, when the White Sea melts, we can leave by boat. Well! We all hope it won't come to that.'

'But why should it? Surely the Tsar needs our trade more than we need his?'

'That's true. And he's such an unpredictable creature that sometimes he does act reasonably.'

I said, 'If we do have to defend ourselves on Rose Island, what happens to our people left behind in Kholmogory?'

'There's a sledge kept ready, and a team of good horses. You can travel fast, in the winter. Our four Englishmen in Kholmogory could reach Rose Island in five hours.'

'But the others?' I asked.

Mr. Taylor said, 'What—the other Englishmen, in Kotlas and Vologda? They keep sledges ready too. Trickier for them, of course.'

'I meant the other people working at Kholmogory.'

'What others? There's only Germans.'

When we reached the rope factory, our compatriots told us that German slaves had grown scarce. The rumour was that some Russian merchants called Stroganov were buying them all. We did, however, find another stonemason—a man outstanding even among Germans for his great height, broad shoulders and glowering silence.

We had so much other marketing to do for Mr. Rowley that we could not start back on the barge that night. We took our new slave to spend the night among the other Germans in the rope factory. 'No need to tie him up,' said Mr. Taylor. 'He's only too delighted to be here, I daresay.'

Indeed, the slaves were not kept prisoner, except as we all were—inside the high spiked wall which guarded the English factory, living quarters and courtyard. The gates were locked; to keep them so there was only the old porter and his wife. The Germans could have broken out. But Mr. Taylor was right; they much preferred us to the wastes of Russia.

Next to the loft where the slaves had to sleep on their pallets was a room with a proper bed in it. This was given to Mr. Taylor and me. I began to undress; he did not.

'The fact is,' he said, 'I've got a . . . another place to go to. All very well for Mr. Rowley to talk. He wasn't born, that man; he was chipped out of ice. . . . Anyway! She's a good girl, though she is a Russian. You won't say anything, will you, Jeremy, when we go back to Rose Island? I know you never said it was me gave you the vodka. I'll do as much for you, perhaps, one day.'

It was afer sundown, but not yet dark. From the window I watched Mr. Taylor cross the courtyard and give a coin to the gatekeeper, who let him out. The night was hot. I was burning. I thought: two years before I finish my apprenticeship. Two years before I earn wages and can keep a woman of my own.

One of the German kitchen maids went to the pump in the courtyard, and filled two buckets. In that milky light I could see the gleam of her hair. I whistled. She looked up at me, then down again, without a sign of interest. But when the buckets were full, and hooked on to the yoke, she did not make for the kitchen, but for the doorway nearest my room. I ran down and met her there. Then I did not know what to say, except, *'Guten Abend.'*

She set the buckets down carefully, so as not to spill any water, and stood rubbing her shoulders where the yoke had marked them. I kissed her. It was not like kissing an English servant girl; there was no wriggling, no twisting away, no threats to tell the master. She only stood, submitting. Her face was nothing much, but her throat was white; I kissed it again and again. Her body felt plentiful and warm.

I began to edge her upstairs, and still there was no resistance. Indeed, she led the way. I followed, scarcely believing my own good luck. She lay quietly on the bed and, when I proved clumsy, showed me what to do.

Through the partition wall I could hear the Germans talking in the loft. I was past caring whether they could hear my novitiate groans and grunts.

The girl did not move with me, nor yet against me. She lay limp. I did not know, in those days, that women ever did otherwise. When I had finished with her she rolled away, shook out her skirts, and left me without a word. I lay for a minute or two in a daze at my unexpected success.

Then I heard, from the courtyard, a sharp slap and a cry. I got up and looked out.

The girl, still carrying the buckets on the yoke, was holding one hand to her face. Confronting her was the new slave, the one we had bought that day. He said, 'Call yourself a German woman? You're a whore!'

She said in a dull, dead voice, 'What can I do? He's one of the masters!'

The shame of that moment made me, for the next few weeks, a very good boy. In the evenings I would borrow a German or Danish Bible from Mr. Rowley (who had studied both languages) and sit comparing some passage with an English Bible. Mr. Rowley had no such thing as a book in Russian. He told me I would have to learn the language as everyone did, by listening.

After supper, when I was absorbed in my books, the other apprentices would grumble at me. I was a wretched companion; why wouldn't I throw dice? I told them that those tedious cubes could do nothing but fall on a one or a six, or some number in between; and when they had done that they could only do it again. Whereas a language has thousands of words, each one different; and there are thousands of languages in the world. They said that was the worst news I could have told them.

One evening, when it was already cold, I was sitting by the stove, reading a Danish Bible. Karen, the Danish women who had spoken to me first, came in to make up the fire, and stayed to look at the Bible. I read to her in her own language, 'Thou shalt not be afraid for the terror by night; nor for the arrow that flieth by day; nor for the pestilence that walketh in darkness, nor for the destruction that wasteth at noonday.'

Karen took the Bible from me, and read aloud, 'A thousand shall fall at thy side, and ten thousand at thy right hand; but it shall not come nigh thee. Only with thine eyes shalt thou behold and see the reward of the wicked.'

On the high bed-shelf above the stove, my companions were, as usual, gambling. One of them dropped his dice on to the floor. He called, 'Hey! Pick that up!'

Karen, who was holding the great Bible as carefully as a baby, did not at once move. He whistled at her, which reminded me how I had whistled at the German girl. Karen made to pick up the dice, but I held her back. 'Come down and get your own lousy dice!' I said.

He asked, 'What for? She too lazy?'

My own shame welled up in me; I was ready to fight him. I shouted, 'Karen's not your dog to whistle at. She might be your mother, or mine. Don't you remember what Mr. Rowley said?'

All the lads began to jeer at me. 'Yes, go and carry tales to Mr. Rowley! We know you're his blue-eyed boy.'

I was thunderstruck, not because they did not like me, but because they thought Mr. Rowley did. What sign had he ever given? I did not stop to argue the point, because Karen was weeping, and I hurried her out of the room.

When we came into the kitchen another Danish woman, the under-cook, Anna, looked up from the fire. Without saying a word, she rose and put her arms round Karen.

'No, no, it's nothing,' Karen said in Danish. 'He's a good lad, but what can he do?'

I said, 'When I'm rich, I'll ransom you and send you home.'

'Home?' said Anna. 'To Oesel, to look at the ashes?'

Karen patted my hand. Then she said, 'You mustn't let your hands get so chapped! What will you do when the real winter comes? You'll have chilblains, and bleeding from the knuckles, and a world of trouble. Let me rub your hands with bear's grease! Anna, have you got a little to spare?'

Anna found some, and we all three sat by the kitchen fire, Karen gently rubbing the grease into my hands. She said, 'I used to do this for my sons. We had cold winters on Oesel.'

Anna said, 'They *seemed* cold to us then.'

'Yes, we grumbled at so many things! When I had to reap the corn in a baking sun, not stopping to straighten my back for a minute even, because the harvest was late.... Or when I went out in the frozen woods to look for kindling.... Or when I tried to fit a patch on a coat that was all patches already .... I used to say, "Why wasn't I born a lady?" Oh, if I had known then how every common action would come back to haunt me! It seems beyond belief, now, that I was ever able to work beside my own husband in our own fields. Or that I looked for kindling in the woods with Anna, and all our other good neighbours.' She looked down at my hands. 'Is it only two years ago I was doing this for my own sons?'

Anna said, 'Magnus will answer for it.'

Karen gave me back my hands. 'Do you feel better now? You're a good boy, only you made me cry, taking my part. It's easier when people shout at me, or say nothing but what work I'm to do next. Then I'm numb. I was numb through all they did to me. I felt nothing.'

'Magnus will feel it,' the other woman said. 'I have cursed him before God, who hears the cry of the widow.'

'What goes round and round in my mind,' Karen said, 'is that there was no sense to it. Even if Magnus wanted to be King of Livonia, why should he trouble us on Oesel? The Isle of Oesel belonged to him already; it was a present from his brother, our King. He thought it wasn't good enough; he left it to go to Moscow; but he could have come back peacefully if he liked. Why did he send those Tartars across the frozen sea? What pleasure was it for him, to see our farms in flames?'

Anna said, 'I have cursed Magnus. He'll die like a beast. His wife will end her days in captivity. If they have children, their children will be murdered. As ours were.'

'My boys tried to fight,' said Karen. 'My Dirk—not fourteen—snatched up the musket, when he saw his father dead. My little Jan stuck a knife in the man who killed him.'

In a dull voice, just like the German girl speaking of me, Anna said, 'My children were all burned to death in the house, while the Tartars pinned me to the ground.'

'I have never questioned the justice of God,' Karen told me. 'Only I wonder sometimes why he did not let me die. He let me be numb a long, long time. I felt nothing when they raped me, and nothing when they made me a drudge at their camp fires, and when they dragged me at the horse's tail, day after day—still nothing! I know all the rivers were overflowing. It must have been the spring when Magnus and his Tartars retreated into Russia. Sometimes, as they dragged me across a river, the water would come up to my neck, and then I would think—oh, God, only a little further to my rest! But it was just as the Bible says—a thousand fell at my side, and ten thousand at my right hand, and still I went on living. They sold me to a peasant. And there's another strange thing! That man bought me—paid out good money—and then got hardly any work out of me. He beat me till I wasn't fit to stand, much less to work.'

'They all do that,' Anna said. 'A Russian puts the pleasure of beating above any thought of money.'

'His wife was just as cruel,' Karen went on. 'Though she was hardly more than a slave herself; he blacked her eyes as well as mine. One day he hit me so hard I fell senseless. I was glad of that afterwards, because it made me miscarry.'

'My baby died when it was born,' said Anna. 'I swear I didn't kill it, though it had a Tartar face.'

'The day I miscarried,' Karen said, 'that Russian woman came into the stable and found me, lying in horse-dung and blood. She went out screaming and calling, and I lay there, thinking, "God is merciful. They know now I can't work, so they'll kill me." Then all the women of the village came running. They picked me up; they washed me; they put a clean shift on me; they laid me in a feather bed. And the woman who had been so cruel fell on her knees, beating her brow against the ground and crying, *"Prosti minya! Prosti!"* In their language that's "Forgive me".'

'I grew so tired of hearing that!' Anna said. 'After every drunken outburst it's that same word: *"Prosti!"* You are supposed to answer that God will forgive. Why do they never think to stay sober and do as God wants in the first place?'

Karen said, 'It was new to me then that they could ask to be forgiven. For the first time I saw that they were human creatures. It was a great misfortune; it made me feel again. Every bruise and every scar I had came to life. I felt for my murdered husband; for my boys. I could not stop weeping, and those women were weeping beside me. When the peasant, my master, came to look for me, they drove him away. They tended me day and night until I was well. But he—my master—got the better of the women; he sold me to a travelling slave-dealer. That's how I came here. And found so many Danes here already! All those young fellows growing the cabbages, just as they used to do on Oesel! And, best of all, my dear Anna, my neighbour, my friend!'

'Do you doubt now why God let us live?' asked Anna. 'We shall be together to see the end of Magnus.'

'So I have been reading this evening,' Karen said. 'God promises we shall see the reward of the wicked. But (if it were God's will) I would rather see it from Heaven.'

'I have so often tried to picture Heaven,' Anna said. 'Hell we know. But Heaven . . . our children will come running to meet us, and our husbands. But shall we really have no cooking to do for them? Or sewing? No floors to scrub? I can't imagine that. What will Heaven look like?'

'Like haymaking time on Oesel?' suggested Karen.

I felt as much a captive as they were. My parents having once apprenticed me, I had no choice but to serve out my time. Two years till my apprenticeship was over, and then seven more years. . . . I said, 'Heaven is a May morning in England.'

3

As if his own words had recalled him to the present, Jeremy now stopped speaking and looked about him. There was no May morning, but the foggy dusk of January. They were lighting their lamps already in Gerrard's Cross.

At the inn they gave us a good room, with a mirror. I scarcely knew the woman I found there. 'Is there so much colour in my cheeks?' I asked my husband.

'Why not?' he said. 'You've been riding all day in the cold.' And then, 'Don't wriggle when I kiss you!'

I said, 'Do you want me to lie limp, like the slave girl?'

'Are you jealous?'

'Jealous? Of something that happened before I was born?'

'God, so it did! How young you are! But now can you understand why I couldn't bear to be told that I'd bought you like a slave? Even if you thought it. . . .'

'I've never, never thought that.'

God! Make me not have thought it! I prayed silently as I kissed him.

He said, 'Bess, you silly girl, you look more beautiful every day.'

Beautiful? Me? I wished he would not talk nonsense. We had to consider seriously how I was to reach the Queen. I said, 'The Queen liked me when I was eleven.'

'You mean she looked at you kindly? That's her trade, Bess—to make people believe she knows them, and will remember them another time.'

'She might possibly remember me. When she stayed at our house, father told her that I was a perfect prodigy in Latin. The Queen asked me if I knew the meaning of: *"Infandum, regina, jubes renovare dolorem"*. I told her, "Beyond all words, O Queen, is the grief thou bidst me revive". Then she asked me if I could go on.'

'And you could, of course.'

'Oh yes, for pages. Only once or twice I hesitated, and she prompted me.'

'Which gave her a chance to show off her own learning. Well done, Bess! If you're as good a courtier now...' He broke off, and then said in earnest, 'My sweet love, it will take more than this to save me.'

I said, 'But after all you've been telling me, how can anyone think you'd betray our Queen for a Russian Tsar? Why, Jeremy, we can find scores of witnesses—guests we've had in the house—to say how you've spoken of Russia. Haven't you always told people that, when you came back to England, you kissed the ground?'

'Well, so I did. And yet, Bess, it's also true that, in my first winter there, I fell in love with Russia.'

He was a whole day—the next day—telling me how that happened.

The last ship that came sailing towards us, that year, was a puzzle. We all watched it, from the top of our new stone walls. We agreed that it was too small to be one of ours. Besides, no English ship would come at this time of year, far into November, with ice already forming round the shores of the White Sea. It was not a Russian boat; it was well constructed, and steered by someone whose proper trade was navigation. It could only be Danish. Yet we knew that the Danes were at daggers drawn with the Russians, and would not venture into their territory without some extraordinary reason.

Then we saw that the ship was heading, not for us, but for the monastery of St. Nicholas. Mr. Rowley frowned. If the Danes were carrying on some trade with Russia, they were infringing our monopoly. It was his duty to the Company to make enquiries at once.

He ordered a rowing boat to be made ready, and then looked at me. Why, I wondered, did my companions think I was his favourite? By this time some of them were calling me not 'blue eyes' but 'bum-boy'. This was all in their fancy; now, as always, Mr. Rowley's brow bent sternly on me, and his voice was harsh. He said, 'You're always chattering Danish. You'd better come along too.'

As we got into the boat we heard the bells of the monastery ringing for joy. They were pealing still, when we arrived. The Danes must have brought some goods that we never carried—or, perhaps, brought the same goods more cheaply. With a stern face, Mr. Rowley knocked at the outer gate.

It was flung open to us, with cries of joy which we could hear even through the bells. 'Welcome, dear friends! Welcome, on our day of rejoicing!'

We were brought into the great hall, where the monks were already gathering for dinner. The abbot came forward to greet us. 'My dear Thomas Thomasovich Rowley! And Yeremy! Forgive us, if this once you are not our guests of honour. We have our great man with us, our famous holy man, the Archimandrite Feodorite.'

The bells at last gave over. We found ourselves kneeling to an old man who was dressed like a simple monk. Why did the abbot and his glittering henchmen humble their cloth of gold before him? I whispered to Mr. Rowley, 'What is an archimandrite?'

He whispered back, 'The head of a big monastery—not much more important than an abbot. This man must be famous for something else.'

At dinner, Mr. Rowley was allowed to sit beside the archimandrite. He asked that I should be placed among the officers from the Danish ship.

My skill in their language only made them suspicious. To my polite enquiries they gave grunts, until I asked, 'Do you mean to stay in Russia long?'

'God forbid!' they all said, and the ship's captain added, 'Off at the next high tide—before the trouble starts.'

'Trouble?' I said. 'But your King's not at war with Russia.'

'A Russian doesn't need to be at war,' said the first mate. The captain asked me, 'Do you know what your friends the Russians did in the Baltic? Have you heard of the Island of Oesel?'

I admitted I had.

'Well then! The people there thought they weren't at war with Russia. Now old Ivan's up to his tricks in Lapland. He won't attack us in Vardö—no peasants for him there, no defenceless women and children. No, he sends this white-bearded villain to outflank us.' He jerked his head at the archimandrite.

'That old man?' I asked.

'The Lapps think he's God. Whenever we ask them or him what he's up to, it's always the same story—he's preaching Christ to the heathen. Those greasy savages bring him their children to be baptised—dipped right into the icy water, Russian fashion. Our Governor told him, if he wants to preach the Russian God, on the Russian side of the frontier, all right. But one step into Danish territory . . .'

I said, 'Where is your frontier? Surely some of those wild parts are claimed by both countries, and Sweden as well.'

'That's what *he* says.' Another glance at the archimandrite. 'But he went too far. He came with his chanting and his holy water to a place where we had a garrison. The soldiers arrested him, sent him to Vardö. Our Governor told them—don't hurt a hair of his head! We don't want a rising among the Lapps. But there's no harm in putting him aboard a boat and seeing him safe home. Let him preach the Gospel to that old heathen Ivan!'

'So that's all you came here for?'

'What would anyone come here for?' asked the first mate. 'Their cooking? What's this they've given us?'

'Elks' brains,' I said. 'A special delicacy, to show you're honoured guests.'

The Danes were at a loss for words, till one of them observed, 'He's not having any. He's got more sense.'

The archimandrite was touching nothing but bread and water. His blue eyes were clear, his wrinkled face as rosy as a child's. I saw him smile, with a sort of radiance, like a man in love. The abbot slackened in his guzzling, the monks in theirs, to gaze at that rapt face. For a moment it seemed possible that their pretended virtues were copies of a real virtue which had not yet lost its power to move them.

Directly dinner was over I reported the Danes' conversation to Mr. Rowley. He frowned and shook his head.

'That's what they want us to think.... Would they really come all this way to deport one old man? I wish we could see inside that ship of theirs.'

I said, 'Sir, we would have a good excuse to go abroad, if we offered them some of our Danish slaves.'

'That's impossible! King Frederik is still refusing to pay any ransom.'

'But sir, if we handed over those two women who work in the kitchen, they could plead with the Governor of Vardö, and he might persuade King Frederik. The King might be more willing to pay ransom, if he knew that those fifteen Danes on our farm were strong young men, likely to make good soldiers.'

'Good sailors, to use against our ships, more like! And why two Danish women? One would be enough.'

'It would be cruel to separate them,' I said. I told him why.

'I knew it!' said Mr. Rowley. 'You're not thinking of the interests of the Company—only of your kitchen friends.'

I said, 'Sir, I confess I took your words to heart, when you told us that these were good Protestant women. When we sit by the fire together reading the Bible, just as my mother reads it at home—'

Mr. Rowley cut me short. 'You talk as if the spring would never come. For six months I can spoil or waste or give away the Company's property. But when the ice melts there's a reckoning. The first ship that reaches us from England takes back my report and accounts, and then what am I to say?'

'Why, sir, that you thought the loss of two kitchen women a small price to pay for seeing what the Danes were after. That under the pretext of asking what room they had for these women, and what provisions, you were able to inspect the ship for signs of illegal trading.'

Mr. Rowley said, 'If ever you meet old Ivan, don't imagine you can get round him!'

He turned his back on me and went up to the Danish captain. The archimandrite, the abbot and the monks had already left the dining hall, sweeping along in state, as if on some holy mission. I knew they were going to sleep. A Russian must be very poor and very hungry, if he stays awake in the afternoon. We made a point of working straight through the hours of daylight, as part of that 'proper pride in being English' which Mr. Rowley was forever urging on us. Yet now I wished I could drop down where I stood, and sleep. I might forget for an hour how lonely I was, how useless, how much Mr. Rowley despised me.

The Danish captain took Mr. Rowley's arm. I heard him say, 'Come aboard and see.'

Mr. Rowley looked round at me. 'What are you standing about for, Jeremy? Take our boat and fetch those women. Hurry! Don't you know it'll be pitch dark in an hour?'

It was dark, and I was glad of the lights at the prow of the Danish vessel, when I rowed towards it with Karen and Anna. They were still dazed, still saying, 'We ought not to go and leave our fellow-countrymen behind.'

'You can plead for them to be ransomed,' I said for the tenth time. 'You can do that better than anyone.'

'So long as we can still be together!' Karen said.

I assured her, 'I told Mr. Rowley that you had to be together, to comfort each other.'

'How young you are, to talk of comfort!' said Karen.

'Comfort?' repeated Anna. 'No. What it is, with us, we never need ask what's the matter.'

As we came alongside the Danish vessel, lanterns were held aloft to guide us. The captain came down the rope ladder, and helped the women to climb it. I heard the sailors cheer them, as they scrambled aboard.

At that moment, while by the light of lanterns I saw the ship spreading its great sails to depart, the monastery bells began the peal anew. Through this music the women's voices came to me, crying, 'God bless you! God reward you!' And from Anna the strange words, 'We shall meet again.'

At the monastery Mr. Rowley told me that the bells were ringing in our honour. We had set free two slaves, and this was a deed pleasing to God.

I said, 'But if the Russians know that it's right to liberate slaves, why do they make so many?'

'Jeremy, your questions.... I suppose you know that you've landed us here for the night.'

'Why, sir, won't the lights on Rose Island be enough to guide us back?'

'They will not. Some of the monks have just come back from fishing; they say that the mouth of the Dvina is choking with ice. There may be a mountain of it floating between here and Rose Island. Did you run into any just now?'

'Yes, now and again something crunched against the boat.'

'We'll have to wait until daylight, when we can see our way round it. That's why the Danes were off in such a hurry; they want to be back north, out of the ice.'

To sail northwards *away* from ice may seem strange. But Vardö, though it is 400 miles nearer the Pole than Rose Island, is open to the winds and currents of the oceans, and the sea there never freezes. Whereas the White Sea, being almost land-locked, forms ice as fast as a lake.

'We have nothing to do until tomorrow morning,' Mr. Rowley said. 'Except eat another of their great greasy meals, and hear them tell us how saintly this archimandrite is, because he doesn't stuff it in like they do. Suffers from the guts-ache, I daresay. They will have it he can walk on air when he likes. Now that would be some entertainment! There's nothing else here. If they had a book to read!'

'Sir, they surely must have a Russian Bible.'

'Oh yes—the book of the four Gospels, encrusted with jewels, in the church. You're supposed to kiss that, not read it. And Jeremy! Don't you trust yourself alone with a monk! They're all the same—persuading young men to their filthy practices...'

Nevertheless, I ventured into the church. It was (as all Russian churches are) a place full of secrets. The altar was behind the iconostasis, a screen covered with paintings of the saints. There were two doors in this iconstasis, and I knew that at certain points in the service the priest would open the doors, giving the congregation a glimpse of the altar. But now the doors were closed, the mysteries hidden. The jewelled Gospels, too, had been locked away. Even the saints, I thought, were hiding something.

In France, the Papist images beckon or smile, or show a grief as near as the artist can make it to human grief. But Russian saints have huge dark eyes that look beyond our natural feelings. Their faces are perfect ovals, crowned with perfect circles of gold. Lifeless, I had always thought them. But, seeing them by candlelight on that black winter afternoon, I thought these might be the faces of people who had known the last extremity of suffering. Like the Danish women, they would never need to ask what was the matter.

Where there was a special place of honour given to St. Nicholas, patron saint of this monastery and of Russia, I drew nearer to look, and stepped on the body of a man.

Each of us began begging the other's pardon. I had not seen this monk prostrated (as their custom is when they pray) full-length on the ground before St. Nicholas. And he had been so rapt in his devotions that he had not seen me. My confusion was increased when the light before the icon showed me that this monk was the archimandrite. I saw, too, that he was barefoot, and his robe threadbare. I stammered out the first thing that came into my head. 'Aren't you cold?'

'If I am,' he said, 'I deserve to be. I have hidden too long in a wilderness of snow. If God had not brought me back by force to the place where my duty lies . . .' He smiled. 'You're English; you don't understand what I say.'

I told him that, though I spoke Russian badly, I could understand a good deal. 'My difficulty is that I can find nothing to read in Russian. I came in here hoping to see a Bible.'

At this he seemed suddenly much younger. He invited me to look at his books.

His room, the best in the monastery, had no chairs, and the bed was a bearskin rug on the top of the stove. But it was richly furnished with manuscripts. 'The Danes let me bring these with me, thank God!' he said. Then he showed me one thing which I had thought never to see—a Russian printed book.

'Yes, the Acts of the Apostles were printed in Moscow,' the archimandrite said. 'But afterwards the printing shop caught fire, and the German printers died in the flames.'

Beautiful as this book was, I took almost more delight in a manuscript New Testament in Greek. It was the first I had seen since I left my dear tutor at home. I began to read it aloud, syllable by syllable.

The archimandrite laughed. 'It should sound more like *this,*' he said, and began to make a sort of gentle music. I listened entranced. But he soon broke off, saying in Russian, 'No, that's not as they pronounce it in Constantinople.'

'Have you really been to Constantinople?'

'Not into the city; it is defiled by the presence of the Turks. I went to the Isle of Sio, where the Patriarch of the Greek Orthodox Church lives now.'

'On a pilgrimage?'

'No, the Tsar sent me. He wanted the Patriarch's particular blessing. His ancestors, you see, were only Grand Dukes of Moscow. Ivan was the first to be crowned Emperor, or Tsar, and I was the first to bring him the blessing of the Church on this new title. This is why the abbot here makes such a stir about me.'

I said: 'I thought it was for your courage in preaching to the wild Lapps.'

'Oh no, no! Where's the courage in that? The Lapps are the gentlest of all God's children, the most hospitable, the most kind. Only they have a silly habit of pretending to cast spells.'

'Our sailors believe they do cast spells, to wreck our ships.'

'Never! Never! All their sorcery is a poor vain attempt to

make spring come a little sooner, autumn last a little longer; to persuade the sun, when it does come, to bring more warmth to their poor country. I told them to praise God for His gift of barren soil, which keeps all greedy and cruel armies away, and leaves them in peace to drink the milk of reindeer.'

'But the Danes and Swedes want their country, surely?

'Only some parts about the coast. What may bring trouble on my poor flock is the hatred of Danes and Swedes and Russians for one another. These Danes, for example, thought I must be still an emissary of the Tsar. Oh! Before supper I must write a letter to the Tsar, to tell him I am safe and well, and need not be made a cause of war! If you will excuse me, my son. . . .'

I could bear to leave him, since he lent me the printed Russian Acts of the Apostles.

Mr. Rowley was pacing up and down the dining hall. He snapped, 'Where in God's name have you been?'

'Finding a book for you,' I said, showing my treasure. He glared at me.

'Who gave you this?'

'I'm afraid, sir, it's only lent, not given. There can't be many of these in all Russia. The printing shop was burned down.'

'Oh, I remember that; I was in Moscow at the time. The bishops did it.'

'The bishops? Why?'

'Because they knew what would become of them, if the people could read this. . . .' He broke off and turned on me with a fierceness which took my breath away. 'Jeremy! You haven't answered my question. Who lent you this book?'

'The archimandrite, sir. His room is packed with books.'

'Then that's where you've been! That old villain lured you into his room! Did he. . . . Did he try to pervert you to his religion?'

'No, sir.'

'They're always trying that. The Russians protect anybody who takes their religion. If a foreigner in Moscow beats his servant, the servant will go running to the church, ask to be dipped in some of their holy water. . . . Then he knows his master can't touch him. Germans have done that, Poles. . . . But, thank God, not one Englishman! Don't ever let them think you'd stoop to that!'

'Of course not, sir.'

'And don't you go to that man's room again! Jeremy, didn't I tell you never to trust a monk?'

I would have laughed, but his eye caught mine with a look so desperate that I dared not.

'Forgive me, sir!' I said. 'But the archimandrite is a very old man. I'm bigger and stronger than he is.' Picking my words carefully, I went on, 'I assure you, sir, I would knock down any man, whoever he was, who made unnatural offers to me.'

Mr. Rowley grunted something about being glad to hear it. Then he flung across the room and sat on the bench by the stove, looking away from me. Did he himself know what was the matter with him? I should not have known, even after that outburst of jealousy, if I had not been prepared by the other boys' teasing.

Disgust was only a part of what I felt. Fear. . . . Yes, a little, for in this lonely place he had almost the power of life and death over me. If I were to tease him, drive him to violence. . . . But why should I do that, when the most part of what I felt was pity? What misery for him, to be in love, to struggle against love with forced frowns and sharp words, and still to betray it so plainly that he was laughed at behind his back!

I wanted to bring him what comfort I honestly could. I said, 'Sir, I hoped that you might help me to read this book. I am afraid it is too hard for me.'

He looked at me with suspicion. But, seeing no mockery in my face, he grunted and made room for me on the bench. Many of the Russian words were as unfamiliar to him as to me; but since he knew the Acts of the Apostles almost by heart, he was able to make them out. Soon his pleasure in teaching me seemed to make him calm. He did not look disturbed, when at supper time the archimandrite insisted that I should sit beside him. 'We must show respect for learning,' the old man said, 'and this boy can read Greek.'

Mr. Rowley sat on his other side, and the archimandrite told us both how he had struggled to learn the Lapp language, which had never before been written. He had made a Lapp alphabet, and had translated the Gospels. 'We have no printing press,' he said. 'But the monks of the Trinity are copying it out.'

'The Trinity?' repeated Mr. Rowley. 'Do you mean the monastery at Pechenga? That huge wooden building our ships pass, a few miles to the Russian side of Vardö? I've always wondered what so many monks were doing, in that lonely place.'

The archimandrite began eagerly to explain how much work there was to do among the Lapps. But the abbot soon broke in, to ask the archimandrite about his letter to the Tsar. How should it be sent? Should Father Vassily carry it? Or Father Nikolai? Or Father Bogdan? (These were his familiars.)

'Who will carry it most quickly?' asked the archimandrite.

'Nobody can carry it quickly,' the abbot said. 'The winter has not hardened.'

He meant that the roads were already clogged with snow, the rivers with ice; and yet neither ice nor snow had made a surface hard enough to bear a sledge. Further south, towards Moscow, there might be nothing but mud.

'If you could send your letter in a month's time . . .' suggested the abbot.

'I must send it now,' said the archimandrite. 'Can we find some strong peasant, who is used to mud?'

'No, no!' cried the abbot. 'It would be disrespectful to the Tsar. We must employ trustworthy, religious men.'

I could see the faces round him shining with desire, at the thought of a visit to Moscow. I myself would have struggled through a thousand miles of mud, to reach those golden domes, those battlements of white stone. But the archimandrite shook his head.

'The court is not a place for a religious man.'

They all nodded gravely, to show their respect for his advice. Then, as if he had not spoken, they agreed that Father Bogdan should go.

While all this was going on I became aware of a trembling. I was not the one trembling; nor anyone beside me. Mr. Rowley was on the other side of the archimandrite; how could I be feeling what he felt? Yet I saw his bloodshot eye fixed on me, and knew I was not mistaken. When there was a pause in the talking I said to the abbot, in my slow and careful Russian, 'Reverend Father, we are trespassing on your hospitality for tonight, and what makes it worse is that I talk in my sleep. My companions on Rose Island complain I am always disturbing them. May I have a room to myself?'

The monks thought it a great joke that I should talk in my sleep. They asked me what I said, and I told them the first thing that came into my head. I had heard a ballad-singer chant it in the market place. 'Moscow is mended with walls of stone. . . .'

They all shushed me, and looked anxiously at the archimandrite. But he seemed as puzzled as I was. I did not know how the song went on; the singer had stopped when he saw me listening.

The abbot said that I must certainly not wake my master with such talk. Of course, of course, I should have a room to myself.

Mr. Rowley said nothing more until we parted for the night. Then he took the Russian Acts of the Apostles from me, saying, 'Let me have this! I shall get no sleep; the bed will be crawling with fleas.'

My bed certainly crawled; yet, in between scratching, I slept. In the morning I was alert and cheerful, till I saw Mr. Rowley's haggard face.

We found the monastery courtyard full of peasants. Amid the clogging snow, out of this country which to our eyes looked unpeopled, people had arisen. Some had been travelling all night, drawn by the pealing of the monastery bells the day before. Each one had brought some present—a lamb, a kid, or, at the least, an earthenware jar of honey. Many of the women carried swaddled, fur-lapped bundles. They held these bundles up, crying out for the holy man to bless their children.

The archimandrite was blessing everyone who came. They at least might have been warm from walking; he stood still, barefoot, on ground powdered with new snow. We had to take our hats off, and, even in the angle of the monastery buildings, the wind howled threats of frostbite to our ears. Yet the archimandrite seemed not to feel the cold.

'My dear children!' he told the crowd. 'Be more gentle to one another, more kind! Remember what a little while we have on earth. Our time is not to be wasted by quarrelling among ourselves. Make peace! Make peace!'

So they did, turning to one another and embracing, and kissing three times over. I found myself grasped in the arms of a strong young peasant, who cried between kisses, *'Prosti minya! Prosti!'*

I had never set eyes on him before, but I answered in the proper form: 'God will forgive.' The tears ran down his cheeks. Mr. Rowley was enduring a similar embrace; to him it might be some pleasure.

The archimandrite followed us down to the boat, carrying the Russian Acts of the Apostles, and insisting that we should take it. We told him it might be some weeks before we could bring it back. He said he would still be there. 'I shall not go to Moscow until I hear from the Tsar. And I'm afraid my letter will not reach him for some time; they have not yet made ready to set out with it.'

As we pushed off, Mr. Rowley muttered, 'To be shut up for the winter among thick-headed monks—that's hard luck, for a man with his brains!'

I took this reluctant praise of the archimandrite as a sign that we were friends. Indeed I believe Mr. Rowley was grateful to me, for keeping myself out of his way the night before. He was, after all, a religious man.

The ice was not in floating mountains, but in a thin crust which our oars broke through. When we came close to Rose Island we saw that in the stillness of our harbour the thin ice was turned to pack ice, which would crush our boat. Mr. Rowley said we must land outside the harbour, on an exposed beach. While I still rowed he stood up in the boat, looking for the best place and guiding me towards it. He was once again my master, giving orders clearly framed for me to follow.

Men came running to pull us ashore. One of them was the

English overseer of the farm. He gasped out a request to Mr. Rowley for permission to horsewhip some of the Danes.

'What for?' Mr. Rowley asked.

'It's only two or three, sir, causing all the trouble. The rest would go back to work at once.'

He was shouting to be heard through the lowing of cows, the desperate bleating of sheep.

'When did they refuse to work?'

'First thing this morning, sir. They wouldn't so much as feed the poor beasts. How can anyone be so cruel? Sir, let me beat it out of them!'

Already the fifteen Danes were approaching us. They took off their hats, and bowed. One of them came forward and said they would start work at once, if Mr. Rowley would tell them what had become of Karen and Anna. Had they been sold to the Russian monks?

Mr. Rowley said softly, 'You see, Jeremy, what you've started?'

I said, 'Sir, they will go back to work if you tell them the truth.'

'And let them know there's a question of their being ransomed?'

'They'll work better, sir, if they have some hope.'

Mr. Rowley snorted. 'I trust you may be right.' He told the Danes the truth, concluding, 'But remember nothing can happen until the spring. So go back to your work, meanwhile. What we all want is to live through the winter.'

They stared at him, still unwilling to believe that some good had come their way at last. Mr. Rowley called me as his witness, and I told them the Danish captain had felt such respect for the women that he himself had helped them aboard. For some reason it was this which moved them. I saw the light of hope as it broke over those young, haggard faces. I knew then that, whatever else I might do in my life, I must bring that look to human faces again.

The Danes went back to work. Soon the lowing and the bleating dwindled into the usual contented farm noises. Mr. Rowley said to the overseer, 'This is a rough life, I know. We can't always be sweet-tempered. But the next time your fingers itch for a horsewhip, remember that if old Ivan attacks us our lives depend on these men. What use is our stone wall, if we have enemies inside it?'

When we were alone again he looked at me sidelong. 'If you know what's good for you, Jeremy, you'll keep out of my way!'

This would have been good advice, if it could have been followed. But for nearly a month we were cut off from the world. The weather kept us almost always indoors. Mr. Rowley made us lads bathe every week, and looked our naked bodies over for the bruise-like marks which are the first signs of scurvy. Mr. Taylor said there was no need; he made each of us eat a piece of raw onion every day, and that would keep us in perfect health. But Mr. Rowley went on looking. He never actually touched me. Yet a dozen times a day I saw his eyes fixed on me.

I read some of the English books in our dining hall—in particular the poems. They seemed all to be by men whose chosen women would not look at them. 'Coyness . . . coldness . . . cruelty. . . .' I wanted to answer the poets back. Suppose the woman, though she thinks you honest, cannot endure the thought of your body touching hers?

Besides the cold, we now had to endure the darkness. We had candles in plenty, Russia being the land of tallow and wax. But it is tedious to cast accounts amid an endless flickering. Towards noonday a milky light would begin to creep through the mica panes. If it did not come, we would know that the panes were drifted thick with snow. Being south of the Arctic Circle, we never quite lost the sun, but even in clear weather we could see it only for a couple of hours every day. Once, peering through the mica, I saw the sun hanging low, and against its redness the black outlines of men, standing where no men could stand. I called Mr. Taylor to look.

'Boys!' he shouted. 'We're free!'

We all put on our furs and went out. It was true; three Russians were standing on the sea. They were catching fish through a hole they had made in the ice, and I saw by the sides of the hole that the ice was more than a foot thick.

Now our horses were led out and harnessed, our sledges packed with merchandise; and there began a constant coming and going between Rose Island and Kholmogory. The eighty miles had become a mere five-hour journey. Even the darkness was not so terrible; moon, stars and the low sun were all reflected in the glittering snow. And if we heard the howling of wolves, we also heard the peal of sleigh bells answering ours, for every peasant seemed to spend the winter skimming to and fro.

We had to be always alert against the cold. The very air was an enemy. We fought it by constant running to and fro; by burrowing amid the bales of cloth on our sleigh journeys; by skating, which is the swiftest and most warming exercise of all. Sometimes for an instant someone would forget. One of our Germans, mending a broken shutter, put a nail between his lips as carpenters do. When he took it away again he took some of his lip with it; the iron had frozen to the flesh.

One comfort was that we had furs in plenty, and could always find more by our own trapping. The very rabbits of that country grow pelts more thick, more silky, than the richest furs from the forests of Germany. The peasant keeps his ears warm in a cap fit for a king.

And their houses are not cold. Once Mr. Taylor took me to the village outside Kholmogory, where his Russian woman lived. Her stove crackled a welcome to us. The hostess bowed; we had to bow, not to her, but to the ikon. The walls were hung with linen towels, as white as the snow outside, and finely bordered with drawn-thread work. I knew that behind these towels the walls were crawling. The cracks were stuffed, not with tow (as ours were on Rose Island) but with moss, which breeds vermin.

The stew the woman put before us was excellent. She was too shy to sit down and eat it with us, but busied herself by the stove.

'She's a good girl,' Mr. Taylor said in English. 'Always makes me hang a veil over the face of the ikon, before I get into bed... She was married, but her husband turned against her, and one fine day he went into the Archangel monastery, leaving her to fend for herself. She was begging for food at the monastery gate when I found her.'

I asked, 'Do you know any more women with husbands in monasteries?'

Mr. Taylor laughed and called in Russian to the woman, 'Nadya! Can you find a pretty girl for Jeremy?'

She asked gravely, 'Does he want to be married, then?'

'Married? No! His parents probably have a bride waiting for him in England.'

'Then he will leave the Russian girl disgraced, as you Englishmen always do.'

'Oh, not always!' Mr. Taylor said. 'We're not all so changeable; you know that.'

He went to the stove and put his arm round her waist. She looked at me, and blushed. *I* blushed, since I knew Mr. Taylor intended to leave her the moment the Company let him go. A month before I would have taken this for granted. But Mr. Rowley's oppressive passion for me had made me feel so much like a woman that I thought, 'There! If we give way to them, that's how they treat us!'

In English Mr. Taylor said, 'I warn you, Jeremy—it's not the husbands you want to worry about. The men here don't give a curse for their wives. Never see 'em before the wedding; too drunk to see much of 'em after. But they think the world of their sisters. If a Russian thinks you're carrying on with his sister, he'll come looking for you with an axe. This one—' he slapped her bottom '—hasn't got a brother, or I'd be for it.'

As soon as I could, I left the house, and went through the village, towards Kholmogory. As I picked my way between the frozen piles of dung, I thought I despised Mr. Taylor for his entanglement. I wanted none of that. Nor would I degrade myself with slave girls; I had learned my lesson there. How much better to lead a life of chastity and hard work!

On the scattered outskirts of the village, a young woman came out of a courtyard gate carrying a live goose. She held it out to me by its wings and said, 'Kill it! Please, your honour, kill it!'

'Why don't you kill it yourself?'

'But I'm a woman. I mustn't kill.'

She pressed the struggling, squawking thing into my hands. I tried to wring its neck, but it defended itself with strong wings, and I could not find the way to hold it as the peasants do. The woman ran and fetched an axe. Then she pinned the goose to the ground so that I could cut off its head. I botched even that simple task; sweating amid the snow, distracted by the creature's cries, I hacked clumsily till at the fourth or fifth attempt I severed the writhing neck.

I had not expected such a spurt of blood. The woman's dress was protected by an apron of sacking, which she now took off and wrapped round the goose's body. I saw by the fading light that my fur coat and gloves were spotted with blood.

The woman said, laughing, that I must be a real gentleman, to make such a pother over killing. Then she asked me into a house exactly like the one I had just left. She made me stand by the stove while she dabbed at the bloodstains with a wet cloth. I was trembling from the killing of the goose, and I trembled all the more at the touch of her hands. I thought they lingered over me. She made me take my coat off, and examined my other clothes for blood. I asked her where her husband was.

'In the drinking shop', she said. 'The Governor of Kholmogory gave him half a rouble for flogging a prisoner, and he won't come home before he's drunk it. Then he'll be fit for nothing but to lie and snore.'

I remembered to ask, 'Have you any brothers?'

'No; I had one, but he died.'

Like the goose, this woman writhed in my arms, but not so as to escape. Soon I found out how much my pleasure was enhanced, when it was the woman's pleasure too. Of course I knew that she was only taking revenge on her drunken husband. I was taking my revenge on Mr. Rowley, who had wanted to turn me into a girl. Yet there was a warmth between us—our complicity in killing, the heaving of the goose's wings, the sudden warm spurt of its blood.

Before I said goodbye, I asked her name. 'Katya,' she said. We agreed that when her husband was away from home she would put a wisp of hay in the latch of the courtyard gate.

I reached the rope factory at almost the same time as Mr. Taylor. There were two people waiting to see us; they had brought a sledge piled with ermine furs enough to adorn all the kings of Europe. Mr. Taylor thrust his hands into the very middle of the pack, because the Russians, in their simple way, never tired of trying to hide their inferior furs inside. But these were not Russians. They had flattish, beardless faces, with black slit eyes. Both were dressed alike, in hooded sealskin tunics laced with leather thongs, sealskin leggings, and rough boots of reindeer hide. Under their tunics they seemed to be wearing some other garment, so thick as to hide their natural shape. It was only when one of them unslung a furry bundle, and opened the front of the tunic to give it suck, that I knew she was a woman.

'The furs are good through and through,' Mr. Taylor said in English. 'That's the best of dealing with *Samoyedy.*'

They caught the last word. In halting Russian they told us that they should not be called *Samoyedy,* which means self-eaters, or cannibals, but *Samiye*—the self-same people of that country where their fathers had lived, since God created the world. While all other peoples had been wandering to and fro, seeking succulent pastures and mines of jewels, they had stayed in their own Siberia, eating the meat and wearing the furs God gave them.

'I'm glad they don't mean to eat *him,*' said Mr. Taylor, in English, nodding at the baby. In Russian he added, 'You're very good fellows, I know, if you do worship elks and bears.'

This too they denied. They believed, they said, in one God, who breathed his breath into every living creature. If, when a man went out hunting in the morning, the first living thing he saw was (for example) a reindeer, then all day long he would worship God through the reindeer. He would praise God for creating such beauty and such grace.

'Bound for Hell, I daresay,' Mr. Taylor said. 'Being heathens. . . . They eat their meat raw, too. Suits them, I suppose; they never get the scurvy. They're good to deal with. They know what we want, and we know what they want.'

In exchange for the princely furs, he handed over a mingle-mangle of iron traps, hunting knives, needles and thread.

'Must we take advantage of their ignorance?' I asked.

'What—you want us to pay what the furs are worth? Where's the profit to the Company in that?'

The gentle Samiye made me think of the archimandrite and his Lapps. When I returned to Rose Island I asked Mr. Rowley's leave to take the Russian Acts of the Apostles back to the old man.

'I know, I know,' said Mr. Rowley. 'You can't endure to spend one minute with . . . with any of your fellow Englishmen.'

But he let me go. And later, when he saw that it would please me, he let me go regularly to the archimandrite, for lessons in grammar.

Most Russians do not know that they have such a thing. The chameleon changes that come over each word are as far beyond their power to explain as they are beyond the power of foreigners to learn. In a general way I had gathered that Russian nouns, like Latin, had three genders and six cases. (Or was it seven cases?) I had an inkling that for each English verb I ought to learn two Russian verbs at least. If I wanted to say 'Go or 'Put' there were countless ways of doing it, and whichever way I chose would be wrong, for reasons which I tried in vain to fathom. My fellow-countrymen had given up trying. 'Just make yourself understood!' Mr. Taylor would say.

My lessons proceeded, as it were, crabwise. Nobody has yet written a Russian grammar. I learned from a Greek one, the archimandrite explaining, page by page, how Russian differed from Greek. If that was not clear he would explain in Latin. Still I was confused.

'Why,' I asked, 'is the word for "dog" feminine, even when it's a male dog?'

'Because a dog is an unclean beast, like a woman.'

'But "goodness" is feminine, and so are "cleanliness" and "honour" and "virtue".'

He smiled. 'I should have thought of that, twenty-five years ago, when I was first put in charge of the monastery at Pechenga. I wanted my monks to be like the monks of Mount Athos, and exclude not only women, but all female animals. They said my rule was making them ill, and so I suppose it was. At Mount Athos they have olive oil, which is edible, whereas we in Pechenga had only seal oil, which is not. My monks told me that they could not live without female reindeer, to give them milk and cheese. But I thought of all things female as unclean. If only I had remembered that it was a woman who first made Russia Christian! And then, under the Tartar yoke, when the men forgot their skill, the women remembered theirs. In Russia whatever a man does is done badly; only what the women do is done well.'

Since he, a Russian, had said so much, I spoke more freely. When I said I had fallen in love with Russia, I meant, rather, with its language. In the dung-puddled market place of Kholmogory, what I saw I hated, but what I heard I loved. The very haggling over prices was a sing-song poetry. The ballads of the people were beyond comparison more beautiful than our doggerel about Robin Hood. And now that I heard Russian spoken by a learned man, I thought it more copious, more elegant than any language in the world. I asked the archimandrite how a people possessing such a language could be content to live in wretchedness, without schools, without doctors, without skilled craftsmen. In a land of timber, they had no carpenters' tools—nor, indeed, any tool but the axe. Their horses went unshod for want of blacksmiths. How did it happen that their only notion of commerce was trickery, their only form of government....

But here I stopped, confused. The archimandrite was, after all, an old friend of Ivan.

He smiled and said, 'You think we are savages. But when you travel from here to Moscow, you will pass along a canal which cuts off a great bend in the River Sukhona. This canal was not made by Englishmen or Germans. Our own Count Gleb Beloser-sky had it cut, more than two hundred years ago. In those days, too, our ancestors began to rebuild the old wooden Kremlin in stone, and it was Russians who did the building then. Then the Tartar invasions came. The Tartars killed so many of our men, and forced so many more to take up fighting as their trade, that there were no crafsmen left alive. I think even the lying and cheating, which you have observed in the market place, come from those days. A conquered people does not speak the truth to its conquerors. We might have kept our ancient virtues, and our ancient skills, if only the Poles and Germans had helped us against the Tartars. Instead, they took advantage of our weakness to attack. Now they call us barbarians, and so we are; but we became so through the barbarous wars we had to fight. We might have forgotten that we were a people, if it had not been for the women. Even in slavery they chanted the old songs and the old stories to their children. It was the Church's doing that we never quite forgot we had a written language. But what kept the Church alive? The devotion of the women! I should have remembered this. Instead I quarrelled with my poor devoted monks, and went to Moscow. That was the beginning of all my misfortunes.'

I begged him then, and often afterwards, to tell me his misfortunes. But he would smile and lead me back into the thickets of grammar.

From these lessons I would come back to Rose Island bursting with curious information, which my companions did not want to hear. I discovered, for example, that the way we wrote our bills and receipts was wrong. We always followed a number by a plural. In English, if you have more than one yard of cloth, you have yards. In Russian, two is followed by a genitive singular 'Two of a yard.' It's the same for three or four. Five is followed by the genitive plural: 'Five of yards'. This goes on till twenty. Twenty-one reverts to the singular; 'Twenty-one yard.' Then: 'Twenty-two of a yard... Twenty-five of yards....' The one thing that cannot follow a Russian number is a simple plural.

Confronted by this insult to his common sense, Mr. Taylor hunted out letters written by Russians, triumphantly showing me numbers followed by plural nouns. I said that what looked like a plural was in fact a genitive singular; in feminine and neuter nouns they may appear the same.

'If they appear the same that's good enough!' said Mr. Taylor.

'But they are differently pronounced,' I protested, while the others laughed.

Ridicule turned to simmering anger when Mr. Rowley took my part. He said that, for the credit of our nation, we ought not to go on speaking a mixture of half-heard phrases and coarse peasant words. From now on we must at least make out all documents in correct Russian. After that I had endless quarrels. 'You're doing it wrong again!' I would shout at some poor confused companion. 'Can't you see it's wrong?' I knew that I was making myself hated, and yet I could not stop. I loved this moody perverse Russian tongue so dearly that a mistake hurt me as a wrong note hurts a musician.

Our quarrels grew more bitter as the winter did. By February we felt that it had always been winter, and always would be. But the cold still made us active. It was late in March, when the wind attacked us with a new venom, that we could no longer force ourselves to run. Once you feel that no exertion can ever make you warm, you move slowly, and then more slowly still. The day seems near when you will stop, and fall into soft snow, and be coffined in hard ice.

In this worst of all Russian seasons, Mr. Rowley gave me good news. 'I've heard from our men in Moscow,' he said. 'They need a young lad there. Somebody who speaks Russian really well. That leaves me no choice but to send you.'

He was watching me closely. I still feel proud that I did not break into a dance of joy. I said, 'Then must I leave you, sir?'

He looked me in the eye and asked, 'Will you mind that?'

I could not quite bring myself to say I would. Passing it off lightly, I said it would not be for ever.

He tried to smile. 'You want to make your fortune; that's natural. You won't be leaving till the summer. Better go and have another lesson from your saint.'

Pregnant with my news, I struggled through soft new snow to the monastery. The whole place was in an uproar; the archimandrite was leaving 'immediately' for Moscow. I cried out, 'Then I shall see you there!'

He smiled and said, 'Perhaps. To God all things are possible.'

I persisted. 'But I shall be there in the summer.'

The archimandrite said, 'We have a custom that before we begin a journey we sit down for a moment. But my friends here are making such a business of the preparations that my sitting down is going to last several hours. I shall be glad of your company.' When we were alone he added, 'It is hard for me to go to Moscow, and not to let one soul on earth know why. Yeremy, I did not want to tell you before, because. . . . There are things it is hard for any Russian to admit to a foreigner. I should have written them down. Here and there in Russia, men banished to monasteries are writing down what they have seen. But a manuscript must be entrusted to someone, and I have nobody . . . that is, nobody I can ask to undergo such danger. Even to hear me speak may be dangerous. Yeremy, if you would rather not. . . .'

I begged him to go on.

4

It was eighteen years ago, in the year 7064 from the creation of the world—by your reckoning, 1556—that I quarrelled with my poor monks in Pechenga. I went to Moscow to put my case before the Church. Soon I found myself imprisoned for heresy. Which was man's injustice, and yet the justice of God, who thus punished me for being stiff-necked and unforgiving. I was liberated through the pleading of Prince Kurbsky, one of the Tsar's generals. He knew me because I had taught him Latin and Greek. It was Prince Kurbsky's doing that the Tsar chose me as his emissary to the Greek Orthodox Church. I felt it a great honour to serve such a noble, such a Christian Tsar. He was then fresh from the conquest of Astrakhan. Kazan had fallen to him four years earlier, and so our Volga had become from end to end a Russian river. The first English merchants were already in Moscow then, making plans to sail down the Volga to Persia. So we knew that the glory of our Tsar had reached the remotest countries. But he seemed still more glorious to those who knew him best. He was young then, and such a humble, earnest Christian that he would go into the belfries and ask the bellringers to teach him their art; he wanted to make music to the glory of God. There was hope, then, in the faces of the people. We all thought the days of darkness and slavery were over.

The Tsar's chief adviser was a man renowned for virtue, Adashev, and a priest, Father Silvester. He was guided, too, by his wife Anastasia. I did not meet her; like any other great lady, she lived in seclusion. But I saw her once in church, and her gentle face made me almost forget my dread of women.

I set out on my journey in the year 7065—or 1557. I had to pass through country constantly raided by the Crimean Tartars. In these peaceful northern parts, Yeremy, you do not feel what it is to be a southern Russian. To be all winter locked in a cold as cruel as any we have here, and not to be glad that the spring comes earlier. Because the warm weather will bring this plague, this whirlwind, this coming of the demons out of Hell. How can the people of the south live as Christians? They scarcely dare even to love their children, since the children may at any moment be carried away into slavery. The Cossacks of the Steppe, who neither plough nor sow, came to ask my blessing with crosses about their necks, to show me that they were Christians. But I could see that they had grown as wild as the Tartars they fought. The people of those parts dare not keep cows or chickens; the Tartars would steal them. So, even when they are at peace, they have no milk or butter or eggs. I ate only dry bread; and I was punished for so often telling my monks that dry bread would not hurt them. The southern bread is poor weak white stuff. I had not known that I could long for any earthly pleasure, so much as I longed then for the black rye bread of the north. God is just; the white bread made me ill.

What with illness, and long roundabout journeys to avoid the Tartars, it was four years before I came back to Moscow.

At first I found nothing but kindness in the Tsar. You see, I had succeeded in my mission. I told the Tsar that the Patriarch had blessed him and hailed him as Emperor—the first Orthodox Emperor since Byzantium fell to the Turks. On hearing this the Tsar embraced me, gave me a sable cloak and a vast sum of money, and offered me a place at court.

The other monks at court told me how much I was needed there. Adashev and Father Silvester had been banished. They had spoken too roughly to the Tsar about his 'little human failings'. Because of their rudeness the Church had lost much of its influence.

Besides, there was a new Tsaritsa. The gentle Anastasia, mother of the Tsar's two sons, had died. Her place was taken by a princess of the Cherkasses. They are a sort of Tartars, who live near the Volga. Of course a Russian Empress cannot be a pagan; she had been received into the Orthodox Church, and baptised under the name of Maria. 'But a Tartar, for all that!' one of the monks whispered. Another leered and said that the girl had nothing good about her but her looks.

These monks who took such an interest in women's looks were near me at the Tsar's dinner table. I saw them laugh like everyone else when the Tsar's musicians played ribald songs. Then the Tsar's buffoons acted a masquerade. Some of them were dressed as women, some as monks, and they performed obscene parodies of the Mass. The Tsar himself, extremely drunk—a new thing with him—applauded the buffoons and urged them on. And the monks too applauded. I sat as if turned to stone, not believing what was before my eyes. Yet it must be real; I had been drinking only water. This was tolerated, because I was a monk. But some unfortunate layman, new to the court, found the Tsar's wine too strong for him and tried to refuse a second cup. The courtiers cried out, 'What's this? Are you trying to call your Tsar a drunkard?' And when he still would not drink they poured the wine over his head. Still the monks laughed. But then, to my relief, all those in holy orders left the table, and a steward whispered to me that I too might go.

Behind me the noise grew louder still. I gladly turned towards the quiet room I had been given. But before I could reach it a lady came and whispered that the Tsaritsa wanted to see me.

Though I was then already old, and known for the strict keeping of my vows, I hesitated to enter the women's quarters at night. But the lady urged me with tears in her eyes. She said,

'There never was a Christian soul who needed comfort more!' So I followed her.

I could not see why the Tsaritsa was called beautiful. Since I was too small to speak, I had been taught to run away from the sight of a Tartar face. I could hardly bear the sight of her black slit eyes, nor the sound of her uncouth accent when she asked me, 'Are you really a famous holy man?'

I said I was not fit to be called holy, and she answered, 'No, you can't be! Because you're doing the same as the other monks!'

I told her sternly that I did not understand her. She said, 'Don't I speak Russian well enough? I've worked so hard to learn it, and to understand your faith! The Metropolitan of Moscow showed me the meaning of the marriage rite. I was to swear to love and obey my husband, and only him, just as a Moslem wife would. But my husband also had to swear—and this is where your faith is better—that he would love me and only me.'

Was this what I had come for? To meet a woman so shameless as to complain to a stranger against her own husband? I said, 'A wife has to bear with her husband's failings, to touch his heart with humility and gentleness.'

'Oh yes! The Metropolitan told me this too. He said I should never raise my voice in rebuke of my husband; the Church would do that. He said the Church was above the greatest kings. He said the Tsar's confessors make him do penance when he sins. Well, so they do—but they encourage him to sin first.'

I repeated (still hating her, still not believing a word she said), 'Encourage him to sin?'

She cried out, 'But you yourself have just done it! You and the other monks left the banquet early, so that he could send for the women.'

Like a flash of lightning, this lit up the dark things I had seen. Though it came from barbaric lips, I knew at once it was true. Yet I would not admit it. I asked her, 'Are you saying that the Tsar's confessors let him keep bad women in his palace?'

'Oh, they are not bad women. Poor, poor creatures! When they can break through their guards they kneel to me, and cling to my dress, and beg me to save them. And I can do nothing! They were captured in war. The Tsar's officers, when they take one of the Baltic towns, pick out girls who seem to be of noble blood, and very young. It is not important that they should be good-looking—only that they should be virgins. The common soldiers are forbidden on pain of death to touch them; it must be the Tsar who takes their virginity. Afterwards he turns them over to his drinking companions. Don't tell me there is good in him! I know there is—there must be—but I can't reach it. I have tried—' She was hiding her face now, and so I had forgotten that it was the face of a Tartar '—I have tried to remind him of his marriage vows. He tells me he has made me pregnant; that should be enough for me. At least I shall be allowed to bear my child. I shall not be made to swallow slippery elm as those poor girls are. Sometimes they die of that.'

I was trying to find words, but my throat was dry. I looked round and saw her ladies listening, and I understood that what made me speechless was terror.

Till then I had not known I was a coward. I had thought a monk's life braver than a soldier's; my profession had led me unarmed through northern wastes, and then through country threatened by the Tartars. Yet now I trembled at the sight of a few women, because they might be spies for Tsar Ivan.

To my shame, the Tsaritsa saw this at once. 'It's the same with the Metropolitan,' she said. 'He's afraid to come and see me now.'

I asked her whether she had a brother to take her part.

'My brother's fighting by the Baltic shore. He's not a Christian; he'd think it right that a king should have the choice of the women captured in war. It's *your* faith which condemns that. But if this faith, which I embraced with all my heart, which I delighted in—if this faith can't find one man with courage to defend it, then I don't want to live in this world. Let me die! The Christian Heaven is open to women as to men; in Heaven I shall find mercy.'

I now began to see in her those virtues which (as you say, Yeremy) are according to our language always feminine. Yet

I could not believe that, in all the court, there was not one Christian besides this woman.

I had to believe it soon. The courtly monks laughed in my face. 'It's your narrow, old-fashioned way of thinking,' they said, 'which has brought the Church into disrepute. Can't you see that these little amusements of the Tsar—oh, yes, if you insist, orgies!—turn out very well for the Church? He's always penitent the morning after. He comes crawling to us for absolution. That's when we can get whatever we like out of him.'

I said, 'Then you can persuade him to live honestly with his wife, and send the foreign girls home.'

One monk laughed. 'Send the whores home? What for? Are they short of whores in those countries?' But another, looking serious, rebuked me. 'Are you completely blind to the interests of the Church? The Tsar has moods when he talks about King Henry of England, and calls him an example to all princes, because he seized the property of the monks. Do you want to see us ruined?'

I said God had punished the monks of the western Church for their sins, and would punish us for ours more terribly, because our faith was better in the first place; we knew more clearly what we ought to do. A penitent, even the Tsar himself, must not be absolved until he showed some serious intention of mending his ways.

At the thought of refusing the Tsar absolution, they made sounds like the whinnying of mares. 'Old Silvester tried that,' said one of them. 'And where is he now?'

I said I would be glad to know where Father Silvester was. I wanted to visit him. At that they turned away without another word.

But Ivan sent for me. This was when he had risen from his afternoon sleep, on the day that followed his orgy. I went ready to rebuke him. What prevented me was that I found him sitting at a table, showing his two sons a manuscript Gospel. The Tsar urged the boys to read aloud to me. The elder boy, a lively child, read fluently, but seemed not to care what he was reading. The younger one could scarcely make out the letters, but stroked with adoring fingers the pictures of saints.

'We shall never make a scholar of young Feodor,' said the Tsar. 'But he loves church bells. I'll teach him the bellringer's art, at least.' And he looked at me with clear, merry eyes. He was not much past thirty, and seemed younger.

Presently he sent the boys away, and again I was about to rebuke him. But he was already showing me a letter. 'You'll enjoy this. It's from your old pupil, Prince Kurbsky. He's leading my armies by the Baltic.'

It was indeed a pleasure for me to read the letter. Prince Kurbsky is one of the few men who can write our language with an eloquence worthy of Cicero. (I recognised some of his favourite phrases from the time when I taught him Latin.) But, if the manner was worthy of ancient Rome, the matter was worthy of a Christian. He was asking the Tsar to show mercy to an enemy.

This was Philipp von Bell, Marshal of a Baltic province, taken prisoner by Kurbsky's men when they captured Ermes, and now sent to Moscow by the Tsar's order. Prince Kurbsky praised von Bell as an honourable fighter, and as a just, wise ruler of his province. The man's talk, wrote Kurbsky, was a continual delight '—which Your Majesty will be able to enjoy, since you speak German.'

The Tsar said, 'What do you think, Father Feodorite? Shall we see what amusement we can find with Landmarschal von Bell?'

I told him that I could not join in the amusement, since I knew no German.

'No more does anyone else here,' answered Ivan, 'though some of them pretend to. Never mind! I myself will be the interpreter.' He called guards to bring in the German, settled himself on his throne, and took his great staff in his hand. The room was at once full of people; I had to give up all thought of a private rebuke. With a clattering of armour and of spears, the prisoner was brought in.

He was a handsome man, very tall and upright, but for several minutes I did not look at him. I could not take my eyes off the soldiers, his guards.

When I had left Moscow, four years earlier, all the soldiers about the Kremlin were our own bearded Russians. Now the men chosen to guard this important prisoner were smooth-cheeked and slit-eyed. If they were not Tartars, they were Cherkasses; and I could see no sign that they had followed their princess into the Christian Church. They wore no crucifixes, and their faces... Was it only my childish dread of a Tartar face that made me see in these the very print of cruelty?

The Tsar was talking a language which, to me, sounded very different from the prisoner's. However, they seemed to understand each other. From time to time the Tsar would fling his court a crumb of explanation.

'Ha! He admits the Germans were defeated because of their drunken habits.... And listen to this! He says they fell into bad habits when they took to the Lutheran heresy. All that smashing of their ikons did them no good.... Ha! He says our men were sent by God. Of course they were! Even a German can see that.... What's he saying now! Something about Attila.... What insolence! He calls us a scourge of God, like Attila the Hun!'

The Tsar struck the stone steps on his throne with his great staff. It rang; and I saw that the staff was tipped with sharp iron.

I looked round. The eyes of all the court were fixed on the angry Tsar. I heard the sharp sucking in of breath. Perhaps it was my own. Only the prisoner continued calm. He said something more, which enraged Ivan so much that he shouted in Russian, 'How dare you slander a Christian army?' Then he spat out something in the prisoner's language. The prisoner made an answer which the Tsar interrupted, leaning forward and laughing in the man's face. 'Ha! You mean our Cherkasses! They were too much for you, eh?' Then all the court laughed at this defeated enemy. When the Tsar stopped laughing, so did everyone else. Then, suddenly, the prisoner began to speak Russian. It was very slow and careful, as if he had rehearsed it. 'You sent a pagan army into a Christian country, which is not the act of a Christian Tsar.'

The Tsar leapt to his feet and shouted, 'To the scaffold!' Instantly the Tartar faces closed round the prisoner, and he was led away. The Tsar still stood shouting, 'At once! Bring me his head at once!'

My cowardice mastered me for a few seconds, which was too long. By the time I was prostrate at the Tsar's feet, begging him to think of Christ's mercy to man, to think at least of his own royal dignity, which ought to be above the killing of an unarmed prisoner—by the time that I had begun to make myself heard, he was already changing his mind. He muttered, 'But the fellow said I wasn't a Christian! ... Oh, very well!' Then, quite calmly, to a courtier standing by, 'Go and tell them not to execute the German.' And, pleasantly, to me, 'Don't excite yourself, Father. It's bad for you, at your age. We've given the fellow a fright, ha? No harm in that. He'll be more careful what he says another time. If what Kurbsky writes is true, we ought to get this German on our side. He'd make a good viceroy for us in Livonia. Shall we make him a viceroy, Father? Would you like that?'

I had half risen when the courtier came back, looking very sheepish. He had been too late; the man's head was already off. Behind him a Cherkass held it up by the hair.

The Tsar said, 'So, when it comes to carrying out my orders, a Cherkass is quicker than a Russian! I'll know in future where to put my trust—anywhere but in my own people.' He prodded the courtier's belly with his iron-tipped staff. 'You're too fat; that's what makes you so slow.' He laughed, and, while the severed head dripped in their midst, the court laughed with him. Then, to the man carrying the head, Ivan said, 'Take that away; he'd never have made a good viceroy.'

If I sank on to the floor, it was from weakness. I was overcome by longing for my gentle Christian Lapps. But the Tsar mistook my prostration for a reproof (which I wish it had been). He cried out, 'What's done is done; you might as well cry over some girl who lost her virginity last night.'

Again the court laughed, and that ugly sound gave me courage. I said quietly, 'I do lament for that as well, and pray for God to change your wicked heart.'

He glowered at me, and then told the courtiers to leave us. When we were alone he dragged me to my feet, and forced me (he was very strong) to sit in his place, on the throne. Then he threw himself down before me.

Ought I to tell you what he said? Why not? It was not under the seal of the confessional. And you should know, Jeremy, what sort of man you English are dealing with in Moscow. A man who could do all I have described, and accuse himself of it aloud, with tears, beating his brow against the stone steps at my feet.

I told him to repent.

'I am repenting. Can't you see that? Isn't there blood on my brow?' He felt it with his finger, to make sure. Then he knocked his head against the steps again, crying out, 'I am not fit to live!' And then with tears, with passion, 'What am I to do? Tell me what I am to do.'

I said, 'The German told you. He is in Heaven among the martyrs; you killed him for telling the truth. Why should Russia show the world a Tartar face? Make peace with your western neighbours, who are Christians of a sort. Lead your armies to the south; deliver your people there from the Tartars.'

'My people in the south ought to fight for themselves.'

'So they do, behind stockades of wood, with bows and arrows. If they had a tenth of the cannons and gunpowder that you have cast away in Livonia, the Tartars would never trouble them again.'

Ivan cried out, 'This is the very voice of old Silvester! A churchman telling me where I'm to fight! "Cast away in Livonia!" Don't you know I've succeeded there? Do you think I'm out of my mind—to give up this war when I'm winning? You think I'm fighting Christians. Let me tell you, half the Poles are Jews. When I took Polotsk, I found the place full of Jews. I didn't leave a single one alive.'

'Did you offer them the choice of baptism?'

'What for? They'd only lick my boots and tell me lies.'

'Did you kill their children too?'

'Of course. They were Jews.'

'The women you capture are Christians; yet you rape them.'

'We have to show these people we're conquerors! What did the Tartars do when they conquered us? Don't you remember the tribute they used to take, on the Maiden's Field, not two miles away from here?'

'The Tartars were pagans. You call yourself a Christian.'

'So I am. When I commit sins, I do penance for them. And then—forgiveness! Christian forgiveness. That's your trade.' He seized the hem of my robe and kissed it. 'Here at your feet, Father Feodorite, I beseech you to forgive me!'

I plucked my robe away and walked into the middle of the room. He cried out: 'But you must forgive me!' He got up and began to follow me about.

'Haven't you heard of Novatian's heresy? Novatian refused to absolve certain sinners, and the Church denounced him as a heretic, because all sins may be forgiven.'

'Yes, if the sinner is repentant.'

'Do you want me to be better than anyone else? Let me tell you, your beloved Kurbsy, when he fancied a common soldier's wife, he took her. Simply took her! And yet the Church absolved him, when he'd done the proper penance. Why him? Why not me?'

I said, 'Your sins are premeditated. You could not commit them if you had not already formed a company of buffoons, given places at court to men you know as worthless, ordered a great search for virgins—who must be rare indeed, in the places your armies have taken. To show penitence you have to dismiss all these people. Send the foreign girls home with dowries and safe-conducts; dissolve your Cherkass army; live by Russia's holy laws. Ask forgiveness from God and from your wife!'

'What wife? I have no wife. My wife was Anastasia, and you let her die.'

'I let her die?'

'The Church did. When she was ill they would do nothing for her. Nothing! A few masses.... But no pilgrimages, no days of prayer and fasting. It's the Church's fault she died. How did they think I could lead a good life without her? This other creature.... She brought me her Cherkass army. She's going to bring me a child. A little Cherkass, ha! With slit eyes.... Do you think that can console me for Anastasia? I miss her all day long, but at nightfall past endurance. It's then I drink, and laugh most loudly at what I most believe. And still the longing grows, till it's a frenzy, and I seize on some white-skinned girl, some tall strong creature who'll fight me with nails and teeth. I let her wrestle, let her think she'll be the first who ever defeated me. Then—pounce! And she's writhing under me, sobbing with pain . . . or pleasure. . . .'

I held up a hand to check this confession, which now sounded more like boasting. He burst out, 'What's the use of telling you these things? You never had a woman in your life. You're like old Silvester; you want to tell me what I should eat or drink, how many times a week I should have my wife. . . . That's over! I'm Tsar; I give orders to the whole snivelling pack of you. Now go! Get out of Moscow!'

I said I would gladly go back to my Lapps, but I did not think his conscience would leave him so easily.

'Conscience! You've been plotting with my so-called wife. Do you think I don't know that? Plotting in the night, behind my back! Yes, go to your Lapps! And don't show your face here again until I give you leave.'

When I reached my room a courtier was already instructing the servants to pack my clothes. I counted out the money Ivan had given me, kept what I would need for my journey to Pechenga, and told the courtier to restore the rest to the Tsar. I left the sable cloak behind, but the Tsar sent it after me. I sold it, and gave the money to the poor.

So, for the sake of a quiet life in the north, I abandoned my duty. I left that wretched girl, the Empress, to bear her fate alone. Of course, I found good reasons . . . I should be powerless, if I stayed in Moscow against the Tsar's decree.

He wrote me a letter afterwards, to say that Prince Kurbsky had sent him another distinguished prisoner. This one he had treated well, and settled on a good estate. For such an act of mercy, the Tsar argued, I ought to mention him in my prayers. But he did not say I should return.

I stayed among my Lapps. News reached me; the Tsaritsa had borne a son, who died. A few years later I heard of her own death. Ivan himself said afterwards that she was poisoned; he did not say by whom. Later he executed her brother.

The Church permitted the Tsar to marry a third time, which is a very rare thing. Our laws against the remarriage of widowers may seem harsh, but we need them here, where the peasants are so apt to beat their wives to death.

That third wife of Ivan's lived only a few days. Then the Church allowed him a fourth wife, which is unheard-of. The Metropolitan, even while he gave permission, decreed a solemn curse against any man who should dare to follow the Tsar's example.

Beside all this, I heard that the Tsar was executing Prince Kurbsky's friends, and their wives and sons. Kurbsky himself ran away to Poland. That was hard for me—to think of my pupil, my friend, as a traitor to Russia. I know that he had to flee, to save his life. But afterwards he fought against his own people, side by side with Poles.

Next, I heard that the Tsar had executed his cousin Vladimir, and Vladimir's wife, and the boys who should have been as dear to him as his own sons. And still I did not see what I ought to do.

I heard of Novgorod—the town the Tartars never conquered, the one town in Russia with honest merchants and good craftsmen. The middle of the river there does not freeze; people were pitched into it from a bridge. If they struggled out they were knocked on the head by Tartars in boats. These victims at least died quickly. But the Tsar was for ever trying new methods of execution. His grandfather kept perfect order in Russia by simply hanging offenders. It was not for the sake of order, but for his own horrible pleasure, that Ivan brought in these cauldrons, these gradual cuts, these slow fires.... And another way, which I, as a Russian, am ashamed to tell you, of making the death agony last fifteen hours, the victim being all the time fully conscious.

On his last day in Novgorod, the Tsar sent his hangmen out to seize one man from every street. They came before him praying that their deaths might be quick. Ivan smiled at them and said, 'You are all forgiven! You have been led astray by traitors, who plotted with Poland. All the blood spilt here is upon their heads. Go home; tell your friends that you have a merciful Tsar. I am at peace with Novgorod!'

It seemed he was at peace with his conscience too; he rode away smiling to Pskov, where he intended to make another blood bath. One man came out to meet him—a wandering holy man, Nikola, who wears nothing but a loin cloth all through the winter. This Nikola hangs his bare body with chains; he is always covered with scars where the metal has frozen to the flesh. He bowed before the Tsar, and offered him a plate of meat.

The Tsar cried out, 'Don't you know I'm a Christian? How dare you offer me meat in Lent?'

Nikola said, 'Yet you drink human blood in Lent.' And even as Ivan raised his whip, Nikola went on, 'If you touch one hair on the head of anyone in this town, your own head will be struck by lightning!'

Ivan looked up and saw thunderclouds overhead. I do not say he heard thunder; when has anyone heard thunder in the winter? Yet for that moment he thought it possible. He did not turn back; he rode on, into Pskov, but not one person there did he touch. And so you see this naked man put all of us to shame—the whole Russian Church, and most of all myself! Because, when you think of us as barbarians, think this too; it could all have been different. I could have made it different, if I had defied the Tsar, and stayed in Moscow. Just as the Princess Olga, a converted pagan, first brought Christianity to Russia, so that converted Cherkass princess might have restored the true faith to the hearts of degenerate Christians, if I had been there to help her. For years now I have prayed for one thing only—that before I die I might appear again before Ivan. God can do all things; he has moved the Tsar to send for me. And God has cured my cowardice. There is no power on earth which can make me betray Him, because for the first time I am not afraid of death. No, not of a death agony which goes on fifteen hours; the natural torments of old age last longer. So I shall beg the Tsar to rule as he ruled once, to end his cruel wars, and to make peace with his former general, Kurbsky.

Now, Yeremy, since we may never meet again, kneel down and let me bless you. May God keep you safe in all your journeys!

5

I would not quarrel with my fellow Englishmen, who say that the Russian faith is a jumble of pagan customs, a mumble of uncomprehended words; and that the jumbling and the mumbling can be no more to God than a saint's name in the mouth of a drunken man. Yet I say God can hear some Russian prayers. And that He heard the archimandrite I am certain, by my safety afterwards in many, many dangers.

The archimandrite's tale of Tsar Ivan made me more curious to see this legendary monster. It was now that ambition began to stir in me, an ambition so enormous that I did not dare confide it to another living soul. I was haunted by the old man's words. 'It could all have been different. I could have made it different.'

Why not? Sir Anthony Jenkinson, unarmed and alone, had by boldly confronting the monster turned him into a man. Already I spoke Russian better than Sir Anthony. I was younger than the archimandrite. All I needed was not to be afraid. I told myself that I was not afraid. Until, one night in April, I was wakened by some loud noise. As I raised my head it was repeated, and I thought I knew it. I shouted, 'It's Ivan! The Russians are attacking us!'

My companions, reluctant to stir, pointed out that nobody had sounded the alarm. I said, 'Perhaps they've killed the watchman.' Then we all heard another, louder crack, and I cried out, 'It is cannon fire! I tell you it's the noise the Spaniards made, when they bombarded Haarlem!'

We began to pull on our clothes as well as we could by the glow of the stove. I told the others we might die if we delayed; Ivan's ammunition might set our sleeping quarters on fire. They said that if I liked to get my cock frostbitten, by going into the wind without my breeches, I might. But they were getting dressed. In fact we all put on our furs. Then we went rushing to the top of the stone wall, to look.

Everything was calm, and lonely, and still. The moon shone quietly on the snow, on the familiar frozen sea. And, while we gaped at the emptiness, the noise came again. Crack! And then a louder crack, and another, sounding right under the wall. If a ghost army were beseiging us, with invisible cannon, at least the noise was real.

'What, lads!' Our night watchman was patrolling the walls. 'Have you come to watch the breaking of the ice? It's early yet; you hear the cracking long before there's anything to see.'

My companions laughed at me a week together. 'Where's your proper pride in being English?' they would say, imitating Mr. Rowley. Yet I found they liked me better. I had been too often right about Russian grammar. It pleased them that this once I had been wrong.

Besides, we were all filled with such a sudden, such a wild rejoicing, as you cannot conceive until you have lived through such a winter. Spring cried its coming to us in the very voice of God. Those thunderous cracks resounded on and on. The snow on the roofs began to murmur a promise, which was fulfilled with gurgling and running and splashing, a music we had forgotten. Then the great broken blocks of ice, pushed northward by the melting of the Dvina, and southward by the currents of the open sea, began to meet and fight and crash together. Sometimes they pushed one another to such a height that I thought they would overtop our stone wall. But the sea, the resurrected, living sea, carried them safe away.

We were once again shut up upon our island, guarded by patrols of moving ice. But this was not a tedious imprisonment. Every day we saw the sun melt the snow a little more, until there were patches of bare earth. While the running water was rejoicing in our ears, we saw that the earth did not rejoice. It was all grey, brown, black, without one blade of green.

'The snow kills the grass, every winter,' Mr. Taylor said. 'But wait a week or two; it'll be up to your knees.'

We celebrated Easter as best we could without a parson. A week afterwards I heard the monastery bells ring for the Russian Easter. The water between us now had only small pieces of ice; some other lads and I crossed in a rowing boat. The monastery courtyard was once more full of peasants. 'Christ is risen!' each peasant said as we greeted him. We answered, 'He is risen indeed!' and then submitted to the necessary kiss.

The monks let us Englishmen stand in the church porch for the Sunday morning service. Their music was as foreign as their words, and yet both words and music spoke to us of resurrection. Just as in England, spring had set the hens to laying. So it seemed right that what the priest took from the altar, and handed out among the people, were hard-boiled eggs.

Afterwards they all went to the graveyard behind the monastery. The people stood in little groups about the graves; and all the women began to scatter something, I thought at first the seeds of flowers. But as I drew near I saw that one woman was crumbling a piece of bread, another the hard-boiled egg she had received from the priest. I asked the girl crumbling the egg, 'Is this the grave of someone belonging to you?'

She said one word that needed no translating, 'Mama'. Then she told me that in the famine, two years before, when all the peasants were eating the barks of trees, her mother had refused to touch the few hares that the family trapped in the woods, or the few fish they caught through a hole in the ice. She had given her share to her children. When at last the English had been allowed to sell their grain to the people, the mother had grown too weak to eat. She died in the very sight of abundance.

From the other peasants I learned that the graveyard was choked with victims of that famine. Not that the way of death made any difference; it is their custom at Easter to scatter crumbs on all graves. Yet now, when I think of Russians, I think of them giving their food to people who have already died of starvation.

When the new grass did come, it came so fast that you might stand and watch it grow. With it came the violets, and then the rosemary. In a tumult of birdsong, I made ready for my journey to Moscow. I could not go until the first ship of the year arrived, bringing letters from home, cloth and gunpowder. (And two new apprentices, before whom we swaggered, saying, 'The women are all right. Never mind their husbands—just watch out for their brothers!') The letters we read again and again till they were tattered; the cloth was partly used to make us new clothes; and the gunpowder was loaded on to barges for Moscow. The Governor of Kholmogory put an armed guard on each barge. I thought I should have to travel on the powder kegs, but the leading barge was laden with letters and supplies for the Englishmen I should find on the way, and food for our journey, and presents for the Tsar. Amid these I slung my hammock. I was the sole English overseer of the whole train of barges. Mr. Rowley could not spare an older man to go with me. I was glad. In England I would have waited ten years, or twenty, for such a position of trust.

Whatever else might explode the gunpowder, the speed of the barges would not. As always in summer, it took us two days to reach Kholmogory. There I visited Katya. In the hay loft, while the wood pigeons crooned outside, she softly echoed their sound, murmuring, *'Lyooblyoo tibya. Lyooblyoo....'* Thinking only how beautiful the word was, I repeated, *'Lyooblyoo'*. Afterwards I felt sorry; what we had both said, she meant. Or so it seemed by her tears, when I told her I was going to Moscow. She sobbed out, 'My husband will kill me!'

'What! Does he know?'

'My mother-in-law told him. She knew because I was looking well and happy. If only I had a brother, he'd defend me right or wrong, but my brother's dead. So my mother-in-law told my husband to kill me. He picked up the axe and began to drag me about by the hair. But he didn't lock the door first; all the village came in. The village elders told him: "We don't want any trouble with the English. The English brought us grain when we were starving. Besides, they're the favourites of the Tsar. Better say nothing about it," they told my husband. "Just give her a good beating and keep her locked up." Then they left me alone with him, but he didn't beat me. He asked me how much money you'd been paying. I said nothing, and he called me a liar. He said the English always give their women money. Some get so much, their husbands can spend all day drinking vodka. He said he'd kill me if I let you in again, and didn't get any money.'

Though I was only an apprentice, and had no wages yet, Mr. Rowley had given me the usual travelling allowance. If he knew I had used some of it to pay Katya.... But if her husband were really to kill her.... If, even, he were to complain to the Company.... The thought was enough to part me from ten roubles. And she was not yet hardened; she did not ask for more.

Sailing south along the Dvina, I sailed into the heat. I had felt hot the previous summer, but then there was always a breeze from the White Sea. Now, inland, the wind that filled the barges' sails was a hot wind. I found the scarcely-setting sun as great an enemy as the cold had been three months before. You can fight cold by making a fire, but how can you fight the sun, or the insects the sun brings forth? I found some relief by swimming in the river. When I had climbed aboard again I would rub myself all over with a decoction of herbs to keep the insects away. And when the afternoon heat grew past all bearing, I would fall asleep.

I understood now why the Russians insisted on sleeping every afternoon. The short night—the dusky time when sunset became sunrise—was too beautiful to spend in sleep. I would sit on the deck of the barge, talking to the boatmen. It was then I learned the Russian words for things belonging to a boat. The boatmen chanted ballads, too; stories of heroes who had fought the

Tartars. They were long, night-long, outlasting every change of the pale sky.

Sometimes the soldiers guarding the gunpowder would begin a song. It would be taken up on the following barge, and then on the next, until from our whole caravan the deep voices trailed over the water. Their ordinary ribald songs would rhyme, as the sayings of the peasants do. But rhyme had no place in the sad, the slowly chanted songs. There was one they sang about a dead soldier.

'The mother weeps like a flowing river,
The sister weeps like a trickling stream,
The widow weeps like the falling dew.
Dew is gone when a new sun rises.'

From the boatmen I heard again that first line which had haunted me in the market place.

'Moscow is mended with walls of stone. . . .'

I begged them to go on, and, now that they knew me, they did. 'Within the walls is the terrible Tsar.'

The song was serious; it had no rhymes. As I listened I found that the common people knew almost everything I knew about their Tsar. The taking of Kazan and Astrakhan, the death of the virtuous Anastasia, the plunge into debauchery, the marriage to the Cherkass princess. . . . Thus far their story tallied with what I had heard. But they went on to blame the Cherkass woman for making their Tsar what they now called him, *'Ivan Grozny'*—Ivan the Terrible. Everything was her fault, even though it had happened after her death. The ballad blamed her for the killing of Prince Vladimir with his wife and sons, the slaughter in Novgorod, even the execution of her own brother.

'But now,' the last verses declared, 'our Tsar is himself again. He has made war against the Germans and the Poles, he has set up King Magnus in Livonia, he has married King Magnus to the daughter of our own Prince Vladimir. . . . He has brought us prisoners upon prisoners, slaves upon slaves!'

Kotlas, where the River Vychegda flows into the Dvina, was our first real town. We reached it in a burning afternoon. I was dead asleep, my head pillowed on my wallet in the only shade I could find on deck. The bump of the barge against the town quay woke me, and I scrambled guiltily to my feet. I knew there was an Englishman called Finch, an agent of the Company, in Kotlas. What would he think, if he saw me sleeping the afternoon away like a Russian?

'Where's the Englishman?' I asked. The boatmen pointed to the largest house on the quayside.

Nobody answered my knock. Yet this was the Company's office, as well as the agent's home; it should have been open for business. I walked round to the back, and found a window open. Through it I could see a bed, and a man covered by a sheet, fast asleep. His red-gold hair looked English, his complexion Russian; he had the bad colour of a vodka drinker. Under the sheet beside him something stirred, and I hastily left the window.

Once Mr. Finch had awakened, and sent his Russian woman back to the kitchen, I found him pleasant enough. 'No point in staying awake here,' he said. 'Even the Stroganovs don't work in the afternoon.'

'The who?'

'Up the side river,' said Mr. Finch, waving sleepily towards the Vychegda. 'You'll meet them. . . . Any letters?'

There was one for him. He read it without excitement. 'My mother's always the same. Read your Bible. . . . Beware of strange women. . . . But for any news of what's going on in the world—'

I told him what my own mother and father had written. 'Sir Humphrey Gilbert, Sir Richard Grenville and others are petitioning the Queen to let them look for the North-East Passage to Cathay. The King of Spain has been sending spies to find out what alloy the English use for casting ship's guns. There's plague in London; the playhouse is closed. The French King's younger brother, the Duc d'Alençon, is in England, asking our Queen to marry him. She's not likely to consent; he's pock-marked, and a Papist. Besides, France is in combustion; the Protestants have taken several towns. They're getting their own back for the massacre of St. Bartholomew.'

'You're lucky, getting a meaty letter like that. Our Queen's never going to marry a Frenchman, surely. You can't trust a Frenchman. The Poles have found that out.'

'The Poles?'

'There are some bits of news we hear first! The Poles chose Prince Henri of France for their king, early this year. I suppose he'd be the elder brother of the one courting our Queen. This Henri spent three months in Poland. Then, the other day, when he was in Cracow, he heard that his brother was dead, which made him King of France. He gave a great ball, slipped away from his guests in the dark, mounted a good horse, and galloped out of Poland. Never wants to see the place again. Don't blame him, if it's anything like here.'

'So the Poles will have to elect another king.'

'Old Ivan thinks they'll choose him. He's pouring out money to bribe the Polish nobles.'

'Does he really think any people on earth will choose him? Why, his own people call him "The Terrible".'

'Ah, yes, but—' Mr. Finch looked very wise. 'You must remember, when they say that, it's only their pride. Have you seen the old man?'

'Not yet.'

'Some of the people round here say he's out of his mind. But then, they're prejudiced. They ran away from Novgorod. When I saw the Tsar last year, in Moscow, he didn't look mad. More . . . cunning. Acting on some design, I shouldn't wonder. When Ivan turns on one of his bosom friends, calls him a traitor, off to the rack, and so forth, does he really mean it?'

'But they really are put on the rack.'

'Oh, yes, and roasted afterwards, or boiled. Sometimes a sharpened wooden stake. . . . But is it for treachery? That's what I question. Ivan surely must have some end in view—keeping his noblemen under, giving more of a chance to the merchants. . . .'

'But I thought he killed mostly merchants, at Novgorod.'

'Well, yes, there. . . . But you wait until you see the Stroganovs!'

When the Stroganovs came, they had trumpeters for heralds; their mounted guards were all tricked out with gold. Yet they themselves were dressed like simple merchants, and they greeted me as an equal.

'It is always a pleasure to meet an Englishman,' said Yakov Stroganov, the elder brother. 'Whichever way we look, we have Englishmen as neighbours. Our saltpits, you know, are on the Vychegda, only half a day's journey from here. A little further up the Vychegda your countrymen are smelting iron.'

'Smelting, poor sods! In this weather!' muttered Finch.

'And we are always glad,' said the younger brother, Grigory, 'to lend our bodyguards and guides to the English, when they go into the wilds, prospecting for iron.'

'If only they didn't find any, they could go home,' said Finch. 'But the stupid bastards always dig up something.'

When we began to drink, it was Finch who drank himself silly. The Stroganovs remained alert. So did their friends and followers, men taller than most Russians, with bright black eyes. These were Cossacks—the people from the south, who, according to the archimandrite, had grown as wild as the Tartars they fought.

The Cossacks questioned me about the English efforts to reach Cathay. I said the mariners who tried it found their way clogged with ice, but they had sailed as far as the mouth of the River Ob. Some thought it possible to go on by sea, but others believed that a strong, well-armed expedition, sailing up the Ob as far as it was navigable, could soon reach Cathay.

''They're quite right,' one of the Cossacks remarked. 'From the Ob to the Great Wall of Cathay is only three months' riding.'

'Three months!' I echoed.

'We took a whole year,' said another Cossack. 'But that was because we did not know the way.'

'Besides,' the first one added, 'we had to wait for the Queen of the Black Mongols to give us a pass. She would know us better another time.'

So I heard the story of the Cossacks' journey to Pekin. The safe-conduct of the Mongol Queen had brought them through a gate in the Great Wall. Then, day after day, they had passed through towns larger than Moscow, built all of stone, their idols covered in gold leaf, their markets perfumed with cloves and cinnamon. As the wealthy Chinese walked about the streets, their servants held little round canopies over their heads, to shield them from the sun.

'Did you see any unicorns?' I asked. They said not one. This made me doubt them; I had always heard that unicorns abounded in Cathay.

Pekin, said the Cossacks, was built of white stone, four-square, in circuit four days' going, cornered with four white towers, very high; and other towers along the wall, white intermingled with blue, and loopholes furnished with ordnance.

I asked them whether they had written all this down. They said yes; in a letter to the Tsar. They added that the Tsar had shown them great kindness, though they had failed in their mission, which was to deliver his messages to the Emperor of Cathay. The Chinese courtiers would not let them enter the gold-roofed inner city of Pekin, since the presents they had brought were not worthy of the Emperor's notice.

'But the English could bring him presents!' Yakov Stroganov said. 'We and the English, working together—we could rebuild the old roads to Pekin, and win back all the trade the Tartars lost!'

As they expounded this plan, the Stroganov brothers looked more and more like Englishmen. Finch was looking more and more like a Russian, as he drowsed forward among the plates. I myself, as I listened to tales of battles against Uzbeks and Bashkirs, was lulled to a drowsy enchantment.

'The Siberian Tartars have never seen gunfire,' said Grigory Stroganov. 'A few cannon could scatter them. And then what trouble would we have with the Samoyedy? We'd soon teach them to eat human flesh and worship elks! All we need . . .' he was now whispering in my ear, '. . . is one of your barges of gunpowder.'

I was at once broad awake. 'They are not *my* barges of gunpowder,' I said. 'Your Tsar has bought them all, and I am charged to deliver them into his hands.'

'But we have the Tsar's commission!' Yakov Stroganov said. He unrolled a parchment, which did indeed bear the Tsar's hand and seal. He was about to roll it up again, when I insisted on reading it.

It empowered the Stroganovs to build fortresses against the raiding Siberian Tartars who came over the Ural Mountains. The Stroganovs might also recruit soldiers, but not peasants belonging to noblemen, or runaway slaves, or known criminals, or highway robbers. (Here the Cossacks laughed.) The Stroganovs were to have the right to mine copper, tin, lead and sulphur wherever, in those north-eastern parts, they might find it. (Not iron; that already belonged to the English.) They might hold their own courts of justice. They were to trade, paying no customs, with the peaceful inhabitants of Siberia. The warlike inhabitants they must pursue and punish.

'So you see,' said Yakov Stroganov, 'we need gunpowder.'

'No doubt,' I said, 'the Tsar will give you some.'

'What!' exclaimed Grigory. 'Take it a thousand miles to Moscow, and then send it a thousand miles back again! It's not our time you're wasting—it's the Tsar's time.'

I said, 'I have to carry out my orders'—and looked at Finch. He continued to snore.

Yakov Stroganov said, 'A number on a sheet of paper—what's that, to a man with your skill in reading and writing? And a young man like you can find some pleasant ways of spending money.'

I shouted, 'You can't bribe an Englishman!' The Stroganovs looked at the sleeping Finch, and smiled.

I tried to shake Finch awake; I pulled that red-gold hair. His eyes remained obstinately shut. I had to find my own way out, and the way I found was that 'proper pride in being English', which had some purpose, after all. Imitating as best I could the tone and manner of Mr. Rowley, I told the Stroganovs I would report their request to the Tsar. The very name was threat enough; these men, the richest in Russia, with their own private army, became obsequious at the merest hint that I might speak unfavourably of them to Ivan. I sailed away safely, with all my barges intact.

When we turned westwards, up the River Sukhona, the wind failed. The boatmen had to walk along the bank, dragging the barges with ropes, as horses would in England. (In Russia, men are cheaper.) We came to the canal which cuts off a bend in the river, near Totma, and I thought of the archimandrite. He was right; the Russians who made this canal, two centuries before, must have had some skill. But their descendants had not the skill to dredge it. We scraped along, bumping against the stony bottom, and at last stuck altogether. Some peasants, who had been lying in wait for this moment, came slowly along the towing path. I had been warned beforehand that I might need their help, and I knew the proper amount to pay. So, when I heard what they were asking, I said, 'Oh, go and rape your mother!'

The bargaining went on for hours. All the boatmen in the barges behind us joined in on my side. At last, when every man present had been many times advised to rape his mother, and accused of having already done so, we settled the price. Then the boatmen laid planks between the boats and the shore, and began to unload their cargoes on to the peasants' backs. I was not allowed to help; the boatmen thought it would bring bad luck if a gentleman were to lift anything. So I had to watch the kegs of gunpowder, and the precious gifts intended for the Tsar, heaved by clumsy hands on to bodies that staggered under the weight. At last the barges were light enough to float, or rather to be dragged along by the boatmen. With how much less labour, I thought, and with how much more profit, all these men might have worked together to deepen the canal!

Coming into Vologda, I saw men making new canals, wider and deeper than I had seen anywhere. They were to carry Ivan's warships, built by English craftsmen. It was a saint's day, and we could scarcely find men to make our barges fast, or sell us food. The whole town had gone to look at the new ships.

I never felt more proud of my own country, than when I saw those twenty ships, each with a different figurehead—lions, dragons, eagles, elephants, unicorns—gilded and silvered and painted with as much delicacy as brightness. Like the Russians, I gaped at the masts, the rigging, the beautifully forged guns. But while the Russians wondered aloud at their hugeness, I wondered how it was possible to make a fighting ship so small. Of course an English warship of the proper size could not be floated on an inland canal. The workmen told me that all the ships had flat bottoms, like barges. They would sail somewhat clumsily, if they came to the open sea, which did not seem likely. Vologda lies four hundred miles from the White Sea, and as far from the Baltic, but that's as the crow flies. By rivers and canals the distance is more than doubled. The great sails, not yet hoisted, would often be useless, for there is no relying on the wind inland. So the ships would have to be dragged, Russian fashion, by men. The weight of the guns would make this very laborious. 'God knows what old Gruff-and-Grum thinks he can do with these,' one shipwright said. 'Better if we'd built them at Yaroslavl, where we make the ships for the Volga trade. You can do a proper job there; the river will take a big ship, with a keel. Still! He would have 'em built here. Done our best, that's all we can do.'

'Gruff-and-Grum's coming to look at 'em soon,' said a carpenter. 'Hope he likes 'em.'

'If he doesn't you know where it's going.'

'Up your arse.'

'Up yours!'

The men laughed more than was warranted by such a commonplace expression. A man who seemed to be in authority emerged from inside one of the ships, and told the workmen that it was no laughing matter.

This man was Thomas Brown, master shipbuilder, not a Company agent but directly employed by the Tsar, and consequently very rich. He invited me to dine at his large house, which was also the home of his workmen. They all ate in common, the food being as much like English food as they could get. Even the cooks and serving men were English.

'My men won't so much as lift a hammer,' Mr. Brown explained 'unless they have English food, English rooms, English beds.... They don't know they're in Russia.'

'We know that all right,' said one of the goldsmiths. 'After that year we spent shut up in this town, with Ivan's men prowling round us like wolves. Mind, Sir Anthony Jenkinson put all that right.'

Mr. Brown shook his head. 'We've been lucky so far. None of those fifteen-hour executions here.'

I said, 'Do the Russians really . . .?'

'Oh, don't you fret yourself,' the goldsmith said. 'Old Gruff-and-Grum knows very well how much he needs us. I've met him, you know. Talked Russian to him, the best I could. I was working on his new crown, and he said, "Take good care of that; you're surrounded by Russians, and they're all thieves." I couldn't help smiling. He asked me why. I told him, "Because you're a Russian yourself." Old Grozny burst out laughing. He said, "Not me! My ancestors were Germans, thank God!" We parted the best of friends.'

The other workmen agreed. 'He'll never treat us like he treats his own people.'

South of Vologda the rivers were shallower, the English canal works incomplete, and the heat still more oppressive than before. I thought we should never reach the next town, Kostroma. Yet we did reach it, and I saw for the first time the River Volga. A great ship was passing along it, and I took off my hat, for the ship, of course, was English. It was on the first part of its two-thousand mile journey to the Caspian Sea, and Persia.

We heard all the bells of Kostroma pealing. This made the boatmen uneasy, because it was not a holiday. Yet it seemed a joyous occasion. The clergy had brought out their ikons and silk banners; the streets were lined with trestle tables, behind which the women stood ready to offer bread and salt; every soul was wearing holiday clothes. Yet nobody smiled; nobody spoke. The women's cheeks were painted red, but I could see that the men's faces were pale.

There was a sound of trumpets, far away, and the people crossed themselves, whispering, 'He is coming!'

All the clergy now gathered in procession, carrying their banners, and made for the gate of the town. I followed. Nobody else did; my boatmen had hidden themselves (under the hatches of the boats, I afterwards discovered).

At the town gate the watchmen let the religious procession go out, but they stopped me. What was I doing, in foreign clothes, and alone? Didn't I know the Tsar was coming? I said I had barges laden with gunpowder for the Tsar, from the Russia Company of England.

At the word 'England' they let me through. I was escorted to the Governor of Kostroma, who was already outside the town, waiting with all his followers for the Tsar's arrival. He embraced me, and asked whether, besides the gunpowder, I had brought the Tsar any presents.

I said, 'I have brought him candlesticks of pure gold, so finely made that the metal seems to be lace, and rolls of a woollen cloth lighter than silk, yet warmer than serge, to make dresses for the Tsaritsa and her ladies. And a mirror framed in silver, as tall as a man.'

'Oh, God be praised!' exclaimed the Governor. 'That will surely please the Tsar.' He sent some of his own servants to the barge to unload these gifts.

Before they could return, we heard the trumpets again. On the road a cloud of dust presaged a great company of horsemen. Then, through the dust, we saw the flash of drawn swords, and the gleam of helmets, and bright silver trappings on black horses. The Tsar's Tartar advance-guard came riding like the centaurs from the fables, man and beast one creature. When they came close they seemed, not to check their horses, but rather to flow aside, as water does on meeting a great rock. In an instant they were motionless, drawn up in two long ranks, between which we must pass to greet the Tsar. The Governor advanced very slowly, as if scarcely able to drag one foot before another. Following him I felt my own blood congeal, as there loomed on either side of me that hideous bodyguard. Looking up at their faces I knew it was true; at one word from their Tsar they would seize a man and put him on a stake sharpened and soaped to enter his bottom; and they would leave him till the weight of his own body made the spike work its way through to liver and heart. Would that really take fifteen hours?

In front of me the Governor and his men sank to their

knees and beat their brows against the ground. I took off my hat and kneeled, as I would to my own Queen. But beside me someone whispered, 'Do you want to kill us all?' Rough hands forced me to abase myself as they did. Though I heard voices, and the noise of horses, I did not dare to take my face off the ground.

I heard one clear, high voice, 'Who is that foreigner?' Then something rapped me sharply on the shoulder. I let out a cry of pain, and looked up.

It was the Tsar's own iron-tipped staff that had struck me, and the Tsar's bony face was grinning into mine. Evidently he had got off his horse to look at me.

I said, 'Great Sovereign, Ivan Vasilyevich, Tsar and Grand Prince of all Russia, of Vladimir, Moscow and Novgorod, Tsar of Kazan, Tsar of Astrakhan, Sovereign of Pskov, Grand Prince of Smolensk, Tver, Yugoria, Perm, Viatka, Bolgar and others; Sovereign and Grand Prince of Novgorod of the Low Country, of Chernigov, Ryazan, Polotsk, Rostov, Yaroslavl, Byelo Ozero; hereditary Sovereign and Master of the Livonian Land of the German Order, of Oodoria, of Obdoria, Kondia and all the Siberian land, and ruler of the northern land, glory to thee!'

'Quite right!' said Ivan, whose eager look showed that he did not find his titles a word too long. 'And where do you come from?'

'England.'

At the magic word he extended his hand for me to kiss. I told him I was travelling with barges full of gunpowder for his army.

He began to question me about my journey, in such a pleasant way that I ventured to remark that the canal near Totma ought to be cut deeper. He told a man who stood near him, 'Make a note of that, Shelkalov!' Shelkalov glowered at me, as he had at Sir Anthony Jenkinson, but the Tsar continued to be gracious. He asked me proudly, 'Did you see my ships at Vologda?' I told him I had. His eyes blazed, and he burst out, 'What traitor showed them to you?'

If Sir Anthony Jenkinson could do it, I could. It was only

that the words would not come out. The twisted face was thrust into mine, the harsh high voice repeating, 'Come on, tell me—which of my traitors dared show my ships to a foreigner?'

I tried to look away. But that meant looking at the faces of the Tsar's attendants. From each ambush of hair and beard, eyes were watching not me but their master. They would be ready, when he ordered the blow. Nearest the Tsar was the worst face of all—a young version of his own. This one smiled, as if knowing that I felt the soaped and sharpened stake already.

Then I saw that a tall man, very young, with beard as downy as my own, was watching me with compassion. Because of this one face, I was able to say quietly, 'May it please Your Majesty, nobody showed your ships to me. It is impossible to pass through Vologda without seeing them; their masts tower above the houses. It was a holiday, and people were flocking to look at the ships. I took the liberty of joining some thousands of your subjects, who were admiring these curious designs.'

'What do you mean—curious designs?'

I described the painted figureheads. That seemed a safe subject; I did not mention the guns.

'A crafty lad—praising the work of his own countrymen!' said the Tsar. He laughed, and the faces round him showed their teeth, which did not make them less terrible. Only the very young man smiled pleasantly, as if glad that I had come so well out of danger.

'You looked carefully at my ships,' the Tsar went on. 'How many did you see?'

I told him—twenty.

'You shall see forty before long. Tell them that, when you go abroad. Tell the foreigners how beautiful they are. Have you ever seen ships more beautiful?'

'Never, Your Majesty.'

'And if you knew what gold and silver ornaments, what tapestries, they are to have inside! Yet some people say that your Queen has the best navy in the world.'

'Your Majesty, she has.'

'Then why have you deceived me? Why did you say mine was the best?'

Still watching that unknown, that single human face, I told the Tsar that though his new ships were indeed the most beautiful in the world, ours were bigger. 'They have to sail the wild seas. Their sides are made thick—so thick that a cannon ball can hardly pierce them.' I began to explain what a sharp keel was, and how it cut its way through the waves as a flat-bottomed boat could not. Ivan took the point.

'Yes, I have studied the ships your countrymen build at Yaroslavl. But I needed those others to sail the inland canals, and come secretly into the Baltic. What a surprise we'll give them there! I'd like to see those great fat German faces, when they look over the walls of Reval, and find we're surrounding them by sea as well as by land!'

I thought. What a surprise he'll give us at Rose Island! We had been so sure he could not attack us in summer, for want of ships. . . .

Now that strong arm was thrown familiarly round my shoulder. The Tsar was of my height; I was no longer among the famine-stunted peasants of the north. Waving aside the Governor of Kostroma, who was trying to present him with a cloth-of-gold kaftan, he walked with me into the town. He seemed scarcely to hear the church bells, or the blessings intoned by the priests. As the women held out bread and salt on white napkins, he would take a pinch here and there, wave, smile, and turn again to me.

'What sort of guns do your ships carry?'

'Every ship carries forty brass pieces of great ordnance, bullets, muskets, powder, chain shot, pikes, defensive armour. . . .' I had to use German words, which are the only words the Russians have for any instrument of modern war.

'*Sprichst du auch deutsch?*' asked the Tsar in delight.

Half in German, half in Russian, he questioned me about the engines which make wildfire. I told him how, in a sea battle, one ship would try to set another afire. If that failed, the attackers would sail as close as might be to the enemy ship, and use grappling irons to draw it closer still. Then sailors and musketeers would leap from one to the other for hand to hand fighting.

'How many?' asked Ivan eagerly. 'How many men?'

I told him, on every royal ship of war a thousand, counting the officers. He asked how they could all be fed. I said there was ample room aboard for flour, oatmeal, beef, bacon, fish, dried peas, butter, cheese, beer, brandy and casks of fresh water.

'And the sails? How many sails?'

'Five or six great sails, besides the small ones. The ship's captain knows when each one is to be spread or furled, or turned as the wind changes, to hold the ship on its course. It is done in a moment, all the sailors moving as one man. When one of our great ships goes out of port, with cables and tackle and masts in perfect order, with sails unfurled, with drummers and pipers and trumpeters playing, then our Queen's arms fly at the masthead on a banner of silk, and the ships of other nations bend and bow.'

Ivan smiled, then suddenly looked round at his courtiers. 'They don't like to hear this,' he said in German. 'And they certainly don't like *me* to hear it.'

Indeed I thought some of those ambushed eyes were darkling, though they ceased to do so when the Tsar turned his head. The Tsar's young replica—his elder son—seemed to find all this tedious. Only the tall man with the friendly face gazed at me as if he could never hear enough. Near him, I now saw for the first time, was a young, beardless boy with the gentle smile of a half-wit, so richly dressed that he could only be the Tsar's younger son, Feodor. The Tsar continued his questioning.

'How many of these great warships does your Queen possess?'

'Forty, Your Majesty.'

'So she could send forty thousand soldiers to a friend! I thought so! Surely she won't begrudge me this help when I need it next!'

I was about to tell him why the Queen could not send soldiers, when we reached the cathedral. Here Ivan let me go, so that the court could range itself in the proper order. From somewhere among the baggage trains came his wife, to kiss his feet and take her place beside him. His two sons followed.

All the court swept past me, and I (being a heretic) stood in the cathedral porch while they gave thanks for their safe arrival. I was glad of the chance to think.

Light-headed at my success in turning away the Tsar's wrath, I had almost blurted out what was wrong with his calculations. What he did not know—and I, thank God, had been prevented from telling him—was that not half the men on a warship were soldiers. And those who were, though they might be fit for a hand-to-hand skirmish, had no notion of a battlefield. Moreover the stores of which I had boasted were ample for short voyages only. The best of ships, in the best of weather, could not reach the White Sea from England in less than a month. With contrary winds it might be two months, or three. Indeed, the Company's captains carried iron rations for four months, because the Danes might be too hostile, or the weather too rough, to permit a landing anywhere on the way. This meant that they could carry very few men.

If I had told Ivan all this, he would have known that he could do what he liked to my countrymen in Russia.

The moment the service was over he came to me again. Now his questions were about the other languages I knew. 'Here am I surrounded by louts who scarcely know their own language. Look at Boris here!' He gave one of his playful iron-tipped taps to the young man whose looks I thought friendly. 'Boris is a good lad in his way. But he can't read or write.'

Bowing low to Boris, I said, 'No doubt he has been too busy serving Your Majesty in other ways.' The friendly face looked at me gratefully.

'Oh, he can fight!' Ivan said. 'He's tall; his men can find him on the battlefield.'

We arrived at the Governor's house, where we were to dine. The Company's presents to Ivan, newly unpacked, were displayed just inside the dining hall. He was delighted, especially with the mirror. He had never before seen himself all at once, from glittering top to red leather toe. He called his courtiers to look at their reflections, but was constantly thrusting them aside to take another look at his own. It was curious to see these bearded men preening themselves like women. (The

Tsaritsa and her ladies had no share in the delights of the mirror; I had already seen them go up the stairs to the women's quarters.)

The Tsar embraced me, and told me to thank my Queen. I explained that these presents were not from the Queen, but from the Company.

'What!' said Ivan. 'Is your country so rich that common merchants can give such presents? Look at this—' he turned to his courtiers '—look what friends I have abroad!'

Throughout dinner he continued to talk to me. 'Have you really been in France? That must be a terrible country! Such cruel people! Why, the French king had a hundred thousand of his own subjects murdered, on the night of St. Bartholomew, only for some difference of religion. Or what they call religion. Papists . . . Lutherans. . . . They're none of them real Christians. The Poles are sorry they chose a French king. They disobeyed me, and God punished them! Let's hope they know better now.'

He called for more wine, and pressed more on me. I pleaded that I was not used to such good wine (which was true) and to my great relief he let me sip instead of swallowing, so long as I would continue to answer his questions.

'Is it true that in England the Queen is Head of the Church?'

'Quite true, Your Majesty.'

'So everyone obeys her—the priests as well!'

'On spiritual matters she listens to the advice of her bishops.'

'Listens to advice—yes, but makes up her own mind! Ha! Tell me, what does she do to people who plot against her?'

'If they rise in open rebellion, like the Papists did in the north, some of them pay a heavy fine, and some are put to death.'

'Surely it's the greater number she puts to death? And cruelly, from what I hear. In your country, traitors are gutted. And castrated, which hardly seems right among Christians. Tell me, have you ever seen that done?'

'No, Your Majesty.'

'Why not? You have a weak stomach, ha?'

'There have not been any traitors executed in London lately, except the Duke of Norfolk, who tried to marry the Queen of Scots. He was beheaded.'

'But his relations? His household servants, who ought to have denounced him? How many of those were executed?'

'None, Your Majesty.'

He laughed in my face. 'You mean you don't want to tell me how many there were! You're very loyal to your Queen. Well! I don't blame you. That's the greatest of all virtues—loyalty. Have you heard of Prince Kurbsky?'

'Your Majesty's former general, who ran away to Poland?'

'That's the man! He sent his servant Vaska to me. This Vaska said shamelessly, in front of everyone: "I have brought you a letter from your exile, Andrey Mihailovich Kurbsky." I was so angry with the wretch, for daring to speak the name of a traitor....'

The archimandrite's voice came back to me. *I shall beg him to make peace with Kurbsky.*

'...so angry at hearing that name,' the Tsar went on, 'that I pinned Vaska's foot to the ground with this iron spike, here, on my staff. I leaned on the staff and said, "Now read me the letter." He never once cried out. His face was as white as that cloth, but he read the letter aloud from beginning to end. Afterwards we tortured him. But not one word would he tell us about Kurbsky's plans. I sent for my old friend Malyuta. He was a good lad, Malyuta. He could get words from the dumb. But this Vaska's only words were in praise of his master. It was a beautiful death. I was moved to tears. I ordered a Mass for Vaska, and prayed to God that He might in His great mercy send me such a servant!'

I succeeded in saying, 'I am sure Your Majesty has many loyal servants.'

He shook his head. His eyes were full of tears. 'Malyuta's dead, God rest him! Laid down his life in battle for me. Well! God takes the best.'

Then he began to talk German, which was agony for me, because he spoke it badly enough when sober, and he was now very drunk. But he lapsed into his own tongue so often that I caught the drift.

'For myself, I take what God sends. But before I die I'd like to see my boys secure. My little boys....'

Watching the gentle, vacant face of Feodor, I could see that he would always be a little boy. But the elder son looked like a full-grown fiend.

'There's one hope,' Ivan went on. 'Your country! England! Your Queen has agreed that if ever I need a friend I can go to her. Sir Anthony Jenkinson brought me a letter to say so. That's why I keep my treasure on an island in the White Sea. Once I'm there, it's not far to Rose Island, and your ships can carry me to England without ever touching the soil of my enemies. It's true, ha? Your ships can sail all that great way without once touching land?'

'Quite true, Your Majesty.'

'My sister, my dear sister—I mean your Queen—has promised me a house and estate in England. She says I'm to pay my own way. Well! I can do that. That would serve my traitors right, eh? Leave them to rule Russia, let them see whether it's easy! Come whenever you like, your Queen wrote; bring your sons, your wife....'

He broke off, looked round and made another effort to speak German. 'My wife, though . . . well, if I'm to believe the priests, my present wife is not my wife at all. No need to bring *her* to England. Because the best thing would be if your Queen and I.... They tell me she's a very beautiful woman.'

I said, 'Your Majesty, Queen Elizabeth has a fair skin, and red-gold hair, and a very noble carriage....'

'Und sie ist eine Jungfrau!' whispered Ivan.

'An undoubted virgin,' I agreed. 'But she is past forty.'

'What!' shouted Ivan. He was once more speaking Russian. 'She is barely thirty!'

'With respect, Your Majesty.... We English celebrate her birthday every year. Last September we drank to it on Rose Island, and she was forty then.'

'Do you mean that a man who has lived in England for more than a year must know the Queen's real age?'

'Yes, Your Majesty.'

'There is treachery somewhere,' muttered Ivan. 'Where's Dr.

Bomel? No, I remember; he's busy. If I have to find myself another doctor. . . .'

His mutterings became incoherent, and seemed directed to his wine, rather than to me. Presently he patted my arm in a friendly way, and rose from the table. Then he lurched off to his afternoon sleep. The Governor, having seen the favour I was in, had made a good room ready for me. The bed was soft and I fell asleep at once, not counting my flea-bites till some hours later, when a servant came to say that the Tsar was asking for me. It seemed he wanted to inspect the kegs of gunpowder on the barges. We strolled beside the river, the Tsar telling me how he had used gunpowder to blow up the Tartars in Kazan. 'We'll give the Poles a taste of this,' he said, 'if they haven't the sense to choose the right king.'

In the gentle mood of evening, the Tsar for the first time spoke to the Governor. Seeing them occupied with their own affairs, I fell back a few paces. It was then that the young man with the friendly face, Boris, asked me, 'Did you say that in your country the barges are pulled by horses?'

Beside Boris walked the gentle, half-witted Prince Feodor. It seemed these two were always together, which puzzled me, since Boris was no half-wit. His talk showed such a wide knowledge of other countries that I could scarcely believe he had it all from hearsay. But people who cannot read often have a wonderful memory. Just as the boatmen could remember their endless ballads, Boris now repeated, word for word, all I had said that morning, and asked me to tell him more.

At supper, everything went as the archimandrite had said. We were entertained by buffoons dressed as women and monks. At a given signal all the real monks rose from the table. So did the half-witted Feodor, and, with him, Boris. The Tsar called to Boris, 'Take good care of my boy! See him safe to bed!' Then he turned to me. 'I have ordered a present for you.'

A man entered, wearing a long robe of a doctor. Before he opened his mouth, I knew he was not a Russian. His nose was too straight for that, his mouth too narrow. As he bowed before the Tsar, he smiled, and the smile gave me a crawling in the flesh. Why should I notice one more wicked face? All round me were drunken, bloodthirsty barbarians. But this was a civilised man, who knew better.

The Tsar said to him in German, 'Eliseus! I have some questions to ask you. But not now. We have more urgent business, ha?'

'The doctor said, 'They are ready for Your Majesty.'

Ivan shouted, 'Bring them in!'

His guards brought in three pale, silent girls, hardly more than children. The Tsar said to me, 'Choose one!'

Each one was plainer than the next. I put my hand out at random, and touched the middle one. She stared at me as if I were demon or ghost.

With a gesture full of paternal kindness, Ivan handed one of the girls to his eldest son. He sat the third girl on his knee and said, 'Now the others!'

The room was at once full of women, some drunk, some extremely drunk. If sober, they could hardly have endured the savage arms which opened for them, the more savage cries of welcome.

These must be the former virgins, no longer of interest to the Tsar. They had the battered look that most whores have. Whereas the three children first brought in were still virgins. To the Tsar's way of thinking this was the best present he could give me—a girl intended for his own bed, or floor. My tastes were different. Like a good many other men, I would rather meddle with a viper than a virgin. My one thought was how I could, without offending Ivan, be rid of the girl.

She was a handful. Before I knew it, she had darted from my side, and hurled herself into the arms of one of the drunken women. Others gathered round; they seemed to be explaining something to her. But the musicians were playing; the men wanted to dance; they lurched in among the women. One man was laying hands on my girl, but I grabbed her away. She struggled, sobbing and crying. I tried to reason with her, first in Russian and then in Danish. 'If only you'll hold your noise, I'll take you to a safe place.'

Ivan was now forcing his own girl to dance, by stabbing at her toes with his spiked staff. So he did not see me drag my girl out of the room.

We were in the courtyard, which was full of soldiers. They had their own whores, and were following their master's example. It was not yet dark; at any moment they might see my girl and seize her. Still clutching her hand, I edged round the courtyard, and then ran for the stairs that led to the women's quarters. The girl was no longer protesting; the sight of the soldiers had silenced her.

At the top of the stairs we were confronted by a closed door. I knocked without result. I tried shaking the door, battering it with my shoulder, and at last kicking it. Still the only noise was the rattling of the bolts that held it shut. I said to the girl: 'If only I could get you to some convent!'

'Oh no!' she said. 'Not a convent! That's where I've been up to now.'

This took my breath away. But where else, after all, would Ivan keep his virgins?

Like Karen and Anna, this girl had been captured in the Tartar raid on Oesel, two and a half years before. She was then eleven. 'The Tartars didn't exactly hurt me,' she said. 'They tied me up and put me in a basket on one side of a horse's saddle. There was another girl on the other side. We bumped along in those baskets for days and days. When they stopped to eat or sleep they would take us out of the baskets, but they always guarded us. We couldn't run away. Then when we came to Moscow they put us into a convent. The nuns are spiteful to us. They've come with us on this journey; they never let us out of their sight. We think they mean to kill us. That is—we did think so, but just now. . . .'

She began to cry. I said, 'Why should the nuns kill you?'

'It's the doctor—Dr. Bomel. He's cruel. Sometimes he comes and takes a girl away, and we don't know where. It's no use her struggling; there are men with muskets. And when we ask the nuns where she's gone, they say, "Ask Dr. Bomel." And when we ask him, he only says, "Ask the Tsar." And whenever a girl annoys one of the nuns, by answering back, or just crying too much, the nun says, "You'll be the next one for Dr.

Bomel." So it can't be anything good he wants them for, can it? And then tonight he picked out three of us at once. And we all thought the men with muskets were going to kill us. Dr. Bomel said, "Mind you put up a good fight! The Tsar likes a fight." We thought the other girls must have died fighting, like men do. But they're alive; I saw some of them tonight. That's what made me cry. Because they didn't look the same. They were so strange—the way they spoke, and their eyes.... Do you think they're under a spell?'

I whispered, 'How old are you? Thirteen?'

'Fourteen, I think.'

I was about to shout curses at the dumb door. But I reflected. The women inside might be frightened of a man's voice. I said, 'If you were to call out.... Do you speak Russian?'

'A bit. We have to speak it, to the nuns.'

'Do you know how to say "Help me"?'

'Pomogi mnye.'

'Say it in the plural: *Pomogitye*. That's more polite. You're talking to the Governor's wife, and the Empress.'

But not, I reflected, to the Cherkass Empress, the woman with the Tartar face who had dared to remind Ivan of his marriage vows. She was dead, and the present Empress cowed, no doubt, by the fear of death.

The girl cried out for help, with a noise more shrill than any man could make, more impossible to bear in silence. Whenever she paused for breath I whispered, 'Go on! Louder!'

At last a woman's voice from inside said, 'Go away! For God's sake go away!'

The girl wailed, 'I'm frightened....'

'Be quiet! You'll have us all in trouble. Who are you?'

The girl answered, 'I am Madelyn van Uxell. My father is a nobleman in Denmark.'

The voice inside said, 'Then you belong to the Tsar.'

I cried out, 'The Tsar gave her to me; she's mine. I can have her protected if I like.'

'Who are you?'

'Jerome Horsey, from England.'

'The Englishman!' There was a great rustling and bustling

inside. At last we heard the bolts being pulled back. A lady appeared; I recognised her as the Governor's wife. Before I could speak she was prostrate before me, beating her brow on the ground.

'Forgive me, forgive me!'

From force of habit I said, 'God will forgive.'

'If we had known it was you . . .' said the Governor's wife. 'My husband said we were not to open the door to anyone, on any pretext, but for you. . . .' She drew the girl inside. Then she turned back to me, whispering, 'I will look after her like my own daughter. Anything for you, after what you've done!'

She was gone, and the door bolted again, before I could ask what she meant.

My sleep was much disturbed by the noise of orgies, and by my own desire to join in. But, if I was frightened by virgins, I was disgusted by drunken women, and so I remained alone, consoled by the thought of my own extraordinary virtue. (Which, it seemed, was greater even than I knew, else why had the Governor's wife been so grateful to me?)

In the morning I was woken early by church bells. The Tsar was going to the cathedral to repent his sins. As he returned I went to bow to him, and saw that his forehead was bruised and bleeding. He advised me to follow his example in doing penance. I said I would. If I had told him that there was no need, he would have known that Madelyn was still a virgin, and therefore, to him, desirable.

Pointing to Boris, the Tsar cried out, 'This is the only one with a good conscience. He sleeps with his wife!' Boris bowed. To me the Tsar whispered, 'Boris is a bit of a milksop, but he makes my Feodor happy. Two illiterates together, ha!'

This, then, was how Boris could live in that court, and remain a human creature. He kept company with the only innocent.

As the court made ready to leave Kostroma, I ventured to ask Boris what I had thought it unwise to ask the Tsar. 'What has become of the Archimandrite Feodorite?'

Boris looked blank. 'I don't know him.'

I described the archimandrite, and said that he had been expecting an audience with the Tsar.

'If he did come,' said Boris, 'I didn't see him. I wasn't there.' Then he smiled. 'I should have liked to tell you something pleasant, my English friend!'

I watched them ride off towards Vologda—the Tartar advance guard, the Tsar, his sons, his courtiers, his soldiers, his baggage train, his wife, the courtiers' wives, their servants, another baggage train, more soldiers, the whores.... How was it possible, I wondered, for the whores to be kept from ever meeting the virgins? Now I could see; they rode with a company of soldiers in between. The virgins were guarded by nuns, and the nuns by Dr. Eliseus Bomel.

When the town's north gate had closed behind the last of the Tsar's rear-guard, the Governor came to me and bowed to the ground. He invited me to return to his house. And there I found the gentry and clergy of the town, waiting for me with gold and jewels which they begged me to accept. I said they were mistaken if they thought that I was the Tsar's favourite, and could do anything for them. They replied that I had saved their lives already.

It seemed that five years before, when the Crimean Tartars were trying to recapture Astrakhan, the Tsar entrusted the command of his army to his first cousin, Vladimir. The army assembled for its journey down the Volga at this very town, Kostroma. The whole town turned out to greet Vladimir. They knew he had been many years in disfavour, and this had seemed to them unnatural—'unbrotherly', as they said. (The Russians often call a first cousin a brother.) Now that they saw Vladimir apparently in favour, they rang their church bells and did him royal honours. 'How could we know that this would displease the Tsar?'

Even before he accused Vladimir of plotting to poison him, the Tsar sent for the then Governor of Kostroma, and all his officials, and put them to death for liking his cousin too much. A few months later Vladimir himself (as they put it) 'was sent by the earthly Tsar to the Tsar of Heaven'.

I asked, 'Is it true that the Tsar executed not only Vladimir and his wife, but their two sons, who were children?'

'That was necessary,' the Governor explained. 'If they had grown up, it would have been their duty to avenge their father.'

All the leading citizens around me nodded gravely. They took it for granted that, if the Tsar killed them, he would have to kill their sons (and their wives, who might be pregnant with more sons). They had all been expecting this, because the Tsar had written to tell them that he had been too lenient with Kostroma; more traitors were flourishing there. He was coming to the town to root out these traitors. The Governor had decided that Kostroma's only hope was to make a great display of loyalty. That was why they had all stood in the streets, the men cheering, the women offering bread and salt. But what had saved them, they believed, was that the Tsar had scarcely noticed they were there. I had kept him entranced with my tales of England. And when at last he did speak to the Governor, it was to point out the barges of gunpowder, and say that these made him feel invincible. He was so strong (Ivan had said) that he could show mercy. He told the Governor to arrange a general amnesty for the town's prisoners.

So as not to offend these grateful townspeople, I accepted their jewels. But I told the Governor privately that if I had (though unwittingly) done him some service, he could repay it best by keeping Madelyn van Uxell in the women's quarters, and seeing that no harm came to her.

He said that his wife was indeed looking after the girl, and would continue to do so for the time being. But he would be glad if I could arrange another hiding place. 'If she stays here, the Tsar might hear of it. . . .'

I thought: What a coward the man is! Then I thought: What a coward I am! I had asked Boris for news of the archimandrite, but had not asked the Tsar. The archimandrite had parted from me with tears, had blessed me, and now I dared not utter his name. What was happening to me? Would I ever before have thought myself noble, because I had refrained from ravishing a girl I did not want?

I told the Governor that when I reached Moscow I would try to arrange a safe place for Madelyn. This (as it proved) was the least of my troubles. In Moscow, amid the golden domes and swinish hovels, I found a little colony of German craftsmen, living in their neat clean houses with their neat clean wives. Ivan was never at war with all Germans, only with those who lived in the Baltic cities. These craftsmen were doing well in Russia, and had good opportunities for travel. On hearing that Madelyn, like themselves, was a Lutheran, they sent one of their number to Kostroma to fetch her. She settled down with an honest German family, who liked her all the better because she cost them nothing. I paid her expenses out of the jewels pressed on me by the Kostroma citizens. This was the beginning of my friendship with the Germans.

My enemies were my fellow-countrymen. Mr. Trumbull, the Company's chief agent in Moscow, had expected a penniless, humble apprentice. By arriving laden with riches, I gave great offence. Worse, I had myself presented the Tsar with gifts which Mr. Trumbull should have handed over with all due ceremony. I had pushed myself forward as if I were some ambassador. Did I really imagine, Mr. Trumbull asked me, that I'd made a good impression on Ivan? He probably hadn't understood a word I'd said. How dared I pretend to speak Russian better than anyone else? And how dared I tell the people of Kostroma that I'd saved their lives? It was taking money on false pretences; if I had a conscience I'd send it all back.

I was then still young enough to believe that all this cavilling would stop when Ivan returned to Moscow, sent for me, made it clear that he delighted in my company, and showered me with costly presents. Did this not prove to my compatriots that I had been telling the truth?

Yes, but it made them bitter. Men who had worked for years in this cruel country, insulted in the streets, cheated by court officials, kept from seeing the Tsar, and not very well rewarded by the Company, saw with rage how everything seemed smooth and easy for me. Nobody accused me, then, of betraying my country. On the contrary, through my favour with Ivan, I gained such benefits for my compatriots that they no longer said a word against me.

But there was always on their part a jealousy, and on mine a knowledge that this jealousy might bring me . . . where it has brought me, Bess, as you now see.

⋞6

In fact I could see nothing, for it had long since grown dark. I knew that we were riding through the City of London. Since we first approached the town our easy guard had grown more attentive to their duty, and were keeping close about us. Now they halted, and made us halt. Torches flared, blinding me. I heard the cry of a sentinel, and the clank of steel. But until we were made to dismount I did not understand that we had reached the Tower.

Jeremy was permitted to kiss me goodbye. He seized me so strongly that I lost my balance. When he at last let me go, I almost fell. Some guards, thinking I was in a faint, held me up. While the gates clanged behind my husband these men spoke gently, directing me to an inn on Tower Hill. I could make no sense of what they said, and would not have found the inn, but my maid led me there.

I passed the night, not in weeping, but in coughing. The damp, smoky, London air had got into my chest. Three or four times my maid, Rosemary, roused herself to bring me cups of water. She said, 'Madam, for God's sake, give over!' Then she tried thumping my back, but that made me worse.

Towards morning I fell asleep. When I woke it was already light, and I jumped out of bed, scolding Rosemary for letting me sleep so long. 'The Tower gates must be open already,' I said. 'Jeremy will be waiting for me.'

So he was, but not impatiently. He had been given a good room, which overlooked the river. He was sitting by the window, telling his manservant, Paul, how to distinguish a flyboat from a cockboat, a carrack from a sloop.

'You can see more here,' he told me, 'than ever you could from that office in Seething Lane.'

I remembered then that Seething Lane was nearby. 'Shall I go there now?' I asked him. 'To the Company's office, to tell them you're in need of witnesses? Or should I go to Cousin Hawtrey first?'

'I've written old Hawtrey a letter,' Jeremy said. 'And I've written to Oulton the lawyer. We can send Paul and Rosemary with the letters.' Then he whispered, 'The first thing you have to do is come to bed.'

'The first thing you have to do,' said Oulton, when he arrived (and he came quickly; we were scarcely dressed again), 'is to ask for a copy of the indictment. You cannot be accused of high treason in general terms. There must be some particular acts. . . . Have you yourself any notion of what they might be?'

Jeremy shook his head. Oulton's broad, powerful face did not look very friendly, and somehow this comforted me. He was a big man, with shoulders fit for a butcher; and, like a butcher advancing to the block, he clearly knew what he was doing. Jeremy would not be able to say that this man was asking silly questions. So I was determined to stay and hear both question and answer, though Oulton appeared somewhat constrained by my presence. He said, 'There is gossip about you in the City. . . .'

'I know,' Jeremy said. 'I'm too rich. I admit I like being rich. But if I did not like it, I could not have helped it, while I was in Moscow. Once it was known I had the Tsar's favour, Russians gave me jewels, furs. . . .'

'As bribes,' put in the lawyer.

'Not bribes. Presents.'

Oulton smiled.

'Surely,' said my husband, 'there is a difference. Mr. Secretary Cecil—I mean Lord Burleigh—is known as the Queen's loyal servant. Nobody imagines that he can be deflected from what he believes is right. And yet people give him such presents that he has built Burleigh House—which is rather bigger than Kimble Manor.'

Oulton shrugged. 'So long as you don't tell him that, when you come before him.'

'But I shall tell him what is also true, that these people did not, as a rule, expect help in any particular instance. They simply thought me such an important person that I ought to be kept in a good temper. The Company did not forbid me to take these presents.'

Oulton said, 'I have heard complaints about your luxury and pride. They say you were not content with the ordinary English living quarters.'

'The Company asked me to use my influence,' Jeremy explained, 'to get them an extra piece of ground, so that the English compound might be enlarged. Part of this ground—with the Company's permission—was set aside for me, and so within the English compound I had my own house, and my own Russian bath house too. That was my luxury. The house was a necessity; I had to entertain the Russian noblemen.'

'And these personal servants you kept. . . .'

'I had to pick my own servants, because I soon found that the Company's cooks and bed-makers were spies for Shelkalov.'

'Is that the Chancellor who hated the English?'

'Yes, and me in particular. He owed the Company a great deal of money, which they despaired of ever seeing again. But when they saw the favour I was in, they asked me to mention the matter to the Tsar. I did, and he roared at Shelkalov: "Do you want to be Tsar of Russia, making your own laws? I'll show you that you can be put in the straightener like anyone else." '

'The straightener?'

'It is a sort of pillory, where people who do not pay their debts are stood in a row and beaten about the shins. When this has been done to the debtors every day for a year, they are allowed to raise the money by selling their wives and children as slaves. The threat is enough to make most people pay up. Shelkalov paid at once. The Company had the money; I had Shelkalov's undying hatred.'

'A good point,' said Oulton, making a note. 'Were there other cases where your position at court helped the Company?'

'It helped them in everything, every day. For example, when they sent me on long journeys, they found I could travel faster than anyone else. I had the Tsar's warrant, and could use post-horses. But always, always, before I showed the warrant, the people at the post houses bowed to me as the Tsar's favourite. As if it had been spread upon the lightning, printed in the clouds, the peasants in the furthest villages knew it at once. And so at every post house my sledge was harnessed first, and with the best horses. When the snow was hard I could travel two hundred miles in a day.'

'Two hundred miles!' repeated Oulton. 'Are you sure? That's from here to York.'

'Between York and London there is more temptation to sleep. We have inns, with beds. The Russian post house is nothing but a bare room with a stove in the middle, and a wooden bench. If you are worn out, you may fall asleep on the bench, but in an hour or two you will wake so stiff and aching that you are glad to go back to your sledge. These post houses were built by Russia's old rulers, the Tartars, who despised comfort and thought only of travelling fast. By one of these journeys, moving swiftly, but frozen and half-starved (the post houses provided as little food as rest) I came to the Baltic port of Narva. Our Company wanted to make more use of this, and less of remote Rose Island. But the way through the Baltic was compassed round about by Russia's enemies. Before I could complete my report on the trading prospects there, I had to go to Oberpalen, and pay court to the "King of Livonia"—Magnus the Dane. He was a tall, fair fellow, who would have been handsome, but that he had a patch over one eye. Magnus did his utmost to treat me well; he wanted to please Ivan, who was his master. But he could not put so much food on his table as you would find at a shopkeeper's house in England. I thought him lucky to have any food at all. Who there was to grow it I could not see; the nearby country was almost a desert. Farmhouses burned down in Ivan's last campaign, five years before, had not yet been rebuilt.

'To show that he was not a Russian, Magnus kept his wife beside him throughout dinner. This was the Russian princess Maria, daughter of that cousin whom Ivan had killed. (A daughter could be allowed to live; it was not her duty to take vengeance.) A child still, she looked uncomfortable in her western clothes, and uncomprehending at a western language —for Magnus insisted on talking to me in German. He explained: "I completed my education at the court of Saxony".

'If he had not much to eat, he had plenty to drink, and his talk soon became perilous. He asked me whether the Tsar ever mentioned him. In fact I had heard Ivan speak with derision of Magnus, his man of straw, but I said no. This too gave offence. "Forgotten all about me, has he?" said Magnus. "I'll give him something to remember me by! He thinks it was my fault we didn't take Reval. I told him from the start—what's the use, I said, when you haven't a ship to your name, what ever is the use of beseiging a town which can be supplied by sea? We, the beseigers—we were the people with nothing to eat. My Russian troops went looking for food round the villages. I told them to buy it. Instead they raped the women, burned the houses, drank the liquor.... If they did find food they were too drunk to bring it back. I was their commander, but a commander needs a second-in-command, officers, sergeants.... I hadn't one man under me who wasn't a savage. As for the Tartars—you know what the Tartars did? When my back was turned they slipped away, and the next thing I knew they were back with prisoners from Oesel. Oesel! My own island! Now everyone thinks I must have ordered that raid. And old you-known-who—he was glad of it. He told me to my face: Now you can't go back to Oesel! You're with me, for good or ill. You're part of Holy Russia! Holy Russia! God! I'll show him. I'll take Reval, if he lends me some of his German mercenaries. They know the meaning of discipline. If I had them, I'd tell the people of Reval—here I am, with good honest Germans like yourselves. No raping, no looting; what we take we pay for. You have me, I'll tell them, if you don't want the Russians. What can they expect, after all? These parts have been fought over by Germans, Lithuanians, Latvians, Poles, Swedes. . . . Now Russia's too strong for all the rest put together. We can all see that. Why should those people in Reval be any different? If only they'll accept Russian rule *through me,* I can make things very comfortable for them. Better me than old Grozny, with his: To the scaffold!"

'Magnus said these last words in Russian, mimicking the wild roling of the eyes, the savage laughter. Then he smirked at his own wit. Beside him the silent girl, his wife, burst into tears. He tried to comfort her in broken Russian, popping little sweetmeats into her mouth. To me he said, "She doesn't like it, when I mimic you-know-who. He killed her father and mother. And her brothers, who were only children. When we went riding through Moscow, after the wedding, the people went mad over her. They were cheering, crying, kissing her feet. . . . Old Grozny thought it was their way of reproaching him. He won't let her show her face in Moscow now; that's why we're stuck out here."

'I spoke a few words in Russian to the little "Queen of Livonia", and she gave me a sort of timid half-smile, so sweet. . . .'

Oulton said, 'Is this to the point?'

Jeremy broke off. He had become so lost in his memories that he looked up at me as if he did not know me. It was a cold day, and he was near the window, but great beads of sweat stood on his brow.

'Your hob-nobbing with kings and queens,' Oulton said, '—did that advance the interests of the Company?'

'Of course.' Jeremy spoke as if thinking of something else.

'And you can bring witnesses—'

'All my best witnesses are dead. If Daniel Sylvester, who was the English ambassador in Moscow then—if he had lived, he could have said what help I gave my country.'

'Then . . . it is only a malicious rumour that you devoted more time to Germans than to your own countrymen.'

'Germans? Oh yes, the captives.' Jeremy seemed to gather his wits again. 'A few months after I came back from the Baltic, war broke out afresh in those parts. The Poles had "disobeyed" Ivan again; instead of choosing him for their king they chose a man almost unknown—Stephan Batory, Prince of Transylvania. Ivan set out to punish them. He still did not take Reval, but he won several other battles, and soon the captives began to reach the slave markets of Moscow. I had to do something for them; it was the gunpowder I brought to Moscow which made them slaves.'

'Even this . . .' Oulton hesitated. 'Well, it has been said that you sought out those who came from wealthy families, and might be ransomed at some profit to yourself.'

'Yes, I heard that said at the time. I could help only those whose languages I spoke. Estonian or Latvian or Finnish peasants—what could I do for them? The Poles I could speak to; I had found out on my travels that their language is like Russian (though both sides deny it). But it's true that most of those I helped were Germans. A good many said that their families would willingly ransom them, if only they had some way of sending a letter. I would find out which agent of the Company was going to Narva, and persuade him to carry the Germans' letters. Then I would find the right court official to give the letters his seal, so that they would not be seized at the frontier. Where there was a ransom paid, and I had some share of it, I used the money to buy and send home other captives. Some of the Germans were simple working men. I told the Tsar they were skilled in their trades; it would be wasteful to sell them to the peasants. He gave them permission to live among the free Germans in Moscow, and build a Lutheran church, and have their own pastor.'

Oulton said, 'But your own countrymen. . . . Did you know a man called Finch?'

'Yes. The Company dismissed him for misconduct.'

'Who denounced him to the Company?'

'Every Englishman who passed through Kotlas, I should think.

He was in the pay of the Stroganovs. I believe it was Mr. Rowley who first complained to the Company.'

'Finch has come back to England now.'

'What! I thought he was dead.'

'Do you mean that you hoped he was dead?' Oulton asked.

'No. I mean I heard he was, and thought it likely.'

'He's been saying that, because of you, he was left stranded in some savage place.'

'He was never stranded! The Company was going to send him home, when he ran away. The next we heard, he was trading on his own account, which is a thing the Company will not stand.'

'Because it undercuts their prices?'

'Of course. Anyone can sell goods cheap, if they're stolen. Even a merchant who comes honestly by his goods can undercut the Company, because the Company bears all the expense of the shipping, the training of apprentices, the upkeep of Rose Island. . . .'

'Yes,' Oulton said. 'I know the Company forbids Englishmen to trade on their own. But can it stop them?'

'It can if they come to Moscow. But this Finch went skulking about in the Ural Mountains. Though the Company had the Tsar's warrant to arrest him, it was impossible for three or four years. After that the matter grew very complicated—'

'Never mind the complications. Whatever was done to Finch was done by the Company, not by you.'

'Yes, that's true.'

'But it was you who went to the Tsar and denounced another Englishman—Dr. Bomel.'

Jeremy looked thunderstruck. 'What! Who says that?'

'Never mind who says it. Is it true?'

'Of all things . . . Eliseus Bomel was not an Englishman. He came from Germany. It's true he spoke English; he was educated at Oxford. And almost hanged in London, for pretending to cast spells. The spells had some success; they made him famous. And so, while he was in the Archbishop's prison at Lambeth Palace, he had a letter from the Tsar inviting him to Moscow. The Archbishop was glad enough to let him go.

When I knew him, Dr. Bomel kept the Tsar's Danish girls, gave them medicines for the pox, and poisons to make them miscarry.'

'This was the famous court brothel,' said Oulton 'which . . .'

He broke off, looking at me. Jeremy smiled. 'Which I was often seen to visit, you were about to say. Yes, but not as my kind countrymen supposed. I had no confidence in the doctor's medicine, and I thought the girls were infectious. Bomel must have thought so too; he was faithful to his English wife Jane. I went to see the Danish girls because I was the only man in Moscow who could speak their language; they relied on me for news. They knew the efforts I made to have them sent home. I never had any success; the Tsar would not let one of them go, even poor little innocent Madelyn, whom I had saved from the brothel. "They'd only spread lies and slanders about me," said Ivan. The girls themselves were doubtful how their families would receive them, now that they were half destroyed with pox and vodka. None of them lived very long. The one whose deathbed I remember best asked me to bring her the German Lutheran pastor. When he came, they could not speak each other's languages; I had to interpret. The girl said that, a little while before, a miscarriage had nearly killed her. Dr. Bomel pretended to be sorry, and told the girl he was her one true friend. She felt so weak and ill that she could not reject a word of sympathy, even from him. He told her he could give her an easier life; she could leave the brothel and live as the kept mistress of one nobleman, provided that she told Dr. Bomel every word of plotting, or treason, or even faint discontent, that the man uttered. The girl thought this a deeper degradation than she had yet been forced to, and she refused. The doctor told her he would make her pay for this, and he dragged her out of bed before she was well. She thought she was being sent to the usual court orgy, but the doctor had arranged a special one, for men whose pleasure was in torturing women. Their mistreatment brought on the bleeding which was now draining her life away. The Lutheran pastor had words for such a deathbed; he was himself a captive, and had comforted more than one German girl who thought herself degraded past redemption. I interpreted his words about God's mercy, and all the while I wondered how many girls had agreed to spy for Bomel. I believe that some of Ivan's most horrible purges of "traitors" had no other cause than this. The nobles thought so, and were glad when it came to the doctor's turn. Bomel had made one mistake. He told Ivan that our queen was a young woman, and a desirable bride. When Ivan learned otherwise . . .'

'When you told him otherwise . . .' Oulton put in.

'Yes! I told him. Is that how the story got about, that I denounced Bomel? I told Ivan the Queen's real age in all innocence, not knowing a word of what Bomel had said. And for five years after that Ivan did not put him to death. He needed him too much. But, once he found another pandering doctor . . .' Jeremy paused. 'No, that sounds as if he calculated.'

'Didn't he? Are you telling me that this Ivan, who for forty years ruled a great country, was a maniac incapable of calculation?'

'There's another thing that made me unpopular with the English in Moscow! When I first arrived there I said (as we did in Rose Island) that Ivan might at any moment order a massacre of the English. I had found my countrymen less and less inclined to admit this, as I travelled deeper into Russia; and in Moscow it was utterly denied. Because, if a massacre did happen there, not one Englishman could escape. The chief agent, Mr. Trumbull, told me that Ivan would never act capriciously. He was a clever man. I said yes, but he was also a monster, and I mentioned Novgorod. Mr. Trumbull pursed his lips, and said, "We do not know all the facts." I said I knew Ivan was mad—not all the time, but now and then. I described his frenzy, when he asked who had showed me his new ships, though he knew it was impossible to pass through Vologda without seeing them. Mr. Trumbull said no doubt I had misunderstood him; my Russian was less good than I pretended. . . . And afterwards, whenever Ivan put out some fantastic story, as that a general who had defeated the Tartars was plotting to let them in, Mr. Trumbull would murmur: "The old man knows what he's doing." He thought Ivan did not believe his own accusations, but used them as a means of keeping his noblemen under. I knew Ivan better than Mr. Trumbull did, though I soon learned not to say so. I believe that in fact he did reason, as in this case of Bomel, whether he could do without this or that man. But that was *before* he accused the man of treachery. *After* he had made the accusation he believed it. He told me, trembling with unfeigned fury, that Bomel had been in league with the Archbishop of Novgorod, to betray him to the King of Poland. This was perhaps the only crime of which Bomel was quite innocent. Since he denied it, he was stretched on a rack, and beaten with wire whips. The Tsar did not himself watch the torture, though it was his custom; he sent his eldest son. Seeing this young devil grinning at him, Bomel confessed—and then went on confessing, when they wanted him to stop. Some of what he said nobody dared to tell the Tsar. It seems he gave a long list of people he had poisoned on Ivan's orders, and then accused the Crown Prince to his face of being a murderer too. He admitted that the plots he had uncovered were his own invention. The Crown Prince was observed to be more thoughtful afterwards.

'Bomel was bound to a stake and roasted between slow fires, which was a usual punishment for traitors. When the executioners thought he was dead, they threw him on a sledge and dragged him through the streets. A crowd went running to see him, and so did I. And as I looked at the scorched thing, its joints pulled awry by the rack, the scars of the wire whips still showing on the burned skin, the thing opened its eyes and whispered, "Oh, Christ, Christ!" '

I cried out, 'But then surely God let him die?'

'Not at once. Afterwards, in a dungeon. Perhaps Christ can have mercy on such a man. He must have had some good in him; his wife mourned—the Englishwoman, Jane. She came to live in the English compound, where she tried to teach the servants our way of cooking. We had all been longing for an Englishwoman, but this one was middle-aged, very plain, and very silent. We would sometimes forget she was there. One day I found her in tears. She had overheard me speak of her husband as a brothel-keeper, and she passionately denied it. I could not dispute the matter with her, knowing that she had watched every moment of that execution. So I begged her forgiveness, and warned the others never to speak of Dr. Bomel. Nor did I ever mention him again till now. Which (now I come to think of it) must be why his true character has been so little known. But if I have to tell all the truth to the Privy Council, I will. I have no reason to feel guilty.'

Oulton asked, 'Is there anything which does make you feel guilty?'

'No. That is . . . I do of course feel guilty of many things. But none of them are breaches of English law.'

'Nevertheless,' Oulton said, 'you will have to tell me about them.' I could have kissed the man.

'Well. . . . One thing I have already told my wife. There were names the Tsar could not bear to hear, and I thought that one of these names would be the Archimandrite Feodorite—my tutor, my dear friend. So I never asked the Tsar directly what had become of him. I asked everyone else in Moscow, but it was as if the earth had swallowed him up. Then, on one of my journeys for the Company, I revisited Rose Island. I did not mean to boast before my old companions, but I could not hide my sables and my velvets.'

'You seem,' remarked Oulton, 'to have done all you could to make your fellow-countrymen love you.'

'There are people so made that they love more than they envy, and Mr. Rowley was one of them. He said, "This is what I dreamed of, when I let you go." And he arranged to send my jewels home to my father for safe keeping. Mr. Rowley's only complaint against me was that he had lost his good Danish farmhands, all ransomed through my intervention. The young men, formerly apprentices with me, who had always been jealous, were now not more so, but less. I had put myself out of their sphere altogether. Besides, they knew my danger. On safe, England-like Rose Island they could feel some pity for me. To Mr. Rowley I was able to say that I never entered Ivan's presence without the fear of death; and that a worse fear was what I myself might do in order to stay alive. I told him how deeply ashamed I felt of not speaking up for the archimandrite. Mr. Rowley said, "You cannot help the man if he is already dead. Go and ask them at the monastery."

'The Abbot of St. Nicholas feasted me on elks' brains. I asked him, "Have you heard any news of the archimandrite?" Father Bogdan, the man who had carried the archimandrite's letter to Moscow, was sitting beside the abbot. They both looked me in the eye, and the abbot said, "What archimandrite?" Father Bogdan began to talk of something else.

'Though this was a long way from Moscow, it was Russia. I knew the language. The archimandrite was dead.'

'But was that your doing?' Oulton asked.

'No. The archimandrite was prepared for martyrdom when he went before Ivan and pleaded for Prince Kurbsky.'

'Then why should you feel guilty?'

'Because I had not the courage to do the same for him.'

'But if you had shown such unseasonable courage as to come between the Tsar and one of his subjects, you might have damaged the interests of the Company.'

'Yes. My fellow-countrymen said so at the time.'

'Good. Now I have a valuable point in your favour. You did not let even your close friendship with a Russian come before English interests.'

'If that is a point in my favour,' Jeremy said, 'I can tell you of a good many other Russians for whom I did not intervene. Monks torn to pieces by bears, men impaled on stakes—'

'You need not tell me about the stakes,' Oulton said.

I asked, 'Did it really take them fifteen hours to die?'

'Sometimes more. And yet the death which haunts my sleep sometimes is the death of an animal. Because we English had opened the way to Persia, the Shah was able to send presents to Moscow. He sent an elephant, which gracefully and slowly folded its front legs and bowed before Ivan. He was delighted. The elephant often repeated this trick, and it always moved me. To see a creature, which could have trampled us all, deliberately choosing to be gentle, looked like a rebuke to mankind. But the climate of Moscow did not suit the elephant; it pined, and grew stiff in its joints. One day Ivan, entertaining foreign guests, sent for it and commanded it to kneel. It stayed upright, for all its keeper's goading. Ivan, angry at this "insult", ordered that the beast should be destroyed. The thing was made a spectacle for the people. The elephant was brought out of the Kremlin to the usual place for executions, in front of St. Vassily's Cathedral. Men bound it with strong ropes, and hacked at it with axes. The trumpeting, the straining at the ropes, the turning of the head from side to side as if the creature looked for friends, the slow crumpling of its giant body as it bled to death. . . . This is what I think of, when I think of Russia. The next day, I bowed before the murderer, and kissed his feet.'

Oulton shrugged. 'If all the Russians were doing the same . . .'

'That's what I told myself. But it's not true of every single Russian. One day in the market place of Moscow I was aware of a filthy face, framed in matted hair, and a pair of mad eyes looking into mine. It was winter, yet this man wore nothing but a tattered skin about his loins, and some chains which had frozen on to his chest. He took a roll off a stall, without paying for it, and the woman keeping the stall bowed to him in gratitude. She said, "Give me your blessing, holy Nikola!" While he munched the roll he held up a hand and blessed her, and the neighbouring stallholders exclaimed at her good fortune. "She has been blessed by Nikola the Holy, Nikola of Pskov!" To them this man's dirt, as much as his nakedness, proved that he denied the flesh; every Russian goes to the bath house for pleasure, at least once a week. Though I could hardly stand the stink I followed the man and asked him, "Is it true that you saved Pskov?" The people round me cried, "Yes, yes; he spoke in the very face of the Tsar; he saved Pskov!" Only the man himself remained silent, until he said suddenly, "Why do you ask me that? Do you think I can save *you*? You do not know what love is, how can you be saved?" '

Oulton looked uneasy. 'Is this true?'

'That I don't know what love is?' Jeremy smiled at me. 'It was true then. Not now.'

'But did he really exist—this naked holy man?'

'Certainly. And he's not the only one. In a country where nobody dares to say what he thinks, there must be some three or four who can say everything. It is an immemorial custom;

Tsar Ivan, whoever else he killed, dared not kill a man who went naked through the winter. Anyone might speak the truth who paid that price.'

Oulton said, 'I begin to see the lines of your defence. You cannot have betrayed our country for a country of cowards and madmen.'

'Cowards and madmen. . . . Yes, I knew plenty of those. But I also knew the young nobleman Boris. I used to spend whole days free of terror, hunting with him. There is no country like Russia for hunting and hawking. We always went with the simpleton, Prince Feodor. In the summer I taught them both to swim, which is a rare accomplishment in Russia. Prince Feodor liked this; he walked heavily, with a slight limp; but he was our equal in the water. Whenever he saw me he would grin from ear to ear and ask me to take him swimming again.

'In his kindness to this lad, Boris was like any other Russian. They all think there is something holy about an idiot. The Tsar himself never spoke an unkind word to his afflicted younger son.

'But Boris, unlike some of his countrymen, did not think there was anything holy about his own ignorance. He had been left an orphan young, and his uncles had neglected his education. He was determined that his younger sister, Irina, should not suffer in the same way. With my help he found a learned monk to teach her Latin, and another to teach her Greek. Then I found a German tutor, a merchant of sober years, who taught her to read printed books from Nuremberg. I myself was not allowed to see her; that, Boris explained, would not be right. A Russian's most holy duty is to protect his sister, by keeping her hidden from young men until she is married. After that the brother protects her from ill-usage by her husband.

'Though he questioned many things, Boris followed the religious customs of his country. When his firstborn son was ill, he carried the boy in his arms to St. Vassily's Cathedral, stripped him naked, and held him over the saint's relics. A priest poured ice-cold holy water into the little mouth. It was winter. The baby died. Boris wept, and so did I, in pity for this barbarian. I could not tell him that he had killed his child.

'But sometimes I did protest. Boris asked me once to help him read something written in his own language. It was one of their tedious "Books of Precedence" which record what office each nobleman has held. Boris pointed to his grandfather's surname, which was the same as his own—Godunov. This much he could read, but he could not make out the context, and he was greatly relieved when I read that his grandfather had held some exalted post at court. The matter was important because, by the rules of precedence, Boris could not hold a position superior to that of a man whose ancestors had been superior to his. In his case, as in most others, the working of the rule was complicated by intermarriage; many people could claim the same ancestor. I told Boris frankly that the system was absurd. I wondered the Tsar did not make away with it. Boris earnestly assured me that the nobility would never tolerate that. Since we were alone I exclaimed, "But they tolerate everything else!" Boris did not answer. But presently he began to ask me questions, in his usual manner. He never said, "Look what happens in Russia!" but always, "Is it true that in England . . . ?" And at last he said, "If I could read my own language, I should still have nothing to read except *this*—" and he pointed to the Book of Precedence. "The things I want to know are all in other languages. So it is better for me to question foreigners. Is it a great nuisance for you to answer me?" I assured him it was a delight. It was Boris who made me remember the archimandrite's words, "It could all have been different." That set me dreaming of what Russians could be like, once they had schools, universities, doctors, laws. . . .'

'I am not concerned with your daydreams,' Oulton said.

'You should be. The worst things I did arose from them.'

'Oh? Do they account for your first return to England, I think in 1580—'

'I set out at the end of 1580,' Jeremy said. 'I arrived in England in February, 1581.'

'Very well. You returned, not as a servant of the Company, but as a secret emissary of the Tsar.'

'The Queen knew that at the time.'

'Nevertheless . . .' began Oulton.

Here we were interrupted. Paul and Rosemary had returned, with old Sir William Hawtrey, my father's cousin. I was so glad to see him that I threw my arms round his neck.

He drew them away and said, very soberly, 'I am sorry to see you here.'

I cried out, 'Oh, but it's all right! Jeremy has an answer for everything.'

All the men were silent. Jeremy looked at me strangely. Was he angry with me? Or sorry I had spoken? Or sorry for something else?

Presently Oulton said, 'Since Sir William is a director of the Russia Company, he had better hear how you returned to England in 1581.'

Jeremy said, 'Yes. I will send my servants to the inn, to bring us all a good dinner, and then I will tell you what happened.'

7

I suppose you will agree it was not my fault that Daniel Sylvester was struck by lightning. Even the Tsar, when he heard of it, said, 'God's will be done!' Afterwards he raged; this blow could not have come at a worse time. Daniel Sylvester was our Queen's ambassador. When the lightning struck him he was at Kholmogory, on his way to ask Ivan what guns and gunpowder he wanted.

Ivan wanted all England could send him, and more. The unknown man the Poles had elected king, Stephan Batory, was proving himself the greatest military leader of our age. He persuaded other countries, which till now had trembled singly before Ivan, to act in unison. Hungarians and Germans marched side by side with Poles; the Swedes harassed Narva so that Ivan could get no supplies by the Baltic. Nor could he send to the Baltic the flat-bottomed ships the English had built for him at Vologda. He sent them down the Volga, to suppress an uprising of the Tartars.

We in Moscow did not know how badly things were going

for Ivan, till we heard that Magnus the Dane, 'King of Livonia', had gone over to Batory.

Ivan's Chancellor, Shelkalov, sent to rally the Russian forces, found himself surrounded by the Polish army, but slipped through their lines in the night, leaving his troops to fight to the death. (Which they did; the Russian gunners, when their ammunition ran out, hanged themselves from their own cannons, rather than surrender.)

Ivan, who punished so many imaginary treasons, did not punish this real treason of Shelkalov. He simply appointed the man to keep order in Moscow. This meant that Shelkalov had to call the people together and tell them that the Poles had taken Polotsk. The women rioted in the market place, crying out, 'Give us back our husbands! Give us back our sons!' But a few Tartar horsemen stilled their cries. The Tsar sent his best fighters to keep a constant watch at those frontier fortresses where Batory was expected to attack. The watchers were disappointed. Marching where no army had ever marched, cutting down trees and undergrowth in their way, laying brushwood paths across the swamps, building bridges across the shrunken summer streams, Batory and his thirty thousand men struck where they were least expected, and took Velikiye Luki. In the autumn of 1580 the Russians beat them back from Smolensk; but when Batory's men retreated it was always in good order, and they could be expected to return in spring.

Later in the same autumn of 1580 the Tsar called his generals to a council of war at his favourite palace in Alexandrov, some fifty miles north-east of Moscow. He summoned me there too; I had no notion why.

I arrived to find the whole court busy with preparations for a wedding. Boris embraced me. There were tears in his eyes. 'Help me!' he whispered. 'Help me to comfort her!'

It was his younger sister, Irina, who was to marry the half-witted Prince Feodor. The Tsar had said so, and there was nothing to be done.

'Don't mistake me!' Boris whispered. 'I love young Feodor. . . .'

He led me to his own part of the palace, and then to the women's quarters.

In the six years I had known Boris, I had three or four times been allowed to meet Masha, his wife. She was the daughter of the terrible 'baby', Malyuta. Yet her face looked kind, and Boris was happy with her. Indeed, their fidelity to each other was almost a miracle, in that court.

The person I had never yet seen was Irina. I might meet her now only because she was as good as married.

Now and again the chubby Russians produce a golden-haired, blue-eyed goddess. I never saw one more perfect than Irina. Because they were secure in their own quarters, the women had not painted their faces, and I was able to see the girl blush as I greeted her. Boris, as if trying to distract a child, told her that he had brought her a man who could tell her stories of countries far away.

I did indeed tell her stories, but soon found that I was not speaking to a child. Her education, which Boris had so often talked of, had made her as bright as she was beautiful. Every word Irina spoke increased my sorrow that she would be sacrificed to a half-wit. I did not mention the matter, till she herself, in the very tone of her brother, asked me, 'Is it true that in England the women choose their own husbands?'

'Not altogether true,' I said. 'Noblewomen, and heiresses to great estates, almost always marry as their fathers wish. Queens may choose for themselves, but . . . our own Queen Mary married the King of Spain, and loved him dearly; and the consequences to our country were so evil that our present Queen has decided not to marry at all.'

This puzzled her. She knew the fact already, but could not imagine how a woman could live unmarried, and yet not be a nun. Could she be safe, in a court full of men?

I said that when the Queen spoke to any man, or even looked at him as if she might speak, he sank at once to his knees. For the first time, Irina smiled. Then she said, 'But a war. . . . How could a woman lead them in a war?'

I told her about the seige of Haarlem. 'A woman called Kenau Simons Hasselaer carried an arquebus, and used it too; and she taught other women to do the same. The Spaniards took Haarlem, but now the Dutch have driven them out again. The free Dutch towns have set up a government, and Kenau Simons Hasselaer is in charge of minting their money.'

As I talked I felt myself once more in the streets of Haarlem. When I was there, it was the fighting which moved me. But now I was more moved to remember what a country it was they fought for. I remembered the clean streets, the cunning patterns of the bricks that paved them, the neat gardens in the courtyards, the well-tended fields within the city walls, the strong defences built against the Spaniards, the strong dykes against the sea. I remembered the Haarlem people boasting that it was one of their citizens, not Gutenberg, who had invented printing. In the midst of the bombardment their presses were turning out books, and every man and woman in the city could read....

How was any Russian to imagine such a country? Yet Irina seemed to imagine it, as I talked. She finished her embroidery before she knew it. Suddenly aware, she looked at her work and said, 'My wedding dress!'—and said no more.

Afterwards Boris thanked me. 'I have not seen her so cheerful since we heard the Tsar's decree.'

The wedding was delayed by many strange accidents, until some people thought it bewitched. But, if there was a curse, it must surely have been on another ceremony which took place a few days earlier, behind locked doors, conducted by a frightened priest who knew that he was breaking the laws of his own church. Ivan himself, before he went to his after-dinner sleep, leaned over his last glass of wine to tell me, 'I was married yesterday.'

As I uttered the proper felicitations, I wondered whether this was his sixth or seventh wife.

Ivan was rambling on. 'I must have had a thousand virgins in my time. Hard work for an old man, ha?' He laughed.

I found myself smiling. I do not mean I smiled because it was in the interests of the Company, in the interests of England, to keep on good terms with the monster. I smiled because, for that moment, I saw things as the monster saw them. A second later I was once again a hypocrite, which is a better thing to be. I knew the difference between right and wrong when I told Ivan, untruthfully, that he did not look old. In fact he looked far more than his real age, which was fifty.

He went on, 'It's time I settled. Time I had more sons. Never would have bastards. Bad thing for a country—bastards disputing a throne. Old Bomel knew how to get rid of them. New fellow's not so good. Bomel must have put a stop to hundreds of bastards . . . thousands. . . .'

I thought: If I stay here much longer, I shall smile at this too.

Ivan was rambling on. 'No Danish girls left. My soldiers don't bring me foreign girls nowadays.' (How could they? They were no longer on foreign soil.) 'And these Russian girls . . . Russian girls have brothers. They're savages, my subjects, you know that? They think more of their sisters than they do of me. All the bodyguards in the world wouldn't keep me alive ten minutes, if I took a Russian girl without marrying her. I have to call it a wedding, whatever the Church calls it. But if you'd seen my real wife, my Anastasia! If only they'd let her live, none of this—' he waved a hand '—none of this would have happened. They poisoned her, you know. The priests . . . the monks . . . Kurbsky. . . .'

The second wife, I remembered, was the Cherkass princess—poisoned, but by whom? The third wife lived only a few days after the wedding. The fourth I had seen at Kostroma. (She was alive still, but repudiated, and imprisoned in a convent.) The fifth died young. The sixth was so beautiful as to make Ivan forget his preference for virgins; she was a widow. But she too was now in a convent.

'And not a single grandchild for me!' complained Ivan. 'There's my eldest boy been married three times. No children! Can Feodor beget any children? God knows!' (Here he piously cast his eyes upwards.) 'God made Feodor as he is; He knows best. The boy never looks at a woman. I'll give him one worth looking at; I can't do more. Must have more sons. . . . Lead a good Christian life. . . .'

Ivan seemed almost asleep over his wine, when he leaped up

suddenly, broad awake and sober. A message had come from Poland. 'Read it aloud!' Ivan said. The messenger, trembling, read a speech which Batory had made to the Polish Assembly. 'It seems that fate has given the whole Russian realm into our hands.'

Ivan's iron-tipped staff rang on the flagstones. 'Yeremy!' he said. 'I'm going to need you.' But then he waved me aside, and sent for his generals.

The wedding of Feodor and Irina took place amid the proper splendour. Feodor, who loved all church ceremonies, wore a blissful smile. His elder brother, Ivan's heir and other self, looked at the bride with interest. I thought no good would come of this; the interest of that young savage could only be of one kind.

Boris was wiping tears away, at the wedding feast. I believe he loved his sister, and felt her tragedy as his own. Yet what he said was, 'At least nobody will dare to dispute my precedence now!'

I whispered, 'Except the relations of the latest Empress.' For a new set of faces had appeared at court; a new family had been ennobled, given court offices, and enriched with estates which had belonged to the relations of Ivan's other wives. The strange thing was to see them swagger as if their good fortune could never change. The new Tsaritsa's brother, Afanassy Feodorovich Nagoy, went out of his way to be offensive to me. 'You're the one who wants to turn us all into Englishmen!' I assured him I had no such intention, but he persisted. 'I know your sort! You want to bring in English customs, English laws. . . .' It is odd how a thick-headed oaf, three parts drunk, may hit the bull's eye. This was in fact what I wanted, though I had never said so, and had no rational hope of it.

'It's your sort,' Afanassy went on, 'who make the young fellows discontented. They're always asking His Majesty for permission to travel abroad. Which they won't get, let me tell you! Well, thank God, there are some of us who love our country. We don't go running to foreign doctors whenever we get a belly-ache. We're proud of being Russians!'

Behind Afanassy I saw Shelkalov, scarcely hiding his triumphant smiles at this attack on me. But was I a person of such importance as to be attacked? Was this diatribe meant, rather, for Boris? I took the first opportunity to tell Boris that I thought so. He said, 'Do you think I have lived so long in this court, without knowing that Shelkalov hates me?' At that moment he looked older than I had seen him. It occurred to me that friendship with Feodor, while it might be the way for a good man to remain good, might also be the way for a clever and ambitious man to seem harmless. But then Boris grinned—that is, looked altogether frank and simple.

'I shall be safe,' he said, 'so long as I have such a friend as you.'

A week later Ivan told me why he had sent for me. He had to send a secret message to England. Since our ambassador's death by lightning, there was nobody to carry it but me.

I faltered that it was already late November; the White Sea would be frozen. Ivan replied, 'You are to go by land.'

How did he think I could get past the Polish army? I murmured, 'No doubt Your Majesty will give me a guide, to show where I may cross the frontier.'

He assured me that there was a place where I could cross in perfect safety. In the same breath he added, 'I won't tell you what the message is; it might be tortured out of you.'

And then he embraced me, with every appearance of affection. 'Let me trust you, at least!' he said. 'I am surrounded with traitors. The Stroganovs have sent a band of armed Cossacks into Siberia. And every man of them a convicted highway robber! You, as an Englishman, tell me—where did the Stroganovs get their gunpowder?'

'Not from the Company!' I said, and thought of Mr. Finch. He would be arrested, if the Stroganovs were, but could Ivan spare men enough to seize them in their Ural fastness?

The court returned to Moscow, and myself with it, and still I was not given the message I was to carry. Ivan told me to keep the matter secret, but I did whisper a warning to Mr. Trumbull, standing well away from the keyhole, and rustling paper as I spoke. Our Russian servants had by now learned some English. But the eavesdropper I feared most was Jane Bomel. She would want to come with me. The poor woman was desperate for England, and the Tsar would not give her a pass.

Mr. Trumbull did not suggest that I should, or could, refuse to do the Tsar's errand. I confess I had one hope I did not tell him. If the jewels I had sent my father were safe, they would be enough to buy me out of the Company's service, and set me up in some business of my own. Then I would never need to see Russia again. True, that would mean parting from a mistress I had then. At the time when the holy man told me I did not know what love was, I thought myself in love with a certain married lady. (Her husband gave little trouble; he lived with a peasant girl from one of his own estates.) I felt some pangs at leaving my mistress; much more at leaving Boris.

Though I could not openly say farewell, I visited all my friends, including the Danish girl, Madelyn. She was now twenty and as plain as ever, but a fine, strong, bustling girl, well trained in household matters by the German woman in whose care she lived. It seemed that a prosperous German merchant (whom I knew) wanted to marry her. 'I'm writing to ask my father's blessing,' she said. 'I'm afraid he'll say it's beneath me, because I'm a noblewoman. But it seems I can't be ransomed, and what use is my noble blood here? Besides, I'll be happy with Heinz, I know I will!'

As my sledge took me home I was kept warm by a glow of righteousness. My past ambitions now seemed foolish; I no longer believed that I could turn the monster into a man. Instead I had felt him turning me into a monster. Yet here and there I still did good, and this was one example.

Ivan summoned me to the Kremlin the next day. He proudly showed me his way of sending his message, a piece of childish cunning. He had ready a little cask of vodka, such as a traveller hangs round a horse's neck. The cask had a false bottom, and in this Ivan placed a letter. (First kissing it, because it was for his 'dear sister', our Queen.) He and Shelkalov sealed up the false bottom, Ivan all the while pointing out to me how skilful he was with his hands. Then he told me that the letter was for the Queen alone; I was not to unseal the cask until I came into her presence. It seemed not to occur to him that (the letter being in Russian) I would have to interpret it, and would therefore know what it said. Perhaps he imagined that our Queen had a wide choice of Russian interpreters.

Ivan also presented me with a wadded coat, carefully frayed to look worn, into which gold coins had been sewn. 'Not roubles!' he said. 'Hungarian ducats, so that nobody will know you've come from Russia.' Unfortunately the coat could not have come from anywhere else.

A nobleman attended me to the frontier. As we skimmed along the frozen road to Novgorod, we overtook a crowd of men trudging along, bound with ropes and guarded by some of Ivan's Tartars. The Tartars told us that these were peasants who had gone into hiding to avoid military service. They were being taken to join the main army near Novgorod.

'Traitors!' exclaimed my companion. He seized the sledge-driver's whip and lashed one of the peasants with it. The man turned, shouting, 'Go and rape your mother!' But, when he saw that it was a nobleman who had struck him, he begged pardon and bowed his head. The Tartars laughed. Our sledge rushed onwards, and myself within it, like one of those peasants, roped and dragged to the war.

Such was our speed that we reached the frontier in three days. But then, because of Batory's gains, the frontier was nearer than before. It was not far west of Pskov that my companion stopped the sledge, had the spare horse unharnessed for me to ride, and demanded a letter from me, to prove to the Tsar that he had brought me as far as he safely could. I would have written him a letter, but I saw soldiers in western clothes emerging from the forest. We had already trespassed on the territory which Batory claimed as his. The sledge turned in a flurry of snow, back to Russia. I rode forward alone, patting my horse's neck, feeling the rope which held the cask of vodka.

The soldiers challenged me in Polish. Doing the best I could in that language, I said that I was a German captive, ransomed out of Russia, and glad with all my heart to have escaped from such a vale of tears. As I said the last part I felt the truth of it, and I wept.

They took me to a castle, where I was questioned and searched, but not roughly. My story seemed likely, and nobody saw anything unusual about the vodka barrel. After two days I was allowed to ride on, with guides to show me the best way to Germany.

They would not take me through Livonia, which was a battlefield still. Russian garrisons were holding out in some of the walled cities, their supply lines much harassed by Polish and Swedish forces guided by Magnus the Dane. (By this time, I thought, he must know his way about these parts.)

Some Swedes led me through the country round Reval, which was now securely theirs. Then they directed me across the ice to an island belonging to the King of Denmark. I asked the name of it. They told me, 'Sarema'. I thought it strange I had not heard of it. But that it was a civilised place I could see; there was a fortress, and a Danish flag flying. That looked safe; Denmark was not at war with Russia. It was not until I reached the fortress that I knew Sarema was the same place as Oesel.

The Danes would not even listen to my story. The Russian coat I was wearing was enough. I was dragged from my horse, bound hand and foot, and brought in a sledge that bumped horribly to the Governor's palace at Arensburg. There I was lodged, not in a dungeon, but in a little servant's bedroom off the kitchen.

I suppose many kitchens, if you look at them too closely, have things that crawl in the flour and about the milkpans; and hens that wander in from the henhouse to peck at the crawling things. Maggots? These were snakes. They came writhing over my bed. The cooks told me, laughing, that they would do me no harm; and indeed I thought there might be worse harm ahead.

All this time I talked. I argued. I demanded my baggage. It was brought to me, after they had searched it. The cask they threw back to me when they had drunk the vodka. I pretended that I could not live without strong liquor, and bribed the cook to fill the cask with brandy. Otherwise, it would have been thought strange that I slept with it under my head. At last I was called before the Governor of Oesel. This crabbed, sour old man sat like a king, amid attendant gentlemen, swordsmen and halberdiers.

I had abandoned my pretence of being a German, and said frankly that I was English. Queen Elizabeth, I said, was at peace with all Christian princes, and certainly with His Majesty of Denmark.

'We have seen this peace of yours,' the Governor said. 'You do no fighting; you look after your own skins; but you are in league with the Muscovite Ivan. You armed the very Tartars who came here, to Oesel.'

I felt my danger all the more as I guessed that the letter in the cask was an appeal for still more arms. But I pleaded that I was myself a simple merchant, apprenticed to the Russia Company before I had a choice.

The Governor asked me my name. Then he asked me to write it down. Suddenly he dismissed me, his voice trembling.

I was taken back among the snakes, but not for long. A very well-dressed young man came to me and said, 'My father would like to see you again.'

This time I was taken into a small, comfortable parlour. The Governor stood there with a letter in his hand. 'It is the same name!' he said. 'The name in my daughter's letter.'

Not believing my luck, I asked, 'Is your daughter Madelyn van Uxell?'

He put his arm about my neck, and burst out crying. The young man said, 'We could not believe what she said in her first letter, because we knew what had happened to all the girls from Oesel. We thought she was trying to spare our feelings.'

'I wrote,' the Governor said, 'I wrote to tell her that she was my dear daughter still, whatever had been done to her. She could tell me all the truth, I said; I would never cease my efforts to bring her home. She answered that, because of her kind English friend, she was living a good honest life among Protestants. Can that really be true?'

'Yes,' I said. 'And she was in good health, only ten days ago.' I told him everything, up to her proposed marriage, which I advised him to approve. The bridegroom was not of suitable rank, I agreed; but he was an honest merchant in good circumstances, and the married state would be some protection to Madelyn. I was about to add that Ivan seldom raped a married woman, unless he had just been watching the execution of her husband. But in this room, with its pretty tiled stove and its rows of silver plates, how could I tell my tales of barbarism? I broke off and said, 'What a beautiful house this is!'

'Yes,' the Governor agreed. 'Magnus built it. He could have lived here in comfort, all his days.... I know why you think that Madelyn will be safer if she is married.'

He now lodged me in his finest room, summoned his tailor to make me a new coat, and called his friends together for a banquet in my honour. I dreaded that, because I could not say a comforting word about anyone else's daughter. (Except that those who had died—by now the majority—had been properly buried by Lutheran rites.)

There was no need for me to say much. Despite Ivan's refusal to release any girl who might spread 'lies and slanders', the truth was known. These Danes no longer thought of saving their daughters, but of avenging them. Their King Frederik had not declared war on Ivan, being hindered by long-standing quarrels with the Swedes. But young Danes by the thousand were volunteering to fight for Batory. They told me so much, at this banquet, of Batory's greatness, of his tolerance to Protestants, his gentleness to women, his humanity to prisoners, that I wished I could fight for him myself. But my road lay elsewhere, and I must make haste. Even the sight of the Governor's library—how long was it since I had seen a real library, with printed books?—did not delay me. I had to reach the west side of the Gulf of Riga by the ice, which freezes later and melts earlier there than on the eastern side. The Governor gave me letters to a Polish general, and his son guided me, over the still solid ice, to a place on the Baltic shore where Batory had triumphed, and there was no more fighting. With good Polish guides, I thought myself past all danger.

I had, however, to pass through a Baltic fortress called Pilten. This meant I had to pay my respects to its master, who was Magnus the Dane. He was fatter now, and redder; his face bulged around his black eye-patch. On catching sight of me, he cried out, 'No! I said No! You can tell Ivan I'm not coming back.'

I reminded him I was an Englishman.

'That's nothing to go by. Ivan's been sending me all sorts. Estonians . . . Finns . . . all with wonderful offers. I'm to get five barrels of gold, and God knows what else, if only I'll put myself in his power again.'

I said he would hear nothing of the kind from me; I was myself too happy to be out of Russia.

'Just come from there, have you?' said Magnus. 'Then you can tell me what old Grozny's up to.' He invited me to dinner.

At the table I was introduced to some Germans. Magnus called them 'very good fellows', and said he had met them 'at the court of Saxony, when I completed my education'. Then he told me, 'I've made it up with my brother, the King of Denmark. He knows it wasn't my fault—what happened at Oesel. He sent me some of his own gentlemen-in-waiting. Yes, and servants, cooks. . . . At last I can have everything done Danish fashion. Especially good Danish liquor!'

Everyone sat down to dinner already drunk, except myself and the Russian princess Maria, Magnus's wife. She was, if possible, more silent than before, her blue eyes wider. But I thought she was now following what Magnus told me in German.

'Old Ivan's more fond of being cruel than he is of winning wars. Look what happened in his last Livonian campaign! I did what I said I'd do; I told the people they could have me and my German mercenaries, who knew how to behave, or they'd get Ivan. Naturally they chose me. Wenden gave in to me without a fight, and Wolmar, and Lenward, and Ronnenburg. . . . At last I really was what Ivan called me—King of Livonia! Wouldn't you think he'd be pleased? No; he came raving and storming; said I'd betrayed him; *he* should have taken those towns. His troops took them over, with such cruelties that every other town decided to resist him to the death. Even if they were outnumbered, even if they had no chance, a walled city that puts up a good fight is a very costly thing to an attacking army. I told him so, to his face. And you know what he did? Put me under arrest. Would have sat me on a stake, but Maria—' for the first time he looked at his wife '—went down on her knees. No! Flat on her face, Russian fashion. She told Ivan, "You took my mother and father; you took my brothers; now you're going to take my husband!" Never believe it of her, would you? Wonder he didn't hand her over to his Tartars. But it seems the old man feels a twinge of remorse, now and then. He let her go. And me! He'd have changed his mind, of course, executed us both, if I hadn't gone over to Batory.'

I asked him to tell me about Stephan Batory.

'Ah! Now there's a sensible fellow. Thinks just as I do. When he comes to some frontier town that's maybe more Russian than Polish, he tells them—you lay down your arms, he tells them, and you'll still have your Russian churches, your Russian customs, your property.... It's only this intolerable tyrant we're fighting, he says, not you! Then he keeps his word. No looting, no rape! Let me tell you—by this time next year, Stephan Batory will be in Moscow!'

I said that on their own soil the Russians might be stubborn, like those gunners near Polotsk who had hanged themselves, rather than be taken prisoner.

'That lot—they had a bad conscience. They know how *they* treat prisoners. But once they see Batory's winning, they'll all be licking his boots. They'd lick anybody's boots. God! If they can stomach Ivan.... Here! Let me tell you something!' Magnus was now very drunk, but still coherent. 'I've seen one of Ivan's noblemen impaled on a stake. You know how they do it? Well! You'd think the man would die at once, but he doesn't. In between screaming, this Russian sent messages to his wife, blessings to his children, and so on. But what I was waiting for was to hear what he thought of Ivan. Because he had nothing more to be afraid of, had he? He was longing for death—praying for it. And yet all he said was, "Long live the Tsar! God bless our beloved Tsar!" '

I suggested, 'Perhaps that was to save his children.'

Magnus shook his head. 'No! He was past pretending. He meant it. They all mean it. Russians! Why, look at this one!'

He turned again to his wife. 'Maria does nothing but snivel and complain that she's not in Holy Russia. Holy arse-holes! Batory's too strong, now, for all the rest put together. Other people can see that; why should she be any different?'

'It's natural she should miss her own country.'

'Natural? To be always whining and praying? I had enough of that with my old mother. She always liked my brother best. Her good boy! Never could stand the sight of me, after I lost this eye. Was that my fault? I was two years old. Can't even remember. It's because of this eye I never could take Reval. The people there called me "the ogre". Do *you* think I look like an ogre?'

He took off his eye-patch, and thrust his face into mine. The socket of the missing eye was not, as I had fancied, red and gaping. The skin had grown over it to a perfect smoothness which I found still more horrible. Yet I stared him out steadily; I had been well trained in Russia. After a moment he laughed, put back his eye-patch, and shouted, 'But it's not true about the webbed feet! My feet are as good as yours!' When I thought he was too drunk to care, I turned to his wife.

'Do you feel homesick for your own country?' I said in Russian.

She responded only with a nod, but Magnus took offence.

'Don't you go speaking to my wife in Russian! Don't you encourage her! What sort of a man are you? Why don't you drink?'

I said I had drunk enough. More would upset my stomach.

'Rubbish!' roared Magnus. 'Good liquor never hurt anybody. You've lived in Russia too long; all you like is vodka. Right! You shall have your vodka. Anna! Bring vodka for all of us!'

The name Anna is common enough, but I turned to look at the serving woman. I could not be sure; her back was towards me, as she went to fetch the vodka.

'There's one thing I won't endure,' declared Magnus, swaying to and fro in his chair. 'One . . . thing . . . I . . . won't endure. . . .' He caught his wife's eye. 'Who do you think you are—my mother's ghost? Don't you dare look at me like that!'

Before I knew it the serving woman was pouring vodka—

an undrinkable quantity of vodka—into my glass. I twisted round, trying to get a good look at her. But Magnus was bawling at me, 'One thing I won't endure is a man who refuses to drink. You think you can keep a cool head and spy on the rest of us. Come on, let me see you pour that vodka down!'

I took one sip of the fiery stuff, which I hated as much then as on my first day in Russia.

'Not like that!' said Magnus. 'Like this!' And he poured his whole glassful straight down his throat. His 'good fellows' cheered, and did their best to imitate him. Since I still would not drink, Magnus tried to force my mouth open and pour in the vodka, calling on his companions to help. His wife was on her feet, trying to restrain him; some of his fellow-drunkards came lumbering to his aid; and I found myself on my back, on the floor, with vodka running over my face, my clothes—but not, thank God, in my guts. The strong serving-woman came between Magnus and me. I knew her voice. 'Are you a prince? And is this how you treat a guest?'

'Guest!' bellowed Magnus, as Anna helped me to my feet. 'He's only an Englishman. Thinks he's a good boy, does he? Looks down on us? And he sells gunpower to Ivan! Well, that won't last, let me tell you! Batory's taken your measure. Next summer, he'll be sending Danes and Swedes into the White Sea, and that'll be the end of your Rose Island!'

I thought: Now he has told me that, how can he let me leave this place alive?

What saved me was that the vodka took its due, and Magnus fell senseless. His wife began begging the servants to carry him away. Anna led me into the kitchen. I found myself, as it were, seven years younger, sitting in front of the fire, tended by Karen and Anna. This, my second strange piece of good fortune, convinced me that I was still guarded by the archimandrite's prayer, however little I had done to deserve it.

Both women kissed me, felt me for bruises, washed the vodka from my face, brought me sweetmeats, and at last, on finding that I truly was not hurt, sat down to talk.

It was not, after all, so strange that they were serving Magnus.

The day that I persuaded Mr. Rowley to send them to Vardö, I knew they would plead for their fellow-countrymen to be ransomed. This they did, and the Governor of Vardö was so much moved that he sent them to King Frederik, who did in fact ransom all the Danes on Rose Island. The King offered Karen and Anna the chance of returning to Oesel. 'But what would we find there, except ashes?' They chose rather to work in the household of King Frederik, who proved a kind master. Then the King made peace with his brother. The old hard, flat note came into Anna's voice, when she said the name, 'Magnus'.

'He has a good heart, our King,' said Karen. 'When Magnus wrote to say that he hadn't a single servant...'

'He didn't say they'd left him because he drank their wages,' Anna put in.

Karen went on, 'King Frederik asked some of us—would we go and be servants to his brother? Of course nobody wanted to go. To leave the best King our country's ever had, and serve a traitor—'

'But then we understood that it was God's will,' Anna said. 'Didn't I tell you so, seven years ago? We shall both see the end of Magnus.'

'He is drinking himself to death,' said Karen.

Anna laughed. 'It's wicked to kill. But it's not wicked to obey my master's orders. When he calls for drink, I pour it out. Let him have plenty!'

'But the most fearful thing,' said Karen, 'is to see him sober. His conscience torments him so that he cannot bear it.'

I said, 'He told me he was not to blame for that raid on Oesel. The Tartars did it without his knowledge.'

'That's what he tells everyone!' exclaimed Anna. 'He knows otherwise. When he saw us, his own people, captives in the hands of the Tartars, did he lift a finger to set us free?'

'The Tartars won't fight,' I said, 'unless they are allowed to take slaves.'

'He must have known that,' replied Anna, 'when he agreed to lead a savage army, for a savage emperor.'

I said I did not think his present companions were much better. There was a murmur of agreement behind me, and I looked round to see that every servant in the place was listening to our conversation. They were all in a mutinous mood, being unpaid. They said these drinking companions had spent every penny Magnus could borrow. At his death, which they all expected shortly, his wife would be a beggar.

From the door a voice broke in, the voice of a Russian lady-in-waiting. 'Your Majesty, he is here!' Everyone bowed, making way for Maria.

'I have been looking for you everywhere,' she said. 'Oh, forgive me, forgive me!'

Karen and Anna smiled at this familiar Russian phrase. I told Maria that I had nothing to forgive her for.

'To think that you should have been forced to take refuge *here!*' she said, looking round in horror. Had she ever before set foot in her own kitchen? 'Please, please come to my apartments.'

It was not the first time that I had been invited to a Russian lady's apartments while her husband lay conveniently drunk. But Maria's invitation was quite innocent. She was seven or eight months pregnant. And she wanted only to say, 'Don't think he's truly like that! His companions.... When we were first married he was so kind! Only sometimes I'm silly; it's my own fault; I cry to be back in Russia. I know that what happens there is horrible, but, you see, it's my country.'

That innocent face.... If I pause now, when I speak of it, you must not suppose I paused then. I thought less of Maria than of my need to leave before Magnus came to. He might remember what he had said about Rose Island.

I rode on to Königsberg and Danzig and so into Germany. At Lübeck I enquired for a merchant whom I had known when he was a captive in Russia. Not only did he make me welcome; the town council voted to honour me. The Burgomaster made a long oration; I was invited to sign the town's book for famous visitors; people I had forgotten came to thank me for setting them free. There was a great banquet, at which I was presented with a silver gilt bowl. It stands on my sideboard to this day. When I found that the bowl was filled with gold and silver coins, I poured them on the table and begged the Burgomaster to give them to the poor. This went contrary to my nature, but I remembered my countrymen who said that I thought only of profitable captives.

Much the same scenes took place at Hamburg. The town council, having heard that I would not accept money, gave me a heavy damask table cloth and two dozen napkins. We use them still at Kimble Manor; they show no signs of wear.

From Hamburg a ship took me in ten days or so to Harwich. It is a dull town, and the country round not beautiful, at least not in a wet, cold February. But there was grass, green grass, though it was winter. I knelt down and kissed the cobblestones of the quay.

When I saw the Queen I had to explain to her why Ivan's letter stank of brandy. Then I had to interpret the letter. It was (after all this trouble) no more than I had supposed; an appeal for help in his wars. He wanted also to be reassured; if all else failed, could he really come to England?

I had my own piece of news to add. Batory was urging the Swedes and the Danes to send their ships into the White Sea, and attack Rose Island. I told the Queen I did not believe the Swedes had ships equal to the task, but the Danes might have.

The Queen asked me what I thought she ought to do. This was the moment I had waited years for. I begged her to help the great Stephan Batory. I said that such a bloodstained monster as Ivan—Here, I remember, the Queen interrupted me. She said that sounded like the writings of the Russian exile Kurbsky, whose denunciations of Ivan were then being printed in most countries of Europe. (Though not in England, where the Russia Company had influence enough to suppress them.) Surely, the Queen said, Kurbsky was exaggerating. I assured her that, on the contrary, his writings fell far short of the truth. He had left Russia before Ivan's worst crimes. I begged her not to send Ivan any gunpowder.

The Queen said . . . But before she said any more she swore me to secrecy. So I could not then report her words to the Company. (Which led to more jealousy against me.) Nor can I report them now, even to save my life, unless Her Majesty gives me leave.

8

'That's another little task for me, then,' Oulton said. 'Besides obtaining a copy of the indictment, and finding some witnesses, I have to ask the Queen to let you quote her own words in a secret conversation.'

'Several secret conversations,' Jeremy said. 'Some of them in the presence of Robert Dudley, Earl of Leicester, who was a good friend to me then. He asked me to buy certain things for him, next time I went to Russia. That was private trading, and against the Company's rules, but the Company never enforced its own rules against its powerful patrons at court. The whole matter led to some correspondence which must be still in the state papers. I hope it is; this is the only way the Earl of Leicester can speak for me, being dead. You can ask for the papers, when you plead for me before the Privy Council.'

'I shall not be allowed to plead for you,' said Oulton. 'You must speak for yourself, if they allow you to speak at all—which they are not bound to do. Once you have pleaded not guilty, they may require you to stand mute, and let the witnesses tell the story.'

Sir William Hawtrey said, 'That's why nobody accused of high treason is ever acquitted.'

I exclaimed, 'Then we might as well be in Russia!'

'Not quite,' replied Oulton. 'Sir William is mistaken; people have been acquitted, once or twice. No doubt this will be another instance. But you will understand'—he turned to Jeremy—'that since I can do no more than prepare you in advance, I must know everything.'

'I have told you everything,' Jeremy said.

'Up to February 1581,' said Oulton, looking at his notes. 'We are now in January 1597. Your account of the last sixteen years must wait; I think I hear the Tower guards coming to turn us out.'

As I kissed Jeremy goodnight, he tasted my salt tears, and said, 'You understand the danger now, poor girl! Forgive me, Bess!'

'What for?' I whispered.

'For marrying you. I did it in all honesty; I thought I could make you happy.'

I was trying to answer him, when they led me away.

Outside the Tower the raw fog made me cough. 'You should send for a doctor,' Oulton said absently. The two men hurried me to my inn as if anxious to be rid of me. But I would not let them go until I had asked them, 'Should I go to the Queen?'

'You!' said Sir William. 'Certainly not.'

Oulton took his arm. 'Wait! Let me look.' He studied my face under the lantern at the inn door. 'You see how she looks now?' he said to Sir William. He might have been a butcher assessing meat. 'White skin, pink cheeks, a gleam of tears.... Not bad, not bad at all. Might do very well at court. Do you know anybody who could introduce her?'

'No!' Sir William said.

'I think you do. The Earl of Essex is not only the Company's patron at court; he's as deep in debt to it as anyone ever has been. He can't refuse it a favour. Lady Horsey ought to see him tomorrow, while he's still in town.'

Sir William looked sour. 'You mean to trouble the Earl in his own house?'

'I will trouble the Devil in Hell,' said Oulton, 'to help a client. But you know the Earl already; it's better if *you* go. Don't write first—simply go, and take her with you. We are, thank God, all subject to the law; but there are things you and I cannot do, which a pretty woman can.'

Oulton then left us, remarking that he had no time to lose. Sir William would have gone home, but I shamed him by weeping at the inn door, until he agreed to have supper with me. All through the meal he was grudging and sour, till I asked him point-blank whether he thought Jeremy guilty, or only certain to be found so.

'Why?' he said. 'What's the difference?'

'When you and my father arranged my marriage, did you think you were giving me to a man capable of treason?'

'That marriage got your father out of a nasty hole. You too! All that property settled on you . . . Jeremy did that to spite the Queen. He knew what was coming to him, and so he made sure his estate should be where the Queen couldn't touch it. It's a great thing for you, Bess. You'll be one of the richest widows in England.'

The inn servants were everywhere. So it was in a whisper that I said, 'You've hunted with Jeremy, dined with him, sat on the magistrates' bench. . . .'

'Yes, why not? This business hadn't come into the open then.'

'But if you thought he wasn't loyal to England—'

'Loyal enough, when he's here, I daresay. And loyal to Russia when he was there. Well and good, so long as he could carry it off. If he can't, he can't. Your snivelling won't help.'

I said, 'It may help, if you take me to the Earl of Essex.'

He tried every possible pretext. The Earl saw petitioners only while he was dressing. The Earl was often listening to music, or composing poetry; he did not hear what the petitioners were saying. If he did trouble to listen, he might not care. If he did care, he might not be able to influence the

Queen. To all this I replied, 'You mean that you are ashamed to be seen with me.'

Sir William opened his mouth, and closed it again. Then he said, 'At least consult your husband first.'

'Time enough to tell him, when I have some good news.'

Though I coughed again that night, I slept much better, and was almost gay when my reluctant, sour old cousin called for me.

We went by water. With its towers fronting the Thames, Essex House was much grander than Kimble Manor. But it was not so comfortable. Petitioners waited in a huge hall with a draught howling through. I expected to wait with them, but Sir William and I were ushered straight into the Earl's dressing room.

For a moment, when I should have greeted him, I gasped. I had never seen anybody so beautiful. His beard of gold, his hair of bronze, his white skin . . . these colours were not more delicate than the grace of his movements, as he came forward in shirt-sleeves to bow to me. I do not mean that I felt any desire to betray my husband. If I had, it would have died in me at the smile which said as plain as words, 'Yes, I am beautiful. Is it any wonder the Queen dotes on me?'

To Sir William the Earl's manner was a mixture of condescension and anxiety. I had seen the same thing in my father, when he greeted social inferiors to whom he owed money. Clearly the Earl had not yet paid for the ruff which a servant was now tying on his neck, nor for the jewelled coat another servant held ready, nor perhaps for the lute being played in the next room. The very secretary who sat with him to make notes—were his wages less than a year behindhand? All the presents the Earl's petitioners brought him could not pay for an establishment like this.

This thought made me bold. I said, 'My husband is in the Tower, accused of high treason. He is so far from being a traitor that he will not reveal certain conversations which the Queen told him to keep secret—not even to his lawyer, not to save his life! Surely she will release him from his vow of secrecy, so that he can defend himself!'

The Earl said he knew Jeremy. 'He tried to teach me Russian once. How many things there are I would like to learn, if only my duties... Tell me, how did he come to be in this trouble?'

When we had been talking for an hour, he said, 'You tell your husband's story so beautifully that you must not fail to tell it to the Queen. I myself will introduce you to Her Majesty.' Then he looked over the secretary's shoulder, to make sure the man had done me justice in his notes.

Coming out of Essex House I said to Sir William, 'You see?' He shook his head, and snorted.

Wind and tide and current all driving together brought our boat to the Tower in a few minutes. The cold made me cough, but when I ran to Jeremy I had breath enough to tell my good news. He seemed rather astounded than pleased.

'The Earl of Essex!' he said. 'You silly girl, why didn't you tell me first?'

'I didn't want to raise your hopes. But Jeremy, he was so kind! He said he knew you.'

'Yes, and I know him. If the Queen came to know as much as I do, he wouldn't be her favourite. He'll invite you to court —yes, if he thinks that'll put you in his power.'

'What for?'

'You know very well what for.'

'But.... He can have beautiful women if he likes.'

'Bess, look in the glass.'

My father had told me no man could want me, and I still half-believed that. Yet I had to admit that the woman in the glass had a full bosom and a slender waist, bright lips, bright cheeks and a clear skin.

'No wonder the Earl was willing to hear my adventures,' Jeremy said, 'when it was you who told them.'

I said, 'This isn't Russia. He can't force me.'

'No, but... you're so simple! And I'm not free to protect you.'

'You do protect me,' I said. 'Your love is protecting me wherever I go.'

After I thought I had made him calm, he sighed. 'Bess... Oulton has not been able to get a copy of the indictment.'

I asked why not.

'Oulton thinks that what is alleged against me may be so secret, or so scandalous, that they will not let a clerk do the copying. If we have to wait until Lord Burleigh finds the time to copy it with his own hand. . . .'

'I can look into that, when I see the Queen.'

'Don't build your hopes too high! You may never see her.'

Indeed I thought I would not. Three weeks passed, during which time Oulton was immersed in the accounts which had led to Jeremy's money troubles with the Company. At last he reported, 'Thank God! There's nothing here you can't explain.' That same day, the Earl sent one of his own boats to bring me to Greenwich Palace. A thick fog hung over the river; my maid Rosemary grumbled at the cold. I huddled her inside my cloak—my sable cloak lined with crimson velvet, my best.

Because of the bad weather, the Queen was taking her daily exercise by walking up and down the long gallery. When the Earl brought me before her she stopped. I curtsied. She looked down at me just as she looked down from her portrait, hanging in my father's hall at Hampden House. Only she had grown more haggard. Her cheeks were worn to the bone, her eyes wide open, as if she held them so by force of will.

I said, *'Infandum, regina, jubes renovare dolorem.'*

'Well, I did not command it,' said the Queen. 'But since you have a sorrowful tale, tell it me.' Then she smiled. 'God's death! Elizabeth Hampden, you did not come here to tell me the fate of Troy.'

I began to tell her Jeremy's trouble.

'Oh, I know,' she said. 'And you are not more anxious than I am to clear his name. It is an insult to my judgment, after all, to tell me that a man I have trusted is a traitor. Yet it has been so, sometimes. . . . In your husband's case, the matters alleged are so grave, and the witnesses appear so credible, that he has to stand his trial.'

I told her that Jeremy did not know what was alleged against him, nor who the witnesses were.

'Has he not seen the indictment?' asked the Queen in surprise. She told the Earl to find out the cause of the delay. Then she turned to me. 'He shall have the right to speak at length before the Council, and whatever else he needs for his defence.'

Already her tone was dismissive. I remembered that I had not said what I came for.

'What he most needs for his defence,' I explained, 'is permission to reveal conversations he had with Your Majesty.'

'Why, how could that help him?' the Queen asked. 'What does he think I said?'

'Your Majesty, he will not tell even me.'

She began to walk briskly again. The Earl kept pace with her, and motioned me to follow. This I did, but not easily, for my persistent cough had left me short of breath. She, a woman of sixty-three, moved like a young girl. After two or three turns up and down the gallery she stopped and said, 'The conversations I had with your husband . . . I have been thinking of them, since this charge was made. What I remember is that, when I saw him first, I thought he was a liar. He told me that Stephan Batory was getting the better of the Tsar. My minister Walsingham had just before informed me that Batory was dead, his army overthrown. Then I found that Walsingham's news had come from a fellow in Ghent—no nearer the battlefield than I was. People expect me to decide right or wrong, peace or war, on such information as that! And the messengers who bring the news are so positive! Young Jeremy Horsey was. He knew what we ought to do. Refuse to sell our gunpowder to the Tsar, send it instead to this Batory—whom he had not met, any more than the fellow from Ghent. Could Batory pay us? I asked. Our merchants in Poland had found them a beggarly people. That might be so, young Horsey admitted; they had nothing like the treasure-houses of the Tsar. But they would surely repay us, when they won. I asked him—was he a grown man, and did not yet know that wars make people poorer? Why does every armourer insist on payment in advance? He urged me then, if I would not help Batory, to send help to neither side.' She smiled. 'I told him he reminded me of my own youth. At my first coming to the throne I forbade our merchants to sell arms to Russia. I said then all Jeremy Horsey could say against it. But I soon found that the Tsar wanted only arms. If we did not trade in them, we should trade in nothing. For myself (I told young Horsey), I did not mind. I could do without the ermine trimming on my gown, and ride out in winter wearing sheepskin instead of sable. I could extinguish these wax candles, and go back to the smoky torches of my father's day. I could find some other place where they grow hemp, and set up some other rope factory, if that were for the good of my country and my people. But you must consider (I told him) that all our wealth is in our shipping, which is also our only defence in war. Am I wrong to glory in our ships, and in our seamen? The men who found the way to the White Sea—should I tell them they may make that voyage as often as they like, but without profit? Should I tell Sir John Hawkins that he may sail the Atlantic, but never with a cargo of ivory and Negroes? No! I have to rule men as they are, not as I would like them to be. If I want ships which can outsail the Spaniards, and seamen who grow constantly more skilled and venturesome, I must let them venture to their own profit. That's what I told young Horsey—'

Here I interrupted the Queen, not of my own accord, by an outburst of coughing. She looked at me sternly, and asked, 'How old are you?' I told her twenty-three.

'And yet you puff and pant as you walk along. What's the matter with you, child?'

'I'm not used to the London air. The fog's been in my chest ever since I came.'

'You should not paint your face to look healthier than you are.'

'It's not paint, Your Majesty. This is my natural colour.'

The winter light was almost gone. She called for candles, and looked more closely at me. 'Where are you staying? . . . I will send my own doctor to you.'

Then, breaking in on my thanks, 'I do indeed hope your husband can clear himself. If he thinks it will help him, he has my leave to repeat all I ever said in his hearing.'

At the turn of the stairs the Earl of Essex was waiting with open arms for my thanks. 'A kiss!' he said. 'Surely that's worth a kiss?'

I curtsied to him and said, 'On behalf of my husband, I thank you with all my heart.'

He laughed, took my hand, and suddenly pulled me through an open door into a little ante-chamber. I looked round for Rosemary, who had been close behind me. She was missing.

'It's time we knew each other better,' said the Earl. He seized me in his arms.

I did not know I still had such a temper. I spat in his face and scratched him, drawing blood. He was so taken aback that he let me go, and I ran back to the staircase.

An old man with a long white beard was coming up the stairs. By the great crowd of servants and petitioners who followed, I knew that he must be William Cecil, Lord Burleigh. I curtsied to him, and made my escape.

At the foot of the stairs I found Rosemary. I asked her angrily, 'Where have you been?'

She said, 'The Earl told me to wait down here.'

'You should have known better than to take any notice.'

'What, madam! Refuse anything to such a handsome fellow!'

She took my reproaches lightly. Happy at seeing the Queen and the great lords, and proud of all she had to tell the inn servants, she did not grumble that night even at my coughing.

When I saw Jeremy I said nothing about the Earl's advances, but told him every word of my conversation with the Queen.

Jeremy asked, 'If she had so much to say, why did she never tell you what the accusations were?'

'It must be something that she herself cannot bear to mention,' Oulton said. 'Have you ever spoken of her . . . disrespectfully?'

'Only to her face.'

'What—when she told you that she must allow the sale of arms to Ivan?'

'Yes. I don't wish to dispute Her Majesty's recollection, but I don't recall she ever made that speech about the ships. Maybe she thought it. But what she said was, first, that I must keep secrecy, because she had promised the Emperor of Germany not to break his blockade on arms to Russia; and, second, that she must send the gunpowder to the White Sea, because the price Ivan had agreed to pay would come to nine thousand English pounds. Gunpowder being a royal monopoly, this would go straight into her treasury. I told her that these were not the acts of a Christian sovereign. For a moment she looked so angry that I remembered the man who had been beheaded for saying the same thing to Ivan. So I went on more soberly, telling the Queen that we might lose even our profits, if the Danes attacked Rose Island. She looked more weary than angry. Then she said, "What I will do is call home the Englishmen who are working for the King of Denmark." It was then she told me that the Danes' good ships were all built by Englishmen.'

'So you learned how profit is made,' Oulton said.

'I learned it in action that summer of 1581, on the return voyage to Rose Island.'

'Despite your views, then, you went back with the gunpowder.'

'I had no choice.'

'What became of your plan to buy yourself out?'

'Oh, that... I asked my father if the jewels I sent from Russia had reached him safely. He said yes; they had come just as the moneylenders were threatening to dispossess him of everything he had. That taught me a lesson—in future I would *be* the moneylender. I did not reproach my parents; after all, I owed my riches to them. When they were already deep in debt, they had hired a tutor to teach me Greek. So it was their doing that I had been able to learn Russian. And some of my money was left. I asked the Company if I might give up to them the house and other property I owned in Moscow, in exchange for my freedom. While they were considering this, the Queen said that they must send me back to Russia. I believe she said she had learned more from me than from all her ambassadors. After that there was no question of the Company releasing me. I had to continue in my duty—which I did, to such effect that at the end of the year the Company paid a dividend of 106 per cent.'

'Remind the Privy Council of that,' Oulton said. 'Most of them are shareholders in the Company.'

9

I shall remind the Privy Council, too, that because of my warning we were prepared for the Danish attack. Instead of the half-dozen ships the Danes expected, we had thirteen.

The Danes did not sail into the White Sea, but lay in wait for us, and caught us as we rounded the North Cape. From a distance we could see on their ships engines to make wildfire. As they sailed nearer we saw their grappling-irons ready. They meant to close with us and board us, though they must surely know that we carried gunpowder. I believe they did not care if they blew themselves to Hell, so long as they took us besides.

Though their ships were good, and built by Englishmen, they were built in the fashion of ten years before. We defended ourselves in a new-fangled way. Our guns were so large, and had so great a range, that we kept the Danes from ever grappling with us. They could not even sail close enough to throw the wildfire. All they could do was fire their own guns, which were nothing like so good as ours. They had one lucky shot, though; they killed an English gunner. I took his place. I should have holed the nearest ship low down, close to the water line. But

I was afraid of aiming too low, and wasting my shot in the water. Instead, I aimed too high, and caught a seaman who was leaning out to retrieve a fallen sail. He went down into clear water, fathoms deep. I do not say for sure I recognised the man; perhaps it only seemed that he was one of the slaves ransomed from Rose Island. Why should my fancy fix on him, and not on his officers, any one of whom might be Madelyn's brother? Surely some were the brothers of less lucky Danish girls. They fought like men avenging intolerable wrongs. Outnumbered and outgunned, they still fired on us, until all their ships were holed, and not one of ours damaged, beyond what we could easily repair. Then they made for their harbour at Vardö. We all cheered and shouted as we saw them turn tail. Yes, myself as well; my blood was up, and I cheered. But when I tried to throw my cap into the air, I found that my right arm hurt. I looked at it and saw blood.

We had a doctor aboard, one Jacob. (Not a Jew; he came from the West Country, where good Christians have these names.) He did up my wound so skilfully that it soon healed. He would not take money, saying that we were both in the service of our country. So I repaid him by teaching him some Russian.

As we sailed past Pechenga, I looked at the monastery of my old friend the archimandrite; and though I had been wounded in a battle I knew I was a coward.

When we landed at Rose Island, we were all heroes; our gunpowder was intact. But when it came to examining our cloth, Mr. Rowley shook his head. 'The quality gets worse every year,' he said. 'Can't they read in London? I've sent them enough complaints. No wonder we've been outflanked by those Hollanders!'

He pointed up the mouth of the Dvina, to the Monastery of the Archangel. There, I now learned, ships were at anchor, flying the Dutch Protestant flag. They had arrived before us, the Danes letting them through unopposed.

'We've been there to ask them what they're up to,' Mr. Rowley said. 'But none of us can speak their sore-throat lingo.

Knowing you, Jeremy, I suppose you're going to tell me you speak it fluently.'

'Not fluently. I did learn some Dutch in Haarlem. Maybe I can remember a little.'

'The look on your face when you say that! You're still the cocky boy who knows it all!'

I found I did know something, when I landed at the Archangel, and heard the voices. The faces, too, seemed familiar. The Dutch always look watchful; they must keep an eye on their country, else it will vanish under the North Sea. They looked at Russia, too, as if it might vanish, and at me as if they wished I would. But when I told them I was one of those Englishmen who had brought supplies to Haarlem, they consented to talk. They were in Russia, they said, with the Tsar's permission, to sell their woollen cloth. And they showed me samples. Their cloth was as good as ours had ever been, and the prices about half. I wrote a letter to tell Mr. Rowley this, concluding, 'We defeated the Danes; can we defeat the Dutch?' Then I sent the letter back by boat, and pressed on by post horses to Moscow.

Ivan gladly paid out the nine thousand pounds. 'Your gunpowder's come just in time,' he said. 'The Poles think they're certain to take Pskov. Now we'll turn the tables on them! What should I do without you, Yeremy?'

The moment I reached Moscow I was rich again. From the Tsar to the humblest petitioner at his gate, Russians overwhelmed me with presents. Dr. Jacob too began to make his fortune. He had done well in England; the Queen consulted him; but never before had his patients paid him in diamonds. The first time he was summoned to the Tsar, I went with him to interpret. Ivan, we found, was not ill; he wanted only to know more about our Queen. Dr. Jacob spoke well of her wisdom; she had followed his advice for her health.

'But her age...' murmured the Tsar. 'She's almost fifty, ha? An old woman.'

Dr. Jacob exclaimed, 'Oh, Your Majesty, you would not say so if you saw her! When there is dancing at court, and the ladies-in-waiting take hands with the young lords, and the

Queen herself leads the dance, then you would swear that she was the most beautiful of all.'

'These ladies-in-waiting—they dance with men, ha?'

I hastily explained that this was the custom of our country. Nobody thought the worse of the ladies.

'Do you mean they are still virgins?' asked Ivan.

'Of course, Your Majesty!' Dr. Jacob exclaimed. 'The Queen's ladies are of spotless reputation.'

Thus far I had acted as interpreter. But Dr. Jacob's Russian was improving; he understood the Tsar's next question. 'Isn't one of these young virgins related to the Queen?'

Before I could interpose, the doctor replied, 'Yes, that's Lady Mary Hastings, a very fine young woman.'

'What on earth possessed you to say that?' I asked him later. 'Do you know what you've done? The Tsar will want to marry Lady Mary Hastings.'

'But he's married already.'

'His Church thinks not. The present Empress is his seventh wife. The Church would never acknowledge an eighth wife either, but, if our Queen did, he might think that good enough. Anything to get Englishmen fighting for him!'

And sure enough Ivan began to talk of sending an ambassador to England. Our gunpowder, it seemed, was not enough. It was not enough, even, that future supplies were assured. Our sea-fight with the Danes (we heard) had made some awkwardness between our Queen and their King, since they were not at war. King Frederik said his men were not his men, but pirates; Queen Elizabeth said her men were simple, peaceful traders. After that, neither mentioned the battle. But, since we had won it, the Danes made concessions. They would in future let us take on fresh water at Vardö. Our ships could carry less water, and consequently more gunpowder.

Yet whatever we brought was too little for the Russians. They were not defeating Batory, only holding him off. Though the Tsar poured out men as he poured out money, he could not re-establish his western frontiers. Narva was gone, and, with it, all Russian trade through the Baltic. And we were blamed. When the general who had raised the siege of

Pskov, one Shuisky, was feasted at court, I heard the old reproach—that the English sent only supplies, never soldiers. I heard it loudest from Shelkalov, and not once did I retort that his main achievement on the field had been to run away.

My most implacable enemy was Afanassy, brother of the Tsaritsa. He thought I wanted to supplant his sister with Lady Mary Hastings. If I were to tell him how it sickened me to think of giving an English girl to Ivan, he might not believe me. Worse, he might. Then he would tell the Tsar that I was plotting to frustrate his purposes.

Amid so many snares, I comforted myself by going to see Madelyn. She was now reigning proudly over her own house, where everything was, as her German husband said, *blitz-sauber*—clean as lightning. Beaming with pride, this German, Heinz Braun, invited me to try the tasty cakes that Madelyn had made. I did not mention our sea-fight, and she, it seemed, had not heard of it. But she made me tell her, over and over, every word of my conversations with her father and brother.

'Poor father!' she said. 'Are there snakes in his kitchen, truly? Since my mother died he's had nobody to keep the servants in order. Oh, if only I were there! Do you think I really shall see them again?'

Heinz told me he had applied for permission to leave Russia. 'Surely, now that Madelyn is my wife, they won't refuse to let her go?' I advised him to be patient; there was fighting, still, at the frontier.

Because of this fighting, Ivan could not send his ambassador to London by land. He had to wait until the White Sea was free of ice—that is, until the summer of 1582. I wanted to go with the ambassador, as interpreter. But Mr. Trumbull refused to release me. I had seen my own country only the year before; there were others more homesick than myself. None of these others could speak Russian as well as I could. When I said so, Mr. Trumbull's mouth tightened. He regarded my perilous journey, my talks with the Queen, as devices to undermine his authority. He tried now to re-establish it by ignoring my advice. He flatly refused to believe that the Tsar wanted an English bride.

Jane Bomel believed it, though. She had heard it where she heard most of Ivan's closely guarded secrets—in the market place. She came to me, this untalkative woman, and said one thing, 'Don't let them bring an English girl here!'

What (I asked her) could I do? I could not send a letter to warn Lady Mary Hastings; the Tsar's officers would intercept it. The Company's chosen interpreter might carry a message by word of mouth, but he was jealous of me, and little inclined either to do me a favour or to believe a word I said.

The day after Ivan's ambassador left Moscow, the Tsar's officers came to the English compound and asked for a thousand roubles in tax. Mr. Trumbull produced the Tsar's own decree, exempting us from all tax. The decree was good, the officers admitted; the decree was perfect, but—the Tsar had changed his mind. His wars had cost more than he had expected.

Mr. Trumbull displayed a sudden affection for me. I was clearly the best person to go to the Kremlin.

The Tsar was not there. He had gone into a monastery, I was told, for a period of meditation. Where, then, was Boris? I found him, to my great surprise, in the apartments of Afanassy, the Tsaritsa's brother. They had been scarcely on speaking terms. Now they were examining a litter of puppies, and Afanassy was urging Boris to choose one for himself. Both listened courteously to my complaint.

Boris told me, 'This is not happening only to the English. People who were assessed to pay a certain sum, according to our old customs, are now being forced to pay treble.'

I asked why.

Afanassy said, 'We are at war.'

I protested, 'The pledged word of a sovereign. . . .'

'Well!' said Boris. 'I will do what I can.'

I asked, 'Can this really be the Tsar's intention? He's sending . . . certain proposals to London. He wants English friends.'

'He wants English soldiers,' corrected Boris. 'He thinks, maybe, that this tax will show how badly he wants them. He has no great hope of the marriage, because he thinks the English may already know that our present Empress is pregnant.'

'What!' I exclaimed. 'Why didn't you tell me this before?'

'His Majesty,' said Boris, 'went into the monastery only this morning.'

I said, 'But if I'd known of the Tsaritsa's pregnancy yesterday—Your ambassador set out then, and our interpreter with him.'

'They will not be travelling very fast,' said Boris. 'If your Mr. Trumbull should wish to send a message—not a written message, of course—Afanassy and I can arrange post horses.'

Afanassy and I. . . . Boris had his own reasons for defending the sanctity of marriage. His power at court had increased since his sister Irina became the wife of Prince Feodor. But such power hung by a thread. Ivan's elder son had been made to divorce two wives in succession because they produced no children. The same fate threatened Irina; she had recently miscarried. Boris was wise, then, to make what friends he could. Smiling, he turned to Afanassy, and said, 'You see, you have misunderstood our dear Yeremy. He wants to prevent this thing as much as we do. Where's Shelkalov?' And, to my astonishment, Shelkalov soon appeared to join this friendly circle. He agreed that the happy news of the Tsaritsa's pregnancy should reach Queen Elizabeth.

'As for the taxes, I cannot help you,' said Shelkalov, looking as if he really regretted this. 'But I will have the collection delayed a little.'

Mr. Trumbull accepted the offer of post horses, and sent a man after the ambassador, ostensibly with medicine for sea-sickness, but really to tell our interpreter about the new taxes, and the Tsaritsa's pregnancy.

Soon we saw that the Tsar's men really were extorting treble taxes from every Russian merchant. 'That madman!' said Mr. Trumbull. 'How does he think people can do business, if they don't know their expenses from one day to the next?'

I smiled at this indignation, coming from a man who had often, as he crossed the market place, stepped calmly over rivulets of blood. Yet now I think that Mr. Trumbull was not altogether wrong. Perhaps it was the uncertainty of the taxes which brought most misery to most people. The richer sort did not rebel against paying; they recouped themselves by oppressing and robbing the poor. It was understood that it should be so; it had happened before, many times. Because of this the hatred between rich and poor was more bitter than I have ever seen it in England, even in a dearth or a plague. Now I heard for the first time what had happened twelve years before, when the Tartars burned Moscow. A Russian merchant told me that he and many others had escaped the flames by standing up to their necks in the river. He was not then rich, he said; the rich were mostly drowned. If they had gold sewn into their clothes, or hidden in their boots, they sank into the mud. Then the poor dragged the bodies out, and took the money. I had thought I could imagine that great fire. But I never imagined that I should hear this man tell me, with a smile on his face, that his fortunes were founded on the night he robbed the dead.

Nor could I have imagined that I would not condemn him. Yet I did not. Any more than I did when I saw him bar every door, for fear of the tax collector, and shutter every window, and still look over his shoulder, before he would show me what furs he had to sell. I thought that he was living as best he could, in that time, at that place.

The Tsar came out of his monastery when the treble tax had been collected. To win back the love of his people, he arranged a spectacle for them. A certain provincial governor had been too open in his bribe-taking. One of the 'gifts' he had extorted was a goose, trussed and stuffed with gold pieces. Now he himself was brought to the place of execution trussed up like a goose. The Tsar sat on the stone platform before St. Vassily's Cathedral. He called to his executioners, 'Can any of you carve a goose?' At his direction, they hacked slices off the man's arms and legs. The Tsar cried out to his victim, 'Do you still think that goose flesh is the best meat?' And then, to the people, 'He meant to eat you up. What would you do, if I were not here to protect you?'

The sport flagged; the victim was no longer shrieking, having become unconscious through loss of blood. Ivan gave the order to cut off his head. As the executioner held it up, the people cheered their smiling Tsar. They did not ask—that is, they did not ask aloud—who kept the money that the bribe-taker had extorted.

Soon Ivan was able to give the people another diversion. In October 1582 a messenger arrived in Moscow. He was one of the Cossacks I had seen years before, when I met the Stroganov brothers. This Cossack, a convicted highway robber, under sentence of death, prostrated himself before the Tsar. He was willing to be executed, he said; but the Tsar ought first to know that the Cossacks, armed and supplied by the Stroganovs, had conquered all Western Siberia. The inhabitants had sworn allegiance to the Tsar. There on the Moscow River was the tribute they had paid—barge after barge piled high with sable and ermine; and some rocks which glittered with seams of gold.

Of course the Cossacks were pardoned. The Stroganovs too were pardoned—or rather, the Tsar seemed simply to forget that he had ever accused them of treason. The people were told, 'God has given a new kingdom to Holy Russia!' The church bells rang out all over Moscow. In the Cathedral of the Assumption in the Kremlin, Prince Feodor himself was among the bellringers. The poor soul was always happy at this one task he could do well.

In the wake of the Cossacks came Mr. Finch. If the Tsar could be so forgiving, he argued, then surely the Company could be friends with him. He had been bravely looking for trade in regions where no Englishman had set foot. He too had sables and gold-bearing rocks to prove it.

Mr. Trumbull sent to Rose Island for Finch's chief accuser, Mr. Rowley. When he came, he showed us a long list of Finch's crimes, beginning, 'Adultery with Russian women.'

The rest of us told him to strike that out. Every Englishman in Moscow had good reason to know that keeping women in seclusion does not make them chaste. When we said so, Mr. Rowley looked in disgust at us all, and especially at me. He was persuaded, however, to go on to his next charges.

'Taking bribes from the Stroganovs . . . supplying them with arms and gunpowder which the Company had already sold to the Tsar. . . . Selling woollen cloth and other property of the Company. . . . Tampering with accounts to cover his defalcations . . . absconding from the Company's custody . . . evading arrest. . . .'

We sent for Finch. His answer to our charges was the Russian proverb: 'God alone is without sin'. Then he told us his adventures in Siberia. None of us, not even Mr. Rowley, could refuse to hear that. What should have been a stern tribunal turned into a dinner, with myself as host. I had wine enough to make Finch eloquent.

'Cossacks! All they know about a gun is how to hold it to the head of some poor traveller. They never could have beaten the Siberian Tartars, if it hadn't been for the Germans.'

'What Germans?' we asked.

'Don't you know that for years the Stroganovs have been buying every German slave they could lay hands on? Some of those fellows are good fighters; they know all about guns. And some are craftsmen; they repaired the Tsar's old ships—you know, that flat-bottomed fleet the English built in Vologda. They were a bit the worse for wear, by the time the Stroganovs bought them. Been patrolling the Volga for years. But the German craftsmen soon had them seaworthy—riverworthy, I mean. The Cossacks went sailing up-river into the Urals, dragged the ships on rollers across the water-sheds, and sailed them down-river into Siberia. The Siberian Tartars had never seen real ships; the figureheads alone made them run away. They came back by night, though, skulking round with their bows and arrows. The Cossacks were blazing away in the dark. Would have used up all their ammunition there and then, if the Germans hadn't persuaded them to wait for dawn. Once the Siberian Tartars were driven off, the other Siberians were easy. The Samoyedy and that sort—what do they know about fighting? All they want is a quiet life. They used to pay their tribute to the Tartars; now they'll do the same for us—I mean, for Yermak.'

We asked him to repeat the name.

'Yermak, the Cossack leader. The Germans call him Hermann. He's not like a Russian. When he captures one of those heathen encampments, there's no looting, no raping . . . The women come round a day or two later, of course, but that's on the sly.'

Seeing the success of his tales, Finch beamed. To me, as to an old friend, he said: 'I hear you've been in England. What's the news? Did the Queen ever marry her Frenchman?'

'No, thank God,' I said.

Finch nodded with drunken solemnity. 'God bless dear old England! God bless our Virgin Queen!'

Mr. Rowley broke in sharply: 'Those rocks you brought back from Siberia . . .'

'Oh, that's real gold,' Finch assured him. 'You test it!'

'We have tested it,' Mr. Rowly said. 'We know the seams are gold. What we want to know is where the rocks come from. Can you guide our prospectors to the place?'

'I could find you guides.'

'We don't want to trust our prospectors to Cossacks, or cannibals. Can you yourself guide them?'

'Oh, come now. You don't mean that I . . . Not myself. Those rocks, you see, didn't come from the place where I was. Not the exact place.'

'How far away?'

'Well, when you ask how far, distances in Siberia . . . The Cossacks, you see, went sailing up and down the River Ob. Took oaths of allegiance from the people they found on the banks. They call that conquering the place. Well, I daresay Yermak *could* have conquered it. Could have kept it for himself, been an independent king. I ask you, would Ivan be any worse off? Wouldn't so much as know—not for a year or two, anyway. But, what it is, the winters are nasty in those parts. Nastier than here. A lot of the Cossacks died of scurvy, that first winter. So did the Germans. Yermak saw he hadn't enough men. If they die at the same rate this coming winter, the Siberian Tartars could finish them off. That's why old Yermak sent this messenger to the Tsar. Humble submission, kiss Your Majesty's feet—and can we please have some of Your Majesty's troops? There's going to be more fighting in Siberia next year. You don't want to send Englishmen into that!'

Mr. Rowley said quietly: 'Where did the rocks come from?'

'Well, as I say, we were on the River Ob. No! I tell a lie. On the Irtish, which flows northwards into the Ob. Or on the Tobol, which flows northwards into the Irtish. One of those rivers. Well! The flat-faced fellows who brought those rocks made out that they came from another river. Not the next river, which is the Yenisei, less than a month's ride from where we were. But somewhere still further east, the Lena. Whether any Christian could find it—'

'It would appear, then,' said Mr. Rowley, 'that the Cossacks are deceiving the Tsar, when they show him those rocks full of gold.'

'Not deceiving him, not to say... The rocks do exist, even if none of us can reach them. That is, the Cossacks are bound to deceive the Tsar, aren't they? They want soldiers.'

We were all silent. Finch blundered on: 'Look, I'm an Englishman. I wouldn't deceive you.'

'We don't intend you should,' said Mr. Rowley. 'You will take some experienced prospectors with you, when you return to Siberia.'

'When I what? You can't mean to send me back there. The winters... And the summers, all swamps and agues...'

'Then you may go to England,' Mr. Rowley said, 'to face the Company's charges.'

Finch preferred to go back to Siberia. But until his departure he avoided us, and lived among the Cossacks—now the darlings of the court.

While the Tsar appeared happy over Siberia, I went to see him about the taxes he claimed from the Company. He would do no more than put them off to a day three months ahead.

'At least,' I pleaded, 'wait until Your Majesty's ambassador in London...'

'Do you dare tell me to wait until next summer? I can't hear from London till then. There's no way in or out of Russia now, except by the White Sea. That's what you English wanted, ha? That's why you stood idle, and let the Poles take all my towns on the Baltic. Do you think I can be comforted by Siberia? A thousand miles of heathen wilderness!'

Then the Tsar shut himself up with envoys from Batory. Boris could do nothing for me. Since the Tsar's return from the monastery, Boris, Afanassy and Shelkalov had quietly drawn apart. Once more they seemed to be deadly enemies, restrained from fighting only by their respect for the Tsar. It was part of their play-acting that, if Boris uttered a word in support of the English, the other two must oppose him.

'So you see,' he told me, 'I had better say nothing. Besides ... The Tsar is suffering terribly.'

'Suffering?'

'If he makes peace with Batory now, we shall have to give up all we ever gained in the west. And yet, how can our lads fight through another winter?'

It was curious to hear Boris, whom I knew as gentle and kind, speak of that savage army as 'our lads'. He went on: 'They're half starved already, and we've had a wretched harvest. My own estate is one of the richest in Russia. It still grows grain, but where are the peasants to gather it in? There are whole villages standing empty. That fool Shuisky—he raised the seige of Pskov, but only by using men as if they were cockroaches. However many the cook stamps on, there's always more in the walls. But there won't always be more peasants!'

I could see for myself every day in the market place that, however the Tsar's officers might threaten, they could not prevent the price of bread from rising. Even before winter set in, the beggars were dying in the streets.

Rejoicings were decreed for the birth of a son to the Tsaritsa. As the child was christened Dmitri, the church bells pealed, but did not drown the mutterings of the people.

Ivan's eldest son, the Crown Prince, became suddenly very popular. When he rode about the streets the people greeted him as 'Our hearts' desire! Our hope!' I wondered at their fantasy. What hope could they have of this bright pupil, whose classrooms had been the orgy and the torture chamber?

I heard a naked holy man crying out in the street that Russia's defeats in war were due to her own wickedness. I went closer to listen, thinking he was right; Ivan's cruelties did indeed invite retribution.

But this was not what the holy man meant. Russia's wickedness was that she tolerated foreigners—men who wore no crosses, who defaced the image of God by shaving away their beards. These foreigners were plotting to bring in a foreign woman, to supplant the Russian Empress who had born the Tsar a son.

The crowd began to look at me, and mutter. They were telling one another that, though I had a beard, I was a foreigner. I must be. I looked well fed. Within my furs, I shivered. For weeks I had not seen the Tsar. Petitioners were no longer bringing me presents. I went back to the English compound and told my countrymen to prepare themselves against a massacre.

We were all armed, when the midnight knocking came at our courtyard gate. What we heard was not the voice of an attacker. There were people crying out, but they cried in German, for pity and for help.

We did not at once dare to open the gates. I climbed up to the top of the courtyard wall, and saw by moonlight men and women standing in tattered, bloodstained rags. These were the survivors of the German colony. As we opened the gates, Heinz Braun come forward, holding in his arms a woman's body. The face was a porridge of blood, but I knew it. Madelyn van Uxell, once delivered from the Tsar, had been raped and murdered by the Tsar's gunners.

Dr. Jacob came to my side, and so did the silent Jane Bomel. Together we saw the dead bodies laid on biers, the living bodies in beds. Many of my countrymen helped willingly, but some were afraid that, by harbouring these victims, we might bring down the same fate on ourselves. They wanted to turn the Germans out. I persuaded them to do nothing until I had been to the Kremlin.

It was not yet dawn, but I found Boris in the great courtyard, about to mount his horse. When I cried out my protest, I felt suddenly very light, as if something had fallen from me. It was my cowardice. For that moment I did not care whether I lived or died. Lights began to appear at the doors and windows; not caring who heard me, I shouted louder.

Boris did not interrupt, but listened with a quiet, intent face.

Another face I saw did interrupt me, a young face, old in wickedness. My cowardice came back. I felt it settle in its accustomed place, low down in my guts.

'Did I hear you say,' asked the Crown Prince, 'that the gunners who did this pretended to have my father's warrant?'

I told him that the Germans were unanimous; when they were roused from their beds and turned out of their houses the captain of the gunners showed them the Tsar's warrant, with the great seal. They were told that their little Lutheran church was a blot on the face of Holy Russia. The gunners burned the church down first, the houses only after they had been looted.

'And the women?' asked the Crown Prince. 'What became of the women?'

'Raped, and some killed as they resisted, and some of the men killed who tried to save them.'

I thought his eyes were narrowing with pleasure. But he burst out: 'This is intolerable! They were solemnly promised our protection. What sort of people will come to Russia now?'

He hurried into the palace. Before I could recover my breath, courtiers began to press round, expressing their sympathy, repeating in other words what the Crown Prince had said. I was in favour, and with the person of whom I had had least hope. Maybe the common people who cheered him in the streets knew more than I did.

When I was alone with Boris I asked him: 'Did the Crown Prince mean what he said?'

'You do not know him,' said Boris. 'Since my sister Irina married Feodor, she has become very friendly with the Crown Prince's wife. He talks to them both sometimes, and Irina has found his talk ... not at all as is generally imagined. He believes now that some things he did without thinking were ... ill-advised. This may be a time of great hope for our country.'

I went back to the English compound with this news. The Germans, once they knew that the Crown Prince was on their side, found courage to bury their dead by their own rites. The

English took it that they would not suffer for harbouring the victims. I am afraid that some of them became kind to the Germans, only when they thought there was nothing to fear.

I was not myself a witness of what happened between the Tsar and his eldest son. The dispute, they say, raged all the morning behind locked doors. Snatches were overheard. It seemed Ivan had lately slapped the face of the Crown Prince's wife, and so terrified her that he made her miscarry, thus frustrating his own desire for grandchildren.. He was on the verge of signing a shameful peace with Batory... He had refused to let the Crown Prince go and fight... In the midst of all these accusations, the massacre of the Germans was forgotten.

And, suddenly, the whole quarrel seemed forgotten. At dinnertime the Tsar was seen to embrace his son. They sat down in peace to eat together, and parted in peace for the afternoon sleep. Father and son met again at supper. The meal went as usual, but, when it was almost over, the Tsar accused his son of ordering post horses by his own authority—'trying to usurp my place!'

Both were fairly drunk, and, in the dispute that followed, the five or six post horses became confused with the slap on the face of the Crown Prince's wife, with Russia's losses in the war, with the imagined plots of the nobility, with the cries of the people when the Crown Prince appeared, and, at last, with the massacre of the Germans. The young man shouted: 'That was not the act of a Christian Tsar!'

Ivan sprang up and advanced upon his son, waving his iron-tipped staff. Boris tried to seize the upraised arm. Ivan turned on him, shouting: 'Traitor! You're taking his part against me!' And he struck at Boris with the staff.

Boris moved his head to dodge the blow, and the iron tip crunched against his shoulder. He fell to the ground. Nobody else dared to stir, as the Tsar attacked his son. Since it was death to bring a sword into the royal presence, the young man was unarmed. He could do no more than try to get the staff out of his father's hand. But Ivan, always very strong, was now fighting for the staff as if it were his empire. He wrested it clear, and brought the iron tip down on the young man's head.

The Crown Prince fell stunned. Feodor ran to him, whimpering: 'Brother! Brother!' The Tsar had raised the staff for another blow, when the sound of the half-wit's weeping stopped him. Now Ivan, like Feodor, sank to his knees, crying out: 'I've killed my son!'

I tell this as I heard it afterwards. We in the English compound knew nothing until that night, when there was more frantic knocking at our gates. Again we feared attack; but the man outside was a courtier whom I knew. He cried out: 'A doctor! For God's sake, an English doctor!'

Dr. Jacob stumbled out of bed, more dead than alive. He had been all day tending the Germans. I went with him, to interpret.

The Crown Prince was conscious, but very pale, and bleeding from a wound in his head. A crowd of courtiers pressed round him. I thought my eyes were mazed by the flickering of candles, when I saw amid the crowd a Lapp dressed in reindeer skins. He seemed to be muttering spells. Further off, monks were chanting prayers.

Ivan hung over his son's bed like a spectre. He seized Dr. Jacob, and hugged the poor man so hard that he had difficulty in getting free to examine his patient.

'He needs peace!' the doctor said. 'A man who has been stunned needs rest and quiet. He must drink nothing but pure water.' But it appeared that the courtiers, eager to help, had already filled the Crown Prince with wine. As for the rest... Though the Tsar now ordered most of the crowd out of the room, he himself kept up a constant wailing, interspersed with pleas for his son's forgiveness.

'God will forgive,' the young man murmured.

But still Ivan cried out: 'I was not myself! I was bewitched! I've been driven out of my mind by the traitors round me. Live, my own dear son, live! I need you so much! We'll fight the traitors together. We'll search them out, put them to the torture... Oh, God! Is this Thy punishment for my sins?'

He was on his knees, and he struck his brow against the flagstones until it bled. Suddenly he sprang to his feet, crying out: 'It's all the fault of the traitors. I am surrounded by traitors! When I attacked you, my own son, not one man tried to stop me.'

'One did,' said the young man. 'Boris.'

'Yes!' Ivan shouted. 'Boris!' He turned to me and seized me in his madman's grip. 'Take this doctor to see Boris. God, God, if Boris were to die! Take him this ring; tell him I know now that he's the only faithful servant I have.'

Boris had death looking out of his eyes, when we found him. Only Masha, his wife, and Irina, his sister, were at his bedside. Prince Feodor was prostrate before the ikon in the corner. I looked round for servants, but there was not one to be seen.

'This patient at least is getting some rest,' said the doctor. He cut away the blood-matted shirt from the shoulder, removed some splinters of bone, and called for water to wash the wound. It was Masha who brought it, with her own hands, and Irina who tore up an old sheet for dressings. From a cradle in the corner came a wailing, and still not one servant appeared. Masha picked up the baby.

'Bring her to me!' said Boris. 'My poor little Ksenia! I'm glad now that she's a girl; she'll be allowed to live. Let me bless her before I die!'

He could not raise his right arm, because Dr. Jacob had bandaged it fast. But he murmured a blessing, as Masha held the baby before him. Then he looked at me. 'Yeremy, if you can, protect my little girl. And try . . . try not to let them put my wife in a convent.'

I made the promises he wanted, but in some bewilderment, for I could not see why he should die of a broken shoulder. I asked the doctor what he thought.

'There's no burning in his brow,' Dr. Jacob said. 'Ask him if he was knocked unconscious.'

I translated this Russian fashion, as: 'Were you without memory?'

'I wish I had been!' said Boris. 'But I remember everything.'

'Then you'll do very well,' the doctor said, 'provided you don't move that shoulder for a week.'

When Boris understood this he said: 'A week! I shall not be allowed to live a week. I raised my hand against the Tsar.'

This was why the servants had fled. I told Boris the Tsar's words. I spoke loudly, so that, wherever the servants were lurking, they should hear. And, sure enough, before I had put the Tsar's ring on Boris's finger, the first pale face appeared in the doorway.

So we left Boris in comfort, and well attended. Irina followed us out, asking timidly: 'Is the Crown Prince really in danger?'

'He hasn't a chance,' replied the doctor, 'if his father won't leave him alone.'

We found the Crown Prince less alone than ever. On one side of the bed Ivan was wailing, and on the other the Crown Prince's wife wailed on a shriller note. Courtiers, monks and magicians had all come back into the room. The patient himself was constantly trying to rise and assure his father of his forgiveness. He was, moreover, trying to drink some vodka, which an old woman assured him would do him good. It was remarkable that he lived another four days.

At his funeral the people wailed; so did the Princess Irina. Prince Feodor wept, having wit enough to know that he was now the heir to an empire he could not rule. The Crown Prince's wife wept with a special bitterness, because she (being childless) was forced to enter a convent.

In the market place that endless ballad which began: 'Moscow is mended with walls of stone . . .' had new verses.

'He is fallen, our morning star,
He is darkened, our clearest light,
He is dead, our beloved prince . . .'

The Tsar's own tears flowed in public and private; remorse tore him from his bed at night and made him wander up and down his palace, howling. Yet, though he clutched out handfuls of his hair, though he stripped off his jewels, though he put on a monk's black habit and talked of becoming a monk, he continued the business of government. He concluded an armistice 'for the time being' with Batory. He entertained a Jesuit envoy and half-promised to reconcile the Russian Church with Rome, if the Pope would send him soldiers.

Then, because it had been his son's dying wish, he rebuilt the houses of the Germans, gave them some compensation for their looted property, and told them that they might meet for worship in the house of their pastor.

Heinz Braun refused the offer of a new house. He asked only for permission to leave Russia. This I obtained for him, and a sum of money besides. Being destitute, he could not refuse the money, but he took it weeping.

'It was greed for money that brought me here,' he said. 'Nobody captured me; nobody dragged me at the horse's tail. I thought I should find riches in Moscow, and marry a nobleman's daughter . . .'

I was once more in favour at court. So were all the English. The time for paying our taxes came and went, yet we were not asked for them. For this I thanked Boris. He was now very powerful. Ivan, in his chastened mood, did not forget that one man alone had tried to prevent him from killing his son. I still believe that Boris, in that moment, was impelled by simple goodness. Yet he could not have done better for his own advantage. He was now in favour both with Ivan and his next heir—with poor Feodor, who had prayed by his bedside when all the servants fled.

'But I am not all-powerful,' Boris told me. 'It was not my doing that the Tsar overlooked your taxes. The fact is . . . He is in earnest now about the English bride.'

I understood. The Tsar had killed one son; he wanted another. The baby Dmitri would not do. The nobles regarded his mother's family as their inferiors. What Ivan wanted was a boy of royal blood, with all England as his bodyguard.

Now that there was peace on the frontier, we could send letters by land. All of us in the English compound were longing to let Queen Elizabeth know that the Tsar had killed his own son. Yet we dared not write the news down. So we mentioned only the birth of a son to the Tsaritsa.

Two months later Mr. Trumbull received a furious letter,

signed by Sir William Hawtrey and the other directors of the Russia Company in London. They said it was not surprising that the Tsar should be demanding taxes from us. We had gone out of our way to incur his displeasure. Some of us had even put about 'lies and slanders' designed to hinder this marriage, which would be of incalculable profit to the Company. The marriage would also advance the cause of the Protestant religion.

'The Protestant religion?' asked Dr. Jacob. 'Don't they know that Lady Mary Hastings would have to be converted to the Russian Church?'

Of course our directors did know this, just as they knew that Ivan had married seven other wives, three of whom were living. But they wrote that they did not believe our news of the Tsaritsa. The Russian ambassador, they said, had assured them that Ivan was already divorced from her. It was a slander, put about by the enemies of Russia, that she had borne a son. 'Even if true,' added the directors, 'it would make no difference.'

It made a difference, however, to Lady Mary Hastings, and to the Queen. In the summer of 1583, the Russian ambassador came back empty-handed. We were happy to see him so. We were less happy to meet the new English ambassador.

The man our Queen had chosen for this task was one highly recommended by the Company, Sir Jerome Bowes. He was a merchant; the directors reasoned that he could understand their interests.

Unfortunately Bowes became puffed up with his new dignity. On the voyage to Rose Island he and Ivan's ambassador were travelling on the same ship. The captain could hardly prevent the two ambassadors from quarrelling at every meal-time, about their precedence. By the time they reached Rose Island, they hated each other so bitterly that they would not travel to Moscow together. Bowes haughtily refused the offer of post horses; he would go by river. This meant that he arrived in Moscow six weeks later than the Russian.

When he reached the English compound we had to kneel to him, because he represented our Queen. But we found it hard to keep a straight face. He was a little fat man, and he sounded like a schoolmaster who cannot control his class.

'What can you fellows have been doing? You've allowed these Russians to get to the point where they show no respect at all. This man they sent to guide me to Moscow along the rivers—he wouldn't let my barge go first.'

'If he was guiding you . . .' I murmured.

'But I'm the ambassador of Her Majesty the Queen!'

'You can see what he's like,' whispered the Company's interpreter. 'He'd put a saint into a passion; God knows what he'll do to old Ivan.'

At supper Bowes asked us: 'This man Rowley, on Rose Island—is he out of his mind? He told me one fantastic tale after another. Says the Tsar killed his own son . . .'

We told him this was true. Bowes turned red, and threatened us all with the Tower, for trying to deceive Her Majesty's ambassador.

The next morning the Tsar sent for Bowes with all imaginable courtesy. Through the October sunshine a glittering nobleman, attended by three hundred of the minor nobility, came riding to the English compound. There was a bodyguard of a thousand gunners, at whose faces I looked, wondering: 'Which of you killed Madelyn?'

The Russians had brought a beautiful gelding, caparisoned with gold, for Bowes to ride. Bowes found fault with this horse; it wasn't so good, he said, as the Spanish horse the nobleman was riding. We told Bowes that, if he rejected the horse, the Tsar would be insulted.

'It's his turn!' said Bowes. 'I've had nothing but insults ever since I left England.'

All this was in English. But his repeated refusals to mount the horse were clear enough to the Russians. They stared at him with bewilderment, and then with anger.

In the end he agreed to ride the gelding if the Tsar's cloth of gold were taken off its back. Bowes's own saddle, and a cloth embroidered with the arms of England, were used instead. When he mounted, I could see the cause of the trouble. Bowes had short legs. If he had used the Russian saddle the grooms would have had to shorten the stirrups, and he could not have borne that. His own saddle was made to fit him. I thought it lucky that he did not know what the people in the streets were shouting as he rode past. 'You dwarf!'

The crowd thought little better of the thirty attendants who rode after Bowes. Each one was carrying a piece of a great silver dinner service, our Queen's present to the Tsar. A wandering holy man cried out: 'Look! There goes Judas, with his thirty pieces of silver.'

We Company men, in our best clothes, rode next, and came in for our share of the insults. One market woman shouted: 'An English wife would be like English cloth—worn out in no time!'

At the Kremlin we were conducted between rows of courtiers, all dressed in cloth of gold. Ivan sat in his throne room, no longer wearing the black monk's habit in which for eleven months he had mourned, but blazing with gems. His three state crowns, of Moscow, Kazan and Astrakhan, each one too heavily jewelled to be worn for long, were displayed on a table. On his head was the ancient 'Hat of Monomakh', a simpler diadem. The iron-tipped staff, with which he had killed his son, was now held by one of the guards. What Ivan held was a sceptre of light-coloured horn, spiralled about with jewels. On each side of his throne stood two young noblemen, holding poleaxes of bright silver. Prince Feodor sat on a lower throne, and round about were all the greatest nobles, in robes of scarlet and gold. Boris, always the tallest and best-looking man at court, was now also the nearest to the throne.

The Tsar himself, though his lean face was lined with an unblest old age, was handsome still. Knowing all that I knew, still I thought: He looks like a great king. His bright eyes darted among us, to find faces he knew. He called me out of the crowd, to be his interpreter. I would rather have gone through another sea battle.

Bowes objected when he found that the Tsar was addressing him by his Christian name.

'He does that to everyone,' I said. 'You'd much better take exception to what he says about the Queen.'

'Why, what's he saying?'

'If you let me get on with it, I'll tell you. He's complaining that the Queen's words of friendship are only words. Why doesn't she ever declare war on his enemies?'

Bowes flared up. 'If he dares to suggest that our Queen doesn't keep her word—'

Translating, I explained Bowes's angry tone by saying that he could not bear any slur on the honour of Queen Elizabeth. Ivan smiled. 'Look!' he said to his courtiers. 'You see how this little man splutters and goes red in the face? That's what you ought to do, if you hear a word said against me!' Then he invited us all to dine with him.

The dinner lasted five hours, and all that time I knew Boris was longing to talk to me. How I knew is hard to explain; we had by now such a feeling for each other that we spoke without words, without a sign that could be noticed even by the bright eyes of Ivan. I hoped Boris, in his turn, would understand that I must be near Bowes. The little man was grumbling at the foreign food. Why had they insulted him by putting on his plate a mess of black jam? 'That's caviare!' I said. I dared not leave him to blunder on alone.

But early the next morning I went to see Boris. He said: 'Yeremy, this promise made to the English...'

'What, the broken promise, about the taxes?'

'No, the other promise. You do know, don't you, that our Church doesn't recognise any marriage after the third?'

'Surely they recognised the Tsar's fourth wife?'

'That was a special dispensation, but the fourth wife—the Kingdom of Heaven be hers—' Boris broke off. 'No! That wife is alive. In a convent, but alive. So there's another good reason!'

I said I did not need any more reasons. 'I've opposed this marriage from the beginning; you know that.'

'But if the English are promised... If they think they are promised that the child of an Englishwoman should inherit the crown—'

I said I had never heard of any such promise. Yet even as

I said it, I thought—he is right. The promise is bound to be made.

'It could never be carried out,' said Boris. 'Never! Our customs, the laws of the Church...'

I said: 'Unless Prince Feodor were to become a monk.'

'He'll never do that. He loves his wife—my sister—so much that he can hardly bear to be one day without her. Feodor, you know, is made like other men. My sister has been pregnant twice. No; he'll never be a monk.'

'Might he perhaps think himself... too gentle to rule Russia?'

I had not seen Boris look so fierce. 'He will be able to rule. He knows well enough how to choose his advisers... his friends...'

Boris, then, was hungering for power. I was glad. When a good man sees his country in distress, what should he hunger for, but power to put things right?

I told Boris that the English wanted to have the taxes taken off, and their trade encouraged. Only to get these things, they would talk of the marriage. But it would never happen.

'Can you answer for your Queen, though?' asked Boris. 'Or for your generals? Are you sure that they won't send an army here, to show the world what they can do?'

I swore him to secrecy. Then I told him that we had no generals, and no real army. 'All our power is at sea.'

'Glory to God!' said Boris, embracing me. 'You have set my mind at rest. If I should ever come to have... more ability to help than I have now, I shall not forget you, my own dear friend.'

It was later, when I was interpreting for Bowes in a secret meeting, that I heard the promise from Ivan's own lips.

'God has punished my sins,' he said, 'by taking to Himself my eldest son. And my Feodor is a good son for a bellringer, not for a Tsar. If God grants me another twenty years, I shall have time to beget and bring up another son. But he must have a mother of royal blood! Where is this Lady Mary Hastings? Why did she not come with you?'

'She has had the smallpox,' Bowes explained. 'Her face is pitted. The Queen has other cousins, better-looking . . .'

'Then bring me one! I swear to you on my sceptre—look, it's a unicorn's horn!—that if I had a son related to the Queen of England, I would change my will and leave my crown to him.'

With some prompting from me, Bowes looked humbly grateful. He promised that this offer should immediately be conveyed to the Queen. With more of my prompting, he obtained a formal and complete remission of all taxes on the Company. And poor Jane Bomel had leave, at last, to go back to England.

Afterwards Boris asked me: 'This Jane Bomel—is she the widow of Dr. Bomel? I never thought of him having a wife. Did she know what he was doing?'

'It seems not.'

Boris laughed. 'He had the strangest way of making friends. He tried to sell me one of his poxy whores—who were spies as well, you know. And when I refused to buy one, he offered me a flask of poison. I told him I didn't want to poison anyone. He said: "Ah! But you will some day. Then you'll think of me, and be grateful." '

Though Ivan would do almost anything for us English, he would not restore our monopoly of trade. That year the Dutch merchantmen in the White Sea had been joined by the French, and even by the Spanish.

'That serves you right,' said Ivan. 'In all these years without a competitor, you've let the quality of your goods go down.'

The calm good sense of this abashed me, and not only because I knew it was true. When Ivan's voice grew harsh with longing for an English bride, when he swore by a sceptre of unicorn's horn, he seemed an old man hag-ridden by fancies. But now, speaking of trade, he showed such a good grasp of the matter that I remembered he was only fifty-four. He might really live another twenty years, and bring up another son in his own image. Then what would become of Boris? Of Russia?

'I'll build a fortress at Archangel Monastery,' said Ivan, 'to keep an eye on these new foreigners. But I must let them come! Their cloth is better than yours.' He called in Shelkalov, who had samples ready to prove it. Bowes stared at him, and asked me: 'Do these barbarians know one cloth from another?'

The word 'barbarian' is understood by every Russian who has met foreigners. It is the insult they mind most. I hastily explained that Bowes was calling the Dutch barbarians. 'Yet they make a fine delicate cloth!' said Ivan.

Bowes began blustering. 'We've had Your Majesty's promise for our monopoly. What are any of your promises worth?' As usual, I toned this down.

What I could not soften was the rudeness of Bowes to the noblemen. They did not understand his words when he said, pointing to St. Vassily's Cathedral: 'What's that—a basket of onions?' But they understood the look, and the tone. I told Bowes that this cathedral had been built by Ivan himself to commemorate his victories over the Tartars. He said: 'You mean the Russians meant it to look like that?'

'They think it's very beautiful. So do I.'

Bowes made such derisive noises that the noblemen thought he was jeering at their religion. They told him the English were 'no Christians'. Those words in Russian are something like our own. Bowes understood, and flew into such a rage that I could with difficulty prevent a fight.

The man who best kept his temper with Bowes was the Tsar. If he heard of any dispute between Bowes and a Russian, it was the Russian he would punish. Great noblemen were flogged, if Bowes complained against them.

None of us in the English compound believed this would last. We tried to explain to Bowes that his manners were bringing us all into danger. The only result was that he soon hated us all as much as he hated the Russians. He demanded a house to himself. The Tsar, still kind, gave him a very fine one, the property of a disgraced nobleman. He went on grumbling. Then he could not keep his thirty attendants in order; one tried to kill another, who fled for safety to my house.

At the beginning of January, 1584, Finch came back from Siberia. 'It's all over,' he said. 'This Yermak was a Russian, after all. He and his bodyguard camped by a river, drank vodka—went to sleep without posting sentinels. Wiped out!

The Siberian Tartars have got the place back, and, believe me, they're welcome.'

We asked what had become of the Company's prospectors.

'I was afraid you'd ask me that,' said Finch.

'You knew we'd ask you,' said Mr. Trumbull.

'Poor fellows!' Finch shook his head. 'Poor fellows! Buried by a Russian priest, I'm afraid. The best I could do for them. It was a sort of Christian burial.'

'What did they die of?'

'I told you—the Siberian Tartars wiped out Yermak's camp. All but one Cossack, who came back to our stockade to warn the rest of us. We'd hardly primed our guns when the Tartars were on us. Drove them away for an hour; they came back in their thousands... When our powder and shot was all spent we took our horses and broke out, those who could still ride. It was dark by then; we escaped.'

I said: 'But you went back the next day.'

'No. Good God, no! Why should I do that?'

'Then when did you bury the bodies?'

'When? ... I told you. We drove the Tartars away for a couple of hours. Maybe three. Didn't waste any time; every man took a spade. The Russian priest was gabbling through the service, as hard as he could go.'

Dr. Jacob asked: 'If you buried them in such a hurry, how did you make sure they were dead?'

'No doubt of it. Poor fellows had their heads blown off.'

'Then how,' pursued Mr. Trumbull, 'could you be sure they were our men?'

'What? Oh, by their clothes, of course.'

I said: 'I thought these Tartars had no firearms.'

'When they wiped out Yermak's camp, they took his powder, his muskets...'

'That would be a powerful musket, to blow off a man's head. Are you sure the Tartars didn't capture a few cannon?'

'Maybe. Well, no. The fact is, one of our cannons blew up.'

'Or maybe,' said Mr. Trumbull, 'you never were in any such battle.'

'I can see that you're all against me,' said Finch.

'The Tsar will be against you too,' I told him, 'if he hears you ran away, when his beloved Cossacks were fighting for him. You know what they do to deserters here? The same as they do to traitors and spies—roast them slowly between two fires.'

Finch took refuge with Bowes, who refused to give him up.

I told the Tsar the bad news from Siberia, adding that I could not vouch for its truth.

Ivan leaned sideways and made a wild groping gesture with his right hand. He was feeling for the iron-tipped staff, which he used to crash down on the steps of his throne when he wanted his courtiers' attention. The guard who held it was not where he should have been; Ivan angrily jerked himself round to look for the man. Then he howled and clutched at his groin.

Men came running to carry him to bed. I went running to fetch Dr. Jacob. By the time I brought him back, Ivan was surrounded by other advisers. The spell-muttering Lapp had proliferated; there were now more than fifty men and women dressed in reindeer skins. They almost outnumbered the monks. Feodor was praying. Boris, Afanassy and Shelkalov, though keeping well apart, all contrived to be near the Tsar's bed. Ivan shouted that everyone should make way for the English doctor. His voice was as loud as ever, but his attempt at a threatening gesture ended in another clutch at his groin. He moaned: 'I have been bewitched.'

Nobody would leave the room. They moved back a little, but still watched what the English doctor was doing to their Tsar. Dr. Jacob now spoke Russian well enough; he made his diagnosis first into the patient's ear and then aloud, to the whole court. 'His Majesty has ruptured himself. He will recover if he lies quite still for a few days. He must have rest and quiet.' The chanting of the monks redoubled; so did the chattering of the Lapps.

Afterwards I said to Dr Jacob: 'Surely rupture is a complaint of labouring men?'

'And of men who move with great violence. If the Tsar knows how to lie still, I shall be surprised.'

Each time he saw the doctor, the Tsar insisted that he was

lying still. 'But you surely don't expect me to stay in one room, all day long?' Every day his attendants carried him to and fro on a litter. I came to the Kremlin during one of these processions. Boris, walking behind the Tsar, beckoned me to follow. With monks holding candles to guide us, we descended into a place where I had never been, Ivan's treasure chamber.

'Set me down here!' said the Tsar. 'Bring me a lodestone, and some needles.' It seemed his attendants had expected the request; they had the things handy. Ivan lectured them on the use of the lodestone in navigation. What he said was perfectly sensible, but the glassy looks of his courtiers told me that he said it every day.

'Look!' Ivan half rose from his litter, to magnetise a needle. 'Look how another needle hangs from this one, and a third, and a whole chain of needles, all miraculously touched by the virtue of this one stone! Now'—he handed back the lodestone and the needles to his attendants—'now, do you think that only this one stone has magic in it? No! God gave every stone its power. Give me a coral! And a turquoise! Look how they lose their colour as they lie on my hand. That's because I'm poisoned with disease. Bring me a diamond! This is the most precious of all. It restrains anger and lust. I very seldom wear it. One little corner of it ground into a drink will poison a horse, let alone a man ... Give me a ruby. A ruby is comforting. It clarifies the blood. An emerald! You must never wear one of these when you go to bed with a woman; it will burst apart when you spend your seed ... The sapphire will surely give me back my strength. Yes! It clears the sight. All these jewels turn pale, foretelling my death, and yet ... Where is my sceptre? Look, because it's unicorn's horn, you may draw a circle with it, on the table ...' Boris took the sceptre and drew the circle. 'Now!' cried Ivan. 'Some spiders!'

The spiders were found in a corner, as they might have been in any other corner of the Kremlin. Ivan watched eagerly as Boris placed two of them inside the circle. Both died. Others were hastily brought in by other hands, and scuttled out of the circle alive.

'Too late!' said Ivan. 'Too late!' He fell back upon his litter, faintly asking to be carried to bed. I followed, wondering whether the spiders had died by magic or a nip from the fingers of Boris.

That night a comet appeared in the sky. Ivan, swaddled in furs, was carried out on the Red Staircase of the Kremlin. He watched the comet blaze above the snowy roofs of Moscow, and cried out: 'This is my sign!'

While the courtiers buzzed with reassurance, I turned and saw the Lapps all standing together on the roof of a house within the Kremlin walls. They seemed to be counting something. That house, I learned, was where the Lapps lived at the Tsar's expense. Every day Shelkalov went to consult them. What they prophesied was kept a secret.

On March the fifteenth 1584 Dr. Jacob was called urgently. I went with him as far as the Tsar's bedroom door, where some dozen courtiers were whispering together. At our coming they fell silent, and remained so all the time that I stood with them, while the doctor was with Ivan. Never before had I seen Russians quiet. If they thought the Tsar was dying, they should have wailed. If they thought there was hope for him, they should have chanted prayers. But where were the monks to lead the chanting?

Dr. Jacob came out and asked them: 'Did any of you see His Majesty make some violent movement?'

Silence. He appealed again. 'His Majesty has had some unusual exertion. Don't any of you know what it was?'

I said in English: 'Why don't you ask the patient?'

'He won't tell me,' Dr. Jacob said. 'Looks as if he tried to jump out of bed—but for what?'

'To kill one last victim?'

The doctor shrugged. 'He's killed himself, this time. There's a piece of gut making a lump in his groin, as big as your fist. And his private parts are swollen, too painful to touch. I've got some opium down his throat, but what a struggle! For all the pain he's in, he didn't want to take it. He said: "I mustn't go to sleep. They'll come and kill me." I told him nobody's going to kill him. Why should they? He can't live more than two days. Three, at the outside.'

'Have you told Boris that?'

'Boris isn't there.'

This was the strangest news of all. I went in search of Boris. His apartments, generally crowded with petitioners, were empty. And, just as when they thought their master marked out for execution, the servants had disappeared. I wandered through the rooms calling to Boris.

He came down the stairs from the women's quarters, quietly, slowly, yet with something in his aspect which took my breath away. I could scarcely falter out my news. He gripped my arm and whispered: 'Three days? Are you sure?'

'The doctor is sure.'

'My sister must hear this.'

In the women's quarters, Princess Irina was clinging to Masha. Both women drew themselves up to receive me, composing tear-swollen faces.

'Tell them!' said Boris.

In the women's presence I did not like to describe the symptoms of the Tsar. I said only when Dr. Jacob expected him to die.

'It will be three days,' Boris echoed. 'Yes, the Lapps told Shelkalov he would die on March the eighteenth. But, Yeremy, you told me something else. Because of his . . . "unusual exertion" the doctor called it . . . he has pain and swelling in the very place where he deserves it most.'

'What!' said his wife. 'Then it was the finger of God.' She turned to Irina. 'Didn't you say so, a moment ago? That when you could not make a sound, your soul cried out to God for deliverance. It was God who kept you innocent, a good wife to Feodor . . .'

Boris motioned her to be quiet.

'It can't be kept a secret,' said Irina. 'I wish it could. Oh Boris, Boris, forgive me!'

'What for?' said Boris gently. '*You* did nothing.'

'I let him trick me. How could I? I've been so much on my guard . . . Oh, Boris, I've always, always thought of the danger to you. When the Crown Prince was in love with me, I never would see him without his wife. And I've never once been alone with the Tsar till this morning, when he pretended to be dying. He said I must send everyone away, and hear his last wish.'

Ivan must be dying indeed, I thought, if he had tried to rape a woman, and failed.

'He tricked me too,' said Boris. 'He told me he'd heard that the guards on the Kremlin wall were sleeping at their posts. He said: "Who can put it right but you? The officers are all your uncles and cousins." There was nothing wrong with the guards, but while I was finding that out—'

'Oh, cunning, cunning!' wailed Irina. 'He pretended his voice had grown faint; I must bend down lower to listen. Then he suddenly clapped his hand over my mouth. That was when my soul cried out to God. And God immediately struck the Tsar, so that he fell back howling with pain. I was delivered. But, because of his howls, everyone came running. They could see the marks of his hand.'

She lifted her face to the light, and I too could see the marks; a separate bruise on her cheek for each iron finger, and a scratch for each nail.

'So I could not keep it secret, what he had attempted. Oh, Boris, I would have kept it secret if I could!'

I knew now why everyone had left Boris. His most holy duty on earth was to avenge the insult to his sister. But nobody wanted to hear him planning to do it.

I said to Irina: 'Your highness, you are avenged already.'

'Oh yes, I think I am,' she said. She turned to Boris: 'Truly, there's no need . . .'

Boris only shook his head. His wife pleaded: 'After all, you are a Christian.'

'A Christian, yes,' he muttered. 'But not a monk.'

I said: 'The Tsar told Dr. Jacob that he was expecting somebody to come and kill him. Is he, perhaps, hoping that you'll make the attempt, and give his bodyguard a pretext for killing you?'

'But why?' said Masha. 'What have we ever done?'

'Nothing but be popular, and powerful.' I turned to Boris. 'Didn't you yourself tell me his complaint against you? You have manned the Kremlin walls with your cousins.'

'I have,' said Boris, 'and they are staying at their posts. Until I need them against his bodyguard.'

'No!' said Irina. 'No! We must not do that. He is going to die in three days, and, meanwhile, we have to consider Feodor.'

Boris looked round. 'Where is Feodor?'

'In our chapel praying,' said Irina. 'He's wounded to the heart. It's his father, after all ... And he understands. There's a great deal that Feodor understands.'

'Very well!' muttered Boris. 'The will of God ... Provided that it is only three days!'

Boris and I found Feodor prostrate before the iconostasis. 'Come!' said Boris. 'Come to your father! He is dying.'

Feodor jumped up and looked from one to the other of us. He asked: 'Then was he in a fever?'

'Yes, delirious,' Boris assured him.

Feodor threw his arms round the neck of Boris, and kissed him three times.

Ivan was full of opium, but still he moaned with every breath he drew. Feodor timidly drew near and asked his father's blessing.

'Bless you, my son!' groaned Ivan. Then he opened his eyes and cried out: 'But my other son! Where is my other son?'

They brought him the baby Dmitri. 'What's that?' he asked. He waved the child aside, moaning: 'Not this! My son, my own, my first-born son! Where have you hidden him? Why won't he come to me? Who stands in the way? Traitors! Traitors!'

Dr. Jacob gave him a stronger dose. All the next day, and the next, he lay stupefied. Then he started up, saying: 'At last you've come! Was it far to travel? Was it very far, son?'

That was on the evening of March the seventeenth. On the morning of the eighteenth he woke without a fever, talking sensibly and saying he would like a bath. When his attendants lifted him into the bath, they saw that the lump in his groin was less. So was the swelling in his private parts. 'I have been bewitched there before,' said Ivan. 'And unwitched again, as you see.'

Boris murmured something to Shelkalov. Then Shelkalov slipped away. 'He's been to see the Lapps,' Boris whispered to me later. 'He told them they would all be burned as false deceiving witches. The day they foretold has come, and the Tsar has recovered.'

'What did they say to that?'

Boris looked at me sidelong. 'They said the day's come, not gone.'

That was the first time I ever doubted his word. It was unlikely that the Lapland witches could know the tale of Caesar and the Ides of March. But Boris knew it, from his learned sister.

Dr. Jacob appeared with a fresh draught of opium. The Tsar was still in the bath house, making a day of it, Russian fashion. The doctor put the glass down on the table. There was then a great coming and going, while the Tsar was carried back to bed. In that press of people, I lost sight of the glass. Or did I look away, for fear I should see who touched it? Certainly it was Dr. Jacob who put it to the Tsar's lips. He said: 'This bath was unwise, Your Majesty.'

'It's done me good,' said Ivan. 'I believe now I could enjoy a game of chess with my friend Boris.'

To show how well he was, Ivan insisted that he himself should set the pieces on the board. But he could not make the king stand upright. It was the careful fingers of Boris that set everything in order.

Ivan smiled at him across the chessboard and said: 'I gave your sister a bit of a fright, ha? All in fun! She knows that. She knows me. A child, that's me. A mischievous little boy . . .'

On those words he choked. His eyes turned in his head, and he fell back.

As Dr. Jacob was bending over the body, Afanassy sidled up to him and whispered something in his ear. The doctor immediately straightened up and said, loud enough for everyone to hear: 'Yes! This was a natural death. I can think of nothing more natural.'

'And yet,' murmured Afanassy, 'His Majesty seemed to be recovering.'

'He seemed so to you, because you know nothing about it. A man cannot recover, once a piece of his guts has become strangulated. There was no witchcraft in it, and no treason; his own violent way of moving killed him.'

Such conversations the doctor was well prepared for. What neither he nor I had expected was the monks' insistence that the Tsar was not dead, only dying. They said he must be made a monk, as all Tsars are before they die. The ceremony is supposed to send them straight to Heaven. The Metropolitan went through it when Ivan had been an hour in Hell.

Dr. Jacob and I had to stand stiffly through interminable prayers. Only Feodor openly lamented his father's death. He seemed more sensible than anyone else in that room. Boris was no longer there. I had not seen him go.

He came back at the moment when the monks' chanting ended, and he did not come alone. He led Irina by the hand. Behind them came Shelkalov, and behind him a great body of armed men. Irina bowed before her husband, and then took her place beside him. Boris bowed to the ground before them both. He cried out: 'Great Sovereign Feodor Ivanovich, Tsar and Grand Prince of all Russia, of Vladimir, Moscow and Novgorod . . .' By this time everyone in the room was prostrate before Feodor. '. . . Tsar of Kazan, Tsar of Astrakhan, Sovereign of Pskov . . .' When he reached the end of the titles, Boris added: 'Your Majesty's bodyguard is here to attend you. The five protectors named in your father's will, of whom I myself am one, await your pleasure.'

Feodor learned slowly, but what he learned he knew. In a perfect imitation of his father's gesture, he held out his hand for Boris to kiss.

When Dr. Jacob and I left the Kremlin, we had to fight our way through the press of lords and bishops, jostling to be the first to kiss the cross and swear allegiance to the new Tsar.

It was now growing dark. In the English compound Jane Bomel reported: 'The people in the market place are saying that Boris poisoned the Tsar.'

'Nonsense!' replied Dr. Jacob firmly.

'They say he had to, when the Tsar tried to rape his sister.'

The doctor said: 'How anyone can dispute my diagnosis—'

'Wait!' I said. 'The people think that Boris killed Ivan, but they think he was in the right.'

'Yes,' Jane Bomel agreed. 'They say he was only doing his bounden duty. But then, if he killed their Tsar, it may be their bounden duty to kill him.'

On hearing this I called my fellow-countrymen together. Armed with pistols and powder, we went back to the Kremlin. Boris gladly accepted us among his personal bodyguard.

We were near him the next morning, when he stood on the Kremlin wall, to confront a great crowd of people. Certainly he had other protectors. While the monks were making their attempt to deny that Ivan was dead, Boris was disposing regiments and cannon. His cousins now guarded all the roads to Moscow, as well as the Kremlin walls. But he could not stay within those walls for ever, if the people were against him. So he faced them, looking unafraid and unarmed. (In fact he wore a breastplate under his kaftan.) Some of the people cried out threats to him, and some were silent, but he greeted them all as friends. He said he knew that they loved Russia, and were ready to defend it with their lives. Not only the men but the women that he saw before him would be ready, like the women of Pskov, to climb upon the ramparts, and beat back the enemies of Russia. 'We are strong. We are strong because of people like you, my friends!'

All this the people had heard before, for Ivan was a master of such arts. But the deep voice of Boris, his gentleness, his beauty, the sudden radiant sweetness of his smile, made every word seem new.

'Our enemies have one hope,' he went on. 'They think we may quarrel among ourselves. That hope is groundless!' Here I heard the first applause. It came, though, from less than half the crowd. 'All the nobility, all the clergy have sworn allegiance to our Tsar, Feodor Ivanovich. Long live the Tsar!'

Nobody could refuse to echo that. But there were some shouts. 'Who really gives the orders?'

Boris firmly said: 'His Majesty the Tsar has ordered that all state prisoners found in the dungeons here should forthwith be pardoned and released. Their names . . .'

Now the crowd was quiet, wanting to hear the names. Instead they heard the pealing of bells. The gate of the Kremlin was opening. The royal bodyguard appeared, and then Feodor, with Irina by his side. They made their way to the stone platform before St. Vassily's where Ivan had been accustomed to stand and watch executions. The crowd cast themselves down in the snow, crying: 'Long live the Tsar!' And then, with more fervour still: 'Long live the Tsaritsa! Long live the blessed, the holy Irina!'

Irina, glowing with happiness, gave alms to the crowd, but that was not why they cheered her. They knew (I gathered from their cries) that she, alone of all the thousand women seized by Ivan, had been miraculously saved. She must be under Heaven's protection—a saint.

On seeing the effect his sister had made, Boris quietly left the wall, and came out (followed by us Englishmen) to do public homage to the Tsar.

Then, one by one, the prisoners came, still wearing fetters, to the platform. There were some bent double by torture, and some who had to be carried. For each one, Boris cried the name aloud. Many of the names were those of once great families; the people knew them well. Some of them I myself knew; they had been arrested for treason on the bare word of Eliseus Bomel. Yet, when he was accused of treason, they were not released. I had thought them dead.

As each prisoner appeared before Feodor, he gave the order for their fetters to be struck away. Those prisoners who could kneel kissed his feet. The people cheered and wept.

I myself was now in tears. Not because Ivan's victims, as they stumbled past me to freedom, screwed up their long-darkened eyes against the glitter of sunlight on snow. No, nor because I smelt the dungeon on them, as if it were the grave.

I was remembering Madelyn. From the moment of seeing her dead, I had known I should never again ease my conscience by doing some small private good. I had been praying for one thing; that since I had become (without my will) the servant of enormous evil, God might give me the means of enormous restitution.

10

'Before you tell us what restitution you made,' Oulton said, 'let's be clear what you did to put your friend into power. You stood by with your pistols primed to guard this Russian.'

'All the English did.'

'At your instigation.'

'Yes. We were not supporting a mutiny. Feodor was the Tsar's only legitimate son. We helped him to succeed to his father's throne in peace.'

'Hm. Tell me, when you confided in Boris that England had no real army, was any third person present?'

'No. If there had been—if the Privy Council knew of it—I would not be afraid. It is not a state secret that we have no army. Every country in Europe knows it. If Ivan did not, it was because he chose to delude himself, imagining an English bodyguard for his imaginary son.'

'For all the help this Boris had from you, was he grateful?'

'He thanked me.'

'Then can you explain why his first act was to put the English ambassador under house arrest?'

'That was for Bowes' own safety.'

'According to Bowes, whenever one of his attendants put his head out of the window, the guards would pelt him with lumps of muddy snow.'

'The common people would have pelted him with something harder. So would the noblemen. Some of them had been flogged because of his complaints against them. Bowes had jeered at their religion. He had called Shelkalov a barbarian to his face.'

'Shelkalov . . . I think he sent a message to Bowes: "The *English Tsar* is dead." '

'Yes, he thought he could send what saucy messages he liked. He had been named in Ivan's will, together with Boris, as a guardian for Feodor. There were three other guardians, but they were men of no great ability. For a few days Shelkalov thought he would have more power than Boris. He had the state papers in his hands; he knew how things were done. And at first Boris treated him with a great show of respect. Shelkalov and the other guardians were present when Boris called me in, to ask what should be done with Bowes.'

'They asked you that?'

'Yes. I said . . .'

Oulton broke in. 'Before you tell me what you said, have you any witnesses?'

'My witness is the fact. Bowes is alive today, and free to spit venom at me. I'm convinced that he's to blame for all my present trouble.'

'Then he must think you urged the Russians to kill him.'

'If they had needed urging! The fact is, I put myself in danger, telling the noblemen to send Bowes back to England with honour and safety. They asked me how I dared to defend a blasphemer. Was I myself laughing at their religion? Did I think they were barbarians? To prove they were not, they turned me out of the room. But Boris afterwards told me in secret what to say. So I went separately to each nobleman, saying that I knew him to be a good Christian, and therefore inclined to mercy. Whatever Bowes had done, I said, he was the ambassador of a great queen. Shelkalov asked me, laughing, what this queen of mine could do to rescue her ambassador. I spoke of the advantages of English friendship, but only Boris really believed in that. When some weeks had passed, Feodor named Boris as Protector of the Realm, with more power than a subject had ever held in Russia. Then Boris felt secure. He arranged that Feodor should receive Bowes for a courteous farewell. Bowes managed to make even this go wrong. In the ante-chamber he refused to give up his sword. It was unheard of to enter the Tsar's presence armed, and so I told him. Some of the noblemen were telling him that what he deserved was to have his legs cut off, and his dwarf's carcase thrown into the river under the window. I did not translate this, but their gestures made it clear, even to Bowes. He took off his sword. Feodor received him courteously, and sent compliments to Queen Elizabeth. Bowes made (through me) a suitable answer, recovered his sword, and was taken back to his house in safety, despite the threats of the crowd. He was ordered to leave Moscow in three days. But he told me he did not know how he was to travel, having spent almost all his money.'

Oulton sat up. 'He asked you for money?'

'And got it.'

'But I thought you were on bad terms. You were harbouring a runaway servant of his, and he was harbouring the Company's runaway, Finch.'

'I hated the sight of Bowes. We all did. But he was an Englishman, and his life was in danger. Not only did I lend him money, I found him thirty carts to transport his baggage, and as many post horses. The Russians gave him a bodyguard to conduct him to Rose Island. The most dangerous part of his journey was the first. Between his house and the city walls of Moscow, the people might overpower his guards. I, with all the Englishmen I could muster, formed an additional bodyguard. We brought him safely to a place ten miles outside Moscow, where I had a tent well prepared with food, wine and mead. We did not say farewell without feasting him as became an ambassador of our Queen.'

'Then what was this trouble between you afterwards?'

'When I came back to Moscow, I begged Boris to send the customary letters and presents after Bowes. This Boris did; his messenger caught up with Bowes at Rose Island. Instead of thanking the messenger, Bowes began to dispute precedence with him. The matter ended with Bowes in such a rage that he threw away the Tsar's letter to the Queen, and slashed to pieces with his sword some valuable sables, a personal gift to the Queen from Boris. On hearing this I could only apologise to Boris, who assured me that he was my friend still.'

'Did you say,' asked Oulton, 'that Finch went back to England with Bowes?'

'Finch went as far as Rose Island with him. They were about to board the ship, when another ship came in sight. It was rolling heavily, as a flat-bottomed river boat will on the open sea. At its prow were the remains of a gilded unicorn, and on its deck the remains of Englishmen. These were the Company's prospectors, whom Finch had led into Siberia. When the news came that Yermak and his camp had been wiped out by the Tartars, Finch took the best horse and fled, leaving his fellow-countrymen to fight alongside the Russians. They really took part in the battle which Finch had described from imagination. When their ammunition ran low, they took to the River Irtish on boats—the same boats I had seen my countrymen building at Vologda, ten years before. They sailed north until the river began to freeze. Then the Englishmen showed the Russians how to lay up the boats for the winter—well greased, and covered from stem to stern in reindeer skins. They built winter quarters, and tried to keep themselves alive by hunting, which is hard when white-furred animals are moving on snow, and the daylight lasts a bare three hours. They might have died of hunger, but some Samoyedy passed that way, and showed them the best places to set their traps. When spring came, it was very sudden; they had frostbite and ague in the same week. The river melted, and the Russians made for a tributary, the Tura, intending to sail up it into the Ural Mountains. From there, they said, they could walk into Russia. The English did not fancy the walk. They sailed north, along the River Ob. They knew (since other

Englishmen had done it) that it was possible to sail from the mouth of the Ob, and south-west along the coast to the White Sea. The Polar ice melts in July, a little. They found a channel through just in time, before the ice began to thicken again, as it does at the end of August.

'Mr. Rowley reproached himself (he told me afterwards) that in the bustle of providing food and beds for these men, he did not at once ask: "Where's Finch?" Finch had recognised the flat-bottomed boat from a distance. Before these unexpected witnesses could land, he had taken a rowing boat, reached the mainland, bought a peasant's horse, and disappeared into the forest. I suppose he made his way to his friends the Stroganovs.'

'Then it was Bowes and his attendants who returned to London.'

'And Jane Bomel.'

'Do you know where she is now?'

'Yes, she has a very pleasant little house on Stepney Green. I went there once to make sure that she had something to live on. She assured me that she was very well off. Apart from that I hardly had a word out of her. Poor woman, she's not very good company.'

'But she may be a very good witness,' Oulton said.

'To what? I still don't know what it is they accuse me of.'

'And yet you think this Bowes is at the bottom of it.'

'Whenever I've heard of a spiteful tale told against me, it's come from Bowes.'

'Then it's none too soon,' said Oulton, 'to look round for witnesses who were in Moscow when you and he were there. Jane Bomel. Who else?'

I suggested: 'This friend of yours in Oxford, Giles Fletcher?'

Jeremy quickly said: 'No! He came later.'

'Mr. Rowley, then?'

Jeremy frowned. 'No. He wasn't in Moscow at the time. He was on Rose Island.'

'But in that case,' I said, 'he must have seen Bowes in his tantrum, when he cut up the sables.'

'Oh, that! The Queen already knows all that.'

'She may require to be reminded of it,' Oulton said.

We had been all day talking, and now the guards came once more to turn us out. Oulton went to the top of the stairs while Jeremy and I kissed goodnight.

'Be a good girl!' Jeremy said. 'Don't go to see any more people without telling me first. Especially not Mr. Rowley.'

'Why not?' I whispered. 'Because I'm not supposed to know he was in love with you?'

Jeremy shook his head. 'He hates me now. The last time I saw him he would scarcely speak to me.'

'But would he tell any lies against you? Surely not?'

Oulton came back. 'Before I go,' he said, 'tell me where this Rowley lives now.'

Jeremy said, 'I don't know.'

'Then I'll have to find out from the Company. We'll send Lady Horsey to see him; she works wonders. Right! I'll see you again when that's done.'

Because Oulton had said it, Jeremy now agreed.

The next morning it was Jane Bomel I went to see first, because I knew where to find her. She had the neatest, prettiest house in Stepney Green. Her maid showed me into a room where every well-waxed panel reflected the warm flickering of the hearth. And yet the mistress of the house, when she appeared, was cold.

'I heard your husband was in the Tower,' she said, and her voice was thick with satisfaction. But she added: 'Lady Horsey, I am sorry for you. I know no harm of you.'

'No harm of him either, surely!' I exclaimed. 'It was his doing you came safely out of Russia.'

'He owed me that.'

How could such a fat person have such a tight little mouth? And what made her spit her words out with such venom? I meant to ask her, but instead I coughed. This tedious cough had followed me out of January into February, and still would not stop. Seeing my distress, she exerted herself so far as to ring for wine.

'You should see a doctor,' she said. 'A good doctor, if there is such a thing nowadays.' While I sipped the wine, she added: 'You'll never find a doctor like my husband.'

Since I had not quite done coughing, there was nothing I need say.

'My husband,' went on the widow of Eliseus Bomel, 'was executed for treason—an innocent man! Your husband should remember that.'

I said Jeremy had told me that Dr. Bomel was quite innocent of the offences for which he died.

'He admits that! Then I hope he's sorry for what he did! Why should my husband—a foreigner in London—why should he have known the right age of our Queen? He made an honest mistake. Was that any reason why Jeremy Horsey should set the Tsar against him?'

'That wasn't Jeremy's fault,' I said.

'Then whose fault was it? Everyone in Moscow loved my husband. The Tsar and the nobleman—they couldn't do enough for him, till Jeremy Horsey came. Money, jewels . . . Do you know, when I came out of Russia, the hem of my petticoat was weighted round with diamonds? Because everything he earned he gave to me. He was the most generous husband who ever breathed. A good doctor and a good man!'

I said: 'But you must have known about the Danish girls.'

'What about them? A doctor has to see every sort of patient. Don't you know that? If you had any sense you'd know. Whatever sort of women he had to attend in his work, it made no difference to our life together.'

I thought: You stupid woman, just because your husband contented you in bed . . . I said aloud, 'But did you never hear that the noblemen thought he was spying, and telling tales to Ivan?'

'Never! Never! If there was any tale-bearing, Jeremy Horsey did it. And went on with it, even after my poor husband was dead. Not one other soul in the English compound ever said a word against my husband.'

'That was because Jeremy warned them not to. Even though it was true—'

'He still says it was true! There! There's the proof that he was to blame for my husband's death!'

'But did you never accuse him of that, in Moscow?'

'I never said a word to him, beyond what was necessary. I needed his help to come safe out of Russia. Not that I cared whether I lived or died. But I knew that, once I was back in England, I could tell people what sort of a man Jeremy Horsey was.'

'Then . . . does this charge of treason come from you?'

'No. That's God's doing, not mine. I would have brought these charges years ago. But the great men in the Russia Company told me I had no proof. All of them in the Company were jealous of my husband. The things they cast up against his memory! Just because he had some difference of opinion with the Archbishop of Canterbury—'

'Isn't it true,' I said, 'that Dr. Bomel was once in the Archbishop's prison?'

'Archbishops! What do they know about medicine?'

I said: 'Listen! Please listen! You were with your husband when the Archbishop let him go, when he went to Russia and entered the service of the Tsar.'

'The Tsar knew how to treat a learned man. He gave us a house to live in—bigger than this.'

'But you didn't think, did you, that the Tsar was a good man?'

'That's not for me to say. I hardly saw him. My husband didn't think the court was a fit place for an Englishwoman.'

'Did he say what made it unfit for you?'

'It was dirty. There were spiders' webs in the corners.'

'Nothing else?'

'How should I know? I minded my own business.'

'But you must have seen the executions.'

'They were nothing to do with me.'

'Mrs. Bomel,' I pleaded, 'you *do* speak Russian? You *did* talk to the people in the market place?'

'What if I did?'

'You must have known that they called their Tsar Ivan the Terrible.'

'What of it?'

'But this was the Tsar at whose court your husband lived for so many years. Do you think he kept the favour of Ivan the Terrible by never doing any wrong?'

'You should ask *your* husband that. Wasn't he a great favourite of Ivan? And what's worse, of that cold-blooded murderer, Boris? Ask Jeremy Horsey what he did for Boris!'

I began: 'He's been explaining . . .' But then, looking at the thin lips in the fat face, I saw that I should get no further. So I got up, and sent for Rosemary, who had been enjoying a more pleasant gossip in the kitchen. As we rode back to London she said: 'Do you know what they call Mrs. Bomel? Mrs. Bumly! Comely Bumly, because of her big backside. Gets herself up very fine of a Sunday, lets everybody see what she gives to the poor at the church door. And at home she's a *miserable* miser. Locks up the maids every night, and then, to make sure, she goes and locks up the larder.'

I laughed. Rosemary went on: 'She's out of her bed half the night, walking to and fro, talking to her dead husband.'

So shall I be soon, I thought, if I don't find witnesses to keep my husband alive.

I rode straight to the Russia Company's office. They received me there with ceremonious courtesy. Mr. Burrough took me into the room where he had, ten years before, refused my father a loan. He asked after Jeremy's health, but seemed scarcely to hear the answer. He wanted only to know what the Queen had said to me. 'Does she seem well disposed towards the Company?'

I said she was, at least, well disposed to Jeremy. 'She will be glad to hear him prove himself innocent.'

'Of course!' replied Mr. Burrough. 'We shall all be glad, if he can.'

I could feel my cheeks flaming at that 'if'. But I had not come to quarrel with this man. I explained that I was going to see every possible witness, and asked him for Mr. Rowley's address.

'Oh, come, come!' he said. 'Why should you trouble—' he

checked himself. 'Why should you trouble *yourself* with such enquiries?'

I said, 'I am not asking you for money.' This was, I knew, a rude word in a place where the thing was always 'capital' or 'interest' or 'an accommodation to a gentleman'. For the pleasure of seeing Mr. Burrough blush, I said it again. 'I don't want money—only the names and addresses of Company men who were with Jeremy in Moscow.'

'Impossible. Why, some of them are still there. We've just had letters the overland way, and we've sent the same messenger back. We shan't be able to write to them again until the White Sea's clear of ice.'

'But some of your agents have retired; some are working in England. If you could tell me where they are—'

'You're asking us to comb through all our books, for twenty years and more. Lady Horsey, we're busy people in this office.'

'And I am busy!' I said. 'With my husband's life.'

A clerk announced the arrival of Mr. Oulton. 'What does he want?' muttered Mr. Burrough.

'The same thing that I want,' I said. 'You had better see us together.'

To Oulton Mr. Burrough was ruder than to me. 'I need hardly remind *you*,' he said, 'that there is a law against interfering with witnesses.'

'Against threatening witnesses,' replied Oulton calmly. 'Do you think that this lady is going about with a cudgel?'

'Oh, if Lady Horsey herself is going to see them . . .'

'I told you all along I was!' I said.

As deftly as a barber extracting a tooth, Oulton got out of Mr Burrough a promise to make a list of all the Company men who had known Jeremy in Russia, with their present whereabouts. For a start, we had the address to which Mr. Rowley had retired.

'But before you go there,' said Oulton, 'have dinner with me.'

I refused, not feeling able to bear the delay. But at my elbow Rosemary was muttering that it was very hard *she* should go hungry, because *my* belly was not empty. Oulton said that his house was round the corner, and his wife always ready for a guest or two. So I consented. Mrs. Oulton, a cosy person, made sympathetic little cluckings when I talked of Dr. Bomel's widow.

'The woman's mad,' I said. 'She won't hear one word against her husband.'

Oulton smiled.

I had to go to Mr. Rowley by boat. He lived at Chelsea, in a house much meaner than Jane Bomel's. I remembered how often I had heard Jeremy say that the pensions the Company paid were too small—'an invitation to dishonesty'.

Mr. Rowley had been honest. His front door led straight into his only living room, and he opened it to me himself. As I entered, a woman came from the kitchen, and I thought: 'Poor man, what a forbidding servant he keeps!' Then he introduced her to me as his wife.

Mrs. Rowley gave me a seat by the fire, and brought me heated wine. Yet, while I sipped it, I was aware of a coldness which neither wine nor fire could reach.

Mr. Rowley was telling me that the house, though small, was just right for the two of them. It had a garden where he and his wife could sit, on summer evenings, and watch the boats go by. In fact it was all he had dreamed of during those Russian writers. 'Quite satisfied...' But while he spoke of satisfaction, his face looked so gaunt and sad that I wondered why he had married. Clearly not for money, still less for beauty or grace. To prove that he was not a lover of boys? Casting about for some way to make him talk naturally, I glanced at his books. 'Is that Virgil's *Aeneid*?' I asked. 'It's a good story for us women, who can't do any real travelling! The Queen knows it almost by heart.'

There was a pause. Then Mrs. Rowley said: 'Her Majesty is very learned in the pagan authors.' She drew herself up and looked at the shelf beside her, which, I now saw, contained the *Institutes* of Calvin.

I thought: Oh, a ranting, canting Puritan! Mr. Rowley, then, had picked her for piety; she might help him to fight his temptations.

I found it hard to talk to them together. Rosemary too

sat with us; in this house there was no other place where she could go.

Yet I had to speak. I explained that Jeremy had been a month in the Tower, and still did not know what the charges were. So he could not say what witnesses he would need.

Mr. Rowley frowned, and was silent.

I told him he was known as an upright man (upright, surely, was a favourite Puritan word) and that I relied on him to speak the truth.

'Yes!' Mr. Rowley said. 'Yes, if they call me as witness, I shall speak the truth.'

I thanked him, but my thanks, I thought, made him uneasy. And the wife was not a woman to do the wifely duty of covering her husband's awkwardness with prattle. My own prattle failed me. I rose to go.

'Have you a boat to take you back?' asked Mr. Rowley.

I said no; I had not kept the one I came in, since Chelsea has a landing stage with boats always waiting for hire.

'They wait,' said Mr. Rowley, 'but without their boatmen. In this weather the men sit drinking in the Jolly Fiddlers. It's a rough place for women; I'll come with you.'

His wife (so far as I could make the woman out) did not dislike his going with Rosemary and me. Yet she did not fetch his cloak for him, nor settle it upon his shoulders, nor kiss him before he went. I thought: Doesn't she know that one temptation can be fought only with a better temptation—with love?

While he led the way along the river bank, holding a lantern, Mr. Rowley said no more than: 'Mind, here's the muddiest part!' But, when we reached the Jolly Fiddlers, he said: 'This place has a parlour for ladies. May I order you some wine?'

So at last I was able to see him alone. (For Rosemary the inn had a wide, welcoming kitchen, full of gossip and beer.)

'I have to tell you,' Mr. Rowley said, 'not to hope that I can help your husband.'

'But you said you would speak the truth!'

'That is what will not help him.'

Keeping my voice quiet, I asked: 'Are you saying Jeremy *has* committed treason?'

'Against this country—no. If I had any evidence of that kind, it would have been my duty to tell the Queen long since. But I know him as a deceiver, a decoy, an accomplice to a crime so monstrous—'

'No!' I said. 'No!'

'Lady Horsey, you look very fine in your sable cloak, with your lining of crimson velvet. Do you know that this blackness was bought with a man's black heart, and this blood colour with blood?'

'It's not true! Jeremy can account for all his money. It's only that people are prejudiced against him.'

'If you knew how I used to be prejudiced *for* him! How I defended him, when all the rest were ready to tear him apart for his cocky ways! If that boy boasts, I used to tell them, it's because he has good grounds for boasting. If only the rest of you (I said) would work as hard, and learn as much! And a good-hearted lad! Or so he seemed. Always fretting over the troubles of some poor creature ... I remember how his eyes filled with tears when he told me that our two kitchen women had seen their children murdered. How could that boy change into . . . Perhaps it was my fault he changed. I sent him to Moscow. God knows I did it for his advantage, not my own. I would as willingly have parted with my sight!'

I said: 'But he still is the boy you knew. He's kind still, and good.'

Mr. Rowley shook his head. 'Ask anyone in the Russia Company—'

I cried out: 'All of them in the Company are jealous of my husband!' The words echoed oddly in my head, until I remembered they were the words of Mrs. Bomel. The look on Mr. Rowley's face was my own baffled, hopeless look.

'Forgive me!' he said. 'Of course you won't believe a stranger against your husband. I would not have said all this, only ... A few days ago I had a shock. The Company's messenger brought letters from Moscow. Tsar Feodor's health is failing, and there's no doubt who means to seize the throne. And ... And the woman in the Troitsky Convent is dead.'

'What woman?'

'She was young to die,' said Mr. Rowley. 'Not yet forty. Whether it was poison, or grief, nobody knows.'

'But who is this woman?'

'A woman who trusted your husband's word. No! Don't ask me. Ask him!'

I could not do that until the following day. All night I told myself that I had no distrust of Jeremy. It was my cough that kept me awake.

How could I doubt him in the morning light? It was a fine, sunny day, though cold. We gave Paul and Rosemary a holiday to see the town. Then we went to bed, enjoying the sweet sinful feel of lovemaking by day. I slept, until Jeremy pulled off the blankets and slapped my bottom.

'The guards will be bringing dinner soon,' he said. 'A lady should eat in her skirt.'

While we were eating I told him only my talk with Jane Bomel.

'I never guessed she felt so bitter,' he said. 'Yet it's natural. Rather than doubt her husband . . .'

After dinner I told him of my visit to the Company, and then, only then, about Mr. Rowley.

Jeremy sat staring into the fire, while I knelt beside him. From time to time he stroked my cheek, saying: 'Poor girl! To run about for me . . . To hear such things . . .' Yet he did not seem perturbed, until I came to the words: 'The woman in the Troitsky Convent is dead.'

Jeremy jerked upright. 'What! Is he sure? How did she die?'

'He says nobody knows whether it was poison or grief.'

Putting his head in his hands, Jeremy began to sob like a child. As if he were my child, I stroked his hair, murmuring, 'Tell me . . . Oh, my dear love, tell me!'

11

But how can I tell you the splendour of that first year, when Ivan was dead, and Russia reborn? Then I began to see on Russian faces that look I had seen long ago on the faces of slaves. Ransom was possible, hope was possible, and freedom ...

Should I tell you how the beautiful Irina moved among the common people? She did it in the only way a Russian lady can, by going on a pilgrimage. Her destination was the Troitsky Convent, forty miles from Moscow, and she went all the way on foot. That showed the people her piety. Also it allowed them to approach her. At every village they brought her their tales of oppression, and she was tireless in listening. She refused all presents, asking the people only to pray that she might have a child.

As the ballad singers tell it, this good Empress came among the Russians like an angel, sent down from Heaven to redress injustice. I knew her, not as an angel, but as a sensible woman. To set right some old injustice makes a good ballad, but what Irina wanted was to prevent new injustice. Rather than carve up a bribe-taker like a goose, she made it possible for officials to live without bribes. She and Boris chose new provincial governors (yes, members of their own family, but where else could they find men to trust?) and increased their pay so that they could maintain the state required of them. They had to keep accounts, which Irina herself inspected. Besides, there was no telling where her pilgrimages would lead her; she might at any moment appear in some remote village, listening to every complaint of the peasants. Knowing this, the governors really did begin to govern better.

She and Boris were always consulting me. One day Irina showed me a manuscript bound in fine leather and clasped with gold. This was the code of laws which Ivan, in his youth, had promulgated. What improvements, asked Irina, could I suggest?

Ivan's laws were a blow to me, not because they were bad, but because they were good. The devil I had known was a fallen angel; and I now saw from what a height he fell. His laws were in some ways better than ours. In Ivan's code a married woman might own property and make a will, which is more than she can in England. It was laid down, even, that her husband must not witness the will, for fear he should browbeat her. These laws had another advantage over ours; they were so simply written that any peasant could understand them. Which is necessary, in a country without lawyers.

The worst thing about the laws, I told Irina, was that nobody knew they existed. In the eleven years since I first came to Russia, I had not heard of them. Nor had the people who most needed their protection. Ivan's code laid it down that a prisoner of war, though he might be made a slave, became a free man on the death of his first master. Where in all the slave markets of Russia had that been whispered? I had visited what passed for a court of justice, and had seen that everything was done according to ancient custom. Since the custom was that a murder should be avenged by the victim's own family, the Tsar's judges concerned themselves little with human life. A husband who had killed his wife might be fined; a master who had killed his servant not punished at all. Yet a poor man who owed a debt to a rich one might be long and horribly tormented in the 'straightener'. A nobleman who got into debt might send a slave to be punished in his place. A rich man could avoid most penalties. On coming into court he would bow to the ikon, and place a lighted candle before it; and sticking to the bottom of the candle was a piece of gold for the judge. This practice was expressly forbidden in Ivan's code; the practice continued while the code lay gathering dust. What had happened when it was first promulgated? Nobody seemed able to remember. Irina said: 'Our people are suspicious of new-fangled ways.'

I remarked that the code was no longer new. People should be shown the yellowed parchment, and reminded that Ivan had written it in the lifetime of his first wife, Anastasia. Her time was remembered as a golden age.

'Make sure that everybody knows these laws,' I said. 'Print them! Send them to every part of Russia! In the remotest places, there are always one or two people who can read.'

Irina, busy already with plans for hospitals and schools, now set a piece of ground aside for a printing press.

In one way she and Boris needed no advice of mine. They had the art (which I wish I possessed) of getting their own way sweetly. Afanassy, for example, they had long since won over. Now, speaking to him as a friend, Boris pointed out the dangers that surrounded Ivan's last son, the baby Dmitri. He was heir presumptive to the throne. It would be unwise to keep him in the same place as the Tsar; any uprising could kill them both at one blow. And Moscow was particularly dangerous, the people being so rebelliously inclined. It would be better for the baby and his mother to live at Ooglich. This town, almost 100 miles from Moscow, is by tradition the place for a widowed empress, and Ivan's will had left it, with all the lands round about, to his last wife. Since Afanassy was her brother, it was his duty to protect her child as if it were his own. Afanassy could not dispute this; it is a rule so absolute among the Russians that in all their fairy tales there is not one wicked uncle. So Afanassy, his brothers and their father found themselves on the road to Ooglich, surrounded by signs of respect, and by a bodyguard which watched their every move. Their places at court were given to cousins of Boris.

Some of Ivan's old favourites kept their places—notably Shelkalov. 'He's better here, where I can watch him,' Boris whispered. 'And I must have a secretary who knows how things are done.' Shelkalov's nature was indeed to be a secretary; he soon lost the cockiness he had shown after Ivan's death. Boris delighted in making this willing servant reverse all the things he had done (just as willingly) for Ivan. More and more forgotten prisoners were found, in remote fortresses, or monastery cells. It was Shelkalov who knew what property had been seized from them; who better than Shelkalov could see to its restoration?

It was his task, too, to make a list of the exiles who had fled into Poland. Even if they had fought against Russia, they were pardoned. So some noblemen who had fought in western armies returned to give Boris the benefit of their experience. Boris was disappointed to learn that the most famous of exiles, Prince Kurbsky, had recently died. 'He knew how to tell the tale!' said Boris. 'I wish he were alive. I wish he could bear witness to what we are doing now.'

It was a time of reconciliation. The only fighting that Boris undertook was the reconquest of Siberia. He sent a small army across the Urals to meet the remnants of the Cossacks, then still struggling up the River Tura. The Cossacks returned with the new forces, routed the Siberian Tartars, and established a new city, Tobolsk. Later they sent to Moscow a prisoner, a Tartar prince, whom Boris treated very honourably. The prince told me that an Englishman with red hair had wandered into his territory, had been captured, and had then escaped. 'But the Samoyedy ate him,' he added with a smile. I thought that was the last I should hear of Finch.

Elsewhere Boris did all he could to put an end to Ivan's old quarrels. Denmark need no longer be an enemy, now that the ravisher of its women was dead. Sweden was at war with Russia still, but there was no real fighting. What Boris wanted most was the friendship of Poland. He listened to every word the returning exiles could tell him about Stephan Batory. This graduate of Padua University, this Hungarian who became Poland's greatest patriot king, might have loved Boris, if they had ever met. But Batory could not believe that a Russian wanted peace. He would do no more than make a two-year armistice. Without waiting even for that, Boris redeemed from slavery thousands of Polish captives, had them fattened up and furnished with new clothes, and sent them back to their country to proclaim his good intentions.

The worst enemies of all, the Crimean Tartars, were quarrelling among themselves. One faction seized power from another, and used that power to send a great prince of theirs on a mission of peace to Boris.

The prince brought his sons with him, and showed us how he trained them. Even the youngest, a boy of eight, got no dinner until he had done his daily target practice, and hit the mark, or nearly. All their education was in horsemanship and war. I asked the Tartars if they never felt the need to put something down on paper. They said they kept some Jews to do that for them.

This prince and his attendants proudly wore patches on their clothes. They kept some ragged edges trailing, too; had they teased them out on purpose? When they made their formal appearance before the Tsar, they brought rich presents—gold and silver things they had looted—but handed them over as if they said aloud: 'Here! If you need such toys . . .' They were tall, broad-shouldered men; not one had a paunch; and when they bowed to Feodor it could be seen how little they were used to bend their backs. Their black slit eyes looked round at us as if they wondered at our chains of gold.

Russia, then, was at peace with almost all foreigners. As for the English—what was there I could ask on their behalf that Boris would not grant? Only our old monopoly of the cloth trade. That was gone for ever, and by our own fault. But since none of Russia's neighbours would sell her gunpowder, we still had that monopoly (which I was glad of, now). Moreover, not one of the Germans and Spaniards and Dutchmen, who now crowded into Moscow, might go beyond it. We English alone had the right to sail down the Volga to Persia.

For myself, my former power and profit seemed a shadow of what I now enjoyed. One thing I still found lacking. I felt it most when I saw Boris playing with his children. He now had a boy as well as a girl. He would pick them up by turns and swing them round his head until he was panting for breath. The girl would grow tired sometimes; the boy never. He would shout: 'Again, Dad, again!'

'I am supposed to be the lord and master of my son,' said Boris. 'But look at me! A pack-horse to him, a slave!'

I told him then how much I envied him. My long love-tangle with a married woman had lately come to an end. Her husband, on the death of his peasant mistress, had returned to her. And she settled back, with evident contentment, into the arms of a drunken savage. How had I come to waste so many years on such a stupid woman? I was now past thirty, the same age as Boris, and I had neither wife nor child.

'We must find you a wife,' said Boris. 'But a good one, a nobleman's daughter! My dear Yeremy, if only I had another sister! ... One thing, though. If you marry a Russian girl, you will have to come into our Church.'

I hesitated—I did hesitate—and then I said: 'Of course. What's the difference? We're all Christians.'

The truth was, I thought I should never go back to England. What could England offer me, to compare with my power in Russia? Not only was I rich; I had a happiness which (they say) can turn poverty into gold. I felt I was doing right. I was helping Boris and Irina to make Russia a civilised country.

And they needed me. For all his winning arts, Boris had enemies. His ascendancy over Feodor's heart was so absolute that to some it looked like witchcraft. The market place was full of spiteful gossip. It could not fasten on the private character of Boris. Now, as always, he was faithful to his wife. He had no taste for feasting, and the poorest peasant drank more. But a determined gossip can always find a pretext. Two hundred and fifty years before, the family of the Godunovs had been founded by a Tartar who turned Christian. The other ancestors of Boris were pure Russian. Yet, when he made his wise attempt to come to terms with the

Crimean prince, his enemies called him 'The Tartar' and professed to see something slit-like in his eyes.

Boris had not chosen his own wife. When he was seventeen or so, Ivan's favourite hangman, Malyuta, had observed his good looks and picked him for a son-in-law. The Tsar had decreed it should be so. Malyuta was killed in a battle, so soon after the wedding that his influence over Boris must have been slight indeed. Yet now people were calling Boris, with a sneer, 'Malyuta's heir'.

When Boris brought back the exiles who had fled from Ivan's cruelty, Shuisky (the general who had raised the seige of Pskov) complained that traitors were being set in authority over loyal Russians. 'The truth is,' Boris muttered, 'he doesn't wan to learn the new ways of making war.'

All this might have been expected. What surprised me was that the Metropolitan of Moscow (a man picked by Ivan for his complete subservience) began to quarrel with Irina. He disliked her plan to set up a printing press. The last one built in Moscow, he said, had been burned down, which was clearly the will of God.

'Why—was it struck by lightning?' asked Irina. 'I know it is blasphemous to put out a fire lit by the finger of God. But the printing press was burned in winter, and perhaps by men.'

The Metropolitan asked whether she intended to print Bibles.

'I hope so,' Irina said. 'The only Russian Bible is printed in Ostrog, and the Poles have Ostrog now. Surely we should print one for ourselves. Why not? This isn't Rome. Our Church has never forbidden laymen to read the Bible. Do you think I am the worse for having read it?'

'He couldn't answer me,' said Irina, laughing, when she reported this conversation.

Within the next few weeks, however, the Metropolitan found an answer. Or thought he had found it, until Boris, with no witness but myself, confronted him.

We were sitting in a little turret room, above the women's quarters of the Kremlin. Boris liked this room in summer; its windows opened on to the green trees that waved among the golden domes. A breeze came to us from the Moscow river. Best of all, we were too high for eavesdroppers.

Boris began talking to the Metropolitan of all that the Church had suffered under Ivan. Monks who had refused to give up monastery treasures were led into an arena, like early Christians, and forced to fight wild bears ... One Metropolitan was deposed; another died mysteriously ... To all this the Metropolitan (who owed his post to his own slavish part in these events) gracely nodded.

'And in those days,' Boris went on, 'the Church was forced into conceding scandalous divorces, followed by unlawful second marriages. Not to speak of third marriages, and fourth ... You must be very glad that this is over.'

The Metropolitan said nothing, but I saw that his hands were trembling.

'The Tartar prince who came to see me,' said Boris, 'had four wives, and, if he wanted to be rid of one, he need only say three times: "I divorce thee." Is that going to be the practice of the Christian Church?'

'My lord,' the Metropolitan faltered out, 'the Church does allow some divorces.'

'It is not going to allow this one,' said Boris.

'No, no, my lord! I never said it would.'

'Your accomplices thought you said so.'

There was a pause.

'Your accomplices,' repeated Boris, 'have even chosen a certain young lady to supplant my sister as Empress.'

'No! That is ... I know nothing about that part.'

'You know everything else. You know, for example, that Feodor loves my sister, and would never divorce her of his own free will. Your accomplices would have to seize Feodor, and force Irina into a convent at the sword's point. Which means that they would first have to kill me. Well! I am only a subject. But to seize your sovereign, hold him prisoner, force him into something against his will—What's the right name for that?'

The Metropolitan was silent.

'Then Yeremy will tell us! He has travelled everywhere. In

England, or France, or Denmark, or Germany, what would they call this design?'

I said: 'In any country it would be treason.'

Boris nodded. 'Only fifteen months ago, a word or a look was called treason. We are not going back to that. Yet there is such a thing as treason, and you have been plotting it.'

'It was not a plot,' pleaded the Metropolitan. 'It was only a matter of ... certain conversations.'

'With whom?'

'Some people are alarmed ... concerned ... because there's no heir to the throne.'

'They have young Dmitri,' sail Boris. 'A fine healthy boy, from what I hear. And what makes you so sure that my sister cannot have a child? She has miscarried twice, but that was when things were happening which might make any woman miscarry. You don't want to go back to those days? No, I am sure none of you do. Not even Shuisky!'

'It was not Shuisky's idea!' exclaimed the Metropolitan.

'You mean it was put into his head by the others, by ...'

The Metropolitan, trembling, begged Boris not to take vengeance on anyone.

'Come, come!' said Boris. 'I do not have noblemen flogged. I do not order venerable churchmen to be knocked on the head and put through a hole in the ice.'

'But, swear! Kiss the cross!'

Boris kissed it solemnly. 'I swear to take no vengeance on anyone for his part in this plot.' Then the Metropolitan told him who Shuisky's accomplices were.

Before they parted, Boris had one more question. 'Your Church was always tolerant of the late Tsar's little human failings. Did it ever recognise his last marriage?'

'Oh no, my lord, never!'

'In that case Dmitri is a bastard. He ought not to be prayed for in the churches, as if he were heir to the throne.'

'As you wish, my lord. As you wish!'

When he had gone, Boris murmured: 'I think I took that in time.'

I asked him how he had found out.

'Shelkalov knows how ... certain things are done.'

I could not insult Boris by asking whether he meant to keep his word. However, he told me of his own accord. 'I lost nothing by that oath. Shuisky's a fool! I wouldn't dirty my hands on him. Besides ... We're going to need all our generals, even the bad ones. Our truce with Poland runs out next year. Batory will refuse to renew it.'

'Are you sure?'

'Well, it looks likely. Do you remember, Yeremy, when you last met Magnus the Dane? You told me he was going to drink himself to death. It seems he did.'

'Did he leave his wife penniless?'

'Quite penniless. That's the trouble. She's Batory's pensioner—his prisoner, rather. He keeps her and her child in a castle in Riga.'

'Poor little Maria!' I said. 'She had trouble enough before.'

'She looked so frightened at her wedding!' said Boris. 'She was only thirteen then. A child's round face, and big blue eyes ... Did she grow into a beauty afterwards?'

'Not a great beauty—not like your sister. But pleasing, and good, and sweet.'

'You liked her?'

'Very much.'

'What I want to know,' said Boris, 'is—why Riga? Why does Batory keep her so near our frontier? He must know that some people thought her father had a better claim to the throne than Ivan.'

I asked: 'Is her child a boy or a girl?'

'Fortunately a girl. But there's no law to say that a Russian Tsar can't be a woman. If Batory persuades Maria to claim the throne ... You've just seen for yourself, Yeremy, that we have people here willing to welcome any sort of pretender.'

I told him how Maria had wept for her own country. 'She would hardly consent to plunge it into war.'

Boris listened with keen interest. 'Do you think, then,' he said, 'that if we offered her a good life here she would come back to Russia?'

'I imagine so. But would the Poles let her go?'

'I hear they let her ride a little way into the country, now and again. Under guard, of course. Guards can be eluded, or overpowered. Provided that the lady herself does not struggle. To bundle a screaming woman into a sledge, and hold her down ... That sort of thing leads to unpleasant accidents. And we can't have the child hurt. No! If she comes back to Russia, it must be of her own free will. Tell her so, Yeremy.'

I had not guessed, until then, that Boris wanted me to be the messenger. Yet who else could be? The Poles, Boris explained, would not let any Russian set eyes on poor Maria. It was not even possible for a Russian to visit Riga, without coming under suspicion as a spy. I would be able to account for my own presence there; I would be on my way back to London, with a letter from the Tsar to my own Queen. This would not expose me to danger, like the letter I had once carried for Ivan. There would be nothing of a military nature in it.

'No doubt I can go to Riga,' I said. 'But who's to open the door of this castle for me?'

'Cardinal Radziwill,' said Boris. 'He's the most important person in those parts. A great enemy of ours; be sure to tell him that you hate the sight of Russia. As for what else you tell him—your reasons for wanting to see Maria—I leave that to you. I know that you'll never be at a loss for words!' He smiled, but at once grew serious again. 'When you do see Maria, tell her the simple truth. Tell her that the tyranny which killed her father and mother and brothers is ended now. Tell her that we have laws. That, if the Poles try to put a pretender on our throne, they'll destroy all our work. Whereas, if she comes back with her little girl—'

Here Boris was interrupted by a scratching, as if a cat were at the door. A voice could be heard: 'Come down, you naughty girl! Come down at once!'

Boris let his little girl in, and called a reassurance to the nurse. He sat playing with the child's brown curls as he went on: 'Promise Maria, in my name—in the Tsar's name—safety and comfort, her father's old estate at Svenigorod, a good allowance, her title as Queen of Livonia . . . What else? Perhaps a new husband. Do you think she wants a husband?'

'He'd have to be very different from her first.'

'Ah! He would be. You and I can find her a good man. What do you say, Yeremy? Do you like her enough?'

'Indeed I do!' I said.

'Then we'll see. The first thing is to bring her back here . . . And Yeremy! Do one thing more for me, when you reach London. Look for a clever doctor, who understands . . . women's matters. There must be a cause for Irina's miscarriages. She'd give every jewel she owns, if she could have a child.'

I left Moscow on August the twentieth, 1585, and travelled with all the splendour befitting a man who carries a letter from an Emperor to a Queen. That is, I was very uncomfortable. I had too many servants to be accommodated in any village we passed. We had to pitch tents, often in drenching rain, while my German servants quarrelled with my Russians. When Boris and Irina had made sure of peace, I thought, I must show them how to build good inns . . .

At the frontier the Poles made me send back my Russian servants. But they let me keep my Germans, treated me courteously, and did not question that I must pass through Riga.

When I paid court to Cardinal Radziwill, his manner was at first austere. Addressing me in stately Latin, he said that, as an Englishman, I was no doubt a heretic. I took this quietly, knowing that the Cardinal had to make a parade of his Catholic devotion. His brother was the leading Protestant of Poland.

Soon he began to speak German, and came to the point. 'Since you are English, why are you carrying a letter for the Tsar?'

I said: 'I have to go to England in any case, on the Company's business, and the Tsar asked me to take his respects to my Queen. He thought his own subjects might not be welcome here, in view of the past.'

'In view of the present, you mean!' said the Cardinal. 'Don't try to tell me the Russians have changed their ways. They're savages—bloodthirsty savages, all of them!'

'My lord,' I said, 'I know a Russian who is a very good and gentle lady—Princess Maria.' (The Poles never spoke of her as Queen of Livonia, since that kingdom was a mere invention of Ivan's.)

I told the Cardinal how Magnus had attacked me, for refusing to drink vodka, and how Maria had tried to restrain him. Embroidering a little, I attributed to her the part which Anna had played in saving me. 'I believe that drunken madman would have killed me otherwise. I would like very much to see the lady, and thank her.'

'Oh, I will pass on your thanks,' the Cardinal said. 'She does not see any visitors, you know; she is in mourning.'

'The mourning for such a husband can hardly last long,' I said. 'I am willing to wait a few days.'

He laughed. I presented him with a handkerchief, richly embroidered in gold. 'Come to supper,' he said. 'Tell me what is happening in Russia.'

I told him, truthfully enough, but not as if I had any part in it. I made myself out a simple merchant, concerned only to buy and sell in peace. This, I explained, was much easier under the new Tsar.

'Under the Tsar's "Protector", I suppose you mean. This Boris Godunov—it's his turn now for the feasting and the women.'

'He shows no interest in either,' I said.

'No interest in women! The man doesn't know what's good for him. What's he after, then?'

'According to himself, the good of his country.'

The Cardinal shook with laugher. 'You're young,' he said. 'I'm so old, I can remember when Ivan was young. He was going to make everything different in Rusia. A pattern of all the virtues ... I wonder how long it will be before Boris turns bloodthirsty.'

A person so prejudiced was not worth trying to convince. I reverted to the subject of Maria.

'She must be much happier now,' I said. 'She has the protection of your King, Stephan Batory.'

'Yes, the King has treated her very generously. She'd have starved ... Seven years!'

'Seven years what, my lord?'

'Seven years of power!' said the Cardinal. 'And Boris will be butchering the people of Novgorod. No—Ivan didn't leave enough there, did he? Some other town, then. Russians! They're all the same.'

'Except Princess Maria,' I persisted.

'She won't have the chance to butcher anyone,' said the Cardinal. 'The lieutenant of the castle sees to that.'

'Then is she a prisoner?'

'Under protection. Under the King's protection. What's that to you?' He leered over his wine. 'Are you in love with her?'

'Perhaps a little,' I said. Carried away with my fancies, I described how I would sing under her windows like Blondel seeking King Richard, or else write poems and pin them to arrows, and shoot them over the battlements of Riga Castle.

The Cardinal was delighted. 'I have just the music for you,' he said.

In a gallery, musicians began to play, and presently the voices of young women rose in songs of courtly love. When the singers paused I exclaimed: 'This is what Russia needs!'

Everyone laughed. 'You're right!' said the Cardinal. 'We have a Russian church here; our King insists that it should be tolerated. I shudder when I hear the noise that comes out of it. That wretched wailing!'

'Sometimes it's beautiful,' I said. 'But it's music for the heartbroken. Your singers make me believe there could be happiness, now and again, on this earth.'

'Oh, my singers are very happy girls.' He called up to the gallery. 'Come down, ladies; we want to thank you in person.'

The women of the Baltic cities look lovely, when they stand in a slave-market wearing rags. Now that I saw them sleek, well-fed, adorned with jewels, they took my breath away. The singing girls were soon joined by other ladies, just as dazzling, until there was one for each chaplain and canon and secretary who sat at the table, and a particularly lovely one for me. An orgy ... But how different from the savage orgies of Ivan! There was not a coarse word spoken, only sweet music, sweet murmurs, and a wooing as courtly as if the outcome were in any doubt. When at last my beauty took a candle, and led me to her bed, the bed smelled sweet. So did she. And she made me feel that this was the most wonderful, the most surprising thing that had ever happened to her.

In the morning it occurred to me to ask what would become of her, if she had a child. She said she already had three; they were put out to nurse with peasants. 'The fourth one isn't yours; don't fret; it's well on the way.'

The next time I saw the Cardinal he was in a religious procession, looking very grave. But he caught my eye as he passed me, and could not keep a straight face. At the first possible moment, I renewed my request to see Maria. The Cardinal wagged his finger at me, and said: 'Greedy!'

'Princess Maria,' I said, 'is above all thoughts of that kind. My pursuit of her is *respectful.*'

'Poor woman!' said the Cardinal. 'She hasn't much to laugh at. She might as well have you.' I presented him with an emerald ring, and he gave me a letter to the lieutenant of the castle.

Very early on a dark morning I found Maria, 'Queen of Livonia', sitting by the open door of a tiled stove, using the red glow as light while she combed her daughter's hair. The child was four years old, and very beautiful. Her hair was golden, and thick, and long. 'What's your name?' I asked her, and she shyly answered: 'Yozhik', which means 'little hedgehog'.

'Oh, you know that's only a nickname!' said her mother. 'Because your hair goes into such tangles, and sticks out all over the place! What are you called *really?*' But the child was too shy to say, and Maria told me. 'She was christened Yevdokia. That was my mother's name.'

I had brought the child a doll from the market place. While she played with it, I asked Maria softly if I might speak to her alone.

'But where?' she said. 'I have only this one room, apart from the bedrooms. And hardly a servant to my name ...'

She had, however, one of the Russian ladies-in-waiting whom I had seen at our last meeting. Maria now confided the child to her, and they sat together by the stove, while Maria led me to the window seat. Trying to compose her round face into a queenly look, she told me to state my business. I asked her to promise that it should be a secret between us. Her only answer was a mistrustful stare. Yet I went on. 'The Tsar, Feodor Ivanovich, your brother'—being Russian, she accepted cousin and brother as the same—'sends you his greetings. He knows how wretchedly you and your daughter are living. He invites you to return to your native country, to live in state there as a queen. The Protector, Boris Feodorovich Godunov, sends his compliments, and his personal promise that everything the Tsar desires for you shall be done.'

'But I don't know these people!' said Maria. 'I hardly know you.'

'Have you forgotten when we met at Pilten?'

'Oh, yes, I remember that day. I wish I could forget.'

'I have always wanted to thank you,' I said, 'for saving my life.'

'Did I do that? No; I'm sure I didn't. All my life I've been trying to save people, and it never was any good. But I remember now—there was a servant ... She told me that you once did something for her. I couldn't make out what; she didn't speak much Russian. But she said you were a good man.'

'What became of her?' I asked.

'Oh, all the Danes went back to Denmark. I wish I could go there too! After all, King Frederik is my brother-in-law; he couldn't treat me like this.' And she looked round at her poorly furnished room.

I said: 'If you went to Russia, they would give you back your father's estate at Svenigorod.'

'Svenigorod! Where I was so happy, till I was nine years old! I could see the mill, and the river ... And Yevdokia could pick mushrooms in the woods there, just as I did ... No, that can't be true. I mustn't be so ready to believe what people tell me. I'm not a child.'

But she looked like one. Her blue eyes opened as innocently as ever; her pink cheeks matched her daughter's. Only her timidity was gone. She no longer had to fear some drunken outburst, every time she opened her mouth. It was wonderful to hear the pent-up thoughts pour out; her longing for Russia, her questions about the changes there . . . She listened as eagerly as she talked, drinking in every word of my answers.

The lieutenant of the castle came to tell her that it was time for her daily ride.

'But I don't want to ride!' Maria wailed, in a mixture of Polish and German. 'I want to talk to this gentleman.'

'The King's orders!' the lieutenant said. 'You are to have fresh air and exercise every day from nine in the morning till eleven.'

'Just because I have a visitor who talks my own language—'

'I want to ride, Mama,' said the little girl.

'Well, I don't!' said Maria, bursting into tears. The child began to cry as well; so did the lady-in-waiting.

'You see what you've done?' the lieutenant said to me. As he hustled me away, Maria cried out: 'Be sure and come again!'

The Cardinal sent for me within the hour. 'What's this?' he said. 'You were supposed to make the woman laugh, not cry.'

'She cried at being forced to part with me.' I presented him with a ruby on a gold chain.

'Oh, very well!' said the Cardinal. 'You may see her again tomorrow, but—you know what I mean—don't be too bold!'

The next day Maria said: 'When I was out riding, I looked at the peasants and thought: Now they've got their harvest in. They're beginning to rebuild their farms. But only just beginning! They've hardly struggled up to their knees from the war. Then I thought: Why are the Poles guarding me so close? It's not because they love me. They mean to make use of my royal blood. Either they'll force me to claim the Russian throne, or they'll force me to marry somebody who claims it. And then all these peasants will see their farms burned again. People will be tied to horses' tails and pulled through rivers.

And perhaps the war will go deep into Russia, and little Svenigorod will be like so many towns—ashes damped with blood!'

I told her, with delight, that she was thinking just as Boris did. 'You'll be able to prevent this war, if you go back to Russia.'

'Yes,' Maria said. 'But then, the Russians may force me into some convent. It's what they do to royal widows.'

'Not now!' I said. 'That's all changed. The widow of Ivan lives royally at Ooglich, on her own estate, with her son.'

'If I could believe it had all changed!' Maria said. 'But I can't believe anyone. I can see that you mean well; you have a kind face, but ... You don't know what happens in Russia. You don't know what Ivan did to me.'

'I do know,' I said. 'He killed your father and mother and brothers, and put you into the Troitsky Convent.'

'But you don't know what happened inside the convent,' Maria said. 'You're a man—you can't imagine a Hell where all the fiends are women. Where the torments are little sharp torments—pinches, and the twisting in of nails, and most often of all a tweak at the hair, because that doesn't leave a mark. And the words are little sharp words. "You're a sly deceitful creature, just like your mother." Or: "You take after your father—always plotting something." At first I used to answer back. I turned on that Mother Superior; I shouted: "And my brothers? What harm did my brothers ever do?" She nearly pulled the hair off my head. She told me: "If your brothers were like you, they were no loss!" After that I learned to be silent. I was nine years old. Sometimes I look at my little hedgehog there, and I think: I looked like her in those days, I know. And I wonder how it was that not a single soul in that convent ever looked at me with pity. I never saw one friendly face. My old nurse had been killed, and all the servants, for "failing to denounce" my father and mother. There was nobody to prevent those fiends, those nuns, from tormenting me every day and every hour of every day. For four years! Then one day suddenly the nuns took me and bathed me, and washed my hair, and dressed me in some clothes I'd never seen. A man called Skuratov had come to look at me.'

'Malyuta Skuratov? The torturer?'

'What could he know about torture? He wasn't a nun. He smiled at me. It was four years since anyone had smiled. Then he said that I would do. What for, nobody told me. But after that the nuns were polite to me, and called me "Your Highness". I was brought to Moscow, and I found the Tsar, who had killed my father and mother, grinning into my face. Do you know what he said? "This is my favourite niece." '

'Boris told me that you looked frightened on your wedding day.'

'I was frightened of doing something wrong, and being sent back to the convent. I didn't mind being married. Magnus was gentle to me in those days. He used to sit and play cat's cradle with me, and pop sweets into my mouth. It was almost like being at home again, with my father. Only Magnus had that ... that weakness ... If I'd been more clever, I'd have thought of a way to save him. But I was useless to him. Useless!'

I said: 'If anyone could have redeemed him, you would have done it, if only by giving him such a beautiful child.'

'He did love her—when he noticed her, you know. When he was in a state to notice anyone.' She was weeping again.

'You've shed enough tears for him,' I said. 'What would you say to another marriage?'

Maria looked up, astonished.

I went on: 'If your brother the Tsar, and Boris Godunov, were to find you a good husband, a man of some note in Russia, a man who hates vodka ...'

'Oh, this is all a dream!' (But a happy dream, by her look.) 'We sit here talking, just as if I could go back to Russia. However much I might want to—'

'You can!' I whispered. 'You can! When you go out for this daily ride of yours, make a habit of playing hide-and-seek with your guards. Tell them it's to amuse the child. Lead her off on long gallops while they lag behind. One day, when the snow's hard, there will be men waiting for you with a sledge.'

Maria said: 'Sometimes I pay the guards to go into a tavern, and leave us alone. But I have so little money!'

'Your brother the Tsar sends you some,' I said. I gave her a purse containing a hundred gold Hungarian ducats. Though I put it into her hand with some discretion. she was so much taken aback that she dropped it. The little girl jumped off the waiting woman's knee, and ran to see what the clatter was. I gave her more gold pieces, for herself.

'Are you sure he won't hurt her?' Maria said, looking at the child. 'This Boris—what kind of a man is he?'

I said: 'When he talked to me about you, he was playing with his own little girl, who is just about this one's age. He said: "We can't have the child hurt." Those were his exact words.'

'And he won't put me in the Troitsky Convent? Because I'd rather die!'

'Live!' I said. 'Live, and be beloved, and happy!'

At that she took my hand, and held it between her hands. The little girl, too, caressed me.

When the lieutenant escorted me away, I found the Cardinal downstairs. 'Did you do better this time?' he asked with a wink.

I said: 'As well as I could hope, with such a virtuous lady.' I presented him with a gold locket.

'We must meet again,' the Cardinal said.

My servants told me that, while I was with Maria, the Cardinal and the lieutenant had been wondering what I found to talk about for so long. Both were uneasy, since I had come out of Russia. But the lieutenant said it must be a simple matter of courtship; I had come wearing fine clothes, like any other young man in love. He wished, he said, that I would take the woman off his hands. The Cardinal warned him that King Stephan Batory would rather lose a hundred thousand dollars than Maria.

Since he knew this, why had he let me see her? I believe he could not resist the pleasure of pandering. His own house was (not to mince words) a brothel, and this could only be because he enjoyed arranging the copulations of others.

As I rode out of the gates of Riga, a young woman ran up to me and pressed a handkerchief in my hand. Tied up inside it was a ring set with small rubies, which I recognised as coming from Maria's hand.

One of my attendants, August Ducker, was a native of Danzig. This free city was now allowing its citizens to make trading journeys into Russia. The Swedes had agreed to let them pass through Narva. As soon as I reached Danzig I wrote a letter to Boris, and sewed letter, handkerchief and ring into the quilting of August Ducker's doublet. Then I sent him back to Russia by way of the Baltic, which had not frozen yet.

So I could go on to England with a clear conscience. My conscience *was* clear; I now felt I was doing right, not only in small ways, but in great. Maria would be happy, and I had prevented a war.

When I passed through Lübeck and Hamburg my old friends, the German merchants, thanked me once again for arranging their ransom. I looked back with a smile on the days when that was the only power I had.

In London I had the thanks of Sir Jerome Bowes. He was so grateful to me, for bringing him out of Russia, that he paid me almost a third of the money he owed me. But this was before I translated Feodor's letter to the Queen.

Lord Burleigh observed how I stumbled and hesitated over this letter. He thought I found the language difficult, and suggested asking the Russia Company for another interpreter. That pricked my vanity; I said I understood the letter perfectly. My difficulty was of another kind. Lord Burleigh said it was nothing to the difficulty I should find myself in, if I were to suppress any part of a letter addressed to the Queen. I knew this was a serious crime—so serious that I was bound to hesitate before accusing another person of it. But I had no choice. Feodor's letter contained the whole story of Bowes and his conduct in Russia, up to the moment when he slashed the sables, and threw the Tsar's letter back into the face of the messenger.

Now you can see why Bowes thinks I carried tales about him to the Queen. In fact what did him the most harm was his own self-defence. Bowes will allow any noise, however stupid and ugly, to come out of his mouth, rather than the word: 'Sorry.' He blustered out that he was glad he had thrown away the letter. Knowing the Russians, he said, and knowing that the Tsar was a half-wit, he had been sure there would be nothing in it worthy of the Queen's attention.

In fact, since Boris dictated the Tsar's letters, they were full of good sense. The one I had carried contained (besides the complaints against Bowes) an explanation of Russia's trading policy, which the Queen was very glad to have. To think that, but for Bowes, she might have had it a whole year earlier, naturally made her angry. She forbade Bowes to come to court.

He had never been very welcome there. But he thought he had. He thought that, but for me, he would have been the Queen's ambassador to countries warmer and more welcoming than Russia. I had robbed him of a splendid future.

So he accused me to the Company of trading on my own account. In fact, as I have explained, I had been trading for the Queen's great favourite, the Earl of Leicester. Since he was the Company's chief patron and friend at court, the Company had always ignored the matter. But they could not ignore a formal, written complaint. I had to appeal to Leicester for protection. He gave it willingly, my dealings on his behalf having been very profitable. But he then had to explain the matter to the Queen. She took it very well. She never allows her courtiers to take bribes which might pervert the course of justice, but she knows that they enrich themselves in other ways. The fact that I had kept the dealings of her favourite a secret did me no harm in the Queen's eyes. On the contrary. She said she had been looking for a man who could keep his mouth shut, and she entrusted me with her reply to the Tsar. So the result of Bowes' malice was that I became ambassador to Russia.

I set sail with royal presents; a huge bull, two lions and their keeper, twelve mastiffs and their trainers, an organ, virginals, musicians and a midwife.

This last was not quite what Boris had ordered. I had

consulted the most eminent doctors I could find about Irina's miscarriages, and they all seemed to think the matter unworthy of them. They said it should be left to a midwife. So I chose one of good reputation, and the Queen wrote a letter recommending her to Irina.

The midwife was seasick all the way to Rose Island. The lions, the mastiffs and the bull were also heard to complain. It was a cold voyage, for I had set out almost too early in the year; and some sharp, dangerous lumps of ice came floating towards us as we entered the White Sea. I did not care, for I was coming home.

There is no law which can punish a man for a design, unacted and unsaid. I never spoke one word of treason. I had not enquired, while I was in England, about the disposition of ships and armies. At the Earl of Leicester's table my questions had been: How can Justices of the Peace be persuaded to serve without pay? If the Lord Lieutenant of the county is a local nobleman, isn't there a risk that his tenants will fight for him, against the Queen? If such a great man does wrong, how can he be brought to justice? Can the Queen collect any tax that Parliament has not passed? I needed to know how a civilised country was governed. Boris needed to know, and I was utterly his man. In my heart and mind I had already cut myself off from England.

But this was all hidden, all within. When we made fast at Rose Island, and Mr. Rowley came aboard, he knelt to me, because I was the English ambassador. In the Queen's name I received the homage of my old master and my old companions. It was a moment which might have turned my head, if my head were not already turned with another ambition. I did not want homage; I wanted power. As adviser to Boris, as husband to a princess, *I could rule Russia.*

Mr. Rowley formally congratulated me on my new post. I said: 'These things are shadows', which passed for modesty, although in fact I meant it.

It was Mr. Rowley's last year at Rose Island. He was to retire, and his place to be taken by Mr. Taylor (who had stayed, after all). I knew the Company would give Mr. Rowley a wretched pension. So, when we were alone, I told him frankly that I felt myself his debtor. He had released me from my duties for those lessons which had made me the first Englishman ever to speak Russian correctly. He had sent me to Moscow to make my fortune, and had never made his own. I begged him to accept a farewell present.

He refused. He had savings, he said, enough to buy a little house. He would be content. 'I shall have the happiness of living in a country with written laws.'

I said Boris was introducing those into Russia.

'Well! He may. I remember when Ivan tried it.'

'What happened?' He shrugged, but I persisted. 'Nobody in Moscow seems to remember. And yet they were good laws.'

'Oh, the laws were good enough. But who was to carry them out? Only Russians. In England, you see, we may not have the best laws; we may not always prevent our judges taking bribes, or keep our J.P.s awake on the bench; but in a general way we trust each other. That's what you can't get here.'

'But with a better system—'

'They still wouldn't trust each other. They'd be mad if they did. Look what's happened to Shuisky! The general, you know. He was in some plot this time last year. It seems Boris made a solemn oath not to arrest him. Now he's been arrested.'

'Perhaps for something else.'

'Oh, I daresay Boris could make out a case for himself, if anybody had the power to question him. I only hope this princess finds Boris more trustworthy.'

'What princess?'

'The one he stole out of Poland. Batory made such an uproar, I thought you'd have heard it in London. The lieutenant of Riga Castle is disgraced. Still! That won't bring her back.'

'Where is she now?'

'In the Kremlin. Where else? Boris would hardly let her out of his sight; she might be a pretender to the throne.'

'But is she safe? Is she well? How's her little girl?'

Mr. Rowley looked at me curiously. 'How did you know she had a little girl?'

I said: 'I saw Princess Maria when her drunken husband was alive.'

'Not since?'

'No.' Even as I said it I thought it a foolish lie; Mr. Rowley might know I had passed through Riga. Indeed, his next words told me that he did.

'The Poles had some story about your seeing her, last autumn, when you were at Riga. After she disappeared they found the tracks of the Russian sledge. But no disturbance in the snow; not a sign that she'd struggled. So she must have made some agreement beforehand. There must have been a messenger—but who?'

'Not me,' I said, looking him in the eye.

'No, Jeremy, of course not. It's only that the Poles will have it they never allowed a Russian to come near her. Which of course I take with a pinch of salt, seeing how slackly she was guarded. I said to Trumbull and our other fellows in Moscow, when they passed on this tattle—you should be more firm, I said. You shouldn't allow such tales to be told of an Englishman. There's not one of us, I said, would advise any human creature, once well out of Russia, to come back!'

I said goodnight to Mr. Rowley with an affection which I still felt, though I wondered how a man of his brains could fail to see that this was a different Russia.

The next morning I gave the necessary orders for the bull, the mastiffs, the lions and the musicians to go by the rivers to Moscow. I myself was to travel by post horses, with no more servants than I needed to maintain my standing as ambassador. Just as I was setting out, a monk rowed over from St. Nicholas, with an urgent message for me. 'Father Bogdan is dying,' he said.

I remembered Bogdan as one of the abbot's familiars. What was his death to me?

'He has something to confess to you,' the monk explained, 'about an old friend of yours.'

I put off my journey, and rowed over to the monastery. The abbot was full of apologies. 'I am sorry, your lordship, to trouble such a great man.'

I said: 'Even Tsars visit the dying.'

Father Bogdan seemed unconscious. But when I spoke he stirred, murmuring: 'Yeremy . . . Yeremy . . .'

The abbot told him that I was on my way to the Tsar, and must now be called: 'My lord'.

'No!' I said. 'I must be prayed for by my Christian name. Perhaps I shall need Father Bogdan's prayers.'

'You will, indeed you will,' whispered the dying man, 'if you are going to Moscow. It's an evil place; it paralysed my soul.'

He would say no more, until we were left alone. Then he croaked out: 'You came here to ask us what had become of the Archimandrite Feodorite. You remember what I said?'

'You said nothing.'

'Nothing. And yet I knew. I was at the court of The Terrible, when the Archimandrite Feodorite begged him to be reconciled with Kurbsky. He told Ivan to his face: "Everything Kurbsky says is true." Ivan called his Tartar guards, who cared nothing for a Christian holy man. "Knock this old wolf on the head," he told them. "Then wait until nightfall, and put him through a hole in the ice." None of us dared utter a word, except Boris.'

'Who?'

'Boris Feodorovich Godunov. He alone was blessed with courage to plead for the archimandrite.'

'You must be mistaken,' I said. I remembered the words of Boris, long ago, when I asked him about the Archimandrite. *I don't know him . . . I didn't see him . . . I wasn't there.*

'How could I be mistaken?' Bogdan groaned. 'I remember Tsar Ivan laughing at Boris, calling him "a bit of a milksop", and then "an illiterate".'

Long ago I myself had heard those words from Ivan. And less than a year ago Boris had told the Metropolitan: 'I do not order venerable churchmen to be knocked on the head and put through a hole in the ice.' The words lodged in my mind because, though I had known of churchmen poisoned, strangled, and torn to pieces by bears, I had never before heard of that particular execution.

Yet I still questioned the story. 'Are you sure it was not someone else who pleaded for the archimandrite?'

'Who else? What other man ever showed such courage, and lived? When I heard that Boris alone had tried to prevent The Terrible from killing his own son, I thought: Who else? I bless Boris. If only he can keep his heart free from the wickedness of Moscow!'

Still I disputed. 'Moscow isn't wicked. Only men can be wicked—not a place.'

'A place may be under a curse,' Bogdan whispered, 'and then the best acts of the best men turn into wickedness.'

I thought this a superstition worse than Papistry. But I could not pursue the point, for Bogdan, with what breath remained, was begging my forgiveness.

'God will forgive,' I said.

Now I had only to say farewell to Mr. Rowley. Without mentioning Boris, I told him how the archimandrite had died.

'Well, we guessed that,' he said. 'You have to guess in this country. You can't always wait until the man you're talking to is on his deathbed, and ready to tell the truth.' He added suddenly: 'I suppose Boris means to reward you for your services.'

I said: 'Yes, I suppose so'—not being able to say that whatever I did for Boris was done with loyalty and love greater than I felt for Queen Elizabeth. So Mr. Rowley, whom I did not see again before his retirement, still believes that I enticed Maria back for money.

My journey to Moscow was as pleasant and swift as dry weather could make it. Only one thing troubled me; in spite of my continual riding, and my natural good health, I could not sleep. Doubts hopped over me like fleas, and were as hard to catch. It was absurd (I told my pillow) to stay awake because Boris had once told me a lie. He had scarcely known me at the time. And he could not have survived Ivan without becoming an accomplished liar. I thought how, to please Ivan, he had pretended to be at daggers drawn with Shelkalov and Afanassy . . . How he had in fact been at daggers drawn with Afanassy, while seeming friendly, getting him and his family a hundred miles from Moscow, a thousand miles from power ... All this was necessary, in ruling Russia. I would do the same. I would (if I knew how to do it so calmly) tell the Metropolitan that young Dmitri was the heir, and a few minutes later that he was a bastard with no claim to the throne.

As it happened I had news of Dmitri. I did not pass through Ooglich, but at Yaroslavl, some fifty miles away, the man who ferried me across the Volga swore that his news was less than two days old. 'This ikon-seller saw the child with his own eyes, and do you know where? Standing in the farmyard, watching them kill a calf. That's what he likes to see, our little prince! A child not four years old! And you know what else I heard? Last winter he was playing in the snow with other boys, and they made snowmen. Our prince was playing with a wooden sword. He went up to the snow people and hacked them about, and stabbed where their hearts ought to be, and shouted: "That's what I shall do when I'm the Tsar!" Oh, he'll be Tsar all right—a real Tsar, like his dad!'

The ferryman's face was beaming with pride. I thought: These are the people Boris has to rule. Is it any wonder ...? Is it any wonder that ...? I delayed finishing the sentence, till I saw him face to face.

And then all was well. Oh, more than well. Not his goodness to me, nor to his country, nor to his country's friends, could I doubt now. He led me to the part of the Kremlin where Ivan's Danish girls had dragged out their miserable lives. It was newly, splendidly furnished. And here, living in state, as 'Queen of Livonia', I found Maria.

She told me that after listening to me she had been tormented by doubts. 'Until I came here, and found such a welcome! They're all so kind—especially Irina. She cried when she saw my pretty little hedgehog; she wants a child of her own so badly. And she deserves one, if ever a woman did! She's wonderful; she knows so many things. I want Yevdokia to grow up like her. Will you find me teachers for Yevdokia? Boris told me you would. And will you come to see me, Yeremy, when my house is ready at Svenigorod? Boris is going to make it as fine as here.'

It was not long before I told her that she should marry again.

'Don't speak of that yet! I'm still in mourning.' Yet she pressed my hand, and looked happy, and told the child to give me a kiss.

I was kissed, too, by the pallid Feodor. His health was failing. 'I haven't been hunting,' he said, 'for such a long time!'

Boris prompted him to thank me for persuading Maria back to Russia.

'She is a sister to me,' said Feodor, who spoke sensibly enough of his affections. 'And her child is a sister's child—as dear to me as my own.'

All these conversations were in private. I could not yet be formally received, because my presents, travelling by the rivers, would take another month to arrive. I told Boris that they included a midwife.

At this he started. 'I had forgotten ...' he said. Then he told me that Irina's modesty had been offended, when he had urged her to consult a German doctor about her barrenness. He was afraid that he would scarcely be forgiven, if she knew he had discussed it with another man—'even with such a friend as you, Yeremy.' He begged me to write a letter telling the midwife to wait at Vologda. 'There are plenty of English to keep her company there.'

I said: 'But the Queen's letter to Irina mentions the midwife.'

'Yeremy, my dear friend—please don't translate that part!'

So I did not. Irina was glad enough to see me, and to hear me read the letter, without that. Elizabeth was a magic name to her. 'Is that really her hand and seal?' she asked, and kissed the letter. 'And is her kingdom quiet? Do all those men still obey her?'

'More than ever,' I said. 'The agents of the Pope have been plotting her murder, and Parliament has formed an association to protect her. Almost every man in England has solemnly sworn to defend her life with his life. What's more, we've all sworn never to acknowledge the pretender for whose benefit the attempt is made. In other words, the Queen of Scots.'

Irina had heard of the Queen of Scots, but wanted to know how she came to be in England.

'She was driven out by her own people,' I said. 'They accused her of murdering her husband. They dragged her through the streets of Edinburgh, shouting: "Burn the hoor." Yet our Queen treated her still as a queen, and her cousin, and believed her story that she had no part in the plots of her followers. And even now, when her plotting has often been proved, the woman has only to write a pathetic letter to set our Queen weeping. Our Queen's advisers are always urging her to have this woman put to death. She ought to consent, for her own safety.'

'What it is to be a queen!' Irina said. 'In this country, where a woman is not allowed to kill even a goose, everyone thinks it's my part to forgive, to reprieve, to show mercy. So I do, but ... Your Queen, from what you say, has forgiven her cousin once or twice already.'

'Oh, again and again! And always found her out in a new plot.'

'Well then, the point comes when a woman must consider—shouldn't I show mercy to thousands of poor people, who will suffer if this one person stirs up rebellion? For all that,' she added quickly, 'as my brother will tell you, Shuisky's death was an accident.'

I had not known that Shuisky was dead.

'You remember the terms of my oath?' said Boris. 'I swore he should not be punished for that plot. He started a new one. Even then, I did no more than order—persuade the Tsar to order—that he should retire to his own estate. The escort of gunners was only to see him safe home ... You know what our people are. Drinking themselves half dead, letting their houses catch fire ... They pulled Shuisky out, but he was already smothered by the smoke.'

Other unpleasant accidents, I found, had happened to Englishmen. One had even been racked, because of a letter passing between two other Englishmen, which described him as a spy. Shelkalov, who knew perfectly well that they meant a Company spy, on the look-out for fraud, had the man arrested and tortured for spying on Russia. Another man was arrested for trying to smuggle out an uncensored letter. My intervention set them both free at once. Yet many of my old companions hated me still. Mr. Trumbull in particular took it very hard that he had to kneel to me in public.

The day of my formal reception at court was a triumph. The people of Moscow crowded to see the bull, which they thought a fabulous creature, so different was it from their own stunted cattle. The lions made them half delighted, half afraid; but they were perhaps more afraid of the keeper, a young Tartar lad. People asked me if England was much troubled by Tartars. I said we had them only for amusement. This lad and his lions had been imported from Venice.

The greatest success of all was the English music. When Irina heard my men playing the organ and the virginals, she ordered all the windows to be opened, so that the people of Moscow could share her delight. They crowded into the Kremlin and stood beneath her windows, listening in utter silence, for hours at a time.

The ladies asked me to translate the songs. 'In a merry May morn . . .' they could understand; May mornings can be merry in Russia sometimes. But what, they asked, was the sad song? I explained that it was about the massacre of the innocents.

'The crime of Tsar Herod!' said Maria. 'My nurse used to tell me I might pray for the souls of all other kings, but not for Tsar Herod, because he killed little children. I thought of that often, after my brothers were killed by Ivan.'

'Don't say that name at nightfall!' murmured a lady-in-waiting.

Maria asked me why one of the songs was half happy and half sad. I told her it was a lullaby sung to a bastard baby. 'Thy father's shame, thy mother's grief . . .' The baby's dancing made the mother happy despite herself, and the song ended: 'Oh soul that thinks no creature harm!'

While we were talking, the golden-haired Yevdokia, the little hedgehog, fell asleep on her mother's lap. Looking down at her, Maria whispered: 'Oh soul that thinks no creature harm!'

When all the formal exchanging of presents was done, Boris

told me that I should have a special present, for bringing Maria back. 'I have three new horses from Persia,' he said. 'Choose one for yourself! And we have discovered the hoard of silver coins Ivan kept for his imagined flight into England. It comes to something like three thousand pounds of your money. If you could bring yourself to accept . . .'

I could. (My own expenses had almost come to that, and I had been forced to borrow from the Company.) But I asked, also, to marry Maria.

'You have my consent already,' Boris told me. 'But there's the Council . . .'

Boris had made away with the five-hour feasts of Ivan's time. Three times a week his Council of Noblemen met, stone-cold sober, at seven in the morning. When they had finished their business they could go home, which they were glad to do, rather than share the frugal dinners of Boris. The business rarely took long. What Boris and Shelkalov proposed they would agree to, knowing themselves to be without education, or any knowledge of what went on outside Russia. They did not protest, even, at the rule that any reprieve or act of mercy was declared to the people as 'due to the intercession of the Tsaritsa Irina and Boris Feodorovich Godunov' whereas any arrest was 'by order of Prince Mtsislavsky and the Council'. Prince Mtsislavsky had a special reason for keeping his mouth shut. It was his sister whom the plotters once chose to supplant Irina.

'One thing they're always ready to dispute, though,' said Boris. 'Those cursed rules of precedence! Do you remember, Yeremy, how you first taught me to think those absurd? I've learned so much from you! And yet I can't abolish those rules! Even Ivan, who could force a man to smile, as he walked past the body of his favourite servant—even Ivan could never make one man serve under another of less noble ancestry. I have to rule—that is, the Tsar has to rule—over men as they are, not as we'd like them to be. And these men think of nothing but precedence. Yeremy, I can't make you a nobleman.'

'I don't want you to.'

'But you want to marry a princess. There's one way. If your own Queen were to make you a nobleman.'

'That's hardly possible.'

'If you had *some* title ... Nobody here would know what it meant.'

I said that I might reasonably expect to be made a knight. 'That's not the same as a nobleman,' I admitted. 'But I should be called "Sir Jerome" and take precedence of all the other Englishmen in Russia.'

This was what we agreed betwen us. I was to return to England the following summer, with magnificent gifts from the Tsar to the Queen. I should also have the Tsar's charter. signed and sealed, granting the Company for all time those rights which in practice it was already enjoying. At such tokens of my success the Queen could not fail to make me a knight. If she needed prompting, my friend the Earl of Leicester would remind her.

In the meantime I spoke of this marriage to Irina. She agreed that there must be some delay. 'I want to see Maria happy,' she said. 'We all love her so much! But ... Even in England, surely, it would be a serious matter to marry a lady with a claim to the throne?'

'Without the Queen's permission—yes. But you see I am asking you, before I make a formal proposal to the lady herself.'

Irina smiled. 'God knows I wish you well. It would all be so simple, if Maria were not the heir! If only I had a child! I did ask Dr. Jacob to find some clever midwife who could advise me, but it's a difficult thing to bring a woman from England.'

I had been to long at that court to show in my face what I felt. If I had blurted it all out at once—that the midwife was even then waiting at Vologda, that Boris was deceiving his own sister—then I would have saved God knows how many lives. But I reflected that Boris might have his reasons. Perhaps he feared that Feodor's child would be like Feodor. He might well pity his sister, for being the wife of a half-wit, and hope to spare her the worse fate of being a half-wit's mother.

The next time I saw Boris, he smiled at me and said: 'Come

with me to Smolensk! I want to show you what we are doing to prevent famine.'

Indeed famine seemed a thing of the past in Russia. As we travelled he showed me why; in the Tsar's name, he was having granaries built in the most fertile districts and great stores put by. Directly he heard of a bad harvest in any part of the country, he would send wagons or barges full of grain. He told me: 'I've decreed—the Tsar has decreed—that peasants must not leave their own villages. We've had too many people wandering the roads, when they ought be growing corn.'

I said it was hard that a peasant should not be able to change a bad master for a good one. 'It's only for the time being,' Boris explained. 'Until we've made sure of our supplies . . .'

At Smolensk he showed me his new fortifications against the Poles. Riding up and down the lines of defence, seeing his constant care for his country's well-being and safety, I somehow could not bring myself to say: 'You told me a lie about a midwife.'

It was at Smolensk we heard that the great Stephan Batory was dead. Until Poland elected a new king, there would be no attack from there. Some of the Russian generals wanted to seize the moment, and storm into Poland. 'They'll never be satisfied,' Boris muttered, 'until they've undone all my work.' He ordered them south, himself leading the way, to strengthen Russia's defences against the Crimean Tartars.

By the time we returned to Moscow Maria was no longer there. 'She has gone to her palace at Svenigorod,' Irina told me.

I asked if I might visit her there.

'Certainly . . . in a little while. She likes you, Yeremy; she told me that if she could bring herself ever to marry again, it would be to you.'

'All we have to do, then,' said Boris, 'is to make sure that the noblemen accept you.'

To this end he did all he could to increase my importance in the eyes of the Russians. He chose me before all his courtiers to ride out of Moscow with him, to see his gerfalcons fly at the wild swans. Only his falconers, and a small bodyguard, rode with us; but we were to meet some 500 noblemen at a clearing in the forest. On the way a ragged friar cried out to us, I thought for alms. But Boris drew rein.

What would great men do without wandering beggars, ballad singers, gypsies—their spies? This friar told Boris that some of the 500 men who were waiting for him in the forest were not his friends. Then the friar glanced at me and whispered something.

Boris did not turn back, only aside. He rode without hurry, pausing now and then to let his falcons fly at some birds they fancied. But he was all the time aiming for a place where we could ford the Moscow River, and so return by a back way. We came safe within the Kremlin walls, but could not reach the palace without passing the crowd of petitioners who thronged the terrace. They set up a cry. Some of them had been trying to reach Boris for two or three days. He rode hastily past them, which would not have mattered if they had been common people. But there were bishops among them, and some great noblemen.

'Those people are not in the forest with your enemies,' I whispered. 'You ought to go back and speak to them.'

Boris glowered at me, and seemed about to say something. I could guess what it might be. The friar's glance had looked like a warning that, if Boris had enemies, it was because he had shown too much favour to me.

Yet he did not say so, but seemed rather to recollect that my advice was good. (As I still think it nearly always was.) He turned, went back to the terrace, received the petitions, and spoke pleasantly to everyone. 'Long live Boris Feodorovich!' they shouted. He said: 'I shall present your petitions to the Tsar.' At this one voice cried out: 'Boris, you have only to say it, and it's done. *You are the Tsar.*'...

Boris did not let his face move. But I heard the sharp and sudden intake of his breath.

Other voices joined in. 'Boris, you are our Tsar! Our real Tsar!'

He might have reproved this as treason. He did not. In that instant I understood why he had changed his mind about the

English midwife. He did not want another life between himself and the crown.

I stood there while his flatterers called him Tsar, and thought: He has no claim. But what claim had Henry Tudor, our Queen's grandfather, to the throne of England? Only that he had won a battle, and that the legitimate heirs were dead.

Feodor's health was poor; he might reasonably be expected to die in a few years. Dmitri might be reckoned a bastard. There was another family with some claim, the Romanovs, but Boris kept them well away from Moscow. Only Maria was free, in blooming health, and an undoubted princess. And her child . . .

That night I suggested to Boris that I might marry Maria at once, in secret, and take her and the child back to England with me. He frowned.

'Let a Russian princess leave Russia? How could I? Have you forgotten all the trouble we had to bring her back?'

'But that was from your enemy's country. If she were in England, she could never be used against you. It's too far.'

'Why, Yeremy, don't *you* mean to come back to us? I thought . . . I hoped you meant to live here always. Was I doing wrong when I showed you our secrets, our fortifications?' His eyes were filled with real tears.

And so, of course, were mine. This man, this liar, led by an ambition on whose burning altar he would sacrifice his own sister, had not lost his power over me. I loved his very ruthlessness.

Yet, from this time on, one part of my mind was always calculating how I could outwit him. Without his consent, had I the slightest hope of bringing Maria and her child out of the heart of Russia? I was an ambassador, and would in other countries have been reckoned inviolate. But this is a notion the Russians have never understood. The most I could do, if I defied Boris, was to die honourably in Maria's defence. Surely there were better ways.

'I swear I will come back,' I told Boris. 'But you in your turn swear that, when I do, I shall marry Maria.'

He kissed the cross. 'You shall marry Maria, if God does not prevent it.'

He added that, until then, it would be better for me not to visit her. I agreed. But Svenigorod is not far from Moscow; I was able to send her messages and presents by my musicians, who were favourites everywhere. (The women in the market place were humming 'Greensleeves'.)

Before my departure, Boris asked if there were any last favour he could do for me. There was one. The Germans of Reval had long been offering to ransom some 500 captives who, they believed, were all in a body together, somewhere in Russia. Shelkalov professed himself unable to remember where Ivan had sent them. Yet as soon as I enquired among the Englishmen who traded along the Volga, I found that they had often seen the Germans. The poor creatures were building a fortress at Nizhny Novgorod. When I told Shelkalov this, he said Russia needed the fortress. I said that the ransom money would be enough to finish the fortress with wage labour. He gave me various answers, none of them to the purpose. Pride would not let him utter the real answer—that a Russian day-labourer was not worth one-tenth of a German slave. Nor did I tell Boris this, when I explained the matter. He was glad to please me so simply; he gave the orders that set the captives free. On their way home they passed through Yaroslavl, just as I was passing through it on my slow and stately progress to Rose Island. The Germans overwhelmed me with gratitude. I was conscious of a familiar feeling. Hearing on every side stories of noblemen who had met the same accident as Shuisky, seeing the printing press not yet built or begun, knowing that the courts were guided still by cruel old customs, tormented with fears for my bride (how could God prevent our marriage, except by her death?) still I could console myself by thinking that I had done one good thing.

Boris was lavish in showing his good will. Along my way to Rose Island, the local governors came out to greet me and feast me. I had a bodyguard of fifty Russians. (Which prevented me, of course, from turning aside to see Maria.)

At Vologda I collected the English midwife. She complained

that I had lured her into Russia with promises of an empress for a patient. Now she had spent more than a year in this barbarous little town, delivering the babies of women seduced by English craftsmen. At every lying-in she had been hindered by filthy old herb women, by sorcerers muttering spells, and by the patients' own attacks of modesty. 'They should have thought of that before, and so I told them.' I could not explain to this woman why Boris had left her stranded. So I took the blame on myself, and tried to quiet her with money.

At the new fortress of Archangel I was feasted sumptuously, and the great guns of the castle saluted me. Russian ships (there were such things now) escorted me to Rose Island, where Dutch and French and English vessels fired their guns in my honour.

And then there came a string of Russian barges, with presents which Boris had sent me for my journey to England.

'Why, Jeremy!' said Mr. Taylor, as he read the bill of lading. 'This is an exercise in your impossible Russian plurals. Sixteen of live oxen, seventy of sheep, six hundred of hens, twenty-one flitch of bacon, two of a swan, two of a crane, three of a young bear—they're not for eating, surely? No, neither is four of a hawk. What, only one wild boar? Your friend's left out a partridge in a pear tree.'

Mr. Taylor then admitted that he envied me. 'Not your fine clothes, Jeremy, nor your favour at court. Just your passage back to England. I meant to go home years ago, only . . . my Nadya gave me a son. I'm daft, I know, but I can't bear to leave the boy, if he is half a Russian!'

My reception by the Queen was as complete a triumph as I could have wished. I found her somewhat sad over the execution of the Queen of Scots. 'It was an accident. My courtiers misunderstood my wishes . . .' But she soon grew cheerful. Besides the usual sables, Boris had sent her the finest Persian silks, the heaviest cloth of gold. She did not consign them to her ladies until she herself had handled them, delighting in their splendour. She was pleased, too, to see the three young bears, and gave them into the keeping of her bear-ward.

It was the terms I made for the Company that pleased her

most. When I translated the letters of Boris, I blushed at the praises of myself, but the Queen said they were all deserved. She looked a long time at the Russian script, asking me to explain the alphabet. Pointing to a sort of E lying on its back she said: 'That's a *sh.*' I could not imagine how she knew. 'They have taken it from Hebrew,' she said, happy at knowing a language I did not. 'I could soon learn this one,' she assured me. She looked round at her courtiers. 'You're all very proud of your French and Latin. And yet here is this great country, our best friend in the world, and not one of you could speak a word there. You!' She pointed to the Earl of Leicester's stepson, Robert Devereux, Earl of Essex. 'You're young enough. Learn Russian!'

I gave young Essex lessons afterwards, and found him bright enough to learn anything he liked, if only he could have done it without work.

Almost at once, the Queen made me a knight. This enraged Bowes; he urged the Company to bring new complaints against me. The midwife too complained. 'What was he up to in Moscow, that he didn't want me to see?' The two Englishmen who had been in prison in Moscow were now in London. They told the Company that, when they knelt before Boris, to thank him for his mercy, Boris replied: 'This is all due to your fellow-countryman, Yeremy.' No doubt he did say this, meaning that their release was due to me. But they perversely maintained that I was to blame for their imprisonment and the torture of one of them.

Mr. Burrough was on better ground when he asked me why I had borrowed the Company's money in Moscow, in a manner somewhat irregular. 'Embezzled' was the word he used. I said I had paid it all back. (As I had, when Boris gave me the three thousand pounds in silver.) But why, Mr. Burrough asked, had a man of my wealth gone borrowing in the first place? I explained that my own capital was out on long-term loans. I had wanted ready cash to maintain my state as ambassador. He asked me why I needed to make a more splendid appearance than any previous envoy. I could not say: 'Because I was wooing a princess, with the intention of taking her religion and staying in her country.' I could only plead the good results of my mission. 'Any more of your brilliant missions, *Sir* Jerome,' spat out Mr. Burrough,' and the Company will be bankrupt.'

The Queen treated all this with contempt. Yet I did not want a bad name in the City, and so I visited the Company's office for one tedious explanation after another. On one of these visits the office was brightened for me by a girl of thirteen, with a straight nose and sad grey eyes, who talked with longing of the Maelstrom and the whales. But, Bess, it was not for love of you I lent money to your dad. It was because the loan had been refused by Mr. Burrough, and by that time I would have called black white, only to contradict him.

It was just after this that the Queen sent for me, and said: 'I believe you speak Danish.'

I knew my life would be more simple if I said: 'No', but my vanity prevented that. What the Queen wanted was that I should go as her ambassador to Denmark, ostensibly to settle a dispute between King Frederik and the English merchants. In secret I was to find out whether the young King of Scots (the son of the beheaded Queen) meant to marry King Frederik's daughter. This part was urgent. 'Be ready within a week,' the Queen said. 'And say nothing to the Company."

I feared this mission might put off my return to Maria. But I could in any case not sail back to Russia till the spring. Meanwhile, if I did well in Denmark, the Queen might give me some new honour—perhaps the Garter—to dazzle the Russians. Feeling like the hero of a fairy tale, who has to win his lady by one ordeal after another, I sailed across wild wintry seas to Copenhagen.

All my business at the Danish court was with great noblemen, whose pride was the more unbending because they were rather poor. I succeeded in settling the grievances of our merchants, and in finding out that King Frederik did indeed intend to give his younger daughter to the King of Scots. But I could not discover what had happened to Karen and Anna. The noblemen were offended that I should ask them about common kitchen women.

When all our business was done, I was invited to King Frederik's table for a farewell dinner. The King asked me how I came to speak Danish. I told him how, as a boy, I used to sit reading the Bible with Karen and Anna.

'Why, yes!' exclaimed the King. 'They told me about you, when they first came from Russia.'

He had not thought it beneath him to question his poor subjects, and (a rarer thing in a sovereign) to remember what they said.

'Karen is dead,' he told me. 'And Anna somewhat crazed in her wits. Poor creature, she sits all day long by the kitchen fire, and grieves. I have given orders that she should not be made to work.'

So, the next day, when I had seen my servants and my baggage go aboard the ship for England, and knew that I had three hours before I sailed, I visited the royal kitchens. Anna was there, but she seemed not to know me. I sat beside her for a while, talking of our last meeting. 'You saved my life.' She only shook her head. 'You came between me and Magnus.'

At that name she looked up, and her bony hand gripped my arm.

'Believe me!' she said. 'Please believe me! If I had known what I know now, I never would have brought down the curse on Magnus.'

I had forgotten the curse. Yet now, fourteen years after, I began to remember.

Anna said: 'God hears the cry of the widow and the fatherless. Remember that! If ever the time comes when you are helpless, utterly in the power of your enemy, be careful what you say. Because God will make every word come true. I said Magnus was going to die like a beast. He died screaming that he was a rat, and a great cat had him in its jaws. I said his wife would end her days in captivity. Poor woman! I did not know her then. Now they've put her into a convent.'

'No!' I said. 'Maria's living on her own estate in Russia.'

'She is in the Troitsky Convent,' Anna said.

Here the other servants broke in. 'Oh yes, sir, this is one of Anna's good days. She's telling the truth. Our King's ambassador has come from Russia, not an hour ago. His grooms have just been drinking here, at the hatch. They told us. There's a dozen of us here that were with Magnus at Pilten; we knew Maria, poor lady! And the little girl. They say the little girl's with her still, in the convent.'

'But the worst is,' Anna said, 'that, when I cursed Magnus, I said his child would be murdered, as my children were.'

I turned on her and shouted: 'Then take off your curse!'

'I have prayed God to take it off. I pray for that every day. Only God says . . .' the wandering look was back on her face '. . . God tells me that he has to show his justice, as well as his mercy. He says the sins of the fathers are visited on the children. Why does God make it so? My children, my children . . . Oh, Karen, my only friend . . .'

I had two hours before my ship sailed. King Frederik agreed to see me at once.

'I have not forgotten,' he said, 'that Maria was my poor brother's wife. But what can I do now? I was ready to take her when she was in the hands of the Poles. Then she ran away to Russia. Poor woman, what fiend prompted her to that?'

I said: 'She thought the Poles would try to make her claim the Russian throne.'

'Some of their generals had that notion. But Batory refused. He was an honourable man. He had just agreed to send her here, you know, when she ran away . . . Now the Russians have the perfect answer to anything we say. She went back to her own country of her own accord. She's chosen to be a nun.'

'Your Majesty, I know the lady. She told me she would rather die than go into a convent. She thinks nuns are worse than torturers. I tell you Boris has imprisoned her!' And forced her to swear celibacy too, I thought. He would say that was God's prevention of our marriage.

'If we had some proof that she was held against her will,' King Frederik said thoughtfully, 'a letter from her, or even the word of some reliable person who had seen her in the convent . . . Then I would make a formal protest.'

That would indeed be something. Boris wanted to be friends with Denmark.

I said: 'I am the only foreigner alive who has the faintest chance of seeing her. And then only if I go at once, before the Russians expect me. Your Majesty, will you give me leave to travel by one of your ships to Lübeck, so that I can go the overland way to Russia?'

'Certainly!' said this best of kings. 'I have not forgotten how you brought my subjects out of slavery. But . . . what will your own Queen say?'

I said: 'She will understand,' though I felt far from sure of that.

In the last hour before the ship sailed, I wrote her a letter. I explained that I had concluded all my business in Denmark, and begged her to believe that I was her true subject still. But I must, on a matter of life and death, go into Russia. I sent this on the ship, with most of my servants and most of my baggage. Then, taking only my German attendant, August Ducker, and one other servant, and what baggage I absolutely needed, I set sail for Lübeck.

My old friends there fitted me out with horses and sledges, and letters of introduction to the German merchants, who maintained their own system of post horses in every town between Lübeck and Smolensk. (I was not going by way of Riga again.) At first my horses had to struggle through snow. But as I neared Poland the roads were all hard ice; I travelled swiftly. The Poles pride themselves on being civilised; unlike the Russians, they learn Latin. That is, they look with reverence at any document in Latin. At the frontier I showed them Queen Elizabeth's letters of credence, and met nobody with learning enough to see that I was accredited to Denmark only. When I came to Russia I needed no letters; my face was recognised by officers who had seen me riding in state beside Boris. They gave me a hearty welcome. Boris had evidently not warned anyone against me. He had not expected me before the spring. So at every post house I was greeted in the old fashion, with bowings and scrapings and the swiftest horses to be found. Within three days of leaving Smolensk, I was knocking at the gate of the English compound in Moscow.

Mr. Trumbull met me with a face of thunder. 'I'm not sure that we ought to have you here.'

'What do you mean? Is that my house over there, or isn't it?'

'I suppose it still is yours,' Mr. Trumbull agreed. 'But we've had this warning from the Company, that you left England in secret to avoid a trial.'

'A what? I did no such thing. *This* is why I left England.' I showed him the Queen's letters.

Mr. Trumbull knew his Latin. 'These are for Denmark.'

'Well? I've done my business there. I've made my report to the Queen. I've come straight on here. Is that a crime?'

'It may be treated as one,' said Mr. Trumbull, 'by your friend Boris.'

I said: 'Not if I see him at once, face to face.'

'What for? To get another agreement signed? He's already broken the one he made with you. They're making us pay the full customs duties again.'

I went into my own house, told my servant to light the stoves, and, without taking off my furs, began to write a letter to Boris.

'That won't do you any good, or us either,' said Mr. Trumbull, who had followed me in uninvited. 'England counts for nothing here now.'

'Why? What's happened?'

'There's a Spanish ambassador in Moscow. His men strut flashing Mexican gold and silver. The ambassador talks all day long about the great fleet his master is preparing against England. This coming year, he says, the Queen of Scots will be avenged, and every Englishman a Catholic, or dead. Shelkalov laps it up.'

Now, just as the stoves began to crackle, the coldness of my house, long empty, struck at me. The coldness of Russia . . . When I had loved it most, I had never imagined a day when England would not be there to go back to. Now I thought of the Spanish army I had seen, moving in perfect order to take Haarlem. England had no real army . . . And Boris knew that. I had told him, long ago.

'Of course, we tell them that our ships will never let them land,' Mr. Trumbull went on. 'But what does that mean to the Russians? They can't even picture a sea-fight.'

'Is Boris in Moscow?'

'God knows! He's always gadding about, persuading some town or other to tell him that he's the real Tsar. Write what you like! It won't do any good.'

Sure enough, my messenger came back from the Kremlin with news that Boris was in Nizhny Novgorod, inspecting the new fortress. I tried to see Irina. She sent back a blank refusal. Through various church dignitaries I asked permission to see Maria. However big the bribe, the answer was the same. She had renounced the world, and wanted no visitors. I could not simply burst into her convent. The whole Troitsky—monastery, convent and ten churches—is enclosed within a stone wall thirty feet high, the gates guarded by soldiers. As for sending her a message—where in Russia could I find anyone to take a message unpleasing to Boris?

I found a man wearing only chains, and a goatskin round his loins. In the market place he was babbling some nonsensical prophecy, but from under his elf-locks his eyes looked at me with more of cunning than madness. I took him home. He said he had murdered his wife, and was going naked, summer and winter, to expiate his crime.

'But God still isn't pleased with me, sir. The fact is, I don't suffer enough. Oh, it's cold, but then, I never did feel the cold much. My wife was always putting more wood in the stove. She made the house too hot for me; that's how we came to quarrel. What used to torment me in those days was hunger. Now everyone offers me food the minute they see me coming. And every house has a corner for me to sleep in. I never do a stroke of work; I say what I like, and nobody can have me flogged. I tell you, sir, it's a comfortable life. Compared with what I'm used to, me being a peasant. I'd be very well contented; but where's my expiation?'

I said: 'I myself have something to expiate.'

'Oh, I know, sir! You're the man they paid to bring Maria Vladimirovna back to Russia.'

This, then, was the common talk of the market place.

'Help me with my expiation!' I said. 'You're a holy man; you can go anywhere. Go to the Troitsky and tell Maria that Yeremy is doing all he can to set her free.'

He was back within a week. 'Did they let you in?' I asked.

'Could they stop me? Sir, I tell you, this is the way to live. Being holy, I blessed the sentries at the outer gate, and they let me in. I blessed the doorkeeper at the convent; he let me in. I went straight into the refectory, where the nuns were having dinner. They all crossed themselves and begged me to accept some of their food. Maria Vladimirovna and her child were sitting in the place of honour, at the right hand of the abbess. The nuns treated her with such respect, they seemed afraid of her. Being holy, I wasn't. I went and whispered your message into her ear. She cried out: "That liar! I curse the day that ever I listened to him." Well, I sat down, ate my fill, blessed them all, made a few prophecies—people expect that—and then, when I was almost out of the door, Maria came running up. She whispered: "Is he sorry for me? Truly?" I said I thought so. She said: "Then tell him I have never taken vows, and never will! I let these nuns know every day what I think of any woman who is fool enough to turn her back on love!" That's all she could say, with the abbess doing her best to overhear.'

I could not reward this man with money, because it was part of his expiation not to carry any. But I stuffed him with his favourite food—a sort of porridge—and asked him to tell Maria that I was writing to the King of Denmark. (So I did, but, by the time my letter reached Copenhagen, King Frederik was dead. His eleven-year-old son, who succeeded him, had not the same interest in the matter.) My holy man set off again for the Troitsky, but he did not return.

I waylaid Irina on one of her pilgrimages. She ignored me, until I cried out in German: 'How can you let that poor princess be kept a prisoner?' Then she turned on me, blazing with anger, and said it was time to tell me what she thought of me. In the narrow space of a church porch, we stood face to face. She said: 'How dare you spread lies and slanders against my brother? It's your doing that people talk of Maria being imprisoned. She has chosen to become a nun.'

'Have you yourself seen her in the convent?' I asked.

'She refuses to see me. She has renounced the world.'

'You mean you have been told that she refuses to see you. If only you would go to the Troitsky, and insist on seeing her—'

'Why should I do any such thing? My brother's word is good enough.'

I said (though I knew better): 'He may be misinformed as well.' Then I told her that Maria was not a nun. 'She has not taken any vows, and never will. She hates all convents, and especially the Troitsky. Did she never tell you that?'

'We all say foolish things when we're young. Once we learn what this world is really like, we're glad enough to renounce it. God knows I should be!' Her face was pale with weariness.

I cried out: 'But Maria loved me!'

'Loved you! What faith could any woman have in you? You knew that the one thing on earth I wanted was a child. You kept the English midwife, who might have helped me, a whole year at Vologda, and never told me she was there!'

'I was tricked into that by your brother, who does not want you to have a child.'

Irina turned her back on me. 'Can you think of nothing better than such a stupid lie?'

Outside the church I found Shelkalov waiting for me. He smiled.

'You don't know our Russian women, Yeremy. They are very pious. They prefer the convent to the marriage bed.'

I said: 'Oh, go and rape your mother!'

In the spring Boris came back to Moscow, but he did not answer my letters. I could not find out whether they reached him.

'What on earth did you expect?' asked Mr. Trumbull. 'You enticed that woman back to Russia, didn't you? You collected your blood money. You surely knew what Boris was going to do. Don't tell me you were such a fool—'

'Yes,' I said. 'Yes, I was such a fool.' I told him, too, that the money Boris gave me I had by now spent in my efforts to reach Maria. Only I never said that I had hoped to marry her.

When the White Sea was open, we heard that the Queen was sending a new ambassador to Russia. Then a letter came to me from the Privy Council, ordering me to return at once.

'I shall obey this,' I told Mr. Trumbull. 'But I must make one last attempt. I'm going to the Troitsky.'

'What for? You'll never get through that stone wall.'

'Even if all I can do is yell at the sentry, she'll hear my voice. She'll know I tried.'

'She'll know that you're crazed in your wits.'

It is forty miles from Moscow to the Troitsky. Half-way there is a post house, but I could not get a fresh horse there, having no warrant from the Tsar. So I continued on my own weary horse, going slower and more slow, till I heard the hoofbeats behind me. I glanced round and saw a company of men. There was something in their look which was not Russian, but this I did not give another thought to. What concerned me was their speed. Their horses were fresh. They must have got them at the post house. Therefore, they had the Tsar's warrant.

For a time I tried to spur on my own poor horse. But it was useless. They were almost on me. In desperation I turned, pulling out both my pistols.

'Come, come!' said an English voice. 'That won't do you any good, you know.'

I cried out: 'Englishmen! Thank God!' I willingly gave up my pistols, as they surrounded me.

Though I had seen some of them before, their leader was a stranger to me. He explained: 'I am Dr. Giles Fletcher.' I bowed. 'The fact is,' he went on, as if apologising, 'I'm Her Majesty's new ambassador to Russia.'

I dismounted, and kneeled to him. He himself dismounted, raised me up, and said very gently: 'I'm sorry to say, Sir Jerome, I have a warrant to arrest you and take you back to England.'

I told him I would gladly face the courts of Queen Elizabeth.

'I give you my word as a gentleman not to escape. But let me go on to the Troitsky Convent! It's only two or three miles more. You can all ride with me.' I began to explain why, but I could hardly get the words out. The face of Dr. Fletcher swam before me; I thought it was my distracted state that made it seem a kind face. While I stammered, a body of gunners, ten times more numerous than ourselves, came from the direction of the Convent. They told us they had orders to shoot anyone who tried to visit the Troitsky without the Tsar's express warrant.

'Ah!' said Dr. Fletcher. 'This is what Mr. Trumbull told me. He said your life would be in danger. That was why we hurried so, to catch you up.'

What was my life to Mr. Trumbull? I suppose he had to save me, as I had saved Bowes, because one Englishman cannot let another be shot by Russians.

All the way back to Moscow, I was in tears. Dr. Fletcher, whose kindness was not a figment of my fancy, tried to comfort me. 'You did all you could,' he said, 'for the poor imprisoned lady.'

But I knew, and was ashamed of knowing, that what made my tears flow was relief. My arrest had probably saved my life, and certainly given me respite from an exhausting struggle.

Dr. Fletcher, having accepted my parole, let me live in my own house as long as he remained in Moscow. He would have let me roam the streets as usual, but neither of us could do that. The Russians restricted our every movement. Rather than prisoner and jailor, we were companions in misfortune, and soon became friends. Dr. Fletcher, a great Greek scholar, was making heroic efforts to learn Russian, and he was glad of my help. He went thoroughly into the Company's charges against me, and concluded that they were absurd. Indeed he was so kind to me that I do not want to face him now. I am ashamed of never having told him that I had intended to marry Maria and abandon my own country. He thought I was moved by a praiseworthy chivalry for a lady in distress. He would have helped me, if he had had access to Boris. But all his presents were sent back to him, with scornful messages. Though it was only a year since I had obtained the charter which 'forever' freed the Company from customs duties, Shelkalov insisted that these duties must be paid. 'While you still have the money to pay them,' he added. 'You'll be asking us for charity soon, when the Spaniards take your country.'

Boris was now hardly ever seen in Moscow. Dr. Fletcher thought he might be in eclipse. Perhaps Shelkalov had seized power? It would have been some comfort to me to think so. But Irina shone in her accustomed glory, and I knew that she would not countenance any supplanter of her brother. Shelkalov was, as he had always been, the man who collected extra taxes while the ruler kept out of sight.

In August, 1588, while Dr. Fletcher and I were beguiling the intolerable heat with study, a letter was brought to him. As he read it, this quietest of men uttered a little cry. I saw tears in his eyes. To my anxious enquiries he said: 'It's just that... All the ships of Spain are scattered.'

'Are you sure?'

'Well, the letter is from a man who should know. Sir Francis Drake.'

'And how many of our ships are lost?'

'Not one. Not so much as a rowing boat.'

I am sure, Bess, when the bells rang out at Hampden Church, to take that news across your gentle hills, they never brought more joy than we felt then.

I immediately translated Sir Francis Drake's letter, and another which he enclosed, giving the text of Her Majesty's address to her troops at Tilbury. Dr. Fletcher had several copies made, and sent them to the Tsar, Irina, Boris and Shelkalov. He kept one Russian copy for reading aloud, which (after some rehearsal with me) he did very creditably.

The Spanish ambassador maintained that such a fleet as his master's could not have been destroyed. The German Emperor's ambassador said that there had been some defeat, but that it would be reversed next year. Half-believing, Shelkalov remitted half the customs duties, and made Dr. Fletcher half an apology. But he insisted that Dr. Fletcher must leave Russia at once, taking me with him, since I was a well-known spy. Dr. Fletcher courteously asked what evidence he had of that.

'Here's the evidence!' cried Shelkalov. He brought out the letter I had written to Mr. Rowley seven years before, when I had visited the Dutch merchants at Archangel. Shelkalov's men had intercepted it, copied it, and made a Russian translation. (Which Dr. Fletcher, to Shelkalov's surprise, carefully read.)

'What the letter shows,' Dr. Fletcher said, 'is that Sir Jerome was doing his duty to the Company. To compare the prices and qualities of cloth is not to spy on Russia.'

'Don't you see the last sentence? "We defeated the Danes; can we defeat the Dutch?" That's just after he'd been brawling with the Danes at sea.'

'Our battles in your seas,' Dr. Fletcher said quietly, 'were all to keep open the trading ways to Russia. The old Tsar thanked us for that.'

'But your Yeremy was plotting a land attack on the Dutch, on Russian soil. Why else did he have that stone wall built round Rose Island?'

Dr. Fletcher kept his temper and obtained a copy of the letter, knowing the good it would do me in England. Shelkalov next accused me of 'an unheard-of insult'. I had told him to go and rape his mother. Again he was baffled by the unexpected Russian fluency of Dr. Fletcher, who pointed out that the insult, though unseemly, was hardly unheard-of. 'I heard yourself use it to another nobleman, you five minutes ago.'

Shelkalov did, however, succeed in keeping Dr. Fletcher from seeing Boris. He was allowed to make his bow before the Tsar, and was then told his mission was at an end.

When Dr. Fletcher and I crossed the Volga at Yaroslavl, the ferryman was full of the news from Ooglich. 'Prince Dmitri's nurse tasted his food before him, and fell dead. Now his mother is preparing his food with her own hands. Can you imagine that? An empress bending over a cooking pot!'

I wish I could say that, when we reached Rose Island, my first feeling was of grief that I could no longer try to help Maria. The truth is that I felt a sharper pain at coming back a prisoner to a place I had left in such grandeur. I knew some of my old companions would be glad of my fall. But there was only a little mischief in Mr. Taylor's grin. 'Never you mind, Jeremy! Up again as high tomorrow, eh?'

'Oh, he will be,' Dr. Fletcher said. 'If I can bring it about.'

I had to rely entirely on him to present my case to the Queen. My old patron, the Earl of Leicester, had died at the very moment when all England was rejoicing for the defeat of the Armada. The Queen, grieving over him, did not forget her other subjects. Dr. Fletcher's eloquence convinced her of my reasons for going to Russia without her leave, which was the one really suspicious thing against me. She sent for the directors of the Russia Company, and told them once for all to drop their charges. I was her servant now, not theirs. If I owed them any money, they must render full accounts. (They did; it took them eight years.)

'But you've been a great fool, Jeremy Horsey,' the Queen told me. 'Why should it lie so heavy on your conscience, that you persuaded a lady to live in her native country?'

'Your Majesty—if it had been any other country but Russia—'

'Is Russia so very different from other places?'

How could I explain this, to a person who had not been there? But the Queen answered her own question. 'The Russians are worse liars than other people, maybe. This Boris has deceived you, and me too, although to me he made his promises in writing. And he deceived this lady, making use of you, so that you feel bound to help her—very well! But God's death, Jeremy! You should have told me all about it first.'

'I did very wrong,' I said humbly, 'not to return from Denmark and confide in Your Majesty. But I thought my only hope of getting into Russia was to go there before I was expected.'

'You could have got into Russia,' said the Queen, 'if I had made you my ambassador again. Why didn't you ask me? I do not like to hear of a princess locked up. It reminds me . . .' I thought for an instant she was going to say: *Of the Queen of Scots*. '. . . of my own youth, when my sister was Queen, and I was in the Tower. Nobody took such a risk for me as you've taken for this lady. The only person who gave me any comfort then was Robert Dudley, who was a prisoner himself.' I knew she meant the Earl of Leicester; I saw the tears come to her eyes when she thought how recently she had lost him. 'People thought I showed him too much favour; but when you have been a prisoner, you know who your friends are. I could forgive him almost anything . . . I shall forgive *you* almost anything, Jeremy Horsey, if you will make one more journey to Russia. You see, I must send someone. I cannot allow my merchants to be insulted and robbed, by any country on earth. And (though I can't interfere between the Tsar and a subject) still, if you can on your own account bring the lady to England, I will gladly receive her.'

I was not such a hero as to be delighted at this. England had never seemed a better place to stay in. I lent more money to Sir Griffith Hampden, and he invited me to Hampden House. I saw the grey-eyed girl who had longed for the Maelstrom, not quite fifteen, but mistress of her father's house, carrying at her waist the keys to dairy and laundry and maltings. As the spring began I rode down a wooded valley beside her, and looked at Kimble Manor, which was mortgaged to me. I knew then that, if I had once wanted to leave England, it was because I had not enough money to lead the life of a country gentleman. Now that I had, I could see that nothing else—not the orgies of Cardinal Radziwill, not all the riches of Russia—could compare with it. To live in Kimble Manor, with a good wife beside me, doing my duty to my tenants and my country . . . I thought of this as the lost soul thinks of salvation. It was not for me. I still thought it was not for me, when Sir William Hawtrey began to hint that young Bess might provide a happy clearance for her father's debts.

'Wait!' I told him. 'Wait until I've made one more journey to Russia.' As I said the words I thought: It is impossible I should come back alive.

I was to carry a letter to the Tsar from the Queen. She had written it with all the pride that she was now entitled to display, as victor over the Armada. She listed the promises the Russians had broken, promises that she had thought irrevocable 'coming from so mighty a prince'. She reminded them of England's power at sea, and how many times it had been used to keep open the way to Russia. Englishmen, she wrote, 'are not a people destitute of a princess'.

When I had translated the letter into Russian, the Queen asked me what I thought of it. I said frankly that every word was justified, and that I should be proud to hand it over, but that, when I had done so, Shelkalov would poison me.

'Sir Francis Drake has given me a very good antidote,' said the Queen. 'It is a juice the Indians of Brazil take from a tree.'

She gave me a flask of it. I took with me also some salad oil (which is good against irritant poisons) and that old remedy, Venice treacle. The Venetians allege that it has two hundred ingredients, and it certainly cures constipation.

I set out in April, 1589, but was windbound in Yarmouth so long that I had my secret wish—to see one more May morning in England. When I did set sail I was driven off course by storms, and then shipwrecked near Emden, which was beseiged at the time. The Emden people took me in by crane over the walls, the gates on the seaward side being rammed up. I was a long time getting a safe-conduct to leave. It was late summer when I came to Poland, where the Queen had instructed me to deal with the complaints of English merchants. The Poles received me well, for now, a whole year after the event, all countries understood that our victory at sea had been complete and overwhelming. Yet negotiations dragged on all the winter, because the Poles could not believe that such a great Queen would concern herself with merchants—'who are only peasants, after all.' I told them that many English merchants were, like myself, gentlemen by birth.

'Gentlemen?' said the Poles. 'Then why are they not in the army?' But we came to terms at last.

Avoiding Riga, I passed through Vilna, which was governed by Prince Christopher Radziwill, the Cardinal's Protestant brother. He wanted his people to believe that I had come to negotiate with him, as if he were an independent prince, and so he entertained me royally. I had to tear myself away from his musicians, his dancing bears, his banquets with unicorns and swans all made of sugar paste. He had the latest Italian fashions in luxury; at his table there was a silver fork laid for each person. It was from these comforts I went into Russia.

The Governor of Smolensk expected me, and had his instructions. I was to proceed, but slowly. Some ten miles west of Moscow they stopped me altogether, and brought me to a bishop's house, where I was kept for some time, so that I should not meet the Polish envoys then in Moscow. (Shelkalov, I heard, was reasoning that since I had come through Poland my main concern must be to plot with Poles against Russia.)

Provided I kept some Russian attendants with me, I was permitted to ride in the forest for exercise. Beggars and ballad singers waylaid me constantly; some of them were messengers from old friends. 'Look out for yourself!' was all they had to tell me. Until one old woman whispered: 'If you go to Svenigorod, the caretaker of Princess Maria's palace will give you news of her.'

I was not far from Svenigorod, and it was now full summer, the ground hard enough to gallop on. I outpaced my Russian attendants, came to Svenigorod and found the 'palace', a sprawling wooden house outside the town. The courtyard gate stood open, the courtyard empty. This was a strange thing in Russia, where the poorest farmhouse is guarded like a fortress against marauders. I hallooed, but got no answer. The caretaker must be sleeping off his vodka; he had not closed the double doors into the house.

I peeped in. The house was unfurnished, but showed signs of being recently repaired and painted. Maria must really have lived here for a time. The little hedgehog had run up and down that hall...

There was a crackle, just like the crackling of a stove, though the only stove that I could see was plainly out. While my mind puzzled over this, my body showed more sense. I jumped backwards, into the courtyard. So I missed the great crossbeam of the door, which crashed near my feet, instead of my head.

Shelkalov was now practised in accidents, and this would have been a good one. Found in a house where I had no business to be...

A week later I was allowed to enter Moscow in state, as ambassador. This time Trumbull did not have the pain of kneeling to me. He had died, I think naturally.

I presented the Queen's letter to the Tsar. Poor Feodor, trembling, handed the letter over to Shelkalov. So far he was playing his well-drilled part. But then he burst into tears. 'Yeremy!' he sobbed. 'Why don't you take me swimming any more? Are we bad friends?' He kissed the cross. 'I swear I never meant to make you angry.'

Shelkalov gave a signal to the attendants, who hurried me out of the room.

Still I was kept from seeing Boris. Until one day, when I heard a great noise of dogs and horses and men going past my house. Boris was riding to the hunt, his little son on a gentle horse beside him. He looked up to my window, and his eyes met mine.

For that instant he was my dear friend, who had begged me to protect his baby girl . . . who had wished he had a sister I might marry . . . who had said: 'I've learned so much from you!'

The same day, at dusk, he sent a man to bring me to a grassy place, almost enclosed within a jutting part of the city wall. We both dismounted. He sent his attendants away, embraced me, and kissed me three times, weeping real tears.

And I loved him still. I told him so. He said: 'Let *their* souls suffer who have come between us!' Then he told me that he could not show me any open favour. 'When those five hundred men waited in the forest to kill me, it was because they'd heard I meant to give Maria to you.'

I said. 'If I can't marry her, then let her go to Denmark!'

He shook his head. 'She can't leave the convent now. She's taken her vows.'

'If she has, it's not of her own free will. Even the papists don't say a vow is binding, if it's imposed by force.'

'What makes you think it was by force?'

'She was in the Troitsky Convent as a child. She told me it was a Hell where all the fiends were women. You yourself once begged me to save your wife from being put into a convent.'

'What a memory you have for idle words!' murmured Boris. And he smiled, so tenderly and sweetly that there seemed not a shadow left between us. 'Let's be good friends again. My sister said we should be. She said that, when she saw you last, she was angrier with you than you deserved.'

This was a puzzle. Irina had come to know the truth about the midwife, and was threatening, perhaps, to tell Feodor...

Boris went on: 'Maria might be released from her vows, if... if you could do one thing for me.'

I waited.

'Young Dmitri's mother is a fool. She's always in debt. Those huge estates at Ooglich are going to ruin. What she needs is advice from a practical man... And while you're there, Yeremy, you can find out what Afanassy and the rest of the family are up to. I'm afraid they've poisoned the boy's mind against me.'

I said: 'You have your own remedy for that. There has never been one soul you could not win, if you set your mind to it.'

'You mean I should bring the boy to Moscow, acknowledge him as heir, spend the time with him that I want to spend with my own boy... That's what Irina says. I thought you'd know better.'

I said: 'Your sister is a good woman.'

'Oh yes—but a woman! Do you know what it means, Yeremy, always to rule through someone else? Everything I do I have to explain twice over, first in terms not painful to a woman's tender feelings, and then in words a simpleton can understand. And what comes next? Am I to rule through a spoiled brat? Do you know what that boy does for pleasure? Beats chickens to death! Why won't you do this one thing for me, Yeremy? I thought once that you would be a tower of strength to me. Was I so much deceived?'

'No,' I said. 'I did mean to live always beside you, take your religion, think only of your country... You prevented that. Nobody but you.'

'Don't put all the blame on me!' said Boris. 'Blame the people I rule over. Their constant plotting, their ingratitude ... When Ivan let whole provinces die of starvation, they died blessing his name. Now that I put food in their mouths, they curse me. There's not one soul I can trust. It's come to this—I have nobody I can send to Ooglich but fools.'

I thought: Yes, the fools who poisoned the nurse instead of the child.

He cried out: 'Why do you look at me like that? I tell you it's the people! The people of Russia forced old Ivan to become The Terrible. Now they're doing the same to me. I swear it's against my will! I never have delighted in blood; you know that.'

'Then don't ask me to go to Ooglich.'

'No, I should not have asked you. Yeremy, forgive me!'

'God will forgive ... Perhaps he will, for what you've done to me. But there's Maria—'

'I will see what can be done for Maria. The Danes have been asking for her again. If I can let her go, I will. Be patient with me, Yeremy! And Yeremy, I am sorry for ... for what happened at Svenigorod. It wasn't my doing. I swear I won't let a hair of your head be touched.'

You might say Boris kept his word. The Russians never touched the hairs of my head. But they incessantly tried to poison the rest of me.

Though a little danger may give a spice to life, to be always in danger is like hearing a bad joke over and over. My butler tasted a present of wine, sent to me from the Kremlin, and fell dead. At that all my Russian servants left me but my cook. She stayed because she was in love with me. I have described my honourable courtship of Maria; now I have to confess that, like other men, I made love less honourably to my cook. She tasted all my food. When the peasants came to sell their vegetables at the kitchen door, she would buy only from those whom she knew. Yet she was caught; she bought a melon and did not see, until she cut it up, that a hole had been bored in the rind. Inside the hole there was a place discoloured; my poor silly girl put her finger on it, and then licked the finger clean. She died before I could put the antidote to her lips. I looked at her twisted, blackened face—it never had been beautiful—and I thought: This is not how the poets fancy a woman who dies for love. I cannot say much for my conduct, but this I will say; I never touched a servant girl again.

After that my man from Danzig, August Ducker, did the tasting. He said that, being a German, he knew how to do things properly. And indeed he took very small quantities, so that he did not die, only broke out in boils and lay writhing.

My laundress cast herself to the ground before me, banging her brow like old Ivan, and confessed that one of Shelkalov's men had paid her to poison me.

Though I saw Boris at meetings of his council, where I reiterated the English complaints, I never again saw him alone. But whenever an attempt to poison me had failed, he sent a message. It was not his doing; it would not happen again... In November, 1590, the message varied a little. I should be safer if I withdrew 'for a little while' to a fine house he had in Yaroslavl.

The time I spent there was the darkest of all my Russian winters. At the beginning of April, 1591, I was accosted in the town by what seemed a skeleton, with a few rags fluttering like the clothes of a man hanged last year. The creature cried out: 'I was imprisoned for you!'

This was my naked holy man, my messenger. I invited him into the house and called for something to eat. He took it with the same precautions as I did; the servant who brought it had to eat some first.

He said: 'Boris may mean to poison me. Already he's done worse than old Ivan. When did Ivan ever dare to lay hands on a man who went naked through the winter? Yet I've been almost three years in the dungeon. I was on my way to Troitsky, with your second message, when some of Ivan's old Tartar guard grabbed hold of me. What does a Christian holy man mean to them? They put me in a dungeon full of rats. I was there till a month ago.'

I had lived so long in Russia that instead of expressing pity I said with suspicion: 'Why did they let you out?'

'They didn't say. "There's your road!"—that's all they said. I looked up to Heaven to ask, was God satisfied? Three years among the rats... Doesn't that make up for the killing of my wife? But God said no; murder is murder; I must suffer a little more before I could live with my wife in Heaven. I thought—what's the quickest way to buy suffering? I went to the Troitsky again, and again they let me in. I found Maria Vladimirovna in the Church of the Assumption, praying by her daughter's grave.'

I did not move or cry out, but let him go on.

'She told me she blamed herself. Some sweetmeats came to her as a present from the Empress Irina. Of course, if the sweetmeats had been from Boris, they would have gone straight into the cesspool. But Maria knew the man who brought them as one of Irina's bodyguard. She said: "I wanted so much to believe they really came from her! I always loved her, and I couldn't understand why she didn't visit me." She tasted the sweetmeats first, of course, before she let the child have them. Well! Some poisons act slowly. It was hours before Maria and the little girl began to vomit. All the nuns who had had a taste were ill too. The grown women recovered; only the little girl died. Maria looked at the grave and said... some song she learned from you. I must try to remember it right. "Oh soul that thinks no creature harm!" She's a nun at last, a real nun. She's taken vows. "What difference does it make now?" she said. "I might as well spend the rest of my life in praying. I must ask God to forgive me, for not taking better care of her." '

I said: 'Does she really blame herself? She would do better to blame me.'

'Sir, she said she forgave you. She said: "He was deceived. We are all deceived when we desperately long to be." You know, sir, I daresay that was why they let me out. Now that she's a real nun, you can't rescue her.'

That same day I wrote a calm, well-reasoned letter to the Privy Council, saying that I had done all I could for English interests in Russia, and would come home as soon as possible.

And I wrote to Boris. I asked for leave to return to Moscow, so that I could sell my household furniture, and collect money owing to me. He sent me promises of money from the Imperial treasury . . . He seemed ready to give me anything, so long as I would leave Russia without looking him in the face.

I was still waiting for the money, and permits for post horses, in May, 1591. A violent knocking on my courtyard gate roused me at midnight, and I tumbled out of bed, seizing my pistols, which I kept always primed nearby. My fifteen servants, also armed, gathered round me, and I climbed up to look over the wall.

By moonlight I recognised Afanassy, the brother of Ivan's last Empress.

'For God's sake, Yeremy!' he cried out. 'Have you an English doctor there?'

'No.'

'Or an antidote for poison?'

'Who's poisoned?'

'My sister. Yeremy, the little boy's been murdered.'

'Prince Dmitri?'

'His throat was cut. The people of Ooglich beat two of the murderers to death. They pleaded for their lives; they cried out that they were obeying orders from Boris. I saved one man from the mob, and put him on the rack. He named Boris. We should have known. They came under pretext of clearing my sister's debts . . . Oh God, my sister! When she saw her son's body she began vomiting. She's been vomiting ever since. Her hair's falling out, and her nails. It must be poison. Yeremy, for the love of Christ, haven't you anything you can give her?'

I still dared not open my gate. But I ran up to my room and fetched some salad oil, and some Venice treacle, and a little phial of the medicine the Queen had given me. 'This one is good against fevers,' I told Afanassy, as I handed it to him over the wall. By this time people were gathering round him, though it was the dead of night. Someone brought him a fresh horse. As he started on his fifty-mile ride to Ooglich, the bells of Yaroslavl began to toll for the murdered Prince Dmitri.

All the tears I had not shed for Maria's child came spilling out of me then. I confess that I was weeping also for myself. I thought it impossible that anyone who knew as much as I knew should come alive out of Russia. There was only one thing left for me to do. On my knees, I prayed God to strike Boris. 'Let all tyrants know that, if they murder children, their own children are not safe.' There I checked my curse, remembering Anna's words—that since I was helpless, utterly in the power of my enemies, God would make every word I said come true. It has not proved so yet; Boris is flourishing; but no doubt God will exact the price of his crimes. And I shall have to pay for my part in them. I believe I shall pay soon.

12

I have put all this down together, though the telling took Jeremy three days. During this time Oulton did not come near us, and our servants were out seeing the sights of London. At night I scarcely heard Rosemary's prattle, except once, when she said that she and Paul had been to a bear-baiting. 'Oh, madam, you should have seen how the bear's claws tore the dogs! He stood up on his hind legs and fought them all, till the blood ran down his old shoulders!'

I said: 'How can you take pleasure in seeing animals tormented?'

'Why, madam, when you go hunting—'

'That's different, you stupid girl. Hunted animals have a chance to escape.'

'If I'm a stupid girl, then so's the Queen. She was there. You should have seen her laugh and clap, when the dogs brought the bear down at last!'

I had known that the Queen loved bear-baiting, but had never before considered that this proved her merciless. What likelihood was there that she would forgive Jeremy, if some forgotten witness now came forward, to tell her that he had once preferred another sovereign?

To Jeremy I said that it was wrong to doubt God's mercy. 'After all, He did bring you safe out of Russia.'

'Yes; Boris had sent the money, and the warrants for post horses, before he knew what a botch his men had made of Dmitri's murder.'

'Which he thought you could have managed better.'

'Which *he* could have managed better, if *I* had been the one torn to pieces by the mob. A foreign agent... plotting with Poles... As things were, Boris had to explain to Feodor, Irina and the Russian people how a cut throat could be an accident. He managed that. It seemed the child had been subject to fits, and his mother, though she knew this, had let him play with a sharp knife... She was put into a convent before she could deny it. People who accused Boris were well answered; their tongues were cut out. More than half the people of Ooglich were dragged from their homes...'

'Like the people of Novgorod.'

'Not quite like; no. Boris did not kill them. His conquests had given him a new punishment. He sent them to Siberia.'

There was one thing I still did not understand. I said: 'From the time you first thought you might marry Maria, till the moment you knew it was hopeless, was at least five years.'

'Nearer six.'

'You have told me what you said and did in all that time, but not once that you loved her.'

'Bess, haven't you understood yet what lies on my conscience? I courted her from ambition. I never loved her. I did not know what it was to love a woman, until I saw you pregnant with my child.'

I thought: Then you did not love me, when we were married. But I let him go on.

'Do you remember, Bess? Once, when I was returning from the hunt, you came to meet me, ambling on the gentle cob the doctor let you ride. I saw your pregnant body; I saw you smile at me. And I was so glad, Bess, that you had come out! I would have been home in half an hour, but it was such a pleasure to spend that half-hour in riding beside you! I knew then that all my life I had been pursuing... Not shadows. That's not it. Sometimes on a Russian road in summer you see a tower ahead. It looks tall and solid, but it's dust, caught up in a sudden whirlwind. Ambition is that whirlwind. It catches up all friendship, loyalty and love, and mingles them into this dance of dust. What I felt for Maria—and I did feel something—was caught up with her being a princess. I thought of her in golden robes beside me, adding lustre to my state.'

'Then you thought of me like that, when we were married. Not that I was a human creature, but that I would be the lady of Kimble Manor.'

'Don't be angry, Bess. That's my nature. I could not think of you without thinking of that life—hunting, hawking, being a Justice of the Peace, perhaps a Member of Parliament... Don't tell me that was wrong! I know it was. That was why I took such care to settle Kimble Manor on you. When all the deeds were signed I felt a hero, to have won such a battle against my nature. I wanted to be praised for that. Instead you burst out: "You've got what you married me for; you're living in Kimble Manor." Which was worse than any lie, being three parts true. I thought then God meant to punish me with an unhappy marriage. But now that I've been so happy with you, I know God must have something else in store.'

'Why can't you believe He has forgiven you?'

'I've often thought of Anna's words. God has to show His justice, as well as His mercy. If Anna was right, if all the helpless captives, dragged here and there, had their curses duly registered in Heaven, then what will fall on Russia? The trouble is that—God's ways being mysterious—it will fall on the innocent. When I finally left Russia, I stayed out of bed, all one sunlit night, waiting for the moment when the ship would pass Pechenga. I thought: The archimandrite's work lives on there; the monks are still translating the Bible into the Lapp language. Russia does hold some seeds of hope...

'The monastery was no longer there. They told us at Vardö

that a Swedish force, thinking the monks were spies, had murdered every one, and burned the monastery to the ground.'

'Are you saying that God punishes the innocent?'

'I wish I knew that. I wish I knew what happened after I left Russia, when Irina surprised everyone by becoming pregnant again, and giving birth to a girl. The baby soon died.'

'Did Boris kill that one too?'

'Boris was a cold-blooded murderer, and yet I think he was too much a Russian to harm his sister's child. No, I think this was God's justice on Irina, who might have done more to restrain her brother's tyranny. As I might . . . Bess, our Felicity is growing more and more like Maria's little girl. When I saw you in the firelight, combing her hair . . . I can't believe that God will let me keep Felicity. That day when you came jumping over the stile to meet me, I thought she had been suddenly struck dead.'

I was opening my mouth to say his fears were foolish, when I found I shared them. I thought of poor old Meg, scarcely able to wake, when the child was taken from the bed beside her. I thought how fast Felicity would run ahead, how recklessly she would pat animals, whether or not they were gentle . . .

But one thing I could say to Jeremy. 'It would not be your fault. If God were to take Felicity, he would be doing only what He does every day. My mother saw five babies buried, and yet the rest of us grew up. Why one should be taken and the other left, we don't know; it happens to the wicked and the good alike.'

He groaned out: 'But if she dies now, without us, because we left her behind . . .'

I said nothing, only I knelt beside him, and put my hand on his brow.

There was a sharp knock at the door, and Oulton came in. He did not greet us, but brought out a rolled parchment, and flung it on the table, saying: 'There's your indictment. And let me tell you, if what it says is true, your only hope is to plead guilty *now*. Then you might be beheaded, instead of the other thing.'

Even before he picked up the indictment, Jeremy asked: 'Who's my accuser?'

'Finch. That is . . . Ever since he came back to England, Finch has been living with Sir Jerome Bowes.'

I was trying to read the indictment over Jeremy's shoulder, but could not make sense of the legal phrases at the beginning. Jeremy pointed to the part which mattered: '. . . that on a day in July, 1582, at his table, talking familiarly as well with Russians as with Englishmen, the said Horsey did utter these words: "Our Virgin Queen is no more a virgin than I am. She enticed the Earl of Leicester to throw his wife downstairs; and so the Earl broke his wife's neck, and became the Queen's lover." '

'Is this all?' Jeremy asked.

'All!' exclaimed Oulton. 'If you said that—'

Jeremy shook his head. 'All these weeks I've been ransacking my soul, fetching out every action or word or thought that could ever be brought against me. After all that, it's this nonsense, a thing I never said . . .'

'But can you prove you never said it?' Oulton asked. 'It's hard to prove a negative. You have to go through every possible occasion.'

'There was only one occasion when Finch dined at my table. I've told you about it. We were investigating his frauds, and his first flight into Siberia, and trying to find out where the gold-bearing rocks came from. There were no Russians present, only Englishmen.'

'And will any of them confirm your story?'

Jeremy thought. 'Trumbull is dead. Jane Bomel wasn't at the table; we used to keep her out when we talked business. Rowley was there. He hates me. And yet he'll tell the truth; he promised my wife . . . I can give you some other names.'

'There's this in your favour,' Oulton said. 'If any of them had heard you say such a thing, fifteen years ago, and not reported it yet, they would themselves be guilty of misprison of treason.'

'Then isn't Finch guilty of that?'

'No. He's been all this time in Siberia. He tells people you kept him there.'

'Rowley can tell the truth about that as well.'

I asked: 'Why wouldn't the Queen tell me, when I saw her, what the accusation was?'

Oulton explained: 'Because this is the slander she can't bear even to speak of. It's been said before, and written. The Jesuits printed a book ... The Queen was obliged to publish a denial. Since then she's forbidden the matter to be mentioned, even for purposes of refutation.'

I said I had never so much as heard the story.

'Oh, it happened long before you were born. Lawyers talk about it still, because it shows the importance of public inquests. Anyone who has been into the matter knows that the Earl did not murder his wife. The inquest established that he was at Windsor, and she died in the Midlands. She was staying with a Mr. and Mrs. Foster; the servants were theirs, not the Earl's. Moreover the Earl had no chance of interfering with witnesses, coroner or jury, because, the moment the Queen heard of his wife's death, she very properly confined him to his house. The inquest was public; the lady's relations were present; doctors examined the body. They said the cause of death was a broken neck, with no other injury and no poison. Would any murderer kill a strong young woman by throwing her downstairs? She would scream and struggle, and then perhaps not die. In this case, it's true, there would have been nobody to hear her screams. The master and mistress of the house were out visiting, and every single servant at a fair. What silenced the slanderers then (although not later) was that all the servants insisted it was the lady herself who made them go to the fair. Two or three tried to stay behind, and she was angry with them. So it seemed like suicide—but then, again, what woman would set out to kill herself on a flight of stairs? She might only have broken her leg. The jury brought in a verdict of misadventure. They were there; they saw the body; they heard the witnesses. And yet people who know nothing about the matter have been tattling ever since ...'

'But if the Queen was proved innocent,' I said, 'why can't she bear to have it mentioned?'

'I suppose because the scandal prevented her from ever marrying the Earl of Leicester.'

'Did she want to?'

'So they say. Why not? He would have made her a good husband.'

Jeremy agreed. 'He was a good counsellor to her, all his life.'

Oulton said: 'If his wife had died naturally—or if people had believed her death was really an accident—the Queen could have lived like other people, with a home and a family. She never forgives anyone who spreads the scandal. Never! This Finch is for it if it's proved that he thought of it, not you.'

'Yes!' Jeremy exclaimed. 'And we can prove that! The man's delivered himself into our hands.' He sat down with Oulton, to make out the list of witnesses.

'We shall have at least a month to prepare our case.' Oulton remarked.

Though it was the middle of the afternoon, I felt suddenly very weary, and asked if I might return to the inn.

'My poor love, I believe you have not been sleeping lately, any more than I have,' Jeremy said. 'Well, now you can rest! I am not going to be hanged, nor quartered neither.'

I found no rest at the inn. A very grand gentleman was asking for me. This was the doctor the Queen had promised to send me, Dr. Strachan.

'I am sorry,' I told him, 'that you should have come here for nothing. When I saw the Queen a month ago I had a cough, but it's gone. I'm perfectly well.'

He said: 'Lady Horsey, I am an obedient subject, and I have the Queen's orders to examine you.' His tone was light, yet he looked very earnestly into my face.

I said: 'It's not convenient; my maid's gone gadding out somewhere . . .'

'I have brought Mrs. Humphreys with me,' he said, 'as I always do when I examine ladies.'

Mrs Humphreys was a midwife-looking woman. Her part, it proved, was to take off my corset, and so arrange my shift that the doctor could listen to my chest without unduly disturbing my modesty.

He seemed a long time listening and tapping. He made me spit into a bowl, and shook his head over that. I saw Mrs. Humphries shake her head, as if in reply.

I said: 'But I can't be really ill. My cheeks are as pink as ever.' I smiled now, because I was looking past them both, into the mirror. My complexion was really dazzling.

Dr. Strachan said: 'A white skin, a perpetual flush in the cheeks . . . You have not always had this colouring?'

'Oh no! When I was a young girl, my complexion was like mud. My nose used to be redder than my cheeks.'

'And you were fatter then?'

'Yes. My waist was too big, and my bosom not big enough, till I had my baby.'

'And before you had your baby, did you cough?'

'Only now and again.'

'Does anyone else cough much, in your family?'

'No. Except my mother, before she died.'

Dr. Strachan said: 'The best advice I can give you is to live in the country.'

'Why, so I do, near Aylesbury. You couldn't find a healthier place.'

'Then go back there now, at once! Don't spend even one more day in this London fog.'

'But I must! I have to stay here at least another month.'

The doctor said: 'My dear child, I am afraid a mortal enemy has taken up his lodging in your lungs.'

I cried out: 'But that's impossible!'

'It's quite certain. Don't be afraid, though; we can keep this enemy at bay. You are not poor? This house of yours has plenty of servants? Well then, go back there at once, live quietly, drink milk instead of wine, eat plenty of butter and eggs, and you may live to be an old lady.'

'But my husband needs me here.'

'Tell your husband that, if you stay in London, your cough will come back. You'll begin to spit blood. The flush in your cheeks will no longer look like health, but like what it is—the mark of a mortal fever. Your elegant shape will turn to the shape of a skeleton. Stay in London, and your next childbirth will kill you. Now, will you tell your husband all this, or shall I?'

'No! Don't tell him. Nobody must tell him!' I began to explain.

Dr. Strachan said: 'Are you sure that he needs you so much? You say he's innocent. Then can't he prove it for himself?'

I said: 'He feels guilty. Not of treason, but of things no human law can punish. He feels he is marked out for the just punishment of God. If he knew of this thing God has visited on me . . . No, that would break his spirit altogether. He would not even try to save his life.'

Dr. Strachan said: 'Then tell him some other reason why you must go back to the country.'

I thought. It would be so easy. I could say that I felt anxious for Felicity, and wanted to go back to her. Jeremy would agree at once. He would never accuse me of desertion.

But it would be desertion, and I could not do it.

I said: 'I will go home the first moment I can. Only for the time being give me some soothing syrup, something I can take just before I go to see him, so that he won't hear me cough.'

At last I got the doctor to agree. 'But I hope,' he said, 'that this man's life is worth your own.'

Surely it was not bad for me to visit all the Privy Councillors, one by one, as I did in the next few weeks. Two or three of them lived outside London, so I had some country air. Lord Burleigh was particularly kind. He knew more than I thought of the Earl of Essex, and his advances to me. Indeed, everybody at court knew, except the Queen. To the delight of his rivals, the Earl had been forced to hide himself, pretending to be ill, until the marks of my nails disappeared from his face. It seemed that this had never happened to him before. Other women melted into his arms, even if their husbands were not in prison. Lord Burleigh promised that Jeremy should not suffer for my fidelity. Since it was the Queen's particular desire, he would have the right to speak, and cross-examine his accuser.

I visited Mr. Rowley, too, and told him of Jeremy's remorse for Maria. 'He did not do this thing for money,' I said. 'He trusted the word of Boris.'

'But then why did he lie to me about it? He must have had some other motive.'

'No,' I said.

I was longing to rest. But there would be time for that, after Jeremy was tried.

When it happened at last, it was almost the end of April. It was not like any other trial for treason. Everything was conducted so as to give Jeremy a fair chance. Oulton was allowed to be present (though not to speak). I was allowed to be present. Jeremy's witnesses waited in an ante-chamber, unthreatened, while Jeremy confronted his accuser.

It was the accuser who looked the more afraid. Finch's tremor, and his ashy skin, may have only been the result of drinking. He was old before his time; above his temples a faded streak, like a dog's piddle in the snow, was the only sign that his hair had once been red. But not even an old man ought to be always turning his head, always looking for prompting to a puffed-up City alderman at his side.

'That's Bowes!' Jeremy told the Privy Council. 'Sir Jerome Bowes, my real accuser.'

Lord Burleigh asked Bowes: 'Do you in fact appear as this man's accuser?'

'I have accused him long ago,' said Bowes, 'of mishandling the Russia Company's money, of slandering me to the Queen—'

'Do you now accuse him of high treason?'

'Well . . . That is . . .'

'Can you of your own knowledge tell us what was said in Moscow in July, 1582?'

'Well, I was there the following year, and I found the defendant very rude and uppish . . .'

'But was he to your knowledge committing high treason?'

Bowes hesitated. Jeremy handed a letter to Lord Burleigh, who looked at the signature and asked Bowes: 'Is this your hand?' Finally Bowes agreed that this was a letter he had written Jeremy, thanking him for the help without which he might not have come alive out of Russia.

'You would not, I suppose,' Lord Burleigh said, 'write such a letter to a man you thought a traitor. Unless you yourself were a traitor.' Bowes remained silent, and Lord Burleigh went on: 'I think we can continue our deliberations without further help from you.'

When Bowes had been escorted from the room, Lord Burleigh told Finch to repeat his testimony. Instead, Finch went into other matters. 'Horsey has admitted, before your lordships, that a Tartar prince told him I had been captured in Siberia. If he had been a true Englishman, he would have tried to save me.'

'You forget,' Jeremy said, 'that the Tartar prince also told me you had been eaten by the Samoyedy. How was I to know they'd spewed you out?'

Finch gave a rambling account of his life among the Samoyedy. Far from eating him, they had given him their own food. Finch had at last made his way to Tobolsk, and lived among the Russians there for some years, as trader and interpreter. But his health began to fail. 'I wanted to see dear old England again before I died.' It had taken him two years to make the journey. 'And when I came here, the Company didn't make me very welcome.'

'Bowes did, though,' Jeremy said. 'You are in Bowes' debt for the very bread you eat. This tale you have told against me is his invention.'

Finch for the first time looked at Jeremy directly. 'No need of that!' he said. 'No need for anyone to tell me; I could see it for myself. It's not what you said or didn't say. *You were a Muscovite*. In your heart you had abandoned your own country.'

Sometimes a thick-headed oaf, three parts drunk, may hit the mark... I saw how deep Jeremy was breathing, while he considered what to say. But Lord Burleigh spoke first.

'Do you, then, stand by your accusation? Did this man, in July 1582, in Moscow, say certain words?'

'It's nearly fifteen years ago,' Finch muttered. 'I know that there was something said about our Virgin Queen.'

'Only by you!' put in Jeremy. 'You said: "God bless our Virgin Queen!" Nobody else made any remark on that.'

'Did you not say: "Amen"?' Lord Burleigh asked.

'No, my lord. The prayer sounded like a blasphemy, in the mouth of this drunken scoundrel, and I think we all preferred to take no notice.'

Jeremy's witnesses were then brought in. Though they did not pretend to remember everything said fifteen years before, they were sure they had never heard Jeremy utter a word against the Queen. Nor had he ever spoken of the Earl of Leicester, except as a kind patron and friend. Mr. Rowley's evidence was particularly telling, for he remembered his own accusations against Finch, and the story of the prospectors left stranded in Siberia. Under his puritan gaze, Finch went to pieces, muttering: 'I suppose I might have been mistaken.'

'Mistaken!' said Lord Burleigh. He and the other lords whispered together for a minute. Then Lord Burleigh rose. 'In the Queen's name!' he said. We all knelt. 'Sir Jerome Horsey' Lord Burleigh went on. 'We remember that you have twice been Her Majesty's ambassador in Russia, where God preserved you through many attempts on your life. Now He has preserved you against another attempt. We find you not guilty.'

Jeremy, still kneeling, thanked the Privy Council. But Finch was on his feet again. 'My lords! You're not letting him go! I tell you he's a Muscovite heart and soul. When I was last in Moscow he threatened me. He said I'd be tied between two fires. He wanted to have me roasted!'

'It is a pity that you were not singed a little,' said Lord Burleigh.

I did not see what happened next, because Jeremy was making his way to me. He said: 'Come on, Bess! We're free.'

So I followed him, and was like him surrounded by all the people who wanted to shake his hand. (Only Mr. Rowley had left quietly, without speaking to either of us.) Jeremy thanked his lawyer and his witnesses; they told him he should rather thank his wife. Which he did, by constantly kissing my cheek. I was ready to sink to the ground from weariness, but his arm around my waist sustained me.

More and more people came crowding into the ante-chamber where we stood. Not until the captain of the guard had shouted three times did anyone make way. Then, squeezed tight against Jeremy, with Oulton on my other side, I saw the guard pass, with Finch, trembling and white, in their midst.

'Where are they taking him?' I asked.

'I suppose to the Marshalsea,' Oulton said. 'That's where they keep the weights.'

'The what?'

'They're going to lay him flat on the floor and put iron weights on his chest.'

I cried out: 'Oh no!'

'Why not?' said Oulton. 'It's the usual punishment for perjury.'

I said: 'But that's horrible. It's like Russia.'

Oulton looked at me with all the scorn of a sensible man for a woman's caprice. 'Don't you understand what this man tried to do to your husband?'

The Privy Councillors were now appearing. 'There's Lord Burleigh!' I said. 'I'll ask him.'

Jeremy's arm round my waist was tight, but I struggled against it. He said angrily: 'Bess!' I whispered in his ear: 'That man told lies, but what he said was true.'

Jeremy snatched his arm away, as if I were red hot. So then I could run forward, to kneel at Lord Burleigh's feet.

'Why, child, what's this?' he said. 'Have you come to thank me? I have done justice, that's all.'

I gasped out: 'Please, please, don't let them put weights on Finch.'

'Why, that's for the judge to decide,' said Lord Burleigh. 'We have done no more than commit him for trial. I suppose the usual punishment will be applied. Why not?'

I said: 'Because . . . because England isn't Russia. So that it should never be Russia.'

'What eyes you have!' said Lord Burleigh. 'You have made

all the counsellors here soft as butter . . . But if you want some particular mercy shown to this man, you must ask the Queen.'

So I did. Jeremy no longer opposed me. 'What can I say?' he asked me sadly. 'You know too much about me now.'

The Queen saw me in a little arbour, sheltered from the winds of spring, where we could talk alone.

'Dr. Strachan gave me a sad report of your health,' she said. 'I hope you will follow his advice.'

'Indeed I shall,' I said. 'Now that my husband is free.' Then I said that, though the false accuser certainly deserved imprisonment, I did not want him crushed under weights. I said I thought that cruel. The word gave offence. The Queen said: 'Do you think I am cruel for the sake of cruelty? If you had my duties . . .'

I thought: Your duty does not take you to the bear-baiting.

Though I had said nothing aloud, she looked at me sharply. 'My sister Mary is remembered as a cruel woman. Believe me, she was far too soft with the people nearest her. They were plotting against her, which I knew, because they tried to involve me in their plots. When I came to the throne they swore loyalty to me. I thought—yes, I know what your loyalty is worth! Well, I have ruled them and their kind these forty years, by fair words mainly, but always in some part by fear. There must always be some who would stir up revolt against me, if they were not afraid of the dungeon and the rack. Look at this man Finch! He tried to spread that slander by the most cunning means—accusing someone else of having said it.'

'But Your Majesty, if I can forgive him . . .'

She cut me short. 'You should forgive him; the harm to you is put right. But not the harm to me! This cruel, stupid rumour, which has plagued me for thirty-seven years, is living still.'

'But it cannot be believed by anyone who knows Your Majesty.'

She shook her head. 'It is believed, and in the face of the clearest evidence. And I shall never be able to establish the truth, because I do not know it myself. Oh, I know I did not kill that woman. No more did Robert—the Earl of Leicester, I mean. When he died'—she was holding her eyes wide open still, and her voice quiet—'when he died I was not with him. I was in St. Paul's, giving thanks for the defeat of the Armada ...' She shook her head briskly. 'But though I did not see him on his death-bed, I had his true confession three years earlier, when he went to command my army in the Netherlands. He was prepared then to die, and he swore to me, as he hoped for Heaven, that though he had never chosen his wife and never loved her, though he had not seen her for a whole year, still he had no notion how she came to die. He thought it must be suicide. What perturbed his conscience was that his neglect might have driven her to it. She might have heard some story ... God knows! I shall never know.'

Whatever I might say would be (I feared) some disrespect. Yet this haggard, lonely old woman had given me back my husband. I wanted very much to give her something. I said: 'When the Countess of Leicester made her servants leave her alone, was that a common thing? Did she like to be alone?'

'No, she hated it. She would always have three or four maids with her, not to do anything, only for company.'

'So this time, she had some powerful reason.'

'The slanderers make out that she expected a messenger from her husband, and wanted to see him alone. When he came, he killed her.'

I thought of my own village. 'A stranger coming up to the gentlefolks' house ... The whole countryside would know. That would come out at the inquest, if nothing else did.'

'Suicide, then. She had been heard praying God to save her from desperation.'

'Your Majesty, there is only one thing which can make any woman jump down a flight of stairs.'

'What's that?'

'One of our maids did it, at Hampden House. Afterwards I heard the cook telling her: "You silly ha'porth; you could have broken your neck. And it never does what it's supposed to do." '

'What is this?' asked the Queen. 'Something the maid knows and the cook knows and you know, and I do not.'

'Your Majesty has led a stainless life. How could you know what a desperate woman will do to bring on a miscarriage?'

'Oh!' said the Queen. And then: 'Impossible! Her husband had not been near her for a year, and—'

She broke off. We looked at each other. I said: 'If the desperation she spoke of were not suicide but some love tangle . . .' I did not dare go on: 'Would it make your mind more easy?' But without asking, I was answered; the Queen's brow cleared. She said: 'If that woman were not a completely innocent victim . . . But you cannot prove what you say. Poor lady, she should not have her name blackened, any more than I should. Never mention this again to anyone!'

I promised I would not. Then the Queen asked: 'What was it you wanted from me?'

'That this Finch, who accused my husband, should not be pressed under weights. Imprisonment is bad enough.'

'Is it?' said the Queen sharply. 'Not for some villains, I assure you! You are a silly girl . . . and yet I can't refuse you, God knows why.'

She gave orders about Finch, and I took my leave. But she called me back to ask: 'Is it really such a common thing among women? To think that I have lived to sixty-three, without being told!'

When Jeremy and I went to the Russia Company's dinner, even Sir William Hawtrey smiled. There was handshaking, clapping, cheering . . . As much, I daresay, as there would have been to celebrate the execution.

Those last few days in London passed me in a daze of weariness. Jeremy took me to see a play about a Jewish moneylender. I was on the moneylender's side, because of Jeremy, and I thought the other people in the play were cruel. Sitting in the open air made me cough. Jeremy grew anxious, but I told him I would be well again once we were home.

As we rode up to Hampden House, my father and brother came out to meet us. I took no pleasure in their welcome. Only my brother's little boy, John Hampden, meant it when he said: 'Uncle Jeremy, I *am* glad to see you!' Sitting on Jeremy's saddle-bow, he begged for a story about Russia.

'Oh, I shall tell you plenty!' Jeremy said. 'And what's more, I'll tell you how to make sure that (as your aunt says) England shall never be Russia.'

We refused to break our journey at Hampden House. I told my father that we were impatient to see Felicity. He did the proper things; he had the bells of Hampden Church rung in our honour.

We rode on, past the great front of Chequers, and then downhill. On our right lay the Happy Valley, and its wild wood full of hares. Trees were budding; some almost in leaf, and some still brown, and at the edges of the wood the hawthorn burst out brightly. Now the sun broke from the clouds and showed us on our left the fertile plain. The young wheat was green, and the young grass, and some fields of oats already a shouting green. We could not yet see Great Kimble, but we heard its church bells answering the bells of Hampden. And then a sound more joyous than the bells; the voice of our Felicity. Old Meg, riding a cob, had brought her to us. Even as I held my darling close I thought: I must be careful how I kiss her; I may be infectious. Jeremy did not know it, but for the past week I had been spitting blood. Yet I smiled as I passed Felicity into his eager arms.

'How well she looks!' he said. 'Has God forgiven me?'

We rode on, down the hill, amid the hawthorn blossoms, the cheering of the people, the pealing of the bells. A May morning in England.

ARCTIC OCE
VARDÖ
PECHENGA
LAPLAND
WHITE SEA
ROSE ISLAND
KHOLM
VOLOC
NARVA
REVAL
DAGÖ
OBERPALEN
NOVGOROD
YAR
OOGLIC
OESEL
ARENSBURG
PSKOV
LIVONIA
WOLMAR
WENDEN
PILTEN
RIGA
LENWARD
GOTHLAND
ÖLAND
VELIKIYE LUKI
SVENIGO
COPENHAGEN
POLOTSK
BALTIC SEA
SMOLENSK
KÖNIGSBERG
VILNA
LÜBECK
DANZIG
HAMBURG
OSTROG
0 100 200 Miles
0 100 200 300 Km